The Retroviridae
Volume 3

THE VIRUSES

Series Editors

HEINZ FRAENKEL-CONRAT, *University of California*
Berkeley, California

ROBERT R. WAGNER, *University of Virginia School of Medicine*
Charlottesville, Virginia

THE VIRUSES: Catalogue, Characterization, and Classification
Heinz Fraenkel-Conrat

THE ADENOVIRUSES
Edited by Harold S. Ginsberg

THE ARENAVIRIDAE
Edited by Maria S. Salvato

THE BACTERIOPHAGES
Volumes 1 and 2 • Edited by Richard Calendar

THE HERPESVIRUSES
Volumes 1–3 • Edited by Bernard Roizman
Volume 4 • Edited by Bernard Roizman and Carlos Lopez

THE INFLUENZA VIRUSES
Edited by Robert M. Krug

THE PAPOVAVIRIDAE
Volume 1 • Edited by Norman P. Salzman
Volume 2 • Edited by Norman P. Salzman and Peter M. Howley

THE PARAMYXOVIRUSES
Edited by David W. Kingsbury

THE PARVOVIRUSES
Edited by Kenneth I. Berns

THE PLANT VIRUSES
Volume 1 • Edited by R. I. B. Francki
Volume 2 • Edited by M. H. V. Van Regenmortel and Heinz Fraenkel-Conrat
Volume 3 • Edited by Renate Koenig
Volume 4 • Edited by R. G. Milne

THE REOVIRIDAE
Edited by Wolfgang K. Joklik

THE RETROVIRIDAE
Volumes 1–3 • Edited by Jay A. Levy

THE RHABDOVIRUSES
Edited by Robert R. Wagner

THE TOGAVIRIDAE AND FLAVIVIRIDAE
Edited by Sondra Schlesinger and Milton J. Schlesinger

THE VIROIDS
Edited by T. O. Diener

The Retroviridae

Volume 3

Edited by

JAY A. LEVY

School of Medicine
University of California
San Francisco, California

PLENUM PRESS • NEW YORK AND LONDON

Library of Congress Cataloging in Publication Data

(Revised for vol. 3)

The Retroviridae.

(Viruses)

Includes bibliographical references and indexes.
1. Retroviruses. I. Levy, Jay A. II. Series.

QR414.5.R48 1992 576′.6484 92-26459

ISBN 0-306-44693-6

A Division of Plenum Publishing Corporation
233 Spring Street, New York, N.Y. 10013

Printed in the United States of America

Contributors

Peter Barry, Department of Pathology, School of Medicine, University of California, Davis, California 95616

Paul R. Bowser, Department of Avian and Aquatic Animal Medicine, College of Veterinary Medicine, Cornell University, Ithaca, New York 14853

Arsène Burny, Molecular Biology and Animal Physiology Unit, Faculty of Agronomy, B5030 Gembloux, and Department of Molecular Biology, University of Brussels, B1640 Rhode-St-Genèse, Belgium

Isabelle Callebaut, Molecular Biology and Animal Physiology Unit, Faculty of Agronomy, B5030 Gembloux, Belgium

James W. Casey, Department of Microbiology, Immunology, and Parasitology, College of Veterinary Medicine, Cornell University, Ithaca, New York 14853

Irvin S. Y. Chen, Departments of Medicine and Microbiology and Immunology, University of California School of Medicine and Jonsson Comprehensive Cancer Center, Los Angeles, California 90024-1736

Louis Droogmans, Department of Molecular Biology, University of Brussels, B1640 Rhode-St-Genèse, Belgium

Michael Endres, Department of Pathology, School of Medicine, University of California, Davis, California 95616

Hung Fan, Department of Molecular Biology and Biochemistry and Cancer Research Institute, University of California, Irvine, California 92717

Patricia N. Fultz, Department of Microbiology, University of Alabama at Birmingham School of Medicine, Birmingham, Alabama 35294

Murray B. Gardner, Department of Pathology, School of Medicine, University of California, Davis, California 95616

Matthew A. Gonda, Laboratory of Cell and Molecular Structure, Program Resources, Inc./DynCorp, National Cancer Institute–Frederick Cancer Research and Development Center, Frederick, Maryland 21702

Patrick L. Green, Department of Microbiology and Immunology, Vanderbilt University School of Medicine, Nashville, Tennessee 37232-2363

Richard Kettmann, Molecular Biology and Animal Physiology Unit, Faculty of Agronomy, B5030 Gembloux, Belgium

Reinhard Kurth, Paul-Ehrlich-Institute, 63225 Langen, Germany

Jo-Ann C. Leong, Department of Microbiology, Oregon State University, Corvallis, Oregon 97331-3804

Dixie L. Mager, The Terry Fox Laboratory, B. C. Cancer Agency, Vancouver, and Department of Medical Genetics, University of British Columbia, Vancouver, British Columbia V5Z 1L3, Canada

Marc Mammerickx, National Institute for Veterinary Research, B1180 Brussels, Belgium

Stephen Norley, Paul-Ehrlich-Institute, 63225 Langen, Germany

Daniel Portetelle, Molecular Biology and Animal Physiology Unit, Faculty of Agronomy, B5030 Gembloux, Belgium

Frederique M. Poulet, Department of Avian and Aquatic Animal Medicine, College of Veterinary Medicine, Cornell University, Ithaca, New York 14853

David A. Wilkinson, The Terry Fox Laboratory, B.C. Cancer Agency, Vancouver, and Department of Medical Genetics, University of British Columbia, Vancouver, British Columbia V5Z 1L3, Canada

Luc Willems, Molecular Biology and Animal Physiology Unit, Faculty of Agronomy, B5030 Gembloux, Belgium

Preface to Volume 3

This third volume in the series on *The Retroviridae* introduces readers to further examples of this important and diverse virus family. A review of retroviruses in chordate and invertebrate species is first covered, with the noteworthy observation on induction of malignancies in snakes and molluscs. These findings reflect the oncogenic potential shared by retroviruses from a variety of different species. Next, the retroviruses of cattle are discussed in terms of two genera, the bovine leukemia-inducing (BLV) and bovine immunodeficiency-inducing (BIV) viruses. Similarly, the subsequent chapters on the primate retroviruses (SIV, SRV, STLV) emphasize the widespread presence of different genera of retroviruses in nature, even in the same animal species.

In a previous chapter in Volume 2, biologic and pathogenic properties of HTLV were discussed. In the present volume, the molecular biology of this human retrovirus is considered. This chapter, along with those on BLV and STLV, indicates why these bovine, primate, and human viruses have been placed within the same genus and how their common biologic and molecular features have added new insights into carcinogenesis.

In regard to cancer induction, a variety of mechanisms by which the Retroviridae can be involved in the oncogenic process are covered in Chapter 7. The important host response to retrovirus infection is then addressed in Chapter 8. This comprehensive analysis of the immune system indicates how several different animal species appear to evoke similar types of reactions to these viruses. These immune responses, both cellular and humoral, can determine whether control of a retrovirus can be achieved, thus preventing disease and assuring survival of the host.

Finally, the overall presence of this virus family in many different species, as reflected in Chapter 1 and covered in other volumes in this series, is underlined by the review on endogenous retroviruses (Chapter 9). Studies dealing with endogenous retroviruses in lower animal species (e.g., the murine and avian species in Volume 1) have suggested that similar agents might be found in human cells. This increasingly important aspect of retrovirology is further supported by the presence of retroviral elements in microorganisms (Volume 1). The recent observations on endogenous viruses in human cells have raised questions on the potential origin of this family of viruses and their role in normal development as well as pathogenic processes.

Readers are recommended to refer to Volumes 1 and 2 in this series for other background information on various aspects of retrovirus infection and pathogenesis. The present volume adds further to our fundamental knowledge on this most interesting virus family.

Jay A. Levy

San Francisco, California

Contents

Chapter 1

Retroviruses of Fish, Reptiles, and Molluscs

Frederique M. Poulet, Paul R. Bowser, and James W. Casey

Chapter 8

Immune Response to Retroviral Infection

Stephen Norley and Reinhard Kurth

CHAPTER 1

Retroviruses of Fish, Reptiles, and Molluscs

FREDERIQUE M. POULET, PAUL R. BOWSER, AND JAMES W. CASEY

I. INTRODUCTION

Neoplasia has long been recognized as a disease of teleosts (Ferguson, 1989). Neoplasms occur not uncommonly in reptiles (Jacobson, 1981). Similarly, five types of neoplasms have been reported worldwide in many species of bivalve molluscs (Brown *et al.*, 1977). Since the documentation of neoplasms in lower vertebrates, tumors have been reported to arise from all organs and cell types having homologous counterparts in higher vertebrate species (Ferguson, 1989; Haddow and Blake, 1933).

The majority of early reports of neoplastic diseases in fish, reptiles, and molluscs represent isolated cases. Recently, however, single occurrences have been replaced by descriptions of epizootics of neoplasms in feral and farm-raised fish, captive reptile, and cultured mollusc populations. In the case of some feral fish populations, the incidence of neoplasms in northern pike and walleye approaches 20 and 30%, respectively. One commercially valuable species, the chinook salmon, is affected by a retrovirus-associated leukemia, termed plasmacytoid leukemia. The incidence of plasmacytoid leukemia in pen-reared salmon can reach 85%. The incidence of neoplasia in bivalve molluscs, specifically the soft-shell clam *Mya arenaria,* ranges from 0 to 64% depending on the collection site and time of year (Brown *et al.*, 1977). These frequencies are the highest recorded for naturally occurring tumors in free-living vertebrate species and molluscs.

This situation likely reflects a real increase in prevalence of neoplasms in some of these species. The expansion of the aquaculture industry, in response to

FREDERIQUE M. POULET AND PAUL R. BOWSER • Department of Avian and Aquatic Animal Medicine, College of Veterinary Medicine, Cornell University, Ithaca, New York 14853. JAMES W. CASEY • Department of Microbiology, Immunology, and Parasitology, College of Veterinary Medicine, Cornell University, Ithaca, New York 14853.

The Retroviridae, Volume 3, edited by Jay A. Levy. Plenum Press, New York, 1994.

destruction of fish natural habitats and escalating demand for food, has been accompanied by apparently new epizootic diseases in these species. It is most probable that artificial rearing conditions, economically feasible under commercial constraints, are responsible for challenging the fragile vital host–pathogen equilibrium. Conversely, the augmented observed prevalence of neoplasms could be attributed to improved monitoring of the health of fish, reptiles, and invertebrates.

Some neoplasms in fish and molluscs, unlike those in homeothermic animals, have the propensity to appear and regress seasonally. This unique feature is not limited to retrovirus-associated neoplasms. The host response to infection, the temperature dependence of viral replication, and the population dynamics resulting from differential response to infectious challenge may all contribute to seasonal appearance of tumors. In that respect, a comparative study of neoplasms of poikilotherm animals opens new research avenues in delineating the regulatory mechanisms of uncontrolled cellular growth.

Some of these epizootic neoplasms of fish, reptiles, and molluscs, as mentioned above, appear to be etiologically associated with retroviruses. Historically, mammalian and avian retroviruses have been instrumental in unraveling key determinants of eukaryotic cellular gene regulation. They have contributed to the most significant advances in oncogenesis and are invaluable tools in contemporary molecular biology. Comparatively, piscine, reptilian, and molluscan retroviruses constitute a relatively recent discovery.

Fish retrovirology is rapidly expanding. There are currently 12 retroviruses associated with spontaneous neoplastic/proliferative diseases of bony fish (Table I). Some of these diseases, particularly those affecting game fish, commercially harvested species, farm-raised fish, or species inhabiting polluted areas, have been studied more extensively than others. For most of them, however, the etiologic relationship rests only on circumstantial evidence. Substantial progress has nevertheless been achieved in some cases by employing contemporary molecular techniques. However, for none of these neoplastic/proliferative diseases has the Henle–Koch postulate been fulfilled. A major impediment partially responsible for delayed progress is the lack of an *in vitro* cell culture system in almost all cases to isolate and propagate piscine retroviruses.

The focus of this chapter is to provide a review of the literature on retrovirus-associated neoplastic/hyperplastic diseases of fish, reptiles, and molluscs. It is emphasized that lower vertebrate and mollusc retrovirology is truly a new and scientifically approachable field which offers unique opportunities to study the evolutionary relationships among Retroviridae and the mechanisms involved in tumorigenesis.

II. RETROVIRAL INFECTIONS IN FISH

A. Epizootiology

Epizootiologic data have long hinted at infectious etiologies for certain fish neoplastic/proliferative diseases (Pilcher and Fryer, 1980). Indeed, an important

TABLE I. Neoplastic or Proliferative Diseases Associated with Retrovirus in Fish

Host species	Disease	Tissue type
Walleye (*Stizostedion vitreum*)	Walleye dermal sarcoma	Mesenchymal
Atlantic salmon (*Salmo salar*)	Sarcoma of the swim bladder	Mesenchymal
Hooknose (*Agonus cataphractus*)	Cutaneous fibroma/fibrosarcoma	Mesenchymal
Angelfish (*Pterophyllum scalare*)	Lip fibroma	Mesenchymal
Walleye (*Stizostedion vitreum*)	Discrete epidermal hyperplasia	Epithelial
Atlantic salmon (*Salmo salar*)	Epidermal papillomatosis	Epithelial
European smelt (*Osmerus eperlanus*)	Spawning papillomatosis	Epithelial
Northern pike (*Esox lucius*)	Smooth type of epidermal hyperplasia	Epithelial
Muskellunge (*Esox masquinongy*)	Smooth type of epidermal hyperplasia	Epithelial
White sucker (*Catostomus commersoni*)	Epidermal papilloma/carcinoma	Epithelial
Chinook salmon (*Oncorhynchus tshawytscha*)	Plasmacytoid leukemia	Hematopoietic
Northern pike (*Esox lucius*)	Hematopoietic neoplasm	Hematopoietic
Muskellunge (*Esox masquinongy*)	Hematopoietic neoplasm	Hematopoietic
Bicolor damselfish (*Pomacentrus partitus*)	Neurofibromatosis-like disease	Nervous

characteristic of most of these retrovirus-associated neoplastic/proliferative diseases is their occurrence at epizootic proportions. Undoubtedly, these diseases in wild fish populations owe their discovery to these epizootics. A notable exception is the cutaneous fibroma/fibrosarcoma of hooknose, which was detected during a survey and had an incidence of 0.7% in the population examined (Anders *et al.*, 1991). In addition, affected fish tend to cluster in several limited geographic areas of their habitat, interspersed with areas populated by healthy fish. This distribution, together with a higher incidence in regions of higher population densities in some cases (Schmale *et al.*, 1983), suggests that these neoplastic/proliferative diseases of fish have an infectious origin and are contagious. Chemical factors apparently play a synergistic role in the epidermal papilloma/carcinoma of white sucker. Indeed, the prevalence of white suckers with neoplasms is greater in polluted sites than in pristine waters (Sonstegard, 1977; Smith *et al.*, 1989a; Hayes *et al.*, 1990).

Three retrovirus-associated neoplastic diseases have appeared as outbreaks in aquaculture situations, where early and adequate monitoring of the fish is facilitated. These are the plasmacytoid leukemia of chinook salmon, epidermal papilloma and sarcoma of the swim bladder of Atlantic salmon. The epidermal papilloma and plasmacytoid leukemia occur also, but less dramatically, in free-living Atlantic (Pilcher and Fryer, 1980) and chinook (Kent, personal communication) salmon.

TABLE II. Summary of Etiologic Evidence Implicating Retroviruses in the Genesis of Fish Neoplastic and Proliferative Lesions[a]

Disease	Electron microscopy				Reverse transcriptase activity		Acellular transmission	Virus purification	Cloning
	Spontaneous lesion	Sucrose gradients	Inoculum	Induced lesion	Spontaneous lesion, cation	Induced lesion			
Cutaneous fibroma/ fibrosarcoma of hooknose	Lentivirus 86–132 nm	ND[b]	NA[c]	NA	ND	NA	ND	ND	ND
Epidermal papilloma/ carcinoma of sucker	Type C[d] 100 nm	ND	ND	NA	1.15–1.16 g/ml ND	NA	Negative	ND	ND
Esocid hematopoietic neoplasm	Types C, B 100 nm	Type C	ND	ND	1.15–1.17 g/ml Mn^{2+}	ND	Successful	Achieved	ND
Smooth type of esocid epidermal hyperplasia	Type C 120–150 nm	ND	NA	NA	ND	NA	ND	ND	ND
Plasmacytoid leukemia of chinook salmon	Type C 110 nm	Type C 110 nm	Type C 110 nm	Type C 110 nm	1.16–1.18 g/ml Mn^{2+}	Present	Successful	SLV[e]	ND
Spawning papillomatosis of European smelt	Type C 88–101 nm 55–76 nm	ND	NA	NA	ND	NA	ND	ND	ND

Damselfish neurofibromatosis-like disease	Absent	Type C 100 nm	ND	ND	1.14–1.18 g/ml ND	ND	Successful	DNFV[f]	ND
Epidermal papillomatosis of salmon	Type C 125–150 nm	ND	NA	NA	ND	NA	ND	ND	ND
Swim bladder sarcoma of salmon	Type C 110 nm	ND	NA	NA	ND	NA	ND	ND	ND
Dermal sarcoma of walleye	Type C 100–135 nm	90 nm	ND	100 nm	1.18 g/ml Mn^{2+}	ND	Successful	WDSV[g]	Achieved
Discrete epidermal hyperplasia of walleye	Type C 120 nm	ND	NA	NA	ND	NA	ND	ND	ND
Lip fibroma of angelfish	Type A 90 nm Type C 150 nm rodlike	ND	ND	NA	ND	NA	Negative	ND	ND

[a] References are given in the text.
[b] ND, not done.
[c] NA, not applicable.
[d] Controversial finding.
[e] SLV, salmon leukemia virus.
[f] DNFV, damselfish neurofibromatosis virus.
[g] WDSV, walleye dermal sarcoma virus.

The etiologic evidence implicating retroviruses in the genesis of fish neoplastic and proliferative lesions is reviewed below and summarized in Table II.

B. Electron Microscopic Evidence

Electron microscopic examination substantiated the suspected infectious etiology of all of these neoplastic and proliferative lesions of fish by revealing retrovirus particles within spontaneous lesions (11 cases), processed tumor material (3 cases), or tumor-derived cell cultures (1 case) from mesenchymal, epidermal, hematopoietic, and neural tissues. All viruses but one, the virus seen in the cutaneous fibroma/fibrosarcoma of hooknose, display the typical morphology of type C oncoviruses (Fig. 1).

Mesenchymal tumors, such as *walleye dermal sarcoma* and *Atlantic salmon swim bladder sarcoma,* contain groups of extracellular and budding type C viral particles. The 100- to 135-nm diameter virus associated with dermal sarcoma was named walleye dermal sarcoma virus (WDSV) by Walker (1969) based on its electron microscopic detection within the tumor. Viruses present in sarcoma of the swim bladder measure 110 nm in diameter (Duncan, 1978). Within the *lip fibroma of angelfish,* several types of particles are either free within the cytoplasm of the neoplastic cells or budding from internal membranes: a 90-nm type A particle, an enveloped, 150-nm, intracytoplasmic form of type C particle, and, in addition, filamentous structures that were identified as rodlike forms within intracytoplasmic vacuoles (Francis-Floyd *et al.*, 1993). Unlike all other previously reported retroviruses associated with tumors of lower vertebrates, in the *fibroma/fibrosarcoma of hooknose,* abundant viral particles are within intracytoplasmic vacuoles of cells identified tentatively as lymphocytes and share morphological resemblance with lentiviruses: 99 nm in diameter, occasional double cores and lateral bodies (Anders *et al.*, 1991).

Within approximately 10% of *the epidermal lesions on the fins of the European smelt,* two types of extracellular retrovirus particles, 88 to 101 or 55 to 76

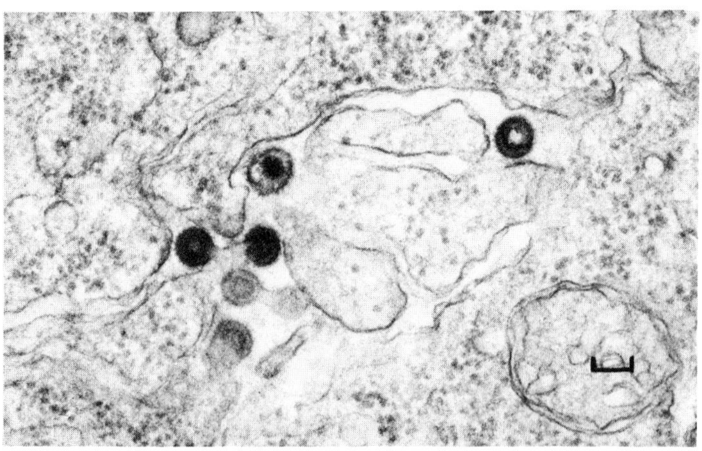

FIGURE 1. Electron micrograph of walleye dermal sarcoma virus (WDSV) in experimentally transmitted neoplasm. Bar = 100 nm.

nm, were observed in addition to large numbers of herpesviruses (Anders, 1989; Anders and Möller, 1985). The extracellular, 80- to 120-nm, type C retrovirus of *discrete epidermal hyperplasia of walleye* differs morphologically from WDSV by its smaller size (80 μm versus 100 μm for WDSV), association with the epithelial Malpighian cells of the epidermis, and less well defined peplomers (envelope surface projections) (Walker, 1969; Yamamoto *et al.*, 1985a,b; Bowser *et al.*, 1988). Intracytoplasmic viruslike particles, 120–150 nm in diameter, and a single extracellular particle are described within the neoplastic cells of *epidermal papillomas of three Atlantic salmon* (Carlisle, 1977). A characteristic feature of *the smooth type of epidermal hyperplasia of northern pike* is the consistent presence of multitudes of type C, 120- to 150-nm, retroviral particles forming dense intercellular aggregates between epithelial cells (Winqvist *et al.*, 1968; Sonstegard, 1976; Yamamoto *et al.*, 1983; Sonstegard and Chen, 1986). The presumptive viral etiology of *epidermal papilloma/carcinoma of white sucker* remains poorly supported: one description of numerous, 100-nm, type C retroviral particles within spontaneous neoplasms (Sonstegard, 1977) has not been confirmed by other investigators (Smith *et al.*, 1989a). Moreover, this description appears to be a publication error, as the same electron micrograph subsequently illustrated the smooth type of epidermal hyperplasia of northern pike (Sonstegard, 1976; Harshbarger and Clark, 1990).

Extracellular and budding, 110-nm, type C retroviral particles, designated salmon leukemia virus (SLV), were detected in *plasmacytoid leukemia of chinook salmon* by ultrastructural examination of neoplastic tissue from spontaneous tumors, experimentally reproduced neoplasms in recipient fish, and sucrose gradient fractions with high reverse transcriptase activity (see below) (Eaton and Kent, 1992). Extensive electron microscopic study failed to reveal conclusive viral particles within spontaneous *hematopoietic neoplasms of the northern pike and muskellunge* (Mulcahy *et al.*, 1970; Brown *et al.*, 1975; Mulcahy, 1963, 1976; Sonstegard, 1976). Suggestive type C viruslike structures, 100 nm, were within the cytoplasm of neoplastic cells of 2 sarcomas out of 17 (Winqvist *et al.*, 1973), in the intercellular spaces of 1 case out of 10, and in the supernatant fluid of cultured tumor cells (Thompson, 1982). However, postmitochondrial fractions prepared by sucrose gradient centrifugation from neoplastic tissue of the northern pike contained characteristic type C retroviruses banding at 1.15 to 1.17 g/ml in sucrose (Papas *et al.*, 1976).

Electron microscopic examination of spontaneous neoplasms of the *bicolor damselfish neurofibromatosis-like disease* did not reveal an infectious agent (Schmale *et al.*, 1983). However, 100-nm, type C particles with a buoyant density of 1.14 to 1.18 g/ml in sucrose were observed within sucrose gradient fractions of media from tumor-derived cell cultures. They were tentatively named damselfish neurofibromatosis virus (DNFV) (Schmale *et al.*, 1993).

C. Reverse Transcriptase

In walleye dermal sarcoma, plasmacytoid leukemia of chinook salmon, esocid hematopoietic neoplasm, and epidermal papilloma/carcinoma of white sucker, tumor tissue homogenates were fractionated by sucrose gradient centrifugation and assayed for reverse transcriptase activity. This activity was detected in all four cases in the fractions banding in sucrose at a buoyant density of 1.18 g/ml for WDSV (Martineau *et al.*, 1991), 1.16 to 1.18 g/ml for SLV (Eaton and Kent, 1992), 1.15 to 1.17 g/ml in esocid hematopoietic neoplasm (Papas *et al.*, 1976; Sonstegard and Chen, 1986), and 1.15 to 1.16 g/ml in white sucker tumor (Sonstegard, 1977). That enzymatic activity was absent from similar preparations of comparable normal tissue. In the walleye, salmon, and northern pike tumors, the reverse transcriptase activity was associated with type C retroviral particles disclosed by electron microscopic analysis of the pooled fractions with high enzymatic activity and a density compatible with that of retroviruses (Papas *et al.*, 1976; Martineau *et al.*, 1991; Eaton and Kent, 1992). The fractions prepared from the epidermal tumors of white sucker were not examined by electron microscopy (Sonstegard, 1977).

Fractions of similar preparations of media from cultured cells derived from the neoplasms of the bicolor damselfish affected with neurofibromatosis-like disease exhibited reverse transcriptase activity and retroviral particles at a density of 1.14 to 1.18 g/ml in sucrose (Schmale *et al.*, 1993).

The esocid hematopoietic neoplasm-associated reverse transcriptase prefers the ribotemplates, poly(rA) · oligo(dT), poly(rC) · oligo(dG), to deoxytemplates (Papas *et al.*, 1976, 1977), uses Mn^{2+} as divalent cation, and has an isoelectric point of 5.5, similar to that of avian myeloblastosis virus reverse transcriptase. It is noteworthy that, unlike mammalian and avian viral enzymes, its optimal temperature is 20°C. Moreover, this fish virus reverse transcriptase retains 82% of its activity at 5°C, yet only 35% at 35°C (Papas *et al.*, 1977). The reverse transcriptase associated with walleye and chinook salmon neoplasms were assayed using Mn^{2+} as the divalent cation and poly(rA) · oligo(dT) as the synthetic ribotemplate (Martineau *et al.*, 1991; Eaton and Kent, 1992).

D. Transmission Studies

Experimental transmission of the disease has been attempted in six cases: walleye dermal sarcoma (Martineau *et al.*, 1990b), chinook salmon plasmacytoid leukemia (Kent and Dawe, 1990, 1993; Newbound and Kent, 1991; Eaton and Kent, 1992), damselfish neurofibromatosis-like disease (Schmale *et al.*, 1993), esocid hematopoietic neoplasm (Brown *et al.*, 1975; Sonstegard, 1975, 1976; Gross, 1983; Sonstegard and Chen, 1986), angelfish lip fibroma (Francis-Floyd *et al.*, 1993), and white sucker epidermal neoplasm (Sonstegard, 1977; Pilcher and Fryer, 1980).

Transmission of dermal sarcoma of walleye was successfully achieved by intramuscular inoculation of walleye fingerlings with 22-μm-filtered, acellular homogenate from sarcomas collected in the spring (Martineau *et al.*, 1990b). Four months after inoculation, an ill-demarcated, irregularly outlined, granular, elevated area appeared on the fingerlings, predominantly at the inoculation site (Fig. 2). These experimentally reproduced neoplasms were identical macroscopically, microscopically, and ultrastructurally to spontaneous tumors of adult fish. Moreover, characteristic type C retroviral particles were detected within the sarcomas of two out of seven tumor-bearing fingerlings by electron microscopy (Martineau *et al.*, 1990b). The optimal temperature for transmission of dermal sarcoma under laboratory conditions was determined to be 15°C (Bowser *et al.*, 1990).

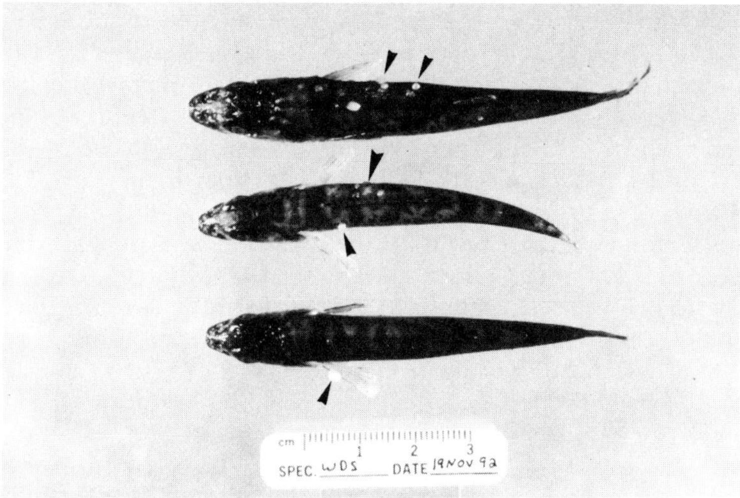

FIGURE 2. Experimental transmission of walleye dermal sarcoma. Walleye fingerlings injected with cell-free filtrate of dermal sarcoma. Note numerous, multifocal nodular dermal sarcomas (arrowheads) on each fish.

Interpretation of the initial transmission studies of chinook salmon plasmacytoid leukemia was complicated by common concurrent infection with *Renibacterium salmoninarum*, the agent of bacterial kidney disease, and/or *Enterocytozoon salmonis*, an intranuclear microsporidium, within the neoplastic cells (Hedrick *et al.*, 1990, 1991; Kent *et al.*, 1990; Morrison *et al.*, 1990). Unequivocal demonstration of a viral etiology has since been accomplished by experimental reproduction of the disease in chinook salmon with a cell-free, 0.22-μm-filtered, kidney and spleen homogenate in the absence of the two other putative causal agents (Kent and Dawe, 1993). Macroscopic, histologic, and hematologic lesions in the inoculated chinook salmon were consistent with plasmacytoid leukemia. Reverse transcriptase activity, associated with type C viral particles banding at 1.16 to 1.18 g/ml in sucrose gradients, was detected in tissues of recipient fish (Eaton and Kent, 1992). Interspecific transmission of the condition from chinook salmon to sockeye salmon, *Oncorhynchus nerka*, Atlantic salmon, *Salmo salar*, and coho salmon, *Oncorhynchus kisutch*, was successful, although the latter was more resistant. The rainbow trout, *Oncorhynchus mykiss*, developed only mild lesions. These mild lesions were, nevertheless, suggestive of plasmacytoid leukemia (Kent and Dawe, 1990; Newbound and Kent, 1991).

Lesions of damselfish neurofibromatosis could be reproduced in both healthy juvenile and adult test homologous fish by inoculation of acellular, 0.22-μm tumor tissue filtrate as well as by injection of tumor cells subcultured *in vitro* (Schmale and Hensley, 1988; Schmale *et al.*, 1993). The induced tumors, identical macroscopically and microscopically to the spontaneous neoplasms, arose along peripheral nerves at the injection site (Schmale and Hensley, 1988).

Esocid hematopoietic neoplasms were transmitted experimentally to healthy homologous fish. Tumors were induced on unaffected muskellunge and pike by natural contact with tumor-bearing fish within the same aquaria (Ljungberg, 1976; Sonstegard and Chen, 1986) and by rubbing tumor tissue onto abraded skin (Sonstegard and Chen, 1986). The neoplasms are transplantable to adult muskellunge and pike and pike fingerlings by inoculation of viable tumor cells (Brown *et al.*, 1975; Sonstegard, 1975, 1976; Ljungberg, 1976; Sonstegard and Chen, 1986). Definite support for a viral etiology was provided by successful cell-free transmission trials with 0.22-μm-filtered tumor homogenates (Mulcahy *et al.*, 1970; Mulcahy and O'Leary, 1970; Brown *et al.*, 1975; Sonstegard, 1975, 1976). However, the outcome of these cell-free transmission experiments varied as material collected during the spring resulted in successful trials, whereas material from late summer yielded negative results. This variation was attributed to seasonal changes in the amount of infectious viral particles in the tumor.

Transfection assays with DNA from northern pike neoplasm into murine (NIH3T3) cells resulted in foci of transformed cells, able to induce tumors in nude mice. Southern blot hybridization analysis of DNA from transfected cells with probes for several oncogenes revealed no homology (Van Beneden *et al.*, 1990). The relationship between the type C retrovirus present in the tumor and the transformed foci has not been addressed.

These transmission experiments using acellular 0.22-μm filtrate from tumor tissue homogenate support the hypothesis that these neoplasms are infectious and have a viral etiology even though the etiological significance of these tumor-associated agents remains unproven. It is only by inducing the typical pathologic changes using the isolated agent or its molecular clones, in the absence of confounding factors, that the etiology will be demonstrated.

Angelfish lip fibromas could not be transmitted to healthy juvenile homologous fish using a 45-μm-filtered, acellular tumor homogenate (Francis-Floyd *et al.*, 1993). Inoculation of viable cells or acellular tumor tissue extracts to test fish failed to induce epidermal papilloma/carcinoma of white suckers. Neither did white sucker neoplasms develop on test fish after contact with affected fish in the same aquaria or after rubbing tumor tissue onto abraded skin (Sonstegard, 1977).

Thus, successful cell-free transmission of a tumor has been documented with four fish viruses (Table II). In the two other cases, further experiments with more concentrated inocula might prove successful.

E. Virus Purification

Four of these presumptive causal agents were partially purified. In all cases they were prepared from particulate fractions and found to band in sucrose gradients at a density compatible with that of retroviruses. The fractions were prepared from tumor tissue homogenates from tumor-bearing fish (Papas *et al.*, 1977; Martineau *et al.*, 1991; Eaton and Kent, 1992) for three retroviruses, WDSV, SLV, and esocid hematopoietic neoplasm-associated virus, and from culture medium from tumor-derived cells cultivated *in vitro* for DNFV (Schmale *et al.*, 1993).

F. Analysis of Viral Proteins

Polyacrylamide gel electrophoresis analysis (PAGE) of purified virus particles from chinook salmon plasmacytoid leukemia disclosed seven polypeptides of apparent molecular mass of 120, 80, 42, 27, 25, 22, and 19 kDa (Eaton and Kent, 1992). Because of lack of cell culture systems and defined antisera, the identity and precursor relationship of these various polypeptides remain unidentified.

G. Analysis of Viral Nucleic Acids

WDSV is thought to be implicated in the genesis of dermal sarcomas of walleye (Martineau *et al.*, 1991). Viral RNA was isolated from sucrose-banded virions, which were purified by sucrose density gradient centrifugation of tumor tissue homogenate, and used to synthesize a cDNA probe. On Southern blot analysis of dermal sarcoma DNA, this cDNA probe recognized an unintegrated, linear, 13-kb viral DNA homologous species in undigested DNA, and two fragments, 5.9 and 6.1 kb, in *Eco*RI-digested DNA (Martineau *et al.*, 1991). No homologous sequences were found in the genomic liver DNA from five normal walleye fingerlings or affected and unaffected adult walleyes and carp. Similarly, the sequences were not detected in the sperm DNA of one unaffected walleye, in bovine lymphoma DNA, and in DNA from several cell lines, including channel catfish ovary (CCO), fathead minnow (FHM), and canine cell lines (Cf2Th) (Fig. 3) (Martineau *et al.*, 1991, 1992).

The WDSV DNA has been cloned and partially characterized. With regard to length, the WDSV genome, 13.2 kb, is closer to that of the Spumavirinae (Loh, 1993) than that of other retroviral subgroups and is the largest one among the

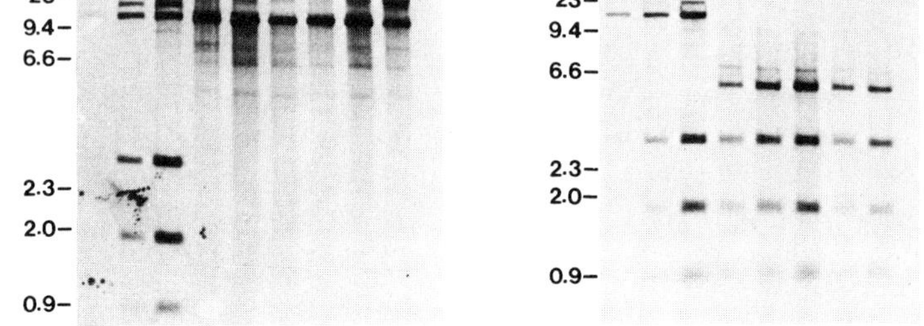

FIGURE 3. Structure of the recombinant clone WDSV-1 and Southern blot analysis of WDSV in tumors. (a) Restriction map of recombinant phage clone (WDSV-1). (b) Southern blot analysis of undigested tumor DNAs. Copy number standards of *Bam*HI-digested WDSV-1 (1, 10, and 50 copies per cell; lanes 1, 2, and 3, respectively) were used. Lanes contain undigested DNA from six different tumors (lanes 4 to 9) and sperm DNA from unaffected walleye (lane 10). (c) Southern blot analysis of *Bam*HI-digested tumor DNAs (lanes 1 to 3, same as for panel b; lanes 4 to 8, same as for panel b except digested with *Bam*HI; lane 9, sperm DNA digested with *Bam*HI). (Reprinted with permission of the *Journal of Virology*.)

Retroviridae. The unintegrated viral DNA has a single-stranded gap region, similar to that found in the unintegrated genome of some lentiviruses and all spumaviruses. In these cases, the gap region corresponds to a duplication of the 3′ polypurine tract, postulated to be an additional primer site for plus-strand DNA synthesis. Within the limits of Southern blot analysis, it was demonstrated that a 13.2-kb, unintegrated, viral DNA species is present in 90% of all dermal sarcomas examined (Martineau *et al.*, 1992). Comparison with copy number standards estimates the number of viral copies in tumors to be 7 to 50 copies per cell assuming that all cells within a tumor are similarly infected (Martineau *et al.*, 1992).

The transcriptional complexity of WDSV was analyzed by Northern blot analysis of polyadenylated mRNA extracted from one individual tumor and from ten pooled neoplasms (Martineau *et al.*, 1992). The viral genome encodes at least four transcripts: a 13-kb transcript, the full-length genome, one of 7.4 kb, presumably the spliced envelope, and at least two smaller transcripts, 2.8 and 1.8 kb, which could possibly encode accessory regulatory proteins. No viral transcripts were detected in liver RNA of a tumor-bearing adult walleye. In conclusion, the sarcomas are the only tissues of tumor-bearing adult walleyes that contain large numbers of viral DNA copies and within which at least four discrete viral transcripts are expressed.

III. DISEASES ASSOCIATED WITH FISH RETROVIRUSES

A. Introduction

Retroviruses are suspected of being etiologically associated with some fish hemic and nonhemic infectious neoplasms/hyperplastic disorders, although causality has not been firmly established for any of these cases. Biological characteristics of retrovirus-associated proliferative diseases in fish are summarized in Table III. Hematopoietic neoplasms of northern pike, muskellunge, and chinook salmon are malignant fatal disorders (Sonstegard and Chen, 1986; Kent *et al.*, 1990). Healed cutaneous lesions are sometimes observed on pike and muskellunge (Sonstegard, 1976; Sonstegard and Chen, 1986). The disorder in muskellunge has a fulminating clinical course and results in the death of 99% of affected fish in 1 year. It is assumed that the evolution of the same disorder in the North American northern pike parallels roughly that of the muskellunge, but is distinguished by a lower mortality rate and frequent apparent regression of cutaneous lesions based on field observations (Sonstegard, 1975, 1976; Sonstegard and Chen, 1986).

Sarcoma of the swim bladder of the Atlantic salmon has a fatal outcome (McKnight, 1978). Neurofibromatosis-like disease of bicolor damselfish is an inexorably progressive, irreversible syndrome with a duration of 1 year (Schmale *et al.*, 1986).

Two neoplasms, cutaneous fibroma/fibrosarcoma of hooknose and epidermal papilloma/carcinoma of white sucker, may occasionally display histological characteristics of malignancy, such as local invasion of adjacent tissues (Smith *et al.*, 1989b; Hayes *et al.*, 1990; Anders *et al.*, 1991), but have never been reported

TABLE III. Biological Characteristics of Retrovirus-Associated Proliferative Diseases in Fish[a]

Disease	Clinical course	Age	Sex	Seasonality	Regression	Cell type associated with retroviral particles
Cutaneous fibroma/fibrosarcoma of hooknose	Locally invasive	Unknown	Unknown	Unknown	Unknown	Lymphocyte
Epidermal papilloma/carcinoma of sucker	Locally invasive	Adults	Unknown	Absent	Demonstrated	Neoplastic Malpighian epithelial cells[b]
Esocid hematopoietic neoplasm	Malignant	Adults, inducible in fingerlings	M = F	Documented in muskellunge, observed in northern pike	Rare	Neoplastic histiocyte
Smooth type of esocid epidermal hyperplasia	Benign	Adults	M = F	Observed	Unknown	Proliferative Malpighian epithelial cell
Plasmacytoid leukemia of chinook salmon	Fatal	Adults, fingerlings	Unknown	Absent	Absent	Neoplastic plasma cells
Spawning papillomatosis of European smelt	Benign	Adults	M = F	Documented	Shedding	Neoplastic epithelial cell of unknown nature
Damselfish neurofibromatosis-like disease	Malignant	Adults, inducible in juveniles	M = F	Absent	Absent	Tumor-derived cell cultures
Epidermal papillomatosis of salmon	Benign	Juveniles	M = F	Documented	Shedding	Neoplastic Malpighian epithelial cell
Swim bladder sarcoma of salmon	Fatal	Ripening fish	M = F	Absent	Absent	Neoplastic fibroblastlike cell
Dermal sarcoma of walleye	Benign	Spontaneous in adults, inducible in fingerlings	M = F	Documented	Demonstrated	Neoplastic fibroblast
Discrete epidermal hyperplasia of walleye	Benign	Adults	Unknown	Observed	Unknown	Proliferative Malpighian epithelial cell
Lip fibroma of angel fish	Benign	Adult females, juveniles	F?	Unknown	Unknown	Neoplastic fibroblast

[a] References are cited in the text.
[b] Controversial finding.

to metastasize. Metastases occur infrequently in fish neoplasms, for unexplained reasons (Ferguson, 1989).

Dermal neoplasms of walleye were termed "sarcomas" based on histological criteria, such as anaplasia and high mitotic index, but they exhibit a benign behavior. Neither local invasion nor metastases have been observed (Walker, 1969; Yamamoto *et al.*, 1976; Martineau *et al.*, 1990a). Lip fibroma of angelfish is a benign, localized neoplasm (Francis-Floyd *et al.*, 1993).

Similarly, some epidermal neoplasms—epidermal papillomatosis of Atlantic salmon (Bylund *et al.*, 1980), spawning papillomatosis of European smelt (Anders, 1989)—and hyperplastic lesions—discrete epidermal hyperplasia of walleye (Walker, 1969), smooth type of epidermal hyperplasia of northern pike and muskellunge (Winqvist *et al.*, 1968; Sonstegard, 1976; Yamamoto *et al.*, 1983)—usually follow a benign clinical course.

B. Seasonality and Regression

A fundamental peculiarity of some neoplastic diseases of poikilotherm animals is their proclivity to regress seasonally. In amphibians, Lucké renal adenocarcinoma of the leopard frog, *Rana pipiens* (Rafferty, 1964; Lucké, 1938), and skin papillomas of the Japanese newt, *Cynops pyrrhogaster* (Asashima *et al.*, 1982), are well-known examples of seasonal regression of neoplasms.

Only six cases of neoplastic/proliferative diseases associated with retroviruses are known to fluctuate seasonally: dermal sarcoma (Bowser *et al.*, 1988) and discrete epidermal hyperplasia of walleye (Walker, 1969; Yamamoto *et al.*, 1985a), spawning papillomatosis of European smelt (Anders and Möller, 1985), epidermal papilloma of Atlantic salmon (Bylund *et al.*, 1980; Pilcher and Fryer, 1980), smooth type of epidermal hyperplasia (Winqvist *et al.*, 1968; Sonstegard, 1976; Yamamoto *et al.*, 1983), and hematopoietic neoplasm of northern pike and muskellunge (Sonstegard, 1976; Pilcher and Fryer, 1980; Sonstegard and Chen, 1986). Actual tumor regression was documented in dermal sarcoma of walleye (Fig. 4) (Bowser and Wooster, 1991) and rare cases of cutaneous hematopoietic neoplasm of muskellunge and northern pike (Sonstegard, 1976; Sonstegard and Chen, 1986). Seasonal shedding of the lesion occurs in spawning papillomatosis of European smelt (Anders and Möller, 1985) and epidermal papilloma of Atlantic salmon (Bylund *et al.*, 1980). Monitoring of white suckers in aquaria has shown that some epidermal neoplasms disappear independently of the season and temperature, although it is unknown whether these lesions slough off or regress (Smith and Zajdlik, 1987; Hayes *et al.*, 1990). The fate of lip fibroma of angelfish and cutaneous fibroma/fibrosarcoma of hooknose is unknown.

The cause of these seasonal variations is most likely multifactorial. They have been attributed in some cases to the host hormonal and immunologic variations, related to the stress of spawning or metamorphosis and subsequent decrease of immune protection. Low temperature alone contributes to immune depression in fish by diminishing cellular immunity and suppressing anamnestic response (Cuchens and Clem, 1977; Rijkers, 1982; Miller and Clem, 1984). In other cases, these variations have been tentatively explained by changes within the replicative cycle of the virus, suggesting a permissive temperature for viral

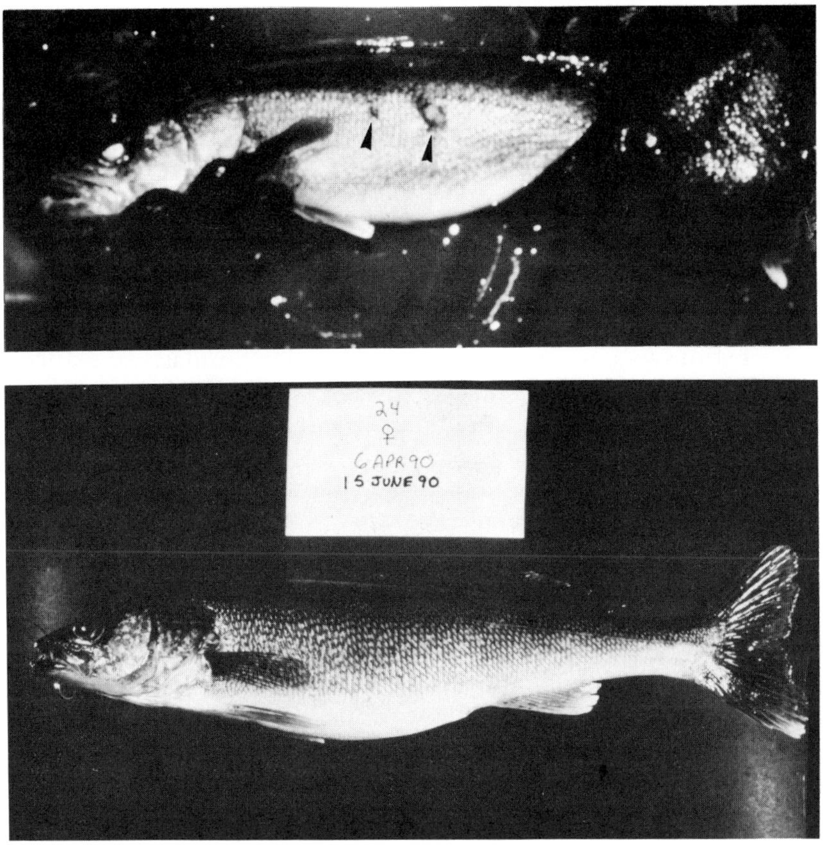

FIGURE 4. Regression of dermal sacoma. (Top) Adult walleye collected in the spring. Two sarcomas are present on the trunk (arrowheads). (Bottom) Same fish in the summer. Both tumors have disappeared.

production (Papas *et al.*, 1976). Seasonality of some infectious diseases and neoplasms in poikilotherms probably depends on the complex interaction between the host immune system and the biology of the agent. Both of these parameters can be affected by seasonal and temperature-related fluctuations.

C. *Agonus cataphractus*–Cutaneous Fibroma/Fibrosarcoma of Hooknose

Previously unreported cutaneous neoplasms of hooknose from the German Wadden Sea are described (Anders *et al.*, 1991) as 3 to 9 mm in diameter, yellow or red-black, well-circumscribed, raised nodules on the skin. Microscopically, the tumors are nonencapsulated and composed of whorling fusiform fibroblasts with interlacing bundles of collagen fibers. Local invasion of the adjacent skeletal muscle was seen in one case. The yellowish tumors differ histologically from the red-colored ones by dense inflammatory infiltrates of lymphocytes, granulocytes, and multinucleated giant cells.

D. *Catostomus commersoni*–Epidermal Papilloma/Carcinoma of White Sucker

Epizootics of epidermal tumors affect the sexually mature North American white sucker in both pristine waters and polluted sites. Higher prevalence of fish with neoplasms and predominance of lip lesions are observed in polluted areas (Sonstegard, 1977). This trend had suggested a possible etiologic role of environmental contaminants. Chemical factors, however, were no longer considered as prerequisites for the development of these neoplasms after Smith and Zajdlik (1987) demonstrated in the laboratory that in a clean environment some epidermal tumors of white suckers regressed, others persisted, and new ones developed.

Three types of epidermal lesions are distinguished grossly by their anatomic location and morphology: one is the epidermal papilloma of the lips, and the two other distinctive types on the body and fins are the discrete, markedly raised focal multinodular papilloma (Fig. 5) and the large, slightly elevated plaque (Smith and Zajdlik, 1987; Smith *et al.*, 1989a,b; Hayes *et al.*, 1990). Lip lesions on both lower and upper lips are frequently facing each other, suggesting that transplantation or infection occurs by direct contact (Smith *et al.*, 1989b). Discrete papillomas arise not uncommonly within a plaque.

All three types are histologically similar (Fig. 6). Localized proliferation of moderately anaplastic Malpighian cells causes a focal thickening of epidermal layers that is sharply demarcated from the normal epidermis by displacement and local hyperplasia of mucous and alarm cells at the margins (Smith *et al.*, 1989a,b). Prominent rete pegs extend interdigitations into the underlying dermis. Diagnosis of rare carcinomas is based on rupture of the basement membrane and local invasion. Metastases are not reported. Morphologic examination of different stages of these epithelial alterations suggests that a gradual progression from hyperplasia to papilloma occurs, and may lead infrequently to squamous cell carcinoma (Smith and Zajdlik, 1987; Smith *et al.*, 1989b; Hayes *et al.*, 1990).

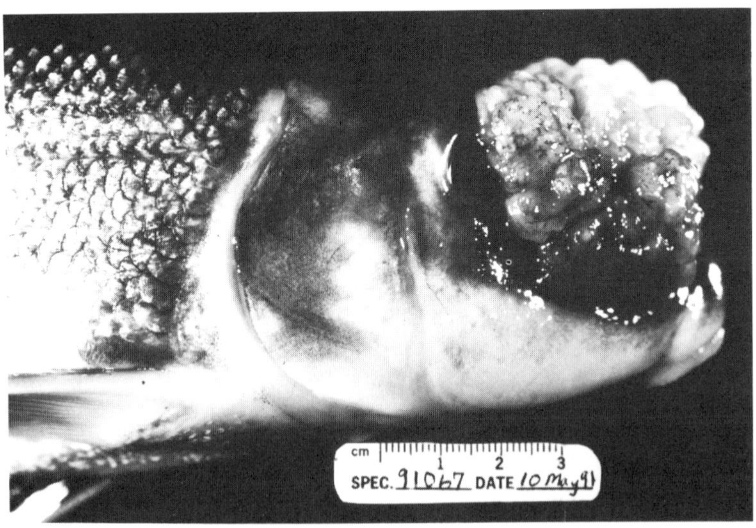

FIGURE 5. White sucker with large papilloma/carcinoma anterior to the eye.

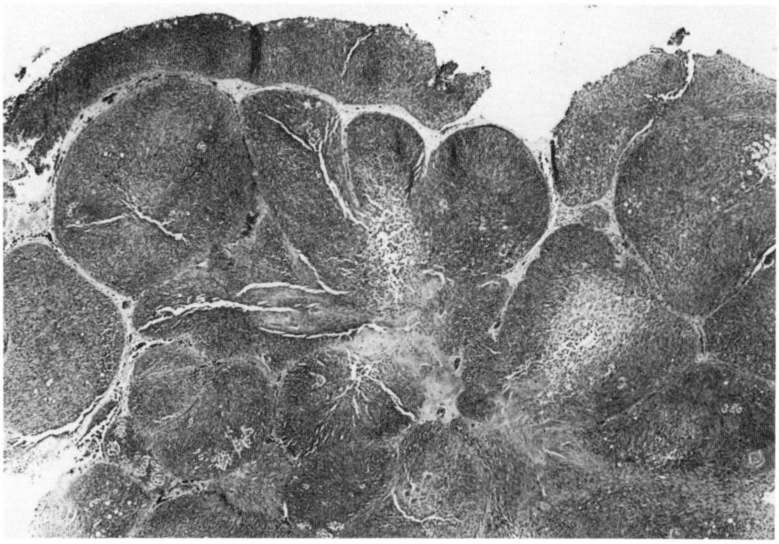

FIGURE 6. White sucker papilloma. This sessile, endophytic neoplasm is well demarcated and forms multiple arborizing nodules of neoplastic epithelial cells within the dermis. The disruption of the overlying normal epidermis in this section is artifactual. HE. ×35.

E. *Esox lucius* and *E. masquinongy*–Hematopoietic Neoplasm

Hematopoietic tumors in fish of the genus *Esox*, northern pike and muskellunge, have been known in North America and Europe for almost one century (Sonstegard and Chen, 1986). The first report of lymphosarcoma in northern pike dates from 1898 (Ohlmacher, 1898). In 1909, Higgins and Hadwen diagnosed a similar neoplasm in muskellunge in Canada (Sonstegard, 1976; Sonstegard and Chen, 1986).

Distinct epizootics of this neoplasm occurred in feral northern pike populations in widely separated geographic areas in North America [Alaska, Alberta, Manitoba, Michigan, New York (Nigrelli, 1943, 1947), Northwest Territories, Ontario, Quebec (Sonstegard, 1976; Sonstegard and Chen, 1986), Wisconsin (Ohlmacher, 1898)] and Europe [France (Sonstegard and Chen, 1986), Finland (Thompson, 1982; Thompson and Hildén, 1988), the Netherlands (Sonstegard and Chen, 1986), Ireland (Mulcahy, 1963), presumably Scotland (Haddow and Blake, 1933), and Sweden (Ljungberg and Lange, 1968; Ljungberg, 1976)]. In contrast, epizootics in muskellunge were observed only in North America (Sonstegard and Chen, 1986). The frequency of esocid hematopoietic neoplasm is the highest recorded for a malignant neoplasm in any species of wild vertebrates: 20.9% of the northern pike population of Michigan, USA, and 16% of the muskellunge population in Ontario, Canada, are affected (Sonstegard, 1975, 1976; Sonstegard and Chen, 1986). Pollution was correlated with more aggressive tumorigenesis (Brown *et al.*, 1975), but since epizootics were also recorded in unpolluted areas (Mulcahy *et al.*, 1970), it is unlikely that chemical factors play a crucial etiologic role.

Since neoplastic conditions from the various geographic regions share biological and epizootiologic similarities, they presumably belong to a single pathological entity. This entity was designated "undifferentiated sarcoma," "lymphoma," "lymphosarcoma," "lymphoblastic lymphoma," and "stem cell lymphoma" based on light microscopic morphologic criteria (Thompson, 1982).

Affected pikes are usually sexually mature. They are caught in salt water in Finland (Thompson, 1982), brackish water in Sweden (Ljungberg and Lange, 1968), and fresh water in North America and Ireland (Mulcahy, 1963, 1976; Thompson, 1982). A sex predilection, 2.5:1 males to females, is noted by Finnish investigators on samplings conducted throughout the year (Thompson, 1982; Thompson and Hildén, 1988). Esocid hematopoietic neoplasm appears to be a lesion of primarily cutaneous origin. Two broad categories of neoplastic conditions, with and without visceral involvement, can be distinguished: in the first group, the North American and Irish pike, extensive metastases occur consistently, whereas in the second group, the Swedish, Scottish, and Finnish cases, the lesions remain confined to the skin.

Anatomically, hematopoietic neoplasm of the North American pike originates almost always as a solitary cutaneous mass on the posterior half of the body (83% of the cases), the fins, flanks, and head (1% of the cases) in order of decreasing frequency (Sonstegard, 1975, 1976; Sonstegard and Chen, 1986). Later, metastases spread to the kidney, spleen, liver, abdominal viscera, and serosa. Cases without cutaneous lesions, in which the primary tumor arises in the kidney, are rare (Nigrelli, 1952; Schlumberger, 1957; Sonstegard, 1976).

In the Irish pike the cutaneous lesions are multiple and involve most frequently the jaws and oral cavity (60% of the cases). Metastases spread to the gills, stomach, kidney, liver, and pancreas (Mulcahy, 1963; Dawe, 1969; Mulcahy *et al.*, 1970; Harshbarger and Dawe, 1973).

In contrast to that extensive visceral involvement, the tumors are restricted to the skin in the Swedish, Scottish and Finnish cases. They arise multicentrically on the posterior aspect of the trunk (Haddow and Blake, 1933; Ljungberg, 1963; Ljungberg and Lange, 1968; Thompson, 1982; Thompson and Kostiala, 1986). The mouth is occasionally involved in the Finnish pike (Thompson, 1982).

The distribution of lesions in the muskellunge resembles that in the North American pike, except that the primary tumor develops always on the skin and prefers the flanks to the fins (Sonstegard, 1976; Sonstegard and Chen, 1986). Leukemia occurs terminally in the muskellunge (Dawe, 1969; Sonstegard, 1975; Sonstegard and Chen, 1986) and the North American (Sonstegard, 1975; Whang-Peng *et al.*, 1976) and Finnish pike (Thompson and Hildén, 1988).

The cutaneous neoplasms consist of round, 0.5 to 5 cm in diameter, soft, bulging, ulcerated white masses variegated with petechial hemorrhages and variable amounts of necrotic tissue (Fig. 7). The tumors are not encapsulated and infiltrate the subjacent skeletal musculature. Affected spleen, kidney, and liver are most often soft and diffusely enlarged. Some reports mention variably-sized metastatic nodules disseminated on the serosa and within the parenchyma of abdominal organs (Ohlmacher, 1898; Nigrelli, 1943, 1947; Sonstegard, 1976).

Histologically, the neoplasms are composed of a homogeneous population of closely apposed, variably sized, round to oval, undifferentiated mononuclear cells. These cells have a high nucleus-to-cytoplasm ratio and are arranged into irregular alveoli, rows aligned along fibers, or in uniform sheets on a delicate, moderately abundant fibrillar stroma (Ljungberg and Lange, 1968; Mulcahy *et al.*, 1970; Winqvist *et al.*, 1973; Sonstegard, 1976; Thompson, 1982). The mitotic

FIGURE 7. Northern pike with hematopoietic neoplasm ventral to the dorsal fin. Reprinted with permission of Iowa State University Press. Kindly provided by Dr. H. W. Ferguson.

rate is variable depending on the tumor or area examined. Necrosis of single cells or of large areas is frequently observed. The neoplasms are nonencapsulated and highly infiltrative. Clusters of tumor cells detach from the irregularly outlined margins and finger into the adjacent normal tissue. The neoplastic mononuclear cells, characterized by indented vesicular nuclei containing one prominent nucleolus, and scant amounts of basophilic cytoplasm, have been tentatively identified by light microscopy as lymphoblasts or stem cells, hence the neoplasm designation.

Finnish investigators have unraveled the cell type of origin of this neoplasm by immunohistochemical, biochemical, and ultrastructural examination of tumors. Although a previous report mentioned that the tumor cells adopt in culture morphologic features of plasma cell differentiation (Thompson, 1982), subsequent studies suggested that they are derived from the histiomonocytic lineage. The evidence supporting this claim is: expression of the intermediate filament protein vimentin (Thompson *et al.*, 1987b), diffuse sodium fluoride-sensitive staining for α-naphthyl acetate esterase (Thompson *et al.*, 1987a; Thompson and Kostiala, 1988), absence of cytoplasmic and surface immunoglobulins (Thompson *et al.*, 1987a; Thompson and Kostiala, 1986, 1988), and ultrastructural characteristics of mononuclear cells. Specifically, these latter characteristics are lack of intercellular junctions and basal lamina, presence of cytoplasmic groups of lysosomes, multitudes of lipid droplets, and scant amounts of rough endoplasmic reticulum (Thompson, 1982; Thompson and Miettinen, 1988). Consequently, based on the tumor's supposed origin from a cutaneous histiocyte, the diagnosis of "histiocytic lymphoma" was proposed to reclassify more appropriately this esocid hematopoietic neoplasm (Thompson and Miettinen, 1988). In other geographic areas, partial identification of the cell type precludes a definite diagnosis. However, similar ultrastructural findings in Swedish and Irish cases (Mulcahy *et al.*, 1970; Winqvist *et al.*, 1973) and positive acid phosphatase staining in Irish cases (Mulcahy *et al.*, 1970) support the possibility of histiomonocytic derivation for these tumors. These studies point toward a single pathological condition with protean manifestations.

Two unusual intracytoplasmic structures, the cylindroid lamella–particle complex, which is reminiscent of the lamella–ribosome complexes found in several types of leukemia and lymphoma in human beings, and the nucleoid body, are described within the neoplastic cells of one North American pike (Dawe *et al.*, 1976). The first structure has also been noted in Irish pike neoplastic cells (Dawe *et al.*, 1976).

It is noteworthy that esocid histiocytic lymphoma is one of the few neoplasms to possess a pathognomic karyotypic signature: minute, sub- and metacentric chromosomes are consistently found in the neoplastic cells of North American and Swedish pike (Ljungberg, 1976; Whang-Peng *et al.*, 1976).

Field and laboratory studies have demonstrated a regular seasonal periodicity of the disease in muskellunge (Sonstegard, 1976; Sonstegard and Chen, 1986). The lowest incidence (0.5%) of fish with neoplasms is seen in midsummer, whereas 10% of fish exhibit early skin lesions in the fall. In early spring, affected fish show more advanced skin lesions and visceral involvement and by late spring about 99% of these fish succumb. Spontaneous regression is seldom observed in muskellunge (Sonstegard, 1975; Sonstegard and Chen, 1986). The clinical course of the disease in northern pike remains unresolved and seasonal periodicity is unclear. Seasonal variations were noted in Swedish (Ljungberg and Lange, 1968) and North American field cases (Sonstegard, 1976), but the distinction between irregular variations and bona fide seasonal pattern could not be made.

F. *Esox lucius* and *E. masquinongy*–Smooth Type of Epidermal Hyperplasia

Two types of epidermal hyperplasia are observed in male and female northern pike and muskellunge from Europe (Winqvist *et al.*, 1968) and North America (Sonstegard, 1976; Yamamoto *et al.*, 1983; Sonstegard and Chen, 1986). The first type of lesion, termed "blue spot disease," consists of granular, pale bluish, opaque plaques of epidermal "giant" hypertrophied cells and is associated with esocid herpesvirus 1 (Yamamoto *et al.*, 1983). The second type of epidermal hyperplasia, which is designated the "smooth type" by opposition to the granular appearance of the herpesvirus-associated lesion, is manifested as multiple, 2- to 3-cm diameter, slightly raised (about 2 mm above the normal skin surface), smooth, translucent whitish plaques on the skin of the body and fins (Winqvist *et al.*, 1968; Yamamoto *et al.*, 1983). They appear comparable to local areas of excessive mucus production. In some cases the plaques are hemorrhagic (Yamamoto *et al.*, 1983). No internal lesions have been detected macroscopically in affected fish.

Microscopically, the smooth type of epidermal hyperplasia is a sharply demarcated, localized proliferation of randomly oriented, undifferentiated, polygonal epithelial cells. Normal stratification of the epidermis is absent. Specialized epidermal cells, such as mucous cells, congregate in a transition zone at the periphery of the lesion (Winqvist *et al.*, 1968; Yamamoto *et al.*, 1983).

G. *Oncorhynchus tshawytscha*–Plasmacytoid Leukemia

An apparently new fatal systemic disease, initially referred to as "marine anemia," emerged in 1987 in cultured chinook salmon and caused high morbidity. Up to 50% mortality was observed in market-size fish in seawater facilities in

British Columbia, Canada, and in 4-month-old freshwater fingerlings in Washington State and California (Harshbarger, 1984; Hedrick *et al.*, 1990; Kent *et al.*, 1990; Morrison *et al.*, 1990). The most prominent clinical sign in fingerlings is gill pallor, whereas adult fish are anemic, lethargic, and darker than normal. Abdominal swelling is frequent and bilateral exophthalmos is noted in 20% of the cases (Kent *et al.*, 1990). On necropsy, there are consistent severe reno- and splenomegaly (Fig. 8), moderate amounts of serous to serosanguinous ascites, thickening of the intestinal wall, disseminated petechial hemorrhages on the viscera, heart, and skeletal muscles, and enlarged and reddened retrobulbar tissue.

Microscopically, the disease is characterized by severe systemic proliferation of plasmacytoid leukemic cells, hence its designation as "plasmacytoid leukemia." Massive numbers of mitotically active, blastlike leukemic cells form dense cellular infiltrates particularly in hematopoietic organs, such as the kidney and spleen, and in the heart sinuses, brain, leptomeninges, gills, skin, retrobulbar tissue, lamina propria of the intestine, pancreas, and liver sinusoids (Figs. 9 and 10). Numerous leukemic cells obliterate small blood vessel lumina and disrupt their endothelium. No mature leukocytes are present. The neoplastic cells are pleomorphic, have an increased nucleus-to-cytoplasm ratio, and exhibit features reminiscent of those of plasma cells, namely, large, lobate to clefted nuclei with one or two nucleoli, moderate amounts of amphophilic cytoplasm, and an occasional juxtanuclear Golgi zone (Kent *et al.*, 1990). The plasma cell lineage of the cell of origin is substantiated by immunohistochemical demonstration of cytoplasmic immunoglobulins and ultrastructural characteristics, such as distended cisternae of rough endoplasmic reticulum containing finely granular material, prominent Golgi apparatus, and eccentric nucleus with heterochromatin clumped along the nuclear membrane (Kent *et al.*, 1990).

H. *Osmerus eperlanus*–Spawning Papillomatosis of European Smelt

Outbreaks of benign epidermal neoplasms have been observed in up to 5.5% of sexually mature male and female European smelt since 1916 (Breslauer, cited

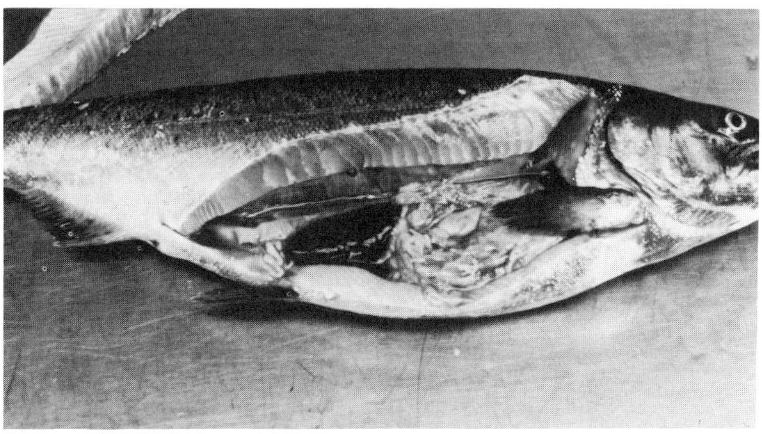

FIGURE 8. Chinook salmon collected during a field epizootic of plasmacytoid leukemia in cage-cultured fish. Note enlarged spleen and kidney. Kindly provided by M. L. Kent.

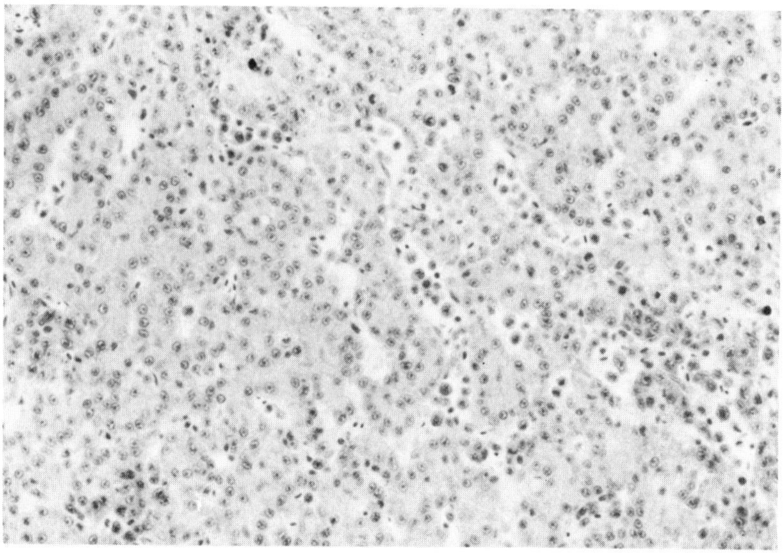

FIGURE 9. Photomicrograph of plasmacytoid leukemia in chinook salmon. Abundant leukemic cells are present within liver sinusoids. Kindly provided by M. L. Kent.

in Schlumberger, 1957). Since the neoplasms occur exclusively during the spring spawning season, this condition has been named "spawning papillomatosis" (Anders and Möller, 1985; Anders, 1989). Two types of smooth, whitish epidermal lesions are identified macroscopically: raised, focal tumors on the fins and head, and large, flat lesions covering extensive areas of the trunk.

Microscopically, within the dermis, the tumors form characteristic, nonencapsulated, well-demarcated, nodular proliferations of epithelial neoplastic cells

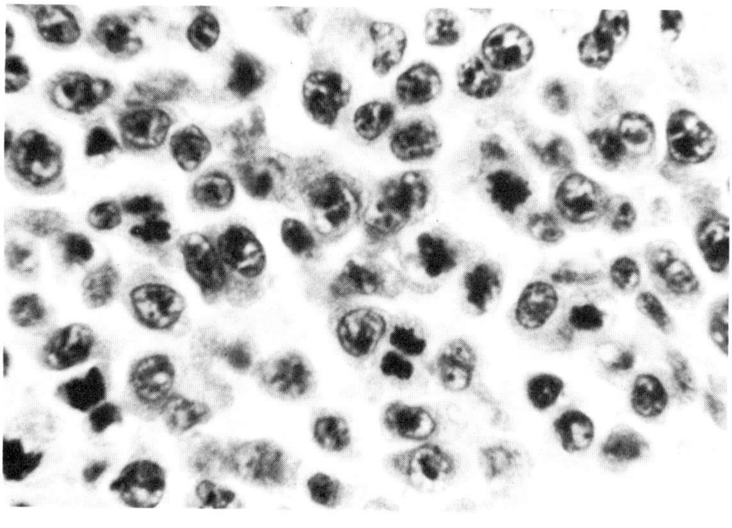

FIGURE 10. Plasmacytoid leukemia in chinook salmon. High magnification of the plasmacytoid leukemic cells within the pancreas. The neoplastic cells have a high nucleus-to-cytoplasm ratio with lobate to clefted nuclei. Numerous mitotic figures are present. Kindly provided by M. L. Kent.

abutting the epidermis. The cell of origin is unknown. Conspicuous Cowdry type A viral inclusion bodies, which represent most likely herpesvirus inclusions, are present in the nuclei of many neoplastic cells (Anders and Möller, 1985; Anders, 1989). In the course of the disease, the neoplastic nodules progress toward the cutaneous surface. In that process, they become gradually encircled by the cutaneous epithelium, which forms an epithelial "capsule" by which they remain attached to the fish body as their basal surface on the dermal side diminishes. At the end of the spawning season, the encircling epithelial layers rupture and the neoplasms detach, leaving ulcerated areas. One case of malignancy has been noted (Anders, 1989).

I. *Pomacentrus partitus*–Neurofibromatosis-like Disease of the Bicolor Damselfish

A neoplastic condition of the bicolor damselfish from southern Florida reefs is designated damselfish neurofibromatosis (DNF) because of several analogies with von Recklinghausen neurofibromatosis, such as multiple neurofibromas, malignant schwannomas, and hyperpigmented epidermal lesions (Schmale *et al.*, 1983, 1986). The incidence of DNF ranges from 0.1 to 23% in various clustered populations in the reefs (Schmale *et al.*, 1986). DNF is a fatal progressive disease manifested in sexually mature fish as multicentric, disseminated Schwann cell neoplasms that originate along peripheral nerves of the skin on the body and fins and in internal organs (Schmale *et al.*, 1983, 1986). The central nervous system is not involved.

Grossly, small, well-delimited, hyperpigmented, cutaneous spots are detected at the first stage of the disease. Progressively, these spots both enlarge to form discrete nodules and increase in number. At the final stage, these multiple nodules coalesce into large, rugose, raised plaquelike areas, which can cover the entire body surface in the most severe cases. Typically, the cutaneous tumors are hyperpigmented while neoplasms from other sites are unpigmented. Both pigmented and unpigmented lesions are highly invasive and destroy the adjacent tissues. Cutaneous neoplasms infiltrate bone and skeletal muscle. Visceral tumors replace the normal architecture of the organ.

Microscopically, early lesions are characterized by gradual irregular thickening of the diameter of peripheral nerves as a result of proliferation of Schwann cells, disorientation of axons, and increased amount of collagen and myelin (Schmale *et al.*, 1983, 1986). The neoplasms are nonencapsulated and form whorls and interlacing bundles of elongate, spindle-shaped cells with palisading parallel nuclei reminiscent of Antoni type A arrangement, intermixed with areas of more rounded, plump, pleomorphic cells.

J. *Salmo salar*–Epidermal Papillomatosis of Atlantic Salmon

Epizootics of a benign, self-limiting disease, epidermal papillomatosis, have been reported since 1968 in cultured and wild Atlantic salmon in fresh and salt water in Finland, Great Britain, Norway, Scotland, Sweden, and the United States (Carlisle and Roberts, 1977; Bylund *et al.*, 1980). During the most serious epizootic, 80% of the fish were affected in some tanks. Males and females are equally susceptible. Age groups display marked differences in incidence (50% in parr, 10% in smolts, and 1.5% in grilse) (Carlisle and Roberts, 1977; Bylund *et al.*, 1980).

Epidermal papillomatosis has a seasonal course. During the summer the papillomas develop as usually multiple, smooth, whitish elevated patches and progressively form confluent plaques up to 4 cm in diameter. The lesions are located preferentially on the dorsal aspect of the posterior region of the body, the caudal fin, the fin bases, and the opercula (Carlisle and Roberts, 1977; Bylund *et al.*, 1980). The disease resolves spontaneously in the fall and winter by detachment of the neoplasms, which leave an ulcerated area that reepithelializes. Low mortality accompanies the uncomplicated cases.

Microscopically, the lesions are characterized by abruptly demarcated proliferations of Malpighian epithelial cells above a convoluted basement membrane, occasionally on a fibrovascular stalk, and delimited at the margins by clusters of mucous cells (Carlisle and Roberts, 1977; Bylund *et al.*, 1980). Moderate mitotic rate, pleomorphism, and anaplasia are present in some cases. There is no rupture of the basement membrane and downward invasion of the subjacent tissues. Infiltration of mononuclear inflammatory cells varies from scant to massive, and is associated with necrosis of large areas of the papilloma prior to its detachment, suggesting an immunologic basis to the seasonal course.

K. *Salmo salar*–Sarcoma of the Swim Bladder of the Atlantic Salmon

A neoplasm of the swim bladder was diagnosed as leiomyosarcoma (McKnight, 1978) or fibrosarcoma (Duncan, 1978) in 4.6% of 1- to 2-year-old male and female Atlantic salmon in Scottish marine farms. Affected fish were in poor physical condition.

Multiple, well-circumscribed, firm, nodular masses to coalescing plaques were distributed along the swim bladder, especially on its rostral aspect. They expanded into the abdominal cavity and encroached on the swim bladder lumen, impairing normal organ function. Neither infiltration of the adjacent tissues nor metastases were detected.

Histologically, the tumors consisted of nonencapsulated but well-circumscribed nodules of interlacing bundles of fusiform neoplastic cells that arose at the junction between the inner circular smooth muscle layer and loose areolar tissue of the swim bladder (McKnight, 1978). Little collagen was present. Mitotic figures abounded. Some nodules had central ischemic necrosis.

L. *Stizostedion vitreum*–Walleye Dermal Sarcoma

An apparently infectious neoplastic disease of walleyes was described by Walker in Oneida Lake, New York (Walker, 1969), and by Yamamoto in Canada (Yamamoto *et al.*, 1976). Walker named this disease "walleye dermal sarcoma." It is manifested as multifocal benign cutaneous tumors on adult fish (Walker, 1969). The disease seasonally affects widespread walleye populations throughout North America and reaches a frequency of 27% in Oneida Lake and 10% in

Canada during the April spawning run (Walker, 1969; Yamamoto *et al.*, 1976, 1985a).

Grossly, adult fish have multiple, randomly distributed, frequently clustered, spherical nodular masses that appear centered on the scales (Fig. 11) (Walker, 1969; Yamamoto *et al.*, 1985a; Martineau *et al.*, 1990a). The tumors are firm, smooth, whitish to pink, and measure on average 1 cm in diameter. The overlying epidermis is commonly ulcerated. A regular seasonal periodicity of tumor prevalence was documented: tumor prevalence is higher during the cold spring and fall months than during the warm summer months (Bowser *et al.*, 1988). Regression of the neoplasms, from spring to summer, has been demonstrated on individually identified adult fish (Bowser and Wooster, 1991).

Microscopically, the neoplasms are nonencapsulated, well-defined nodular masses that expand superficially within the stratum spongiosum layer of the dermis and abut the epidermis (Fig. 12) (Martineau *et al.*, 1990a). They arise multicentrically from apparently the upper surface of the scales (Walker, 1969; Bowser *et al.*, 1988). Adjacent tumors coalesce. The overlying cutaneous epithelium is often ulcerated. Neoplasms vary from highly cellular to densely fibrous sarcomas. Within most tumors, compact whorls of oval to fusiform neoplastic mesenchymal cells are separated by variable amounts of intercellular collagenous or osteoid matrix. The mitotic rate is usually low. Neither local invasion nor metastases have been observed. The neoplastic cells have moderately abundant, occasionally vacuolated, pale basophilic cytoplasm and oval, vesicular nuclei with a single conspicuous nucleolus. The accompanying inflammatory infiltrate is mainly composed of lymphocytes.

Ultrastructurally, the neoplastic cells are elongated and possess numerous long interdigitating cytoplasmic processes, which course among the interstitial collagen fibers. Prominent rough endoplasmic reticulum and large mitochondria occupy a large part of the cytoplasm. Occasional lipid droplets are present (Walker, 1969; Martineau *et al.*, 1990a).

M. *Stizostedion vitreum*–Discrete Epidermal Hyperplasia

Two different epidermal hyperplastic lesions occur in the spring on adult walleyes (Walker, 1969; Yamamoto *et al.*, 1985b). The overall incidence of both

FIGURE 11. Adult walleye with numerous dermal sarcomas.

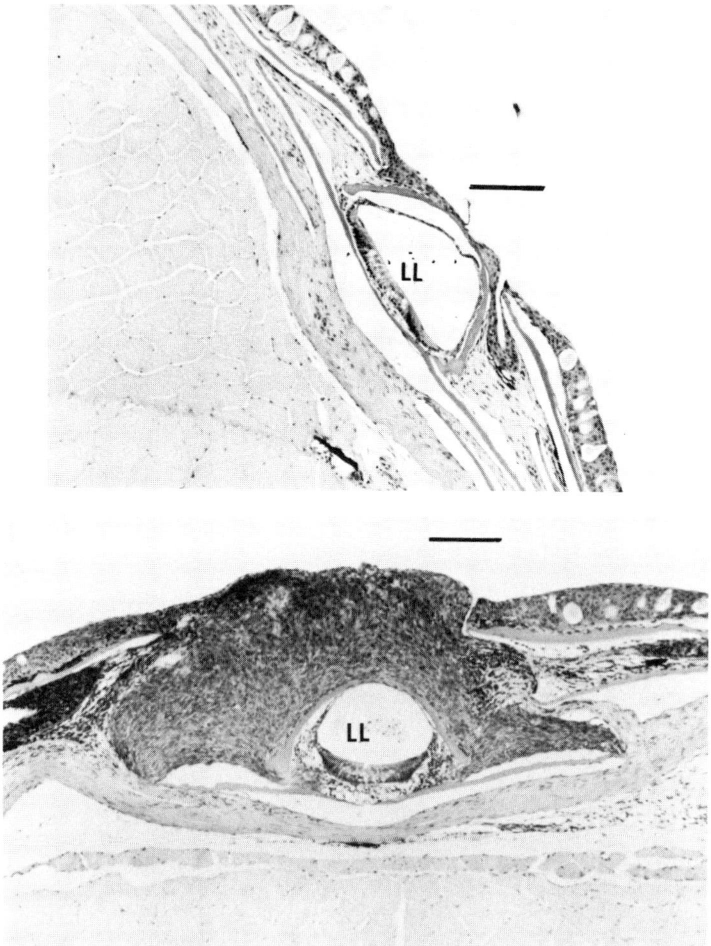

FIGURE 12. Photomicrograph of walleye dermal sarcoma. (Top) Walleye fingerling inoculated with cell-free filtrate of normal walleye skin. Normal appearance of the lateral line region (LL). (Bottom) Experimentally transmitted dermal sarcoma on a walleye fingerling inoculated with tumor extract. A nonencapsulated but well-circumscribed mesenchymal neoplasm is located in the region of the lateral line (LL). HE. Bars = 100 μm.

types of epidermal hyperplasia ranges from 0 to 20% (Yamamoto *et al.*, 1985b). The first lesion, diffuse epidermal hyperplasia, is associated with Percid herpesvirus 1 (Kelly *et al.*, 1983). The second lesion, discrete epidermal hyperplasia, is characterized by usually multiple, broad, flat, translucent, sharply delimited plaques of thickened epidermis that cover large areas of the body, head, and fins (Fig. 13). It is most commonly located on the tail fin (Walker, 1969; Pilcher and Fryer, 1980; Yamamoto *et al.*, 1985a). In a given fish population, the discrete form of epidermal hyperplasia is observed more frequently that the diffuse one (Kelly *et al.*, 1983; Yamamoto *et al.*, 1985a,b).

Microscopically, these discrete epidermal plaques consist of localized hyperplasia of Malpighian cells, demarcated abruptly from the normal adjacent epidermis by conspicuous margins containing abundant mucous cells (Fig. 14). Mitotic

FIGURE 13. Adult walleye with two lesions of discrete epidermal hyperplasia (arrows).

figures are numerous (Yamamoto *et al.*, 1985b). The subjacent dermis is unaltered. Neither epidermal papilloma nor carcinoma has been reported (Walker, 1969).

N. *Pterophyllum scalare*–Lip Fibroma of Angelfish

Solitary or multiple benign neoplasms occurred on the lips of numerous juvenile and adult female angelfish from three different facilities (Francis-Floyd *et al.*, 1993). Grossly, white to pinkish, 3- to 9-mm, multinodular, firm masses arise at

FIGURE 14. Photomicrograph of discrete epidermal hyperplasia of walleye. The hyperplastic lesion is sharply demarcated from the normal epidermis. Abundant rete pegs are formed within the lesion. HE. ×35.

the mucocutaneous junction near the midline of the lower or upper lip. Neoplasms on both lips are frequently located in direct opposition, suggesting that the neoplastic cells spread by direct contact. The overlying labial epithelium is usually unremarkable, except for rare ulcerations. The cut surfaces of the tumors consist of dense, whitish, fibrous tissue; occasionally, the tumors contain variably-sized fluid-filled cavities.

Microscopically, these neoplasms are nonencapsulated, well-circumscribed nodules of elongated neoplastic fibroblasts arranged in whorls in a dense collagenous stroma. Loose matrix was seen in one case. Neither local invasion of adjacent tissues nor metastases are described.

IV. FISH RETROVIRUSES IN CELL CULTURE

A. Introduction

Most of the piscine retroviruses described above are not adapted to cell culture and therefore have not been propagated *in vitro*. One exception is DNFV. Schmale *et al.* (1993) were able to isolate DNFV from culture medium of subcultured tumor-derived cells from neurofibromatosis-affected damselfish. In all other cases where *in vitro* systems have been initiated using either a variety of established fish cell lines or primary cultures of neoplastic/hyperplastic cells, attempts have been unsuccessful: dermal sarcoma (Yamamoto *et al.*, 1985a) and discrete epidermal hyperplasia (McKnight, 1978; Yamamoto *et al.*, 1985a) of walleye, spawning papillomatosis of European smelt (Anders, 1989), epidermal papilloma/carcinoma of white sucker (Sonstegard, 1977; Pilcher and Fryer, 1980), plasmacytoid leukemia of chinook salmon (Kent *et al.*, 1990), epidermal papilloma of Atlantic salmon (Carlisle, 1977), smooth type of epidermal hyperplasia of northern pike (Yamamoto *et al.*, 1983), and hematopoietic neoplasm of northern pike and muskellunge (Sonstegard, 1976).

To circumvent the current inability to isolate the putative causal agent, molecular cloning of the virus and generation of infectious viral clones represent the most promising alternative to demonstrate causality. There are currently four culture-adapted piscine retroviruses available. Three are spontaneously released in established cultures from Asian freshwater fish species (Frerichs *et al.*, 1991) and the fourth is apparently endogenous to the *Xiphophorus* fish (Petry *et al.*, 1992).

Fish-derived cell lines do not require cultivation at 37°C. Unlike avian and mammalian cell lines, piscine cell lines will grow at a wide variety of temperatures, ranging from 10 to 30°C. The optimum temperature for growth of a specific cell line usually reflects the temperature of the environment in which the fish, from which the cell line was originally derived, lived.

B. Retroviruses in Cell Lines Derived from Asian Freshwater Fish Species

Spontaneous *in vitro* production of type C retroviral particles occurred in four cell lines derived from three Asian freshwater fish species. Two of the infected cell lines, SSN-1 and SSN-3, originated from the striped snakehead fish

(*Ophicephalus striatus*), the third cell line, CP, was derived from the climbing perch (*Anabas testudineus*), and the fourth one, SGP, from the snakeskin gourami (*Trichogaster pectoralis*) (Frerichs *et al.*, 1991).

The spontaneous release of 85- to 90-nm, type C retroviral particles from all four cell lines was discovered by electron microscopic examination of the established monolayers. In addition, reverse transcriptase assay of cell culture medium revealed high levels of Mn^{2+}-dependent enzymatic activity using poly(rA) · oligo(dT) template-primer at room temperature. Peaks of this reverse transcriptase activity banded at 1.16 g/ml in sucrose gradient fractions. Cell culture supernatant fluids, inoculated into bluegill fry (BF-2) monolayers, induced in all four cases marked cytopathic effects that paralleled levels of reverse transcriptase activity.

A fifth cell line, SSN-2, also derived from the striped snakehead fish, showed neither electron microscopic nor enzymatic evidence of retroviral infection.

It is not known whether these retroviruses represent endogenous or exogenous agents or elements. Their association with disease in these fish species also remains to be determined.

C. *Xiphophorus* Retrovirus

Perhaps the most thoroughly studied tumor system in fish is the malignant melanoma of the *Xiphophorus* hybrid. In this tumor a dominant oncogene (Tu), recently renamed Xmrk, encodes a receptor tyrosine kinase and is regulated by tumor suppressor genes (R), which are linked and/or not linked to Tu (Wittbrodt *et al.*, 1989). Tu derepression results in the development of melanomas and occurs by genetic segregation through successive interspecies introgressive backcrossings of the platyfish (*Xiphophorus maculatus*) and the swordtail (*Xiphophorus helleri*), which lead to the elimination in the progeny of the nonlinked regulatory R genes.

Retrovirus of *Xiphophorus* fish was first suggested by ultrastructural findings (Kollinger *et al.*, 1979; Perlmutter and Potter, 1987). Some extracellular, 100-nm, type C retroviruses were found spontaneously within the normal kidneys of embryos from inbred platyfish (*X. maculatus*), but large numbers of these viral particles could be induced in the embryos' kidneys by treatment of the female with 5-bromodeoxyuridine during embryologic development (Perlmutter and Potter, 1987). Intracytoplasmic, 100-nm, type C and B retroviruslike particles were detected within *N*-methyl-*N*-nitrosourea-induced neuroblastomas after 5-bromodeoxyuridine treatment of tumor-bearing hybrid fish (Kollinger *et al.*, 1979).

Extensive characterization of a retrovirus in *Xiphophorus* fish was carried out with one cell line, BsT, derived from whole-body melanoma tissue of a particular genotype of platyfish (Petry *et al.*, 1992). The BsT cell line spontaneously released after the 40th passage 100-nm, morphologically immature, type C retroviral particles with a buoyant density of 1.15 to 1.17 g/ml. Culture supernatant fluids expressed a Mn^{2+}-dependent reverse transcriptase activity using the template-primer poly(rA) · oligo(dT). Five proteins were visualized after SDS-PAGE and Western blotting: 180, 120, 70, 65, and 28 kDa. Antiserum against FeLV p27 cross-reacted with the 70-, 65-, and 28 kDa proteins. Southern blot hybridization analysis using probes synthesized in an endogenous reverse transcriptase reaction showed homologous sequences in the DNA of the BsT cell line and of the platyfish and swordtail. These results suggested that this *Xiphophorus* retrovirus could be endogenous. The same probe recognized three transcripts, 8.5, 4.2, and 1.5 kb, within the RNA of BsT cells.

Two findings lend support to a possible relationship between this *Xiphophorus* retrovirus and feline leukemia virus (FeLV). The first is serological cross-reactivity between their reverse transcriptases: antiserum against FeLV reverse transcriptase induced a 40% inhibition of *Xiphophorus* retrovirus reverse transcriptase. The second is the detection of FeLV homologous sequences in the DNA of the BsT cell line and the genomes of the platyfish and swordtail using an FeLV long terminal repeat (LTR)–gag probe.

These observations indicate that an endogenous retrovirus, similar to type C retroviruses, is apparently present in *Xiphophorus* fish. The unequivocal relationship of this virus to other type C retroviruses, however, awaits cloning and sequencing analysis. Similarly, whether this endogenous retrovirus plays a role in the induction of tumors in *Xiphophorus* fish or whether it represents an innocuous bystander that has been fortuitously activated remains to be determined.

V. RETROVIRUSES OF REPTILES

The two well-documented current examples of reptilian retroviruses occur in snakes: Russell's viper retrovirus (VRV) (Zeigel and Clark, 1971) and corn snake retrovirus (CSRV) (Lunger *et al.*, 1974). Although these two retroviruses are related to reptilian neoplasms by the fact that they arose after long-term *in vitro* cultivation of neoplastic cells or upon inoculation of established cell lines by tumor tissue homogenate, evidence for disease etiology is lacking. Neither *in vivo* transmission experiments nor *in vitro* assessment of transforming capacity of these retroviral isolates have been performed. Contrary to most piscine ones, the two reptilian retroviruses can be propagated in cell culture systems. These two viral isolates were originally obtained from animals kept in zoos or private collections. Outbreaks of viral diseases in reptiles are likely to be caused by stressful situations such as crowding or introduction of new reptiles into healthy collections.

A. Russell's Viper Retrovirus

VRV originates from a late passage (48th to 52nd) of cultured spleen cells from an Asian pit viper, *Vipera russelli* (Zeigel and Clark, 1969, 1971). This viper had a large mass diagnosed as edematous myxofibroma anterior to the heart (Zeigel and Clark, 1971). Attempts were made to cultivate *in vitro* tissues from the primary tumor, liver, heart, spleen, kidney, pancreas, and salivary gland. Only the spleen formed a continuous cell line, which was named VSW. Electron microscopic examination of the spleen-derived cell line VSW revealed spontaneous production of 108-nm, type C retrovirus particles (Zeigel and Clark, 1969, 1971). No virus was found either within the neoplasm or within tissues of the tumor-bearing viper. These studies constitute the first example of *in vitro* cultivation of a type C retrovirus from a poikilotherm. Later, histological examination of the visceral organs disclosed the widespread presence of numerous neoplastic emboli and metastases from the primary tumor (Zeigel and Clark, 1971). Based on the presence of these metastases in the spleen and common ultrastructural features between VSW cells and neoplastic cells from the primary mass and visceral metastases, it was suggested that the VSW spleen-derived cell line, which expresses VRV, originated from a metastasis of the myxofibroma (Zeigel and Clark, 1971).

Early preliminary analysis of VRV reverse transcriptase and structural proteins indicated that this retrovirus differed from the avian and murine viruses. VRV reverse transcriptase has an apparent molecular mass of 109 kDa, intermediate between the avian and murine retrovirus reverse transcriptases (Twardzik *et al.*, 1974). Both Mg^{2+} and Mn^{2+} can serve as divalent cations although Mg^{2+} is preferred. The optimum temperature for this enzyme ranges from 35 to 40°C. The virions, banding at 1.16 g/ml, contain a 70 S RNA species that sediments with the 70 S RNA of murine leukemia virus (Gilden *et al.*, 1970).

The structural polypeptides of VRV have been analyzed by SDS-PAGE (Menko *et al.*, 1976). Approximately 16 polypeptides, 3 major and 13 minor ones, were visualized, but the identity and precursor relationships were not established. The major structural protein, 24 kDa, has been assessed for antigenic relatedness to several other retroviral core proteins. Limited reciprocal cross-reactivity was observed with Mason-Pfizer (MPMV) retrovirus (Andersen *et al.*, 1979), whereas none was obtained with other retroviral core proteins, including murine and avian ones (Gilden *et al.*, 1970; Andersen *et al.*, 1979). Nucleic acid relationships between VRV, type C and type D retroviruses were not found using DNA-driven Cot analysis (Andersen *et al.*, 1979). However, greater than 70% homology was detected between a VRV cDNA probe, synthesized in an endogenous reverse transcriptase reaction, and DNA from normal viper cells. This result indicates that VRV is endogenous to Russell's viper. The number of VRV copies within viper cell DNA is estimated at 15 to 20 (Andersen *et al.*, 1979). Since VRV is an endogenous retrovirus, it may have been fortuitously activated in late-passaged VSW cells and may not be directly involved in the induction of the original myxofibroma.

B. Corn Snake Retrovirus

CSRV was first observed on electron micrographs of an embryonal rhabdomyosarcoma from a mature female corn snake (*Elaphe guttata*) (Lunger *et al.*, 1974). The corn snake, affected with rapidly growing, multifocal subcutaneous tumors, was kept in a community cage with five other corn snakes originally collected in New Jersey. One snake in that cage had previously died with similar multiple subcutaneous swellings, suggesting that transmission of an infectious agent might have occurred. Morphologically, CSRV appeared as a typical type C retrovirus. Budding and immature virus particles, approximately 105 to 115 nm in diameter, were observed in the rhabdomyosarcoma and spleen of the tumor-bearing animal. No particles were seen in the spleens of tumor-free corn snakes.

Although initial attempts to establish a cell line from the tissues of the affected corn snake failed (Lunger *et al.*, 1974), infection of a rattlesnake fibroma cell line (C-89) and of early passage rattlesnake heart (CH) and kidney (CK) cells has been accomplished with tumor tissue homogenate (Clark *et al.*, 1979). CRSV purified from the cell cultures has been subjected to comparative analysis of its polypeptides with VRV polypeptides. Of five major CSRV polypeptides, some (12, 13, and 24 kDa) display similar molecular mass as in VRV but no antigenic cross-reactivity. Two molecules are not observed in VRV: the major core protein, with an apparent molecular mass of 24 kDa, and a glycoprotein of 72 kDa. Transmission experiments and more detailed nucleic acid analysis of these ophidian isolates have not been reported.

C. Reptilian Type C Retrovirus Particles and Malignancies

In addition to the two ophidian retroviruses described above, several type C viruslike particles associated with lymphoid leukemia/lymphosarcoma in diverse species of snakes make a growing list of case reports that suggest a possible etiologic relationship between these viral particles and the neoplasms (Jacobson *et al.*, 1980; Zachiesche *et al.*, 1988). The results, however, rely solely on morphologic characteristics determined by electron microscopic examination of the tumors.

In one case of lymphosarcoma of a California king snake (*Lampropeltis getulus californiae*), intranuclear and intracytoplasmic viruslike particles were described in the neoplastic cells (Jacobson *et al.*, 1980). The suggestion that these inclusions are related to retroviruses is based on their morphologic similarity to nucleocapsids.

Electron microscopic analysis of leukemic cells in various tissues of a four-lined chicken snake (*Elaphe obsoleta quadrivittata*) with lymphoblastic leukemia revealed 80- to 100-nm, type C virus particles (Zachiesche *et al.*, 1988).

VI. RETROVIRUSES OF MOLLUSCS

The pathologic basis of diseases in invertebrate species has only recently been addressed. Study of invertebrate diseases, and their possible relation to pollution and significance for public health, is rapidly expanding partially because of environmental concern and commercial interests (Perkins, 1993). Neoplasms in molluscs have been reported worldwide in 15 different species of bivalve molluscs, namely, oysters, clams, and mussels, since 1887 (Brown *et al.*, 1977).

The soft-shell clam, *Mya arenaria*, has been particularly studied (Peters, 1988). Two neoplastic conditions, a hematopoietic neoplasm referred to as "disseminated sarcoma" and a gonadal neoplasm, occur in adult male and female feral soft-shell clams inhabiting both polluted and pristine sites. The incidence of these two conditions combined ranges from 0 to 64% (Brown *et al.*, 1977). Disseminated sarcoma is an enzootic malignant neoplasm characterized by widespread tissue and vascular invasion by mitotically active, anaplastic cells tentatively identified as hemocytes. Although the disease is usually fatal, remission was noted in approximately 12% of the cases (Cooper *et al.*, 1982). Study of seasonal fluctuation in disease prevalence documented two peaks, one in late spring and another in late fall (Cooper *et al.*, 1982).

Disseminated sarcoma of *M. arenaria* appears etiologically associated with a newly discovered molluscan retrovirus. Virions were purified by sucrose gradient centrifugation of neoplastic tissue from affected clams (Oprandy *et al.*, 1981). The visible band, at a density of 1.17 to 1.18 g/ml, was collected and found to contain abundant 120-nm, type B retroviral particles when examined by electron microscopy. Transmission experiments, with 0.45-μm-filtered viral fractions purified on sucrose gradients and in which the presence of viral particles was confirmed by electron microscopic examination, provide the most convincing evidence that disseminated sarcoma of *M. arenaria* has a retroviral etiology. Test clams developed the neoplasm 2 months postinoculation (Oprandy *et al.*, 1981). In addition, two-stage serial transmission resulted in neoplasm induction in test clams using virus purified as described above from one experimentally infected clam that had developed disseminated sarcoma.

Additionally, 5-bromodeoxyuridine treatment of nonneoplastic clams was followed by both replication of a similar if not identical retrovirus in clam tissue and development of disseminated sarcomas (Oprandy and Chang, 1983). This retrovirus, induced by 5-bromodeoxyuridine treatment, reproduced the disease in healthy clams. Whether 5-bromodeoxyuridine treatment induced an endogenous retrovirus or activated an exogenous one remains to be determined. Mg^{2+}-dependent reverse transcriptase activity has been found associated with purified virus particles (Oprandy and Chang, 1983; Medina *et al.*, 1993). The enzyme is active at 6°C and inactive above 25°C.

VII. PERSPECTIVE AND CONCLUSION

The scope of this review encompasses the neoplastic/proliferative diseases of possible retroviral etiology that affect lower vertebrates and molluscs. It focuses on fish systems, which have provided the most information. Piscine, reptilian, and molluscan retrovirology is a field in its infancy, especially when compared with its avian and mammalian counterparts. Much of the work suggesting a retroviral etiology for a specific disease is based on transmission with filterable agents as inocula. Similar approaches to establish retroviruses as tumorigenic agents were employed almost 90 years ago (Payne, 1992). To date, no piscine retrovirus has been successfully cultivated *in vitro* and subsequently shown to reproduce the disease in a susceptible species. This experimental deficit has played a major role in preventing further analysis of these viruses, especially of their molecular genetics.

As experimental models, fish systems have been used to extend our understanding of basic biological processes in diverse disciplines, such as neurobiology, endocrinology, developmental biology, toxicology, and carcinogenesis (Powers, 1989). Fish are amenable to extensive genetic analysis and transgenic manipulation (Streisinger *et al.*, 1981; Maclean *et al.*, 1987; Liu *et al.*, 1990). In carcinogenesis, some fish tumors appear to have a retroviral etiology. In this regard, the study of these agents could provide new insights into the mechanism of oncogenesis particularly at different temperatures and in various animal species. This peculiarity may enable the temporal and spatial dissection of potential viral and cellular factors that trigger neoplastic transformation and growth. These factors thus identified might later be found to play a central role in cell cycle regulation. Our current view of the retroviral world is limited, to a great extent, to the agents that can be readily propagated *in vitro*. We await similar success with the viruses discussed in this review so that the origin, evolutionary relationships, and modes of disease induction can be evaluated.

The recent molecular genetic information on one fish retrovirus moreover suggests that the current taxonomy (Coffin, 1992) may soon need revision in order to accommodate piscine retroviruses. Whether these viruses fit into an existing subgenus or require the creation of a new one awaits future description. This review will, hopefully, stimulate further investigation into this important field.

ACKNOWLEDGMENTS. We thank Dr. V. M. Vogt for reviewing the manuscript. The research of J.W.C. was sponsored in part by Public Health Service grant CA-53623A from the National Cancer Institute. P.R.B. was funded in part by grant 93-03355 through the National Research Initiative Competitive Grants Program of the United States Department of Agriculture. Photographs were kindly provided by M. L. Kent (plasmacytoid leukemia of chinook salmon) and Dr. H. W. Ferguson (hematopoietic neoplasm of northern pike).

VIII. REFERENCES

Anders, K., 1989, A herpesvirus associated with an epizootic epidermal papillomatosis in European smelt (*Osmerus eperlanus*), in: *Viruses of Lower Vertebrates* (W. Ahne and E. Kurstak, eds.), p. 184, Springer-Verlag, Berlin.

Anders, K., Hilger, I., and Möller, H., 1991, Lentivirus-like particles in connective tissue tumours of fish from German coastal waters, *Dis. Aquat. Org.* **11**:151.

Anders, K., and Möller, H., 1985, Spawning papillomatosis of smelt, *Osmerus eperlanus* L., from the Elbe estuary, *J. Fish Dis.* **8**:233.

Andersen, P. R., Barbacid, M., Tronick, S. R., Clark, H. F., and Aaronson, S. A., 1979, Evolutionary relatedness of viper and primate endogenous retroviruses, *Science* **204**:318.

Asashima, M., Komazaki, S., Satou, C., and Oinuma, T., 1982, Seasonal and geographical changes of spontaneous skin papillomas in the Japanese newt *Cynops pyrrhogaster*, *Cancer Res.* **42**:3741.

Bowser, P. R., and Wooster, G. A., 1991, Regression of dermal sarcoma in adult walleyes, *J. Aquat. Anim. Health* **3**:147.

Bowser, P. R., Wolfe, M. J., Forney, J. L., and Wooster, G. A., 1988, Seasonal prevalence of skin tumors from walleye (*Stizostedion vitreum*) from Oneida Lake, New York, *J. Wildl. Dis.* **24**:292.

Bowser, P. R., Martineau, D., and Wooster, G. A., 1990, Effects of water temperature on experimental transmission of dermal sarcoma in fingerling walleyes, *J. Aqua. Anim. Health* **2**:157.

Brown, E. R., Dolowy, W. C., Sinclair, T., Keith, L., Greenberg, S., Hazdra, J. J., Beamer, P., and Callaghan, O., 1975, Enhancement of lymphosarcoma transmission in *Esox lucius* and its epidemiologic relationship to pollution, *Comp. Leuk. Res.* **43**:245.

Brown, R. S., Wolke, R. E., Saila, S. B., and Brown, C. W., 1977, Prevalence of neoplasia in 10 New England populations of the soft-shell clam (*Mya arenaria*), *Ann. N.Y. Acad. Sci.* **298**:522.

Bylund, G., Valtonen, E. T., and Niemelä, E., 1980, Observations on epidermal papillomata in wild and cultured Atlantic salmon *Salmo salar* L. in Finland, *J. Fish Dis.* **3**:525.

Carlisle, J. C., 1977, An epidermal papilloma of the Atlantic salmon. II. Ultrastructure and etiology, *J. Wildl. Dis.* **13**:235.

Carlisle, J. C., and Roberts, R. J., 1977, An epidermal papilloma of the Atlantic salmon. I. Epizootiology, pathology and immunology, *J. Wildl. Dis.* **13**:230.

Clark, H. F., Andersen, P. R., and Lunger, P. D., 1979, Propagation and characterization of a C-type virus from a rhabdomyosarcoma of a corn snake, *J. Gen. Virol.* **43**:673.

Coffin, J. M., 1992, Structure and classification of retroviruses, in: *The Retroviridae*, Vol. 1 (J. A. Levy, ed.), pp. 19–50, Plenum Press, New York.

Cooper, K. R., Brown, R. S., and Chang, P. W., 1982, The course and mortality of a hematopoietic neoplasm in the soft-shell clam, *Mya arenaria, J. Invertebr. Pathol.* **39**:149.

Cuchens, M. A., and Clem, L. W., 1977, Phylogeny of lymphocyte heterogeneity. II. Differential effects of temperature on fish T-like and B-like cells, *Cell. Immunol.* **34**:219.

Dawe, C. J., 1969, Neoplasms of blood cell origin in poikilothermic animals. A review, *Natl. Cancer Inst. Monogr.* **32**:7.

Dawe, C. J., Banfield, W. G., Sonstegard, R., Lee, C. W., and Michelitch, H. J., 1976, Cylindroid lamella-particle complexes and nucleoid intracytoplasmic bodies in lymphoma cells of northern pike (*Esox lucius*), *Prog. Exp. Tumor Res.* **20**:166.

Duncan, I. B., 1978, Evidence for an oncovirus in swimbladder fibrosarcoma of Atlantic salmon *Salmo salar* L, *J. Fish Dis.* **1**:127.

Eaton, W. D., and Kent, M. L., 1992, A retrovirus in chinook salmon (*Oncorhynchus tshawytscha*) with plasmacytoid leukemia and evidence for the etiology of the disease, *Cancer Res.* **52**:6496.

Ferguson, H. W., 1989, *Systemic Pathology of Fish: A Text and Atlas of Comparative Responses in Diseases of Teleosts,* Iowa State University Press, Ames.

Francis-Floyd, R., Bolon, B., Fraser, W., and Reed, P., 1993, Lip fibromas associated with retrovirus-like particles in angel fish, *J. Am. Vet. Med. Assoc.* **202**:427.

Frerichs, G. N., Morgan, D., Hart, D., Skerrow, C., Roberts, R. J., and Onions, D. E., 1991, Spontaneously productive C-type retrovirus infection of fish cell lines, *J. Gen. Virol.* **72**:2537.

Gilden, R. V., Lee, Y. K., Oroszlan, S., Walker, J. L., and Huebner, R. J., 1970, Reptilian C-type virus: Biophysical, biological, and immunological properties, *Virology* **41**:187.

Gross, L., 1983, Tumors, leukemia and lymphosarcomas in fish, in: *Oncogenic Viruses,* p. 103, Pergamon Press, Elmsford, N.Y.

Haddow, A., and Blake, I., 1933, Neoplasms in fish: A report of six cases with a summary of the literature, *J. Pathol. Bacteriol.* **36**:41.

Harshbarger, J. C., 1984, Pseudoneoplasms in ectothermic animals, *Natl. Cancer Inst. Monogr.* **65**:251.

Harshbarger, J. C., and Clark, J. B., 1990, Epizootiology of neoplasms in bony fish of North America, *Sci. Total Environ.* **94**:1.

Harshbarger, J. C., and Dawe, C. J., 1973, Hematopoietic neoplasms in invertebrate and poikilothermic vertebrate animals, *Bibl. Haematol.* **39**:1.

Hayes, M. A., Smith, I. R., Rushmore, T. H., Crane, T. L., Thorn, C., Kocal, T. E., and Ferguson, H. W., 1990, Pathogenesis of skin and liver neoplasms in white suckers from industrially polluted areas in Lake Ontario, *Sci. Total Environ.* **94**:105.

Hedrick, R. P., Groff, J. M., McDowell, T. S., Willis, M., and Cox, W. T., 1990, Hematopoietic intranuclear microsporidian infections with features of leukemia in chinook salmon *Oncorhynchus tshawytscha, Dis. Aquat. Org.* **8**:189.

Hedrick, R. P., Groff, J. M., and Baxa, D. V., 1991, Experimental infections with *Enterocytozoon salmonis* Chilmonczyk, Cox, Hedrick (Microsporea): An intranuclear microsporidium from chinook salmon *Oncorhynchus tshawytscha, Dis. Aquat. Org.* **10**:103.

Jacobson, E. R., 1981, Neoplastic diseases, in: *Diseases of the Reptilia,* Vol. 2 (J. E. Cooper and O. F. Jackson, eds.), p. 429, Academic Press, New York.

Jacobson, E. R., Seely, J. C., and Novilla, M. N., 1980, Lymphosarcoma associated with virus-like intranuclear inclusions in a California king snake (*Lampropeltis getulus californiae*), *J. Natl. Cancer Inst.* **65**:577.

Kelly, R. K., Nielsen, O., Mitchell, S. C., and Yamamoto, T., 1983, Characterization of Herpesvirus vitreum isolated from hyperplastic epidermal tissue of walleye, *Stizostedion vitreum vitreum* (Mitchill), *J. Fish Dis.* **6**:249.

Kent, M. L., and Dawe, S. C., 1990, Experimental transmission of a plasmacytoid leukemia of chinook salmon, *Oncorhynchus tshawytscha, Cancer Res.* **50**(Suppl.):5679.

Kent, M. L., and Dawe, S. C., 1993, Further evidence for a viral etiology in plasmacytoid leukemia of chinook salmon *Oncorhynchus tshawytscha, Dis. Aquat. Org.* **15**:115.

Kent, M. L., Groff, J. M., Traxler, G. S., Zinkl, J. G., and Bagshaw, J. W., 1990, Plasmacytoid leukemia in seawater reared chinook salmon *Oncorhynchus tshawytscha, Dis. Aquat. Org.* **8**:199.

Kollinger, G., Schwab, M., and Anders, F., 1979, Virus-like particles induced by bromodeoxyuridine in melanoma and neuroblastoma of *Xiphophorus, J. Cancer Res. Clin. Oncol.* **95**:239.

Liu, Z., Moav, B., Faras, A. J., Guise, K. S., Kapuscinski, A. R., and Hackett, P. B., 1990, Development of expression vectors for transgenic fish, *Bio/Technology* **8**:1268.

Ljungberg, O., 1963, Report on fish diseases and inspection of fish products in Sweden, *Bull. Off. Int. Epizootiol.* **59**:111.

Ljungberg, O., 1976, Epizootiological and experimental studies of skin tumours in northern pike (*Esox lucius* L.) in the Baltic Sea, *Prog. Exp. Tumor Res.* **20**:156.

Ljungberg, O., and Lange, J., 1968, Skin tumours of northern pike (*Esox lucius* L.). I. Sarcoma in a Baltic pike population, *Bull. Off. Int. Epizootiol.* **69**:1007.

Loh, P. C., 1993, Spumaviruses, in: *The Retroviridae*, Vol. 2 (J. A. Levy, ed.), pp. 361–397, Plenum Press, New York.

Lucké, B., 1952, Kidney carcinoma in the leopard frog: A virus tumor, *Ann. N.Y. Acad. Sci.* **54**:1093–1109.

Lunger, P. D., Hardy, W. D., Jr., and Clark, H. F., 1974, C-type virus particles in a reptilian tumor, *J. Natl. Cancer Inst.* **52**:1231.

McKnight, I. J., 1978, Sarcoma of the swimbladder of Atlantic salmon (*Salmo salar* L.), *Aquaculture* **13**:55.

Maclean, N., Penman, D., and Zhu, Z., 1987, Introduction of novel genes into fish, *Bio/Technology* **5**:257.

Martineau, D., Bowser, P. R., Wooster, G., and Forney, J. L., 1990a, Histologic and ultrastructural studies of dermal sarcoma of walleye (Pisces: *Stizostedion vitreum*), *Vet. Pathol.* **27**:340.

Martineau, D., Bowser, P. R., Wooster, G. A., and Armstrong, L. D., 1990b, Experimental transmission of a dermal sarcoma in fingerling walleyes (*Stizostedion vitreum vitreum*), *Vet. Pathol.* **27**:230.

Martineau, D., Renshaw, R. R., Williams, J. R., Casey, J. W., and Bowser, P. R., 1991, A large unintegrated retrovirus DNA species present in a dermal tumor of walleye *Stizostedion vitreum*, *Dis. Aquat. Org.* **10**:153.

Martineau, D., Bowser, P. R., Renshaw, R. R., and Casey, J. W., 1992, Molecular characterization of a unique retrovirus associated with a fish tumor, *J. Virol.* **66**:596.

Medina, D. J., Paquette, G. E., Sadasiv, E. C., and Chang, P. W., 1993, Isolation of infectious particles having reverse transcriptase activity and producing hematopoietic neoplasia in *Mya arenaria*, *J. Shellfish Res.* **12**:101 (Abstract).

Menko, A. S., Sokol, F., Clark, H. F., and Tan, K. B., 1976, Structural and enzymatic characterization of viper C-type virus, *Arch. Virol.* **50**:125.

Miller, N. W., and Clem, L. W., 1984, Temperature-mediated processes in teleost immunity: Differential effects of temperature on catfish in vitro antibody responses to thymus-dependent and thymus-independent antigens, *J. Immunol.* **133**:2356.

Morrison, J. K., MacConnell, E., Chapman, P. F., and Westgard, R. L., 1990, A microsporidium-induced lymphoblastosis in chinook salmon *Oncorhynchus tshawytscha* in freshwater, *Dis. Aquat. Org.* **8**:99.

Mulcahy, M. F., 1963, Lymphosarcoma in the pike, *Esox lucius* L., (Pisces; Esocidae) in Ireland, *Proc. R. Ir. Acad.* **63**:103.

Mulcahy, M. F., 1976, Epizootiological studies of lymphomas in northern pike in Ireland, *Prog. Exp. Tumor Res.* **20**:129.

Mulcahy, M. F., and O'Leary, A., 1970, Cell-free transmission of lymphosarcoma in the northern pike *Esox lucius* L. (Pisces; Esocidae), *Experientia* **26**:891.

Mulcahy, M. F., Winqvist, G., and Dawe, C. J., 1970, The neoplastic cell type in lymphoreticular neoplasms of the northern pike, *Esox lucius* L, *Cancer Res.* **30**:2712.

Newbound, G. C., and Kent, M. L., 1991, Experimental interspecies transmission of plasmacytoid leukemia in salmonid fishes, *Dis. Aquat. Org.* **10**:159.

Nigrelli, R. F., 1943, Causes of diseases and death of fishes in captivity, *Zoologica* **28**:203.

Nigrelli, R. F., 1947, Spontaneous neoplasms in fishes. III. Lymphosarcoma in *Astyanax* and *Esox*, *Zoologica* **32**:101.

Nigrelli, R. F., 1952, Virus and tumors in fishes, *Ann. N.Y. Acad. Sci.* **54**:1076.

Ohlmacher, A. P., 1898, Several examples illustrating the comparative pathology of tumors, *Bull. Ohio Hosp. Epileptics* **1**:223.

Oprandy, J. J., and Chang, P. W., 1983, 5-Bromodeoxyuridine induction of hematopoietic neoplasia and retrovirus activation in the soft-shell clam, *Mya arenaria*, *J. Invertebr. Pathol.* **42**:196.

Oprandy, J. J., Chang, P. W., Pronovost, A. D., Cooper, K. R., Brown, R. S., and Yates, V. J., 1981, Isolation of a viral agent causing hematopoietic neoplasia in the soft-shell clam, *Mya arenaria*, *J. Invertebr. Pathol.* **38**:45.

Papas, T. S., Dahlberg, J. E., and Sonstegard, R. A., 1976, Type C virus in lymphosarcoma in northern pike (*Esox lucius*), *Nature* **261**:506.

Papas, T. S., Pry, T. W., Schafer, M. P., and Sonstegard, R. A., 1977, Presence of DNA polymerase in lymphosarcoma in northern pike (*Esox lucius*), *Cancer Res.* **37**:3214.

Payne, L. N., 1992, Biology of avian retroviruses, in: *The Retroviridae*, Vol. 1 (J. A. Levy, ed.), pp. 299–404, Plenum Press, New York.

Perkins, F. O., 1993, Infectious diseases of molluscs, in: *Pathobiology of Marine and Estuarine Organisms*, p. 255, CRC Press, Boca Raton, Fla.

Perlmutter, A., and Potter, H., 1987, Retrovirus-like particles in embryonic kidney tissue of the platyfish, *Xiphophorus maculatus*, *J. Exp. Zool.* **243**:125.

Peters, E. C., 1988, Recent investigations on the disseminated sarcomas of marine bivalve molluscs, *Am. Fish. Soc. Spec. Publ.* **18**:74.

Petry, H., Petry, K., Schmidt, M., Hunsmann, G., Anders, F., and Lüke, W., 1992, Isolation and characterization of a retrovirus from the fish genus *Xiphophorus*, *Virology* **188**:785.

Pilcher, K. S., and Fryer, J. L., 1980, The viral diseases of fish: A review through 1978. Part 2: Diseases in which a viral etiology is suspected but unproven, *CRC Crit. Rev. Microbiol.* **8**:1.

Powers, D. A., 1989, Fish as model systems, *Science* **246**:352.

Rafferty, K. A., 1964, Kidney tumors of the leopard frog: A review, *Cancer Res.* **24**:169.

Rijkers, G. T., 1982, Kinetics of humoral and cellular immune reactions in fish, *Dev. Comp. Immunol.* Suppl. **2**:93.

Schlumberger, H. G., 1957, Tumor characteristics for certain animal species. A review, *Cancer Res.* **17**:823.

Schmale, M. C., and Hensley, G. T., 1988, Transmissibility of a neurofibromatosis-like disease in bicolor damselfish, *Cancer Res.* **48**:3828.

Schmale, M. C., Hensley, G. T., and Udey, L. R., 1983, Neurofibromatosis, von Recklinghausen's disease, multiple schwannomas, malignant schwannomas. Animal model: Multiple schwannomas in bicolor damselfish, *Am. J. Pathol.* **112**:238.

Schmale, M. C., Hensley, G. T., and Udey, L. R., 1986, Neurofibromatosis in the bicolor damselfish (Pomacentrus partitus) as a model of von Recklinghausen neurofibromatosis, *Ann. N.Y. Acad. Sci.* **486**:386.

Schmale, M. C., Aman, M. R., and Gill, K. A., 1993, A retrovirus associated with neurofibromatosis in the bicolor damselfish, *Proc. Am. Assoc. Cancer Res.* **34**:545 (Abstract).

Smith, I. R., and Zajdlik, B. A., 1987, Regression and development of epidermal papillomas affecting white suckers, *Catostomus commersoni* (Lacépède), from Lake Ontario, Canada, *J. Fish Dis.* **10**:487.

Smith, I. R., Baker, K. W., Hayes, M. A., and Ferguson, H. W., 1989a, Ultrastructure of Malpighian and inflammatory cells in epidermal papillomas of white suckers *Catostomus commersoni*, *Dis. Aquat. Org.* **6**:17.

Smith, I. R., Ferguson, H. W., and Hayes, M. A., 1989b, Histopathology and prevalence of epidermal papillomas epidemic in brown bullhead, *Ictalurus nebulosus* (Lesueur), and white sucker, *Catostomus commersoni* (Lacépède), populations from Ontario, Canada, *J. Fish Dis.* **12**:373.

Sonstegard, R., 1975, Studies of the etiology and epizootiology of lymphosarcoma in northern pike (*Esox lucius*) and muskellunge (*Esox masquinongy*), *Comp. Leuk. Res.* **43**:242.

Sonstegard, R. A., 1976, Studies of the etiology and epizootiology of lymphosarcoma in Esox (*Esox lucius* L. and *Esox masquinongy*), *Prog. Exp. Tumor Res.* **20**:141.

Sonstegard, R. A., 1977, Environmental carcinogenesis studies in fishes of the Great Lakes of North America, *Ann. N.Y. Acad. Sci.* **298**:261.

Sonstegard, R. A., and Chen, T. T., 1986, A review of lymphosarcoma of muskellunge and northern pike, *Am. Fish. Soc. Spec. Publ.* **15**:47.

Streisinger, G., Walker, C., Dower, N., Knauber, D., and Singer, F., 1981, Production of clones of homozygous diploid zebra fish (*Brachydanio rerio*), *Nature* **291**:293.

Thompson, J. S., 1982, An epizootic of lymphoma in northern pike, *Esox lucius* L., from the Åland islands of Finland, *J. Fish Dis.* **5**:1.

Thompson, J. S., and Hildén, M., 1988, Population characteristics of true histiocytic lymphoma in northern pike, *Esox lucius* L., from the Åland islands of the Baltic sea, *Aqua Fennica* **18**:85.

Thompson, J. S., and Kostiala, A. A. I., 1986, Immunological classification of a pike lymphoma, *Vet. Immunol. Immunopathol.* **12**:421.

Thompson, J. S., and Kostiala, A. A. I., 1988, Cutaneous tumor of northern pike, *Esox lucius* L. II. Expression of surface and cytoplasmic immunoglobulins and alpha-naphtyl acetate esterase by mononuclear cells from lymphoid tissues and tumor, *Dev. Comp. Immunol.* **12**:75.

Thompson, J. S., and Miettinen, M., 1988, Ultrastructural pathology of cutaneous tumours of northern pike, *Esox lucius* L., *J. Fish Dis.* **11**:47.

Thompson, J. S., Kostiala, A. A. I., and Miettinen, M., 1987a, Cutaneous tumour of northern pike *Esox lucius* L.: Evidence for a monocytic neoplasm, *J. Fish Biol.* **31**(Suppl. A):167.

Thompson, J. S., Virtanen, I., and Lehto, V.-P., 1987b, Intermediate filaments in normal tissues and lymphomas of northern pike, *Esox lucius* L., from the Åland islands of Finland, *J. Comp. Pathol.* **97**:257.

Twardzik, D. R., Papas, T. S., and Portugal, F. H., 1974, DNA polymerase in virions of a reptilian type C virus, *J. Virol.* **13**:166.

Van Beneden, R. J., Henderson, K. W., Blair, D. G., Papas, T. S., and Gardner, H. S., 1990, Oncogenes in hematopoietic and hepatic fish neoplasms, *Cancer Res.* **50**:5671s.

Walker, R., 1969, Virus associated with epidermal hyperplasia in fish, *Natl. Cancer Inst. Monogr.* **31**:195.

Whang-Peng, J., Sonstegard, R. A., and Dawe, C., 1976, Chromosomal characteristics of malignant lymphoma in northern pike (*Esox lucius*) from the United States, *Cancer Res.* **36**:3554.

Winqvist, G., Ljungberg, O., and Hellstroem, B., 1968, Skin tumours of northern pike (*Esox lucius* L.). II. Viral particles in epidermal proliferations, *Bull. Off. Int. Epizootiol.* **69**:1023.

Winqvist, G., Ljungberg, O., and Ivarsson, B., 1973, Electron microscopy of sarcoma of the northern pike (*Esox lucius* L.), *Bibl. Haematol.* **39**:26.

Wittbrodt, J., Adam, D., Malitschek, B., Mäueler, W., Raulf, F., Telling, A., Robertson, S. M., and Schartl, M., 1989, Novel putative receptor tyrosine kinase encoded by the melanoma-inducing *Tu* locus in *Xiphophorus*, *Nature* **341**:415.

Yamamoto, T., MacDonald, R. D., Gillespie, D. C., and Kelly, R. K., 1976, Viruses associated with lymphocystis disease and dermal sarcoma of walleye (*Stizostedion vitreum vitreum*), *J. Fish. Res. Board Can.* **33**:2408.

Yamamoto, T., Kelly, R. K., and Nielsen, O., 1983, Epidermal hyperplasias of northern pike (*Esox lucius*) associated with herpesvirus and C-type particles, *Arch. Virol.* **79**:255.

Yamamoto, T., Kelly, R. K., and Nielsen, O., 1985a, Morphological differentiation of virus-associated skin tumors of walleye (*Stizostedion vitreum vitreum*), *Fish Pathol.* **20**:361.

Yamamoto, T., Kelly, R. K., and Nielsen, O., 1985b, Epidermal hyperplasia of walleye, *Stizostedion vitreum vitreum* (Mitchill), associated with retrovirus-like type-C particles: Prevalence, histologic and electron microscopic observations, *J. Fish Dis.* **19**:425.

Zachiesche, W., Konstantinov, A., Ippen, R., and Mladenov, Z., 1988, Lymphoid leukemia with presence of type C virus particles in a four-lined chicken snake: *Elaphe obsoleta quadrivittata*, in: *Erkrankungen der Zootiere. Verhandlungsbericht des 30 Internationale Symposiums über die Erkrankungen der Zoo- und Wildtiere* (R. Ippen and H. D. Schröder, eds.), p. 275, Akademie-Verlag, Berlin.

Zeigel, R. F., and Clark, H. F., 1969, Electron microscopic observations on a "C"-type virus in cell cultures derived from a tumor-bearing viper, *J. Natl. Cancer Inst.* **43**:1097.

Zeigel, R. F., and Clark, H. F., 1971, Histologic and electron microscopic observations on a tumor-bearing viper: Establishment of a "C"-type virus-producing cell line, *J. Natl. Cancer Inst.* **46**:309.

CHAPTER 2

Bovine Leukemia Virus

RICHARD KETTMANN, ARSÈNE BURNY, ISABELLE CALLEBAUT, LOUIS DROOGMANS, MARC MAMMERICKX, LUC WILLEMS, AND DANIEL PORTETELLE

I. INTRODUCTION

In bovine pathology, research performed between 1871 and 1877 by Siedamgrotzky was apparently the first attempt to study bovine leukosis (BL) and characterize it as a clinical entity. At the beginning of the 20th century, pathologists interested in this disease observed that BL is a herd disease of probable infectious origin. In 1957, Bendixen initiated an epidemiological study of BL in Denmark (Bendixen, 1963), and documented the existence of an enzootic disease [enzootic bovine leukosis (EBL)]. EBL is a chronic contagious disease that evolves over an extended period of time (1–8 years), with tumors eventually developing in only a small number of affected animals. He also defined a group of sporadic manifestations [sporadic bovine leukosis (SBL)], which are not contagious or linked to the presence of a virus.

Evidence to support the infectious nature of EBL became more and more evident and accepted, and prompted many virological studies (acellular transmissions, electron microscopy) in the 1960s. In 1969, Miller *et al.* observed viral particles in short-term cultures of leukocytes from EBL cattle with persistent lymphocytosis (PL). These particles were called "bovine leukemia virus" (BLV) after demonstrating that infected biological material (e.g., blood, milk) was capa-

---

RICHARD KETTMANN, ISABELLE CALLEBAUT, LUC WILLEMS, AND DANIEL PORTETELLE • Molecular Biology and Animal Physiology Unit, Faculty of Agronomy, B5030 Gembloux, Belgium. ARSÈNE BURNY • Molecular Biology and Animal Physiology Unit, Faculty of Agronomy, B5030 Gembloux, and Department of Molecular Biology, University of Brussels, B1640 Rhode-St-Genèse, Belgium. LOUIS DROOGMANS • Department of Molecular Biology, University of Brussels, B1640 Rhode-St-Genèse, Belgium. MARC MAMMERICKX • National Institute for Veterinary Research, B1180 Brussels, Belgium.

The Retroviridae, Volume 3, edited by Jay A. Levy. Plenum Press, New York, 1994.

ble of inducing leukosis in cattle and sheep (Mammerickx, 1970, 1972; Olson *et al.*, 1972).

A few years later, Miller and Olson (1972) demonstrated the presence of antibodies directed against a BLV internal antigen in EBL-affected cattle, thus opening the route to serological studies. Later, Van Der Maaten *et al.* (1974) and Van Der Maaten and Miller (1976) succeeded in adapting BLV to bovine fetal cells, thus deriving permanently infected cell lines [fetal lamb kidney (FLK)]. Subsequently, it was shown that BLV is a retrovirus exogenous to the bovine species (Kettmann *et al.*, 1976, 1978; Deschamps *et al.*, 1981). In this review, we emphasize new developments in BLV research during the past 6 years and refer the reader to previous reviews for more historical aspects (Burny *et al.*, 1980, 1984, 1985, 1990; Ghysdael *et al.*, 1984; Burny and Mammerickx, 1987).

II. GENOME AND GENE PRODUCTS

The BLV proviral genomic structure has been described in detail by Rice *et al.* (1984, 1985, 1987a), Sagata *et al.* (1985a), and Weiss *et al.* (1984, 1985). Analysis of purified BLV virions by SDS-polyacrylamide gel electrophoresis and gel filtration has resolved four major nonglycosylated proteins (p24, p15, p12, p10) and two glycosylated components (gp51 and gp30) (Fig. 1). The precursors of these structural proteins have also been identified both *in vivo* and by *in vitro* synthesis (Ghysdael *et al.*, 1979; Bruck *et al.*, 1984a; Yoshinaka *et al.*, 1986). It is apparent that: (1) the gag polyprotein contains the viral structural proteins p15 (109 aa), p24 (215 aa), and p12 (69 aa); (2) a protease, p14 (195 aa), is coded by an open reading frame overlapping the *gag* gene to the left and the *pol* gene to the right; *gag, prot,* and *pol* genes are in three different reading frames; (3) the *env* gene codes for a 72-kDa env precursor (Pr 72^{env}: 515 aa including the signal peptide) and is cleaved to form the two glycosylated envelope proteins gp51 (268 aa) and gp30 (214 aa); (4) two overlapping open reading frames (ORFs), located between *env* and the 3′ LTR (long terminal repeat) code for the Tax (p34; 309 aa) and Rex (p18; 156 aa) proteins (Fig. 1).

The two gag polyprotein precursors of 66 kDa (also referred to as 70 kDa by some authors) and 44 kDa are synthesized in BLV-infected cells, in reticulocyte cell-free lysates programmed with BLV 38 S RNA, and in frog oocytes microinjected with the same RNA message. Pr 66 is the precursor of: (1) p15, the myristylated fragment derived from the NH_2-terminus; (2) p24, the major core protein; (3) p12, the RNA interacting fragment; and (4) p14, the protease, which is probably synthesized via a frameshift suppression by a lysine-specific tRNA of the gag terminal codon (Yoshinaka *et al.*, 1986). Pr 44 is the same except that it lacks the p14 protease. The protein, p10, is an additional cleavage product that is found in purified BLV preparations and is an NH_2-terminal polypeptide of the gag polyprotein. It is also myristylated like the matrix-associated (MA) p15, the larger and more common antigen derived from that region of the precursor. MA p15 forms a complex specifically with dimer RNA from the U5-5′ *gag* region which is presumed to contain the BLV packaging signal (Katoh *et al.*, 1991). RNA dimerization is mediated by the region harboring U5, the primer binding site (PBS), and 30 bases immediately downstream of PBS (Katoh *et al.*, 1993).

The reverse transcriptase-endonuclease protein (852 aa) is made as a 145-kDa

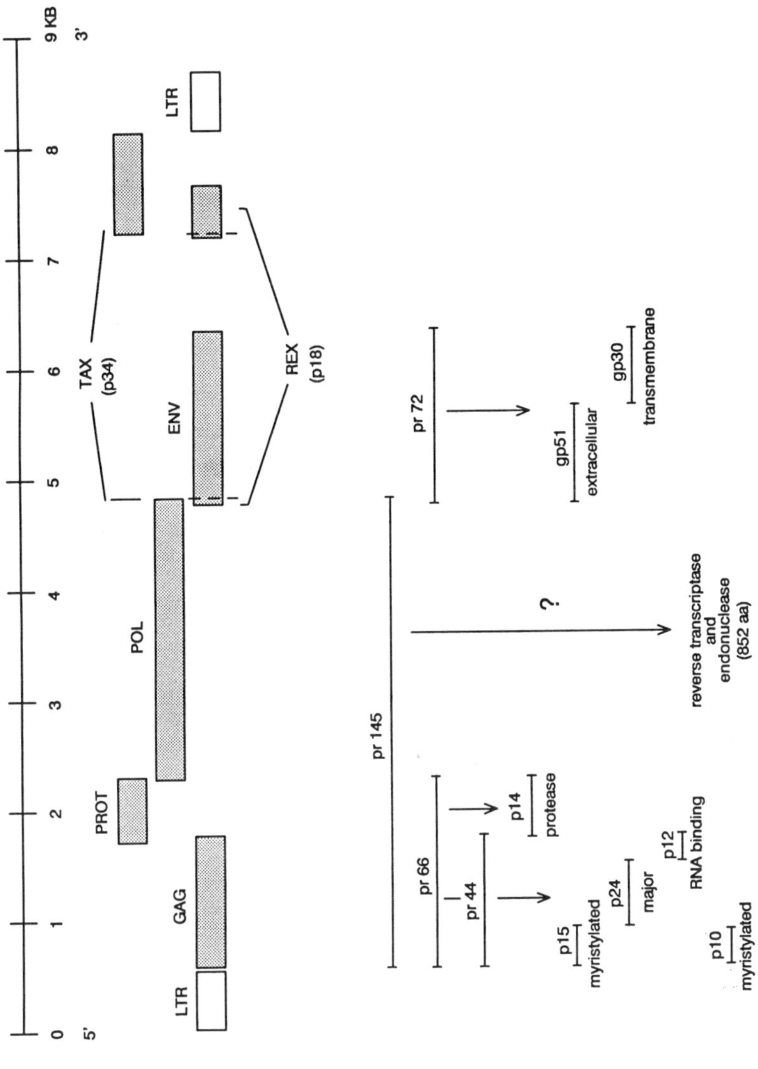

FIGURE 1. Genomic structure of BLV provirus and translation products of viral genes.

gag–pol precursor. Pr 145 contains all of the tryptic peptides of Pr 70^{gag}. It is present as an elongation product of Pr 70^{gag} in BLV-infected cells and in heterologous protein-synthesizing systems programmed with BLV 38 S RNA (Ghysdael *et al.*, 1979).

Pr 72^{env} is the precursor of gp51, the external envelope glycoprotein, and gp30, the transmembrane component. Pr 72^{env} overlaps the right-hand side of pol by a stretch of 17 amino acids and is cleaved into its two components at the Arg-Val-Arg-Arg sequence found at the carboxyl-end of gp51. The unglycosylated homolog Pr 72^{env} is a 47-kDa polypeptide, which is not processed further.

The complete nucleotide sequence of the 8714-bp BLV provirus revealed the presence of a region of 1817 bp (called the X region) between the *env* gene and the 3′ LTR (Rice *et al.*, 1984; Sagata *et al.*, 1985a). This region, also found in the HTLV-I and -II genomes (Seiki *et al.*, 1983; Shimotohno *et al.*, 1985), contains several ORFs. A subgenomic mRNA of 2.1 kb containing two overlapping ORFs has been identified in BLV-infected cells. This mRNA is generated by a complex splicing mechanism resulting in juxtaposition of the 5′ end of the *env* gene and two overlapping X-ORFs (Mamoun *et al.*, 1985; Rice *et al.*, 1987b). Translation of this mRNA can yield at least two distinct proteins depending on which initiation codon is used (Sagata *et al.*, 1985b,c; Rice *et al.*, 1987b). The BLV long open reading frame (LOR) on the X region encodes a protein (apparent molecular weight 34,000–38,000) referred to in this review as $p34^{tax}$ (Tax) that is predominantly located in the nucleus of infected cells (Sagata *et al.*, 1985c). The short open reading frame (SOR) product was identified as a nuclear phosphoprotein, referred to here as $p18^{rex}$ (Rex), with an apparent molecular weight of 18,000–19,000 (Rice *et al.*, 1987b). Both proteins are recognized by sera from cattle or sheep in the tumor phase of infection (Yoshinaka and Oroszlan, 1985; Willems *et al.*, 1988; Powers *et al.*, 1991). In addition to *tax* and *rex*, BLV and HTLV may have the potential to express other small ORFs in the X region (Rice *et al.*, 1984, 1987b; Sagata *et al.*, 1985a; Seiki *et al.*, 1983; Shimotohno *et al.*, 1985). These ORFs have been termed XBL-III and XBL-IV for BLV and pXI and pXII for HTLV. Proteins encoded by these genes would have to be translated from mRNAs that are spliced differently from the mRNA for the Tax and Rex proteins. Recently, complex splicing was shown for HTLV and BLV (Furukawa and Shiku, 1991; Ciminale *et al.*, 1992; Berneman *et al.*, 1992; Koralnik *et al.*, 1992; Alexandersen *et al.*, 1993). Four alternatively spliced mRNAs encoding potential new regulatory proteins have been recently identified in BLV-infected cattle. Interestingly, among them, the presence of the GIV mRNA seems correlated to the development of persistent lymphocytosis (Alexandersen *et al.*, 1993) but is not required for *in vivo* infection (Willems *et al.*, 1993b).

Because of its genome structure, proviral nucleotide sequence, and the size and amino acid sequence of the structural and nonstructural viral proteins, BLV is an obvious relative of the human T lymphotropic viruses (HTLV-I and -II) and the simian T lymphotropic virus (STLV-I) (Weiss *et al.*, 1984; Gallo and Wong-Staal, 1990). These four viruses replicate and act via *trans*-activation of the enhancer–promoter regions of both their own LTRs and cellular genes, and they can induce diseases with similar pathologies, characterized by the absence of chronic viremia, a long latency period, and a lack of preferred proviral integration sites in the tumors.

Oroszlan *et al.* (1982) have shown that the major core proteins of BLV and HTLV-I share significant homology in both their NH_2- and COOH-terminal regions. In preliminary studies, anti-HTLV-I sera failed to cross-react with BLV p24 in a competitive RIA; however, hyperimmune goat anti-HTLV-I serum does react with BLV p24 upon denaturation (Oroszlan *et al.*, 1984). Similarly, using a competitive ELISA, Onuma *et al.* (1987) observed competition between an anti-BLV p24 monoclonal antibody and some human anti-HTLV-I sera only after denaturation of the antigen. Using Western blotting, Maruyama *et al.* (1989) also attempted to study the cross-reactivity between BLV and HTLV-I. Finally, using sera from selected infected hosts and nondenatured virion proteins, Zandomeni *et al.* (1991) showed that BLV, HTLV-I, HTLV-II, and STLV-I share an antigenic determinant(s) that distinguishes them from other retroviruses. The presence of antibodies against this determinant(s) is found only in infected hosts that have high titers of antibodies against the major *gag* proteins.

III. TRANSMISSION OF BLV

The data presented above suggest that BLV is essentially transmitted as a cell-associated infectious agent. BLV is found worldwide, and in temperate climates the virus is spread mainly through iatrogenic transfer of infected lymphocytes. In warm climates and areas heavily populated by hematophagous insects, there are indications of insect-borne propagation of the virus. BLV transmission has been the subject of many field observations and experimental trials (Table I). Cases of natural infection are documented in cattle, sheep, capybara, and water buffalo. Infection can be experimentally transmitted to goats, pigs, rabbits, rhesus monkeys, chimpanzees, and buffalos. There is considerable documentation to support the fact that horizontal transmission is most common (Burny *et al.*, 1980; Burny and Mammerickx, 1987). Since infected cells are the best potential vehicles for infectious BLV particles, the concentration of BLV-positive cells in the blood or other body fluids plays a major role in the success or failure of BLV transmission. As few as 926 lymphocytes (more than 50% containing proviral information) from a cow with PL have been shown to be capable of transmitting infection to serologically negative sheep (Mammerickx *et al.*, 1987).

Recently, Willems *et al.* (1992c) successfully infected sheep through the intradermal inoculation of proviral DNA (100 μg) mixed with a cationic liposome solution. Seroconversion occurred 1 to 2 months after injection. Thus, a cloned BLV provirus is capable of inducing infection in the absence of other cellular factors, underlining the fact that the provirus encodes all of the genes required for its own infectivity. This *in vivo* transfection technique is particularly useful for studying genetically modified proviruses (Willems *et al.*, 1993b) (see Section VI) and avoids the use of mixed viral populations and BLV-infected cells as the source of infectious material.

Recombinant BLV vectors have been designed for the analysis of virus infectivity *in vitro* (Ban *et al.*, 1989; Derse and Martarano, 1990; Milan and Nicolas, 1991), and most cell lines from a variety of species appear to be susceptible to BLV infection (Table I). Thus, the receptor(s) for BLV seems to be widely expressed on various species and cell types.

TABLE I. BLV Infectious Potential

In vitro			*In vivo*[a]			
				Sero-conversion[b]		
Species	Cell culture origin	Infectious BLV material used	Species	A	B	Clinical symptoms
Bat	Adult lung[c,d]	WT[e]	Capybara	+		– (?)[f]
Cat	Lung	Recomb.	Chicken		+	
Cow	Kidney	Recomb.	Chimpanzee		+	+
	Lymphocytes[c]	WT	Cow	+	+	+
Dog	Primary osteogenic sarcoma	Recomb.	Goat		+	+
Goat	Fetal lung	Recomb.	Guinea pig		+	–
Human	Embryonal rhabdomyosarcoma	Recomb.	Macaque		+	–
	Epithelioid carcinoma	Recomb.	Pig		+	–
	Neural[g]	WT	Rabbit		+	+
Monkey	Kidney	Recomb.	Rat		+	–
Mouse	Embryo fibroblast	Recomb.	Sheep	+	+	+
Rabbit	Lung	Recomb.	Water buffalo	+	+	– (?)
Rat	Spleen	WT	Human[h]	–		–
Sheep	Fetal kidney[c,i]	WT				
	Fetal lung	Recomb.				
	Lymphocytes	Recomb.				

[a] From Burny *et al.* (1980, 1984), Burny and Mammerickx (1987), Altanerova *et al.* (1989b; 1990), Wyatt *et al.* (1989).
[b] A, after natural infection; B, after experimental infection.
[c] Led to well-known BLV-producing cell lines.
[d] Graves and Ferrer (1976).
[e] WT, wild-type BLV; recomb., recombinant BLV (from Derse and Martarano, 1990, and/or Milan and Nicolas, 1991).
[f] (?), not followed for a long enough time.
[g] Altaner *et al.* (1989).
[h] Not observed to date.
[i] Van Der Maaten and Miller (1976).

IV. DIAGNOSIS AND CONTROL OF ENZOOTIC BOVINE LEUKOSIS

A. Diagnosis

EBL must be distinguished from SBL, a disease that occurs primarily in young animals. Although lesions in SBL have a similar appearance to those in EBL (frequently B-lymphocyte-derived tumors) (Parodi, 1987), no virus has been found as an etiologic agent of this disease. Thus, diagnosis of EBL can be confirmed by a positive laboratory test for BLV. The various tests employed have been reviewed by Burny *et al.* (1980) and Mammerickx *et al.* (1985).

Because of the absence of viremia and the importance of the anti-gp51 response, investigators developed a test for the early detection of BLV infection that was based on serologically screening for anti-gp51 antibodies. The agar-gel immunodiffusion test (AGID) has been widely used to identify infected animals as a means of helping to eradicate the disease under field conditions. Since an ELISA for detecting BLV infection was introduced, some technical modifications have improved the specificity and sensitivity of the test (Portetelle and Mammerickx, 1987). In particular, ELISA tests involving one or two monoclonal antibodies (MAbs) appear to be the most practical, sensitive, and specific means presently available for the large-scale detection of BLV infection in samples containing low amounts of anti-BLV antibodies (individual milk, bulk milk, and pooled sera). MAbs have been used in several ways: (1) to prepare gp51-coated plates, (2) as a competitor of bovine anti-gp51 antibodies, or (3) as a specific reagent for the detection of bovine anti-gp51 (Florent *et al.*, 1988; Portetelle *et al.*, 1989c; Ban *et al.*, 1990; Klintevall *et al.*, 1991).

Recently, anti-p24 MAbs have been produced, and these will be particularly useful for the development of an ELISA that is capable of distinguishing between infection (animals with anti-p24 antibodies) and immunization (using a gp51 vaccine).

In addition, batteries of anti-gp51 and anti-p24 MAbs have been used for the development of extremely sensitive sandwich ELISAs used to quantify BLV expression in peripheral blood lymphocyte (PBL) cultures (Portetelle *et al.*, 1991) or multiple site ELISAs for the identification of BLV gp51 variants (Bruck *et al.*, 1984a; Portetelle *et al.*, 1989a). For EBL eradication programs, the PBL-ELISA may also be useful, along with serodiagnosis of BLV infection, by providing a simple means of identifying those seropositive cattle that are most likely to transmit infection. Thus, PBL-ELISA could provide a basis for culling strategies designed to limit the spread of infection within a herd (Cowley *et al.*, 1992).

In recent years, recombinant DNA technology has been widely used to produce large quantities of viral proteins in either prokaryotic or eukaryotic systems for their use as diagnostic reagents or in synthetic vaccines (see Section VIII). Different portions of the BLV *env* and p24 genes have been expressed both in bacteria (Platzer *et al.*, 1990; Ulrich *et al.*, 1990, 1991b; Siakkou *et al.*, 1990; Ban *et al.*, 1992) and in yeast (Dumont *et al.*, 1989; Legrain *et al.*, 1989). However, their potential use for diagnostics has been disappointing particularly for gp51 since the yeast-expressed protein is only partially glycosylated as heterodisperse molecules, which are poorly recognized by sera from BLV-infected cows (Legrain *et al.*, 1989).

The polymerase chain reaction (PCR) is currently regarded as the most useful and rapid method for detecting BLV proviral sequences (Gaudi *et al.*, 1990; Ballagi-Pordany *et al.*, 1992) as well as viral RNA expression *in vivo* (Sherman *et al.*, 1992). In order to provide a more sensitive and efficient method for the diagnosis of BLV infection in naturally infected cattle and in sheep, PCR can be followed by an enzyme-linked assay (PCR-ELA) as described by Naif *et al.* (1992). This strategy should permit the identification of infected animals at very early stages of infection. PCR also provides an attractive and alternative method to the conventional multiple site ELISA for studying the intrastrain variability of BLV after the intradermal injection of a cloned BLV provirus in sheep (Willems *et al.*, 1993a). Current data suggest there is an extremely low level (or absence) of intrastrain variability for BLV *in vivo*, which is in sharp contrast with the high mutation rates observed in the lentivirus family. The variability of the BLV *env* gene derived from seven BLV variants that were isolated in different regions of the world was also found to be limited (less than 6%) (Mamoun *et al.*, 1990). Similar observations have been made for HTLV-I (Malik *et al.*, 1988; Gessain *et al.*, 1992).

B. Control

Few countries have carried out seroepidemiological surveys of BLV infection in their entire cattle population. Generally, the national prevalence is estimated by testing a part of the population. Prevalence is considered to be the percentage of positive herds (containing at least one seropositive animal) at any given moment. However, these estimations are of little interest since many parameters can influence the result; e.g., the incidence of BLV infection appears to be related to herd size and cattle breed, the number of BLV responders is variable within a given positive herd (from 1 to 100%), some areas in a country are more highly affected than others, and the different serological tests used do not have equal sensitivity and specificity.

Over the past 5 years (from 1989 to 1993), a program has been initiated in Belgium to eradicate EBL. This program involves:

1. Detection of the BLV-infected herds by ELISA tests on pooled sera (one-third of the entire cattle population per year during the first 3 years of the program)
2. Identification of the infected animals within the positive herds by AGID tests on individual serum samples
3. Elimination of all BLV-positive animals
4. Testing of the remaining animals at 4-month intervals until seropositive animals have been entirely eliminated

An infected herd is considered to be EBL-free if negative test results are obtained on three successive occasions and if newly imported animals originate from other EBL-free herds. The results suggest that the incidence in Belgium is 0.94% (from 0.45 to 2.42% according to the area). This figure is very low relative to other countries. In 1993, after completion of the eradication program, Belgium is now considered to be an EBL-free country. This was achieved after testing more than

75,000 herds (about 2×10^6 animals) and "cleaning" 707 infected herds (Knapen *et al.*, 1990, 1991, 1993).

V. VIRUS–CELL INTERPLAY

The BLV envelope is composed of the membrane glycoproteins gp51 and gp30 and is directly involved in virus infectivity (Fig. 1). This has been demonstrated by the ability of polyclonal sera and certain MAbs directed against gp51 to inhibit some biological activities and BLV-induced cell fusion. The gp51 is the external subunit which contains the cellular receptor-binding domain while gp30 anchors the envelope both to the membrane of infected cells and in virions.

Studies designed to demonstrate the importance of the gp51–gp30 complex in the virus–host interplay and to identify, at the molecular level, the specific regions involved in protective antigenic properties, membrane fusion, and receptor recognition activities have been performed.

A. The Immunological Importance of gp51

All available data indicate the following important roles for BLV gp51 in the immunological response of naturally or experimentally infected animals:

1. BLV gp51 is the first viral antigen to elicit an immunological reaction in the newly infected host. Anti-gp51 antibodies are detectable 2 to 5 weeks after the initiation of experimental infections. Antibodies against gp51 appear earlier than antibodies raised against the major internal protein p24 (Bex *et al.*, 1979; Mammerickx *et al.*, 1980) and are present at higher titers. These findings suggest that some of the viral proteins, and in particular gp51, are potent antigens. The presence of anti-BLV antibodies remains throughout the life of an infected animal and is a reliable marker for viral presence.
2. Anti-gp51 antibodies, produced after natural or experimental BLV infection in cattle and sheep, display distinct antiviral activities: they can neutralize virus infectivity (pseudotypes)* (Zavada *et al.*, 1979) and syncytium-inducing activity† (Guillemain *et al.*, 1978) (Fig. 2), and elicit a strong cytolytic effect on BLV-producing cells in the presence of rabbit complement (Portetelle *et al.*, 1978).
3. Lambs or calves immunized passively with purified immunoglobulins derived from the serum of infected sheep (Kono *et al.*, 1986), the colostrum of

* The VSV–BLV pseudotypes are infectious virions, containing the genome and the internal proteins of the vesicular stomatitis virus (VSV) and the surface glycoprotein of BLV. The VSV–BLV pseudotypes can be utilized in a highly accurate *in vitro* assay for measuring the efficiency of BLV-specific neutralizing antibodies (Zavada *et al.*, 1979).

† One peculiar property of BLV-producing cells (effector cells) is to induce syncytia in mixed cultures with indicator cells such as mouse sarcoma virus-transformed cat cells (CC81 cell line). This syncytium-inducing activity, called early polykaryocytosis by its discovers, occurred as soon as 24 hr after initiation of the culture, and is specifically neutralized by antibodies to BLV gp51 (Guillemain *et al.*, 1978).

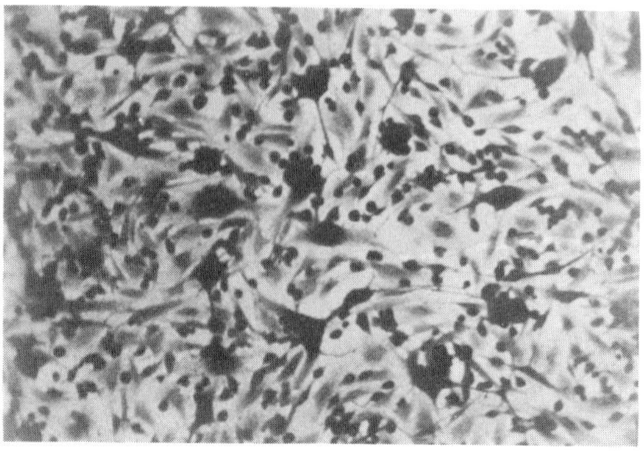

FIGURE 2. Syncytium induction by the BLV-producing cell line (FLK) cocultivated with indicator CC81 cells. ×400.

infected sheep (Mammerickx *et al.*, 1980) or cows (Lassauzet *et al.*, 1989), can successfully resist an infectious challenge provided they have sufficiently high anti-gp51 antibody titers.

In order to characterize regions of gp51 that potentially could act as an effective BLV subunit vaccine or as an efficient diagnostic probe, MAbs, synthetic peptides, and more recently molecular modeling have been used to precisely identify crucial epitopes involved in the biological activities of BLV gp51.

B. Molecular Dissection of gp51 with Monoclonal Antibodies

Mouse MAbs directed against gp51 have identified eight different epitopes referred to as A, B, B′, C, D, D′, E, F, G, H (Bruck *et al.*, 1982a). Three of these, F, G, and H, are associated with viral activities (infectivity and syncytium induction) (Bruck *et al.*, 1982b). The processing of Pr 72^{env} and the biological activity of F, G, H are both conformation-dependent and glycosylation-dependent phenomena (Bruck *et al.*, 1984a). The gp51 sites recognized by neutralizing antibody vary in antigenicity among BLV isolates from the same and different geographical origins (Bruck *et al.*, 1984a; Portetelle *et al.*, 1989a; Mamoun *et al.*, 1990; Coulston *et al.*, 1990). However, none of the currently known variants were simultaneously lacking F, G, and H, suggesting that loss of these three epitopes probably means loss of infectivity. Molecular analysis revealed that sequence variations among widely different isolates are limited to point mutations resulting in substitutions of less than 14 amino acids. Furthermore, no variation in the number and the position of glycosylation sites and cysteine residues has been found, stressing the uniformity of selection pressures exerted on BLV. Limited proteolytic hydrolysis of gp51 and a careful analysis of gp51 amino acid sequence variation indicate that the F, G, and H epitopes are located at the NH_2-moiety of the protein (Portetelle *et al.*, 1989a). F, G, and H are the most accessible epitopes found on intact BLV virions and in immunostimulating complexes (ISCOMs). ELISA tests based on either BLV virions disrupted with a nonionic detergent or purified gp51 as the antigen in a sandwich assay and where the same MAb was chosen as both the binding MAb and the conjugate, suggested the existence of oligomeric complexes for the native env structure where F, G, H are maximally reactive. The antigenicity of these oligomers may therefore be relevant in the design of diagnostic procedures and subunit BLV Env vaccines (Portetelle *et al.*, 1990).

C. Molecular Dissection of gp51 with Synthetic Peptides

A total of 31 synthetic peptides, covering 78% of the BLV gp51 polypeptide backbone, have been employed by our group to precisely identify epitopes recognized by MAbs and to characterize the most efficient B- and T-cell epitopes on this antigen. This is a particularly valuable approach for understanding the antigenic structure of gp51 and for designing a vaccine containing several protective epitopes and at the same time avoiding potentially harmful structures.

A strategy using ELISA tests based on synthetic peptides and/or the corresponding antipeptides, the native BLV gp51, or heterologous *env* gene fragments produced in *E. coli,* was used to localize the epitopes previously defined by MAbs. Epitope B was mapped to amino acids (aa) 181 to 192, B′ to aa 195 to 205, D and D′ to aa 218 to 237, and A to aa 249 to 260 (Portetelle *et al.*, 1989b; Callebaut *et al.*, 1991; Ban *et al.*, 1992) (Fig. 3).

Four epitopes identified recently by Platzer *et al.* (1990) and Ulrich *et al.* (1991a), and referred to as I, K, L, and M, are unrelated to epitopes A to H. Epitope I (recognized by MAb 14; Portetelle *et al.*, 1989b) is located between aa 57 and 64; K, between aa 170 and 217; and M, at the COOH-terminus of gp51 between aa 217 and 268. The conformational epitopes C and L have not yet been located. A MAb against epitope L demonstrated significant competition with sera from infected cattle for gp51 binding, as was previously observed for MAbs against F, G, H (Bruck *et al.*, 1984b). The amino acid residues which are potentially a part of these determinants have been identified by comparing the epitope phenotype with the aa sequence from several BLV isolates (Portetelle *et al.*, 1989a; Mamoun *et al.*, 1990).

The most striking inhibition of syncytium formation was found with rabbit antibodies directed against the peptides 64–73, 98–117, and 177–192. Based on the predicted positions for epitopes F and H, it is interesting to note that: (1) peptide 64–73 is adjacent to a mutation site that may be linked to the F phenotype (residue 62), (2) antisera raised against peptide 64–73 compete with MAb F for

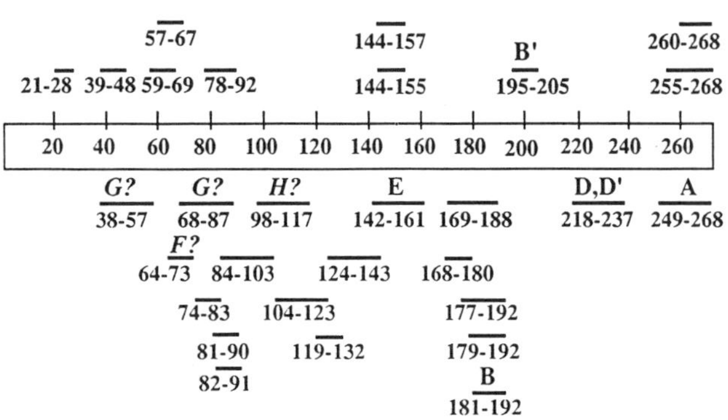

FIGURE 3. Mapping of the gp51 epitopes using synthetic peptides (numbers) and the monoclonal antibody recognition sites (letters) on the polypeptide backbone of BLV gp51 (′, overlapping site; ?, supposed location).

gp51 recognition in an ELISA, (3) residue 111 in peptide 98–117 may belong to the H phenotype, and (4) peptides 98–117 and 177–192 include two of the three loops that are thought to be located at the top of the gp51 head (as defined by modeling) and delineate the putative gp51 binding region for the cell receptor (see Section V.D). It is worth noting that antibodies directed against the region corresponding to peptide 64–73 in the HTLV-1 gp46 sequence [the alignment was performed using the hydrophobic cluster analysis (HCA) method (Callebaut *et al.*, 1994b) are also neutralizing (Palker *et al.*, 1992).

Neutralization of VSV–BLV pseudotypes by antisera directed against peptides 39–48, 78–92, 144–157, and 177–192 has been reported (Callebaut *et al.*, 1993). The peptides 39–48 and 78–92 contain residues 40, 41, 88, which may belong to the G phenotype, and peptide 144–157 of BLV gp51 corresponds to the region in the HTLV-1 envelope glycoprotein involved in neutralization (Matsushita *et al.*, 1986; Ralston *et al.*, 1989; Callebaut *et al.*, 1991). Interestingly, VSV–BLV pseudotypes are resistant to neutralization by two syncytium-inhibiting antibodies raised against peptides 64–73 and 98–117, and it seems likely that the epitopes involved on the pseudotype particles are not accessible to the antipeptide antibodies. Similar instances of failure to neutralize pseudotypes have been reported for anti-HIV-gp41 MAbs that inhibit syncytium formation (Dalgleish *et al.*, 1988). Because antisera against the 177–192 region harbor neutralizing activities while antisera against peptide 169–188 do not, the finding illustrates the importance of one or several residue(s) of the tetrapeptide WAPE (189–192) in the infection mechanism.

Some helper T-cell determinants have also been defined by sera from gp51-immunized mice and BLV-infected cattle. Although none of the peptides tested behaved as a "universal" helper T-cell epitope, peptide 98–117 stimulated proliferation of T cells from BALB/c mice and three infected cows, while peptide 169–188 strongly stimulated T-cell proliferation from one infected cow. Experiments performed with three overlapping peptides from the latter region (177–192, 179–192, 181–192) underline the particular importance of residue(s) 177 (P) and/or 178 (D) in the T-helper epitope (Callebaut *et al.*, 1993).

In another experiment, a complete series of 26 overlapping peptides, each 20 amino acids in length, were tested for their ability to elicit PBL proliferation and cytotoxic T-lymphocyte responses (Gatei *et al.*, 1993a). Peptides 51–70 and 61–80 produced a proliferative response from uninfected animals only (both sheep and cattle). Such results were not observed in our own experiment (Callebaut *et al.*, 1993). On the other hand, cytotoxic T-lymphocyte activity was associated only with peptides 121–140 and 131–150 and was found only in BLV-infected sheep. These results identify, for the first time, cytotoxic T-lymphocyte ($CD8^+$) epitopes on BLV gp51.

D. Molecular Modeling of gp51

Knowledge about the structural organization of the BLV envelope is essential to our understanding of virus infectivity, i.e., receptor binding and the dynamics that follow this event, which are certainly dependent on conformation. This information may also help to explain properties of neutralizing antibodies that will help in the design of an effective vaccine.

The particular abundance of glycosides on viral Env proteins has severely limited their ability to be crystallized, and only one 3-D structure, that of influenza A hemagglutinin, has been solved (Wilson *et al.*, 1981; Wiley and Skehel, 1987). The general organization of hemagglutinin is comparable with retroviral glycoproteins: it is divided into an extramembrane subunit (EM) bearing the receptor binding site and a transmembrane subunit (TM) carrying a hydrophobic "fusion" peptide. However, influenza A does not belong to the Retroviridae family and hemagglutinin does not have any obvious sequence homology (>25% —the limit of classical alignment methods) with retroviral Env glycoproteins. The well-established HCA method (Lemesle-Varloot *et al.*, 1990; Woodcock *et al.*, 1992) is a particularly efficient and sensitive means of detecting structural homologies in spite of weak sequence identities. It has been suggested that the gp30 TM as well as the NH_2- and COOH-terminal portions of the gp51 EM share structural characteristics with the equivalent influenza A hemagglutinin segments, which correspond to the stem of the complex. However, the folding of the central part of gp51 differs from that of the hemagglutinin head. The gp51 head is thought to fold into an antiparallel β structure, based on modeling from the immunoglobulin-like α 3 domain of HLA A2 (Bjorkman *et al.*, 1987), and appears to be organized into trimeric structures (Callebaut *et al.*, 1994b). The major information highlight from this model is that the two loops, against which the two most efficient anti-gp51 neutralizing antibodies are directed, are in close proximity at the top of the domain and could represent a potential site for receptor binding (Fig. 4).

The transmembrane glycoprotein, gp30, is a highly glycosylated polypeptide of 214 aa. It has six cysteine residues, four of which are conserved among all type C viruses, suggesting a crucial pattern of disulfide bonding. It does not seem likely that S–S bridges between gp51 and gp30 serve to link the two proteins after

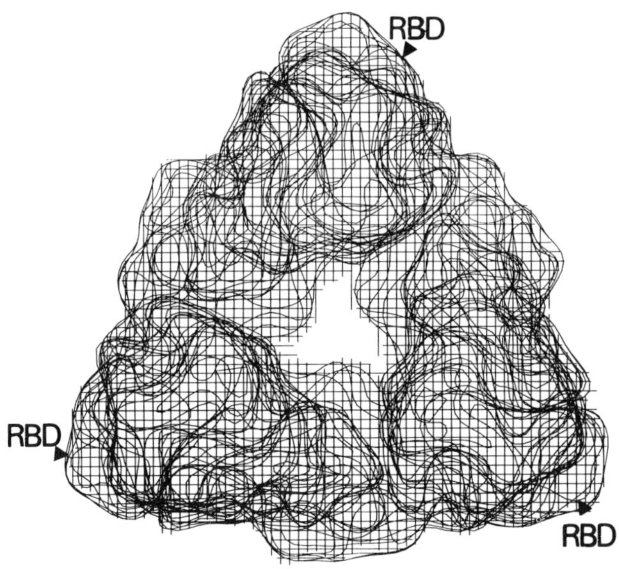

FIGURE 4. Trimeric organization of the globular domains of the gp51 head represented by the surface accessible to the solvent (Conolly's surface). The receptor binding domains (RBD) defined with antipeptide antibodies are indicated.

their proteolytic cleavage from the precursor. gp51 is readily lost from virions during purification procedures, which would favor this hypothesis (Rice *et al.*, 1984, 1985, 1987a; Sagata *et al.*, 1985a; Schultz *et al.*, 1984). Nearest-neighbor relationships between the components of BLV virions were investigated using chemical cross-linking and led to a structural model of the virus particle (Uckert *et al.*, 1984).

E. Virus–Cell Fusion

Enveloped viruses, like BLV, make use of their envelope glycoproteins for fusion with the target cell membrane, particularly because hydrophobic fusion peptides are located in the NH_2-terminal portion of the transmembrane glycoprotein. Brasseur *et al.* (1988; Brasseur, 1991) have suggested that the fusion peptides may act as "sided insertional helices" and insert obliquely into the membrane, thus leading to its destabilization.

In order to test this hypothesis, modified BLV glycoproteins, expressed by vaccinia virus recombinants, were examined in a syncytium-formation assay for their fusogenic capacities (Vonèche *et al.*, 1992a,b). This analysis revealed that: (1) both of the BLV envelope glycoproteins are necessary for cell fusion; (2) insertion of the NH_2-terminal segment of gp30 (the fusion peptide) into the lipid bilayer in an oblique orientation, results in fusogenic capacities that are higher than insertion in a perpendicular or parallel orientation; (3) replacement of the BLV fusion peptide with its simian immunodeficiency virus (SIV) counterpart does not modify the fusogenic capacity of the BLV glycoprotein; and (4) the segment of gp51 between residues 19 and 27 probably adopts an amphipathic β-strand structure that can interact with either the fusion peptide or another region of gp30. Theoretical analysis suggests that these peptides are divided into two regions: one is highly hydrophobic and inserts into the lipid bilayer in α-helical oblique orientation and the other, which is particularly rich in small residues such as alanine, glycine, serine, and threonine, confers mobility to the polypeptide chain. Obliquity and mobility are thus probably responsible for destabilization of the membrane (Callebaut *et al.*, 1994a).

Finally, the ability of some di- and tripeptides to inhibit cell fusion by BLV has been investigated (Vonèche *et al.*, 1993). Unexpectedly, hydrophobic and/or positively charged peptides strongly enhance syncytium formation. These observations can be correlated with theoretical studies predicting the stability of a peptide at the lipid–water interface and its destabilizing properties in a lipid layer, and they suggest that the enhancer peptides interact with the membrane of BLV-producing and/or indicator cells. This enhancement could improve the sensitivity of retroviral detection tests based on syncytium formation.

F. BLV Receptor

Several classes of viral receptors have been studied. A receptor for the human immunodeficiency virus, as well as other human viral receptors such as those for poliovirus and rhinovirus, belong to the immunoglobulin superfamily. Other retroviral receptors have been found to contain amino acid and phosphate permeases, such as the ecotropic murine leukemia virus receptor which contains a basic amino acid transporter and the gibbon ape leukemia virus receptor which has a phosphate permease component homologous to that of *Neurospora crassa.*

Using a lambda gt11 cDNA library derived from BLV permissive bovine cells (MDBK cell line), a cDNA (BLVRcp1) was isolated and characterized. BLVRcp1 encodes a polypeptide that represents part of the putative BLV receptor and selectively binds the envelope glycoprotein gp51 (Ban *et al.*, 1993a). BLVRcp1 codes for a protein fragment of 729 aa; the missing piece of this protein has recently been identified using the BLVRcp1/5′ cDNA (Ban *et al.*, 1993b) to generate an ORF that encodes a putative receptor of 849 aa. This ORF contains a signal sequence, an extracellular domain including the BLV gp51 binding region, a transmembrane region, and a putative intracellular cytoplasmic domain. This molecule appears to be of the EGF-receptor structural type.

No homology with sequences in the nucleotide or amino acid data bases could be found other than a strong homology of approximately 250 bp with Venter's tag sequences corresponding to an unidentified gene from a human brain library. Southern blot analysis revealed that the putative BLV receptor is widely distributed among mammals, which suggests that it plays an important role in the physiology of many cell types. Elucidating the mechanism by which the virus is confined to a few cell types *in vivo* awaits more detailed analysis.

VI. VIRUS EXPRESSION

A. *In Vivo* Studies

BLV expression *in vivo* is thought to be transcriptionally blocked since neither virus particles nor viral proteins or RNA have been readily detected in freshly isolated PBL or tumor cells (Baliga and Ferrer, 1977; Kettmann *et al.*, 1980a, 1982). Even transcriptionally competent proviruses are silent in BLV-infected tumor cells (Van den Broeke *et al.*, 1988). This phenomenon was initially attributed to the presence of a blocking factor in the plasma of BLV-infected cattle (Gupta and Ferrer, 1982). It seems likely that host factors suppress and/or stimulate BLV expression, with their relative concentration in the animal acting as an important determinant of susceptibility or resistance to the development of PL and lymphosarcoma in infected cattle (Zandomeni *et al.*, 1992). Using *in situ* hybridization, the presence of viral RNA has only been detected in 1/2000 to fewer than 1/500,000 of freshly isolated PBL from clinically normal BLV-infected sheep (Lagarias and Radke, 1989). Thirty-eight BLV-induced tumors (16 bovine, 22 bovine) were examined by the same technique, and in about 35% of the tumors, an average of 1/5000 cells contained viral RNA (Fig. 5) (Van den Broeke, personal communication). However, it cannot presently be excluded that these positive cells are nonneoplastic BLV-infected PBL that have infiltrated the tumor. Using the sensitive polymerase chain reaction after reverse transcription (RT-PCR), the presence of the *tax/rex* 2.1-kb mRNA was found in 3/12 normal seropositive animals, in 6/6 animals with PL, and in 5/8 tumors but not in BLV seronegative animals (Jensen *et al.*, 1991). The fact that some BLV-induced tumors were negative for the 2.1-kb mRNA species suggests that expression of

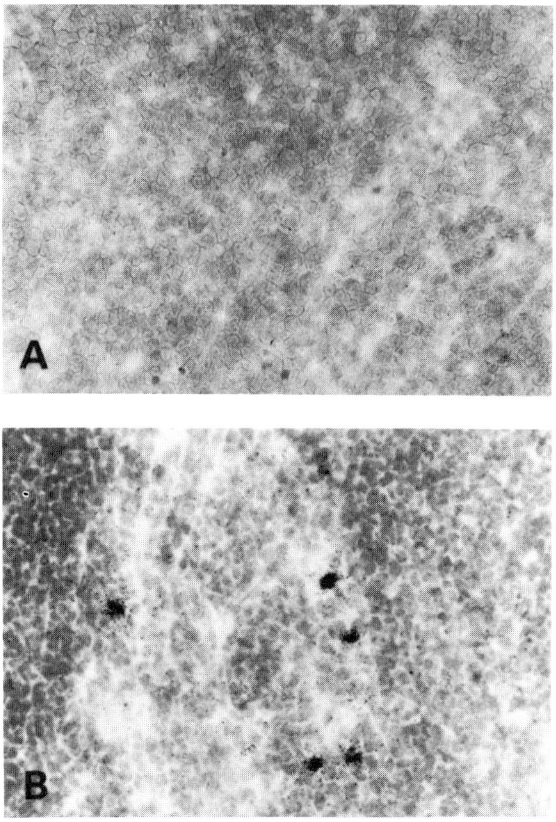

FIGURE 5. Hybridization to RNA in cytological preparations with ^{35}S-labeled antisense RNA corresponding to the 3′ region of the BLV genome. (A) Normal thymus, (B) EBL tumor. ×200.

p34^{tax} is not necessary to maintain the neoplastic state as previously proposed (Kettmann *et al.*, 1982, 1985).

It has recently been reported that BLV transcripts can be amplified by RT-PCR from total or cytoplasm-enriched bovine lymphocyte RNA from all BLV-infected cells but only a few aleukemic animals. Using primer pairs that differentiate the spliced transcripts (genomic, envelope, and *tax/rex*), it appears that a trend toward the exclusion of both genomic and envelope RNAs, with retention of the *tax/rex* message, occurs as the leukemia/lymphosarcoma develops (Haas *et al.*, 1992). The constant presence of *tax/rex* mRNA in the cytoplasm of cells from cows at all stages of infection led these authors to propose that persistent low-level expression of either of these gene products may be important to the pathogenesis of BLV infection.

Taken altogether, the minimal *in vivo* expression of structural genes could be viewed as a unique strategy designed to evade immunosurveillance and permit the persistent and low-level infection that is characteristic of BLV. This could also explain the extremely low genetic variability of BLV relative to the HIV/SIV lentivirus group (Willems *et al.*, 1993a).

Direct inoculation of a cloned BLV provirus in sheep has recently permitted the study of viral infectivity of genetic mutants *in vivo* (Willems *et al.*, 1993b). Three BLV variants cloned from BLV-induced tumors and 12 *in vitro*-modified proviruses were isolated and analyzed for viral expression in cell culture. These proviruses were then inoculated in sheep to assess viral infectivity *in vivo* (see

Section III). Out of three variants cloned from BLV-induced tumors (344, 395, and 1345), one (344) was found to be infectious *in vivo*. This particular provirus was used to engineer 12 BLV mutants. A hybrid between the 5′ region of the complete but noninfectious provirus 395 and the 3′ end of mutant 344 was found to be infectious *in vivo*, suggesting that the *tax/rex* sequences were defective in virus 395. As expected, several regions of the BLV genome appear to be essential for viral infection: the *prot*, the *pol*, and the *env* genes. Even discrete modifications in the fusion peptide located at the NH_2-terminus of the transmembrane gp30 glycoprotein, that do not impair its fusogenic capacity, destroy infectious potential. In contrast, mutations and deletions in the X3–X4 region present between the *env* gene and the 3′ *tax/rex* region do not interfere with viral infection *in vivo*. The function of this region is unknown, and could provide a target regarding the introduction of foreign sequences.

B. *In Vitro* Studies

1. Transactivation

a. *The Transactivator Protein p34^{tax}*

Transcriptional initiation signals for retroviruses lie within the LTR sequences that flank the integrated provirus (Fig. 1). To study these signals, various cell lines have been transfected with recombinant plasmids carrying the bacterial chloramphenicol acetyl-transferase (CAT) gene coupled to the BLV LTR. Transient expression of CAT activity was found only in the established BLV-producer cell line derived from fetal lamb kidney cells (FLK; Van Der Maaten and Miller, 1976) and in BLV-infected bat lung cells (Derse *et al.*, 1985; Rosen *et al.*, 1985). Enhancer activity of the BLV LTR sequences was apparent only in virus-producing cells (Derse and Casey, 1986; Rosen *et al.*, 1986a). It has been suggested that the major determinant for efficient functioning of BLV transcriptional elements is the presence of BLV-induced (viral or cellular) transacting factors because CAT gene expression directed by a plasmid containing the BLV LTR was only observed in BLV-infected cells (Rosen *et al.*, 1985).

Transient expression of a cDNA encompassing the X region ORFs (see Section II) demonstrated that a product encoded by this region can activate, *in trans*, gene expression directed by the BLV LTR (Rosen *et al.*, 1986b; Willems *et al.*, 1987a). Infection of fibroblasts with recombinant murine retroviruses exclusively expressing p34, provided further evidence that the X-LOR gene product (p34^{tax}, Tax) specifically transactivates the BLV LTR (Willems *et al.*, 1987b). Similar observations were obtained using other recombinant plasmids expressing the same product (Derse, 1987). Synthesis of functional p34^{tax} in a recombinant baculovirus system demonstrated that the p34^{tax} product is phosphorylated in insect SF9 cells and requires mammalian cellular factors for transactivation (Chen *et al.*, 1989).

Attempts to identify the region(s) responsible for the functional activity of the transactivator protein by mutational analysis (amino acid deletions or insertions) have been unsuccessful because most of the modifications abrogated transactivation (Willems *et al.*, 1989; Sakurai *et al.*, 1991), suggesting that BLV p34^{tax}

is under heavy evolutionary constraints. By constructing hybrids between a well-defined region of Tax and the specific DNA binding portion of the yeast GAL4 transactivator protein (amino acids 1 to 147), the Tax domain between amino acids 157 and 197 was identified as an activating region (Willems *et al.*, 1991). This segment is approximately located in the middle of $p34^{tax}$ and is globally neutral (a net charge of zero). This putative Tax domain contains 24% leucine residues, which are potentially involved in heterologous protein interactions.

b. The Cis-Acting Response Elements and Cellular Activators

Sequences upstream of the TATA element in the BLV LTR (Couez *et al.*, 1984) were analyzed for their ability to respond to $p34^{tax}$ using the CAT transient expression assay (Gorman *et al.*, 1982). Deletion experiments suggested there were multiple *cis*-acting elements present that respond to $p34^{tax}$ activation (Derse and Casey, 1986; Derse, 1987; Katoh *et al.*, 1989). Transactivation of the BLV LTR by $p34^{tax}$ requires the presence of a 75-bp element upstream of the viral promoter in the LTR U3 region. This fragment was identified as a cell-specific "enhancer" because its activity is restricted to cells productively infected with BLV and contains one or more response elements that are sensitive to p34 expression. Nucleotide sequence analysis shows that this *cis*-acting element contains two 21-bp repeats that are centered around position 148 and located 123 bp upstream of the RNA start site. A third, related element was found at position −48 upstream of the RNA cap site and is apparently poorly active in the absence of the upstream sequences (Fig. 6). These three repeats are very similar, with respect to their nucleotide sequence and spatial array, to elements found in the LTRs of HTLV-I and -II (see chapter 6).

All of these elements contain the common 8-bp core sequence, TGACGTCA, which has been proposed to be a cyclic AMP (cAMP)-responsive element (CRE) (Montminy *et al.*, 1986) or a binding site for the cellular transcription factor, ATF (Lee *et al.*, 1987). The mutually exclusive transactivation of the BLV and HTLV promoters by their cognate Tax proteins suggests that the shared CRE represents the functional center for a larger response element whose specificity is determined by the flanking nucleotides (Derse, 1987). Adenovirus-5 E3 and E4, c-fos, and somatostatin regulatory regions containing a CRE/ATF element have been reported to exhibit responsiveness to $p34^{tax}$ in a CAT transient expression assay (Katoh *et al.*, 1989).

A lambda gt11 cDNA library, constructed with mRNA isolated from a BLV-induced tumor, was screened with an oligonucleotide corresponding to the 21-bp

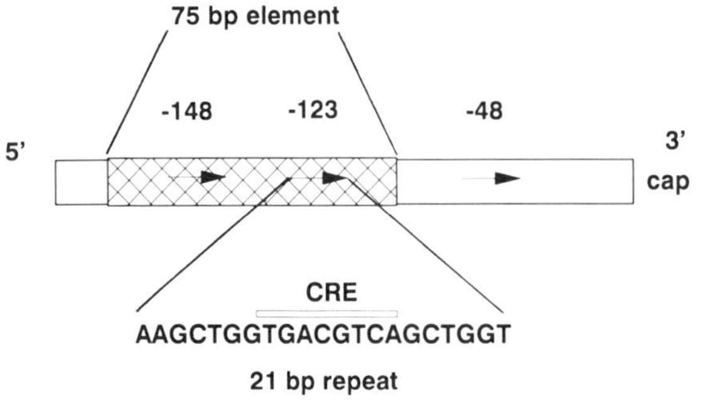

FIGURE 6. U3 region of the BLV LTR.

tax activation response element, and permitted the identification of the bovine CREB2 protein (Willems *et al.*, 1992a). In the absence of p34^{tax}, both the CREB2 protein and the cAMP-dependent protein kinase A can activate the BLV LTR. Thus, CREB2 might be a key element for regulating BLV expression, and play a role in the depression of proviral expression in the absence of Tax and regulation of Tax-induced *trans*-activation during the replicative cycle.

c. *Trans-Regulation of mRNA Processing: The Role of p18^{rex}*

The BLV 2.1-kb mRNA encoding p34^{tax} is translated, in an alternate reading frame, to produce the nuclear phosphoprotein p18^{rex}. Co-transfection experiments using mutated BLV proviruses in combination with BLV-X region expression plasmids indicate that p18 is required to regulate, *in trans*, the accumulation of full-length (genomic) and single-spliced (*env*) viral mRNAs. In contrast, synthesis of the double-spliced X mRNA is not influenced by p18 expression. As a consequence, immediately after its induction viral transcription would predominantly yield the doubly spliced X mRNA. Translation of this message and the subsequent accumulation of p34^{tax} and p18^{rex} would result in both the amplification of viral transcription and a switch to the processing events that permit expression of the structural genes. Large regional deletions and substitutions of proviral sequences have located the elements essential for p18 regulation in the 3′ LTR. Furthermore, sequences within a 250-nucleotide region between the AATAAA signal and poly(A) site were found to be essential for efficient mRNA 3′-end processing and responsiveness to p18^{rex} regulation (Derse, 1988).

2. Transformation

The BLV genome is devoid of a typical oncogene (Deschamps *et al.*, 1981). All BLV-induced tumors are clonal and contain minimally part of a provirus that is randomly integrated in the host genome (Kettmann *et al.*, 1980b, 1982, 1983; Grégoire *et al.*, 1984). All deleted proviral copies in tumor cells have been shown to preserve the X region, stressing its probable important role in tumorigenesis. Considering that the presence of virus is mandatory during the initial steps of the process and apparently dispensable later on (Kettmann *et al.*, 1982, 1985; Van Den Broeke *et al.*, 1988), we can speculate that transient expression of viral functions can lead to the permanent expression of critical cellular genes. Two possibilities must be considered: (1) either the transformation process rests entirely on regulatory mechanisms without DNA alterations (transactivation of host functions; possible mechanisms by which a gene may be switched on by a transient signal and remain on after that signal has disappeared) or (2) alterations in the cellular genome make the cell more susceptible to progression toward the neoplastic state (through mutations, deletions, amplifications, translocations, visible chromosomal abnormalities, and so forth). Aneuploidy and structural modifications have been reported in BLV-induced lymphosarcoma and are observed in cultured tumor cells from cow and sheep in the tumor phase where karyotypic analysis shows that profound rearrangement of one X chromosome is a common event (Popescu, personal communication). The major question that arises is whether chromosomal abnormalities are a primary event in tumorigenesis involving oncogenes, growth factor receptor genes, differentiation genes, or tumor suppressor genes, or a secondary step in the transformation process and reflect an adaptation of the cancer cells to tissue culture conditions or to abnormalities in metabolic pathways.

It is generally accepted that BLV $p34^{tax}$ acts as a positive regulator of viral transcription. A current hypothesis for the induction of leukemogenesis suggests that the transactivation of cellular gene(s) is the first step in transformation. The identity of these genes and their mode of action remain to be determined. Transcriptional activation of the interleukin-2 (IL-2) gene and its receptor (IL-2R) permits the preferential proliferation of the HTLV-infected T cells by an autocrine mechanism (Inoue *et al.*, 1986), although subsequent steps are required to stabilize this activation state (see chapter 6).

In the BLV system, $p34^{tax}$ has been shown to immortalize primary rat embryo fibroblasts (REF) in culture. Furthermore, Tax, in cooperation with the Ha-ras oncogene, can induce foci *in vitro* (Fig. 7) that will develop into tumors *in vivo* (Willems *et al.*, 1990). Thus, BLV Tax can be classified as a member of the immortalizing oncogene subgroup, with other proteins such as adenovirus E1A, SV40 large T, mutated p53 and *myc* gene products. It has recently been reported that Tax, mutated in its putative zinc finger structure, may completely lose its transactivating activity but maintain its transforming capacity (Willems *et al.*, 1992b). Thus, transactivation of the BLV LTR and target cell transformation appear to be independent functions of Tax involving different functional domains of the protein. One could conjecture that Tax must form a complex with a still unknown tumor suppressor gene in order to induce full malignancy.

3. *In Vitro* Culture of BLV-Infected Lymphocytes

BLV was initially identified using short-term cultures of B lymphocytes from cattle in PL (Miller *et al.*, 1969). Virion proteins and virus particles can be detected in some cells as early as 3 to 6 hr after the initiation of these cultures, but are most abundant after 24 to 48 hr in culture (Stock and Ferrer, 1972; Baliga and Ferrer, 1977; Djilali *et al.*, 1987). BLV expression is most easily detected in cultured PBL from cattle with PL (Cockerell and Rovnak, 1988) or sheep with elevated blood leukocyte numbers and an inverted B/T-lymphocyte ratio (Cornil *et al.*, 1988). The virus is found infrequently in cultured PBL from hematologically normal BLV-infected animals (Cockerell and Rovnak, 1988). Rare mononuclear cells, rapidly isolated from the peripheral blood (PBMCs) of asymptomatic sheep with latent infection, contained low amounts of BLV RNA as determined using *in situ* hybridization. After their separation and culture, PBMCs from blood were shown to produce BLV in as many as 1 in 2000 cells. BLV transcripts increased in a biphasic manner in these cultured cells (Lagarias and Radke, 1989).

Cellular transcriptional factors, induced by lymphocyte activation, could initiate low-level transcription from viral promoters and thus lead to the synthesis of a small amount of the viral transactivators, which could then amplify transcription (Derse, 1988; Willems *et al.*, 1992a). The fetal bovine serum present in the culture medium was identified as the principal inducer of viral transcription, an event that occurs in early B-cell activation. Further treatment of PBMCs with T-cell [concanavalin A (Con A)] or B-cell [lipopolysaccharide (LPS), pokeweed mitogen (PWM)] activators increases not only the number of cells containing viral RNA but also the level of viral transcripts within each cell. This finding suggests

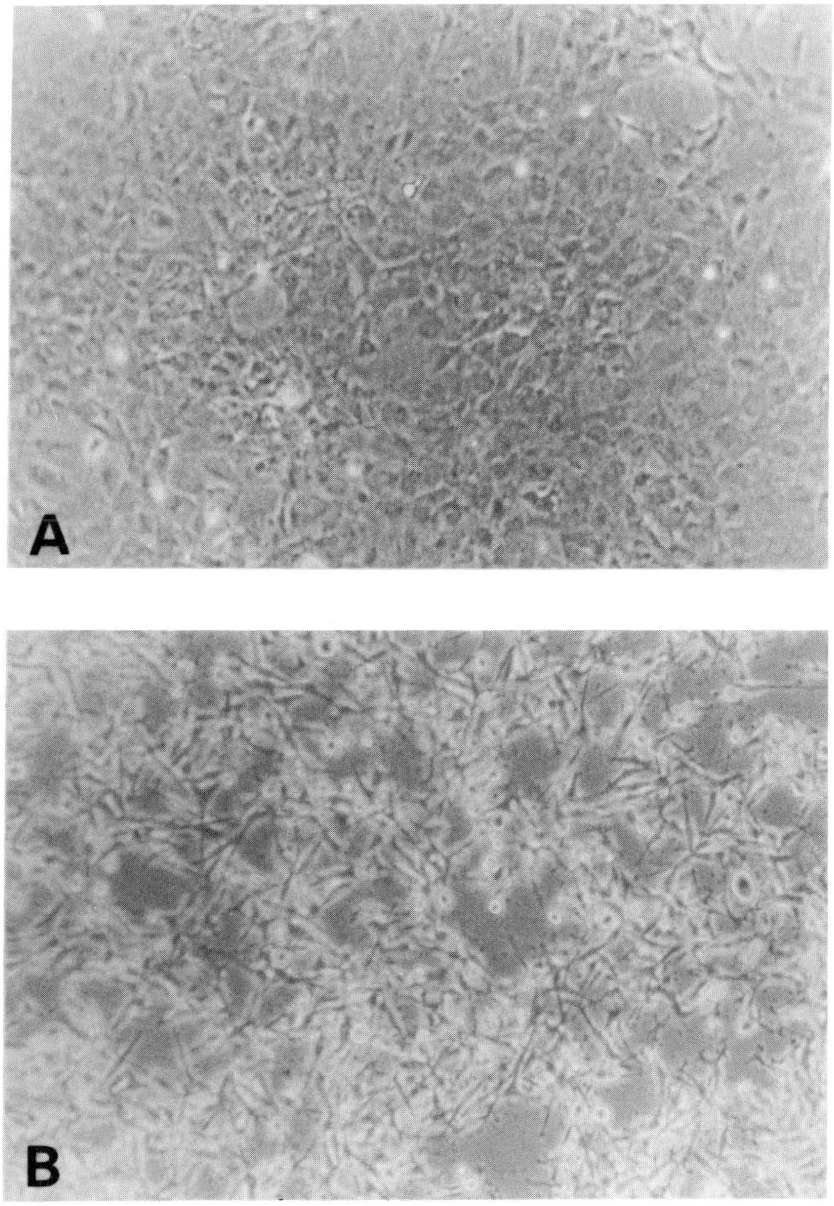

FIGURE 7. (A) Tax-immortalized primary rat embryo fibroblasts (REF) in culture. (B) Tax/Ha-ras-transformed REF cells. ×400.

that immune activation of latently infected cells could induce viral expression in the host (Lagarias and Radke, 1989). Some investigators believe that T and B cells cooperate for BLV expression in ovine lymphocytes (Cornil *et al.*, 1988; Jensen *et al.*, 1990). It also seems likely that T-cell factors alone may enhance BLV expression in B lymphocytes. Alternatively, BLV-infected sheep have been found to possess a plasma factor that can specifically suppress the *in vitro* expression of virus in ovine lymphocyte cultures (Cornil and Levy, 1989) as previously observed by Gupta and Ferrer (1982) in BLV-infected cattle.

Recently, Powers and Radke (1992) demonstrated that transcription can be activated as early as 30 min after PBMCs from a BLV-infected sheep are placed in cultured. This expression is characterized by early and late stages, each distinguished by the unique pattern of cytoplasmic RNAs present. In early expression, the cytoplasmic viral RNA is exclusively the 2.1-kb *tax/rex* transcript, although all of the transcripts are present in the nucleus. Early expression rapidly gives way to late expression, where all of the viral transcripts accumulate in the cytoplasm. The transition from early to late expression requires new protein synthesis, and probably reflects an essential role of viral $p18^{rex}$ in this transition. These results demonstrate the existence of staged viral expression in lymphocytes naturally infected with BLV and validate the data derived using the *in vitro* model of BLV gene expression based on transfection experiments in nonlymphoïd cells (Derse, 1988).

VII. PATHOGENESIS INDUCED BY BLV

A. Overview

BLV particles were first observed by Miller *et al.* (1969) in short-term cultures of PBL from BLV-infected animals in PL. Numerous attempts to detect the virus in body fluids from PL animals and in tumors thought to be of the EBL type failed until it was discovered that viremia can only be monitored during the first 10–12 days postinfection prior to the appearance of permanent antiviral neutralizing antibodies. The continued presence of anti-BLV antibodies in infected animals indicates the existence of a constant antigenic stimulation, which is mediated via either viral proteins and/or virus particles produced by B lymphocytes and possibly other cell types. Considering the long-term progression of BLV infection, cattle fall into three major groups (Fig. 8). The first and largest of these groups (about 60%) include those animals that develop a persistent infection and humoral immune response but are normal in every other respect. The second group, representing perhaps 30 to 35% of all BLV-infected cattle, develop PL. This lymphocytosis results from expansion of the B-lymphocyte population. Some of these B cells carry the BLV provirus but others do not, and it has been suggested that the latter population is related to immune expansion in response to the BLV infection. Expression of a spliced mRNA encoding a potential new regulatory protein seems correlated to development of PL (Alexandersen *et al.*, 1993). A third, and much smaller, group (less than 5%) include those animals that eventually develop lymphosarcoma or localized lymphoïd tumors (Figs. 9 and 10). The role of the bovine major histocompatibility complex (BoLA) in BLV infection and disease progression has been examined in recent years. It appears that resistance and susceptibility to subclinical progression of EBL are associated with the different BoLA-A locus alleles found in various breeds of cattle (Lewin and Bernoco, 1986; Lewin *et al.*, 1988; Leveziel, 1989).

BLV is not only the etiological agent of bovine leukosis but also the rare natural occurrence of ovine leukosis. Although less than 5% of BLV-infected cattle go on to develop tumors, it has been shown that nearly all experimentally BLV-infected sheep progress to and die in the tumor phase of the disease (Mammerickx *et al.*, 1976). Infection, once established, is lifelong. The progression of

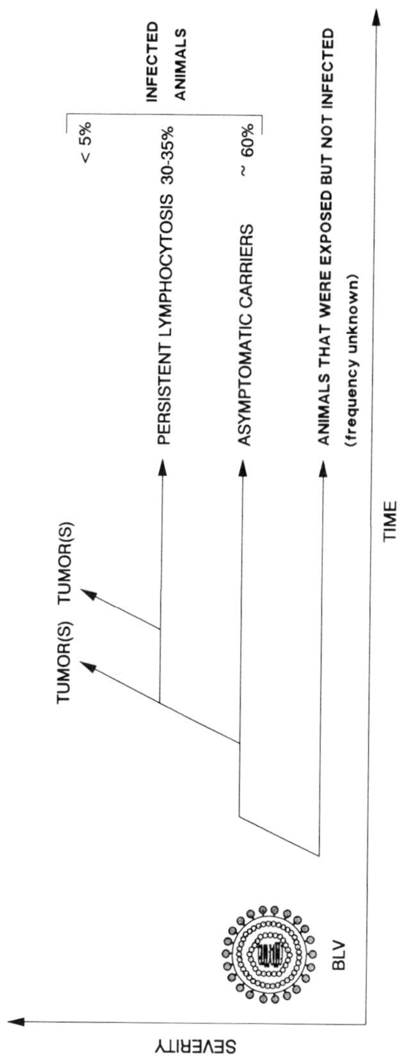

FIGURE 8. BLV-induced pathogenesis.

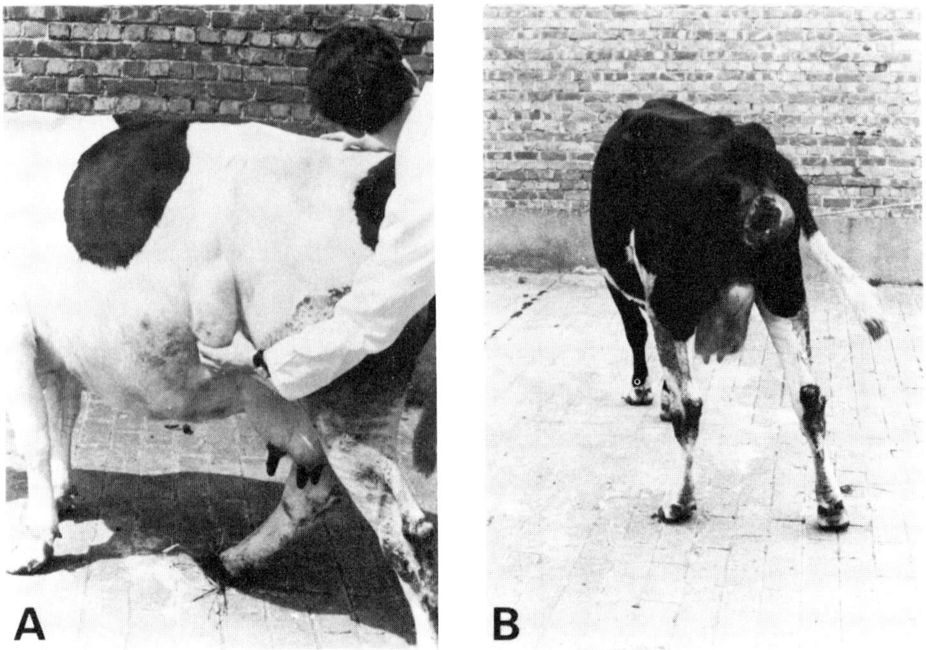

FIGURE 9. Tumor hypertrophy of the precrural lymph node (A) and lymphoid tumor infiltrating the vulva (B).

infection is linked to the infectious dose that seems to play a role in latency before the onset of the neoplastic phase (Mammerickx *et al.*, 1987, 1988).

Furthermore, when BLV is injected into rabbits, it induces profound perturbations in the white cell compartment (neutropenia, lymphopenia) together with diarrhea, and wasting syndrome which is followed by death at a time when the antibody titer has dropped to zero (Burny *et al.*, 1985; Altanerova *et al.*, 1989a; Wyatt *et al.*, 1989).

B. Cooperation between B and T Lymphocytes

Cows with PL have an elevated number of surface immunoglobulin (sIg)-bearing B lymphocytes (Muscoplat *et al.*, 1974). Nylon-adherent bovine cells highly enriched for sIg-positive B lymphocytes have been found to be the main producer of BLV in short-term cultures (Paul *et al.*, 1977). Phytohemagglutinin (PHA) stimulates BLV expression in bovine PBMCs, and this effect is reportedly independent of both its mitogenic effect and T-cell help (Chatterjee *et al.*, 1985; Jensen *et al.*, 1990). Mitogen stimulation does not appear to be required for *in vitro* BLV expression in BLV-infected cattle (Stock and Ferrer, 1972; Driscoll *et al.*, 1977; Jensen *et al.*, 1990). In sheep experimentally infected with BLV, PHA has been reported to either be necessary for (Djilali *et al.*, 1987; Lévy *et al.*, 1987), to markedly enhance (Cornil *et al.*, 1988), or to be dispensable regarding (Jensen *et al.*, 1990) the induction of BLV expression in cultured PBMCs. We can speculate that the induction of viral expression is T-cell independent but T lymphocytes may produce factors that enhance BLV expression in cultured lymphocytes.

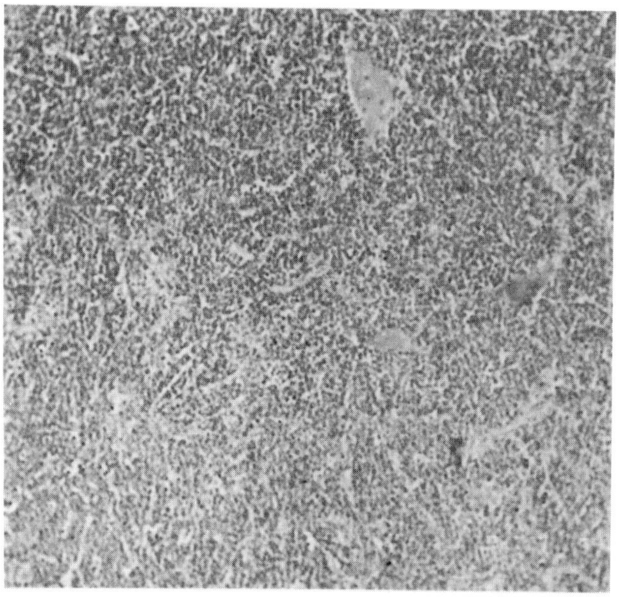

FIGURE 10. Thin section of a sheep tumoral prescapular lymph node. H & E. ×100.

In cattle, it has been shown that BLV can also infect a subpopulation of T lymphocytes in the peripheral blood of infected aleukemic animals (Williams *et al.*, 1988). It has been suggested that infection of this cell population may alter T-cell function and thus facilitate the transition from the aleukemic to PL stage of EBL. In another study, integrated proviral DNA sequences were found integrated in helper and cytotoxic/suppressor T lymphocytes after their *in vitro* cultivation in the presence of recombinant human IL-2 and either PHA, PWM, or Con A (Scott *et al.*, 1991). The ability of BLV provirus to enter and integrate into the genome of both major subpopulations of T lymphocytes, as well as to evoke an increase in the absolute number of T cells in animals with PL (Williams *et al.*, 1988) suggests that the T-lymphocyte population is also afflicted by BLV infection. This deregulation of specific T-cell-derived cytokine synthesis during infection may have an important influence on disease progression. Since $CD5^+$ B cells from cattle in PL proliferate in response to IL-2 (Matheise *et al.*, 1992), T cells may well be involved in the induction of proliferation in a manner similar to that occurring in MAIDS (Mosier *et al.*, 1987).

C. Markers Present on the Surface of the B-Cell Target

BLV-infected B cells from cows with PL and the majority of B cells from BLV-infected sheep with hematological disorders, express high levels of MHC class II, surface IgM, CD5 but also CD11b and CD11c normally borne by cells belonging to the myeloid lineage. It seems that BLV-infected cells of animals in PL are activated cells (Depelchin *et al.*, 1989; Letesson *et al.*, 1990; Matheise *et al.*, 1992). The reason underlying the CD5 T-cell marker's expression on sIg^+ B cells from BLV-infected animals is not understood, but three possible explanations have been suggested: (1) BLV could have tropism for a minor subpopulation of normal CD5$^+$ B cells that clonally expand as a consequence of viral integration; (2) CD5 expression might be linked to either a direct or indirect transactivating effect of the integrated provirus; or (3) the CD5$^+$ phenotype *in vivo* could result from IgM cross-linking of antigens with repetitive epitopes or stimulation with anti-Ig such as that observed for some murine or human B cells (Kantor, 1991; Ying-Zi *et al.*, 1991). Whatever the explanation, it appears that in cattle, at least, CD5 expression is not specifically related to BLV infection, as a dramatic increase in CD5$^+$ B cells has also been observed following a challenge with *Trypanosoma congolense* (Williams *et al.*, 1989). This particular marker, which is frequently found on various mouse B-cell tumors and on human B-CLL cells, is also apparent in some autoimmune diseases (Hayakawa and Hardy, 1988).

Flow cytometric analysis of three BLV-infected bovine tumor cell lines determined that although they apparently belong to the B-cell type, some of them lack detectable surface immunoglobulin. However, all of these lines express both CD5 marker and some myeloid surface markers (Letesson *et al.*, 1991).

Seventeen cases of BLV-induced bovine lymphoid tumors were examined by Heeney and Valli (1990) and determined to be of the B-cell lineage based on their immunoglobulin gene rearrangements. Immunohistochemical studies on frozen fixed sections of bovine lymphosarcoma using a pan-T MAb revealed that they all lacked detectable T cells. Although one tumor failed to react with MAbs directed against either T- or B-cell determinants, all others were positive for various B-cell markers. The most frequent phenotype was Ia$^+$, cIgM$^+$, or sIgM$^+$ with an occasional concurrent appearance of the IgG isotype and sporadic cases of cells positive for terminal deoxynucleotidyl transferase (TdT$^+$). However, the conclusion derived from studying these 16 cases of EBL was that BLV-induced tumors are composed of mature B cells expressing a spectrum of μ, γ_1, or γ_2 probably due to isotype class switching, which is characteristic of mature B cells.

Immunological studies using a highly adsorbed antiserum demonstrated that tumor cells from cattle with EBL also express a tumor-associated antigen (TAA), which is distinct from the BLV-encoded antigens (Onuma and Olson, 1977; Okada *et al.*, 1983). Aida *et al.* (1985, 1993a,b) produced 13 MAbs, identifying three types of TAA expressed on EBL tumors: common TAA, partially common TAA, and individually distinct TAA. One of these MAbs, c143, reacted strongly with tumor cells from individual cows with EBL but did not recognize cells from normal cattle or cells expressing the BLV-encoded proteins. Further examination, however, revealed that c143 TAA is weakly expressed on lymphoid tissues from BLV-free normal cattle of various ages (Okada *et al.*, 1987; Reinacher *et al.*, 1989) and on PBL from several BLV-infected cattle where there was no evidence of tumors (Onuma *et al.*, 1985; Aida *et al.*, 1987; Suzuki *et al.*, 1987). Using flow cytometry, Aida *et al.* (1992) recently confirmed that c143 TAA was expressed on PBL from BLV-free normal cattle. This antigen is composed of two glycoproteins, with molecular weights of 32,000 and 34,000 and the difference between them due to the oligosaccharides. These glycoproteins have been found to be serine phosphorylated on BLV-transformed bovine B lymphocytes but not on normal cells (Aida *et al.*, 1992), and a possible role for phosphorylation in BLV-induced leukemogenesis is being investigated. In BLV-infected cattle, the proportion of PBMCs expressing the c143 TAA increases during the progression of EBL, and BLV-infected cattle without lymphosarcoma show increased proportions of c143

MAb-positive cells coexpressing IgM, CD5, B1 low, and B2 but not CD2, CD4, or CD6. Furthermore, most of the c143 MAb-positive tumors from cattle with EBL appear to be monoclonal and derived from a single B cell that can be divided into two types: c143 TAA$^+$, CD5$^+$, B1 low$^+$, B2$^+$, and sIgM$^+$ *or* sIgM$^-$ (Aida *et al.*, 1993a,b). Further study including cloning and sequencing analysis is now required to clearly define the structure and function of the c143 TAA.

D. The Relevance of Cytokines in BLV-Infected B-Cell Growth

Very little data are available concerning the cytokine responsiveness of BLV-infected B cells. Recently, Matheise *et al.* (1992) found that B cells from animals in PL can proliferate in response to IL-2 (MLA144). This observation is in agreement with the activated state of these cells revealed by their surface markers. IL-6 receptors have also been found on the surface of several bovine and ovine cell lines derived from BLV-induced tumors (Droogmans *et al.*, personal communication). Interestingly, by screening a BLV-induced tumor cDNA library several IL-6 cDNAs were isolated, suggesting that this B-cell growth and differentiation factor is expressed in the tumor cells (Droogmans *et al.*, 1992a). Cloning the bovine IL-6 gene promoter upstream of the bacterial CAT gene (Droogmans *et al.*, 1992b) permitted the studies of the influence of BLV infection on IL-6 promoter activity. No interference was observed between BLV expression and IL-6 gene transcription (L. Droogmans, personal communication). A potential role for BLV infection in the expression of IL-6 receptors remains unknown.

VIII. VACCINATION TRIALS

Surveys of published vaccination trials have been reviewed by Miller (1986) and Portetelle *et al.* (1993), and they reveal that although the work is encouraging, negative data have also been obtained (Table II). The first attempts to provide immune protection against BLV infection were carried out by Miller and Van Der Maaten in 1978, using chemically inactivated BLV virions that were prepared from the cell culture supernatant of BLV-infected FLK cells as a vaccine. This limited study was repeated with less success in 1983, and in 1984, Miller *et al.*, found that antigen treated with glycosidases also failed to prevent infection. Another experiment conducted by Pätrascu *et al.* (1980) obtained better results using inactivated whole virus in an oil adjuvant to stimulate the immune response, and they reported complete protection in 18 out of 20 animals after challenge.

In 1984, Onuma *et al.* vaccinated sheep with several different BLV preparations: gp51 and p24 antigens purified from the virus, glutaraldehyde-fixed FLK cells chronically infected with BLV, and sheep fibroblasts transformed with BLV (SF-28 cells) were used as the inoculum. Animals that were vaccinated with the p24 antigen or the SF-28 cells and did not produce antibodies to the glycoprotein antigen, were not protected against a challenge. In contrast, sheep vaccinated with either viral glycoproteins or FLK cells made antibodies to glycoproteins which appeared to confer protection. Two other vaccination trials were performed by Parfanovich *et al.* (1983) and by Kono *et al.* (1986). The first group used

TABLE II. Vaccine Studies in Sheep and Calves

Vaccine type	Material	Adjuvant[a]	Animal	Challenge[b]	Protection[c]	Reference
Challenge following active immunization with inactivated-BLV or BLV-subunit material						
Inactivated virus[d]						
AEI or BEI	Antigens from culture fluid of FLK cells	AH	Calves	Cattle lymphocytes	3/4	Miller and Van Der Maaten (1978)
BEI	Idem		Cattle	Cattle lymphocytes	2/12	Miller *et al.* (1983)
EI	Antigens from culture fluid of NBL or FLS cells	IFA	Calves	Cattle lymphocytes	18/20	Pätrascu *et al.* (1980)
Formaldehyde	Purified BLV	AH	Cattle	Active BLV	8/8	Parfanovitch *et al.* (1983)
			Cattle	Cow blood	8/8	Parfanovitch *et al.* (1983)
			Sheep	Active BLV	13/13	Parfanovitch *et al.* (1983)
			Sheep	Cow blood	14/14	Parfanovitch *et al.* (1983)
Tween 80	Purified BLV	Iscom	Sheep	Cow blood	0/2	unpublished results
Triton X-100	Antigens from culture fluid of FLK cells	CFA	Sheep	Sheep blood	4/8 then 0/8	Kono *et al.* (1986)
Subunit proteins						
gp51-enriched (AEI treated)	Antigen from culture fluid of FLK cells	CFA	Sheep	Cow lymphocytes	3/3	Onuma *et al.* (1984)
p24-enriched (AEI treated)	Antigen from culture fluid of FLK cells	CFA/none	Sheep	Cow lymphocytes	0/3	Onuma *et al.* (1984)
Glycosidase-treated	BLV gp51-enriched antigen	AH	Calves	Steer lymphocytes	2/8	Miller *et al.* (1984)
Low-pH-treated gp51	Purified BLV	Iscom	Calves	?	?	Merza *et al.* (1991)
Challenge following active immunization with cellular material						
Killed BLV-infected cell						
Glutaraldehyde	FLK cell line	CFA/none	Sheep	Cow lymphocytes	3/3	Onuma *et al.* (1984)
Killed BLV-transformed cell						
Glutaraldehyde	FLK cell line	CFA/none	Sheep	Cow lymphocytes	3/3	Onuma *et al.* (1984)
Killed BLV-transformed cell						
Glutaraldehyde	SF-28 cell line	CFA/none	Sheep	Cow lymphocytes	0/3	Onuma *et al.* (1984)

Killed BLV-induced tumorous lymph node					
Cellular extract		CFA	Calves	Cattle blood	0/10	Ristau *et al.* (1987b)
		None	Sheep	Cattle blood	8/14	Ristau *et al.* (1987a)
		None	Sheep	Cattle blood	5/6	Ristau *et al.* (1987a)
Plasma membrane		None	Sheep	Cattle lymphocytes	9/9	Ristau *et al.* (1987a)
Live cell						
Non-BLV-infected lymphoid cell line	BL-3 cell line	None/AH	Calves	Cattle lymphocytes	8/9	Theilen *et al.* (1982, 1984)
	BL-3 cell line	AH	Calves	Cow lymphocytes	0/8	Ristau *et al.* (1989)
	BL-20 cell line	None	Sheep	Cow lymphocytes	3/4 (challenged at 1 month) 1/4 (challenged at 4 months)	Roberts *et al.* (1982)
	BL-20 cell line	None	Calves	Cow lymphocytes	0/6	Roberts *et al.* (1984)
Producing BLV gp51, gp30, p24	NP-2 cell line	None	Heifers	BLV-producing VP-1 cells	3/3	Altaner *et al.* (1991)
				Purified BLV	2/2	Altaner *et al.* (1991)
Vaccine studies in sheep: challenge following active immunization with recombinant material						
Epitope approach						
Recombinant HBC Ag	aa 56–103 of gp51	?	?	?	?	Borisova *et al.* (1989)
Live recombinant virus						
Vaccinia-env	BLV env-gene	None	Sheep	Cow blood	7/8	Portetelle *et al.* (1991)
Live and killed vaccinia env (+β-propiolactone for the boosters)	BLV env-gene	None	Sheep	Cow blood	0/8	Ohishi *et al.* (1990, 1991)
Vaccinia-env	Truncated BLV env-gene	None	Sheep	Cow blood	0/10	Gatei *et al.* (1993b)
Vaccinia-env						
P7.5 promoter	BLV env-gene	None	Sheep	Cow blood	6/6	Gatei *et al.* (1993b)
PFE/L promoter	BLV env-gene	None	Sheep	Cow blood	4/6	Gatei *et al.* (1993b)

[a] AH, aluminum hydroxide; IFA, incomplete Freund's adjuvant; CFA, complete Freund's adjuvant.
[b] See references for more details [number of vaccinations, schedule, dose of infected material for challenge, etc].
[c] No. protected/No. tested.
[d] AEI, *N*-acetylethylenimine; EI, ethylenimine; BEI, binary ethylenimine.

virus purified on sucrose gradients, inactivated by different methylated amino acids and aluminum hydroxide as an adjuvant. These authors concluded that vaccinated cattle were completely protected against challenge, although this finding is considered questionable because electron microscopy was used to determine positive infections. The second group used BLV purified from FLK cell supernatants by ammonium sulfate precipitation which was subjected to further treatment with 0.1% Triton X-100. Sheep were injected with the immunogenic Freund's complete adjuvant, and 2 weeks after the last injection, the animals were challenged with BLV-infected sheep lymphocytes. These experiments clearly suggest that the induction of a neutralizing antibody response is an important parameter for the short-term protection against BLV infection, if sheep attain sufficiently high anti-gp51 titers. They would also suggest that this protection does not last longer than a few months.

Attempts have also been made using nonviral immunogens. Theilen *et al.* (1982, 1984) used the BLV-free lymphoblastoid cell line BL-3 (derived from an SBL case) as a source of antigen, but the results obtained were inconsequential (unpublished results in Miller, 1986). Another lymphoid cell line, BL-20 (also derived from an SBL case), was used by Roberts *et al.* (1982) for the immunization of sheep. Although a slight protective effect was observed shortly after vaccination, a subsequent study in cattle demonstrated no protection (Roberts *et al.*, 1984). Cellular extracts or plasma membranes obtained from lymph nodes of BLV-infected cattle or BL-3 cells were used by Ristau *et al.* (1987a) to immunize sheep and calves. BLV infection was not apparent in sheep but was apparent in calves using bovine BLV-infected lymphocytes as a challenge inoculum (Ristau *et al.*, 1987b, 1989).

Tests have also been performed to determine whether live heterologous mammalian cells could serve as an immunogen for cattle. Non-virion-producing clonal ovine cells, NP-2 line, have been tested as an immunogen (Altaner *et al.*, 1991). The NP-2 cells synthesize only the BLV *env* gene products (gp51 and gp30) and the main structural protein p24. Based on their data, they concluded that this approach induces the long-term presence of virus-neutralizing antibodies which can protect the animals against a BLV challenge.

The use of the ISCOM technology for presenting gp51 has been shown to induce a considerably higher response in mice than purified gp51 (Merza *et al.*, 1989). In an initial trial, whole BLV was used in the preparation of ISCOMs. Immunoprotection by BLV-ISCOMs was not obtained in sheep since seroconversion occurred after the challenge with a high dose of infected lymphocytes (Portetelle, unpublished data). In the ISCOMs, hydrophobic interactions hold the viral protein and the ISCOM matrix together. The purified gp51, which lacks an accessible hydrophobic region, seems to overcome this problem by exposure of the hidden hydrophobic areas after low-pH treatment of the molecule. Indeed, when purified gp51-ISCOMs were inoculated into four calves, they elicited a high antibody response (Merza *et al.*, 1991). It remains to be shown whether their immunogenicity protects animals against a challenge infection.

Expression of the core antigen of hepatitis B virus (HBC Ag) in *E. coli* leads to the highly efficient synthesis of capsids that are morphologically and immunologically indistinguishable from viral core particles. The usefulness of recombinant HBC Ag as a carrier for foreign oligopeptide sequences has been proposed and special vectors have been constructed to test the expression and exposure of oligopeptides such as aa 56–103 of gp51 (Borisova *et al.*, 1989); however, information is not yet available concerning the immunogenicity of this construct in rabbits, sheep, or cows.

The recombinant vaccinia virus appears to be particularly efficient for producing the gp51 antigen in its native configuration, with accurate addition of the saccharide moiety and without further extraction, purification, or denaturation of the oligomeric structures. Several recombinant vaccinia viruses containing part of (gp51 alone) or the complete (gp51 and gp30) BLV *env* gene have been constructed and used to study immunogenicity in rabbits or to confer protection against experimental BLV challenge in sheep. Considering the worldwide distribution of BLV and the variability of the BLV envelope, three isolates have been used in these constructions: an Australian provirus isolated from a sheep tumor (Kumar *et al.*, 1990; Gatei *et al.*, 1993b), a Belgian provirus from the bovine tumor T15-2 (Portetelle *et al.*, 1991), and a Japanese provirus from the bovine tumor BLV-1 (Ohishi *et al.*, 1988). The BLV *env* gene was cloned into the hemagglutinin HA locus of vaccinia virus, except for the Australian isolate where the thymidine kinase (TK) locus was used.

On the one hand, high neutralizing antibody titers seem to confer parallel protection against BLV infection in various animals (Portetelle *et al.*, 1991). On the other hand, sheep inoculated with recombinants containing the complete *env* gene showed a CD4 response to peptide 51–70 that could be correlated with protection as determined by antibody response, PCR, and infectivity of PBMCs in recipient animals, but this response disappeared 4 months postchallenge; no CD4 response was recorded in animals inoculated with recombinants containing a truncated *env* gene and these animals were infected upon challenge (Gatei *et al.*, 1993b).

Moreover, consistent inhibition of BLV growth has been observed in sheep inoculated with an Env recombinant although this protection was not sufficient to prevent BLV infection after challenge (our observations; Ohishi *et al.*, 1990). Ohishi *et al.* (1988, 1991) observed that the recombinant induces a strong delayed-type hypersensitivity response and stimulates helper T cells without inducing detectable anti-gp51 antibodies, suggesting that the neutralizing titer is not the crucial parameter but reflects the overall efficiency of the protective response. Cell-mediated immune responses are thus assumed to play a major role in protective immunity, and if vaccination induces this protective immunity, it may be sufficient to: (1) suppress partially the growth of BLV (subtotal immunity) in carrier animals independent from neutralizing antibody titers (Ohishi *et al.*, 1991, 1992), or (2) suppress strongly BLV infection (abortive infection) as defined by transient early virus recovery 6 weeks after infection and subsequent absence of residual provirus 4 months postinfection and for the remainder of the trial (16 months) (Gatei *et al.*, 1993b).

IX. CONCLUSIONS

Although BLV infection was first described as a clinical entity in the 19th century, the majority of our knowledge has been obtained in the past 25 years.

In terms of the clinical aspects, some countries, with a low incidence of BLV infection, have successful eradication programs based on the detection of infected animals by specific and sensitive immunological reagents. However, in regions with a high prevalence of EBL, vaccination appears to be the most reasonable approach to solve this problem. Subunit vaccines based on recombinant DNA technology appear to be promising. Increased knowledge of BLV infectivity mechanisms and the identification of both B- and T-cell epitopes on the surface of the envelope gp51 glycoprotein should permit us to improve the design of an effective vaccine.

Another exciting perspective in BLV research appears to be the study of leukemogenesis. Molecular biology has permitted the identification of genes that are potentially involved in the transformation process, i.e., *tax*, GIV. In an exciting new approach, the *in vivo* transfection of infectious proviral DNA in sheep has provided the means to study the role of these genes in the induction of tumor formation. This makes EBL an excellent system for testing the role of regulatory gene sequences on cellular proliferation and transformation *in vivo*.

The BLV model is not only rewarding within itself but also provides a valuable animal model for understanding certain aspects of human HTLV-induced diseases.

ACKNOWLEDGMENTS. The work performed in the authors' laboratory was helped financially by the *Fonds Cancérologique de la Caisse Géñerale d'Epargne et de Retraite*, the Belgian *Service de Programmation de la Politique Scientifique* (SPPS), and the Belgian National Fund for Scientific Research (FNRS).

I.C. is Research Assistant, R.K. is Research Director, and L.W. is Research Associate of the FNRS. L.D. is a Fellow of the FNRS-Télévie.

The authors warmly thank Dr. K. Willard-Gallo for a critical review of the manuscript and B. Chavée and Y. Braet for preparing all documents.

X. REFERENCES

Aida, Y., Onuma, M., Ogawa, Y., Mikami, T., and Izawa, H., 1985, Tumor-associated antigens on bovine leukemia virus-induced bovine lymphosarcoma identified by monoclonal antibodies, *Cancer Res.* **45**:1174.

Aida, Y., Onuma, M., Kasai, N., and Izawa, H., 1987, Use of viable-cell ELISA for detection of monoclonal antibodies recognizing tumor-associated antigens on bovine lymphosarcoma cells, *Am. J. Vet. Res.* **48**:1319.

Aida, Y., Okada, K., Ohtsuka, M., and Amanuma, H., 1992, Tumor-associated Mr 34,000 and Mr 32,000 membrane glycoproteins that are serine phosphorylated specifically in bovine leukemia virus-induced lymphosarcoma cells, *Cancer Res.* **52**:6463.

Aida, Y., Okada, K., and Amanuma, H., 1993a, Phenotype and ontogeny of cells carrying a tumor-associated antigen that is expressed on bovine leukemia virus-induced lymphosarcoma, *Cancer Res.* **53**:429.

Aida, Y., Okada, K., and Onuma, M., 1993b, Antigenic regions defined by monoclonal antibodies on tumor-associated antigens of bovine leukemia virus-induced lymphosarcoma cells, *Leuk. Res.* **17**:187.

Alexandersen, S., Carpenter, S., Christensen, J., Storgaard, T., Viuff, B., Wannemuehler, Y., Belousov, J., and Roth, J., 1993, Identification of alternatively splicing mRNAs encoding potential new regulatory proteins in cattle infected with bovine leukemia virus, *J. Virol.* **67**:39.

Altaner, C., Altanerova, V., Ban, J., Niwa, O., and Yokoro, K., 1989, Human cells of neural origin are permissive for bovine leukemia virus, *Neoplasma* **36**:641.

Altaner, C., Ban, J., Altanerova, V., and Janik, V., 1991, Protective vaccination against bovine leukaemia virus infection by means of cell-derived vaccine, *Vaccine* **9**:889.

Altanerova, V., Ban, J., and Altaner, C., 1989a, Induction of immune deficiency syndrome in rabbits by bovine leukemia virus, *AIDS* **3**:755.

Altanerova, V., Portetelle, D., Kettmann, R., and Altaner, C., 1989b, Infection of rats with bovine leukemia virus: establishement of a virus-producing rat cell line, *J. Gen. Virol.* **70**:1929.

Altanerova, V., Ban, J., Kettmann, R., and Altaner, C., 1990, Induction of leukemia in chicken by bovine leukemia virus due to insertional mutagenesis, *Arch. Geschwulstforsch.* **60**:89.

Baliga, V., and Ferrer, J., 1977, Expression of the bovine leukemia virus and its internal antigen in blood lymphocytes, *Proc. Soc. Exp. Biol. Med.* **156**:388.

Ballagi-Pordany, A., Klintevall, K., Merza, M., Klingeborn, B., and Belak, S., 1992, Direct detection of bovine leukemia virus infection: Practical applicability of a double polymerase chain reaction, *J. Vet. Med.* **39**:69.

Ban, J., First, N., and Temin, H., 1989, Bovine leukemia virus packaging cell line for retrovirus mediated gene transfer, *J. Gen. Virol.* **70**:1987.

Ban, J., Gieciova, E., Orlik, O., and Altaner, C., 1990, Use of monoclonal antibodies in an ELISA for the diagnosis of bovine leukaemia virus infection, *J. Virol. Methods* **30**:79.

Ban, J., Czene, S., Altaner, C., Callebaut, I., Krchnak, V., Merza, M., Burny, A., Kettmann, R., and Portetelle, D., 1992, Mapping of sequential epitopes recognized by monoclonal antibodies on the BLV external glycoprotein expressed in E. coli by means of antipeptide antibodies, *J. Gen. Virol.* **73**:2457.

Ban, J., Portetelle, D., Altaner, C., Horion, B., Milan, D., Krchnak, V., Burny, A., and Kettmann, R., 1993a, Isolation and characterization of a 2.3-Kbp cDNA fragment encoding the binding domain of the bovine leukemia virus cell receptor, *J. Virol.* **67**:1050.

Ban, J., Horion, B., Altaner, C., Burny, A., Portetelle, D., and Kettmann, R., 1993b, Isolation of the missing 5′-end of the encoding region of the bovine leukemia virus cell receptor gene, *Arch. Virol.*, submitted for publication.

Bendixen, H. J., 1963, *Leukosis Enzootica Bovis. Diagnostik, Epidemiologi, Bekaempelse,* Carl F. Mortensen, Copenhagen.

Berneman, Z., Gartenhaus, R., Reitz, M., Blattner, A., Manns, A., Hanchard, B., Ikehara, O., Gallo, R., and Klotman, M., 1992, Expression of alternatively spliced human T-lymphotropic virus type I pX mRNA in infected cell lines and in primary uncultured cells from patients with adult T-cell leukemia/lymphoma and healthy carriers, *Proc. Natl. Acad. Sci. USA* **89**:3005.

Bex, F., Bruck, C., Mammerickx, M., Portetelle, D., Ghysdael, J., Cleuter, Y., Leclercq, M., Dekegel, D., and Burny, A., 1979, Humoral antibody response to bovine leukemia virus infection in cattle and sheep, *Cancer Res.* **39**:1118.

Bjorkman, P., Saper, M., Samraoui, B., Bennett, W., Strominger, J., and Willey, D., 1987, Structure of the human class I histocompatibility antigen, HLA-A2, *Nature* **329**:506.

Borisova, G. P., Berzins, I., Pushko, P., Pumpen, P., Gren, E., Tsibinogin, V., Loseva, V., Ose, V., Ulrich, R., Siakkou, H., and Rosenthal, H., 1989, Recombinant core particles of hepatitis B virus exposing foreign antigenic determinants on their surface, *FEBS Lett.* **259**:121.

Brasseur, R., 1991, Differentiation of lipid-associated helices by use of three-dimensional molecular hydrophobicity potential calculations, *J. Biol. Chem.* **266**:16120.

Brasseur, R., Cornet, B., Burny, A., Vandenbranden, M., and Ruysschaert, J. M., 1988, Mode of insertion into a lipid membrane of the N-terminal HIV peptide segment, *AIDS Res. Hum. Retroviruses* **4**:83.

Bruck, C., Mathot, S., Portetelle, D., Berte, C., Franssen, J. D., Herion, P., and Burny, A., 1982a, Monoclonal antibodies define eight independent antigenic regions on the bovine leukemia virus (BLV) envelope glycoprotein gp51, *Virology* **122**:342.

Bruck, C., Portetelle, D., Burny, A., and Zavada, J., 1982b, Topographical analysis by monoclonal antibodies of BLV-gp51 epitopes involved in viral functions, *Virology* **122**:353.

Bruck, C., Rensonnet, N., Portetelle, D., Cleuter, Y., Mammerickx, M., Burny, A., Mamoun, R., Guillemain, B., Van Der Maaten, M., and Ghysdael, J., 1984a, Biologically active epitopes of bovine leukemia virus glycoprotein gp51: Their dependence on protein glycosylation and genetic variability, *Virology* **136**:20.

Bruck, C., Portetelle, D., Mammerickx, M., Mathot, S., and Burny, A., 1984b, Epitopes of BLV glycoprotein gp51 recognized by sera infected cattle and sheep, *Leuk. Res.* **8**:315.

Burny, A., and Mammerickx, M., 1987, *Enzootic Bovine Leukosis and Bovine Leukemia Virus,* Nijhoff, The Hague.

Burny, A., Bruck, C., Chantrenne, H., Cleuter, Y., Dekegel, D., Ghysdael, J., Kettmann, R., Leclercq, M., Leunen, J., Mammerickx, M., and Portetelle, D., 1980, Bovine leukemia virus: Molecular biology and epidemiology, in: *Viral Oncology* (G. Klein, ed.), pp. 231–289, Raven Press, New York.

Burny, A., Bruck, C., Cleuter, Y., Couez, D., Dekegel, D., Deschamps, J., Ghysdael, J., Kettmann, R., Marbaix, G., Mammerickx, M., and Portetelle, D., 1984, Leukemogenesis by bovine leukemia virus, in: *Mechanism of Viral Leukemogenesis* (J. Goldman and O. Jarrett, eds.), pp. 229–260, Churchill Livingstone, Edinburgh.

Burny, A., Bruck, C., Cleuter, Y., Couez, D., Deschamps, J., Ghysdael, J., Gregoire, D., Kettmann, R., Mammerickx, M., Marbaix, G., Portetelle, D., and Willems, L., 1985, Bovine leukemia virus, a versatile agent with various pathogenic effects in various animal species, *Cancer Res.* **45**:4578.

Burny, A., Cleuter, Y., Kettmann, R., Mammerickx, M., Marbaix, G., Portetelle, D., Van Den Broeke, A., Willems, L., and Thomas, R., 1990, Bovine leukemia: Facts and hypotheses derived from the study of an infectious cancer, in: *Retrovirus Biology and Human Disease* (R. Gallo and F. Wong-Staal, eds.), pp. 9–32, Dekker, New York.

Callebaut, I., Burny, A., Krchnak, V., Gras-Masse, H., Wathelet, B., and Portetelle, D., 1991, Use of synthetic peptides to map sequential epitopes recognized by monoclonal antibodies on the bovine leukemia virus external glycoprotein, *Virology* **185**:48.

Callebaut, I., Voneche, V., Mager, A., Fumiere, O., Krchnak, V., Merza, M., Zavada, J., Mammerickx, M., Burny, A., and Portetelle, D., 1993, Mapping of B-neutralizing and T-helper cell epitopes on the bovine leukemia virus external glycoprotein gp51, *J. Virol.*, **93**:5321.

Callebaut, I., Tasso, A., Brasseur, R., Burny, A., Portetelle, D., and Mornon, J. P., 1994a, Common prevalence of alanine and glycine in mobile reactive centre loops of serpins and viral fusion peptides. Do prions possess a fusogenic peptide? *J. Comput. Aided Mol. Des.*, In press.

Callebaut, I., Portetelle, D., Burny, A., and Mornon, J. P., 1994b, Structural and functional features of bovine leukemia virus gp51 head domain. Towards a general model for oncovirus envelope glycoproteins. *Eur. J. Biochem.*, in press.

Chatterjee, R., Gupta, P., Kashmiri, S., and Ferrer, J., 1985, Phytohemagglutinin activation of the transcription of the bovine leukemia virus genome requires de novo protein synthesis, *J. Virol.* **54**:860.

Chen, G., Willems, L., Portetelle, D., Willard-Gallo, K., Burny, A., Ghysen, D., and Kettmann, R., 1989, Synthesis of functional bovine leukemia virus (BLV) p34^{tax} protein by recombinant baculoviruses, *Virology* **173**:343.

Ciminale, V., Pavlakis, G., Derse, D., Cunningham, C., and Felber, B., 1992, Complex splicing in the human T-cell leukemia virus (HTLV) family of retroviruses: Novel mRNAs and proteins produced by HTLV type I, *J. Virol.* **66**:1737.

Cockerell, G., and Rovnak, J., 1988, The correlation between the direct and indirect detection of bovine leukemia virus infection in cattle, *Leuk. Res.* **12**:465.

Cornil, I., and Levy, D., 1989, *In vivo* inhibition of bovine leukemia virus (BLV) expression, *Leukemia* **3**:159.

Cornil, I., Delon, P., Parodi, A., and Levy, D., 1988, T–B cell cooperation for bovine leukemia virus expression in ovine lymphocytes, *Leukemia* **2**:313.

Couez, D., Deschamps, J., Kettmann, R., Stephens, R., Gilden, R., and Burny, A., 1984, Nucleotide sequence analysis of the long terminal repeat of integrated bovine proviral DNA and of adjacent viral and host sequences, *J. Virol.* **49**:615.

Coulston, J., Naif, H., Brandon, R., Kumar, S., Khan, S., Daniel, R., and Lavin, M., 1990, Molecular cloning and sequencing of an Australian isolate of proviral bovine leukemia virus DNA: Comparison with other isolates, *J. Gen. Virol.* **71**:1737.

Cowley, J., Molloy, J., Dimmock, C., Walker, P., Bruyeres, A., and Ward, W., 1992, Infectivity of bovine leukaemia virus infected cattle: An Elisa for detecting antigens expressed *in vitro* cultured lymphocytes, *Vet. Microbiol.* **30**:137.

Dalgleish, A., Chanh, T., Kennedy, R., Kanda, P., Clapham, P., and Weiss, R., 1988, Neutralization of diverse HIV-1 strains by monoclonal antibodies raised against a gp41 synthetic peptide, *Virology* **165**:209.

Depelchin, A., Letesson, J., Lostrie, N., Mammerickx, M., Portetelle, D., and Burny, A., 1989, Bovine leukemia virus (BLV) infected B cells express a marker similar to the CD5 T-cell marker, *Immunol. Lett.* **20**:69.

Derse, D., 1987, Bovine leukemia virus transcription is controlled by a virus-encoded trans-acting factor and by cis-acting response elements, *J. Virol.* **61**:2462.

Derse, D., 1988, Trans-acting regulation of bovine leukemia virus mRNA processing, *J. Virol.* **62**:1115.

Derse, D., and Casey, J., 1986, Two elements in the bovine leukemia virus long terminal repeat that regulate gene expression, *Science* **231**:1437.

Derse, D., and Martarano, L., 1990, Construction of a recombinant bovine leukemia virus vector for analysis of virus infectivity, *J. Virol.* **64**:401.

Derse, D., Caradona, S., and Casey, J., 1985, Bovine leukemia virus long terminal repeat: A cell type specific promoter, *Science* **227**:317.

Deschamps, J., Kettmann, R., and Burny, A., 1981, Experiments with cloned complete tumor-derived bovine leukemia virus information prove that the virus is totally exogenous to its animal species, *J. Virol.* **40**:60.

Djilali, S., Parodi, A., and Levy, D., 1987, Bovine leukemia virus replicates in sheep B-lymphocytes under a T-cell released factor, *Eur. J. Cancer Clin. Oncol.* **23**:81.

Driscoll, D., Baumgartner, L., and Olson, C., 1977, Concanavalin A and the production of bovine leukemia virus antigen in short-term lymphocyte culture, *J. Natl. Cancer Inst.* **58**:1513.

Droogmans, L., Cludts, I., Cleuter, Y., Kettmann, R., and Burny, A., 1992a, Nucleotide sequence of bovine interleukin-6 cDNA, *DNA Sequence* **2**:411.

Droogmans, L., Cludts, I., Cleuter, Y., Kettmann, R., and Burny, A., 1992b, Nucleotide sequence of the bovine interleukin-6 gene promoter, *DNA Sequence* **2**:115.

Dumont, J., Legrain, M., Portetelle, D., Brasseur, R., Burny, A., and Hilger, F., 1989, High yield expression of the bovine leukemia virus (BLV) p24 major internal protein in Saccharomyces cerevisiae, *Gene* **79**:219.

Florent, G., Delgoffe, J. C., and Zygraich, N., 1988, Detection of antibodies to bovine leukemia virus in bovine milk samples with an ELISA involving two monoclonal antibodies, *Vet. Microb.* **18**:89.

Furukawa, K., and Shiku, H., 1991, Alternatively spliced mRNA of the pX region of human T lymphotropic virus type I proviral genome, *FEBS Lett.* **295**:141.

Gallo, R., and Wong-Staal, F., eds., 1990, *Retrovirus Biology and Human Disease*, Dekker, New York.

Gatei, M., Good, M., Daniel, R., and Lavin, M., 1993a, T-cell responses to highly conserved CD4 and CD8 epitopes on the outer membrane protein of bovine leukemia virus: Relevance to vaccine development, *J. Virol.* **67**:1796.

Gatei, M., Naif, H., Kumar, S., Boyle, D., Daniel, R., Good, M., and Lavin, M., 1993b, Protection of sheep against bovine leukemia virus (BLV) infection by vaccination with recombinant vaccinia viruses, expressing BLV envelope glycoproteins: Correlation of protection with CD4 T-cell response to gp51 peptide 51–70, *J. Virol.* **67**:1803.

Gaudi, S., Ponti, W., Agresti, A., Meneveri, R., Malcovati, M., Bonizzi, L., Poli, G., Amato, A., and Ginelli, E., 1990, Detection of bovine leukaemia virus (BLV) infection by DNA probe technology, *Mol. Cell. Probes* **4**:163.

Gessain, A., Gallo, R., and Franchini, G., 1992, Low degree of human T-cell leukemia/lymphoma virus type I genetic drift *in vivo* as a means of monitoring viral transmission and movement of ancient human populations, *J. Virol.* **66**:2288.

Ghysdael, J., Kettmann, R., and Burny, A., 1979, Translation of BLV virion RNAs in heterologous protein synthesizing systems, *J. Virol.* **29**:1087.

Ghysdael, J., Bruck, C., Kettmann, R., and Burny, A., 1984, Bovine leukemia virus, *Curr. Top. Microbiol. Immunol.* **112**:1.

Gorman, C., Moffat, L., and Howard, B., 1982, Recombinant genomes which express chloramphenicol acetyltransferase in mammalian cells, *Mol. Cell. Biol.* **2**:1044.

Graves, D., and Ferrer, J., 1976, *In vitro* transmission and propagation of the bovine leukemia virus, in monolayer cell cultures, *Cancer Res.* **36**:4152.

Grégoire, D., Couez, D., Deschamps, J., Heurtz, S., Hors-Cayla, M.-C., Szpirer, J., Szpirer, C., Burny, A., Huez, G., and Kettmann, R., 1984, Different bovine leukemia virus-induced tumors harbor the provirus in different chromosomes, *J. Virol.* **50**:275.

Guillemain, B., Mamoun, R., Levy, D., Astier, T., Irgens, K., and Parodi, A. L., 1978, Early polykaryocytosis inhibition: A simple quantitative *in vitro* assay for the detection of bovine leukemia virus infection in cattle, *Eur. J. Cancer* **14**:811.

Gupta, P., and Ferrer, J., 1982, Expression of bovine leukemia virus genome is blocked by a non immunoglobulin protein in plasma from infected cattle, *Science* **215**:405.

Haas, L., Divers, T., and Casey, J., 1992, Bovine leukemia virus gene expression *in vivo*, *J. Virol.* **66**:6223.

Hayakawa, K., and Hardy, R., 1988, Normal, autoimmune and malignant $CD5^+$ B cells: The LY-1 lineage? *Annu. Rev. Immunol.* **6**:197.

Heeney, J., and Valli, V., 1990, Transformed phenotype of enzootic bovine lymphoma reflects differentiation-linked leukemogenesis, *Lab. Invest.* **62**:339.

Inoue, J., Seiki, M., Taniguchi, T., Tsuru, S., and Yoshida, M., 1986, Induction of interleukin 2 receptor gene expression by $p40^x$ encoded by human T-cell leukemia virus type 1, *EMBO J.* **5**:2883.

Jensen, W., Sheehy, S., Fox, M., Davis, W., and Cockerell, G., 1990, *In vitro* expression of bovine leukemia virus in isolated B-lymphocytes of cattle and sheep, *Vet. Immunol. Immunopathol.* **26**:333.

Jensen, W., Rovnak, J., and Cockerell, G., 1991, *In vivo* transcription of the bovine leukemia virus tax/rex region normal normal and neoplastic lymphocytes of cattle and sheep, *J. Virol.* **65**:2484.

Kantor, B., 1991, The development and repertoire of B-1 cells (CD5 B cells), *Immunol. Today* **12**:389.

Katoh, I., Yoshinaka, Y., and Ikawa, Y., 1989, Bovine leukemia virus trans-activator $p38^{tax}$ activates heterologous promoter with a common sequence known as a cAMP-responsive element or the binding site of a cellular transcription factor ATF, *EMBO J.* **8**:497.

Katoh, I., Kyushiki, H., Sakamoto, Y., Ikawa, Y., and Yoshinaka, Y., 1991, Bovine leukemia virus matrix-associated protein MA (pl5): Further processing and formation of a specific complex with the dimer of the 5′-terminal genomic RNA fragment, *J. Virol.* **65**:6845.

Katoh, I., Yasunaga, T., and Yoshinaka, Y., 1993, Bovine leukemia virus RNA sequences involved in dimerization and specific gag protein binding: Close relation to the packaging sites of avian, murine, and human retroviruses, *J. Virol.* **67**:1830.

Kettmann, R., Portetelle, D., Mammerickx, M., Cleuter, Y., Dekegel, D., Galoux, M., Ghysdael, J., Burny, A., and Chantrenne, H., 1976, Bovine leukemia virus: An exogenous RNA oncogenic virus, *Proc. Natl. Acad. Sci. USA* **73**:1014.

Kettmann, R., Burny, A., Cleuter, Y., Ghysdael, J., and Mammerickx, M., 1978, Distribution of bovine leukemia virus proviral DNA sequences in tissues of animals with enzootic bovine leukosis, *Leuk. Res.* **2**:23.

Kettmann, R., Marbaix, G., Cleuter, Y., Portetelle, D., Mammerickx, M., and Burny, A., 1980a, Genomic integration of bovine leukemia provirus and lack of viral RNA expression in the target cells of cattle with different response to BLV infection, *Leuk. Res.* **4**:509.

Kettmann, R., Cleuter, Y., Mammerickx, M., Meunier-Rotival, M., Bernardi, G., Burny, A., and Chantrenne, H., 1980b, Genomic integration of bovine leukemia provirus: Comparison between persistent lymphocytosis and lymph node tumor form of enzootic bovine leukosis, *Proc. Natl. Acad. Sci. USA* **77**:2577.

Kettmann, R., Deschamps, J., Cleuter, Y., Couez, D., Burny, A., and Marbaix, G., 1982, Leukemogenesis by bovine leukemia virus: Provirus DNA integration and lack of RNA expression of viral long terminal repeat and 3′ proximate cellular sequences. *Proc. Natl. Acad. Sci. USA* **79**:2465.

Kettmann, R., Deschamps, J., Couez, D., Claustriaux, J.-J., Palm, R., and Burny, A., 1983, Chromosome integration domain for bovine leukemia provirus in tumors, *J. Virol.* **47**:146.

Kettmann, R., Cleuter, Y., Gregoire, D., and Burny, A., 1985, Role of the 3′ long open reading frame region of bovine leukemia virus in the maintenance of cell transformation, *J. Virol.* **54**:899.

Klintevall, K., Näslund, K., Svedlund, G., Hajdu, L., Linde, N., and Klingeborn, B., 1991, Evaluation of an indirect ELISA for the detection of antibodies to bovine leukaemia virus in milk and serum, *J. Gen. Virol. Med.* **33**:319.

Knapen, K., Kerkhofs, P., and Mammerickx, M., 1990, Eradication de la leucose bovine enzootique en Belgique: bilan après une première année de dépistage sur un tiers du cheptel national, *Ann. Med. Vet.* **134**:557.

Knapen, K., Kerkhofs, P., and Mammerickx, M., 1991, Eradication de la leucose bovine enzootique en Belgique: bilan après une deuxième année de dépistage sur un second tiers du cheptel national, *Ann. Med. Vet.* **135**:491.

Knapen, K., Kerkhofs, P., and Mammerickx, M., 1993, Eradication de la leucose bovine enzootique en Belgique: bilan du dépistage de masse réalisé sur l'ensemble du cheptel national en 1989, 1990 et 1991, *Ann. Med. Vet.* **137**:197.

Kono, Y., Arai, K., Sentsui, H., Matsukawa, S., and Itohara, S., 1986, Protection against bovine leukemia virus infection in sheep by active and passive immunization, *Jpn. J. Vet. Sci.* **48**:117.

Koralnik, I., Gessain, A., Klotman, M., Lo Monico, A., Berneman, Z., and Franchini, G., 1992, Protein isoforms encoded by the pX region of human T-cell leukemia/lymphotropic virus type I, *Proc. Natl. Acad. Sci. USA* **89**:8813.

Kumar, S., Andrew, M., Boyle, D., Brandon, R., Lavin, M., and Daniel, R., 1990, Expression of bovine leukaemia virus envelope gene by recombinant vaccinia viruses, *Virus Res.* **17**:131.

Lagarias, D., and Radke, K., 1989, Transcriptional activation of bovine leukemia virus in blood cells from experimentally infected asymptomatic sheep with latent infection, *J. Virol.* **63**:2099.

Lassauzet, M., Johnson, W., Thurmond, M., and Stevens, F., 1989, Protection of colostral antibodies against bovine leukemia virus infection in calves on a california dairy, *Can. J. Vet. Res.* **53**:424.

Lee, K., Hai, T., Siva Raman, L., Thimmappaya, B., Hurst, H., Jones, N., and Green, N., 1987, A cellular protein, activating factor, activates transcription of multiple E1A-inducible adenovirus early promoters, *Proc. Natl. Acad. Sci. USA* **84**:8355.

Legrain, M., Portetelle, D., Dumont, J., Burny, A., and Hilger, F., 1989, Biochemical and immunological characterization of the bovine leukemia virus (BLV) envelope glycoprotein (gp51) expressed in Saccharomyces cerevisiae, *Gene* **79**:227.

Lemesle-Varloot, L., Henrissat, B., Gaboriaud, C., Bissery, V., Morgat, A., and Mornon, J. P., 1990, Hydrophobic cluster analysis: Procedures to derive structural and functional information from 2D representation of protein sequences, *Biochimie* **72**:554.

Letesson, J., Mager, A., Mammerickx, M., Burny, A., and Depelchin, A., 1990, B cells from bovine leukemia virus (BLV)-infected sheep with hematological disorders express the CD5 T-cell marker, *Leukemia* **4**:377.

Letesson, J., Van Den Broeke, A., Cleuter, Y., Mager, A., Delcommenne, M., Mammerickx, M., Burny, A., and Depelchin, A., 1991, FACS analysis of BLV infected cell lines with monoclonal antibodies to B cells and to monocytes-macrophages, *Vet. Immunol. Immunopathol.* **27**:207.

Leveziel, H., 1989, BoLA et leucose bovine, *Ann. Rech. Vet.* **20**:395.

Lévy, D., Kettmann, R., Marchand, P., Djilali, S., and Parodi, A., 1987, Selective tropism of bovine leukemia virus (BLV) for surface immunoglobulin-bearing ovine B-lymphocytes, *Leukemia* **1**:463.

Lewin, H., and Bernoco, D., 1986, Evidence for BoLA-linked resistance and susceptibility to subclinical progression of bovine leukaemia virus infection, *Anim. Genet.* **17**:197.

Lewin, H., Wu, M., Stewart, J., and Nolan, T., 1988, Association between BoLA and subclinical bovine leukemia virus infection in a herd of Holstein–Friesian cows, *Immunogenetics* **27**:338.

Malik, K., Even, J., and Karpas, A., 1988, Molecular cloning and complete nucleotide sequence of an adult T-cell leukemia virus/human T cell leukemia virus type I (ATLV/HTLV-I) isolate of Caribbean origin: Relationship to other members of the ATLV/HTLV-I subgroup, *J. Gen. Virol.* **69**:1695.

Mammerickx, M., 1970, Sur l'utilisation du mouton pour les expériences sur la leucose bovine, *Exp. Anim.* **3**:285.

Mammerickx, M., 1972, La transmission verticale et horizontale de la leucose bovine enzootique. Premiers résultats après cinq ans d'expérimentation, *Ann. Med. Vet.* **116**:647.

Mammerickx, M., Dekegel, D., Burny, A., and Portetelle, D., 1976, Study on the oral transmission of bovine leukosis to sheep, *Vet. Microbiol.* **1**:347.

Mammerickx, M., Portetelle, D., Burny, A., and Leunen, J., 1980, Detection by immunodiffusion—and radioimmunoassay—tests of antibodies to bovine leukemia virus antigens in sera of experimentally infected sheep and cattle, *Zentralbl. Veterinaermed. Reihe B* **27**:291.

Mammerickx, M., Portetelle, D., and Burny, A., 1985, The diagnosis of enzootic bovine leukosis, *Comp. Immunol. Microbiol. Infect. Dis.* **8**:305.

Mammerickx, M., Portetelle, D., De Clercq, K., and Burny, A., 1987, Experimental transmission of enzootic bovine leukosis to cattle, sheep and goats: Infectious doses of blood and incubation period of the disease, *Leuk. Res.* **11**:353.

Mammerickx, M., Palm, R., Portetelle, D., and Burny, A., 1988, Experimental transmission of enzootic bovine leukosis in sheep: Latency period of the tumoral disease, *Leukemia* **2**:103.

Mamoun, R., Astier-Gin, R., Kettmann, R., Deschamps, J., Rebeyrotte, N., and Guillemain, B., 1985, The pX region of the bovine leukemia virus is transcribed as a 2.1 kb mRNA, *J. Virol.* **54**:625.

Mamoun, R., Morisson, M., Rebeyrotte, N., Busetta, B., Couez, D., Kettmann, R., Hospital, M., and Guillemain, B., 1990, Sequence variability of bovine leukemia virus env gene and its relevance to the structure and antigenicity of the glycoproteins, *J. Virol.* **64**:4180.

Maruyama, K., Fukushima, T., and Mochizuki, S., 1989, Cross-reactive antibodies to BLV and HTLV in bovine and human hosts with retrovirus infection, *Vet. Immunol. Immunopathol.* **22**:265.

Matheise, J., Delcommenne, M., Mager, A., Didembourg, C., and Letesson, J., 1992, $CD5^+$ B cells from bovine leukemia virus infected cows are activated cycling cells responsive to interleukin 2, *Leukemia* **6**:304.

Matsushita, S., Guroff, M., Trepel, J., Crossman, M., Mitsuya, H., and Broder, S., 1986, Human monoclonal antibody directed against an envelope glycoprotein of human T-cell leukemia virus type 1, *Proc. Natl. Acad. Sci. USA* **83**:2671.

Merza, M., Linne, T., Hoglund, S., Portetelle, D., Burny, A., and Morein, B., 1989, Bovine leukaemia virus ISCOMs: Biochemical characterization, *Vaccine* **7**:22.

Merza, M., Söber, J., Sundquist, B., Toots, I., and Morein, B., 1991, Characterization of purified gp51 from bovine leukemia virus integrated into ISCOM: Physicochemical properties and serum antibody response to the integrated gp51, *Arch. Virol.* **120**:219.

Milan, D., and Nicolas, J.-F., 1991, Activator-dependent and activator-independent defective recombinant retroviruses from bovine leukemia virus, *J. Virol.* **65**:1938.

Miller, J., 1986, Bovine leukemia virus vaccine, in: *Animal Models of Retrovirus Infection and Their Relationship to AIDS* (L. A. Salzman, ed.), pp. 421–430, Academic Press, New York.

Miller, J., and Olson, C., 1972, Precipitating antibody to an internal antigen of the C-type virus associated with bovine lymphosarcoma, *J. Natl. Cancer Inst.* **49**:1459.

Miller, J., and Van Der Maaten, M., 1978, Evaluation of an inactivated bovine leukemia virus preparation as an immunogen in cattle, *Ann. Rech. Vet.* **9**:871.

Miller, J., Miller, L., Olson, C., and Gillette, K., 1969, Virus-like particles in phytohemagglutinin-stimulated lymphocyte cultures with reference to bovine lymphosarcoma, *J. Natl. Cancer Inst.* **43**:1297.

Miller, J., Van Der Maaten, M., and Schmerr, M. J. F., 1983, Vaccination of cattle with binary ethylenimine-treated bovine leukemia virus, *Am. J. Vet. Res.* **44**:64.

Miller, J., Van Der Maaten, M., and Schmerr, M., 1984, Vaccination with glycosidase-treated glycoprotein antigen does not prevent bovine leukemia virus infection in cattle, in: *Fifth International Symposium on Bovine Leukosis, Tübingen, October 19–21, 1982* (O. C. Straub, ed.), pp. 507–513, Commission of the European Communities Report EUR 8471, Brussels.

Montminy, M., Sevarino, K., Wagner, J., Mandel, G., and Goodman, R., 1986, Identification of a cyclic-AMP-responsive element within the rat somatostatin gene, *Proc. Natl. Acad. Sci. USA* **83**:6682.

Mosier, D., Yetter, R., and Morse, H., 1987, Functional T lymphocytes are required for a murine retrovirus-induced immunodeficiency disease (MAIDS), *J. Exp. Med.* **165**:1737.

Muscoplat, C., Johnson, D., Pomeroy, K., Olson, J., Larson, V., Stevens, J., and Sorensen, D., 1974, Lymphocyte surface immunoglobulin: Frequency in normal and lymphocytotic cattle, *Am. J. Vet. Res.* **35**:593.

Naif, H., Daniel, R., Cougle, W., and Lavin, M., 1992, Early detection of bovine leukemia virus by using an enzyme-linked assay for polymerase chain reaction-amplified proviral DNA in experimentally infected cattle, *J. Clin. Microbiol.* **30**:675.

Ohishi, K., Maruyama, T., Shida, H., Nishimali, J., Miki, K., Sagata, N., Ikawa, Y., and Sugimoto, M., 1988, Immunogenicity of a recombinant vaccinia virus expressing envelope glycoprotein of bovine leukaemia virus, *Vaccine* **7**:428.

Ohishi, K., Suzuki, H., Yamamoto, T., Maruyama, T., Sugimoto, M., Ikawa, Y., Numakunai, S., Okada, K., and Ohshima, K. I., 1990, Consistent inhibition of the growth of BLV in sheep inoculated with the vaccinia–retrovirus env gene recombinant, in: *Vaccines 90*, pp. 413–417, Cold Spring Harbor Laboratory Press, Cold Spring Harbor, N.Y.

Ohishi, K., Suzuki, H., Yamamoto, T., Maruyama, T., Miki, K., Ikawa, Y., Numakunai, S., Okada, K., Ohshima, K., and Sugimoto, M., 1991, Protective immunity against bovine leukaemia virus (BLV) induced in carrier sheep by inoculation with a vaccinia virus–BLV env recombinant: Association with cell-mediated immunity, *J. Gen. Virol.* **72**:1887.

Ohishi, K., Suzuki, H., Yasutomi, Y., Onuma, M., Okada, K., Numakunai, S., Ohshima, K., Ikawa, Y., and Sugimoto, M., 1992, Augmentation of bovine leukemia virus-specific lymphocyte proliferation responses in ruminants by inoculation with BLV *env*-recombinant vaccinia virus: Their role in the suppression of BLV replication, *Microbiol. Immunol.* **36**:1317.

Okada, K., Onuma, M., Numakunai, S., Kagawa, Y., Minamino, K., Ito, T., Kobayashi, Y., Morimoto, N., Morita, H., and Ohshima, K., 1983, Tumor-associated antigen detected by complement-dependent antibody cytotoxicity test and immunofluorescence test in enzootic bovine lymphosarcoma, *Jpn. J. Vet. Sci.* **45**:195.

Okada, K., Sakaguchi, I., Numakunai, S., Aida, Y., Onuma, M., and Ohshima, K., 1987, Existence of antigens cross-linking to tumor associated antigens of enzootic bovine leukosis in normal tissues of the bovine fetus, *Jpn. J. Vet. Sci.* **49**:373.

Olson, C., Miller, L., Miller, J., and Hoss, H., 1972, Transmission of lymphosarcoma from cattle to sheep, *J. Natl. Cancer Inst.* **49**:1463.

Onuma, M., and Olson, C., 1977, Tumor-associated antigen in bovine, and ovine lymphosarcoma, *Cancer Res.* **37**:3249.

Onuma, M., Hodatsu, T., Yamamoto, S., Higashihara, M., Masu, S., Mikami, T., and Izawa, H., 1984, Protection by vaccination against bovine leukemia virus infection in sheep, *Am. J. Vet. Res.* **45**:1212.

Onuma, M., Aida, Y., Okada, K., Ohshima, K., Kawakami, Y., and Izawa, H., 1985, Usefulness of monoclonal antibodies for detection of enzootic bovine leukemia cells, *Jpn. J. Cancer Res.* **76**:959.

Onuma, M., Tsukiyama, K., Ohya, K., Morishima, Y., and Ohno, R., 1987, Detection of cross-reactive antibody to BLV p24 in sera of human patients infected with HTLV, *Microbiol. Immunol.* **31**:131.

Oroszlan, S., Sarngadharan, M., Copeland, T., Kalyanaraman, V., Gilden, R., and Gallo, R., 1982, Primary structure analysis of the major internal protein p24 of human type C T-cell leukemia virus, *Proc. Natl. Acad. Sci. USA* **79**:1291.

Oroszlan, S., Copeland, T., Kalyanaraman, V., Sarngadharan, M., Schultz, A., and Gallo, R., 1984, Chemical analysis of human T-cell leukemia virus structural proteins, in: *Human T-cell Leukemia/Lymphoma Virus* (R. Gallo, M. Essex, and L. Gross, eds.), pp. 101–110, Cold Spring Harbor Laboratory Press, Cold Spring Harbor, N.Y.

Palker, T., Riggs, E., Spragion, D., Muir, A., Scearce, R., Randall, R., McAdams, M., McKnight, A., Clapham, P., Weiss, R., and Haynes, B., 1992, Mapping of homologous, amino-terminal neutralizing regions of human T-cell lymphotropic virus type I and II gp46 envelope glycoproteins, *J. Virol.* **66**:5879.

Parfanovich, M., Zhdanov, V., Lazarenko, A., Nomm, E., Simovart, Y., Parakin, V., and Lesmesh, V., 1983, The possibility of specific protection against bovine leukaemia virus infection and bovine leukaemia with inactivated BLV, *Br. Vet. J.* **139**:137.

Parodi, A. L., 1987, Pathology of enzootic bovine leukosis. Comparison with the sporadic form, in: *Enzootic Bovine Leukosis and Bovine Leukemia Virus* (A. Burny and M. Mammerickx, eds.), pp. 15–49, Nijhoff, The Hague.

Pätrascu, I. V., Goman, S., Sandu, I., Stiube, P., Munteanu, I., Coman, T., Ionescu, M., Popescu, D., and Mikailescu, D., 1980, Specific protection against bovine leukemia virus infection conferred on cattle by the Romanian inactivated vaccine BL-VACC-RO, *Rev. Roum. Med. Virol.* **31**:95.

Paul, P., Pomeroy, K., Castro, A., Johnson, D., Muscoplat, C., and Sorensen, D., 1977, Detection of bovine leukemia virus in B-lymphocytes by the syncytia induction assay, *J. Natl. Cancer Inst.* **59**:1209.

Platzer, C., Siakkou, H., Sober, J., Kopp, J., Scheve, E., and Rosenthal, S., 1990, Identification of an immunodominant region on the isolated bovine leukaemia virus (BLV) major envelope protein gp51 by monoclonal antibodies presumably not exposed during natural BLV infection, *Acta Virol.* **34**:255.

Portetelle, D., and Mammerickx, M., 1987, ELISA, a highly sensitive and specific method for the diagnosis of bovine leukemia virus infection, in: *Enzootic Bovine Leukosis and Bovine Leukemia Virus* (A. Burny and M. Mammerickx, eds.), pp. 201–217, Nijhoff, The Hague.

Portetelle, D., Bruck, C., Burny, A., Dekegel, D., Mammerickx, M., and Urbain, J., 1978, Detection of complement-dependent lytic antibodies in sera from bovine leukemia virus-infected animals, *Ann. Rech. Vet.* **9**:667.

Portetelle, D., Couez, D., Bruck, C., Kettmann, R., Mammerickx, M., Van Der Maaten, M., Brasseur, R., and Burny, A., 1989a, Antigenic variants of bovine leukemia virus (BLV) are defined by amino acid substitutions in the NH_2 part of the envelope glycoprotein gp51, *Virology* **169**:27.

Portetelle, D., Dandoy, C., Burny, A., Zavada, J., Siakkou, H., Gras-Masse, H., Drobecq, H., and Tartar, A., 1989b, Synthetic peptides approach to identification of epitopes on bovine leukemia virus envelope glycoprotein gp51, *Virology* **169**:34.

Portetelle, D., Mammerickx, M., and Burny, A., 1989c, Use of two monoclonal antibodies in an ELISA test for the detection of antibodies to bovine leukemia virus envelope glycoprotein gp51, *J. Virol. Methods* **23**:211.

Portetelle, D., Burny, A., Desmettre, P., Mammerickx, M., Paoletti, E., and Zavada, J., 1990, Development of a specific serological test and an efficient subunit vaccine to control bovine leukemia virus infection, *Dev. Biol. Stand.* **72**:81.

Portetelle, D., Limbach, K., Burny, A., Mammerickx, M., Desmettre, P., Riviere, M., Zavada, J., and Paoletti, E., 1991, Recombinant vaccinia virus expression of the bovine leukemia virus envelope gene and protection of immunized sheep against infection, *Vaccine* **9**:194.

Portetelle, D., Callebaut, I., Bex, F., and Burny, A., 1993, Vaccination against animal retroviruses, in: *Progress in Vaccinology,* Vol. 4 (R. Pandey, S. Hoglund, and G. Prasad, eds.), pp. 87–138, Springer-Verlag, Berlin.

Powers, M., and Radke, K., 1992, Activation of bovine leukemia virus transcription in lymphocytes from infected sheep: Rapid transition through early to late gene expression, *J. Virol.* **66**:4769.

Powers, M., Grossman, D., Kidd, L., and Radke, K., 1991, Episodic occurrence of antibodies against the bovine leukemia virus rex protein during the cause of infection in sheep, *J. Virol.* **65**:4959.

Ralston, S., Hoeprich, P., and Akita, R., 1989, Identification and synthesis of the epitope for a human monoclonal antibody which can neutralize human T-cell leukemia/lymphotropic virus type 1, *J. Biol. Chem.* **264**:16343.

Reinacher, M., Thurmond, M. C., Onuma, M., Portetelle, D., Picanso, J., and Theilen, G. H., 1989, Immunohistological demonstration of virus and tumor associated antigens in tissues in experimental and spontaneous bovine leukemia virus (BLV) infection, *Vet. Immunol. Immunopathol.* **22**:223.

Rice, N., Stephens, R., Couez, D., Deschamps, J., Kettmann, R., Burny, A., and Gilden, R., 1984, The nucleotide sequence of the env gene and post-env region of bovine leukemia virus, *Virology* **138**:82.

Rice, N., Stephens, R., Burny, A., and Gilden, R., 1985, The *gag* and *pol* genes of bovine leukemia virus: Nucleotide sequence and analysis, *Virology* **142**:357.

Rice, N., Stephens, R., and Gilden, R., 1987a, Sequence analysis of the bovine leukemia virus genome, in: *Enzootic Bovine Leukosis and Bovine Leukemia Virus* (A. Burny and M. Mammerickx, eds.), pp. 115–144, Nijhoff, The Hague.

Rice, N., Simek, S., Dubois, G., Showalter, S., Gilden, R., and Stephens, R., 1987b, Expression of the bovine leukemia virus X region in virus-infected cells, *J. Virol.* **61**:1577.

Ristau, E., Beier, D., Wittmann, W., and Klima, F., 1987a, Schutz von schafen gegen eine BLV-infecktion durch vakzinierung mit tumorzelle oder tumorzellpräparaten aus lymphknoten von leukosekranken rindern, *Arch. Exp. Vet. Med. Leipzig* **41**:185.

Ristau, E., Beier, D., and Wittmann, W., 1987b, Verlauf der infektion mit bovinem leukosevirus (BLV) bei kälbern nach applikation von zellextrakt aus lymphknotentumoren von BLV-infizierten rindern, *Arch. Exp. Vet. Med. Leipzig* **41**:323.

Ristau, E., Wittmann, W., Starick, E., and Kluge, K., 1989, Bovine leucosis virus challenge infection of calves following application of BL-3 cells, *Arch. Exp. Vet. Med. Leipzig* **43**:155.

Roberts, D., Lucas, M., Sands, J., and Wibberley, G., 1982, Protection against bovine leukosis infection in sheep with the BL-20 bovine lymphoblastoid cell line, *Vet. Immunol. Immunopathol.* **3**:635.

Roberts, D., Lucas, M., Sands, J., Wibberley, G., Spooner, R., and Oliver, S., 1984, Further studies on the use of the BL-20 cell line as a vaccine against bovine leukosis virus infection, in: *Fifth International Symposium on Bovine Leukosis, Tübingen, October 19–21, 1982* (O. C. Straub, ed.), pp. 481–492, Commission of the European Communities Report EUR 8471, Brussels.

Rosen, C., Sodroski, J., Kettmann, R., Burny, A., and Haseltine, W., 1985, Trans-activation of the bovine leukemia virus long terminal repeat in BLV-infected cells, *Science* **227**:320.

Rosen, C., Sodroski, J., Kettmann, R., and Haseltine, W., 1986a, Activation of enhancer sequences in type II human T-cell leukemia virus and bovine leukemia virus long terminal repeats by virus-associated trans-acting regulatory factors, *J. Virol.* **57**:738.

Rosen, C., Sodroski, J., Willems, L., Kettmann, R., Campbell, K., Zaya, R., Burny, A., and Haseltine, W., 1986b, The region of bovine leukemia virus genome encodes a trans-activator protein, *EMBO J.* **5**:2585.

Sagata, N., Yasunaga, T., Tsuzuku-Kawamura, J., Ohishi, K., Ogawa, Y., and Ikawa, Y., 1985a, Complete nucleotide sequence of the genome of bovine leukemia virus: Its evolutionary relationship to other retroviruses, *Proc. Natl. Acad. Sci. USA* **82**:677.

Sagata, N., Yasunaga, T., and Ikawa, Y., 1985b, Two distinct polypeptides may be translated from a single spliced mRNA of the X genes of human T-cell leukemia and bovine leukemia viruses, *FEBS Lett.* **192**:37.

Sagata, N., Tsuzuku-Kawamura, J., Nagayoshi-Aida, M., Shimuzu, F., Imagawa, K., and Ikawa, Y., 1985c, Identification and some biochemical properties of the major XBL gene product of bovine leukemia virus, *Proc. Natl. Acad. Sci. USA* **82**:7879.

Sakurai, M., Taneda, A., Nagoya, H., and Sekikawa, K., 1991, Construction and functional characterization of mutants of the bovine leukemia virus trans-activator protein $p34^{tax}$, *J. Gen. Virol.* **72**:2527.

Schultz, A., Copeland, T., and Oroszlan, S., 1984, The envelope proteins of bovine leukemic virus: purification and sequence analysis, *Virology* **135**:417.

Scott, M., Thurmond, D., Dunn, S., Osburn, B., and Scott, J., 1991, Integrated bovine leukosis proviral DNA in T helper, and T cytotoxic/suppressor lymphocytes, *J. Gen. Virol.* **72**:307.

Seiki, M., Hattori, Y., Hirayama, Y., and Yoshida, M., 1983, Human adult T-cell leukemia virus: Complete nucleotide sequence of the provirus genome integrated in leukemia cell DNA, *Proc. Natl. Acad. Sci. USA* **80**:3618.

Sherman, M., Garth, D., Ehrlich, D., Ferrer, F., Sninsky, J., Zandomeni, R., Dock, N., and Poiesz, B., 1992, Amplification and analysis of specific DNA and RNA sequences of bovine leukemia virus from infected cows by polymerase chain reaction, *J. Clin. Microbiol.* **30**:185.

Shimotohno, K., Takahashi, Y., Shimizu, N., Gojobori, T., Golde, D., Chen, I., Wiwa, M., and Sugimura, T., 1985, Complete nucleotide sequence of an infectious clone of human T-cell leukemia virus type II: An open reading frame for the protease gene, *Proc. Natl. Acad. Sci. USA* **82**:3103.

Siakkou, H., Ulrich, R., Uelze, A., Möhring, R., and Rosenthal, S., 1990, Immunological characterization of BLV proteins synthesized in Escherichia coli, *Acta Virol.* **34**:256.

Stock, N., and Ferrer, J., 1972, Replicating C-type virus in phytohemagglutinin-treated buffy-coat cultures of bovine origin, *J. Natl. Cancer Inst.* **48**:985.

Suzuki, M., Okada, K., Ohshima, K., Numakunai, S., Aida, Y., and Onuma, M., 1987, Application of complement-dependent antibody cytotoxicity test for diagnosis of enzootic bovine lymphosarcoma, *Jpn. J. Vet. Sci.* **49**:841.

Theilen, G., Miller, J., Higgins, J., Rupanner, R., and Garret, W., 1982, Vaccination against bovine leukemia virus infection, *Curr. Top. Vet. Med. Anim. Sci.* **15**:547.

Theilen, G., Rupanner, R., Miller, J., Murray, S., Yah, H., and Garret, W., 1984, Continuing studies using non-viral antigen to protect cattle against bovine leukemia virus infection, in: *Fifth International Symposium on Bovine Leukosis, Tübingen, October 19–21, 1982* (O. C. Straub, ed.), pp. 493–498, Commission of the European Communities Report EUR 8471, Brussels.

Uckert, W., Wunderlich, V., Ghysdael, J., Portetelle, D., and Burny, A., 1984, Bovine leukemia virus (BLV)—A structural model based on chemical crosslinking studies, *Virology* **133**:386.

Ulrich, R., Siakkou, H., Platzer, C., Bossmann, H., Möhring, R., Wiedmann, M., Bähring, S., and Rosenthal, S., 1990, Synthesis of bovine leukemia virus antigens in Escherichia coli, *Arch. Exp. Vet. Med. Leipzig* **44**:909.

Ulrich, R., Pushko, P., Kozlovskaya, T., Siakkou, H., Pumpen, P., Dreilina, D., and Platzer, C., 1991a, Precise localization of the epitope of major BLV envelope glycoprotein, *Acta Virol.* **35**:302.

Ulrich, R., Bähring, S., Siakkou, H., and Rosenthal, S., 1991b, Expression of bovine leukaemia virus antigens fused to MS2 polymerase in E. Coli, *Acta Virol.* **35**:391.

Van Den Broeke, A., Cleuter, Y., Chen, G., Portetelle, D., Mammerickx, M., Zagury, D., Fouchard, M., Coulombel, L., Kettmann, R., and Burny, A., 1988, Even transcriptionally competent proviruses are silent in bovine leukemia virus-induced sheep tumor cells, *Proc. Natl. Acad. Sci. USA* **85**:9263.

Van Der Maaten, M., and Miller, J., 1976, Replication of bovine leukemia virus in monolayer cell culture, *Bibl. Haematol.* **43**:360.

Van Der Maaten, M., Miller, J., and Boothe, A., 1974, Replicating type-C virus particles in monolayer cell cultures of tissue from cattle with lymphosarcoma, *J. Natl. Cancer Inst.* **52**:491.

Vonèche, V., Portetelle, D., Kettmann, R., Willems, L., Limbach, K., Paoletti, E., Ruysschaert, J. M., Burny, A., and Brasseur, R., 1992a, Fusogenic segments of bovine leukemia virus and simian immunodeficiency virus are interchangeable and mediate fusion by means of oblique insertion in the lipid bilayer of their target cells, *Proc. Natl. Acad. Sci. USA* **89**:3810.

Vonèche, V., Callebaut, I., Kettmann, R., Brasseur, R., Burny, A., and Portetelle, D., 1992b, The 19–27 amino acid segment of gp51 adopts an amphiphilic structure and plays a key role in the fusion events induced by bovine leukemia virus, *J. Biol. Chem.* **267**:15193.

Vonèche, V., Callebaut, I., Lambrecht, B., Brasseur, R., Burny, A., and Portetelle, D., 1993, Enhancement of bovine leukemia virus induced syncytia formation by di- and tripeptides, *Virology* **192**:307.

Weiss, R., Teich, N., Varmus, M., and Coffin, J., 1984, *RNA Tumor Viruses*, Vol. 1, Cold Spring Harbor Laboratory Press, Cold Spring Harbor, N.Y.

Weiss, R., Teich, N., Varmus, H., and Coffin, J., 1985, *RNA Tumor Viruses*, Vol. 2, Cold Spring Harbor Laboratory Press, Cold Spring Harbor, N.Y.

Wiley, D., and Skehel, J., 1987, The structure and function of the hemagglutinin membrane glycoprotein of influenza virus, *Annu. Rev. Biochem.* **56**:365.

Willems, L., Bruck, C., Portetelle, D., Burny, A., and Kettmann, R., 1987a, Expression of a cDNA clone corresponding to the long open reading frame (XBL-I) of the bovine leukemia virus, *Virology* **160**:55.

Willems, L., Gegonne, A., Chen, G., Burny, A., Kettmann, R., and Ghysdael, J., 1987b, The bovine leukemia virus p34 is a transactivator protein, *EMBO J.* **6**:3385.

Willems, L., Chen, G., Burny, A., and Kettmann, R., 1988, Expression in bacteria of β-galactosidase fusion proteins carrying antigenic determinants of two X gene products of bovine leukemia virus, *Leukemia* **2**:1.

Willems, L., Chen, G., Portetelle, D., Mamoun, R., Burny, A., and Kettmann, R., 1989, Structural and functional characterization of mutants of the bovine leukemia virus transactivator protein p34, *Virology* **171**:615.

Willems, L., Heremans, H., Chen, G., Portetelle, D., Billiau, A., Burny, A., and Kettmann, R., 1990, Cooperation between bovine leukemia virus transactivator protein and Ha-ras oncogene in cellular transformation, *EMBO J.* **9**:1577.

Willems, L., Kettmann, R., and Burny, A., 1991, The amino acid (157–197) peptide segment of bovine leukemia virus $p34^{tax}$ encompass a leucine-rich globally neutral activation domain, *Oncogene* **6**:159.

Willems, L., Kettmann, R., Chen, G., Portetelle, D., Burny, A., and Derse, D., 1992a, A cyclic AMP-responsive DNA-binding protein (CREB2) is a cellular transactivator of the bovine leukemia virus long terminal repeat, *J. Virol.* **66**:766.

Willems, L., Grimonpont, C., Heremans, H., Rebeyrotte, N., Chen, G., Portetelle, D., Burny, A., and Kettmann, R., 1992b, Mutations in the bovine leukemia virus Tax protein can abrogate the long terminal repeat-directed transactivating activity without concomitant loss of transforming potential, *Proc. Natl. Acad. Sci. USA* **89**:3957.

Willems, L., Portetelle, D., Kerkhofs, P., Chen, G., Burny, A., Mammerickx, M., and Kettmann, R., 1992c, *In vivo* transfection of bovine leukemia provirus into sheep, *Virology* **189**:775.

Willems, L., Thienpont, E., Kerkhofs, P., Burny, A., Mammerickx, M., and Kettmann, R., 1993a, Bovine leukemia virus, an animal model for the study of intrastrain variability, *J. Virol.* **67**:1086.

Willems, L., Kettmann, R., Dequiedt, F., Portetelle, D., Vonèche, V., Cornil, I., Kerkhofs, P., Burny, A., and Mammerickx, M., 1993b, *In vivo*, infection of sheep by bovine leukemia virus mutants, *J. Virol.*, **67**:4078.

Williams, D., Barta, O., and Amborski, G., 1988, Molecular studies of T-lymphocytes from cattle infected with bovine leukemia virus, *Vet. Immunol. Immunopathol.* **19**:307.

Williams, D., Newson, J., and Naessens, J., 1989, Humoral immune responses in trypano tolerant, and susceptible breeds of cattle following secondary challenge with Trypanosoma congolense, *Immunobiology* **4**:152.

Wilson, I., Skehel, J., and Wiley, D., 1981, Structure of the hemagglutinin membrane glycoprotein of influenza virus at 3A resolution, *Nature* **289**:366.

Woodcock, S., Mornon, J. P., and Henrissat, B., 1992, Detection of secondary structure elements in proteins by hydrophobic cluster analysis, *Protein Eng.* **5**:629.

Wyatt, C., Wingett, D., White, J., Buck, C., Knowles, D., Reeves, R., and Magnuson, N., 1989, Persistent infection of rabbits with bovine leukemia virus associated with development of immune dysfunction, *J. Virol.* **63**:4498.

Ying-Zi, C., Rabin, E., and Wortis, H. H., 1991, Treatment of murine CD5$^-$ cells with anti-Ig, but not LPS, induces surface CD5: Two B-cell activation pathways, *Int. Immunol.* **3**:467.

Yoshinaka, Y., and Oroszlan, S., 1985, Bovine leukemia virus post-envelope gene coded protein: Evidence for expression in natural infection, *Biochem. Biophys. Res. Commun.* **131**:347.

Yoshinaka, Y., Katoh, I., Copeland, T., Smythers, G., and Oroszlan, S., 1986, Bovine leukemia virus protease: Purification, chemical analysis, and *in vitro* processing of gag polyproteins, *J. Virol.* **57**:826.

Zandomeni, R., Carrera-Zandomeni, M., Esteban, E., and Ferrer, J., 1991, The trans-activating C-type retroviruses share a distinct epitope(s) that induces antibodies in certain infected hosts, *J. Gen. Virol.* **72**:2113.

Zandomeni, R., Carrera-Zandomeni, M., Esteban, E., Donawich, W., and Ferrer, J., 1992, Induction and inhibition of bovine leukaemia virus expression in naturally infected cells, *J. Gen. Virol.* **73**:1915.

Zavada, J., Cerny, L., Zavadova, A., Bozonva, J., and Alstein, A. D., 1979, A rapid neutralization test for antibodies to bovine leukemia virus with the use of rhabdovirus pseudotypes, *J. Natl. Cancer Inst.* **62**:95.

CHAPTER 3

The Lentiviruses of Cattle

MATTHEW A. GONDA

I. NATURAL HISTORY OF BOVINE LENTIVIRUSES

The bovine immunodeficiency virus (BIV) is a horizontally transmitted, nononcogenic, pathogenic retrovirus and member of the lentivirus genus (Fig. 1). It was discovered in the late 1960s during the intensive search for the cause of bovine leukemia/lymphosarcoma. BIV was initially isolated in the southern United States from a pregnant cow with clinical signs of disease consisting of lymphocytosis, lymphadenopathy, encephalitis, progressive weakness, and emaciation (Van Der Maaten *et al.*, 1972). Little work was performed on BIV in the succeeding 15 years because of the general lack of interest in nononcogenic retroviruses. It was not until the discovery that the human immunodeficiency virus (HIV) is a lentivirus (Gonda *et al.*, 1985, 1986) that there was a renewed interest in the biology of BIV (Gonda *et al.*, 1987).

The lentiviruses are a morphologically, biologically, and genetically distinct genus of retroviruses that are associated with immunodeficiencies and chronic, inflammatory, organ-specific diseases of various severity (Haase, 1986; Gallo and Nerurkar, 1989; Gallo, 1991; Levy, 1993; Gonda, 1994). Chronologically, BIV was the third distinct lentivirus discovered (Van Der Maaten *et al.*, 1972), being preceded by equine infectious anemia virus (EIAV) (Vallée and Carré, 1904) and the ovine lentiviruses which include visna, maedi, and progressive pneumonia viruses (Sigurdsson *et al.*, 1957, 1960; Sigurdsson and Pálsson, 1958; Kennedy *et al.*, 1968). To date, there are seven groups of lentiviruses that are distinguished based on host specificity and include isolates from humans, monkeys, cattle, cats, goats, sheep, and horses (Gonda, 1994).

Generally, lentiviruses infect the natural host and, occasionally, closely related species. Overall, the genomes of these seven distinct lentivirus groups are highly divergent in their nucleotide sequences, which has complicated analyses of

---

MATTHEW A. GONDA • Laboratory of Cell and Molecular Structure, Program Resources, Inc./DynCorp, National Cancer Institute–Frederick Cancer Research and Development Center, Frederick, Maryland 21702.

The Retroviridae, Volume 3, edited by Jay A. Levy. Plenum Press, New York, 1994.

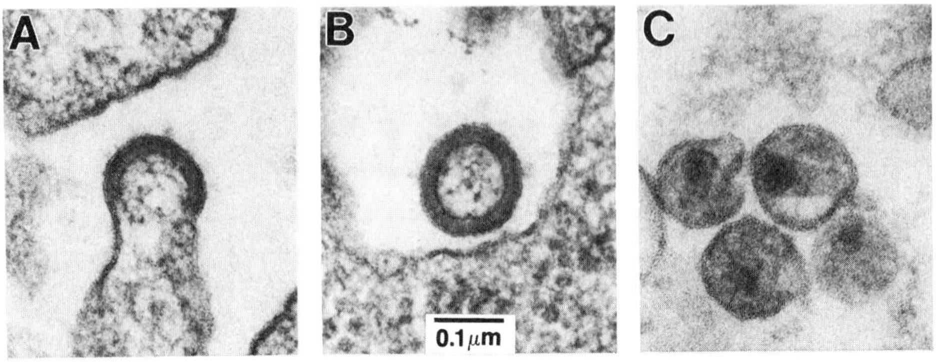

FIGURE 1. Electron microscopy of BIV. (A) Budding virus. (B) Immature extracellular virus. (C) Mature extracellular virus.

their genetic relationship. Nevertheless, the reverse transcriptase (RT) domain of the *pol* gene of retroviruses is one region that is highly conserved and upon which evolutionary relationships can be drawn (Gonda *et al.*, 1985, 1986, 1987; Garvey *et al.*, 1990). Phylogenetic analyses based on comparisons of amino acids encoded by the RT domain of *pol* for various retroviruses, including at least one member of each of the separate lentivirus groups, have been made (Fig. 2). These analyses demonstrate that the lentiviruses cluster together as a genetically defined group among the retroviruses whose RTs have a preference for Mg^{2+} cations. Moreover, based on comparisons of this segment of RT, BIV is a distinct entity among the lentiviruses that diverged from a common progenitor early in the evolution of this taxonomic group, and it is relatively equidistant from all of the other lentiviruses. Thus, BIV did not have its origins in any of the presently known lentivirus groups.

The importance of BIV infection to the health of cattle has not been firmly established but most infections appear to be without overt signs of disease in the absence of appropriate environmental cofactors. An understanding of BIV pathogenesis has in part been hampered by sparse information on the worldwide prevalence of BIV infections. However, serologic data do exist for BIV infections in North American cattle (Amborski *et al.*, 1989; Black, 1990; Whetstone *et al.*, 1990; Cockerell *et al.*, 1992; Jacobs *et al.*, 1992; Luther *et al.*, in preparation). BIV is more frequently encountered in the United States in southern than northern locales. About 4% of randomly selected cattle sera from the southern United States are positive for antibodies to BIV. The frequency of seropositivity in individual herds can be considerably higher. Average frequencies of 68 and 40% for dairy and beef herds, respectively, have been found in the Louisiana area (Luther *et al.*, in preparation). The higher frequency of BIV in dairy cattle may be the result of herd management practices or the extended productive life (and risk of exposure) of dairy animals relative to beef animals. BIV also has been serologically detected in the Netherlands, Switzerland, Canada, Costa Rica, Venezuela, and New Zealand (Horner, 1991; Horzinek *et al.*, 1991; H. Lutz, J. Bonilla, and R. Walder, personal communications), which suggests that infections in cattle are worldwide.

BIV R-29 is the single representative serotype of the bovine lentivirus group. At present, all of the serologic information on the presence of BIV infections has

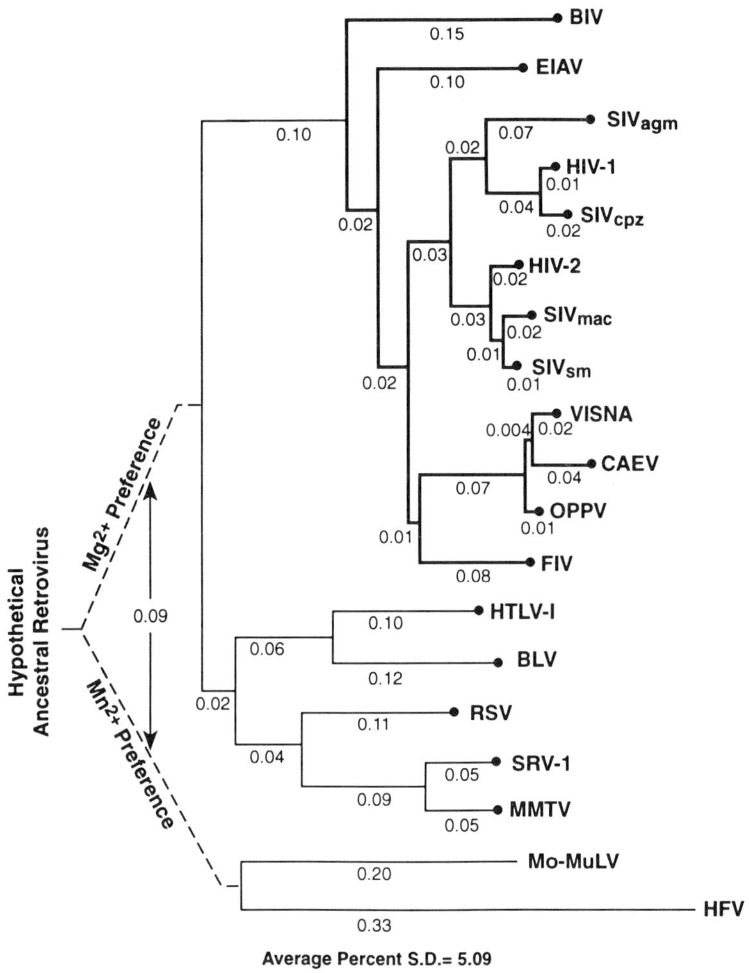

FIGURE 2. Evolutionary relationship of BIV. Fitch–Margoliash phylogenetic tree of retroviral relationships based on the highly conserved 90 amino acids from the RT domain of the *pol* gene as described (Gonda *et al.*, 1986, 1987; Gonda, 1988; Garvey *et al.*, 1990). Retroviruses used are visna virus, OPPV (ovine progressive pneumonia virus), CAEV (caprine arthritis encephalitis virus), SIVsm, SIVcpz, SIVmac, SIVagm (simian immunodeficiency viruses), HIV-2, HIV-1, EIAV (equine infectious anemia virus), BIV, FIV (feline immunodeficiency virus), HTLV-I (human T-cell lymphotropic virus type I), BLV, RSV (Rous sarcoma virus), SRV-1 (simian AIDS retrovirus), MMTV (mouse mammary tumor virus), Mo-MuLV (Moloney murine leukemia virus), and HFV (human foamy virus). Sequences were obtained from Garvey *et al.* (1990) and GenBank. Branch lengths are in units of $-\log M$, where M is the frequency of matching residues. The tree was rooted with HFV and Mo-MuLV as the outgroup taxons because they consistently had the lowest alignment scores and because their RTs preferentially use Mn^{2+} rather than Mg^{2+} as cofactor. The average standard deviation of the tree was 5.09%. (From Gonda, 1992, with permission of Current Science, Ltd., London.)

been obtained using the original R-29 isolate (Van Der Maaten *et al.*, 1972) as antigen. Only recently have new field isolates of BIV become available, and these appear serologically to be highly related to BIV R-29 (Suarez *et al.*, 1993; Oberste *et al.*, in preparation). Nonetheless, emerging data suggest that variants of BIV, which are antigenically distinct from BIV R-29 and demonstrate immunologic cross-reactivity only with the major capsid (CA) protein (p26), are present in the eastern United States and Europe (Horzinek *et al.*, 1991; Gonda *et al.*, 1994). This finding in cattle is not unexpected considering that there are several species of true cattle and that a large number of divergent simian immunodeficiency virus (SIV) and feline immunodeficiency virus (FIV) variants have been recently described for monkeys and cats (Johnson and Hirsch, 1992; Olmsted *et al.*, 1992). It is not known whether BIV evolved with domestic cattle or was acquired more recently from infection with a closely related species. Domestic cattle (*Bos taurus*) are found worldwide and are one of seven species of the genus *Bos* that are of Old World origins. The other species, known by their common names, include Brahma, gauers, bantengs, yaks, gayals, and kouprey, and other than the Brahma, are more restricted in location. There also are several other closely related ungulate species including the bison and buffalo. Both Brahma and bison have been bred into the *Bos taurus* background to improve certain physical qualities of the breeding stock. It is not yet known whether any of these closely related species harbor a lentivirus infection. It will be important to isolate and characterize putative new variants of BIV so as to investigate the presence of lentivirus infections in these distinct ungulate species, which may provide clues as to the origins and possible reservoirs of BIV.

II. THE VIRION AND ITS REPLICATIVE PROPERTIES

Although several new field strains of BIV recently have been obtained (Suarez *et al.*, 1993; Oberste *et al.*, in preparation), at present, all work on the molecular biology and pathogenesis of BIV is based on the original R-29 laboratory strain (Van Der Maaten *et al.*, 1972) and several complete functional proviral molecular clones of BIV R-29 (Braun *et al.*, 1988; Garvey *et al.*, unpublished data).

A. Structure of the Virion and Protein Locations

The BIV morphology and morphogenesis resemble closely those of the other lentiviruses (Fig. 1). The particle can be dissected into two major components: the viral envelope and the core (Fig. 3). The viral envelope is comprised of the lipid bilayer with associated cellular proteins, the Env glycoproteins, and the matrix (MA) protein derived from the NH_2-terminus of the Gag precursor by proteolytic cleavage. The NH_2-terminus of the MA remains associated with the viral envelope and is largely inside the virion. The surface (SU) glycoprotein, gp100, resides on the outside of the membrane and electrostatically interacts with the extracellular domain of the transmembrane (TM) glycoprotein, gp45. gp45 has a cytoplasmic tail and traverses the membrane at a very hydrophobic domain in the TM protein; this integral interaction of gp45 and the lipid bilayer serves to anchor the gp100–gp45 complex to the viral envelope (Garvey *et al.*, 1990). The Env glycoproteins of lentiviruses are thought to interact as oligomers (Pinter *et al.*, 1989; Schawaller *et al.*, 1989; Chakrabarti *et al.*, 1990; Earl *et al.*, 1990; Weiss *et al.*, 1990; Moss *et al.*, 1991), and BIV Env proteins readily form dimeric and multimeric structures (Rasmussen *et al.*, 1992). The core structure in mature virions, which can be purified away from the envelope by mild detergent extraction, contains the diploid polyadenylated genomic RNA. The RNA molecules are be-

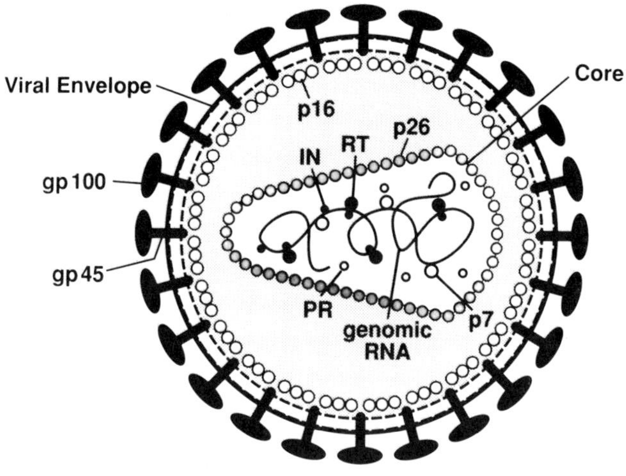

FIGURE 3. Schematic representation of the mature BIV virion and location of the structural gene products. Proteins that make up the viral envelope are gp100, gp45, and p16 (MA). Proteins that are found in the viral core are p26 (CA), p7 (NC), PR, RT, and IN. The genome-length viral nucleic acid is also found in the viral core. (From Gonda *et al.*, 1994, with permission of Elsevier Science Publishers, Amsterdam.)

lieved to be held together near the 5′ ends in a structure that includes the $tRNA^{Lys}_{1,2}$ molecules that function as replicative primers. The viral RNA is probably coated with nucleocapsid (NC) and encased in the rod- or cone-shaped core that is structured based on interactions between CA molecules. The catalytic *pol* gene products are also found in the core but in an abundance lower than the Gag proteins. This is reflective of the relatively low frequency of the −1 translational frameshift event (Varmus, 1988; Jacks *et al.*, 1988).

B. Genome Organization

Multiple full-length, infectious proviral molecular clones of BIV R-29 have been cloned, sequenced, and extensively characterized (Braun *et al.*, 1988; Garvey *et al.*, 1990, and unpublished data). The prototypic BIV R-29 127 genome contains 8482 nucleotides (nt) and 8960 base pairs (bp) in the forms of the viral RNA and proviral DNA, respectively, and encodes at least nine genes (Garvey *et al.*, 1990; Gonda, 1992). As is the case with all other replication-competent retroviruses, the BIV genome (Fig. 4) contains the structural genes, *gag*, *pol*, and *env*, flanked by long terminal repeats (LTRs) on either end. The central region of the

FIGURE 4. Genome organization of BIV. The structural genes identified in the BIV genome are *gag*, *pol*, and *env* and the nonstructural accessory genes are *vif*, *tat*, *rev*, *vpy* (Y), *vpw* (W), and *tmx*. (Adapted from Gonda *et al.*, 1990b, with permission of the editors and Elsevier Science Publishers, Amsterdam.)

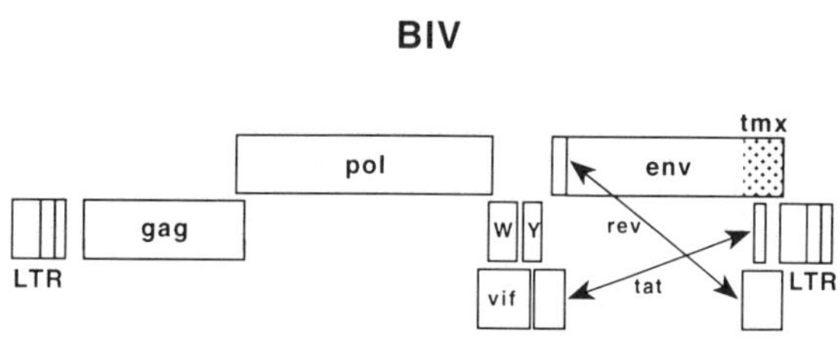

genome, between and overlapping the *pol* and *env* open reading frames, contains genes and/or coding exons for the nonstructural accessory proteins termed *vif* (viral infectivity factor), *tat* (*trans*-activator of transcription), and *rev* (regulator of virus expression) and for two potential proteins termed *vpw* and *vpy*, of unknown function. Although BIV has no apparent open reading frame equivalent to the *nef* gene of primate lentiviruses, cDNA libraries created from BIV-infected cell cultures contain abundant copies of a ninth gene, termed *tmx* for gene encoding transmembrane protein X, which consists of a single coding exon and is found in the 3′ end of the genome (Greenwood *et al.*, in preparation).

C. Transcripts

BIV has a complex transcription map that has been deduced from Northern blot analyses, splice junction mapping, and characterization of cDNAs. It reflects the intricacy of the genome (Fig. 5). At least five size classes (8.5, 4.1, 3.8, 1.7, and 1.4 kb) of BIV-specific RNA are transcribed and processed in infected cells (Oberste *et al.*, 1991). All viral transcripts initiate at the start of R (+1) in the LTR and contain an untranslated common leader sequence and terminate at the polyadenylation site located in the 3′ LTR sequences. The *gag* and *gag–pol* encoded precursor proteins are translated from the full-length, 8.5-kb transcript. In keeping with other characterized lentiviruses (Jacks *et al.*, 1988; Varmus, 1988; Henderson *et al.*, 1992), translation of *pol*, which overlaps *gag* but is in a different reading frame, is predicted to occur via a −1 frameshift near the 3′ terminus of the *gag* gene (Battles *et al.*, 1992). The *vif/vpw* and *env* transcripts are singly spliced mRNAs of 4.1 and 3.8 kb, respectively. Thus, the *gag*, *gag–pol*, *vif*, *vpw*, and *env* transcripts are predicted to contain the putative BIV Rev-responsive element (RRE), which facilitates transport of singly spliced messages from the nucleus to the cytoplasm. Preliminary data have confirmed the Rev responsiveness of the *gag*, *gag–pol*, and *env* transcripts (Oberste *et al.*, 1993). The *tat* and *rev* transcripts are multiply spliced and consist of three exons. The use of alternative splice acceptor sites for the second exon in *rev* and the third exon in *tat* creates heterogeneous splicing patterns, and therefore *tat* and *rev* mRNAs exist in more than one form (Liu *et al.*, 1992; Oberste *et al.*, 1993; Pallansch *et al.*, submitted). Splicing of exon 2 to exon 3 in these mRNAs removes the putative RRE. Thus, *tat* and *rev* are early messages and are transcribed soon after infection of a naive cell. The recently characterized *tmx* mRNA is predicted to be less than 1.4 kb in length. The single coding exon of *tmx* begins at the same splice acceptor site as the third exons of *tat* and *rev* and terminates in the 3′ LTR. The *tmx* message is also an early mRNA as its splicing pattern removes the predicted RRE (Greenwood *et al.*, in preparation).

D. Structural Gene Products

The BIV Gag precursor polyprotein (Pr53) is translated from the full-length mRNA (Fig. 6) and initiates formation of the morphologically recognizable virus particle beneath the plasma membrane (Rasmussen *et al.*, 1990). The Pr53 polyprotein is the minimal essential viral information needed to produce a budding

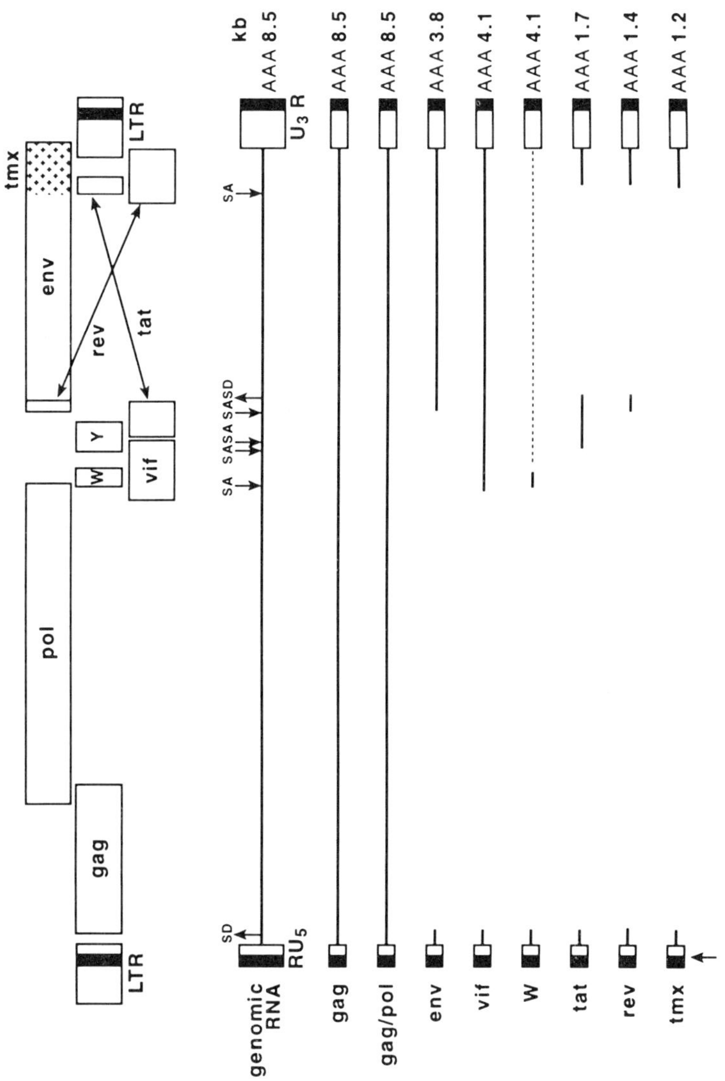

FIGURE 5. Transcriptional map of BIV. (From Gonda, 1992, with permission of Current Science, Ltd., London.)

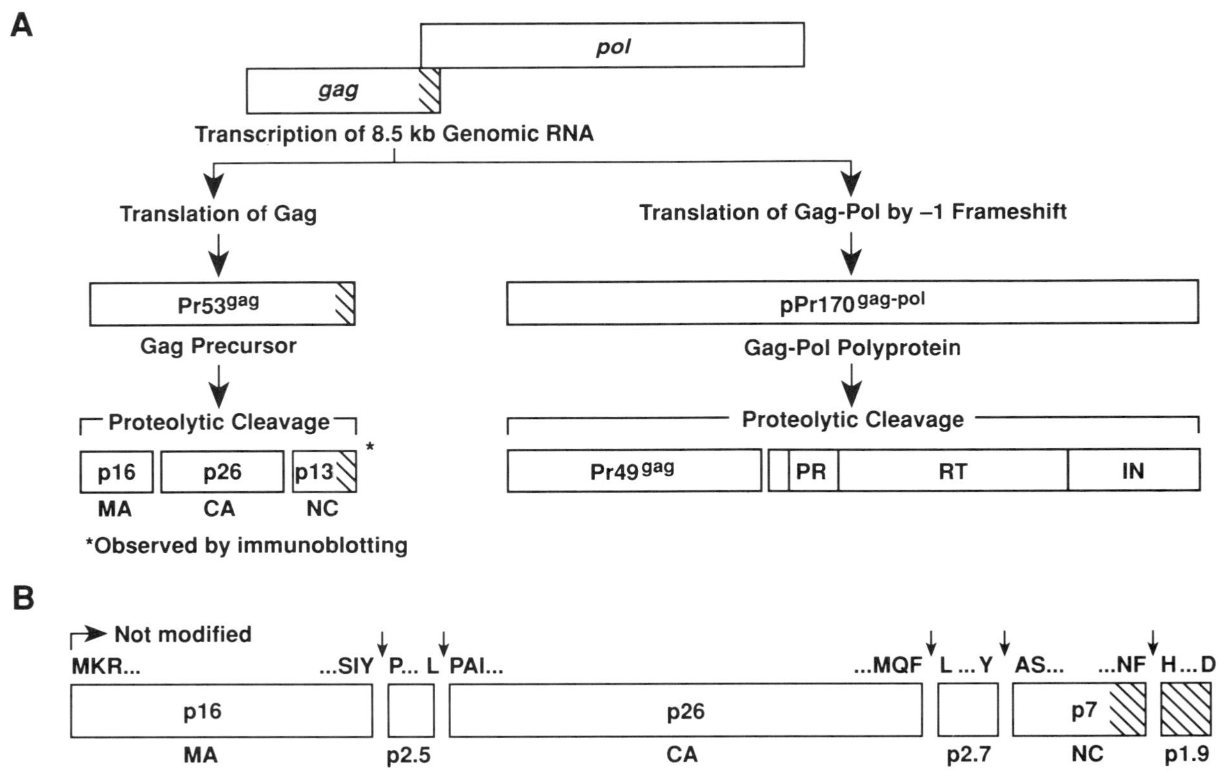

FIGURE 6. Processing of the BIV Gag and Gag–Pol precursors. The processing of the BIV Gag precursor by the viral PR was determined experimentally by immunologic methods (A) and amino acid sequence analysis (B). The processing of the BIV Gag–Pol precursor (A) is that predicted from analyses performed with other lentiviruses. (From Gonda *et al.*, 1994, with permission of Elsevier Science Publishers, Amsterdam.)

TABLE I. BIV Structural Proteins[a]

Gene	Protein location	Protein	Molecular mass (kDa)	
			Predicted	Observed
gag	nt316–1743	Gag precursor	53.4	53
	aa1–126	MA	14.6	16
	127–148	p2.5	2.5	p3
	149–367	CA	24.6	26
	368–392	p2.7	2.7	N.D.
	393–458	NC	7.3	13[b]
	459–476	p1.9	1.9	N.D.
pol	—	Gag–Pol polyprotein precursor	168.0	170
	nt1581–4739	Pol polyprotein	120.7	N.D.
	aa51–143	PR	10.6	N.D.
	144–699	RT	72.2	N.D.
	700–771	?	8.0	N.D.
	772–1053	IN	32.4	N.D.
env	nt5415–8126	Env precursor	102.3[c]	145[d]
	aa1–555	SU	62.1[c]	100[d]
	556–905	TM	40.2[c]	45[d]

[a] From Garvey *et al.* (1990) unless otherwise noted. N.D., not determined; aa, amino acid; nt, nucleotide; MA, matrix; CA, capsid; NC, nucleocapsid; PR, protease; RT, reverse transcriptase; IN, integrase; SU, surface glycoprotein; TM, transmembrane glycoprotein. (Table adapted from Gonda, 1992, with permission of Current Science, Ltd., London.)
[b] The p7 NC protein migrates as a p13 on gel electrophoresis.
[c] Unglycosylated backbone.
[d] Glycosylated.

particle, and it also serves to incorporate the genomic viral RNA into the forming virus particle. Soon after budding, Pr53 is cleaved by the viral protease (PR) into the major mature Gag proteins, MA, CA, and NC, p16, p26, and p7, respectively, during maturation of the immature virus particle (Fig. 6, Table I). The Gag products have been extensively characterized immunologically (Battles *et al.*, 1992) and by amino acid analysis (Tobin *et al.*, submitted). In contrast to all other lentiviral MA proteins studied to date, the NH_2-terminus of the BIV MA does not appear to be modified by fatty acids as, during amino acid sequencing, it is accessible to the Edman degradation reaction and retains the NH_2-terminal methionine. It is an enigma as to how the NH_2-terminus of BIV Pr53 associates with or inserts into the plasma membrane to facilitate virus particle formation. The NC protein contains two cysteine arrays that resemble zinc-finger motifs. It is believed that the NC domain of the Gag precursor facilitates incorporation of genomic viral RNA into the virion, presumably through the interaction of the zinc-finger motifs and the viral RNA. The NC migrates as a 13-kDa protein in denaturing gel electrophoresis at pH 8.8. The retarded gel mobility of the NC protein may be a result of either its low pI (6.8) or the formation of NC dimers. Dimerization may be precipitated by an additional cysteine residue present between the two zinc-finger motifs; this cysteine is not present in other lentiviruses. In addition to the major Gag proteins, MA, CA, and NC, several short peptides result from PR cleavage of Pr53 during particle maturation (Fig. 6, Table I) (Tobin *et al.*, submitted). Peptides of 22 and 25 residues are cleaved from the MA–CA (p2.5) and CA–NC (p2.7) junctions, respectively. An 18-amino-acid peptide (p1.9) is removed from the COOH-terminus of Pr53 and contains a PSAPP amino acid sequence that is conserved among the COOH-terminal proteins of other lentiviruses (e.g., HIV p6) (Henderson *et al.*, 1992; Tobin *et al.*, submitted). It is noteworthy that many minor but highly immunogenic alternate cleavage products arise during processing of BIV Gag and Env proteins and are detectable in Western blots. Many of these proteins have mobilities in gel electrophoresis that are identical in size to many of the major structural gene products (Battles *et al.*, 1992).

The Gag–Pol precursor polyprotein (Pr170), also translated from the primary transcript by a −1 frameshift (Fig. 6), is incorporated into the virus particle in a 1:10 ratio with the Gag precursor (Battles *et al.*, 1992). Unlike Pr53, Pr170 is not essential for formation of the virus particle (Rasmussen *et al.*, 1990). The viral PR cleaves Pr170 into at least three mature catalytic proteins PR, RT, and integrase (IN) simultaneous with Pr53 processing (Fig. 6). The RT has both polymerase and RNase activity. The BIV RT, which exhibits an Mg^{2+} cofactor curve almost identical to that of the HIV-1 RT, has been functionally identified (Gonda *et al.*, 1990a; Kashanchi *et al.*, 1991). The RT and other *pol* gene products, however, have not been immunologically identified with monospecific antisera, but the putative products of these *pol* domains have been predicted from translations of the DNA sequence (Table I). The BIV *pol* gene may encode a fourth product of 75 amino acid residues that resides between RT and IN in the region assigned dUTPase activity in EIAV, FIV, caprine arthritis encephalitis (CAEV), and ovine lentiviruses (Elder *et al.*, 1992; Threadgill *et al.*, 1993). The sequence in this region of *pol* in BIV (Garvey *et al.*, 1990) is not related to proteins with dUTPase activity and thus is of unknown function.

The BIV Env precursor polyprotein (gPr145) is heavily glycosylated (Table I); over 40% of its empirically derived molecular weight can be attributed to carbohydrate modifications (Rasmussen *et al.*, 1992). The Env precursor is processed into the mature SU and TM glycoproteins, gp100 and gp45, respectively (Table I) (Rasmussen *et al.*, 1992). After cleavage, the SU is associated with the TM glycoprotein by noncovalent bonds, and it is shed readily during virus maturation or by procedures used to isolate virions (Battles *et al.*, 1992). By analogy with other lentiviruses, the SU and TM glycoproteins contain the primary virus receptor and fusion domains, respectively.

E. Nonstructural Accessory Gene Products

The six BIV nonstructural accessory genes are *tat, rev, vif, vpw, vpy,* and *tmx* (Fig. 4). Based on genome locations, conserved traits in the predicted products, and/or functional studies, they appear to be analogous to the primate lentivirus *tat, rev, vif, vpr, vpu, vpx,* and *nef* genes, although there is limited sequence similarity between analogous BIV, HIV, and SIV nonstructural accessory genes or products (Garvey *et al.*, 1990). The predicted and observed sizes, where known, of the BIV nonstructural accessory genes are shown in Table II.

The BIV Tat protein (p14) is a potent *trans*-activator that can enhance virus gene expression as much as 50- to 200-fold in some cell lines (Liu *et al.*, 1992; Carpenter *et al.*, 1993; Pallansch *et al.*, submitted; S. Fong, unpublished data). Two species of BIV *tat* cDNA have been found: one uses only the first coding exon (exon 2) and is translated into a polypeptide of 103 residues and the second uses the first 98 residues from the first coding exon (exon 2) and 10 residues from

TABLE II. Putative BIV Accessory Gene Proteins[a]

Gene	Location of coding exons	Protein	Molecular mass (kDa)	
			Predicted	Observed
vif	nt4601–5194	Vif	22.8	N.D.
vpw	nt4729–4890	Vpw	6.6	N.D.
vpy	nt5089–5328	Vpy	9.5	N.D.
tat-1[b]	nt5228–5536	Tat-1	11.7	14
tat-2[b]	nt5228–5521 nt8056–8085	Tat-2	12.2	14
rev[c]	nt5415–5540 nt7637–8068	Rev	20.7	23
tmx	nt7650–8126	Tmx	18.0	19[d]

[a] The predictions of BIV accessory gene products are from Garvey *et al.* (1990) unless otherwise noted. N.D., not determined; nt, nucleotide. (Table adapted from Gonda, 1992, with permission of Current Science, Ltd., London.)

[b] Two forms of *tat* cDNAs have been found. One encodes Tat-1 by translating 103 amino acids from exon 2. The other encodes Tat-2 by translating the first 98 amino acids from exon 2 and 10 amino acids from exon 3, as observed in HIV-1 (Liu *et al.*, 1992; Pallansch *et al.*, submitted).

[c] *rev* cDNAs consist of three exons. Only the second and third exons are used to encode the Rev protein. Calculation for predicted molecular mass was obtained from translation of *rev* exons 2 and 3 contained in *rev* cDNA (Oberste *et al.*, 1993).

[d] A second form of Tmx migrates at a molecular mass of 18 kDa.

the second coding exon (exon 3) to produce a protein of 108 amino acids (Pallansch *et al.*, submitted). The two-exon form of Tat found in BIV parallels the organization present in HIV-1 (Gonda and Oberste, 1992). Also, as in HIV-1, the *trans*-activating function of the BIV Tat protein resides in exon 2 (Seigel *et al.*, 1986; Liu *et al.*, 1992; Pallansch *et al.*, submitted). Biochemical and immunoelectron microscopy studies have shown that BIV Tat is a phosphoprotein that localizes to the nucleus and nucleoli of infected cells (Pallansch *et al.*, submitted).

The BIV Rev protein (p23) is a 186-residue phosphoprotein that also localizes to the nucleus and nucleoli of infected cells (Oberste *et al.*, 1993). It is encoded by the second and third exons of a three-exon message that is expressed early in the infection cycle. The first coding exon (exon 2) is in the same reading frame as the *env* coding region and uses the first 42 amino acids of the *env* translation (Oberste *et al.*, 1991; Rasmussen *et al.*, 1992). Complementation studies have shown that BIV Rev positively regulates the expression of unspliced and singly spliced viral transcripts, but has no effect on the expression of doubly spliced messages such as *tat* (Oberste *et al.*, 1993). Regulation of splicing in BIV, like in HIV, is believed to occur through a *cis*-acting RRE, a lentiviral RNA sequence that binds Rev and permits RNA export from the nucleus to the cytoplasm (Emerman *et al.*, 1989; Felber *et al.*, 1989). The RRE of BIV is presumably encoded within the sequence for the extracellular domain of the TM envelope protein (nt 5541 to 7636). This region is contained within all unspliced and singly spliced BIV transcripts, but not within any known multiply spliced mRNAs (Oberste *et al.*, 1991; Pallansch *et al.*, submitted).

The functions of the remaining accessory genes *vif*, *vpw*, *vpy*, and *tmx* are not known in detail, although some information is available about the structure and gene expression patterns of *vif* and *tmx*. *vif* is found in the center of the genome overlapping but in a different reading frame than *pol*, and it is predicted to encode a protein of 23 kDa. BIV Vif has little sequence similarity with the predicted Vif proteins of other lentiviruses, although there are two well-conserved sequence motifs that suggest a conservation of function (Oberste and Gonda, 1992). Proviruses in which stop codons have been introduced into *vif* appear to replicate less efficiently, and virus produced from Vif-deficient proviruses appears to be less infectious (Gonda, unpublished data). Tmx (p19) is a recently characterized nonstructural accessory protein that is transcribed by a unique singly spliced mRNA. Translation of *tmx* mRNA begins 14 nt from the splice junction of the common leader and the *tmx* coding exon at the first AUG that is in frame with the *env* gene TM glycoprotein coding sequence. Thus, Tmx shares primary structure with the carboxyl 40% of the TM protein. Although Tmx localizes to the cytoplasm, it is also an abundant protein in the virus particle. It is unclear whether all of the Tmx found in the virion is translated from the 1.3-kb *tmx* transcript or whether a form of this protein results from enzymatic processing of the TM glycoprotein as in EIAV (Rice *et al.*, 1990). BIV *tmx* mRNA has also been found in the spleen and lymph nodes of BIV-infected cattle, and an EIAV *tmx*-like mRNA has been found in tissues from EIAV-infected horses, reinforcing the importance of *tmx* in the virus life cycle (Beisel *et al.*, 1993; Greenwood *et al.*, in preparation). The relative placement of *tmx* in the BIV genome is similar to that of *nef* in primate lentiviruses, suggesting that the two genes may have common functions; however, in contrast to Tmx, the primate Nef is not found in the virion. The biological effect of BIV Tmx on replication and pathogenesis remains to be elucidated (Gonda, 1992).

F. Infection Cycle

Since BIV is an exogenous lentivirus, its replication cycle is predicted to be similar to that of HIV and other horizontally transmitted retroviruses (Gonda and Oberste, 1992; Gonda, 1992, 1994). BIV particles consist of two positive-sense, single-stranded, protein-encapsidated RNA genomes surrounded by an envelope that is derived from the host cell membrane during budding. Replication initiates when glycoproteins on the surface of free virus bind to a specific host cell surface receptor, resulting in the direct fusion of the viral envelope with the plasma membrane (Stein *et al.*, 1987; Maddon *et al.*, 1988). Fusion leads to the release of genomic RNA and mature *pol* gene products into the cytoplasm, and transcription of genomic RNA into double-stranded DNA is catalyzed by the action of the viral RT. The double-stranded viral DNA, termed the provirus, is transported to the nucleus and integrated at apparently random locations in the host genome in a reaction facilitated by the viral IN. The integrated provirus has two fates: it can remain transcriptionally silent or it can be actively transcribed by the host RNA polymerase II using promoter elements present in the viral 5′ LTR. As discussed above and further below, the virally encoded Tat protein can substantially enhance viral gene expression. The appearance in the cytoplasm of genome-length viral transcripts derived from an actively expressed provirus, and singly and multiply spliced subgenomic viral messages derived from the primary transcript, is facilitated by the concerted action of the cellular splicing machinery and another virally encoded protein, Rev. In these functions, BIV Tat and Rev resemble similar proteins encoded by primate lentiviruses.

Genomic and subgenomic viral messages are translated on ribosomes in the cytoplasm. Precursors for the viral Gag and Gag–Pol proteins assemble on the inner surface of the cell plasmalemma and, in the budding process, incorporate viral genomic RNA. The immature viral particles released from the cell are decorated with SU and TM glycoproteins. These particles undergo a maturation event in which Gag-related precursors are cleaved into functional subunits by the viral PR. Mature particles can initiate the infection cycle again by binding to a naive cell expressing the BIV receptor.

G. Viral LTR and Factors That Modulate Virus Expression

As with other lentiviruses, the regulation of BIV replication is dependent on viral and cellular *trans*-acting factors that interact with the viral LTR (Liu *et al.*, 1992; Pallansch *et al.*, 1992, submitted; Carpenter *et al.*, 1993). The BIV LTR contains U3, R, and U5 elements, with the U3 region containing signals for transcription activation, initiation, and termination (Fig. 7).

The BIV LTR contains transcription factor binding sites, including NF-κB, HRE, ATF/CRE, Sp1, AP-1, and AP-4, that are present in some eukaryotic promoters, as well as a retroviral core enhancer. It is noteworthy that Sp1, HRE, ATF/CRE, NF-κB, AP-1, and AP-4 are present in only single copies, whereas in more virulent lentiviruses with relatively high *in vitro* replication rates, such as HIV-1 and visna virus, there are two or more of each of the most essential sites (Hess *et al.*, 1989; Tong-Starksen and Peterlin, 1990; Ross *et al.*, 1991; Andrésson *et al.*, 1993; Chang *et al.*, 1993; Gonda, 1994). The transcription factor binding sites and the retroviral core enhancer in the BIV LTR are more dispersed and distal from the TATA box relative to that of other lentiviruses. For example, in BIV the single NF-κB site is 309 bp from the TATA box, whereas in HIV-1 (where there are two NF-κB sites) the distance is only 56 bp from the proximal NF-κB site. Deletion analysis using LTR-reporter gene constructs has shown that the AP-1–retroviral core enhancer–AP-4 region is absolutely essential for LTR promoter function and several of the other transcription factor binding sites may contribute in regulating virus gene expression. These phenomena are highly dependent on the species and tissue origin of the cell line used (Pallansch *et al.*, submitted).

The virally encoded BIV Tat protein is a strong positive activator of tran-

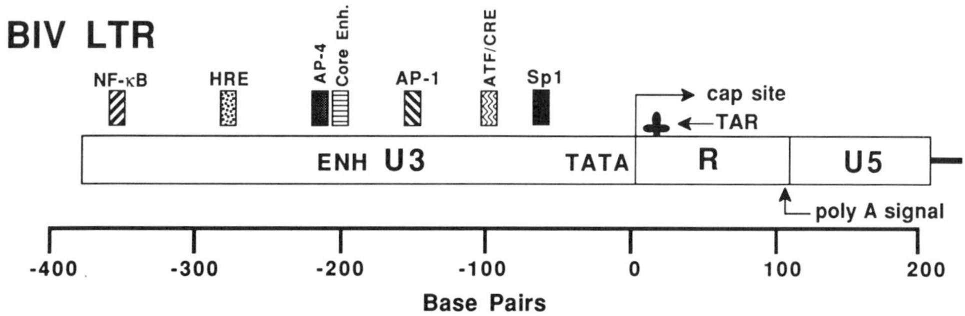

FIGURE 7. Structure of the BIV LTR. (From Gonda *et al.*, 1994, with permission of Elsevier Science Publishers, Amsterdam.)

scription (Liu *et al.*, 1992; Carpenter *et al.*, 1993; Pallansch *et al.*, submitted). Most of our understanding of lentiviral Tat *trans*-activation comes from the study of HIV-1 Tat (Rosen and Pavlakis, 1990; Jeang *et al.*, 1991). In HIV-1, Tat appears to act by binding a 59-nt RNA stem-loop structure termed the *trans*-activation responsive element (TAR) that is present at the 5′ termini of all viral transcripts. Recent studies suggest that binding of Tat to TAR allows the protein to physically contact and stimulate transcription by RNA polymerase II complexes present on the viral LTR (Selby and Peterlin, 1990). The BIV LTR R region also encodes a TAR, an RNA predicted to have strong secondary structure. Deletion analysis has shown that the BIV TAR sequence, like that of HIV-1, is immediately downstream from the transcription initiation site (nt +1 to +20). Although the TAR sequences in BIV are predicted to form a stem-loop structure as do those of HIV-1, there is no obvious similarity in primary sequence or secondary structure (Garvey *et al.*, 1990; Carpenter *et al.*, 1993; Pallansch *et al.*, submitted). Nevertheless, the Tat proteins of BIV and HIV-1 are capable of *trans*-activating each other's LTRs, although the level of *trans*-activation is less than in the homologous system (Liu *et al.*, 1992; Pallansch *et al.*, submitted). The latter suggests that the Tat proteins of BIV and HIV-1 can bind to each other's TAR sequence to activate transcription. Alternatively, the Tat proteins of HIV-1 and BIV may be able to indirectly activate each other's LTRs by activating specific cellular transcription factors in a TAR-independent manner. For example, HIV-1 Tat has been recently demonstrated to indirectly activate the cytomegalovirus promoter (Kim and Risser, 1993).

H. Genetic Diversity and Virus Variability

The availability of DNA sequences for several proviral molecular clones of BIV has permitted an analysis of virus variability (Braun *et al.*, 1988; Garvey *et al.*, 1990, and unpublished data). For example, when looking at the Gag, Pol, and Env structural protein precursors, using percent amino acid homology as a measurement of sequence variability, BIV is only 27 and 28% related to HIV and SIV, respectively, whereas HIV and SIV share 51% amino acid homology. Using the same type of analysis for individual structural gene protein precursors, the Env proteins are the least conserved. BIV is 20% related to either HIV or SIV, and HIV and SIV share 38% homology. Sequence comparisons between two clones, BIV R-29 106 and BIV R-29 127 (Garvey *et al.*, 1990), derived from a single tissue culture isolate (Braun *et al.*, 1988), have shown that a substantial amount of sequence variability can exist. When all substitutions, deletions, and additions are considered, the overall genetic variability between these two clones is 1.7%, with 75% of the substitutions occurring in the SU-coding region of the *env* gene. This genetic diversity and the highly variable nature of the *env* gene sequence have been observed for other lentiviruses, particularly HIV (Hahn *et al.*, 1986; Starcich *et al.*, 1986; Payne *et al.*, 1987; Saag *et al.*, 1988; Gonda *et al.*, 1989). Recent DNA sequence analyses of new field isolates of BIV have also demonstrated that substantial genetic variation exists between different field strains (Suarez *et al.*, 1993; Oberste *et al.*, in preparation). Very likely, genetic diversity in BIV and other lentiviruses plays a role in their ability to evade immune selection pressures, increase replication rate and pathogenicity, and to adapt to new host niches. For example, they could acquire modifications in cryptic or degenerate transcription factor binding sites in the LTR and make them biologically relevant as observed in closely related visna virus strains (Andrésson *et al.*, 1993).

III. CELLS SUSCEPTIBLE TO BIV INFECTION

A. *In Vitro* Tropism and Cytopathogenicity

BIV has a broader host cell range than the primate lentiviruses as it infects cell cultures of bovine, rabbit, and canine, but not primate origin. *In vitro*, BIV replicates in fibroblastlike cells and is cytopathic in most, causing syncytium formation and cell death (Fig. 8). Primary cultures of bovine embryonic spleen, thymus, lung, kidney, testes, brain, choroid plexus, and synovial membrane tissues are infectible (Gonda *et al.*, 1987) as are the established cell lines, embryonic rabbit epithelium (EREp), canine thymus (Cf2Th), and several bovine cell lines

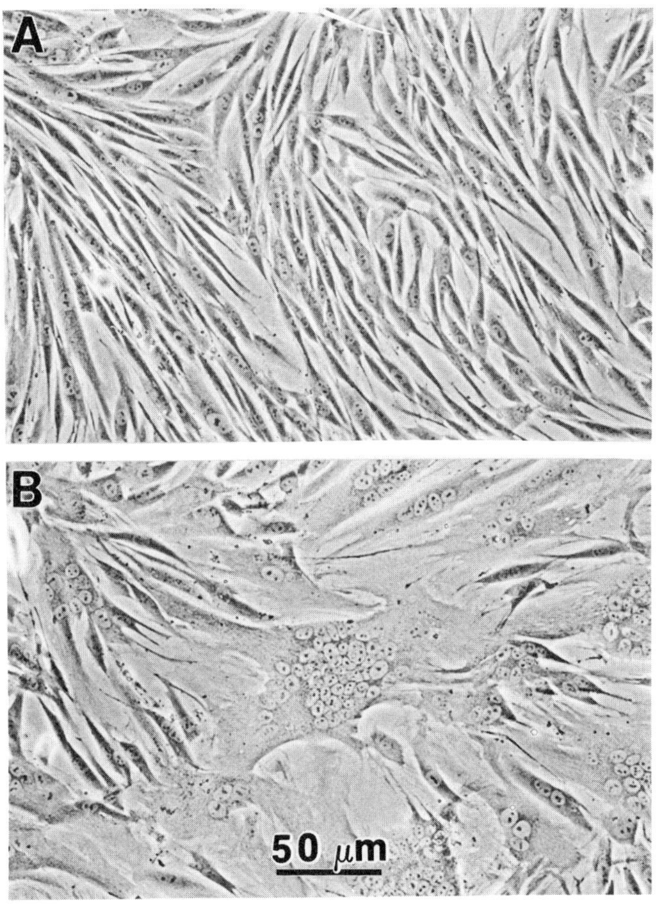

FIGURE 8. Induction of syncytia in BIV-infected EREp cells. (A) Uninfected EREp cells. (B) BIV-infected EREp cells. (From Gonda *et al.*, 1994, with permission of Elsevier Science Publishers, Amsterdam.)

(Gonda *et al.*, 1990a,b; Gonda, 1992; Pifat *et al.*, 1992). Although most cultures succumb to the cytopathic effects of BIV gene expression, the Cf2Th cell line can maintain a long-term productive infection (Bouillant *et al.*, 1989; Gonda *et al.*, 1990a,b) and, when properly maintained, continuously infected cultures can supply infectious BIV for laboratory investigations.

B. *In Vivo* Tropism

The *in vivo* targets of BIV infection appear to be more restricted than those *in vitro*. By polymerase chain reaction (PCR) amplification, BIV DNA is most often detected in DNA obtained from spleen, lymph node, peripheral blood mononuclear cells (PBMCs), lung, and brain extracts of infected rabbits and cattle (Pifat *et al.*, 1992; Oberste *et al.*, in preparation). BIV RNA has been localized in PBMCs from cattle by *in situ* hybridization (Carpenter *et al.*, 1992), and BIV protein has been observed in atypical blastic splenic rabbit macrophages by immunohistochemical techniques (Pifat *et al.*, 1992). Moreover, BIV can be rescued from monocyte-enriched cultures of PBMCs from infected cattle but not from lymphocyte-enriched cultures, and monocyte-enriched cultures but not lymphocyte-enriched cultures can be infected with BIV.

These results suggest that the predominant *in vivo* targets of BIV infection are most likely cells of the monocyte/macrophage lineage (Carpenter *et al.*, 1992; Onuma *et al.*, 1992; Pifat *et al.*, 1992). In this regard, BIV resembles visna and the caprine lentiviruses (Narayan *et al.*, 1993). However, recent studies with new field isolates of BIV suggest that the tropism may vary and be broader than monocytes and macrophages, as observed with HIV-1 (Dalgleish *et al.*, 1984; Klatzmann *et al.*, 1984; Gendelman *et al.*, 1989; Gartner and Popovic, 1990; Meltzer *et al.*, 1990; Gonda, unpublished data).

C. Receptors

The CD4 molecule has been identified as the primary cellular receptor of the human and simian immunodeficiency viruses (Dalgleish *et al.*, 1984; Klatzmann *et al.*, 1984; Maddon *et al.*, 1988). A 50-kDa sheep cellular protein (Crane *et al.*, 1991) and a sheep MHC class II molecule (Dalziel *et al.*, 1991) have been implicated as the receptor for ovine lentiviruses. The BIV SU glycoprotein (gp100) is presumed to attach to target cells via a specific cell surface receptor. The BIV cellular receptor has not been identified. However, the observations that BIV primarily replicates in cells of the immune system *in vivo* and that other lentiviruses use leukocyte cell differentiation antigens as receptors suggest that the BIV receptor may be a protein of the cell differentiation series found on leukocytes.

IV. PATHOGENESIS AND IMMUNOBIOLOGY

A. Infection in Natural Host

The importance of BIV infections to the health of cattle has been difficult to assess in domestic settings. The reasons for this difficulty are that cattle are production animals with a high turnover rate, and consequently a short life span. Moreover, many are dually infected with BLV and BIV, obscuring the contributions of BIV to disease. Since most BIV infections in the United States are encountered in the South, studies on the effects of natural BIV infection are limited to observations of these herds. As an example, several relevant observations were made in a long-term study (>4 years) of a Louisiana dairy herd with a high frequency (>80%) of BIV infection (Luther *et al.*, unpublished data). The cattle had various signs of disease including lethargy, mastitis, chronic pododermatitis unresponsive to medical intervention, pneumonia, mycotic abomasitis, lymphadenopathy, lymphosarcoma, and injection site myositis with abscessation (Fig. 9). Many cattle exhibited neural lesions on histopathologic examination. These lesions included slight to moderate lymphocytic meningitis and lymphocytic peri-

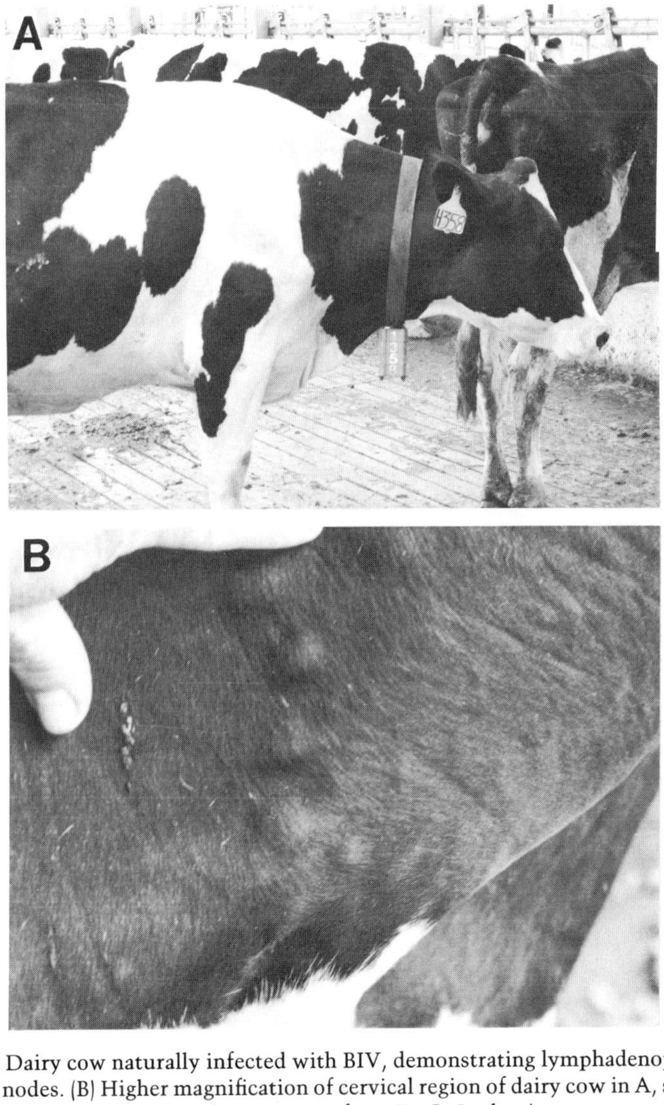

FIGURE 9. (A) Dairy cow naturally infected with BIV, demonstrating lymphadenopathy in the prescapular lymph nodes. (B) Higher magnification of cervical region of dairy cow in A, showing swollen prescapular lymph nodes. (Photographs courtesy of Dr. D. G. Luther.)

vascular cuffing. In some animals, there were focal mononuclear cell accumulations in areas of neuronal degeneration and focal lymphocytic infiltration of the neuropil associated with enlarged astrocytes, and other areas of the neuropil had diffuse microglial activation reflective of the severity of the encephalitis. The nonsuppurative meningoencephalitis was similar to the brain lesions in cattle described earlier (Van Der Maaten *et al.*, 1972). The various maladies observed resulted in high mortality and cull rates. In addition, pathogenesis may be exacerbated by the stress associated with dairy farming and other environmental factors. There is a need for concurrent and retrospective serologic studies of BIV infections where clinical information on the health status and maintenance of animals is well documented. The experimental study of BIV infections in natural settings in which infected and noninfected animals are segregated will provide further relevant data as to BIV transmission and pathogenesis.

Lentiviruses induce chronic lifelong infections in their hosts (Gonda, 1994). There is usually an acute stage of active viral replication followed by an asymptomatic phase. Experimental infections with BIV R-29 (Van Der Maaten *et al.*, 1972; Carpenter *et al.*, 1992; Onuma *et al.*, 1992) and molecular clones derived from it (Braun *et al.*, 1988; Oberste *et al.*, in preparation) indicate that this is the case for BIV. Additional confirmatory information has come from the study of several new field isolates (Suarez *et al.*, 1993; Oberste *et al.*, in preparation). The initial pathogenesis of BIV in cattle is similar to that observed for HIV in humans. There is a persistent, generalized lymphadenopathy and, in some studies, a transient lymphocytosis has been observed (Van Der Maaten *et al.*, 1972; Carpenter *et al.*, 1992; Onuma *et al.*, 1992; Suarez *et al.*, 1993). In the early stages of experimental BIV infection of cattle (<6 months), virus can be readily rescued from PBMCs and BIV DNA can be found in these cells. This virus production is accompanied by a strong increase in antibody titer (acute phase) which declines to and remains at a lower level after the first 6 months. The serological response to virus replication during the acute stage results in the production of antibodies to a spectrum of Gag and Env precursors and processed products (Fig. 10) whose titers fluctuate over the course of infection. Virus expression during the asymptomatic phase can be stimulated by transfusion of blood from an uninfected donor cow, which results in a sharp rise in BIV-specific antibody titers. Although there were no overt signs of disease within the short duration (<2 years) of these studies (Oberste *et al.*, in preparation), two major types of histopathological lesions were observed in experimentally infected cows. One was a follicular hyperplasia of germinal centers in lymph nodes (Fig. 11A). The second was a nonsuppurative perivascular cuffing in the cerebrum indicative of a viral encephalitis (Fig. 11B).

B. Infection in Experimental Hosts

Although the host range of lentiviruses is generally restricted to closely related species, BIV has been shown to infect rabbits (Pifat *et al.*, 1992; Van Der Maaten and Whetstone, 1992) and possibly sheep (Whetstone *et al.*, 1991). Rodents do not appear to be infectible (Gonda, 1992). Rabbits can be infected with cell-associated or cell-free virus inocula and may provide insights into disease mechanisms and virus–host interactions of BIV. Rabbits respond to the virus by mounting a rapid (within weeks), high-titered (ELISA titer > 1:25,000), and per-

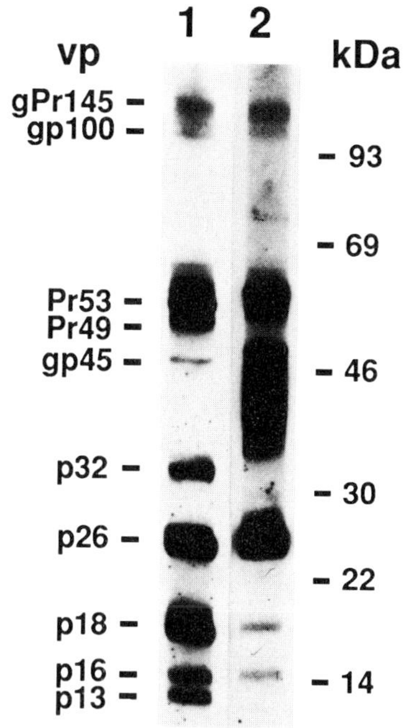

FIGURE 10. Diagnostic Western immunoblot of BIV proteins. Purified BIV virions were resolved on an 8 to 16% gradient polyacrylamide gel and transferred onto an Immobilon-P membrane. Antisera used were from a naturally infected bovine (lane 1) and a rabbit experimentally infected with BIV (lane 2). BIV proteins (vp) gPr145, gp100, Pr53, Pr49, gp45, p32, p26, p18, p16, and p13 are designated on the left; molecular mass markers are indicated on the right side. The lower molecular mass BIV protein bands (p18, p16, p13) each contain multiple Gag- or Env-related proteins with similar sizes that result from the primary or alternate cleavage of precursors (Battles *et al.*, 1992). (From Battles *et al.*, 1992, with permission of the American Society for Microbiology, Washington, D.C.)

sistent (for the life of the animal) humoral immune response (Pifat *et al.*, 1992). Antibodies to both Gag and Env proteins are elicited (Fig. 10). A mild lymphadenopathy is seen early in the infection, but there are no apparent clinical signs of disease. Virus can be rescued from PBMCs, spleen, and lymph nodes, and these and other immune tissues as well as brain are the easiest to confirm as sites of BIV replication by PCR amplification of BIV sequences. In addition, *in situ* immunohistochemical staining of splenic macrophages with BIV-specific serum suggests that macrophages are the active site of virus replication in rabbits (Pifat *et al.*, 1992).

To study the biological effects of virus gene expression in controlled *in vivo* experiments, a number of transgenic mouse lines have been made with complete functional BIV proviruses (Pallansch *et al.*, in preparation). Two lines develop antibodies to BIV proteins and exhibit substantial clinical disease, including dermal lesions and a neurologic syndrome. Pathogenesis was observed in several organs including brain, thymus, spleen, lymph nodes, and skin. One line exhibited early mortality (50%) attributable to meningoencephalitis. Virus gene expression in these animals was low and tissue-specific. It could only be detected by PCR, was strongest in brain tissue, and generally correlated with the severity of the pathology observed in the affected organs. The neuropathology observed in BIV proviral transgenic mice is similar to the meningoencephalitis observed in cattle.

C. Immunobiology

BIV infection of cattle results in a strong host response (Gonda *et al.*, 1987; Whetstone *et al.*, 1990, 1991; Battles *et al.*, 1992; Carpenter *et al.*, 1993; Suarez *et*

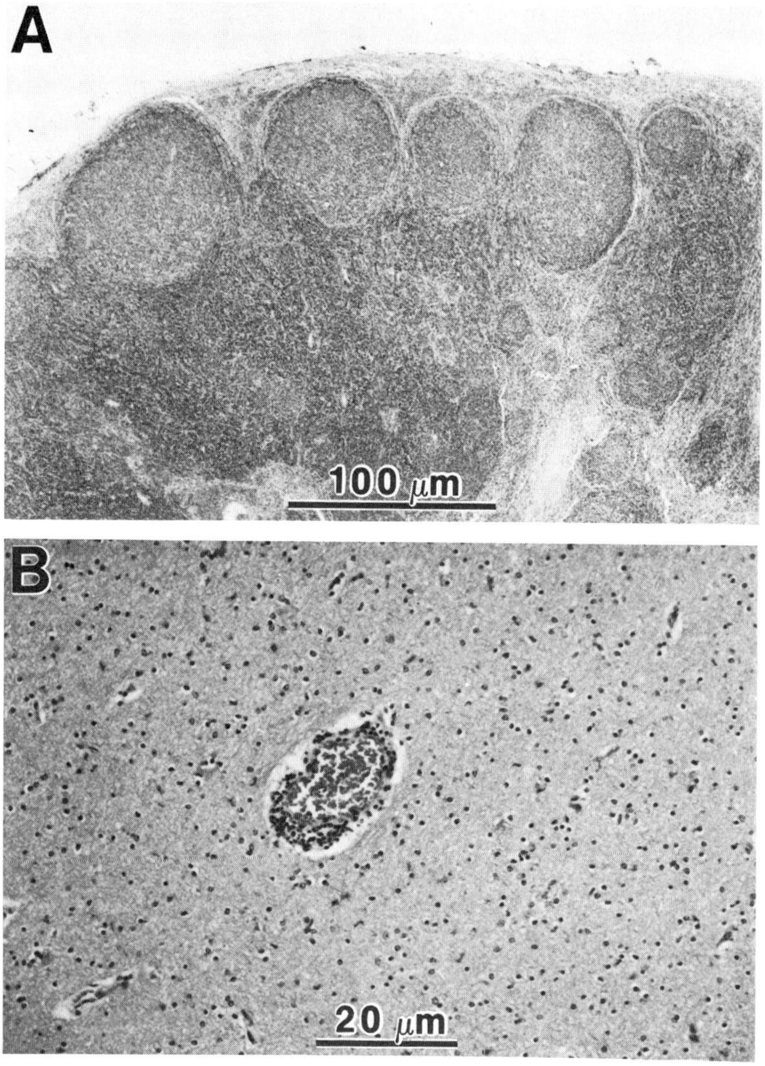

FIGURE 11. Pathology of tissues from cattle experimentally infected with BIV. (A) Follicular hyperplasia of germinal centers of a subcutaneous lymph node showing numerous enlarged germinal centers. (B) Perivascular cuffing of mononuclear cells in the cerebrum. (From Gonda *et al.*, 1994, with permission of Elsevier Science Publishers, Amsterdam.)

al., 1993; Oberste *et al.*, in preparation). Within the first month after BIV infection, there is a humoral response to Gag and Env proteins. It is not clear when a response to the nonstructural accessory proteins begins. There are no reports on the bovine cellular immune response (e.g., development of cytotoxic T lymphocytes, antibody-dependent cell cytotoxicity) to BIV antigens in infected animals. However, immune dysfunction of immune cells from experimentally infected cattle has been investigated. Onuma *et al.* (1992) have shown that BIV infection of cattle reduces the responsiveness of various important monocyte functions without a change in CD4/CD8 ratios; there was also a slight delay in humoral antibody response to mouse antigen. Other studies that examined the effect of

BIV infection on immune function have demonstrated either mild or no immunosuppression on the basis of lymphocyte blastogenesis tests, neutrophil function tests, mononuclear subset analysis, and histopathological changes (Martin *et al.*, 1991; Carpenter *et al.*, 1992; Flaming *et al.*, 1993).

The immunobiology of BIV infections has also been studied by coinfecting rabbits with BIV and BLV. The infection of rabbits by BLV can only be achieved with great difficulty (Wyatt *et al.*, 1989). BIV appears to potentiate infection of rabbits by BLV when it is introduced into rabbits prior to or after BLV inoculation (Gonda *et al.*, unpublished data). Rabbits coinfected with BIV and BLV exhibited a strong and persistent humoral response to both viruses indicating active replication. In contrast, rabbits in this study given BLV alone have not seroconverted to BLV. The mechanism whereby BIV potentiates the replication of BLV in rabbits is unknown. It is possible that BIV perturbs the development of a protective immune response of rabbits to BLV infection or stimulates the growth of cells permissive to BLV infection. The immune dysfunction described for monocytes in BIV-infected cattle and the apparent replication of BIV in rabbit macrophages suggest that BIV interferes with the rabbits' ability to mount a cellular immune response to BLV, perhaps through dysfunction of monocytes and macrophages.

V. CONCLUSIONS

At present, a considerable amount of knowledge has accumulated about the molecular biology of BIV. BIV has the most complex genome of the nonprimate lentiviruses and contains a series of accessory genes that are used to positively regulate virus gene expression in a fashion similar to that exhibited by the primate lentiviruses. This similarity in BIV and primate lentivirus accessory gene function may assist in efforts to elucidate the role of accessory genes in lentivirus replication and pathogenicity. The comparative study of divergent nonprimate lentiviruses, such as BIV, may assist in our understanding of HIV-induced disease mechanisms.

The early stage of infection by ungulate lentiviruses is in many ways analogous to that of HIV infection in humans. The common features of the ungulate lentivirus–host interaction include infection of the immune cells, prolonged periods of clinical latency (months to years) with restricted virus replication and mutable genomes that lead to genetic variability (primarily in the *env* sequence). In addition, chronic inflammatory disease is present that is usually slow in onset, involves multiple organs, and, when the brain is affected, is often accompanied by a substantial neuropathology (Haase, 1986; Narayan and Clements, 1989; Gonda, 1994). BIV infection of cattle does not produce acute clinical disease in the absence of cofactors, and thus is typical of most other ungulate and simian lentiviruses with reduced pathogenicity, which have naturally evolved during virus–host adaptation. This finding contrasts with more acutely pathogenic lentiviruses, such as SIV_{pbj} (Dewhurst *et al.*, 1990) and visna virus strain K1772 (Andrésson *et al.*, 1993). They have recently mutated to forms that induce a rapid clinical disease and swift death for the host. It will be important to determine the factors that upregulate BIV expression and enhance pathogenicity.

A great deal of effort has gone into developing small-animal models for HIV infection. SIV infection in macaques offers many advantages in terms of an animal model for AIDS (Desrosiers and Ringler, 1989). Transgenic mice also have played an important role in understanding the pathogenesis of HIV gene expression (Leonard *et al.*, 1988; Kopp *et al.*, 1993) and have aided our understanding of the pathogenic effects of BIV gene expression *in vivo*. BIV transgenic mice have a prominent neuropathology and chronic skin lesions similar to those observed in cattle (Van Der Maaten *et al.*, 1972; Oberste *et al.*, in preparation). The study of these mice may provide clues on the induction of inflammatory central nervous system lesions by lentivirus infection in the absence of opportunistic infections, as seen in HIV (Vazeux *et al.*, 1990). Moreover, it will be important to better understand the immunobiology of ungulate lentiviruses and the role of the immune system in defense and disease induction. BIV also infects rabbits (Pifat *et al.*, 1992), indicating that rabbits may be a useful animal model for studying lentiviral replication. The replication of BIV in immune cells, its induction of chronic infections in spite of a strong host immune response in small laboratory animals, and its shared antigenic, genetic, and pathologic features, all suggest that BIV infections could provide useful tools for testing antiviral strategies relevant to HIV and AIDS.

ACKNOWLEDGMENTS. The author thanks the many colleagues who over the years have contributed to this work. I am especially indebted to G. Tobin and S. Fong for reading the manuscript and for the many helpful suggestions, G. Serig for aid in preparation of the manuscript, and S. Rivard for preparation of graphics.

VI. REFERENCES

Amborski, G. F., Lo, J. L., and Seger, C. L., 1989, Serological detection of multiple retroviral infection in cattle: Bovine leukemia virus, bovine syncytial virus and bovine visna virus, *Vet. Microbiol.* **20**:247.

Andrésson, Ó. S., Elser, J. E., Tobin, G. J., Greenwood, J. D., Gonda, M. A., Georgsson, G., Andrésdóttir, V., Benediktsdóttir, E., Carlsdóttir, H. M., Mäntylä, E. O., Rafnar, B., Pálsson, P. A., Casey, J. W., and Pétursson, G., 1993, Nucleotide sequence and biological properties of a pathogenic proviral molecular clone of neurovirulent visna virus, *Virology* **193**:89.

Battles, J. K., Hu, M. Y., Rasmussen, L., Tobin, G. J., and Gonda, M. A., 1992, Immunological characterization of the *gag* gene products of bovine immunodeficiency virus, *J. Virol.* **66**:6868.

Beisel, C. E., Edwards, J. F., Dunn, L. L., and Rice, N. R., 1993, Analysis of multiple mRNAs from pathogenic equine infectious anemia virus (EIAV) in an acutely infected horse reveals a novel protein, Ttm, derived from the carboxy terminus of the EIAV transmembrane protein, *J. Virol.* **67**:832.

Black, J. W., 1990, Bluetongue and bovine retrovirus committee report, in: *Proceedings of the 93rd Annual Meeting of the U.S. Animal Health Association*, pp. 150–152, Carter Printing Co., Richmond.

Bouillant, A. M., Ruckerbauer, G. M., and Nielsen, K. H., 1989, Replication of the bovine immunodeficiency-like virus in diploid and aneuploid cells: Permanent latent and virus-productive infections *in vitro*, *Res. Virol.* **140**:511.

Braun, M. J., Lahn, S., Boyd, A. L., Kost, T. A., Nagashima, K., and Gonda, M. A., 1988, Molecular cloning of biologically active proviruses of bovine immunodeficiency-like virus, *Virology* **167**:515.

Carpenter, S., Miller, L. D., Alexandersen, S., Whetstone, C. A., Van Der Maaten, M. J., Viuff, B., Wannemuehler, Y., Miller, J. M., and Roth, J. A., 1992, Characterization of early pathogenic effects after experimental infection of calves with bovine immunodeficiency-like virus, *J. Virol.* **66**:1074.

Carpenter, S., Nadin-Davis, S. A., Wannemuehler, Y., and Roth, J. A., 1993, Identification of transactivation-response sequences in the long terminal repeat of bovine immunodeficiency-like virus, *J. Virol.* **67**:4399.

Chakrabarti, S., Mizukami, T., Franchini, G., and Moss, B., 1990, Synthesis, oligomerization, and biological activity of the human immunodeficiency virus type 2 envelope glycoprotein expressed by a recombinant vaccinia virus, *Virology* **178**:134.

Chang, L.-J., McNulty, E., and Martin, M., 1993, Human immunodeficiency viruses containing heterologous enhancer/promoters are replication competent and exhibit different lymphocyte tropisms, *J. Virol.* **67**:743.

Cockerell, G. L., Jensen, W. A., Rovnak, J., Ennis, W. H., and Gonda, M. A., 1992, Seroprevalence of bovine immunodeficiency-like virus and bovine leukemia virus in a dairy cattle herd, *Vet. Microbiol.* **31**:109.

Crane, S. E., Buzy, J., and Clements, J. E., 1991, Identification of cell membrane proteins that bind visna virus, *J. Virol.* **65**:6137.

Dalgleish, A. G., Beverley, P. C. L., Clapham, P. R., Crawford, D. H., Greaves, M. F., and Weiss, R. A., 1984, The CD4 (T4) antigen is an essential component of the receptor for the AIDS retrovirus, *Nature (London)* **312**:763.

Dalziel, K. G., Hopkins, J., Watt, N. J., Dutia, B. M., Clarke, M. A. K., and McConnell, I., 1991, Identification of a putative cellular receptor for the lentivirus visna virus, *J. Gen. Virol.* **72**:1905.

Desrosiers, R. C., and Ringler, D. J., 1989, Use of simian immunodeficiency viruses for AIDS research, *Intervirology* **30**:301.

Dewhurst, S., Embretson, J. E., Anderson, D. C., Mullins, J. I., and Fultz, P. N., 1990, Sequence analysis and acute pathogenicity of molecularly cloned SIVSMM-PBj14, *Nature (London)* **345**:636.

Earl, P. L., Doms, R. W., and Moss, B., 1990, Oligomeric structure of the human immunodeficiency virus type 1 envelope glycoprotein, *Proc. Natl. Acad. Sci. USA* **87**:648.

Elder, J. H., Lerner, D. L., Hasselkus-Light, C. S., Fontenot, D. J., Hunter, E., Luciw, P. A., Montelaro, R. C., and Phillips, T. R., 1992, Distinct subsets of retroviruses encode dUTPase, *J. Virol.* **66**:1791.

Emerman, M., Vazeux, R., and Peden, K., 1989, The *rev* gene product of the human immunodeficiency virus affects envelope-specific RNA localization, *Cell* **57**:1155.

Felber, B. K., Hadzopoulou-Cladaras, M., Cladaras, C., Copeland, T., and Pavlakis, G. N., 1989, rev protein of human immunodeficiency virus type 1 affects the stability and transport of the viral mRNA, *Proc. Natl. Acad. Sci. USA* **86**:1495.

Flaming, K., Van Der Maaten, M., Whetstone, C., Carpenter, S., Frank, D., and Roth, J., 1993, Effect of bovine immunodeficiency-like virus infection on immune function in experimentally infected cattle, *Vet. Immunol. Immunopathol.* **36**:91.

Gallo, R. C., 1991, Human retroviruses: A decade of discovery and link with human disease, *J. Infect. Dis.* **164**:235.

Gallo, R. C., and Nerurkar, L. S., 1989, Human retroviruses: Their role in neoplasia and immunodeficiency, *Ann. N.Y. Acad. Sci.* **567**:82.

Gartner, S., and Popovic, M., 1990, Macrophage tropism of HIV-1, *AIDS Res. Hum. Retroviruses* **6**:1017.

Garvey, K. J., Oberste, M. S., Elser, J. E., Braun, M. J., and Gonda, M. A., 1990, Nucleotide sequence and genome organization of biologically active proviruses of the bovine immunodeficiency-like virus, *Virology* **175**:391.

Gendelman, H. E., Orenstein, J. M., Baca, L. M., Weiser, B., Burger, H., Kalter, D. C., and Meltzer, M. S., 1989, The macrophage in the persistence and pathogenesis of HIV infection, *AIDS* **3**:475.

Gonda, M. A., 1988, Molecular genetics and structure of the human immunodeficiency virus, *J. Electron Microsc. Tech.* **8**:17.

Gonda, M. A., 1992, Bovine immunodeficiency virus, *AIDS* **6**:759.

Gonda, M. A., 1994, Molecular biology and virus-host interactions of lentiviruses, *Ann. N.Y. Acad. Sci.* **724**:22.

Gonda, M. A., and Oberste, M. S., 1992, AIDS—The human immunodeficiency virus: Molecular and structural aspects of its biology, in: *Control of Virus Diseases*, (E. Kurstak, ed.), pp. 3–31, Dekker, New York.

Gonda, M. A., Wong-Staal, F., Gallo, R. C., Clements, J. E., Narayan, O., and Gilden, R. V., 1985, Sequence homology and morphologic similarity of HTLV-III and visna virus, a pathogenic lentivirus, *Science* **227**:173.

Gonda, M. A., Braun, M. J., Clements, J. E., Pyper, J. M., Wong-Staal, F., Gallo, R. C., and Gilden, R. V., 1986, Human T-cell lymphotropic virus type III shares sequence homology with a family of pathogenic lentiviruses, *Proc. Natl. Acad. Sci. USA* **83**:4007.

Gonda, M. A., Braun, M. J., Carter, S. G., Kost, T. A., Bess, J. W., Jr., Arthur, L. O., and Van Der Maaten, M. J., 1987, Characterization and molecular cloning of a bovine lentivirus related to human immunodeficiency virus, *Nature (London)* **330**:388.

Gonda, M. A., Boyd, A. L., Nagashima, K., and Gilden, R. V., 1989, Pathobiology, molecular organization, and ultrastructure of HIV, *Arch. AIDS Res.* **3**:1.

Gonda, M. A., Oberste, M. S., Garvey, K. J., Pallansch, L. A., Battles, J. K., Pifat, D. Y., Bess, J. W., Jr., and Nagashima, K., 1990a, Development of the bovine immunodeficiency-like virus as a model of lentivirus disease, *Dev. Biol. Stand.* **72**:97.

Gonda, M. A., Oberste, M. S., Garvey, K. J., Pallansch, L. A., Battles, J. K., Pifat, D. Y., and Nagashima, K., 1990b, Contemporary developments in the biology of the bovine immunodeficiency-like virus, in: *Animal Models in AIDS* (H. Schellekens and M. Horzinek, eds.), pp. 233–255, Elsevier, Amsterdam.

Gonda, M. A., Luther, D. G., Fong, S. E., and Tobin, G. J., 1994, Bovine immunodeficiency virus: Molecular biology and virus–host interactions, *Virus Res.* **32**:155.

Haase, A. T., 1986, Pathogenesis of lentivirus infections, *Nature (London)* **322**:130.

Hahn, B. H., Shaw, G. M., Taylor, M. E., Redfield, R. R., Markham, P. D., Saladhuddin, S. Z., Wong-Staal, F., Gallo, R. C., Parks, E. S., and Parks, W. P., 1986, Genetic variation in HTLV-III/LAV over time in patients with AIDS or at risk for AIDS, *Science* **232**:1548.

Henderson, L. E., Bowers, M. A., Sowder, R. C., II, Serabyn, S. A., Johnson, D. G., Bess, J. W., Jr., Arthur, L. O., Bryant, D. K., and Fenselau, C., 1992, Gag proteins of the highly replicative MN strain of human immunodeficiency virus type 1: Posttranslational modifications, proteolytic processings, and complete amino acid sequences, *J. Virol.* **66**:1856.

Hess, J. L., Small, J. A., and Clements, J. E., 1989, Sequences in the visna virus long terminal repeat that control transcriptional activity and respond to viral *trans*-activation: Involvement of AP-1 sites in basal activity and *trans*-activation, *J. Virol.* **63**:3001.

Horner, G. W., 1991, Serologic evidence of bovine immunodeficiency-like virus and bovine syncytial virus in New Zealand, *Surveillance* **18**:9.

Horzinek, M., Keldermans, L., Stuurman, T., Black, J., Herrewegh, A., Sillekens, P., and Koolen, M., 1991, Bovine immunodeficiency virus: Immunochemical characterization and serological survey, *J. Gen. Virol.* **72**:2923.

Jacks, T., Power, M. D., Masiarz, F. R., Luciw, P. A., Barr, P. J., and Varmus, H. E., 1988, Characterization of ribosomal frameshifting in HIV-1 *gag–pol* expression, *Nature (London)* **331**:280.

Jacobs, R. M., Smith, H. E., Gregory, B., Valli, V. E. O., and Whetstone, C. A., 1992, Detection of multiple retroviral infections in cattle and cross-reactivity of bovine immunodeficiency-like virus and human immunodeficiency virus type 1 proteins using bovine and human sera in a Western blot assay, *Can. J. Vet. Res.* **56**:353.

Jeang, K.-T., Chang, Y.-N., Berkhout, B., Hammarskjöld, M. L., and Rekosh, D., 1991, Regulation of HIV expression: Mechanisms of action of Tat and Rev, *AIDS* **5**(Suppl. 2):S3.

Johnson, P. R., and Hirsch, V. M., 1992, Genetic variation of simian immunodeficiency viruses in nonhuman primates, *AIDS Res. Hum. Retroviruses* **8**:367.

Kashanchi, F., Liu, Z. Q., Atkinson, B., and Wood, C., 1991, Comparative evaluation of bovine immunodeficiency-like virus infection by reverse transcriptase and polymerase chain reaction, *J. Virol. Methods* **31**:197.

Kennedy, R. C., Eklund, C. M., Lopez, C., and Hadlow, W. J., 1968, Isolation of a virus from the lungs of Montana sheep affected with progressive pneumonia, *Virology* **35**:483.

Kim, Y. S., and Risser, R., 1993, TAR-independent transactivation of the murine cytomegalovirus major immediate-early promoter by the Tat protein, *J. Virol.* **67**:239.

Klatzmann, D., Champagne, E., Chamaret, S., Gruest, J., Guetard, D., Hercend, T., Gluckman, J.-C., and Montagnier, L., 1984, T-lymphocyte T4 molecule behaves as the receptor for human retrovirus LAV, *Nature (London)* **312**:767.

Kopp, J. B., Rooney, J. F., Wohlenberg, C., Dorfman, N., Marinos, N. J., Bryant, J. L., Katz, S. I., Notkins, A. L., and Klotman, P. E., 1993, Cutaneous disorders and viral gene expression in HIV-1 transgenic mice, *AIDS Res. Hum. Retroviruses* **9**:267.

Leonard, J. M., Abramczuk, J. W., Pezen, D. S., Rutledge, R., Belcher, J. H., Hakim, F., Shearer, G., Lamperth, L., Travis, W., Fredrickson, T., Notkins, A. L., and Martin, M. A., 1988, Development of disease and virus recovery in transgenic mice containing HIV proviral DNA, *Science* **242**:1665.

Levy, J. A., 1993, Pathogenesis of human immunodeficiency virus infection, *Microbiol. Rev.* **57**:183.

Liu, Z. Q., Sheridan, D., and Wood, C., 1992, Identification and characterization of the bovine immunodeficiency-like virus *tat* gene, *J. Virol.* **66**:5137.

Maddon, P. J., McDougal, J. S., and Clapham, P. R., 1988, HIV infection does not require endocytosis of its receptor, *Cell* **54**:865.

Martin, S. J., O'Neill, T. P., Bilello, J. A., and Eiseman, J. L., 1991, Lymphocyte transformation abnormalities in bovine immunodeficiency-like virus infected calves, *Immunol. Lett.* **27**:81.

Meltzer, M. S., Nakamura, M., Hansen, B. D., Turpin, J. A., Kalter, D. C., and Gendelman, H. E., 1990, Macrophages as susceptible targets for HIV infection, persistent viral reservoirs in tissue, and key immunoregulatory cells that control levels of virus replication and extent of disease, *AIDS Res. Hum. Retroviruses* **6**:967.

Moss, B., Earl, P. L., and Doms, R. W., 1991, Transport and assembly of the HIV-1 envelope protein, in: *Retroviruses of Human AIDS and Related Animal Diseases* (M. Girard and L. Valette, eds.), pp. 177–180, Pasteur Mérieux, Paris.

Narayan, O., and Clements, J. E., 1989, Biology and pathogenesis of lentiviruses, *J. Gen. Virol.* **70**:1617.

Narayan, O., Zink, M. C., Gorrell, M., Crane, S., Huso, D., Jolly, P., Saltarelli, M., Adams, R. J., and Clements, J. E., 1993, The lentiviruses of sheep and goats, in: *The Retroviridae*, Vol. 2 (J. A. Levy, ed.), pp. 229–255, Plenum Press, New York.

Oberste, M. S., and Gonda, M. A., 1992, Conservation of amino-acid sequence motifs in lentivirus Vif proteins, *Virus Genes* **6**:95.

Oberste, M. S., Greenwood, J. D., and Gonda, M. A., 1991, Analysis of the transcription pattern and mapping of the putative *rev* and *env* splice junctions of bovine immunodeficiency-like virus, *J. Virol.* **65**:3932.

Oberste, M. S., Williamson, J. C., Greenwood, J. D., Nagashima, K., Copeland, T. D., and Gonda, M. A., 1993, Characterization of bovine immunodeficiency virus *rev* cDNAs and identification and subcellular localization of the Rev protein, *J. Virol.* **67**:6395.

Olmsted, R. A., Langley, R., Roelke, M. E., Goeken, R. M., Adger-Johnson, D., Goff, J. P., Albert, J. P., Packer, C., Laurenson, M. K., Caro, T. M., Scheepers, L., Wildt, D. E., Bush, M., Martenson, J. S., and O'Brien, S. J., 1992, Worldwide prevalence of lentivirus infection in wild feline species: Epidemiologic and phylogenetic aspects, *J. Virol.* **66**:6008.

Onuma, M., Koomoto, E., Furuyama, H., Yasutomi, Y., Taniyama, H., Iwai, H., and Kawakami, Y., 1992, Infection and dysfunction of monocytes induced by experimental inoculation of calves with bovine immunodeficiency-like virus, *J. AIDS* **5**:1009.

Pallansch, L. A., Lackman-Smith, C. S., and Gonda, M. A., 1992, Bovine immunodeficiency-like virus encodes factors which *trans* activate the long terminal repeat, *J. Virol.* **66**:2647.

Pallansch, L. A., Greenwood, J. D., Williamson, J. C., Rasmussen, L., Ennis, W. H., Mentzer, A. F., Nagashima, K., Copeland, T. D., and Gonda, M. A., Structural and functional studies of cDNAs and proteins encoded by the *tat* gene of bovine immunodeficiency virus, submitted.

Pallansch, L. A., Lackman-Smith, C. S., Mikovits, J. A., Ruscetti, F. W., and Gonda, M. A., Functional analysis and mapping of promoter elements in the bovine immunodeficiency virus long terminal repeat, submitted.

Payne, S. L., Fang, F.-D., Liu, C.-P., Dhruva, B. R., Rwambo, P., Issel, C. J., and Montelaro, R. C., 1987, Antigenic variation and lentivirus persistence: Variations in envelope gene sequences during EIAV infection resemble changes reported for sequential isolates of HIV, *Virology* **161**:321.

Pifat, D. Y., Ennis, W. H., Ward, J. M., Oberste, M. S., and Gonda, M. A., 1992, Persistent infection of rabbits with bovine immunodeficiency-like virus, *J. Virol.* **66**:4518.

Pinter, A., Honnen, W. J., Tilley, S. A., Bona, C., Zaghouani, H., Gorny, M. K., and Zolla-Pazner, S., 1989, Oligomeric structure of gp41, the transmembrane protein of human immunodeficiency virus type 1, *J. Virol.* **63**:2674.

Rasmussen, L., Battles, J. K., Ennis, W. H., Nagashima, K., and Gonda, M. A., 1990, Characterization of virus-like particles produced by a recombinant baculovirus containing the *gag* gene of the bovine immunodeficiency-like virus, *Virology* **178**:435.

Rasmussen, L., Greenwood, J. D., and Gonda, M. A., 1992, Expression of bovine immunodeficiency-like virus envelope glycoproteins by a recombinant baculovirus in insect cells, *Virology* **186**:551.

Rice, N. R., Henderson, L. E., Sowder, R. C., Copeland, T. D., Oroszlan, S., and Edwards, J. F., 1990, Synthesis and processing of the transmembrane envelope protein of equine infectious anemia virus, *J. Virol.* **64**:3770.

Rosen, C. A., and Pavlakis, G. N., 1990, Tat and Rev: Positive regulators of HIV gene expression, *AIDS* **4**:499.

Ross, E. K., Buckler-White, A. J., Rabson, A. B., Englund, G., and Martin, M. A., 1991, Contribution of NF-κB and Sp1 binding motifs to the replicative capacity of human immunodeficiency virus type 1: Distinct patterns of viral growth are determined by T-cell types, *J. Virol.* **65**:4350.

Saag, M. S., Hahn, B. H., Gibbons, J., Li, Y., Parks, E. S., Parks, W. P., and Shaw, G. M., 1988, Extensive variation of human immunodeficiency virus type-1 *in vivo*, *Nature (London)* **334**:440.

Schawaller, M., Smith, G. E., Skehel, J. J., and Wiley, D. C., 1989, Studies with crosslinking reagents on the oligomeric structure of the env glycoprotein of HIV, *Virology* **172**:367.

Seigel, L. J., Ratner, L., Josephs, S. F., Derse, D., Feinberg, M. B., Reyes, G. R., O'Brien, S. J., and Wong-Staal, F., 1986, Transactivation induced by human T-lymphotropic virus type III (HTLV-III) maps to a viral sequence encoding 58 amino acids and lacks tissue specificity, *Virology* **148**:226.

Selby, M. J., and Peterlin, B. M., 1990, Trans-activation by HIV-1 Tat via a heterologous RNA binding protein, *Cell* **62**:769.

Sigurdsson, B., and Pálsson, P. A., 1958, Visna of sheep. A slow, demyelinating infection, *Br. J. Exp. Pathol.* **39**:519.

Sigurdsson, B., Pálsson, P. A., and Grímsson, H., 1957, Visna, a demyelinating transmissible disease of sheep, *J. Neuropathol. Exp. Neurol.* **16**:389.

Sigurdsson, B., Thormar, H., and Palsson, P. A., 1960, Cultivation of visna virus in tissue culture, *Arch. Gesamte Virusforsch.* **10**:368.

Starcich, B. R., Hahn, B. H., Shaw, G. M., McNeely, P. D., Modrow, S., Wolf, H., Parks, E. S., Parks, W. P., Josephs, S. F., Gallo, R. C., and Wong-Staal, F., 1986, Identification and characterization of conserved and variable regions in the envelope gene of HTLV-III/LAV, the retrovirus of AIDS, *Cell* **45**:637.

Stein, B., Gowda, D., Lifson, R., Penhallon, R., Beusch, K., and Engelman, E., 1987, pH independent HIV entry into CD4-positive T-cells via virus envelope fusion to the plasma membrane, *Cell* **49**:659.

Suarez, D. L., Van Der Maaten, M. J., Wood, C., and Whetstone, C. A., 1993, Isolation and characterization of new wild-type isolates of bovine lentivirus, *J. Virol.* **67**:5051.

Threadgill, D. S., Steagall, W. K., Flaherty, M. T., Fuller, F. J., Perry, S. T., Rushlow, K. E., Le Grice, S. F., and Payne, S. L., 1993, Characterization of equine infectious anemia virus dUTPase: Growth properties of a dUTPase-deficient mutant, *J. Virol.* **67**:2592.

Tobin, G. J., Sowder, R. C., III, Bryant, D. K., Hu, M. Y., Battles, J. K., Fenselau, C., Henderson, L. E., and Gonda, M. A., Amino acid sequence analysis of the proteolytic cleavage products of the bovine immunodeficiency virus Gag precursor polypeptide, submitted.

Tong-Starksen, S., and Peterlin, B. M., 1990, Mechanisms of retroviral transcriptional activation, *Semin. Virol.* **1**:215.

Vallée, H., and Carré, H., 1904, Nature infectieuse de l'anémie du cheval, *C.R. Acad. Sci.* **139**:331.

Van Der Maaten, M. J., and Whetstone, C. A., 1992, Infection of rabbits with bovine immunodeficiency-like virus, *Vet. Microbiol.* **30**:125.

Van Der Maaten, M. J., Boothe, A. D., and Seger, C. L., 1972, Isolation of a virus from cattle with persistent lymphocytosis, *J. Natl. Cancer Inst.* **49**:1649.

Varmus, H., 1988, Retroviruses, *Science* **240**:1427.

Vazeux, R., Cumont, M. C., and Montagnier, L., 1990, HIV replication in the central nervous system of AIDS patients, in: *Modern Pathology of AIDS and Other Retroviral Infections* (P. Racz, A. T. Haase, and J. C. Gluckman, eds.), pp. 69–81, Karger, Basel.

Weiss, C. D., Levy, J. A., and White, J. M., 1990, Oligomeric organization of gp120 on infectious human immunodeficiency virus type 1 particles, *J. Virol.* **64**:5674.

Whetstone, C. A., Van Der Maaten, M. J., and Black, J. W., 1990, Humoral immune response to the bovine immunodeficiency-like virus in experimentally and naturally infected cattle, *J. Virol.* **64**:3557.

Whetstone, C. A., Van Der Maaten, M. J., and Miller, J. M., 1991, A Western blot assay for the detection of antibodies to bovine immunodeficiency-like virus in experimentally inoculated cattle, sheep, and goats, *Arch. Virol.* **116**:119.

Wyatt, C. R., Wingett, D., White, J. S., Buck, C. D., Knowles, D., Reeves, R., and Magnuson, N. S., 1989, Persistent infection of rabbits with bovine leukemia virus associated with development of immune dysfunction, *J. Virol.* **63**:4498.

CHAPTER 4

Simian T-Lymphotropic Virus Type I

PATRICIA N. FULTZ

I. HISTORICAL AND EPIDEMIOLOGIC PERSPECTIVES

Counterparts of the known human retroviruses have been shown to naturally infect various nonhuman primate species of African and Asian origin. These retroviruses belong to three distinct retrovirus groups (Coffin, 1992): (1) the HTLV/BLV group (formerly included among the oncovirus family), which includes the type C human T-cell leukemia/lymphoma viruses (HTLV) types I and II (Sugamura and Hinuma, 1993) and the simian T-lymphotropic viruses (STLV) type I; (2) the lentiviruses, comprised of the human immunodeficiency viruses (HIV) types 1 and 2 and the simian immunodeficiency viruses (SIV) (see Chapter 5); and (3) the spuma or foamy viruses, of which there are closely related human and simian counterparts (Mergia *et al.*, 1990; Flugel, 1991). All of these viruses share common regulatory features and similar genomic organizations and encode the basic retroviral genes, *gag*, *pol*, and *env*. In addition, viruses of all three retrovirus groups establish long-term persistent infections, but they induce very different disease syndromes. The lentiviruses, HIV-1 and HIV-2, are the causative agents of acquired immunodeficiency syndrome (AIDS), a disease which develops in from 2 to 10 years in at least 50% of infected persons and is pandemic throughout the globe (Mann and Chin, 1988). Of the type C viruses, HTLV-I is associated not only with adult T-cell leukemia/lymphoma (Poiesz *et al.*, 1980; Hinuma *et al.*, 1981; Yoshida *et al.*, 1982), which develops in only a small percentage of individuals after many years of infection, but also with neurological disorders known as tropical spastic paraparesis or HTLV-I-associated myelopathy (Gessain *et al.*, 1985; Bhagavati *et al.*, 1988; Jacobson *et al.*, 1988). In contrast, neither HTLV-II nor the spumaviruses have been conclusively linked with specific diseases.

PATRICIA N. FULTZ • Department of Microbiology, University of Alabama at Birmingham School of Medicine, Birmingham, Alabama 35294.

The Retroviridae, Volume 3, edited by Jay A. Levy. Plenum Press, New York, 1994.

The simian counterparts of HTLV-I, known collectively as STLV-I, are the subject of this review. The discussion of seroprevalence to STLV-I among various nonhuman primate populations refers to STLV-I and HTLV-I interchangeably in the context of seropositivity to viral antigens; this reflects the fact that HTLV-I antigens were employed in the analyses. Because of the close genetic and antigenic relationship between STLV-I and HTLV-I isolates, the serologic studies, in general, provide reliable estimates for the prevalence of infection in the various species tested.

A. Discovery of STLV-I

Following the initial isolation and identification of HTLV-I in 1980 (Poiesz *et al.*, 1980), test reagents were generated, and areas of endemicity were identified in seroepidemiologic studies (Levine *et al.*, 1988). Endemic areas of infection were found in the Caribbean, southeastern United States, selected areas in South America, large portions of Africa, and in southern Japan, where other early HTLV-I isolates were obtained (Hinuma *et al.*, 1981; Yoshida *et al.*, 1982). The subsequent demonstration that peripheral blood lymphocytes from Japanese macaques (*Macaca fuscata*) could be transformed by cocultivation with irradiated HTLV-I-infected human MT2 cells (Miyoshi *et al.*, 1982a) led to the speculation that these animals might be naturally infected with a closely related virus. Indeed, the first tangible evidence for the existence of HTLV-I-related simian viruses came from cultures of lymphoid cells from HTLV-I-seropositive Japanese macaques in which type C retroviral particles were detected by electron microscopy (Miyoshi *et al.*, 1982b). Several species of nonhuman primates with natural habitats in both Asia and Africa were then found to have serum antibodies that cross-reacted with HTLV-I antigens (see below).

Conclusive evidence that nonhuman primates were infected by viruses highly related to HTLV-I was provided when several investigators established cell lines from seropositive monkeys and chimpanzees (*Pan troglodytes*) (Komuro *et al.*, 1984; Yamamoto *et al.*, 1984; Guo *et al.*, 1984; Tsujimoto *et al.*, 1985). Genomic DNA from these cell lines was subjected to Southern blotting with probes that encompassed the entire HTLV-I genome or represented specific genes. These studies showed that the nonhuman primate viruses were highly related to HTLV-I throughout the length of their genomes, but restriction enzyme patterns indicated that the various isolates were not identical (Guo *et al.*, 1984; Komuro *et al.*, 1984; Tsujimoto *et al.*, 1985). Nucleotide sequence analysis ultimately revealed 90 to 95% homology with HTLV-I, depending on the species from which the virus was isolated (Inoue *et al.*, 1986; Watanabe *et al.*, 1986), and thus justified classification of these simian retroviruses in the same genus (see below). These studies also showed that STLV-I, like HTLV-I, was an exogenous retrovirus for the various host species. Furthermore, using type-specific synthetic peptide epitopes in the central regions of HTLV-I or -II external glycoproteins (gp46 and gp52, respectively), Rudolph *et al.* (1991) showed that seropositive monkeys were indeed infected with an HTLV type I-like rather than a type II-like virus.

B. Seroprevalence of STLV-I Infection

1. Seroprevalence among Animals in Their Native Habitats

Multiple studies undertaken to determine the seroprevalence of STLV-I infection in various nonhuman primates, particularly Old World monkeys and apes, have been conducted using fresh and frozen serum samples from both captive and feral monkeys and apes (Table I). In studies conducted in different regions of Japan among more than 40 troops of Japanese macaques (*M. fuscata*), about 25% (range, 5% to 95%) of the animals were seropositive for antibodies cross-reactive to HTLV-I (Miyoshi *et al.*, 1983a; Hayami *et al.*, 1984). Dracopoli *et al.* (1986) surveyed 535 serum samples from feral African green monkeys in different areas of East Africa. The overall seroprevalence for antibodies to STLV-I was 33% (178/535), but the frequencies ranged from 4% to 44%, depending on the region from which the serum samples were collected. Among other populations of animals living in their native habitats, Ishikawa *et al.* (1987) found a seroprevalence of 13% (44/334) for monkeys and apes in Indonesia and 18% (41/233) for monkeys and great apes in Africa. In Indonesia, 7 (all *Macaca spp.*) of 11 different species of monkeys, but only 1 (a siamang, *Hylobates syndactylus*) of 22 apes had antibodies to STLV-I. Among the African species, seropositive animals were identified in 6 of 15 species tested, which included 1 of 12 gorillas (*Gorilla gorilla*) and none of 52 chimpanzees. On both continents some of the animals tested were wild-caught and some were housed in zoos, research institutions, or private homes. Because the number of individual animals (i.e., serum samples) from many of the species was low (range, 1 to 83), not all of the results of these studies represent an accurate estimate of the proportion of animals or num-

TABLE I. STLV-I Seroprevalence among Animals in Their Native Countries[a]

Species	Country	Seroprevalence (%)
Macaca fuscata	Japan	853/3041 (28)
Macaca sp.	Indonesia	43/301 (14)
Presbytis sp.	Indonesia	0/11
Hylobates sp.	Indonesia	1/18 (6)
Orangutan	Indonesia	0/4
Cercopithecus sp.	Kenya	158/361 (44)
	Ethiopia	11/137 (8)
	Uganda	25/85 (29)
	Gabon	0/19
	Ghana	0/9
	Cameroon	17/32 (53)
Cercocebus sp.	Gabon	0/6
Erythrocebus patas	Cameroon	2/14 (14)
Papio sp.	Cameroon	0/16
Papio c. anubis	Kenya	3/5 (60)
Papio sphinx	Gabon	2/20 (10)
Pan troglodytes	Gabon	0/52
Gorilla gorilla	Gabon	1/12 (8)

[a] Numbers are compiled from data discussed and referenced in the text.

ber of species infected by STLV-I, either among captive animals or animals in the wild. Of interest is the finding that none of 125 captured African green monkeys (*Cercopithecus sp.*) on Barbados were seropositive (Blakeslee *et al.*, 1985). The ancestors of these animals had originally been transported from Africa to the island in the 17th century, but the reason for the lack of infection is not known.

2. Seroprevalence among Animals Housed in the United States

Several studies have assessed the seroprevalence of antibodies to STLV-I in captive Asian and African Old World monkeys and apes housed in zoos and research institutions in the United States (Table II). The overall consensus, with few exceptions, was that the extent of STLV-I infection in these animals was low. Lowenstine *et al.* (1986) found 7.4% (39 of 526) of animals tested to exhibit seroreactivity to HTLV-I antigens. These 39 seropositive animals represented 15 different species and included chimpanzees, baboons, and various African and Asian monkeys. Likewise, in a survey of sera from four different *Macaca* species at the New England Regional Primate Center, Daniel *et al.* (1988) identified only 3.2% (27 of 848) of the macaques as having antibodies to STLV-I.

Two exceptions to the apparent low level of STLV-I infection among captive monkeys and apes in the United States have been reported. First, Mone *et al.* (1992) tested 329 animals, representing five species of baboons, from a colony of 3200 and reported that approximately 45% of the animals were seropositive. Second, a survey of a large breeding colony and a smaller group of African sooty mangabeys (*Cercocebus atys*) housed at two different U.S. primate centers (both groups were originally part of the same colony) showed that 33% (45 of 138) and

TABLE II. Seroprevalence of Antibodies to STLV-I among Nonhuman Primates Housed in the United States[a]

Species	No. tested/No. positive	Percent
Macaca sp.	27/848	3
	13/107	12
Total	**40/955**[b]	**4**
Cercopithecus sp.	**14/185**	**8**
Cercocebus atys	2/22	9
	14/32	44
	45/138	33
Total	**61/192**	**32**
Baboons (*Papio* sp.)	6/37	16
	152/329	46
Total	**158/366**	**43**
Hylobates sp.	**0/22**	
Presbytis sp.	**0/42**	
Colobus sp.	**0/38**	
Pan sp.	**3/8**	**38**
Gorilla gorilla	4/10	40
	0/26	
Total	**4/36**	**11**

[a] Numbers are compiled from data discussed and referenced in the text.
[b] Bold numbers are totals for a given species.

44% (14 of 32), respectively, of the animals were seropositive for antibodies cross-reactive with HTLV-I (Murphey-Corb *et al.*, 1986; Fultz *et al.*, 1990). These percentages are in close agreement with the overall seroprevalence of infection among African green monkeys in the wild, as reported by Dracopoli *et al.* (1986). In an unrelated earlier study, Blakeslee *et al.* (1987) reported that serum samples from 40% (4 of 10) of the gorillas at one primate center were reactive with HTLV-I antigens.

3. Seroprevalence among New World Monkeys

Limited data are available regarding possible natural infection of New World monkeys with an STLV-I-like virus. Miyoshi *et al.* (1983a) and Hunsmann *et al.* (1983) failed to identify any HTLV-I cross-reactive antibodies in serum samples from 11 different species of New World monkeys. More recently, Andrada-Serpa *et al.* (1989) tested sera from 44 wild-caught *Cebus apella* monkeys in Amazonia; none had antibodies to indicate infection by an STLV-I/HTLV-I-related virus. Although New World monkeys do not appear to be infected in the wild, experimental oral transmission of HTLV-I into common marmosets (*Callithrix jacchus*) and squirrel monkeys (*Saimiri sciureus*) has been reported (Kinoshita *et al.*, 1986; Nakamura *et al.*, 1986).

Thus, it appears that natural STLV-I infections are confined to Old World species of monkeys and apes indigenous to both Asia and Africa. This finding contrasts with natural infections with various SIV subgroups of animals in the wild, which appear to be restricted to nonhuman primates of African origin (Lowenstine *et al.*, 1986; Ohta *et al.*, 1988). Of interest, however, are recent serologic data which indicate that some New World simian species may be naturally infected with an STLV-II-like virus (T. M. Folks, personal communication).

C. Epidemiology and Transmission of Infection

In some of the above-referenced studies, seropositivity for STLV-I antigens was reported as a function of sex and/or age at time of serum collection. These analyses revealed a consistent picture, irrespective of whether the animals surveyed were feral or captive or of Asian or African origin. The frequency of infection among both feral African green monkeys (Dracopoli *et al.*, 1986; Ishikawa *et al.*, 1987) and captive macaques (Miyoshi *et al.*, 1983a), sooty mangabeys (Fultz *et al.*, 1990), and baboons (Mone *et al.*, 1992) increased with age. Similar findings were also documented for cynomolgus macaques (*M. fascicularis*) and maura monkeys (*M. maura*) in Indonesia (Ishikawa *et al.*, 1987) and for a colony of hamadryas baboons (*Papio hamadryas*) in the Soviet Union (Voevodin *et al.*, 1985).

Most of these investigators also reported no difference in the percentages of males and females that were infected with STLV-I, especially if one included only sexually active animals in the analysis. This latter point is important because simian females, in general, become sexually active at an earlier age than males. Dracopoli *et al.* (1986) and Mone *et al.* (1992), however, did report a higher proportion of STLV-I seropositive females than males, in all age groups of Kenyan African green monkeys and captive baboons, respectively. Since the prevalence of infection with STLV-I increases with age, it appears as though the primary mode of transmission is sexual. This conclusion is (1) supported by the finding that none of 27 mangabeys younger than 1 year old in a captive breeding colony had antibodies to STLV-I (Fultz *et al.*, 1990), and (2) consistent with documentation that HTLV-I is transmitted sexually as well as by blood transfusion, needle sharing, and from mother to child via milk (Blattner, 1989).

There are insufficient data to ascertain whether STLV-I is naturally transmitted via breast milk, blood, or saliva, although one family study of Japanese monkeys did suggest the virus could be maternally transmitted (Miyoshi *et al.*, 1983a). Since HTLV-I infections can be acquired via transfusion (Sato and Okochi, 1986), it is probable that exposure to infected blood during fighting or from open wounds is a potential mechanism for transmission of STLV-I. HTLV-I proviral sequences have been detected in lymphocytes isolated from saliva of infected persons (Miyoshi *et al.*, 1992). Thus, it is also possible that STLV-I is present in saliva, but there are no data that document transmission of either STLV-I or HTLV-I by this route. It is apparent, however, that both viruses, which are highly cell-associated (Tsujimoto *et al.*, 1985; Sato and Okochi, 1986), are primarily transmitted via infected cells and not as cell-free virus particles. Thus, the potential exists for horizontal transmission of STLV-I with any body fluid that contains lymphocytes.

II. STLV-I-ASSOCIATED DISEASE AND PATHOLOGY

Although there has been no definitive demonstration that STLV-I causes neoplastic diseases, strong correlations between STLV-I infection and the development of malignant leukemia/lymphomas in several species of nonhuman primates have been reported. Homma *et al.* (1984) tested serum from 75 healthy macaques and 13 macaques with malignant lymphomas for the presence of antibodies cross-reactive with HTLV-I. Of those with lymphoproliferative diseases, 85% (11 of 13) were seropositive, whereas only 7% of the healthy controls had antibodies. Additional evidence for a causal relationship between STLV-I infection and the development of leukemia/lymphoma was provided by Voevodin *et al.* (1985) in their study of a colony of hamadryas baboons in which outbreaks of malignant lymphomas had occurred since the early 1970s. Serum from 98% (57 of 58) of baboons with lymphoma and 45% of 177 healthy baboons in this colony had antibodies that cross-reacted with HTLV-I antigens. In contrast, the seroprevalence of antibodies to HTLV-I in two separate, lymphoma-free groups of baboons at this same institution was only 5 and 8%.

The above circumstantial observations were strengthened by a report from Sakakibara *et al.* (1986). These investigators initially studied neoplastic cells from a spontaneous malignant lymphoma in an African green monkey. These cells were similar in morphology to the pleomorphic cells observed in human patients with adult T-cell leukemia/lymphoma (ATLL), had antigens that reacted specifically with antibodies to HTLV-I, and phenotypically were $CD2^{+}$ $CD8^{+}$ $CD4^{-}$ $CD25^{+}$. This work was extended by Tsujimoto *et al.* (1987) who isolated chromosomal DNA from primary peripheral blood lymphocytes from 31 STLV-I-seropositive African green monkeys and analyzed it for STLV-I proviruses. Of the 31 infected monkeys, 1 animal was leukemic and 5 were preleukemic; in these 6 cases the STLV-I provirus was monoclonally integrated. In contrast, in cells from the apparently healthy animals, the proviruses reflected polyclonal integration. This strong association between monoclonal integration of the provirus and ATLL-like disease indicates that STLV-I is leukemogenic. It further suggests that STLV-I integrates into the genome at random, and that secondary events occur in one or a few cells such that these transformed cells proliferate continuously and ultimately overgrow other STLV-I-infected cells.

An association between STLV-I infection and development of leukemia/lymphoma has also been described in other species of nonhuman primates. A wild-caught olive baboon (*Papio anubis*) maintained in captivity for 10 years developed spontaneous diffuse malignant lymphoma with a leukemic phase, anemia, and thrombocytopenia (McCarthy *et al.*, 1990). Similarly, non-Hodgkin's lymphoma, with marked lymphadenopathy (Fig. 1), was described in an STLV-I-seropositive baboon (Mone *et al.*, 1992). Analysis of a continuous cell line established from this animal revealed STLV-I-encoded Gag proteins by immunoblot and monoclonal integration of an STLV-I provirus. As with most STLV-I-associated neoplasms, the leukemic cells were pleomorphic and medium- to large-sized with a single, highly lobulated nucleus (Fig. 2), closely resembling neoplastic cells in human ATLL. Subsequently, 27 additional cases of non-Hodgkin's lymphomas in STLV-I-seropositive baboons were identified in the same domestic colony of 3200 baboons (Hubbard *et al.*, 1993).

A few cases of lymphoproliferative disease in African monkeys dually infected with STLV-I and SIV have been reported. In one U.S. breeding colony of sooty mangabey monkeys, three instances of disease have been reported (McClure *et al.*, 1992). T-cell leukemia, derived from $CD8^+$ T lymphocytes and

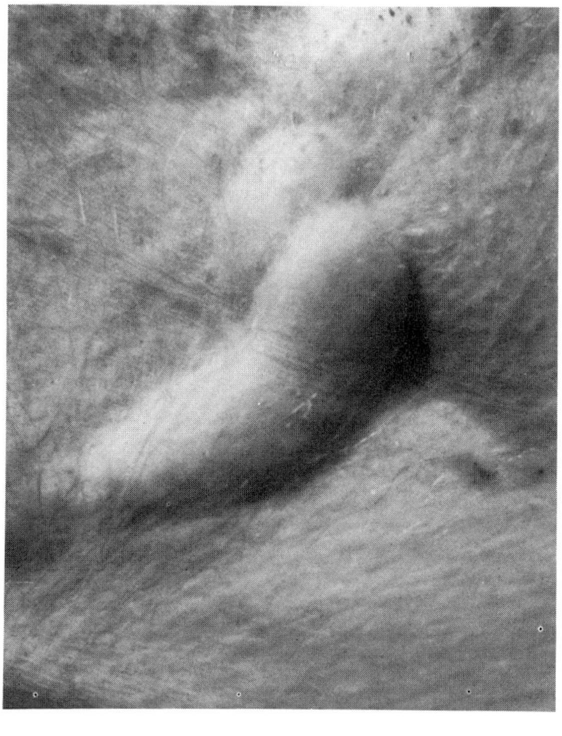

FIGURE 1. Marked lymphadenopathy of the inguinal lymph nodes in an STLV-I antibody-positive baboon (*Papio cynocephalus anubis*) with non-Hodgkin's lymphoma of T-cell lineage. The largest node measured approximately 5 cm × 2 cm. Courtesy of Dr. Gene Hubbard, Southwest Foundation for Biomedical Research.

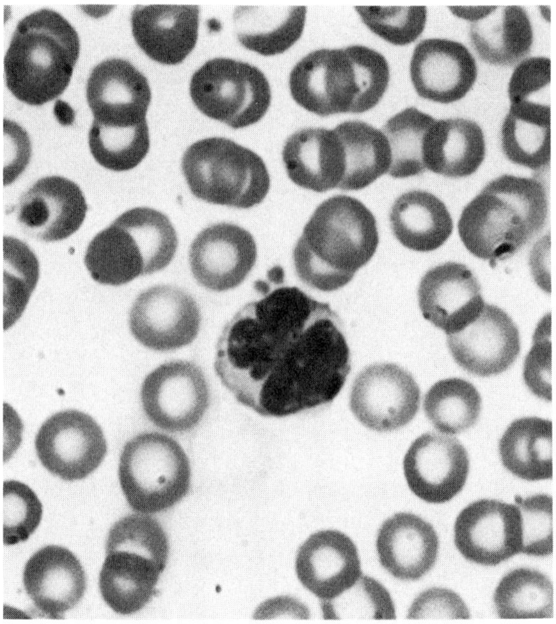

FIGURE 2. Typical hyperlobulated neoplastic lymphocyte with abundant basophilic cytoplasm in the peripheral blood of an STLV-I antibody-positive baboon with ATLL-like disease. Wright's stain; magnification, ×1000. Courtesy of Dr. Jonathan Allan, Southwest Foundation for Biomedical Research.

accompanied by generalized skin lesions, lymphadenopathy, and hepatosplenomegaly, was diagnosed in a captive, 14-year-old female sooty mangabey that was coinfected naturally with STLV-I and SIV. In addition, a second dually infected female mangabey had chronic lymphocytosis with lymphoid interstitial pneumonia and pulmonary syncytial giant cells. A third adult female mangabey developed disseminated malignant lymphoma with multinucleated giant cells in the lungs, liver, and bone marrow. Although 33 and 57% of the mangabeys in this colony of more than 140 animals are naturally infected with STLV-I and/or SIV, respectively, these were the only cases of leukemia/lymphoma identified. Diseases that might be related solely to SIV infection have occurred rarely. In sooty mangabeys, therefore, coinfection with these two retroviruses does not appear to potentiate disease routinely, as has been suggested for coinfection with HIV and HTLV-I (Bartholomew *et al.*, 1987).

At another U.S. facility, both lymphoproliferative and immunodeficiency diseases were reported (Traina-Dorge *et al.*, 1992) in an African green monkey that was estimated to be more than 10 years old. This naturally infected animal had symptoms and lesions characteristic of SIV infection in macaques, including only 5% $CD4^+$ lymphocytes in peripheral blood and numerous multinucleated giant cells containing lentivirus particles in multiple tissues. In addition, Southern blot analysis of neoplastic lymphoid tissue revealed clonal integration of STLV-I, suggesting that this virus was the causative agent of the neoplasm. Like sooty mangabey monkeys infected with SIVsmm, African green monkeys infected with SIVagm generally do not develop AIDS-like disease. In the present case, however, induction of disease symptoms characteristic of AIDS may have been enhanced by coinfection with STLV-I.

Among the great apes, lymphoma or chronic fatal disease associated with STLV-I infection has only been documented in gorillas. In one report a captive gorilla that subsequently died developed a T-cell lymphoma, which was comprised of large cells with lobulated nuclei (Lee *et al.*, 1985). In a second report (Blakeslee *et al.*, 1987), three seropositive gorillas in one facility died of similar symptoms, which included chronic diarrhea, gastroenteritis, weight loss of 17 to 38%, shigellosis, and periodic episodes of anemia and lymphopenia. In contrast to other cases of disease associated with STLV-I infection, no lymphoproliferative disease was reported in these animals. Whether STLV-I was actually the causative agent for this syndrome is unknown.

III. PROPERTIES OF STLV-I

A. Physical Properties

Electron microscopic examination of chronically infected cell lines established from several species of nonhuman primates revealed typical mature type C retroviral particles, which varied from 80 to 150 nm in diameter and were identical morphologically to HTLV-I (Miyoshi *et al.*, 1982b; Ishikawa *et al.*, 1987). The particles had an electron-dense nucleoid surrounded by an outer membrane which, like other retroviruses, is acquired as the virus buds from the cell (Fig. 3).

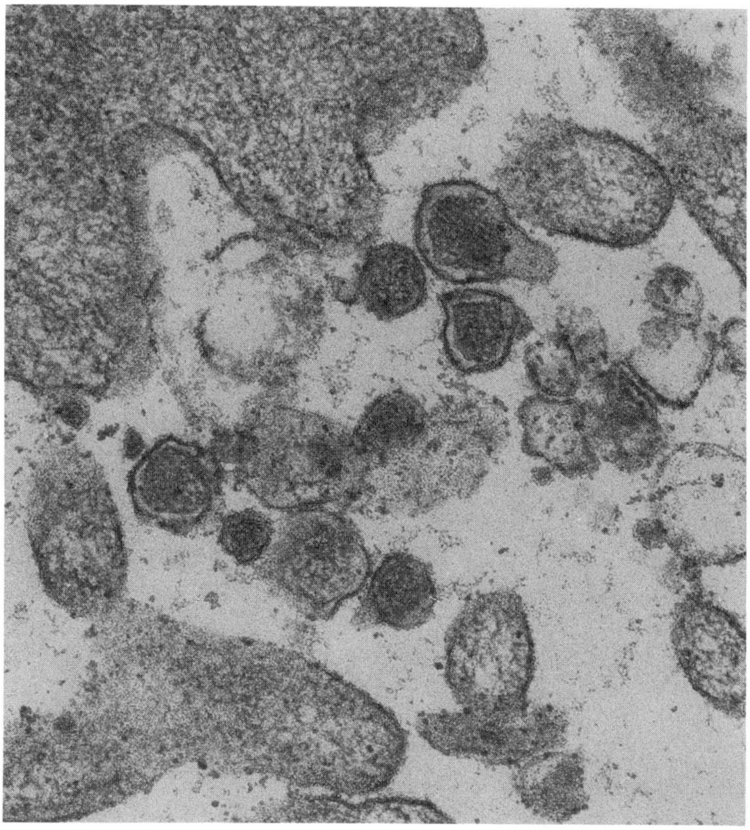

FIGURE 3. Electron micrograph of STLV-I virions produced by a continuous T-lymphocytic cell line from a naturally infected sooty mangabey monkey; magnification, ×70,000 (P. N. Fultz).

STLV-I isolates from different Asian and African monkeys and from a chimpanzee were shown to have a single-stranded genome of 8.5 kb (Komuro *et al.*, 1984), the same as that of HTLV-I. In addition, the RNA-dependent DNA polymerase, or reverse transcriptase enzyme, encoded by STLV-I viruses has a preference for Mg^{2+} over Mn^{2+} (Tsujimoto *et al.*, 1985), a property shared not only with HTLV-I (Poiesz *et al.*, 1980) but also with the lentiviruses, HIV and SIV (Hoffman *et al.*, 1985).

B. Biological Properties

Because the STLV-I viruses have not been studied extensively, there is only limited information on their biological properties. Like HTLV-I, STLV-I is usually transmitted *in vitro* via cell–cell interactions and not as free virus. Thus, most STLV-I isolates exist as persistently infected, continuous cell lines (Table III). Two consequences of cell-to-cell transmission are: first, only small numbers of virus particles are released into culture supernatants, as evidenced by low or undetectable levels of cell-free reverse transcriptase activity; and second, cells chronically infected with STLV-I form syncytia, i.e., large multinucleated giant cells, with uninfected cells. Fusion occurs at neutral pH, leads to cell death, and probably is mediated via interaction between the STLV-I surface glycoprotein on infected cells and the viral receptor on uninfected cells.

Although the cellular receptor for HTLV-I, HTLV-II, and STLV-I has not been identified, all of these primate type C retroviruses do have a common receptor, which is distinct from the CD4 receptor utilized by the T-lymphotropic lentiviruses, HIV and SIV. This property was demonstrated by Sommerfelt and Weiss (1990) using syncytial interference assays and is consistent with the earlier finding that STLV-I and HTLV-I can transform both human and simian lympho-

TABLE III. Isolates of STLV-I Obtained from Nonhuman Primates

Species	Common name	References[a]
Asian monkeys		
Macaca fuscata	Japanese monkey	1–3
M. nemestrina	Pig-tailed macaque	1, 3
M. arctoides	Stump-tailed macaque	1, 3
M. cyclopis	Formosan monkey	1, 3
M. radiata	Bonnet monkey	1, 3
M. fascicularis	Cynomolgus monkey	4
M. maura	Maura monkey	4
African monkeys		
Cercopithecus aethiops	African green monkey	1, 3
C. mitis	Sykes' monkey	4
Cercocebus atys	Sooty mangabey monkey	5
Papio cynocephalus	Baboon	6
Apes		
Pan troglodytes	Chimpanzee	1, 3
Hylobates syndactylus	Siamang	4

[a] References: 1, Komuro *et al.* (1984); 2, Yamamoto *et al.* (1984b); 3, Tsujimoto *et al.* (1985); 4, Ishikawa *et al.* (1987); 5, Fultz and Su (unpublished); 6, Guo *et al.* (1984).

cytes *in vitro* (Miyoshi *et al.*, 1983b). Cell lines established from STLV-I-infected animals generally form rosettes with sheep red blood cells, i.e., they are T lymphocytes, and all express the IL-2 receptor, CD25, for which the anti-TAC monoclonal antibody is specific. Phenotypically, the majority of STLV-I-transformed T-cell lines characterized to date had either the CD4 or CD8 cell surface markers (Ishikawa *et al.*, 1987), whereas most (but not all) HTLV-I-transformed cells carry the CD4 marker associated with T helper/inducer activity (Hattori *et al.*, 1981). Although human B cells do not appear to be infected with HTLV-I (Richardson *et al.*, 1990), 2 of 27 non-Hodgkin's lymphomas in baboons were of B-cell lineage (Hubbard *et al.*, 1993).

Because of limited biological characterization of STLV-I isolates and the overall close genetic homology of these viruses to HTLV-I, at the present time, specific properties of STLV-I isolates must be inferred from studies with HTLV-I. Although both STLV-I and HTLV-I isolates are considered to be T-cell tropic, HTLV-I isolates have been shown to infect a wide variety of nonlymphoid primary cells and established cell lines derived from different species (Sugamura and Hinuma, 1993). These findings suggested that the receptor for these viruses is broadly distributed. This conclusion has been confirmed by studies which show that HTLV-I-transformed T cells can form syncytia with cells from many species and that pseudotypes formed between vesicular stomatitis virus (VSV) and HTLV-I [VSV(HTLV-I)] can infect a wide range of cells (Clapham *et al.*, 1984; Nagy *et al.*, 1984). Clapham *et al.* (1984) also demonstrated that antisera from an STLV-I-infected chimpanzee and African green monkey inhibited syncytium formation by HTLV-I and neutralized VSV(HTLV-I) pseudotype viruses.

Replication in cell lines derived from various species is more restricted. Cell types shown to be infectable by and to support replication of HTLV-I include human osteogenic sarcoma (HOS) cells (Clapham *et al.*, 1983), primary human endothelial cells (Ho *et al.*, 1984; Hoxie *et al.*, 1984), primary glial cells (Saida *et al.*, 1988), promyelocytic leukemic cells (Hiramatsu *et al.*, 1986), and human cancer cells of gastric origin (Akagi *et al.*, 1988). Although transmission of HTLV-I and STLV-I to various cell lines is achieved *in vitro* more readily by cocultivation of persistently infected cells, peripheral blood mononuclear cells and some cell lines, such as Molt4, HUT78, and HOS, can be infected by cell-free virus (Clapham *et al.*, 1983; Fan *et al.*, 1992). HOS cells also support replication of STLV-I. This fact was demonstrated by Inoue *et al.* (1986) who transfected HOS cells with an infectious molecular clone of STLV-I isolated from a pig-tailed macaque (Pt-STLV-I). In this same study, it was also shown that STLV-I could be transferred to a B lymphoblastoid cell line (Ball-1 cells) by cocultivation of these cells with HOS/Pt-STLV-I cells. Thus, the limited available information appears to support the assumption that STLV-I has a host range similar to that of HTLV-I.

A property of HTLV-I that probably is important in the pathogenesis of the virus is its ability to activate resting lymphocytes (Gazzolo and Duc Dodon, 1987). Furthermore, this activation of lymphocytes to proliferate occurs in the absence of accessory cells and appears to be mediated through the CD2 molecule (Duc Dodon *et al.*, 1989). Associated with spontaneously proliferating lymphocytes from HTLV-I-infected persons is constitutive production of interleukin-6 and tumor necrosis factor-α (Lal and Rudolph, 1991). It would be of interest to define such properties for STLV-I, but studies such as these have not been done, perhaps because of the shift in emphasis from the HTLV/STLV viruses to HIV and SIV in the 1980s.

Continuous cell lines chronically infected with both STLV-I and SIVsmm have been established from a sooty mangabey naturally infected with both viruses (P.N.F., unpublished). By radioimmunoprecipitation of cell lysates, these cells were shown to produce high levels of both SIV p27 *gag* and STLV-I p24 *gag* antigens. However, only one glycoprotein, with an apparent molecular weight slightly less than that of SIVsmm, was identified unequivocally. Furthermore, electron microscopy revealed that these cultures contained large numbers of extracellular retroviral particles with either a lentivirus or type C virus morphology. Phenotypically, these cells are $CD2^+$ $CD4^-$ $CD8^-$ $CD20^-$ $CD25^+$; however, it is possible that the CD4 protein is downregulated or cannot be detected on the cell surface because of production of and subsequent complex formation with the SIV envelope glycoprotein (Hoxie *et al.*, 1988). Cloned cell lines that express only STLV-I antigens have been generated by limiting dilution and cocultivation with lymphocytes from uninfected pig-tailed macaques and sooty mangabeys. Cultures of transformed T cells infected with only STLV-I produce less extracellular viral particles than those infected with both SIV and STLV-I (P.N.F., unpublished), which suggests that coinfection with SIV may potentiate STLV-I expression.

Molecular cloning and nucleotide sequence analysis of the sooty mangabey STLV-I (sm-STLV-I) are in progress (see below). These continuous simian lymphoid cell lines may prove useful for studying (1) the influence of STLV-I on infection with SIV or other simian viruses and (2) replicative and biological properties of STLV-I and SIV isolates in simian cells.

C. Molecular Properties

1. Genomic Analysis

The initial studies that led to the identification of STLV-I in multiple nonhuman primate species, as well as those to define clonality of leukemic cells from affected monkeys, utilized probes generated from molecular clones of HTLV-I. The results indicated that STLV-I and HTLV-I were highly related throughout the genome, including the *env* and *pX* regions (Komuro *et al.*, 1984; Guo *et al.*, 1984; Tsujimoto *et al.*, 1987). Komuro *et al.* (1984) established cell lines from five different species of nonhuman primates, which included three *Macaca* species, an African green monkey (*Cercopithecus aethiops sabaeus*), and a chimpanzee. Cytogenetic analysis revealed that most of the cell lines were diploid, but some showed varying degrees of aneuploidy, including hexaploidy. Southern blot analyses with HTLV-I probes specific for either *gag, env, pX,* or LTR regions or representative of the entire genome showed that STLV-I viruses from all species exhibited a high degree of homology with HTLV-I and that the genetic map of the proviruses was probably identical to that of HTLV-I (Fig. 4). Of interest was the observation that the viruses from the three Asian macaque species had very similar, if not identical, restriction enzyme maps while those from the African monkey and chimpanzee were distinct from each other and from the macaque isolates.

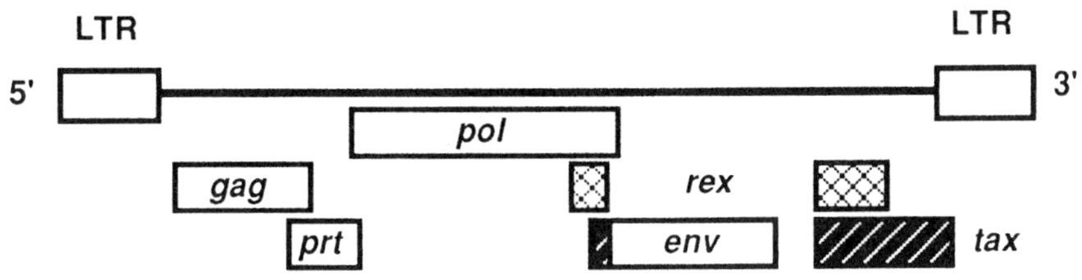

FIGURE 4. Coding regions of a typical STLV-I/HTLV-I provirus.

This apparent difference was subsequently confirmed by nucleotide sequence analyses. Watanabe *et al.* (1985) and Inoue *et al.* (1986) compared nucleotide sequences in the 3′ half (*env, pX,* LTR) and in the putative protease coding region, respectively, of HTLV-I with those of an STLV-I isolate from a pig-tailed macaque (*M. nemestrina*) (Pt-STLV-I); both studies revealed approximately 90% homology in all regions. Like HTLV-I isolates, the LTR of Pt-STLV-I had several unique characteristics: (1) a long R region of 227 bases which can form a stem-loop structure; (2) a poly(A) signal located 274 bases upstream of the poly(A) site; and (3) three 21-base direct repeats on the 5′ side of the TATAA box. Overall, the Pt-STLV-I LTR was encoded in 745 nucleotides and had a 10-base deletion in U3 relative to HTLV-I. The *env* gene of Pt-STLV-I also had characteristics of HTLV-I: (1) a 5-nucleotide overlap of the *env* and *pol* genes; (2) 20 hydrophobic amino acids at the NH_2-terminus; and (3) five possible glycosylation sites. Finally, the *pX* coding region of Pt-STLV-I contains 1071 nucleotides, including 79 at the 5′ end of the 3′ LTR, also like HTLV-I (Seiki *et al.*, 1983).

The nucleotide sequence results were extended to include a comparison with the LTR sequences of STLV-I isolates from an African green monkey and a chimpanzee (Watanabe *et al.*, 1986). From this limited analysis, it appeared that HTLV-I, irrespective of the geographical location where it was isolated, was more closely related to the African STLV-I isolates, which were equally distinct from each other. Furthermore, Pt-STLV-I from an Asian monkey was more distantly related both to HTLV-I isolates and to STLV-I strains from Africa (Table IV). Restriction enzyme mapping and Southern blot analysis of HTLV-I isolates from the African countries of Ghana and Gabon showed that the Ghana isolate closely resembled other HTLV-I isolates from Japan and the United States, whereas the isolate from Gabon appeared to be more closely related to a chimpanzee STLV-I (Fukasawa *et al.*, 1987).

The more distant relationship of Pt-STLV-I from other isolates is supported by the isolation, molecular cloning, and sequence analysis of the *env* gene of a

TABLE IV. Percent Nucleotide Sequence Homology in the LTR of Various Primate Oncornaviruses[a]

	HTLV-I	Pt-STLV-I	AG-STLV-I	CH-STLV-I
HTLV-I	99	90	94.7	94.7
Pt-STLV-I		>99	90	90
AG-STLV-I			>99	94.8

[a] Adapted from data in Watanabe *et al.* (1986). Pt, pig-tailed macaque isolates; AG, African green monkey isolates; CH, chimpanzee isolates.

virus obtained from African sooty mangabeys, sm-STLV-I (see above; P.N.F. and L. Su, unpublished). Comparison of nucleotide sequence data from sm-STLV-I with that of the Pt-STLV-I and several HTLV-I isolates showed that sm-STLV-I was more closely related to the HTLV-I strains than to the Pt-STLV-I (Fig. 5). This relationship was also true for partial nucleotide sequence data from a new STLV-I isolate from a tantalus monkey in the Central African Republic (Saksena *et al.*, 1993). The completion of the entire nucleotide sequence of the provirus and the generation of an infectious molecular clone of sm-STLV-I are in progress.

In an ongoing study, determination and comparison of the nucleotide sequences of the gp21 transmembrane protein derived from the *env* gene of 15 STLV-I isolates, all from different species of Asian and African nonhuman primates, indicate that STLV-I varies in this gene among all species tested (Genoveffa Franchini, personal communication). The genetic variation, however, is much less than that seen among different HIV and SIV isolates from different species (Myers *et al.*, 1992).

2. Protein Analysis

The above studies demonstrated that the genomic organization of STLV-I is the same as that of HTLV-I, namely, LTR–*gag*–*pol*–*env*–*pX*–LTR. In addition, the proteins encoded by these genes have the same or similar apparent molecular weights as the corresponding HTLV-I proteins (see Chapter 6). Yamamoto *et al.* (1984b) compared structural gene products of HTLV-I and STLV-I from a Japanese macaque and could detect no differences in the p19 *gag* proteins. Furthermore, Rudolph *et al.* (1991) showed by peptide-based immunoassay that 58 of 60 STLV-I-seropositive monkey sera reacted with a COOH-terminal region of p19 of HTLV-I. Because of the close genetic homology between these human and simian retroviruses, a monoclonal antibody to the p24 capsid protein was generated that recognizes both STLV-I and HTLV-I, thus demonstrating the presence of a common epitope. This monoclonal antibody was used to develop an antigen

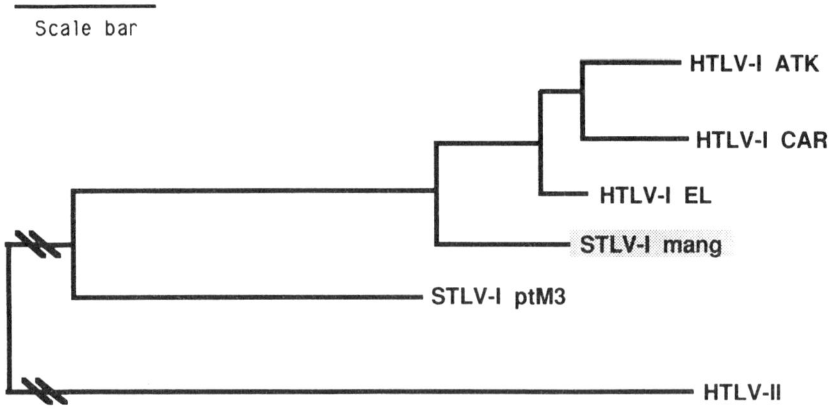

FIGURE 5. Minimum length phylogenetic tree based on 1447 total sites in the *env* gene, of which 555 sites were varied. The tree was rooted on the HTLV-II sequence. Horizontal distances are proportional to single nucleotide changes and, read from the scale bar, represent 1.7% nucleotide changes; vertical distances are for clarity. Courtesy of Drs. Kersti MacInnes and Gerald Myers, Theoretical Division, Los Alamos National Laboratory.

detection assay (Toedter *et al.*, 1992). This assay enhances the ability to detect both HTLV-I and STLV-I in cultures, which generally exhibit very low or no reverse transcriptase activity. Although both viruses have antigenically identical p24 *gag* proteins, these proteins can be distinguished by their isoelectric points (Yamamoto *et al.*, 1984b).

More variation was seen in the envelope glycoproteins, which were shown to differ in their apparent molecular weights (HTLV-I, gp68 precursor versus two precursors, gp57 and gp50, for STLV-I) and to be distinct serologically. In contrast, Inoue *et al.* (1986) found the molecular weights of the glycoprotein precursors of HTLV-I and an STLV-I isolate from a pig-tailed macaque to be very similar. Using computer algorithms for structural motifs, Lal and Griffis (1991) predicted T- and B-cell epitopes from the amino acid sequences for the *env* gene of HTLV-I and the Pt-STLV-I isolate. The finding that six of seven B-cell epitopes and six of seven T-cell epitopes predicted for HTLV-I had 100% sequence identity with corresponding sequences in STLV-I Env reinforces the close genetic relationship of these two viruses. However, since isolates from different monkey species are not identical, variations are to be expected depending on the particular isolates used for comparison.

Although no definitive studies have been reported, it is reasonable to assume that replication and regulatory aspects of STLV-I are very much the same as those for HTLV-I and are likely to involve similar *tax* ($p40^x$) and *rex* gene products, both of which are essential for HTLV-I replication (Green and Chen, 1990) (see Chapter 6). In fact, 89% amino acid identity in $p40^x$ was observed between certain STLV-I and HTLV-I isolates (Watanabe *et al.*, 1985). This finding strongly suggests that these proteins have similar functions. In addition, monoclonal antibodies to the $p40^x$ *trans*-activator gene product of HTLV-I showed differential reactivity to *tax* proteins in cell lines infected with various STLV-I isolates (Tanaka *et al.*, 1992). The STLV-I *tax* proteins were more variable in their apparent molecular masses than the HTLV-I *tax* protein, ranging from 36 kDa for Pt-STLV-I to 41 kDa for a chimpanzee STLV-I isolate.

IV. ORIGIN OF STLV-I

Widespread infection by HTLV-I and STLV-I, along with HTLV-II, suggests that it is probable that these viruses evolved centuries ago from a common progenitor. Because of the close genetic homology of these three groups of viruses, they have been referred to collectively as the primate T-cell lymphoma/leukemia viruses (PTLV) (Saksena *et al.*, 1992). The limited nucleotide divergence among HTLV-I isolates and the general finding that isolates from the same region are more closely related to each other than to isolates from more distant regions (Paine *et al.*, 1991), have led to the assumption that these differences can be used (1) to estimate times of divergence and (2) to speculate on the origins of the viruses. The proviral sequences of the prototype HTLV-I and HTLV-II isolates vary by 38% (Seiki *et al.*, 1983; Shimotohno *et al.*, 1985), whereas the greatest divergence among HTLV-I and STLV-I isolates is about 10 to 12% (Watanabe *et al.*, 1985; Inoue *et al.*, 1986; P.N.F. and L. Su, unpublished). Thus, it is clear that

HTLV-II diverged first from the ancestral virus, and that at a somewhat later time, the Pt-STLV-I lineage was established.

As noted above, STLV-I strains appear to infect Old World nonhuman primates, but not New World species. In contrast, no HTLV-II-like viruses, i.e., STLV-II, have been detected in Old World nonhuman primates. If the existence of a simian STLV-II is confirmed (T. Folks, personal communication) among New World monkeys, it would suggest that STLV-II diverged from the progenitor PTLV after separation and formation of distinct African and South American continents. Furthermore, since all human populations evolved from an African ancestor (Cavalli-Sforza *et al.*, 1988), finding STLV-II in New World monkeys would strongly support the concept that HTLV-II entered the human population by cross-species transmission of STLV-II, and by analogy, cross-species transmission and adaptation of STLV-I to humans resulted in HTLV-I. These conclusions, however, are merely speculative at this time.

That there are distinct differences in endemic areas of STLV-I and HTLV-I infection in Japan (Hayami *et al.*, 1984) and in the Caribbean (Blattner *et al.*, 1982; Blakeslee *et al.*, 1985) makes it highly probable that monkeys are not *now* a reservoir for HTLV-I. Because most HTLV-I isolates were more closely related to STLV-I from African nonhuman primates than to the Indonesian (Asian) macaque isolates, it was suggested previously that HTLV-I originated in Africa (Saxinger *et al.*, 1984). However, with the isolation and sequence data now available for the new variants of HTLV-I from Papua New Guinea, more recently it was proposed that the PTLV family originated in the Indo-Malay region, then spread to Africa, and from there to the Americas and Japan (Saksena *et al.*, 1992; Sherman *et al.*, 1992). Because there is no apparent independent phylogenetic segregation between HTLV-I and STLV-I isolates, it is possible that cross-species transmission between nonhuman primates and man occurred not only over long periods but also on different continents. Analysis of additional isolates from more nonhuman primate species and more diverse human populations may clarify the issue.

V. SUMMARY AND CONCLUSIONS

Seroepidemiologic data indicate that STLV-I infection is widespread among different species of Old World monkeys and apes in Asia and Africa, but not among New World monkeys. In addition, the natural history of these viruses is similar to that of HTLV-I infection of humans, with the possible exception of the induction of immunodeficiency, which has been reported only in animals dually infected with STLV-I and SIV. Because STLV-I infection of nonhuman primates can result in the development of malignant lymphomas and leukemia, it could serve as an animal model for retrovirus-induced neoplastic disease in humans. The limitation of such a model is that the frequency of lymphoproliferative disease in infected animals appears to be low, just as in HTLV-I-infected humans. Experimental infection of macaques and squirrel monkeys with HTLV-I has not resulted in disease (Yamamoto *et al.*, 1984a; Nakamura *et al.*, 1986). However, by using various STLV-I isolates (not HTLV-I, as in previous studies) and manipulating the system, e.g., by experimental cross-species infections, it may be possible to identify an STLV-I that is more aggressive in a particular simian species.

Potentially useful information might be gained by developing coinfection of monkeys with STLV-I and SIV as a model system for studying the *in vivo* interaction of HTLV-I and HIV. Because of increasing rates of infection by both of these human retroviruses in defined populations, e.g., intravenous drug abusers (Lee *et al.*, 1991), it is possible that new disease syndromes could emerge or progression of disease could be potentiated in coinfected individuals.

Another area in which a simian STLV-I model system might prove of value is in the development of new therapies and vaccines. In fact, a preliminary study in which pig-tailed macaques were immunized with HTLV-I antigens and challenged with an STLV-I-infected cell line has been reported (Dezzutti *et al.*, 1990). The vaccinated animals developed antibodies that recognized and prohibited syncytium formation by both viruses, and appeared to be partially protected, although the data provided no clear-cut conclusions. However, this study could serve as a guide to facilitate future vaccine development, if such a need appears warranted.

It is clear that retroviruses cause disease, the full spectrum of which has not been defined. The similarities and differences (e.g., regulation of virus expression, natural history of infection) between the human onco- and lenti-retroviruses can be exploited to gain a better understanding of factors that influence the development of specific disease syndromes. Utilization of the simian counterparts, such as STLV-I and SIV, can only facilitate that understanding.

VI. REFERENCES

Akagi, T., Yoshino, T., Motoi, M., Takata, H., Yano, S., Miyoshi, I., Oka, T., and Ohtsuki, Y., 1988, Isolation of virus-producing transformants from human gastric cancer cell line, HGC-27, infected with human T-cell leukemia virus type I, *Jpn. J. Cancer Res.* **79**:836.

Andrada-Serpa, M. J., Tosswill, J., Schor, D., Linhares, D., Dobbin, J., and Pereira, M. S., 1989, Seroepidemiologic survey for antibodies to human retroviruses in human and non-human primates in Brazil, *Int. J. Cancer* **44**:389.

Bartholomew, C., Blattner, W., and Cleghorn, F., 1987, Progression to AIDS in homosexual men co-infected with HIV and HTLV-I in Trinidad, *Lancet* **2**:1469.

Bhagavati, S., Ehrlich, G., Kula, R. W., Kwok, S., Sninsky, J., Udani, V., and Poiesz, B. J., 1988, Detection of human T-cell lymphoma/leukemia virus type I DNA and antigen in spinal fluid and blood of patients with chronic progressive myelopathy, *N. Engl. J. Med.* **318**:1141.

Blakeslee, J. R., Sowder, W. G., and Baulu, J., 1985, Wild African green monkeys of Barbados are HTLV negative, *Lancet* **1**:525.

Blakeslee, J. R., McClure, H. M., Anderson, D. C., Bauer, R. M., Huff, L. Y., and Olsen, R. G., 1987, Chronic fatal disease in gorillas seropositive for simian T-lymphotropic virus I antibodies, *Cancer Lett.* **37**:1.

Blattner, W. A., 1989, Human T-lymphotropic viruses and diseases of long latency, *Ann. Intern. Med.* **111**:4.

Blattner, W. A., Kalyanaraman, V. S., Robert-Guroff, M., Lister, T. A., Galton, D. A. G., Sarin, P. S., Crawford, M. H., Catovsky, D., Greaves, M. F., and Gallo, R. C., 1982, The human type-C retrovirus, HTLV, in blacks from the Caribbean region, and relationship to adult T-cell leukemia/lymphoma, *Int. J. Cancer* **30**:257.

Cavalli-Sforza, L. L., Piazza, A., Menozzi, P., and Mountain, J., 1988, Reconstruction of human evolution: Bringing together genetic, archaeological, and linguistic data, *Proc. Natl. Acad. Sci. USA* **85**:6002.

Clapham, P., Nagy, K., Cheingsong-Popov, R., Exley, M., and Weiss, R. A., 1983, Productive infection and cell-free transmission of human T-cell leukemia virus in a nonlymphoid cell line, *Science* **222**:1125.

Clapham, P., Nagy, K., and Weiss, R. A., 1984, Pseudotypes of human T-cell leukemia virus types 1 and 2: Neutralization by patients' sera, *Proc. Natl. Acad. Sci. USA* **81**:2886.

Coffin, J. M., 1992, Structure and classification of retroviruses, *in* The Retroviridae, Volume 1, (J. A. Levy, ed.), pp. 19–49, Plenum Press, New York.

Daniel, M. D., Letvin, N. L., Sehgal, P. K., Schmidt, D. K., Silva, D. P., Solomon, K. R., Hodi, F. S., Jr., Ringler, D. J., Hunt, R. D., King, N. W., and Desrosiers, R. C., 1988, Prevalence of antibodies to 3 retroviruses in a captive colony of macaque monkeys, *Int. J. Cancer* **41**:601.

Dezzutti, C. S., Frazier, D. E., and Olsen, R. G., 1990, Efficacy of an HTLV-I subunit vaccine in prevention of a STLV-I infection in pig-tailed macaques, in: *Developments in Biological Standardization*, Volume 72, (International Association of Biological Standardization, eds.), pp. 287–313, S. Karger AG, Switzerland.

Dracopoli, N. C., Turner, T. R., Else, J. G., Jolly, C. J., Anthony, R., Gallo, R. C., and Saxinger, W. C., 1986, STLV-1 antibodies in feral populations of east African vervet monkeys (*Cercopithecus aethiops*), *Int. J. Cancer* **38**:523.

Duc Dodon, M., Bernard, A., and Gazzolo, L., 1989, Peripheral T-lymphocyte activation by human T-cell leukemia virus type I interferes with the CD2 but not with the CD3/TCR pathway, *J. Virol.* **63**:5413.

Fan, N., Gavalchin, J., Paul, B., Wells, K. H., Lane, M. J., and Poiesz, B. J., 1992, Infection of peripheral blood mononuclear cells and cell lines by cell-free human T-cell lymphoma/leukemia virus type I, *J. Clin. Microbiol.* **30**:905.

Flugel, R. M., 1991, Spumaviruses: a group of complex retroviruses, *J. Acq. Immune Def. Syn.* **4**:739.

Fukasawa, M., Tsujimoto, H., Ishikawa, K.-I., Miura, T., Ivanoff, B., Cooper, R. W., Frost, E., Delaporte, E., Mingle, J. A. A., Grant, F. C., and Hayami, M., 1987, Human T-cell leukemia virus type I isolates from Gabon and Ghana: Comparative analysis of proviral genomes, *Virology* **161**:315.

Fultz, P. N., Gordon, T. P., Anderson, D. C., and McClure, H. M., 1990, Prevalence of natural infection with SIVsmm and STLV-I in a breeding colony of sooty mangabey monkeys, *AIDS* **4**:619.

Gazzolo, L., and Duc Dodon, M., 1987, Human T-lymphotropic virus, type I, is a direct activator of resting T lymphocytes, *Nature* **326**:714.

Gessain, A., Barin, F., Vernant, J. C., Gout, O., Maurs, L., Calender, A., and de The, G., 1985, Antibodies to human T-lymphotropic virus type I in patients with tropical spastic paraparesis, *Lancet* **2**:407.

Green, P. L., and Chen, I. S. Y., 1990, Regulation of human T cell leukemia virus expression, *FASEB J.* **4**:169.

Guo, H.-G., Wong-Staal, F., and Gallo, R. C., 1984, Novel viral sequences related to human T-cell leukemia virus in T cells of a seropositive baboon, *Science* **223**:1195.

Hattori, T., Uchiuama, T., Toibana, T., Takatsuki, K., and Uchino, H., 1981, Surface phenotype of Japanese adult T cell leukemia cells characterized by monoclonal antibodies, *Blood* **58**:645.

Hayami, M., Komura, A., Nozawa, K., Shotake, T., Ishikawa, K., Yamamoto, K., Ishida, T., Honjo, S., and Hinuma, Y., 1984, Prevalence of antibody to adult T-cell leukemia virus-associated antigens (ATLA) in Japanese monkeys and other nonhuman primates, *Int. J. Cancer* **33**:179.

Hinuma, Y., Nagata, K., Hanaoka, M., Nakai, M., Matsumoto, T., Kinoshita, K.-I., Shirakawa, S., and Miyoshi, I., 1981, Adult T-cell leukemia: Antigen in an ATL cell line and detection of antibodies to the antigen in human sera, *Proc. Natl. Acad. Sci. USA* **78**:6476.

Hiramatsu, K., Masuda, M., and Yoshikura, H., 1986, Mode of transmission of human T-cell leukemia virus type I (HTLV-I) in a human promyelocytic leukemia HL60 cell, *Int. J. Cancer* **37**:601.

Ho, D. D., Rota, T. R., and Hirsch, M. S., 1984, Infection of human endothelial cells by human T-lymphotropic virus type I, *Proc. Natl. Acad. Sci. USA* **81**:7588.

Hoffman, A. D., Banapour, B., and Levy, J. A., 1985, Characterization of the AIDS-associated retrovirus reverse transcriptase and optimal conditions for its detection in virions, *Virology* **147**:326.

Homma, T., Kanki, P. J., King, N. W., Hunt, R. D., O'Connell, M. J., Letvin, N. L., Daniel, M. D., Desrosiers, R. C., Yang, C. S., and Essex, M., 1984, Lymphoma in macaques: Association with virus of human T lymphotropic family, *Science* **225**:716.

Hoxie, J. A., Matthews, D. M., and Cines, D. B., 1984, Infection of human endothelial cells by human T-cell leukemia virus type I, *Proc. Natl. Acad. Sci. USA* **81**:7591.

Hoxie, J. A., Haggarty, B. S., Bonser, S. E., Rackowski, J. L., Shan, H., and Kanki, P. J., 1988, Biological characterization of a simian immunodeficiency-like retrovirus (HTLV-IV): Evidence for CD4-associated molecules required for infection, *J. Virol.* **62**:2557.

Hubbard, G. B., Mone, J. P., Allan, J. S., Davis III, K. J., Leland, M. M., Banks, P. M., and Smir, B., 1993, Spontaneously generated non-Hodgkin's lymphoma in twenty-seven simian T-cell leukemia virus type 1 antibody-positive baboons (*Papio* species), *Lab. Anim. Sci.* **43**:301.

Hunsmann, G., Schneider, J., Schmitt, J., and Yamamoto, H., 1983, Detection of serum antibodies to adult T-cell leukemia virus in non-human primates and in people from Africa, *Int. J. Cancer* **32**:329.

Inoue, J.-I., Watanabe, T., Sato, M., Oda, A., Toyoshima, K., Yoshida, M., and Seiki, M., 1986, Nucleotide sequence of the protease-coding region in an infectious DNA of simian retrovirus (STLV) of the HTLV-I family, *Virology* **150**:187.

Ishikawa, K., Fukasawa, M., Tsujimoto, H., Else, J. G., Isahakia, M., Ubhi, N. K., Ishida, T., Takenaka, O., Kawamoto, Y., Shotake, T., Ohsawa, H., Ivanoff, B., Cooper, R. W., Frost, E., Grant, F. C., Spriatna, Y., Sutarman, Abe, K., Yamamoto, K., and Hayami, M., 1987, Serological survey and virus isolation of simian T-cell leukemia/T-lymphotropic virus type I (STLV-I) in nonhuman primates in their native countries, *Int. J. Cancer* **40**:233.

Jacobson, S., Raine, C. S., Mingioli, E. S., and McFarlin, D. E., 1988, Isolation of an HTLV-1-like retrovirus from patients with tropical spastic paraparesis, *Nature* **331**:540.

Kinoshita, K., Yamanouchi, K., Ikeda, S., Morita, S., Amagasaki, T., Soda, H., Ichimaru, M., Moriuchi, R., Katamine, S., Miyamoto, T., and Hino, S., 1986, Oral infection of a common marmoset with human T-cell leukemia virus type-I (HTLV-I) by inoculating fresh human milk of HTLV-I carrier mothers, *Gann* **76**:1147.

Komuro, A., Watanabe, T., Miyoshi, I., Hayami, M., Tsujimoto, H., Seiki, M., and Yoshida, M., 1984, Detection and characterization of simian retroviruses homologous to human T-cell leukemia virus type I, *Virology* **138**:373.

Lal, R. B., and Griffis, K. P., 1991, Predictive B- and T-cell linear epitopes in structural proteins of HTLV-I, HTLV-II, and STLV-I, *AIDS Res. Hum. Retroviruses* **7**:663.

Lal, R. B., and Rudolph, D. L., 1991, Constitutive production of interleukin-6 and tumor necrosis factor-α from spontaneously proliferating T cells in patients with human T-cell lymphotropic virus type-I/II, *Blood* **78**:571.

Lee, H. H., Swanson, P., Rosenblatt, J. D., Chen, I. S. Y., Sherwood, W. C., Smith, D. E., Tegtmeier, G. E., Fernando, L. P., Fang, C. T., Osame, M., and Kleinman, S. H., 1991, Relative prevalence and risk factors of HTLV-I and HTLV-II infection in US blood donors, *Lancet* **1**:1435.

Lee, R. V., Prowten, A. W., Satchidanand, S. K., and Srivastava, B. I. S., 1985, Non-Hodgkins' lymphoma and HTLV-I antibodies in a gorilla, *N. Engl. J. Med.* **312**:118.

Levine, P. H., Blattner, W., Clark, J., Tarone, R., and Maloney, E. M., 1988, Geographic distribution of HTLV-I and identification of a new high risk population, *Int. J. Cancer* **42**:7.

Lowenstine, L. J., Pedersen, N. C., Higgins, J., Pallis, K. C., Uyeda, A., Marx, P., Lerche, N. W., Munn, R. J., and Gardner, M. B., 1986, Seroepidemiologic survey of captive old-world primates for antibodies to human and simian retroviruses, and isolation of lentivirus from sooty mangabeys (*Cercocebus atys*), *Int. J. Cancer* **38**:563.

McCarthy, T. J., Kennedy, J. L., Blakeslee, J. R., and Bennett, B. T., 1990, Spontaneous malignant lymphoma and leukemia in a simian T-lymphotropic virus type I (STLV-I) antibody positive olive baboon, *Lab. Anim. Sci.* **40**:79.

McClure, H. M., Anderson, D. C., Gordon, T. P., Ansari, A. A., Fultz, P. N., Klumpp, S. A., Emau, P., and Isahakia, M., 1992, Natural simian immunodeficiency virus infections in nonhuman primates, *Top. Primatol.* **3**:425.

Mann, J. M., and Chin, J., 1988, AIDS: A global perspective, *N. Engl. J. Med.* **319**:302.

Mergia, A., Shaw, K. E. S., Lackner, J. E., and Luciw, P. A., 1990, Relationship of the *env* genes and the endonuclease domain of the *pol* genes of simian foamy virus type I and human foamy virus, *J. Virol.* **64**:406.

Miyoshi, I., Taguchi, H., Fujishita, M., Yoshimoto, S., Kubonishi, I., Ohtsuki, Y., Shiraishi, Y., and Akagi, T., 1982a, Transformation of monkey lymphocytes with adult T-cell leukemia virus, *Lancet* **1**:1016.

Miyoshi, I., Fujishita, M., Taguchi, H., Matsubayashi, K., Miwa, N., and Tanioka, Y., 1983a, Natural infection in non-human primates with adult T-cell leukemia virus or a closely related agent, *Int. J. Cancer* **32**:333.

Miyoshi, I., Yoshimoto, S., Fujishita, M., Taguchi, H., Kubonishi, I., Niiya, K., and Minezawa, M., 1982b, Natural adult T-cell leukemia virus infection in Japanese monkeys, *Lancet* **2**:658.

Miyoshi, I., Taguchi, H., Yoshimoto, S., Fujishita, M., Mitani, M., Yamashita, M., and Ohtsuki, Y., 1983b, Transmission of Japanese monkey type C virus to human lymphocytes, *Lancet* **2**:166.

Miyoshi, I., Takashi, S., Iwahara, Y., Ishii, K., Kubonishi, I., and Taguchi, H., 1992, Excretion of HTLV-I in saliva [letter], *J. Am. Med. Assoc.* **267**:236.

Mone, J., Whitehead, E., Leland, M., Hubbard, G., and Allan, J. S., 1992, Simian T-cell leukemia virus type I infection in captive baboons, *AIDS Res. Hum. Retroviruses* **8**:1653.

Murphey-Corb, M., Martin, L. N., Rangan, S. R. S., Baskin, G. B., Gormus, B. J., Wolf, R. H., Andes, W. A., West, M., and Montelaro, R. C., 1986, Isolation of an HTLV-III-related retrovirus from macaques with simian AIDS and its possible origin in asymptomatic mangabeys, *Nature* **321**:435.

Myers, G., MacInnes, K., and Korber, B., 1992, The emergence of simian/human immunodeficiency viruses, *AIDS Res. Hum. Retroviruses* **8**:373.

Nagy, K., Weiss, R. A., Clapham, P., and Cheingsong-Popov, R., 1984, Biological properties of human T-cell leukemia virus envelope antigens, in: *Human T-Cell Leukemia/Lymphoma Virus* (R. C. Gallo, M. E. Essex, and L. Gross, eds.), pp. 121–131, Cold Spring Harbor Laboratory, Cold Spring Harbor, N.Y.

Nakamura, H., Tanaka, Y., Komuro-Tsujimoto, A., Ishikawa, K., Takadaya, K. I., Tozawa, H., Tsujimoto, H., Honjo, S., and Hayami, M., 1986, Experimental inoculation of monkeys with autologous lymphoid cell lines immortalized by and producing human T-cell leukemia virus type-I, *Int. J. Cancer* **38**:867.

Ohta, Y., Masuda, T., Tsujimoto, H., Ishikawa, K., Kodama, T., Morikawa, S., Nakai, M., Honjo, S., and Hayami, M., 1988, Isolation of simian immunodeficiency virus from African green monkeys and seroepidemiologic survey of the virus in various non-human primates, *Int. J. Cancer* **41**:115.

Paine, E., Garcia, J., Philpott, T. C., Shaw, G., and Ratner, L., 1991, Limited sequence variation in human T-lymphotropic virus type 1 isolates from North American and African patients, *Virology* **182**:111.

Poiesz, B. J., Ruscetti, F. W., Gazdar, A. F., Bunn, P. A., Minna, J. D., and Gallo, R. C., 1980, Detection and isolation of type C retrovirus particles from fresh and cultured lymphocytes of a patient with cutaneous T-cell lymphoma, *Proc. Natl. Acad. Sci. USA* **77**:7415.

Richardson, J. H., Edwards, A. J., Cruickshank, J. K., Rudge, P., and Dalgleish, A. G., 1990, In vivo cellular tropism of human T-cell leukemia virus type 1, *J. Virol.* **64**:5682.

Rudolph, D. L., Keesling, S. S., Lerche, N., Yee, J. A., and Lal, R. B., 1991, Dominance of HTLV type I-specific antibody responsiveness in Old World monkeys, *AIDS Res. Hum. Retroviruses* **7**:721.

Saida, T., Saida, K., Funauchi, M., Nishiguchi, E., Nakajima, M., Matsuda, S., Ohta, M., Ohta, K., Nishitani, H., and Hatanaka, M., 1988, HTLV-I myelitis: Isolation of virus, genomic analysis and infection in neural cell cultures, *Ann. N.Y. Acad. Sci.* **540**:636.

Sakakibara, I., Sugimoto, Y., Sasagawa, A., Honjo, S., Tsujimoto, H., Nakamura, H., and Hayami, M., 1986, Spontaneous malignant lymphoma in an African green monkey naturally infected with simian T-lymphotropic virus (STLV), *J. Med. Primatol.* **15**:311.

Saksena, N. K., Herve, V., Sherman, M. P., Durand, J. P., Mathiot, C., Muller, M., Love, J. L., LeGuenno, B., Barre-Sinoussi, F., Dube, D. K., and Poiesz, B. J., 1993, Sequence and phylogenetic analyses of a new STLV-I from a naturally infected tantalus monkey from Central Africa, *Virology* **192**:313.

Saksena, N. K., Sherman, M. P., Yanagihara, R., Dube, D. K., and Poiesz, B. J., 1992, LTR sequence and phylogenetic analyses of a newly discovered variant of HTLV-I isolated from the Hagahai of Papua New Guinea, *Virology* **189**:1.

Sato, H., and Okochi, K., 1986, Transmission of human T-cell leukemia virus (HTLV-I) by blood transfusion: Demonstration of proviral DNA in recipients' blood lymphocytes, *Int. J. Cancer* **37**:395.

Saxinger, W., Blattner, W. A., Levine, P. H., Clark, J., Biggar, R., Hoh, M., Moghissi, J., Jacobs, P., Wilson, L., Jacobson, R., Crooks, R., Strong, M., Ansari, A. A., Dean, A. G., Nkrumah, F. K., Mourali, N., and Gallo, R. C., 1984, Human T-cell leukemia virus (HTLV-1) antibodies in Africa, *Science* **225**:1473.

Seiki, M., Hattori, S., Hirayama, Y., and Yoshida, M., 1983, Human adult T-cell leukemia virus: complete nucleotide sequence of the provirus genome integrated in leukemia cell DNA, *Proc. Natl. Acad. Sci. USA* **80**:3618.

Sherman, M. P., Saksena, N. K., Dube, D. K., Yanagihara, R., and Poiesz, B. J., 1992, Evolutionary insights on the origin of human T-cell lymphoma/leukemia virus type I (HTLV-I) derived from sequence analysis of a new HTLV-I variant from Papua New Guinea, *J. Virol.* **66**:2556.

Shimotohno, K., Takahashi, Y., Shimizu, N., Gojobori, T., Golde, D. W., Chen, I. S. Y., and Sugimura, T., 1985, Complete nucleotide sequence of an infectious clone of human T-cell leukemia virus type II: an open reading frame for the protease gene, *Proc. Natl. Acad. Sci. USA* **82**:3101.

Sommerfelt, M. A., and Weiss, R. A., 1990, Receptor interference groups of 20 retroviruses plating on human cells, *Virology* **176**:58.

Sugamura, K., and Hinuma, Y., 1993, Human retroviruses: HTLV-I and HTLV-II, in: *The Retroviridae,* Volume 2, (J. A. Levy, ed.), pp. 399–435, Plenum Press, New York.

Tanaka, Y., Masuda, M., Yoshida, A., Shida, H., Nyunoya, H., Shimotohno, K., and Tozawa, H., 1992, An antigenic structure of the *trans*-activator protein encoded by human T-cell leukemia virus type-I (HTLV-I), as defined by a panel of monoclonal antibodies, *AIDS Res. Hum. Retroviruses* **8**:227.

Toedter, G., Pearlman, S., Hofheinz, D., Blakeslee, J., Cockerell, G., Dezzutti, C., Yee, J., Lal, R. B., and Lairmore, M., 1992, Development of a monoclonal antibody-based p24 capsid antigen detection assay for HTLV-I, HTLV-II, and STLV-I infection, *AIDS Res. Hum. Retroviruses* **8**:527.

Traina-Dorge, V., Blanchard, J., Martin, L., and Murphey-Corb, M., 1992, Immunodeficiency and lymphoproliferative disease in an African green monkey dually infected with SIV and STLV-I, *AIDS Res. Hum. Retroviruses* **8**:97.

Tsujimoto, H., Komuro, A., Iijima, K., Miyamoto, J., Ishikawa, K., and Hayami, M., 1985, Isolation of simian retroviruses closely related to human T-cell leukemia virus by establishment of lymphoid cell lines from various non-human primates, *Int. J. Cancer* **35**:377.

Tsujimoto, H., Noda, Y., Ishikawa, K., Nakamura, H., Fukasawa, M., Sakakibara, I., Sasagawa, A., Honjo, S., and Hayami, M., 1987, Development of adult T-cell leukemia-like disease in African green monkey associated with clonal integration of simian T-cell leukemia virus type I, *Cancer Res.* **47**:269.

Voevodin, A. F., Lapin, B. A., Yakovleva, L. A., Ponomaryeva, T. I., Oganyan, T. E., and Razmadze, E. N., 1985, Antibodies reacting with human T-lymphotropic retrovirus (HTLV-1) or related antigens in lymphomatous and healthy hamadryas baboons, *Int. J. Cancer* **36**:579.

Watanabe, T., Seiki, M., Tsujimoto, H., Miyoshi, I., Hayami, M., and Yoshida, M., 1985, Sequence homology of the simian retrovirus genome with human T-cell leukemia virus type I, *Virology* **144**:59.

Watanabe, T., Seiki, M., Hirayama, Y., and Yoshida, M., 1986, Human T-cell leukemia virus type I is a member of the African subtype of simian viruses (STLV), *Virology* **148**:385.

Yamamoto, N., Hayami, M., Komuro, A., Schneider, J., Hunsmann, G., Okada, M., and Hinuma, Y., 1984a, Experimental infection of cynomolgus monkeys with a human retrovirus, adult T-cell leukemia virus, *Med. Microbiol. Immunol.* **172**:57.

Yamamoto, N., Okada, M., Hinuma, Y., Hirsch, F. W., Chosa, T., Schneider, J., and Hunsmann, G., 1984b, Human adult T-cell leukemia virus is distinct from a similar isolate of Japanese monkeys, *J. Gen. Virol.* **65**:2259.

Yoshida, M., Miyoshi, I., and Hinuma, Y., 1982, Isolation and characterization of retrovirus from cell lines of human adult T-cell leukemia and its implications in the disease, *Proc. Natl. Acad. Sci. USA* **79**:2031.

CHAPTER 5

The Simian Retroviruses
SIV and SRV

MURRAY B. GARDNER, MICHAEL ENDRES, AND PETER BARRY

I. INTRODUCTION: SIMIAN RETROVIRUSES THAT CAUSE SIMIAN AIDS

Two genera of simian retrovirus cause fatal immunosuppressive disease in macaques: the simian immunodeficiency lentiviruses (SIV) and the type D simian retroviruses (SRV). Simian retroviruses belonging to the exogenous oncovirus, e.g., simian T-lymphotropic virus (STLV), and spumavirus genus, e.g., simian foamy virus, are described in other chapters (Chapter 4, this volume; Loh, 1993). The nonhuman primate retroviruses have been the subject of several other recent reviews (Gardner and Luciw, 1989, 1992; Desrosiers and Ringler, 1989; Desrosiers, 1990; Schneider and Hunsmann, 1988; McClure, 1991).

In 1982, soon after the recognition of AIDS in humans, an immunodeficiency syndrome remarkably similar to the human disease was observed in macaque species at several U.S. primate centers (for review, see Gardner *et al.*, 1988). Over the next several years it was determined that not one but two different subfamilies of simian retrovirus were capable of causing fatal immunosuppressive disease in macaques. Both syndromes have been called simian AIDS (SAIDS). Therefore, it is important to specify the retrovirus (SRV or SIV) responsible for the SAIDS under consideration.

The first SAIDS retrovirus to be discovered was a type D retrovirus closely related to the previously isolated Mason–Pfizer monkey virus (MPMV). MPMV was isolated in 1970 from a naturally occurring breast carcinoma in a captive rhesus macaque (Chopra and Mason, 1970). This virus group is now known to be indigenous and widespread in Asian macaques. A serosurvey of captive macaques in U.S. primate centers in the mid-1970s found that about 25% had antibodies to MPMV (Fine and Arthur, 1981). Experimental transmission of MPMV into naive

MURRAY B. GARDNER, MICHAEL ENDRES, AND PETER BARRY • Department of Pathology, School of Medicine, University of California, Davis, California 95616.

The Retroviridae, Volume 3, edited by Jay A. Levy. Plenum Press, New York, 1994.

macaques at that time failed to induce any tumors, but did cause a fatal immunosuppressive disease (Fine *et al.*, 1975). However, this finding attracted little attention because it preceded the recognition of AIDS in humans and investigators at that time were looking for an etiologic agent for cancer. The simian type D retrovirus subgroup consists of at least five serotypes (SRV 1–5) that are indigenous to Asian macaques (Marx *et al.*, 1985). Type D retroviruses have not been found conclusively in humans (Gardner *et al.*, 1977). SRV is apparently nononcogenic but is responsible for almost all (>99%) of the spontaneous cases of SAIDS in U.S. primate centers or other primate facilities (Gardner *et al.*, 1988; Daniel *et al.*, 1988a). SRV-SAIDS is infectiously transmitted, primarily by saliva and biting (Lerche *et al.*, 1986), and is quite prevalent in captive macaques; it represents a major practical problem and must be screened for (along with STLV) and eliminated in any SIV macaque studies. Fortunately, vaccines are quite effective against SRV infection and disease, at least under experimental conditions (Hu *et al.*, 1989; Marx *et al.*, 1986; Brody *et al.*, 1992) (see Sections XV and XVII).

The second subfamily of simian retrovirus capable of causing SAIDS is a lentivirus, serologically related to the human immunodeficiency virus (HIV). This virus, first isolated from several immunodeficient macaques in 1985, was initially called simian T-lymphotropic virus strain III (STLV-III), and now SIV_{mac} (Letvin *et al.*, 1985). Subsequently, other SIV isolates were obtained from immunosuppressed or lymphomatous macaques at the Washington, Tulane, and California primate centers. These isolates and their derived infectious molecular clones are closely related to, but distinguishable from, SIV_{mac}. All of the SIV isolates from macaques are thought to be the result of inadvertent spread of the virus from naturally infected sooty mangabeys (see below) or a closely related species cohoused in various U.S. primate centers (Desrosiers *et al.*, 1989b; Hirsch *et al.*, 1989b). The suspected but unproved routes of SIV transmission from African monkey species to Asian macaques are by fighting and biting and by investigators doing serial transmission of tissue extracts or blood in search of other infectious agents. The fatal immunodeficiency disease induced in captive macaques (Letvin *et al.*, 1985) by these particular SIV strains constitutes the version of SAIDS that now serves as the major animal model for AIDS in humans. SIV is the animal lentivirus most closely related to HIV, about 50% related to HIV-1 and 75% related to HIV-2 (for review, see Gardner and Luciw, 1988). It is noteworthy that SIV is carried as a harmless infection in its natural host, certain species of African nonhuman primates, and is not indigenous in macaques (Section III).

This chapter will describe the existing SIV strains (Table I), compare the natural history, pathology, and pathogenesis of SIV in the natural and experimental hosts, and summarize the status of research on AIDS vaccine and antiviral therapy using the SIV macaque model. Following the same format, the SRV macaque model will then be reviewed more briefly.

II. PRIMATE FAMILY: SPECIES, HABITAT, EVOLUTION

A. Retroviruses of Old World and New World Nonhuman Primates

Retroviruses are mainly found in Old World vertebrates, i.e., animals from the continents of Europe, Africa, and Asia, in contrast to "New World" (Ameri-

TABLE I. SIV Isolates

Species	Virus designation	Reference
Macaca		
Rhesus monkey (*M. mulatta*)	$SIV_{mac\text{-}1}$	Daniel *et al.* (1985)
Pigtailed macaque (*M. nemestrina*)	$SIV_{mne\text{-}1}$	Benveniste *et al.* (1986)
Stumptailed macaque (*M. arctoides*)	$SIV_{stm\text{-}1}$	Lowenstine *et al.* (1992)
Cynomolgus monkey (*M. fascicularis*)	$SIV_{fas\text{-}1}$	Daniel *et al.* (1988a)
Cercopithecus		
African green monkey		
Vervets (*C. pygerythrus*)	$SIV_{agm\text{-}ver}$	Ohta *et al.* (1988)
Grivets (*C. aethiops*)	$SIV_{agm\text{-}gri}$	Daniel *et al.* (1988b)
Tantalus (*C. tantalus*)	$SIV_{agm\text{-}tan}$	Allan *et al.* (1990a)
Green (*C. sabaeus*)	$SIV_{agm\text{-}sab}$	Kraus *et al.* (1989), M. C. Muller *et al.* (1993)
Sykes monkey (*C. mitis*)	SIV_{syk}	Emau *et al.* (1991)
Cercocebus		
Sooty mangabey (*C. atys*)	SIV_{sm}	Murphey-Corb *et al.* (1986), Fultz *et al.* (1986), Lowenstine *et al.* (1986), Marx *et al.* (1991)
White-crowned mangabey	SIV_{wcm}	Tomonaga *et al.* (1993)
Papio		
Mandrill (*P. sphinx*)	SIV_{mnd}	Tsujimoto *et al.* (1988)
Ape		
Chimpanzee (*Pan troglodytes*)	SIV_{cpz}	Peeters *et al.* (1989; 1992)

cas) animals that are largely free of these viruses. Based on the continental drift theory and fossil data, Africa was probably the site of origin of the New World monkeys (Ciochon and Chiarelli, 1980). The dispersal from Africa to South America probably occurred during the late Eocene epoch, around 40–60 million years ago when these continents were in closer proximity. The stepwise evolution of nonhuman primates is shown in Fig. 1. Table II outlines representative species of Old World and New World monkeys.

Prosimians, which comprise 41 species, are found only in the Old World, primarily in Asia and on the island of Madagascar (Napier and Napier, 1986). The only retrovirus ever found in prosimians is an endogenous type C virus (TRV-1) of the tree shrew (Flugel *et al.*, 1978). New World primates, consisting of 51 extant species, are thought to have evolved about 60 million years ago, from prosimian predecessors migrating from Africa. Old World primates diverged about 40 million years ago from New World primates and currently consist of 77 species of monkeys and 13 species of apes (Napier and Napier, 1986). Among the apes, gorillas and chimpanzees live only in Africa; orangutans and gibbon apes live only in Asia. Major features of Old World monkeys that tend to distinguish them from New World monkeys include diurnal activity, larger size, terrestrial behavior, nails instead of claws, absence of a prehensile tail, and a thinner longer nose accompanied by a lesser importance of the sense of smell.

The only retroviruses that have been isolated from New World monkeys are an endogenous (i.e., inherited) type C virus in the owl monkey (Todaro *et al.*, 1978c) and an endogenous type D virus in the squirrel monkey (Heberling *et al.*, 1977). Exogenous (i.e., nongenetic) type C or type D retroviruses have not been identified in any New World monkeys. Among Old World simians, endogenous type C viruses were isolated in the 1970s from baboons (Africa) (Benveniste *et al.*,

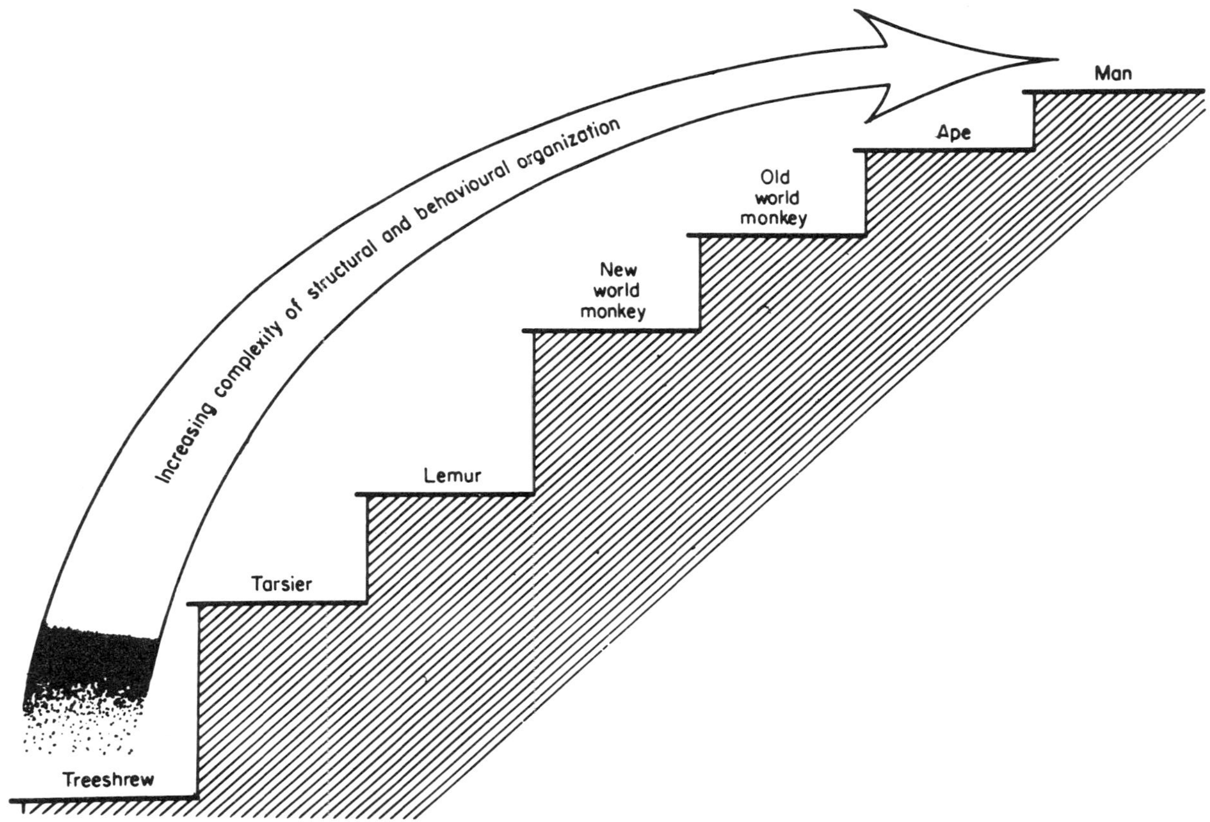

FIGURE 1. A staircase of primate evolution illustrating the morphological and behavioral grades found among living primates. The height of each step reflects the "distance" between each grade. (From Napier and Napier, 1986.)

TABLE II. Classification and SIV Status of *Cercopithecus* Monkeys (Guenons)[a]

Genus	Subgenus	Superspecies	Species	Common name (No. of subspecies)	SIV status[b]
Cercopithecus	*Allenopithecus*		*nigroviridis*	Allen's Swamp monkey	
	Miopithecus	*talapoin*	*talapoin*	Talapoin	+
	Erythrocebus		*patas*	Patas (4)	+
	Cercopithecus	*diana*	*diana*	Diana (2)	+
			dryas	Dryas	
			salongo	Salongo	
		aethiops	*aethiops*	Grivet (4)	+
			pygerythrus	Vervet (14)	+
			sabaeus	Sabaeus	+
			tantalus	Tantalus (3)	+
		lhoesti	*lhoesti*	L'Hoest's	+
			preussi	*Preuss'*	
			solatus	*Solatus*	
			neglectus	*DeBrazza's*	+
			hamlyni	*Hamlyn's*	+
		cephus	*ascanius*	Redtail (5)	+
			cephus	Moustached (2)	
			erythrogaster	Red-bellied	
			erythrotis	Red-eared (3)	
			petaurista	Spot-nosed (2)	
		mona	*campbelli*	Campbell's (2)	
			mona	Mona	
			pogonias	Crowned (3)	
			wolfi	Wolf's (4)	
		nictitans	*nictitans*	Greater white-nosed (3)	
			mitis	Blue (8)	+
			albogularis	Sykes' (12)	+

[a] Classification scheme has been adapted from Lernould. (Redrawn from Allan, 1992.)
[b] Other SIV$^+$ genera include sooty mangabeys (SIV$_{smm}$), mandrills (SIV$_{mnd}$), chimpanzees (SIV$_{cpz}$), colobus, and in rare instances, baboons.

1974) and macaques (India) (Todaro *et al.*, 1978b) and exogenous type C viruses were isolated from gibbon apes (Southeast Asia) (Kawakami *et al.*, 1972). In 1978, an endogenous type D virus was isolated from a langur monkey (India) (Todaro *et al.*, 1978a). This virus was thought to be the evolutionary precursor of the exogenous MPMV-related type D retroviruses of macaques. In the 1980s, many Old World monkeys from both Africa and Asia were shown to harbor STLV-1 (Tsujimoto *et al.*, 1985), the simian counterpart of the human T-leukemia virus, strain 1 (HTLV-1) (Sugamura and Hinuma, 1993).

In the last 7 years, SIV, the closest animal lentivirus relative of HIV-1, has been found to be highly prevalent in species of African monkeys of the genera *Cercopithecus* (including African green monkeys) and *Cercocebus* (including sooty mangabeys) (Allan *et al.*, 1990a) (see Section VII). SIV infection appears particularly prevalent in members of the guenon family (Fig. 2) of the genus *Cercopithecus*. By contrast, SIV does not occur naturally in Asian apes or macaques. All but one of the 17 species of macaques live in Asia; the exception is the Barbary macaque (*Macaca sylvanus*) that habitates the Rock of Gibraltar in Morocco (Napier and Napier, 1986). The type D retroviruses are indigenous in Asian macaques and have not been isolated from any African monkey species (Lowenstine *et al.*, 1986; Ohta *et al.*, 1988).

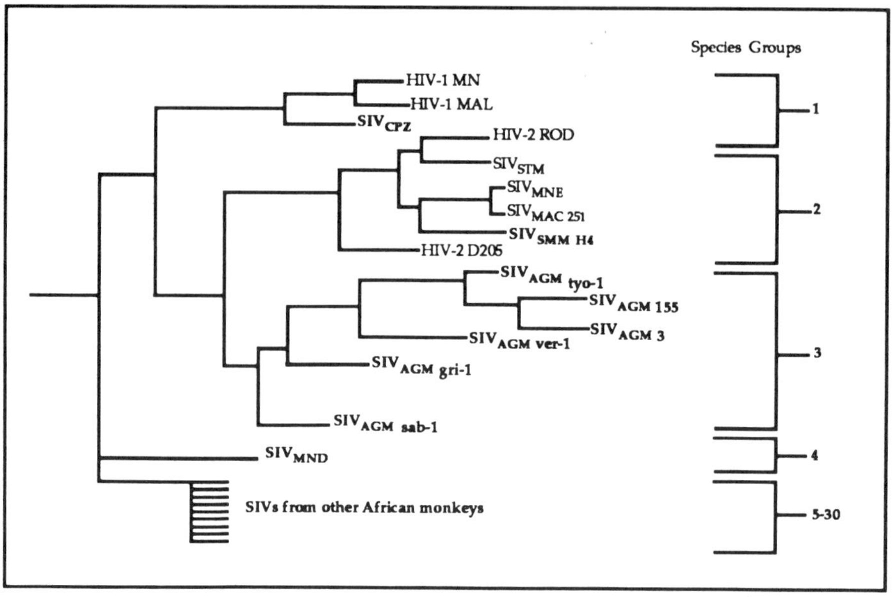

FIGURE 2. Phylogenetic tree of the family of primate immunodeficiency viruses. The tree is based on analysis of the protein sequence encoded by the *gag* gene for each virus. Each branch represents the evolutionary steps toward new species. Boldface type indicates viral species that infect nonhuman primates in the wild. In each species group (numbered on right), there is a naturally occurring virus that probably produced all others. In the first two groups, nonhuman primate viruses have apparently crossed species to infect humans and other non-African monkey species. (Redrawn from Allan, 1992.)

B. Importation and Quarantine

In the past 40 years, numerous African monkeys and apes were imported to zoos and to primate centers for biomedical research. In the realm of infectious diseases, macaques have been particularly useful for developing vaccines against poliovirus (Paul, 1971) and tuberculosis (Good, 1968) and sooty mangabeys were used for research on leprosy (Gormus *et al.*, 1988). African apes and monkeys and Asian macaques were extensively utilized in the 1970s for attempted transmission of atypical viruses that caused spongiform encephalopathies such as the kuru and Jakob–Creutzfeldt agents (Gajdusek, 1990). SIV may have been inadvertently transmitted from African to a few Asian monkeys during these experiments, although this possibility was never proven.

Since the mid-1970s, nonhuman primates have not been imported from Africa or India. Some macaques still come from Indonesia, China, or the island of Mauritius in the Indian Ocean, and some African green monkeys come from the Caribbean islands of Barbados and St. Kitts. Between June of 1990 and September of 1991, 130 shipments consisting of about 11,000 cynomolgus macaques, 900 rhesus macaques, and 500 African green monkeys entered 19 primate quarantine facilities in the United States. A mortality of about 2% was associated with the shipping process (Schrier *et al.*, 1992). Fortunately, most of the macaques needed for AIDS and other research purposes can now be bred in captivity in the United States. An exception is the pigtailed macaque (*M. nemestrina*), now being imported from Indonesia because of its somewhat unique susceptibility to acute infection with HIV-1 (Agy *et al.*, 1992).

Newly imported monkeys are kept in quarantine for 1–3 months, TB skin-tested every 2 weeks, examined for fecal and blood-borne parasites, especially malaria, and vaccinated against measles. TB skin test-positive monkeys are sacrificed for confirmation of infection and exposed animals are placed on a 90-day quarantine. Animals are also screened for antibodies to filovirus and to STLV, SIV, and type D retroviruses (Lairmore *et al.*, 1990). Because many type D virus-infected animals are seronegative (Lerche *et al.*, 1991), virus isolation from peripheral blood mononuclear cells (PBMC) is also done on all animals. For AIDS research, NIH has sponsored an AIDS Animal Model Program, including chimpanzee breeding and research at five sites and the formation of six specific-pathogen-free (SPF) colonies of macaques that are bred free of STLV, type D retroviruses, and simian type B herpesviruses. Included in this program is a Simian Retrovirus Reference Laboratory (UC Davis) for serologic screening of nonhuman primates and animal handlers. This surveillance mechanism recently detected two instances of SIV seroconversion in laboratory personnel exposed to SIV (Khabbaz *et al.*, 1992a,b) (see Section XI).

C. Herpes B Virus

B virus is a neurotropic herpesvirus that is ubiquitous and relatively nonpathogenic in macaques, just like herpes simplex virus in humans (Kalter and Heberling, 1988). Almost 100% of rhesus macaques are seropositive for B virus by $2\frac{1}{2}$ years of age (Weigler *et al.*, 1990). Therefore, screening for B virus is not a routine quarantine procedure. However, humans exposed to monkey bites can become infected with B virus and develop fatal encephalitis (Weigler *et al.*, 1990). Fortunately, this infection does not often occur. About 12,000 monkey bites are reported each year in animal handlers and only about 1% of these cause infection with B virus (Julia Hilliard, personal communication). From 1934 to 1993, some 32 cases of human infection were recognized with 26 deaths resulting from encephalitis (Kalter and Heberling, 1988; Scinicariello *et al.*, 1993). Four deaths have occurred in the past 5 years. Importantly, 7 cases of human infection identified in recent years have been effectively controlled by early diagnosis and continued treatment with acyclovir. The diagnosis of B virus infection in humans is based on a rise in antibodies and detection of the virus at the bite site by culture and/or PCR (Scinicariello *et al.*, 1993). These assays are done on request at a B-Virus Diagnostic Lab at the Southwest Foundation for Biomedical Research in San Antonio, Texas (Julia Hilliard, personal communication).

III. SEROEPIDEMIOLOGY: GLOBAL SCALE

Simian lentiviruses (i.e., SIV family) appear to occur exclusively in African monkey species, mostly in the *Cercopithecus* (guenon family) (Fig. 2) and *Cercocebus* genera (Lowenstine *et al.*, 1986; Ohta *et al.*, 1988; Allan, 1992) (Fig. 3). On the basis of seroepidemiological data, it appears that as many as 30 distinct SIV strains may be harbored in their African monkey hosts. Among Old World monkeys, lentiviruses have been isolated from African green monkeys (*Cercopithecus aethiops*), sooty mangabeys (*Cercocebus atys*), white-crowned mangabeys,

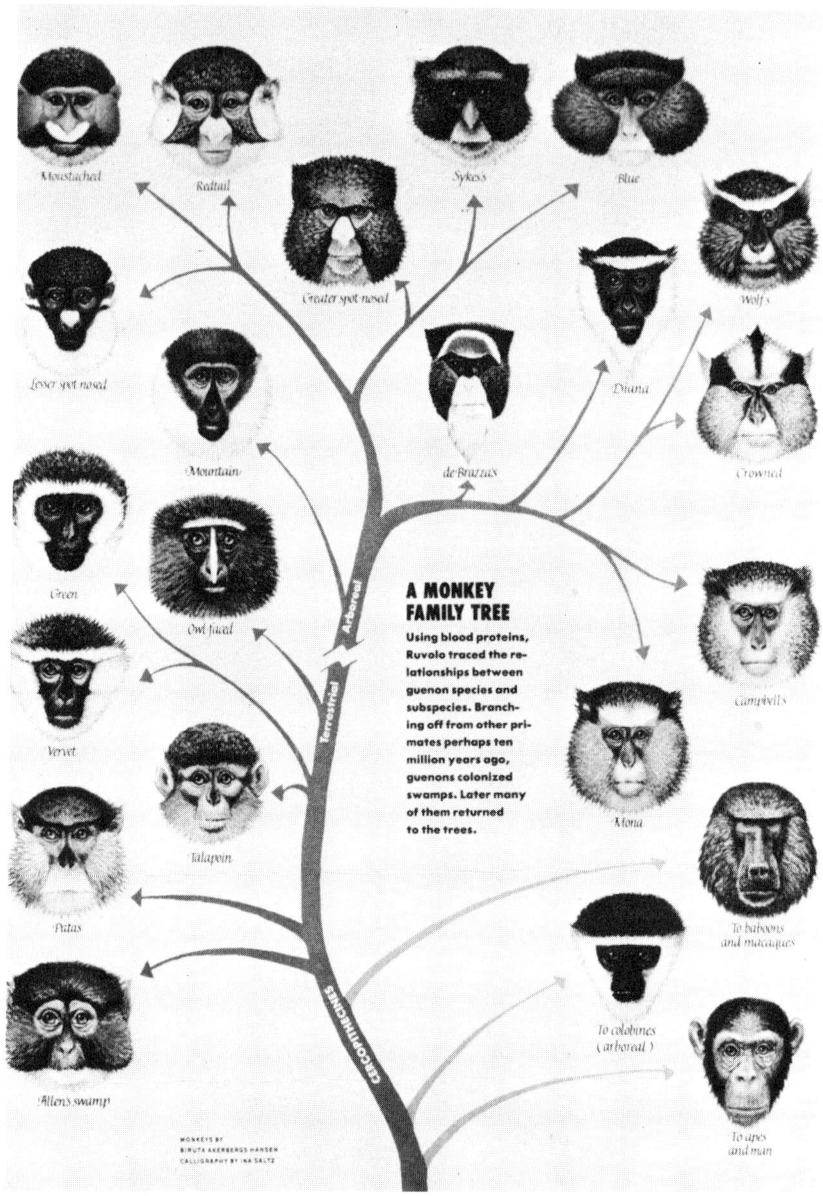

FIGURE 3. Guenon monkey family tree. (From G. Maranto, in *Discover*, Nov. 1986, p. 100, with permission.)

mandrills (*Papio sphinx*), Sykes monkeys (*Cercopithecus mitis*), and chimpanzees (*Pan troglodytes*) (Table I) (see below). A phylogenetic tree of the current family of primate immunodeficiency viruses is shown in Fig. 2.

Based on serologic cross-reactivity with SIV or HIV, many additional primate species harbor related lentiviruses. These species include colobus monkeys (Lowenstine *et al.*, 1986), yellow baboons (Kodama *et al.*, 1989a), talapoins (Lowenstine *et al.*, 1986), Hamlyn's monkeys (Nicol *et al.*, 1989), DeBrazza's monkeys (Ohta *et al.*, 1988), l'Hoesti monkeys (Nicol *et al.*, 1989), Diana monkeys (Lowenstine *et al.*, 1986), Kolb's guenon (Lowenstine *et al.*, 1986), Dent's monkeys (Ohta *et al.*, 1989), redtail monkeys (Novembre *et al.*, 1993a), tantalus monkeys (Muller *et al.*, 1993), and blue and highland Sykes monkeys (Emau *et al.*, 1991; Novembre *et al.*, 1993a). Most of these monkeys (at least 13 species) belong to the genus *Cercopithecus* (total of 27 species), as do African green monkeys (Table II).

Apparently SIV infection is very infrequent in baboons. Only 2 of 279 sera from baboons in their natural habitat (Tanzania, Ethiopia) were seropositive to SIV and virus could not be isolated from these monkeys (Kodama *et al.*, 1989a). More recently, 2 of 106 baboons from Kenya tested SIV-seropositive by ELISA and Western blot, but virus isolation attempts were negative (Novembre *et al.*, 1993a). By contrast, screening of over 2000 macaques from the wild has revealed no evidence of infection with SIV (Daniel *et al.*, 1988; Ohta *et al.*, 1988; Benveniste *et al.*, 1986; Lowenstine *et al.*, 1986). Over 1200 captive macaque sera were also seronegative for SIV (Lairmore *et al.*, 1990). Of these sera, 22% were seropositive for STLV-I and 8.4% were seropositive for type D retrovirus. Antibodies cross-reactive with SIV, SRV and HTLV-1 were detected in 20–30% of baboons imported from Ethiopia. These findings suggest that wild caught baboons are infected with retroviruses only partially related to human primate viral isolates (Benveniste *et al.*, 1993a).

The seroprevalence, as best as is known, of SIV in various African monkey species held in captivity in primate facilities or zoos or in the wild is presented in Table III. Among all African nonhuman primates tested, African green monkeys (AGM) comprise the largest reservoir of SIV in the wild. Four subspecies of AGM are distinguished: vervets (*C. pygerythrus*) found from southern Ethiopia to South Africa; grivets (*C. aethiops*), limited to Ethiopia and the Sudan; tantalus monkeys (*C. tantalus*) from the Central African Republic; and sabaeus monkeys (*C. sabaeus*) from West Africa (Fig. 4). From 20 to 50% of AGM (*C. aethiops*) in Kenya, Ethiopia, South Africa, and Senegal have antibodies to SIV (Kanki *et al.*, 1985b; Ohta *et al.*, 1988). Seropositives included all four subspecies of AGM. Sera from 28 of 67 (42%) wild-caught, healthy AGM (vervet species) from Kenya

TABLE III. SIV Seroprevalence in African Monkey Species

Species	Approximate seroprevalence[a]
African green monkeys[b,c]	36%
Sooty mangabeys	57%
Mandrills	15%
Baboons (other than mandrills)	<1%
Sykes monkeys	34%
Talapoins	70%
Colobus	<10%
Chimpanzees	<1%
Orangutans	0
Gorillas	0

[a] Based mainly on survey of captive animals in zoos and primate centers and to a lesser extent on animals trapped in the wild in Africa.
[b] All four subspecies of AGM.
[c] AGM on Caribbean islands are SIV seronegative.

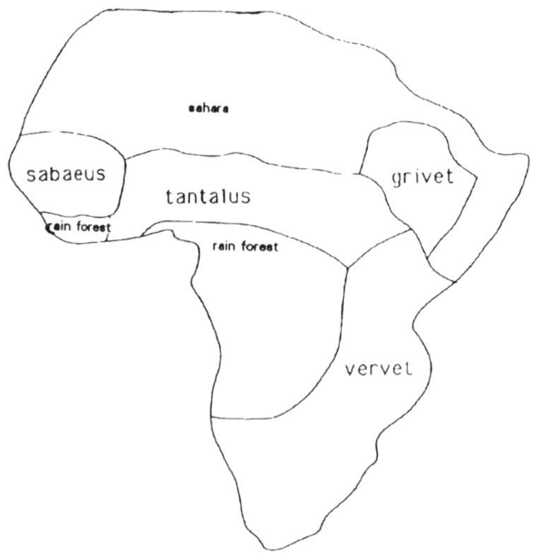

FIGURE 4. Geographic distribution of the four AGM species: the grivet (*C. aethiops*), tantalus monkey (*C. tantalus*), vervet (*C. pygerythrus*), and sabaeus monkey (*C. sabaeus*).

contained antibodies to SIV_{mac} envelope and core antigens. At the Institute of Primate Research in Kenya, the seroprevalence of SIV in AGM was 30% (Otsyula, personal communication). About 10% of these animals were coinfected with STLV. In Japan, SIV was isolated from cryopreserved PBMC from 16 of these seropositive animals from Kenya (Ohta *et al.*, 1988) (Section V).

In a retrospective survey, 10% (6 of 60) of sera collected from AGM in Africa between 1957 and 1962 were positive for SIV_{mac} antibodies and this seroprevalence increased to 42% (14 of 33) in such sera collected in Africa between 1980 and 1985 (Hendry *et al.*, 1986). In another large serologic survey of 855 nonhuman primates from 32 Old World and New World monkey species and apes, about 60% of the AGM sera had antibodies to SIV_{mac} (Kanki *et al.*, 1985b). Antibodies to SIV_{agm} core proteins were not detected in sera of SIV_{agm}-infected AGM (Schneider *et al.*, 1987). In a colony of AGM (grivets and vervets) at Bowen Gray School of Medicine, 53 of 152 (35%) animals were seropositive and SIV was isolated from 6 of these animals (Krugner-Higby *et al.*, 1990). A serologic survey of 526 captive Old World monkeys and apes, representing 50 species and 20 genera, living in U.S. zoos, for antibodies cross-reactive to HIV-1 core, revealed that, while infection with SIV was not widespread (~4%) in captive Old World monkeys, cross-reactive HIV-1 core antibodies were indeed present in monkeys from five different species (sooty mangabeys, talapoins, mandrills, guenons, colobus) at three zoos (Lowenstine *et al.*, 1986). SIV strains were subsequently isolated from PBMCs from 4 of these seropositive sooty mangabeys and 1 seropositive mandril housed in a single cage at one zoo. However, SIV could not be isolated from PBMC of 11 of the seropositive talapoins or 2 seropositive colobus monkeys, suggesting that these animals might have recovered from the infection or were latently infected.

Most recently, 30 of 80 (37%) wild-caught tantalus monkeys from the Central African Republic and 131 of 325 (40%) of feral sabaeus monkeys from Senegal were found seropositive to SIV (Muller *et al.*, 1993). SIV isolates were recovered from 4 of the seropositive tantalus monkeys and from 4 seropositive sabaeus monkeys (see Section V).

In contrast to the AGM from Africa, several hundred AGM sera from Caribbean islands, as mentioned above, were all negative for SIV antibodies (Lowenstine *et al.*, 1986; Daniel *et al.*, 1988b). Also in contrast to the monkey sera from Africa, sera from 106 wild AGM in Barbados were entirely free of STLV antibodies (Blakeslee *et al.*, 1985). Since the Caribbean AGM are descended from animals brought to the West Indies on slave-trading vessels in the mid-1700s (Denham, 1987), it appears possible that SIV and STLV might have been introduced into the African monkey population at some time during the past 200 years. However, in view of the considerable divergency of SIV strains isolated from AGMs in Africa (see below), it appears more likely that SIV has been present in African monkeys for many centuries, and that the absence of SIV in their Caribbean descendants represents a sampling artifact.

Sooty mangabeys (SM) (*Cercocebus* genus) are extensively infected with SIV in captivity (Murphey-Corb *et al.*, 1986; Fultz *et al.*, 1986; Lowenstine *et al.*, 1986). Although SIV infection of SM does occur in their native habitat (West Africa), the prevalence of natural infection in SM appears low in that only 1 of 15 pet mangabeys in Liberian households was seropositive (Marx *et al.*, 1991). Further studies are also required to ascertain the prevalence of SIV infection in SM, mandrills, Sykes monkeys, and other African monkey species in the wild. SIV infection appears to be very infrequent in feral or captive chimpanzees in that only two SIV-seropositive chimpanzees have been found among several hundred tests. No SIV-seropositive gorillas have been found (Lowenstine *et al.*, 1986; Peeters *et al.*, 1989; Marx *et al.*, 1991). Further surveys may well detect SIV infection in yet other African nonhuman primate species. This information is needed to establish the prevalence and genetic diversity of SIV in these wild populations of nonhuman primates and to resolve the questions concerning coevolution of these viruses and their hosts including the possible cross-species transmission of SIV to humans (Allan 1992) (see below).

The type D viruses are indigenous in various Asian macaque species and have not yet been isolated from nonhuman primates in Africa. However, based on serology, African talapoin monkeys in a zoo (Lowenstine *et al.*, 1986) and in the wild (Ilyinskii *et al.*, 1991) appear to be infected with a type D virus. Type D viruses are an important cause of AIDS-like disease in captive macaques held in primate facilities throughout the world. The type D viruses are covered in Section XVII.

IV. SIV ISOLATES FROM MACAQUES

A. The SIV_{mac} Strain

In early 1985, a new simian retrovirus related to HIV was isolated from three immunodeficient and one lymphomatous rhesus monkey at the New England Regional Primate Center (NERPC) (Daniel *et al.*, 1985). Three of these animals were part of a lymphoma transmission study, and in contrast to all of the other macaques with immunodeficiency disease at this center, they were not infected with the type D retrovirus (SRV-1). Although three of these four monkeys were seropositive for STLV-I, an etiologic association of lymphomas with this oncovirus could not be proven at the molecular level. Surprisingly, all four immunosuppressed monkeys had antibodies reactive with HIV-1 proteins, and a lentivirus, SIV_{mac}, was subsequently isolated by coculturing PBMC of these four animals with human T lymphocytes following the same procedure used for isolation of HIV-1 from human PBMC. SIV_{mac} is the prototype SAIDS-inducing lentivirus in macaques (Letvin *et al.*, 1985). Biologic and molecular clones of the SIV_{mac} strain (Table IV) have been most extensively utilized for studies on the pathogenesis of SIV-induced SAIDS in macaques and for AIDS vaccine research (Sections XIV and XV).

1. Cellular Host Range

The prototype SIV_{mac251} strain was infectious for human $CD4^+$ but not $CD8^+$ lymphocytes, and its ability to infect macaque and human lymphocytes could be blocked with monoclonal anti-CD4 antibodies (Kannagi *et al.*, 1985). SIV_{mac} did not replicate in PBMC cultures from chimpanzee, squirrel monkey, and cotton-topped tamarin. SIV_{mac} also differed from HIV-1 in its apparent inability to grow in EBV-transformed human B lymphocytes, the different range of nonhuman T cells that supported its growth, and its less striking cytopathicity for macaque and human T lymphocytes. In addition, sera from HIV-1-infected human subjects which neutralize HIV-1 *in vitro* did not inhibit SIV_{mac} infection *in vitro* and a monkey serum which neutralized SIV_{mac} did not affect HIV-1 replication *in vitro.* It is noteworthy that SIV_{mac} induced expression of class II major histocompatibility complex (MHC) antigens on human infected target T-cell lines *in vitro* (Kannagi *et al.*, 1987). MHC-II antigens were also detected on the envelope of virus particles derived from these cells (Kannagi *et al.*, 1987), a prophetic finding in view of the important role shown several years later for the MHC antigens in immune protection achieved with inactivated whole SIV vaccines grown in such human T cells (Arthur *et al.*, 1992b) (Section XV).

2. Serologic Relationships

The SIV_{mac} virus was antigenically more closely related to HIV-2 than to HIV-1 in keeping with the closer genetic relationship of SIV_{mac} with HIV-2 than

TABLE IV. Molecular Clones of SIV_{mac}, SIV_{sm}, and SIV_{agm}

Clone	Source	Infectious *in vitro*?	Infectious *in vivo*?	Induction of AIDS?	Reference
SIV_{mac251}	Rhesus	Yes	Yes	No	Naidu *et al.* (1988)
SIV_{mac142}	Rhesus	Yes	No	No	Naidu *et al.* (1988a)
SIV_{mac239}	Rhesus	Yes	Yes	Yes	Kestler *et al.* (1990)
$SIV_{mac251(1A11)}$	Rhesus	Yes	Yes	No	Marthas *et al.* (1989)
SIV_{mneC18}	Pigtailed	Yes	Yes	Yes	Overbaugh *et al.* (1991)
SIV_{smH3}	Mangabey	Yes	Yes	Yes	Johnson *et al.* (1990b)
SIV_{smH4}	Mangabey	Yes	Yes	Yes	Johnson *et al.* (1990b)
$SIV_{agm\text{-}TYO\text{-}1}$	African green monkey (vervet)	Yes	Yes	No	Fukasawa *et al.* (1988)
SIV_{agm155}	African green monkey (vervet)	Yes	Yes	No	Johnson *et al.* (1990b)
SIV_{agm3}	African green monkey (vervet)	Yes	Yes	No	Baier *et al.* (1990b)
$SIV_{agm385.266}$	African green monkey (vervet)	Yes	ND[a]	ND	Li *et al.* (1989b)

[a] ND, not done.

with HIV-1 (see next paragraph) (Kanki *et al.*, 1985b). By immunoblot, human antisera specific for HIV-1 detected mainly core proteins of SIV_{mac} while human HIV-2 antisera reacted with both Env and core proteins of SIV_{mac} (Fig. 5). By the more sensitive radioimmunoprecipitation (RIP) assay of SIV_{mac}-infected cells, using human serum samples positive for HIV-1 antibodies, specific proteins similar in size to the major Env (gp160 and gp120) and Gag (p55 and p24) proteins of HIV-1 were detected (Daniel *et al.*, 1985). Monoclonal antibodies to HIV-1 p24, the major core protein, also immunoprecipitated a 24-kDa protein in lysates of cells infected with the macaque virus, and disrupted macaque virus proteins specifically competed with binding of HIV p24 to goat HIV-1 antisera. However, in an RIP competitive assay for the Gag proteins, the macaque virus showed only a partial relationship to HIV-1, presumably because of a lesser affinity of the SIV Gag antigens to HIV-1 antisera. Moreover, two-dimensional peptide maps showed only 21–27% similarity in the main core polypeptides (p19 and p24) of both viruses (Jurkiewicz *et al.*, 1986).

3. Genomic Relationship to HIV-1 and HIV-2

To examine the genetic relationship of SIV_{mac} and HIV-1, an HIV-IIIB-9-kb probe (BH10 clone) was hybridized in Southern blots to DNA fragments from the SIV-infected HUT-78 human T-cell line (Daniel *et al.*, 1985). No detectable cross-hybridizing sequences were found, even under low stringency (6 × SSC, 30% formamide at 37°C with washing in 2 × SSC at 55°C). The same negative result was found using subclones of the HIV *gag* and *pol* regions. These results indicated that there was ≤75% sequence homology between SIV_{mac} and HIV-1.

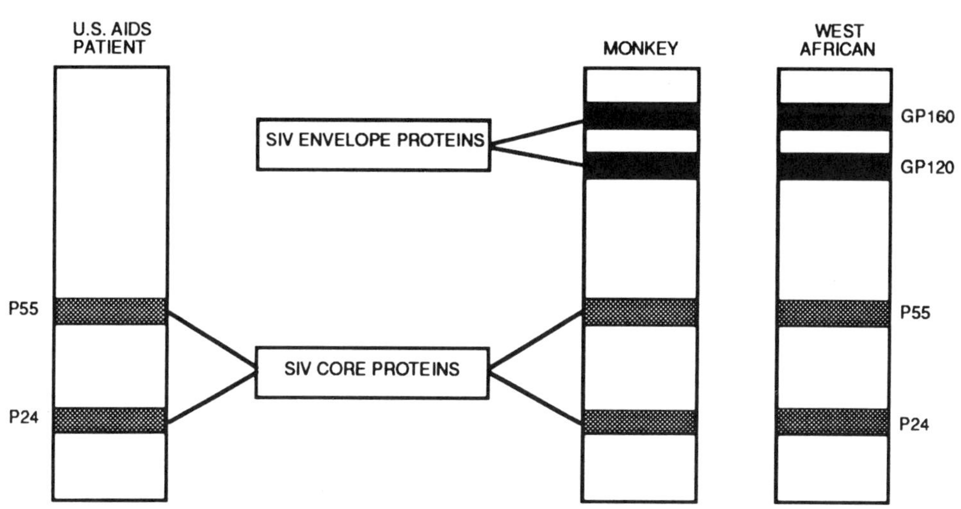

FIGURE 5. Serologic relationship of SIV_{mac}, HIV-1, and HIV-2. HIV-2 was discovered by serological testing with SIV proteins. The blood of U.S. AIDS patients had antibodies only to the core proteins, in keeping with the roughly 50% relation of SIV to HIV-1. As one might expect, SIV-infected monkeys had antibodies to both core and envelope proteins of SIV. People in high-risk groups in West Africa turned out also to have antibodies that reacted with both the core and envelope proteins, indicating that they were infected by a human virus more closely related to SIV than to HIV-1. The virus is now designated HIV-2.

Later sequence comparisons showed that these viruses are about 50% related overall. SIV_{mac} was shown to be more closely related to HIV-2, about 75% sequence homology overall (Chakrabarti *et al.*, 1987), but the divergence of the sequences of SIV_{mac} and HIV-2 was greater than that observed between HIV-1 isolates. Thus, these human and monkey viruses, although immunologically related, are quite distinct, suggesting that one could not be a recent progenitor of the other. This low degree of sequence conservation also argues against transmission of SIV to humans in recent times (Desrosiers, 1986).

4. Molecular Clones

Several infectious molecular clones of SIV_{mac} have been derived and characterized for their *in vitro* and *in vivo* biologic properties (Naidu *et al.*, 1988) (Table IV). One of the clones (SIV_{mac239}) is pathogenic *in vivo* (Kestler *et al.*, 1990a). Another clone, $SIV_{mac1A11}$, is nonpathogenic (Luciw *et al.*, 1992; Marthas *et al.*, 1989). Recombinants of these clones are presently being used to define the viral determinants of pathogenesis (Marthas *et al.*, 1993). Features of SAIDS induced by the various biologic and molecular clones of SIV will be discussed in Section XIII.

B. The SIV_{mne} Strain

Following the isolation of SIV_{mac}, another closely related lentivirus was recovered in 1986 from the stored cultured lymph node of a pigtailed macaque (*M. nemestrina*) that had died of lymphoma in 1982 at the University of Washington Medical Lake Primate Facility (Benveniste *et al.*, 1986). At that time this was the only monkey at the Washington center from which SIV could be isolated and a serologic survey done there on several hundred animals was completely negative for SIV. By contrast, infection of macaques with the type D virus, SRV-2 strain, was already highly prevalent at this center (Benveniste *et al.*, 1986). SRV-2 was also isolated from the stored frozen lymph node from which SIV was recovered. SIV_{mne} replicated to high titers in the human HUT-78 T cells and to a lower level in baboon and macaque PBMC. SIV_{mne} produced persistent viremia and eventual SAIDS (66–87 weeks after inoculation) in rhesus and pigtailed macaques but did not infect baboons (Benveniste *et al.*, 1988). By immunologic and amino acid sequence analysis, SIV_{mne} was partially related to but clearly distinct from HIV-1 and HIV-2 and closely related to SIV_{mac} (Benveniste *et al.*, 1986) (Fig. 2). Subsequent nucleic acid sequence analysis confirmed that SIV_{mne} was very similar (97–98% *gag* homology) to SIV_{mac} and somewhat more distantly related (87–88% *gag* homology) to SIV from sooty mangabeys (SIV_{sm}) (Benveniste *et al.*, 1989) (see below). A single cell clone (E11S) of SIV_{mne} was derived that produces high titers of the virus and provides a rich source of the envelope glycoprotein (gp120). This cell clone contains only two SIV_{mne} proviral DNAs; virus harvested from this biologic clone (E11S) has been used for vaccine studies described in Section XV. A molecular clone (C18) of SIV_{mne} is infectious and pathogenic for pigtailed macaques (see below) (Overbaugh and Rudensey, 1992).

C. The SIV_{stm} Strain

In a retrospective study, based on seroepidemiology, pathology, virus isolation, and virus transmission, an epizootic of SAIDS and lymphoma in stumptailed macaques (*M. arctoides*) that began in 1976 at the California Primate Research Center (CPRC) (Holmberg *et al.*, 1985) was determined to have been caused by SIV (Lowenstine *et al.*, 1992). A lentivirus, designated SIV_{stm}, was isolated in human HUT-78 T cells at the CPRC from a rhesus macaque inoculated with the stored homogenate of a lymph node from a seropositive stumptailed macaque from this outbreak and this virus was pathogenic for rhesus monkeys. Two surviving, infected stumptailed macaques from the CPRC sent to the Yerkes Primate Center in 1981 remained infected, but asymptomatic, for at least 7 years prior to death. After this time, the virus was found to be disseminated throughout the Yerkes stumptailed macaque colony and responsible for many fatal cases of SAIDS (Novembre *et al.*, 1992). SIV_{stm} was thus associated with two spontaneous outbreaks of SAIDS in macaques at different primate centers 15 years apart. Heretofore, in the different primate centers, SIV had been identified only in a few individual macaques and all outbreaks of SAIDS had been caused by SRV (for summary, see Gardner *et al.*, 1988). The mode of transmission of SIV_{stm} is uncertain, as is the natural mode of transmission of SIV in general (see below). However, in a recent experiment the horizontal transmission of SIV_{stm} among group-housed male rhesus has been documented and the route of transmission appears to be via biting and fighting during the stage of acute viral infection (Jennings *et al.*, 1992).

The origin of SIV in the stumptailed macaques is unknown. Cross-species spread is possible since at least one known SIV-infected stumptail was housed in 1976 in close proximity to sooty mangabeys, some of which were seropositive for SIV. A history of shipments in the 1970s of stumptails and sooty mangabeys mixed together also suggests the potential for transmission by contaminated needles, tattooing instruments, or direct animal-to-animal contact.

Molecular analysis of SIV_{stm} indicates that it is highly divergent from other members of the SIV/HIV-2 group (Fig. 2). SIV_{stm} has only 83–84% nucleotide sequence homology in the *gag* region with the other SIV macaque isolates (SIV_{mac}, SIV_{mne}) whereas the latter are 97–98% homologous to each other (Khan *et al.*, 1991a). Yet, SIV from sooty mangabey (SIV_{sm}) has 87–88% identity in the *gag* region with SIV_{mac} and SIV_{mne}. It is noteworthy that SIV_{stm} has a slightly higher degree of sequence homology (74–79%) with HIV-2 than do the macaque and sooty mangabey SIV strains, which are 71–76% homologous to HIV-2. Nucleotide sequence analysis in the *env* region corroborates the relatedness seen in the *gag* region by indicating that SIV_{stm} is most divergent from the other macaque SIVs and is the closest SIV relative to HIV-2 (Khan *et al.*, 1991b). An evolutionary tree analysis based on nucleotide variation in the *gag* and *env* regions suggests that SIV_{stm} diverged from the SIV_{sm}/SIV_{mac} group at least 30 years ago and is equidistant from these SIV strains (Novembre *et al.*, 1992) (Fig. 2), whereas SIV_{sm} and SIV_{mac} could have diverged as recently as about 15 years ago. By similar analyses, SIV and HIV-2 could have diverged somewhat more than 30 years ago, but the time interval is not conclusive. The data thus suggest that SIV_{stm} is the oldest SIV of the SIV/HIV-2 group reported to date; its sequence is the closest representative of the hypothetical ancient progenitor to the SIV/HIV-2 group of lentiviruses. An infectious molecular clone of SIV_{stm} has not yet been derived.

D. The SIV_{fas} Strain

This SIV strain was isolated from a seropositive cynomolgus macaque (*M. fascicularis*) at the NERPC and has not as yet been further characterized (Daniel *et al.*, 1988a).

V. SIV ISOLATES FROM AFRICAN MONKEY SPECIES

A. Sooty Mangabey (SIV_{sm})

In addition to macaques, SIV isolates were recovered in 1985–1987 from healthy sooty mangabeys housed at two U.S. primate centers (Murphey-Corb *et al.*, 1986; Fultz *et al.*, 1986), at a U.S. zoo (Lowenstine *et al.*, 1986), and more recently (1990) from a pet sooty mangabey in West Africa (Marx *et al.*, 1991) and feral African white-crowned mangabey monkeys (Tomonaga *et al.*, 1993).

1. The $SIV_{smDelta}$ Strain

At the Delta (now Tulane) Primate Center in Covington, Louisiana, 13 of 44 mangabeys had antibodies cross-reactive with HIV-1, and a lentivirus called $SIV_{smDelta}$ was isolated from cultured PBLs of several such seropositive sooty mangabeys (Murphey-Corb *et al.*, 1986). SIV_{sm} was also isolated from rhesus macaques with SAIDS and B-cell lymphoma, the result of prior inoculation with homogenates of a cutaneous lepromatous leprosy lesion from a presumably seropositive sooty mangabey from the Yerkes Primate Center (Baskin *et al.*, 1986). A rhesus herpesvirus, related to but distinct from human EBV, was recovered as well from cultured lymphoma cells from two of the immunosuppressed macaques (Rangan *et al.*, 1986). $SIV_{smDelta}$ was cytopathic for human $CD4^{+}$ cells and induced syncytial giant cells *in vitro* and *in vivo*. *In vivo* transmission of $SIV_{smDelta}$ in rhesus macaques induced fatal SAIDS within 2 months to 2 years (Baskin *et al.*, 1988). Sera from infected macaques and mangabeys reacted in Western blot with $SIV_{smDelta}$ core (p26) and envelope (gp110) proteins. Based on the degree of cross-reactivity with HIV and SIV_{mac} antigens and the apparent size of the viral proteins, $SIV_{smDelta}$ appeared closely related to but distinct from SIV_{mac}. Nucleic acid sequence comparisons later demonstrated that SIV_{sm} and SIV_{mac} were indeed closely related, showing about 88% homology in the *gag* region. Infectious molecular clones of $SIV_{smDelta}$ have been derived but not yet shown to be pathogenic.

2. The $SIV_{smYerkes}$ Strain

The same or a very closely related virus to $SIV_{smDelta}$ was isolated from PBMC of 14 of 15 randomly selected healthy sooty mangabeys housed at the Yerkes Regional Primate Center in Atlanta, Georgia (Fultz *et al.*, 1986). Forty mangabeys from Yerkes had been sent to the Delta Regional Primate Center in 1979 for experimental leprosy studies and these were almost certainly the source of $SIV_{smDelta}$. The Yerkes mangabey colony was established in 1969 with wild animals caught in West Africa and most of the existing 100 animals were born in captivity. Retrospective serosurvey indicated that SIV infection was widespread in the Yerkes mangabey colony before 1981. The $SIV_{smYerkes}$ grew well in stimulated human PBMC and caused a decrease in the number of $CD4^{+}$ T cells coincident with the rise in reverse transcriptase activity. Some $SIV_{smYerkes}$ isolates also replicated well in certain human T-cell lines (e.g., HUT-78). None of the mangabey sera contained substantial neutralizing antibody against the indigenous $SIV_{smYerkes}$. Immunologic and molecular hybridization data indicated that $SIV_{smYerkes}$ was quite distinct from HIV-1 and was virtually identical to $SIV_{smDelta}$. The $SIV_{smYerkes}$ strain, like $SIV_{smDelta}$, was also strongly pathogenic for rhesus monkeys (McClure *et al.*, 1989). Two infectious pathogenic molecular clones (H3, H4) of $SIV_{smYerkes}$ were derived from a rhesus macaque that died of SAIDS after infection with $SIV_{smYerkes}$ (Johnson and Hirsch, 1992) (Table IV).

3. $SIV_{smPBj14}$

In the course of experimental transmission of SIV isolates from sooty mangabeys at the Yerkes Primate Center, a virus pool from lymphoid tissues of a macaque (PBj) infected for 14 months was found to be associated with acute disease and death (Fultz *et al.*, 1989). Although the virus, called PBj14, was serologically indistinguishable from $SIV_{smYerkes}$, it induced death within 8 days in nine of nine pigtailed macaques (*M. nemestrina*). It also induced death acutely in SIV-seronegative but not in SIV-seropositive sooty mangabeys. Serum from macaque PBj neutralized the parental $SIV_{sm\text{-}9}$ but not $SIV_{smPBj14}$ suggesting that the variant may have been generated by immune selection. Previous infection of pig-tailed macaques with the parental $SIV_{sm\text{-}9}$ protected against the acute disease on subsequent inoculation of $SIV_{smPBj14}$. Cynomolgus macaques were equally susceptible to the PBj variant, but rhesus macaques were less vulnerable to the acute lethal disease and the survivors developed the chronic SAIDS syndrome (see below). The increased susceptibility of pigtailed and cynomolgus macaques was thought to be related to enhanced virus replication resulting from the preactivation of their $CD4^{+}$ cells from indigenous microbial pathogens other than retroviruses.

Death from PBj14 is caused by bloody diarrhea, dehydration, and electrolyte imbalance (Lewis *et al.*, 1992b). The major pathology reveals extensive lymphoid hyperplasia, especially prominent in gut-associated lymphoid tissue, without mucosal ulceration. Proliferating lymphoblasts are either $CD4^{+}$ or $CD8^{+}$ phenotype. SIV antigen and lentivirus particles are detected in the gut-associated lymphoid tissue but not in the gut epithelial cells. Accompanying this lymphoid hyperplasia is a marked increase in plasma levels of interleukin-6 (IL-6) and tumor necrosis factors (TNF); these lymphokines may be involved in the pathogenesis of the acute diarrhea. An identical syndrome has also been induced by a molecularly cloned virus derived from the PBj14 biological clone (Dewhurst *et al.*, 1990).

SIV_{smPBj} differs from its parent virus, $SIV_{sm\text{-}9}$, not only in pathogenicity but also in multiple (at least eight) *in vitro* properties. Compared with $SIV_{sm\text{-}9}$, $SIV_{smPBj14}$ infection gives higher yields of virus in primary PBMC from normal pigtailed macaques and mangabey monkeys and is cytopathic for mangabey $CD4^+$ lymphocytes (Fultz *et al.*, 1989). Moreover, unlike the parental virus, the variant replicates with equal efficiency in resting and in mitogen-stimulated mangabey $CD4^+$ lymphocytes and induces both $CD4^+$ and $CD8^+$ lymphocytes to proliferate. This proliferative response is not observed with human or chimpanzee PBMC. The ability to activate simian PBMC to proliferate is apparently specific for the acutely lethal variant and is associated with a viral protein but not necessarily with the virus's ability to replicate well in macrophages (Rosenberg *et al.*, 1991; Novembre *et al.*, 1993b). Activation is observed only on incubation with concentrated $SIV_{smPBj14}$, even heat-inactivated, and not with other viruses. The *in vivo* and *in vitro* observations suggest that the ability of $SIV_{smPBj14}$ to replicate in resting cells and to induce lymphoid cell proliferation may contribute to the extreme virulence of this virus variant.

Sequence analysis of a pathogenic molecular clone of $SIV_{smPBj14}$ and parental $SIV_{Yerkes\text{-}9}$ shows a high degree of conservation and only 36 scattered amino acid changes, primarily in the envelope (Dewhurst *et al.*, 1990). Duplication of the NF-κB site in the LTR of PBj14 may contribute to the lethality but only in concert with other subtle changes in the genome (see Section XIV). Apparently, a complex set of viral determinants are involved in PBj pathogenesis (Novembre *et al.*, 1993b; Hirsch *et al.*, 1993c). Comparison of sequential viruses obtained from the same animal between time of infection with $SIV_{sm\text{-}9}$ and isolation of $SIV_{smPBj14}$ revealed that the transition to the acutely lethal virus occurred between 6 and 10 months after infection (Fultz and Tao, 1993). Although the possibility that similar lethal HIV variants exist or may yet be generated must be considered, the unusual phenotype naturally acquired by $SIV_{smPBj14}$ probably represents a rare event that has not as yet been observed in any other experimental SIV infections.

4. $SIV_{smCalif}$

At the California Primate Center, another lentivirus was isolated from PBMC of four of seven healthy HIV-1-seropositive sooty mangabeys housed in a single exhibit at a midwestern U.S. zoo (Lowenstine *et al.*, 1986). This virus was highly pathogenic for rhesus macaques. Based on NH_2-terminal amino acid sequencing of the major *Gag* protein (p24), this virus appeared closely related, although not identical, to the sooty mangabey SIV isolates from the Tulane and Yerkes primate centers and to the macaque SIV isolates from the New England and Washington primate centers (L. Henderson, personal communication). This virus has not been molecularly cloned or sequenced. The proposed nomenclature for the three separate SIV sooty mangabey isolates from the Tulane, Yerkes, and California primate centers is $SIV_{sm\text{-}1}$, $SIV_{sm\text{-}2}$, and $SIV_{sm\text{-}3}$, respectively (Schneider and Hunsmann, 1988).

5. SIV_{smLib}

Another SIV isolate (SIV_{smLib}) was obtained from a sooty mangabey in Liberia, West Africa (Marx *et al.*, 1991). This animal was the only seropositive one among 15 healthy pet sooty mangabeys tested living in households in this area. The virus was isolated at the CPRC from cryopreserved PBMC cocultured with the human CEMX174 cell line (a T–B hybrid cell line) and also from similarly cocultured rhesus PBMC following inoculation *in vivo* of the cryopreserved PBMC. A single rhesus monkey has been persistently infected with this virus for over 1 year with no evidence of illness. Serologically, SIV_{smLib} was closely related to but serologically distinguishable from SIV_{mac} and $SIV_{sm\text{-}3}$. For genetic analysis, 1.2 kbp of the *pol* gene of $SIV_{smLib\text{-}1}$, $SIV_{sm\text{-}1}$, and $SIV_{sm\text{-}2}$ were amplified by PCR, cloned, and sequenced (Marx *et al.*, 1991). Within the SIV_{sm} group, Pol amino acid sequence identity ranged from 83 to 90%, indicating an extensive genetic diversity of SIV_{sm} isolates. The amino acid sequences of the SIV_{sm} group also had 87 to 92% identity to this region of SIV_{mac}, and 81 to 90% identity to HIV-2. Indeed, some SIV_{sm} isolates were as close genetically to some HIV-2 isolates as to other SIV_{sm} isolates and some HIV-2 isolates were as close genetically to some SIV_{sm} isolates as to other HIV-2 isolates. SIV_{sm}, SIV_{mac}, and HIV-2 thus form a single subgroup of primate lentiviruses, every bit as divergent among themselves as that observed in other SIV groups (SIV_{agm}, SIV_{mdr}) (see below) and among HIV-1 strains. The similarity between SIV_{sm} and HIV-2 isolates and the colocation of the natural habitat of sooty mangabeys and the HIV-2 endemic area (Fig. 4) suggest, but do not prove, cross-species transmission of SIV_{sm} from mangabeys as the source of HIV-2 in humans (Hirsch *et al.*, 1989b; Desrosiers *et al.*, 1989b). As mentioned, SIV_{sm} is also suspected of being the source of the several SIV isolates from captive macaques.

6. SIV_{wcm}

Several SIV isolates, designated SIV_{wcm}, were obtained from white-crowned mangabey monkeys which are indigenous in West Africa (Tomonaga *et al.*, 1993). Unexpectedly, serologic and partial sequence data indicated that this virus is more closely related to SIV_{agm} isolates than to SIV_{sm} from sooty mangabeys. These results suggest that SIV_{agm} (*Cercopithecus* genus) and SIV_{wcm} (*Cercocebus* genus) may be members of an SIV group that is genetically distinct from SIV_{sm} from sooty mangabeys (*Cercocebus* genus).

B. African Green Monkeys (SIV_{agm})

The first reported SIV_{agm} isolates from Kenya and a putative human retrovirus isolate (HTLV-IV) (Kanki *et al.*, 1985a) were later shown to be a laboratory contaminant with SIV_{mac} (Kornfeld *et al.*, 1987; Franchini *et al.*, 1987b; Kestler *et al.*, 1988; Hahn *et al.*, 1987). The first authentic AGM SIV strain, $SIV_{agm\text{-}TYO\text{-}1}$, was isolated in 1987 in Japan from PBMC of a healthy seropositive AGM (vervet) from Kenya (Ohta *et al.*, 1988). Sixteen separate bona fide SIV isolates were recovered from seropositive AGM vervets that had been imported to Japan from East Africa between 1978 and 1983. SIV_{agm} had no cytopathic effect on cultured AGM PBMC but was cytopathic for human $CD4^+$ T cells (Molt 4, HUT-78). Binding of $SIV_{agm\text{-}TYO\text{-}1}$ gp130 micelles to the CD4 molecule on Molt 4 clone 8 cells was inhibited by several anti-CD4 monoclonal antibodies directed against the OKT4a epitope as well as with soluble recombinant CD4 (Polzien *et al.*, 1992). It was noted that the envelope glycoproteins of $SIV_{agm\text{-}TYO\text{-}1}$ and $SIV_{sm\text{-}3}$ were better preserved during pelleting of the virus than seen with HIV-1 or SIV_{mac} suggesting structural similarities between the envelopes of the respective viruses (Schneider *et al.*, 1987).

After 1988, SIV isolates were obtained from many seropositive AGM by researchers in Japan, Germany, France, and the United States. As noted in Section III, the AGM are divided into four phenotypically and geographically distinct species, commonly termed vervets (*C. pygerythrus*), grivets (*C. aethiops*), tantalus (*C. tantalus*), and green (*C. sabaeus*). A different name is thus indicated for SIV from each AGM species, i.e., $SIV_{agm\text{-}ver}$, $SIV_{agm\text{-}gri}$, $SIV_{agm\text{-}tan}$, and $SIV_{agm\text{-}sab}$ (Allan *et al.*, 1990a). Most of these SIV isolates were from grivet and vervet monkeys exported from eastern Africa in recent years but some were from tantalus and sabaeus from central and western Africa (Fig. 4). Several SIV strains (e.g., $SIV_{agm385,266}$) were isolated at the NERPC from PBMC and monocyte/macrophage adherent cells of three seropositive feral vervet monkeys captured in the wild near Nairobi, Kenya (Daniel *et al.*, 1988b). These SIV_{agm} isolates exhibited a surprising degree of genetic variability, much more than seen between individual HIV-1, HIV-2, or SIV_{mac} isolates (Li *et al.*, 1989a,b). At the Southwest Foundation for Biomedical Research in San Antonio, Texas, SIV was isolated from grivet, vervet, and sabaeus monkeys that were part of a large AGM colony housed at Bowman Gray Medical School in North Carolina (Allan *et al.*, 1990a). Comparisons of a PCR-amplified *pol* gene fragment among these different SIV isolates ($SIV_{agm\text{-}gri,\ ver,\ sab}$) revealed extreme species-specific diversity (30–35%) at the nucleotide and amino acid levels, exceeding the variability among any other group of primate lentiviruses (Johnson *et al.*, 1990; Allan *et al.*, 1991; Fomsgaard *et al.*, 1991). In Frankfurt, Germany, 15 SIV_{agm} strains were isolated from a colony of AGM (grivets), initially from Ethiopia (Kraus *et al.*, 1989). One of the isolates (SIV_{agm3}) was molecularly cloned and characterized (Baier *et al.*, 1989, 1990b). The isolates grew well in some human T-cell lines and bound to the primate analogue of the CD4 receptor. Although these $SIV_{agm\text{-}gri}$ isolates were fairly similar to one another, they exhibited the genomic variability typical of lentiviruses.

Several SIV_{agm} strains have been molecularly cloned and completely sequenced (Table IV) and several others have been molecularly cloned, partially sequenced, or mapped by restriction enzymes. The genomic organization of SIV_{agm} was fundamentally similar to that of other primate lentiviruses; however, SIV_{agm} does not have the *vpr* gene, unlike SIV_{mac}, HIV-1, and HIV-2; like SIV_{mac} and HIV-2, SIV_{agm} also has no *vpu* gene (see Section IX). Initial comparisons of the SIV_{agm} isolates showed a greater amino acid sequence divergence (50–60%) from HIV-1 than seen with SIV_{mac} and SIV_{sm}. Indeed, the newly isolated SIV_{agm} strains appeared almost as distinct from HIV-2 as from HIV-1. However, considerable heterogeneity exists among the SIV_{agm} strains derived from the four AGM subspecies from different geographic regions of Africa with some of the isolates appearing closer to HIV-1 while others more closely resemble HIV-2.

Mutational analysis of the genes of an infectious DNA clone of $SIV_{agm\text{-}TYO\text{-}1}$ revealed a functional relatedness to HIV-1. Indeed, most of the genes could be exchanged between the viruses as judged by transfection assays (Shibata *et al.*, 1990c). Three SIV_{agm} isolates from Kenya showed only 80–87% amino acid identities in a region of *pol* and conservation of only 3–5 restriction sites among about 40 mapped across the genome (Li *et al.*, 1989a,b). Apparent size differences of the major core protein (p24), and the *pol* and *env* gene products as well, and a high degree of divergence in the *rev* and *tat* genes characterized the SIV isolates from the different subspecies. Comparison of SIV_{agm3} with $SIV_{agm\text{-}TYO\text{-}1}$ nucleotide sequences revealed an unusually high degree of variability in *pol* in relation to *gag* and *env*. However, hypervariable genes in *env* were relatively restricted in comparison with HIV-1 and HIV-2 (Baier *et al.*, 1990b). The relatively high genome variability suggests that SIV_{agm} has been present in its natural host for a longer time than has HIV been present in humans. Nevertheless, these SIV_{agm} isolates are still more closely related to one another than to any of the HIV-1, HIV-2, or other SIV groups, characterized to date (Desrosiers, 1990a). Experimental transmission studies with $SIV_{agm\text{-}TYO\text{-}1}$ showed that the virus caused persistent infection for at least 42 weeks in seronegative AGM and cynomolgus macaques, but was nonpathogenic in both hosts (Honjo *et al.*, 1990). Nor have the other SIV_{agm} isolates, with possibly one exception (Gravell *et al.*, 1989), been pathogenic for macaques (Johnson and Hirsch, 1992). Recently, however, a strain of $SIV_{agm\text{-}ver}$, pathogenic for pigtailed macaques, has been characterized and a possibly pathogenic molecular clone ($SIV_{agm9063}$) derived (Dapolito *et al.*, 1993).

At the Pasteur Institute and Georgetown University, SIV isolates were recently obtained from the tantalus species of AGM trapped in the Central African Republic and Uganda, respectively, and four additional SIV isolates were recovered from feral sabaeus species of AGM in Senegal (M. C. Muller *et al.*, 1993; Hirsch *et al.*, 1993a). Based on antigenic properties and partial *env* sequences, the tantalus SIV were the most divergent among SIV_{agm} in feral AGM in Africa. The species specificity of SIV for each of the four AGM species supports the coevolution of these viruses and their natural hosts and suggests that SIV transmission seldom, if ever, occurs among separate AGM subspecies in the wild.

C. SIV Isolates from Sykes Monkeys (SIV_{syk})

One hundred wild-caught or colony-bred Sykes monkeys (*Cercopithecus mitis*) collected between 1986 and 1988 at the Institute for Primate Research in Kenya were screened by ELISA for antibodies cross-reactive with HIV-2 by EIA (Emau *et al.*, 1991). Fifty-nine of these were confirmed positive by RIP analysis with cell lines persistently infected with either HIV-2, SIV_{sm}, or SIV_{agm}. PBMC from six seropositive and three seronegative animals were sent to Yerkes Primate Center in Georgia where a lentivirus, designated SIV_{syk}, was isolated from five of six seropositive monkeys. However, in four cases isolation was possible only after depletion of $CD8^+$ lymphocytes and coculture with PBMC from seronegative Sykes monkeys. These findings suggest that $CD8^+$ lymphocytes, as in HIV infection (Walker *et al.*, 1986), can suppress virus replication and recovery in SIV-infected Sykes monkeys (Emau *et al.*, 1991). The apparent inability of SIV_{syk} to infect PBMC from humans or other nonhuman primates is unique since other SIV and HIV strains are able to establish infections in cells of primate species other than the natural host. However, one of the SIV_{syk} isolates was eventually adapted by serial passage to human CEMX174 cells. SIV_{syk} also differed substantially from other SIV isolates in that sera from positive Sykes monkeys immunoprecipitated env antigens from HIV-1 as well as from HIV-2, SIV_{sm}, and SIV_{agm}. Preliminary sequence data from the *gag* region indicated that SIV_{syk} was distinct from all other SIV and HIV strains.

These findings indicate a high prevalence of natural infection in Sykes monkeys in Kenya with a lentivirus that appears unique among SIV strains with respect to its restricted host range and antigenic cross-reactivity with the Env of HIV-1, HIV-2, and other SIV strains.

Three species of macaques (rhesus, pigtailed, and cynomolgus) were inoculated i.v. with the uncloned, cell-free SIV_{syk}. Macaques of all three species became infected based on seroconversion and virus isolation from PBL (Hirsch *et al.*, 1993a). However, virus isolation became inconsistent after the first few months of infection. The animals remain healthy with no decrease in $CD4^+$ cells at 10 months.

A full-length infectious molecular clone of SIV_{syk} was generated by PCR amplification (Hirsch *et al.*, 1993a). The genome organization of SIV_{syk} is similar to that of other primate lentiviruses. Like SIV_{agm} and SIV_{mnd} (see below), it lacks *vpr* and *vpx*. A unique feature of SIV_{syk} is the absence of an NF-κB binding site in the LTR.

D. SIV Isolates from Mandrills (SIV_{mnd})

A strain of SIV (SIV_{mnd}) was isolated by coculturing PBMC from two wild-caught seropositive, healthy mandrills (*Papio sphinx*) in Gabon, West Africa, with Molt 4 clone 8 human T cells (Kikukawa *et al.*, 1986) which are highly susceptible to HIV-1 infection and cytopathology (Tsujimoto *et al.*, 1988). These two animals were the only two seropositives among 16 mandrills tested from this region. The virus induced cytopathic effects in the human $CD4^+$ cells. Immunoblot analysis and cross hybridization experiments revealed that SIV_{mnd} was sufficiently distinct from the other main HIV and SIV groups to represent yet another group of primate lentiviruses. Nucleotide sequence analysis of the *pol* gene suggests that SIV_{mnd} has evolved from a common ancestral primate lentivirus just before the divergence of HIV-1 from HIV-2 (Fig. 2). The sequences responsible for efficient replication of SIV_{mnd} in monocyte/macrophages were mapped to the *tat* gene (Sakai *et al.*, 1992a,b). Another strain of SIV_{mnd} was recently isolated from PBMC of a seropositive mandrill at a U.S. zoo, but it has yet to be characterized (Lerche, personal communication).

VI. SIV ISOLATES FROM CHIMPANZEES

Infection by a lentivirus was detected in 1989 by serologic studies of 2 infant wild chimpanzees (*Pan troglodytes*) in Gabon, West Africa. A virus, referred to as $SIV_{cpz\text{-}gab}$, was isolated from one of these animals (Peeters *et al.*, 1989). Only these 2 chimpanzees out of 83 tested were seropositive against HIV-1, indicating that $SIV_{cpz\text{-}gab}$ is not widely dispersed in this region. Indeed, 250 chimpanzees caught in West Africa over the past 15–20 years and several hundred chimpanzees held in captivity throughout the world were also all seronegative. An infectious proviral clone of $SIV_{cpz\text{-}gab}$ was obtained from $SIV_{cpz\text{-}gab}$-infected human lymphocytes. Sequence analysis revealed that the $SIV_{cpz\text{-}gab}$ genome was more closely related to HIV-1 than to any of the simian lentiviruses (Huet *et al.*, 1990). The overall genomic organization was the same as that of HIV-1 but more divergent than any

HIV-1 sequence yet reported in that several regions of $SIV_{cpz\text{-}gab}$ (e.g., *vpu*) were more than 50% divergent with respect to $HIV\text{-}1_{LAI}$. Functionally, the *tat* and *nef* gene products of $SIV_{cpz\text{-}gab}$ were identical to HIV-1 (Sakuragi *et al.*, 1992). The presence of an HIV-1-specific *vpu* gene and the absence of a *vpx* gene, common to HIV-2 and most SIV strains, showed that $SIV_{cpz\text{-}gab}$ is the first nonhuman lentivirus to have the same genetic structure as HIV-1. The *vpu* gene product, however, was particularly divergent (64%) from $HIV\text{-}1_{LAI}$. A phylogenetic tree analysis (Fig. 2) indicated that $SIV_{cpz\text{-}gab}$ branched before any of the HIV-1 strains sequenced to date. Thus, although HIV-1-like viruses are present in chimpanzees, it is not possible to conclude that $SIV_{cpz\text{-}gab}$ was the immediate precursor to HIV-1.

A second HIV-related virus, $SIV_{cpz\text{-}ant}$, was recently isolated from PBMC of an HIV-1-seropositive, healthy, 6-year-old, wild-capture chimpanzee living in captivity in Belgium (Peeters *et al.*, 1992). This animal had not been exposed to HIV or SIV nor did it have contact with human blood products. Neutralizing antibody titers against $SIV_{cpz\text{-}ant}$ were low or absent in this animal. The virus showed a restricted host range capacity on different continuous cell lines and did not replicate in human monocytic cell lines. $SIV_{cpz\text{-}ant}$ core and envelope proteins had higher molecular weights than $SIV_{cpz\text{-}gab}$ and more HIV-1-positive human sera reacted with the envelope proteins of $SIV_{cpz\text{-}gab}$ than with $SIV_{cpz\text{-}ant}$. Sequence analysis of PCR-amplified, cloned products revealed a 70% nucleotide homology for the *pol* and 55% homology for the *env* genes between $SIV_{cpz\text{-}gab}$ and $SIV_{cpz\text{-}ant}$. These results suggest that natural infection of chimpanzees with an HIV-1-related virus, although apparently infrequent, may not be as uncommon as previously thought. The diversity of the two chimpanzee isolates and their different geographical origin suggest that these animals have not been recently infected with SIV.

Over a period of 33 months, nine sequential sera and nine sequential $SIV_{cpz\text{-}ant}$ strains were obtained at 3 to 4-month intervals. Little or no neutralizing antibody was detected to virus isolates obtained at the same time the serum was collected. However, over time, neutralizing antibody titers increased to high levels against $SIV_{cpz\text{-}ant}$ isolates obtained at earlier time points. No mutations were found in the V3 loop, suggesting that the neutralization escape mutants had mutations outside the V3 loop (Peeters *et al.*, 1993). Similar findings were previously reported in HIV-1-infected chimpanzees (Nara *et al.*, 1990) and in goats and sheep naturally infected with their lentiviruses (Narayan and Clements, 1989).

VII. ORIGIN OF SIV AND HIV

As reviewed above, the SIV constitute a family of naturally occurring lentiviruses indigenous in certain simian species living in the wild in Africa. In their natural African simian host, these viruses apparently cause no disease. To date, five species of African nonhuman primates that harbor genetically distinct subgroups of SIV have been identified: (1) African green monkeys (*Cercopithecus aethiops*) (SIV_{agm}); (2) sooty mangabeys (*Cercocebus atys*) (SIV_{sm}); (3) mandrills (*Papio sphinx*) (SIV_{mnd}); (4) Sykes monkeys (*Cercopithecus mitis*) (SIV_{syk}); and (5) chimpanzees (*Pan troglodytes*) (SIV_{cpz}) (Table I). Except for the relationship of

SIV_{cpz} to HIV-1, individual virus isolates of a single species are more closely related to one another than to members of a different group.

Within each group, however, the genetic heterogeneity (80–100%) of the conserved Pol amino acid sequences between individual SIV isolates is equal to or greater than that seen between different HIV-1 isolates. Based on such nucleotide sequence differences, computer-generated phylogenetic trees have been constructed and the divergence time of HIV-1, HIV-2, and SIV has been estimated to be as recent as 40–280 years. It is widely believed that HIV has emerged within only the last 50 years (Myers and Pavlakis, 1992). It is also possible that HIV-1, HIV-2, and current-day SIVs emerged within the last several centuries from earlier progenitor lentiviruses or recombinational events between ancestral simian viruses related to SIV_{mnd}, SIV_{sm}, or SIV_{agm} (Schneider *et al.*, 1987). Alternatively, HIV-1 and HIV-2 may have been confined for a long time (centuries) to a few isolated human populations and only emerged recently in epidemic form. The marked degree of heterogeneity of SIV isolates between and within species suggests that SIV coevolved with African monkey species for many centuries (Allan, 1992). The phylogenetic tree of the family of primate immunodeficiency viruses as modified from Myers and Pavlakis (1992) is shown in Fig. 2. Perhaps natural selection over a long period of time accounts for the current-day nonvirulence of these viruses for their natural host whereas this adaptation process has not had sufficient time to take hold in the HIV-infected human population in which the virus has emerged and spread relatively recently.

The genetic and phenotypic differences among the various subgroups of SIV suggest that SIV is not transmitted between these monkey species. However, partial sequence comparisons of different SIV_{agm} isolates suggest the possible transmission of SIV in the wild among different AGM species (Hahn *et al.*, 1993). Surprisingly, isolates of SIV_{sm} are more closely related (80–90%) to several SIV isolates from captive macaques (SIV_{mac}) and to various HIV-2 isolates (70–80%), than they are to SIV_{agm}, SIV_{mdr}, SIV_{syk}, SIV_{cpz}, or HIV-1 (40–50%). Therefore, SIV_{sm}, SIV_{mac}, and HIV-2 appear to belong to the same lentivirus subgroup. Not surprisingly, then, HIV-2 isolates or PCR-amplified sequences, virtually indistinguishable from SIV_{sm}, have been detected in humans in Africa (Gao *et al.*, 1992; Dietrich *et al.*, 1989). Since macaque monkeys in their natural habitat do not harbor SIV, and since sooty mangabeys and HIV-2 share the same geographic localization to West Africa, it has been proposed that sooty mangabeys or closely related monkeys may have been the source for cross-species infection of macaques in captive colonies and of humans in West Africa (Hirsch *et al.*, 1989b). Ample opportunity exists for contact spread of SIV from infected sooty mangabeys to humans in Africa. The recent seroconversion to SIV of two laboratory workers exposed to this virus and the isolation of SIV from one of these individuals further supports the possible transmission of SIV from monkeys to man under natural conditions (Khabbaz *et al.*, 1992a,b; Khabbaz *et al.*, 1994; Essex 1994). Seroepidemiologic surveys addressing this question of cross-species transmission have not been reported from Africa.

SIV_{cpz} is more closely related to HIV-1 than to SIVs or HIV-2, and therefore might be a "missing link" to the origin of HIV-1 (Huet *et al.*, 1990). However, its genetic divergence from other HIV-1 strains indicates that SIV_{cpz} represents a distinct subtype which is probably not the immediate precursor of HIV-1. Thus, the possible origin of HIV-1 and HIV-2 from recent (decades) cross-species infection with SIV_{cpz} or SIV_{sm}, respectively, although plausible, remains to be established. In this scenario, AIDS would be a zoonotic disease. We can conclude that, at this time, the origin of the AIDS virus remains uncertain (Essex and Harbin, 1988). Characterization of further isolates of HIV-1, HIV-2, and SIV-related lentiviruses from humans and nonhuman primate species in Africa or elsewhere may yet fill in the missing branches in the evolutionary tree of the primate immunodeficiency viruses. It is also possible that other HIVs will arise from this large, genetically diverse primate lentivirus reservoir. These evolutionary issues are covered in another chapter in this series (Myers and Pavlakis, 1992).

VIII. VIRUS MORPHOLOGY

Although a detailed morphological analysis of the different SIV isolates has not been described, the structural features of SIV, taken as a whole, are typical of the lentivirus group of viruses (Munn *et al.*, 1985; Marx *et al.*, 1988; Gelderblom, 1991). Ultrathin section and negative-staining electron microscopy (EM) reveal irregularly hexagonal, mature virus particles 125–160 nm in diameter with eccentric nucleoids and electron-dense cores, primarily conical in shape (Ringler *et al.*, 1989a; Lowenstine *et al.*, 1986; Grief *et al.*, 1989) (Fig. 6). The cores, 110 nm in length, vary in diameter from 25 to 50 nm and are frequently surrounded by a 7 to 10-nm outer shell of irregular structure (Grief *et al.*, 1989). Surrounding this core is an outer envelope with spoke-shaped glycoproteins, 6 nm in length and 10 nm in diameter, prominently projecting from the surface (Grief *et al.*, 1989). Mature extracellular virions, which bud only from the plasma membrane, contain more of the glycoprotein projections than do intracisternal or intravacuolar forms of the virus (Banapour *et al.*, 1991). Mature virions are sometimes con-

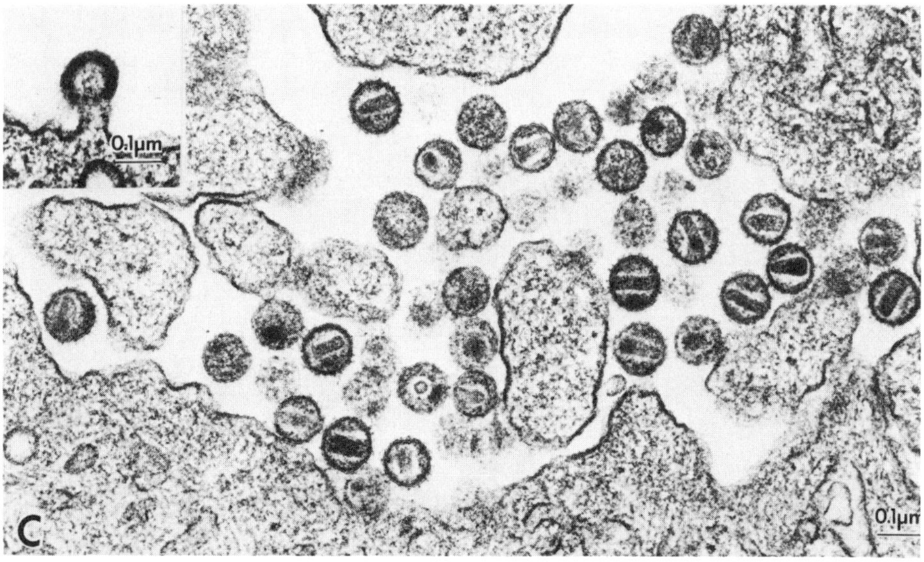

FIGURE 6. Electron micrograph of SIV virus particles (×80,000). [Thanks go to R. Munn (UC Davis) for his help with the electron microscopy.]

tained within cytoplasmic vacuoles of macrophages (Fig. 6) but are only seen budding from the plasma membrane of T cells. Budding virus is characterized by a broad C-shaped crescent bud on a smooth plasma membrane (Lowenstine *et al.*, 1986).

IX. GENOMIC ORGANIZATION

In general, the genetic organization of most SIV strains is identical to HIV-2, although there are some exceptions (e.g., SIV_{cpz}) (see below). Proviral genomes are composed of two identical long terminal repeats (LTR) flanking coding regions for proviral structural proteins, the prototypical *gag, pol,* and *env* genes (Fig. 7). Most SIV isolates code for accessory gene functions specified by the *vif, tat, rev, vpr, vpx,* and *nef* open reading frames. Their functions are described below. The *vpx* gene is found also in HIV-2 but not in HIV-1 isolates, whereas the *vpu* gene is found in HIV-1 but not HIV-2 isolates. SIV_{cpz} contain a *vpu*-like gene although there is less than 40% conservation of amino acid sequence with HIV-1. It has also been suggested that the Vpx protein of SIV_{agm} is closer in amino acid sequence to HIV-1 *vpr* than to SIV/HIV-2 *vpx* (Tristem *et al.*, 1992).

A. LTR

The LTR of SIV is divided into the U3, R, and U5 regions found in all retroviral LTRs (Luciw and Leung, 1992). Within the 5′ LTR, signal sequences in U3

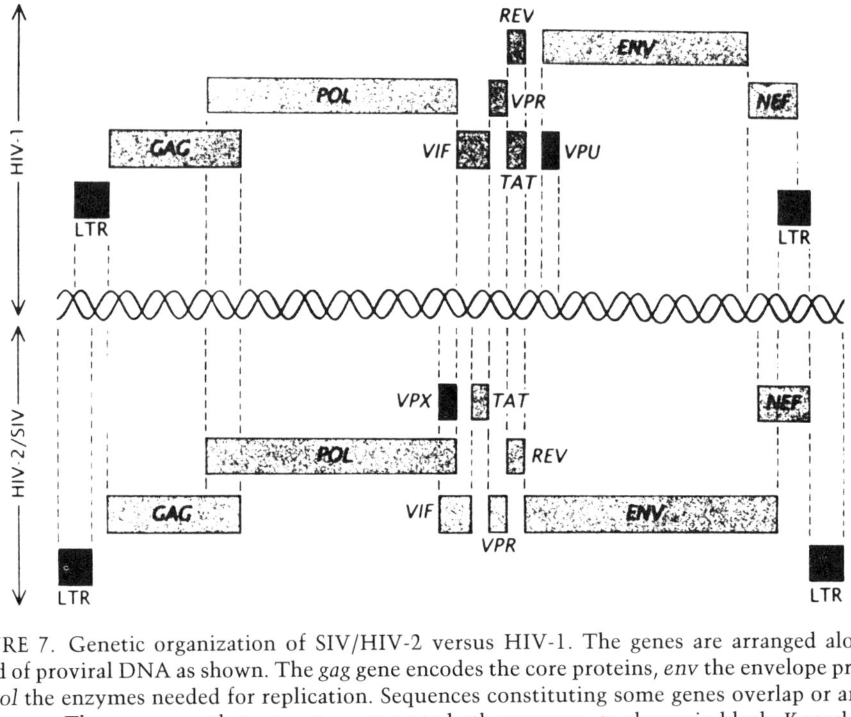

FIGURE 7. Genetic organization of SIV/HIV-2 versus HIV-1. The genes are arranged along the strand of proviral DNA as shown. The *gag* gene encodes the core proteins, *env* the envelope proteins, and *pol* the enzymes needed for replication. Sequences constituting some genes overlap or are non-contiguous. The two genes that are not common to both genomes are shown in black. Knowledge of their function might help to show why HIV-1 causes lethal disease but HIV-2 may not.

regulate the frequency and start site of transcription, which begins at the U3/R boundary. Included within U3 are the TATA sequence, which positions the transcription initiation start site, and binding sites for different cell-encoded proteins involved in transcription regulation. These include the Sp-1 and NF-κB protein binding sites, the latter of which are believed to be involved in a T-cell-specific activation of the provirus (Anderson and Clements, 1991, 1992; Arya, 1988). Responsiveness to T-cell activation signals is a phenotype of HIV-1 (Tong-Starksen *et al.*, 1990) and efficient replication in primary macrophages (Ballas *et al.*, 1993). The R region also contains a *cis*-acting sequence, the *trans*-acting responsive (TAR) element, which is an essential regulatory motif mediating *trans*-activation by the virus-encoded Tat protein (discussed below). In its transcribed form, the TAR element can potentially form a stable secondary structure which is vital for an efficient response to Tat. The secondary structure for SIV TAR is more similar to HIV-2 TAR than to HIV-1 TAR. The R region also specifies the polyadenylation site for SIV RNA. When present within the 3′ LTR, the sequence AAUAAA specifies the addition of poly(A) to the end of viral transcripts.

B. Gag

The Gag precursor protein is translated from a genome-length transcript which encodes all of the SIV open reading frames, a feature common to all retroviruses. As part of the maturation process for Gag, the SIV_{mne} Gag and Gag–Pol precursor polyproteins are proteolytically cleaved by the virus-encoded protease into six mature proteins, p16, p28, p2, p8, p1, and p6 (as arranged in the precursor protein in an amino-to-carboxyl order) (Fig. 8). The amounts of p16, p28, p2, and p8 are essentially equimolar, while lower amounts of p1 and p6 are observed (Henderson *et al.*, 1988a). This result most likely reflects a function of the mechanism of translation of the Gag and Gag–Pol precursor proteins. For many retroviruses, including SIV/HIV, the Gag coding region overlaps with Pol such that the Pol coding region is out of frame with the Gag. Translation of the Pol open reading frame is probably achieved via a −1 ribosomal frameshifting that occurs as the ribosomes translate the carboxyl end of the Gag coding region (Jacks *et al.*, 1988). At a certain probability, ribosomal slippage will generate the Gag–Pol

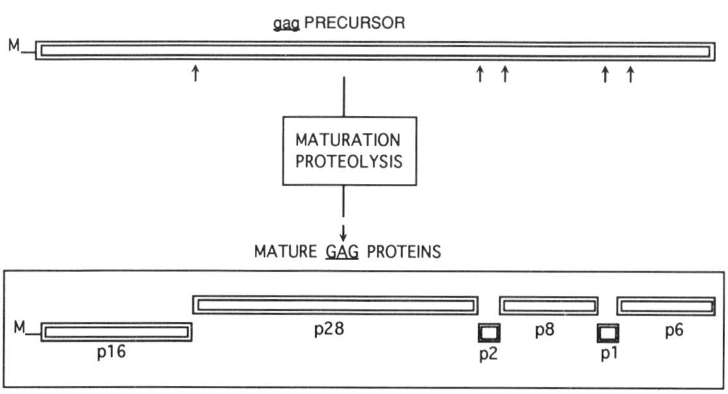

FIGURE 8. Proteolytic cleavage pattern for SIV/HIV Gag precursor.

precursor. Accordingly, Gag coding regions downstream of the ribosomal frameshift site, in this case p1 and p6, will be underrepresented compared with upstream Gag regions, i.e., p16, p28, p2, and p8 (Henderson *et al.*, 1988a). The processing pattern is similar to both HIV-1 and HIV-2, including the amino acids surrounding the proposed protease cleavage sites (Henderson *et al.*, 1988a; Jurkiewicz *et al.*, 1986). The processing of the Gag precursor into six mature proteins distinguishes the SIV/HIV group of viruses from all other retroviruses. The major core protein, p28, is partly expressed on the surface of infected cells (Nishiuo *et al.*, 1992).

p16 codes for the matrix (MA) protein which remains inside the mature virus particle in association with the envelope. p28 forms the capsid (CA) structure surrounding viral RNA, which, in turn, is complexed with the p8 nucleocapsid (NC) protein. SIV NC are similar to the other lentiviruses in containing two copies of the "cys-array," a short amino acid motif essential for packaging of viral RNA, which also influences infectivity. Other retroviruses contain either only one copy of the cys-array, or one copy of the cys-array and a second modified copy. There are no known functions for p2, p1, and p6, although all three are found in the core in SIV/HIV isolates.

Among the SIV_{sm}, SIV_{mac}, and SIV_{mne} isolates, the overall amino acid sequence homology for the Gag proteins is between 84 and 93%. This extent is similar to the degree of homology to HIV-2 isolates. However, there is only a 50–55% degree of homology with SIV_{agm} and slightly less homology to HIV-1. SIV_{agm} is equidistant in terms of Gag protein homologies to HIV-1 as are the other SIV isolates. This result is consistent with the notion that SIV_{agm} forms a distinct line of SIV.

Retroviral Gag precursor proteins are covalently modified at the amino-terminus, and of the processed proteins, the MA protein bears this modification. Most retroviruses, including SIV/HIV, have a covalent attachment of myristic acid ($C_{16}H_{31}O$–) to the amino-terminus of the Gag precursor. Attachment to the penultimate glycine in the Gag precursor first requires the removal of the methionine initiation codon. After processing, myristic acid remains a covalently attached modification. Baculovirus expression of the SIV_{mac} gag precursor results in spontaneous assembly of virus-like particles, which are myristylated, and which bud from the cell into the culture supernatant (Delchambre, 1989). If a mutation in the N-terminal glycine is introduced, myristylation is inhibited. In this case, intracellular particles are assembled, but do not bud from the cell (Delchambre, 1989).

C. Pol

The Pol coding region of SIV is believed to code for three proteins (the viral enzymes) of total molecular mass 119,600 Da: protease (PR), reverse transcriptase (RT), and integrase (IN) (Prakash *et al.*, 1992), and overlaps out of frame with the very carboxyl end of the Gag precursor. The proteins are translated from a full-length genomic transcript. Translation is most likely achieved by a −1 ribosomal frameshifting on the transcript to generate the Gag–Pol precursor polyprotein. This mechanism of ribosomal frameshifting is found in some other retroviral groups (Luciw and Leung, 1992).

SIV PR is a 99-amino-acid polypeptide (11.6 ± 0.3 kDa) which functions as a dimeric aspartyl protease (Grant *et al.*, 1991). SIV_{mac} PR has 50% amino acid identity and 73% amino acid similarity with HIV-1 PR. Both PR react with similar kinetics. The tertiary structure of SIV PR is close to that of HIV-1 PR with an Asp-Thr-Gly amino acid motif of each monomer subunit forming part of the active site. Protein unfolding studies have determined the conformational and chimeric stabilities of HIV-1 and SIV proteases (Grant *et al.*, 1992). Active protease is believed to be contained within the capsid structure of the mature virion (Henderson *et al.*, 1990). PR cleaves both the Gag and Pol precursor proteins into the mature proteins (Debouck, 1991).

RT encodes reverse transcriptase and the RNase H activities, both essential elements in the reverse transcription of genomic RNA into the double-stranded DNA form of the viral genome. SIV_{agm} RT, the best characterized RT, is 120–130 kDa and enzymatic activity resides within a heterodimeric form. One subunit is 64 kDa and the other is 48–50 kDa (Lüke *et al.*, 1990; Baier *et al.*, 1990a). Both subunits express RT activity, although enzymatic activity is greater with the larger subunit. Like HIV-1 RT, SIV RT is Mg^{2+}-dependent and functions on both poly(rA)-oligo(dT) and poly(rC)-oligo(dG) as template–primer pairs (Lüke *et al.*, 1990; Prasad *et al.*, 1990). Optimum concentrations for both mono- and divalent cations, as well as enzyme kinetics, are similar to those of HIV-1 RT (Lüke *et al.*, 1990). SIV RT is equally as sensitive as the RT from HIV-1 and HIV-2 to the chain terminating compounds azidothymidine and dideoxycytidine, as well as other compounds (Baier *et al.*, 1990a; Prasad *et al.*, 1990; Lüke *et al.*, 1990). Purified recombinant SIV_{agm} RT has an *in vitro* error rate 2- to 20-fold less than that observed with purified recombinant HIV-1 RT (Manns *et al.*, 1991). The difference in accuracy is dependent on whether balanced or unbalanced nucleotide pools are used in the assay. It is not known if the fidelity for other SIV RT is greater than HIV-1 RT.

In addition to RT activity, the RT heterodimer possesses RNase H activity. RNase H degrades the RNA in RNA–DNA hybrids during reverse transcription. SIV RNase H activity has not been characterized in detail. The ratio of the level of RT activity to the level of RNase H activity, expressed as units of enzymatic activity per milligram of protein, is 2.5 for both virion-associated RT and purified recombinant RT (Baier *et al.*, 1990a).

The integrase protein facilitates integration of linear double-stranded viral DNA into the cellular genome. To date no characterization of SIV IN has been reported. However, when IN function is deleted from a replication-competent clone of SIV_{mac}, replication can still occur, although the level of mutant virus is two- to threefold less than the parental, wild-type clone (Prakash *et al.*, 1992). The IN-deficient provirus fails to integrate in infected cells.

D. Env

The SIV envelope gene product, like that of HIV-1 and HIV-2, is synthesized as a large precursor protein which is processed into two smaller mature proteins which are essential for virus replication. SIV Env precursor is translated as a 160-kDa glycosylated protein (Veronese *et al.*, 1989; Kanki *et al.*, 1986). The unglycosylated form of the protein is 80 kDa (Kodama *et al.*, 1991). SIV_{agm} engineered to contain a premature termination codon within the extracellular glycoprotein domain is noninfectious *in vitro* (Shibata *et al.*, 1990b).

The signal peptide is a cleavage product from processing at a cysteine residue at amino acid position 22 and a threonine residue at position 23 (Veronese *et al.*, 1989; Myers *et al.*, 1992). The cysteine residue is conserved in SIV_{mac}, SIV_{smm}, and HIV-2 isolates and in almost all HIV-1 isolates (Myers *et al.*, 1992). The threonine residue is not conserved. The signal peptide is slightly shorter in SIV_{mac}, SIV_{smm}, and HIV-2 than in HIV-1 (Myers *et al.*, 1992).

SIV_{mac} gp160 is also cleaved between residues 526 and 527 to yield the external glycoprotein (gp120) and the transmembrane protein (gp32) (Veronese *et al.*, 1989; Planelles *et al.*, 1991). The primary function of gp120 is the binding to the cellular CD4 protein (Planelles *et al.*, 1991). The amino acid sequence around the processing site between gp120 and gp32 (. . . RNKR/GVFV . . .) is highly conserved among SIV_{mac}, SIV_{smm}, SIV_{stm}, and HIV-2 isolates, and less conserved with HIV-1 isolates (Khan *et al.*, 1991b; Myers *et al.*, 1992). The gene coding region for molecularly cloned SIV_{mac} gp32 contains an in-frame termination codon absent in other isolates of SIV (Khan *et al.*, 1991b; Hirsch *et al.*, 1989). As a result, SIV_{stm} and SIV_{smm} code for larger transmembrane proteins (gp40) (Khan *et al.*, 1991b). The truncation of the transmembrane protein is an apparent artifact of virus propagation in human cell tissue culture (Hirsch *et al.*, 1989a; Cranage *et al.*, 1989; Kodama *et al.*, 1989b) (discussed below). Potential glycosylation sites within the transmembrane protein (N-X-T or N-X-S; X = any amino acid) are highly conserved between SIV_{mac}, SIV_{smm}, and HIV-2 (Kodama *et al.*, 1991).

The amino-terminus of gp32/40 plays a critical role in mediating membrane fusion between the virus and cellular membrane (I. Martin *et al.*, 1991; Bosch *et al.*, 1989; Ritter *et al.*, 1993). Introduction of mutations within the 15 terminal amino acids of SIV_{mac} gp32 that increased hydrophobicity or truncation of this cytoplasmic domain increased the ability of the transmembrane protein to induce syncytia (Bosch *et al.*, 1989; Spies *et al.*, 1994). Polar or charged amino acid mutations reduced syncytium formation. The fusion domain of HIV-1 is similarly located.

As with HIV-1 (Earl *et al.*, 1990), the SIV_{mac} envelope gp160 forms stable homodimers and other higher-order oligomers prior to posttranslational cleavage (Doms *et al.*, 1990). For HIV-1 the oligomer structure is retained after processing of gp160 to gp120 and gp41, and the amino-terminal 129 amino acids of gp41 mediate the formation of oligomers (Earl *et al.*, 1990). Recombinant SIV_{mac} gp120 forms little or no higher-order structures in the absence of gp32/40 (Planelles *et al.*, 1991). When both HIV-1 and SIV_{mac} envelope genes are coexpressed within cells, heterodimers were formed between HIV-1 and SIV envelope proteins (Doms *et al.*, 1990). Recombinant SIV gp120 may be secreted by alternate pathways with or without its transitory appearance on the cell surface (Spies and Compans, 1993). Although SIV_{mac} and HIV-1 retain only 56% amino acid homology within the amino-terminus of gp41, a functional domain has evidently been conserved to facilitate the formation of higher-order heterostructures (Doms *et al.*, 1990). The function of oligomerization may be to allow binding of the envelope to multiple CD4 molecules, enhancing viral binding and fusion (Earl *et al.*, 1990).

E. Tat

SIV *tat* open reading frames code for a small protein (potentially 97–130 aa) and are present within all SIV isolates. The *tat* gene product is an essential element for viral replication. When a recombinant SIV_{agm} provirus, deleted of Tat function, was transfected into human cell lines, a low but measurable amount of viral RNA and cell-free reverse transcriptase activity was detected (10% of that observed with parental provirus) (Shibata *et al.*, 1990d). The virus, however, failed to replicate when inoculated into susceptible cell lines, leading to the conclusion of a vital role for Tat in the replicative cycle. Tat is an essential gene product for replication of both HIV-2 (Shibata *et al.*, 1990a) and HIV-1 (Dayton *et al.*, 1986; Fisher *et al.*, 1986).

The *tat* gene product of HIV-1 is the best characterized of the SIV/HIV group. Although the mechanism is not fully understood, Tat increases the frequency of transcription initiation (Rice and Mathews, 1988; Laspia *et al.*, 1989) by apparently binding to TAR RNA and interacting with components of the transcription complex (Berkhout *et al.*, 1990; Selby and Peterlin, 1990). Tat eliminates a block to transcription elongation on the HIV-1 LTR in a manner that may not be independent of the mechanism of transcription initiation (Kao *et al.*, 1987; Laspia *et al.*, 1989; Selby *et al.*, 1989). In addition, evidence exists which points to a second, posttranscriptional role for Tat, at least in some cell types (Barry *et al.*, 1991; Braddock *et al.*, 1989, 1990; Cullen, 1986; Parkin *et al.*, 1988; Rosen *et al.*, 1986; Wright *et al.*, 1986). Tat function is dependent on cellular factors which regulate its transactivation (Barry *et al.*, 1991).

While SIV Tat is not as well studied, its mechanism of function appears similar to HIV-1 Tat. There is little amino acid conservation between SIV and HIV-1 Tat outside of a cysteine-rich core and a second, basic region (Colombini *et al.*, 1989). Both domains are important for the ability of HIV-1 Tat to function (Colombini *et al.*, 1989). The Cys-rich core is present within SIV_{mac}, SIV_{smm}, and SIV_{agm}, while the highly basic region is prominent in just SIV_{mac} and SIV_{smm} (Myers *et al.*, 1992). Like HIV-1 Tat, SIV_{mac} Tat stimulates expression from the SIV promoter. In transient expression assays in human cell lines, in which the chloramphenicol acetyltransferase gene (CAT) was under the transcriptional control of the SIV LTR, CAT activity was dramatically stimulated when the SIV LTR/CAT construct was cotransfected with SIV *tat*, compared with transfection in the absence of *tat* (Viglianti and Mullins, 1988). Increases in SIV mRNA levels were observed as well. Increases in RNA and protein expression are hallmarks of HIV-1 and HIV-2 Tat-mediated transactivation (Arya, 1988; Jakobovits *et al.*, 1988).

SIV LTR/CAT is as fully responsive to transactivation by HIV-1 Tat as it is to transactivation by its autologous transactivator (Viglianti and Mullins, 1988; Arya, 1988). Yet, the converse is not true. HIV LTR/CAT is only partially transactivated by SIV Tat, compared with transactivation by HIV-1 Tat. Similar results have been observed for reciprocal transactivation analyses with the HIV-1 and HIV-2 Tat and LTR/CAT plasmids (Arya, 1988). These results lead to the suggestion that the HIV-1 and SIV/HIV-2 Tat proteins function by similar, although not identical, mechanisms. This conclusion is strengthened by observations using deletion mutations of HIV-1 and SIV *tat*. *tat* transcripts are the product of complex splicing events, and the coding region is contained within two exons (discussed below) (Unger *et al.*, 1991; Viglianti *et al.*, 1990; Colombini *et al.*, 1989; Robert-Guroff *et al.*, 1990). HIV-1 Tat deleted of the second exon coding region is as fully competent in transactivation, as is the full-length construct. SIV Tat deleted of the second exon transactivates the SIV LTR/CAT at a level only 5% of that observed with full-length SIV Tat (Viglianti and Mullins, 1988). Thus, there is a functional distinction between the SIV_{mac} and HIV-1 Tat proteins.

In HIV-1 Tat there is a precise arrangement and position of amino acids involving three to four basic amino acids flanking an original residue, which are critical for RNA binding in the bulge region of TAR and for transactivation (Tao and Frankel). Significantly, one element critical for electrostatic interactions is absent in SIV Tat, and this absence accounts for differences in transactivation between HIV and SIV Tat (Tao and Frankel, 1993).

The importance of Tat in the infection cycle of SIV_{agm} may be lower than in the other SIV/HIV isolates. Basal promoter strength of four SIV_{agm} isolates was compared with that of HIV-1, HIV-2, SIV_{mac}, and SIV_{mnd} in transient CAT expression assays (Sakuragi *et al.*, 1991). CAT activity, in the absence of any viral transactivator, was substantially higher for the four SIV_{agm} LTR than for the other LTR when assayed in various human and monkey cell lines. When the level of transactivation by Tat was assayed, the SIV_{agm} LTR were less responsive to transactivation by Tat, autologous or heterologous, than the HIV-1, HIV-2, and SIV_{mac} LTR. The SIV_{mnd} LTR also had a low response to Tat transactivation, but unlike SIV_{agm} LTR, had a low basal level of expression. These results suggest that as SIV_{agm} evolved, its replicative cycle became less dependent on transactivation by Tat (Sakuragi *et al.*, 1991). Part of this may have been compensated by a higher basal promoter activity. Although the proposed secondary structure of SIV_{agm} TAR differs from those found in other LTR, it is not known if this structure confers the apparent difference in gene regulation.

F. Rev

SIV Rev are small proteins (84–107 amino acids), slightly smaller than HIV-1 Rev proteins (115–118 aa). Like Tat, the Rev coding region is derived from two exons. Within the amino half of Rev, there is approximately 50% amino acid conservation between SIV_{smm}, SIV_{mac}, SIV_{agm}, HIV-2, and HIV-1. In the carboxyl half, there is less than 10% amino acid conservation (Cheng *et al.*, 1990a,b). As with HIV-1 and HIV-2 (Shibata *et al.*, 1990), Rev is absolutely essential for virus production. SIV_{agm} virus deleted of Rev function failed to produce detectable levels of virus after transfection of mutated proviral DNA into human cell lines (Shibata *et al.*, 1990a,b). Infection of susceptible $CD4^+$ cells with cell-free supernatant from these same transfected cells also failed to give rise to either a cytopathic effect or detectable reverse transcriptase activity.

Control of SIV/HIV expression is a complex process of transcriptional and posttranscriptional regulation that still requires resolution. The Rev protein of HIV-1 appears to be the key component in shifting the pattern of viral protein expression from nonstructural to structural (Malim *et al.*, 1989a,b). Rev functions through a *cis*-acting element, the Rev responsive element (RRE) (discussed below), in the Env transcript, to facilitate transport of unspliced transcripts from the nucleus to the cytoplasm (Malim *et al.*, 1989b). These unspliced transcripts code for the viral structural proteins.

In transient expression assays, SIV Rev facilitated the expression of unspliced Env message and Env protein. Cheng *et al.* (1990a,b) used an expression plasmid with an SIV_{smm} subgenomic fragment containing either the Env coding region, flanked by both exons of Rev (rev^+ env^+), or the Env region and only the second exon of Rev (rev^- env^+), to assay Rev function. Only when Rev was expressed were unspliced *env*-containing message and Env protein synthesized. Similar to studies with SIV Tat, SIV and HIV-1 Rev function via similar, but not identical, mechanisms. HIV-1 Rev mutations could not be complemented by Rev from HIV-2, SIV_{agm}, or SIV_{mnd} (Sakai *et al.*, 1990, 1991). Conversely, mutations in HIV-2 and SIV *rev* genes could be partially complemented by HIV-1 *rev*. An SIV *rev* mutant was complemented with high efficiency by HTLV-1 *rex* (Krohn *et al.*, 1993). In addition, SIV_{mnd} Rev mutations could not be complemented by either SIV_{agm} or HIV-2 Rev, whereas SIV_{agm} Rev mutations could be partially complemented by SIV_{mnd} and HIV-2 Rev. HIV-1 Rev acts in a sequence-specific manner through the RRE, which can potentially form a very stable RNA structure (Malim *et al.*, 1989b). The RRE of HIV-1, HIV-2, SIV_{agm}, and SIV_{mnd} all are capable of forming highly structured motifs (Dillon *et al.*, 1990; Sakai *et al.*, 1991). All structures, however, are distinct from each other. The mechanisms underlying the differential abilities of *rev* complementation await further investigation.

G. Nef

The SIV *nef* gene codes for an approximately 29-kDa protein, the function of which remains controversial. The *nef* gene product is not essential for viral replication *in vitro*, but the *nef* gene is conserved in all isolates of SIV, HIV-1, and HIV-2. The role of Nef with regard to viral replication *in vitro* is in some dispute. Some reports point to Nef acting as a negative regulator of viral transcription and replication in tissue culture cells, while other reports fail to note such a regulatory capacity for Nef. Niederman *et al.* (1991) and Binninger *et al.* (1991) observed that proviral clones of SIV_{mac} with mutations in *nef* (insertions, deletions, and premature stop codons) replicated with faster kinetics than the unmutated parental clone. This change included increased RNA and protein expression. These reports are consistent with some reports for HIV-1 (Luciw *et al.*, 1987; Niederman *et al.*, 1989, 1992). For HIV-1 there is evidence that virus isolated from patients at an advanced stage of disease is more refractory to downregulation by Nef than virus isolated from individuals at asymptomatic stages (Cheng-Mayer *et al.*, 1989).

Other reports fail to note any downregulation by SIV Nef. Elimination of Nef function in molecular clones of SIV_{mac} (Kestler *et al.*, 1991) and SIV_{agm} (Shibata *et al.*, 1990a,b) does not impair viral replication, compared with the parental, wild-type provirus. Several infectious molecular clones of SIV_{mac} and HIV-2 have mutated *nef* genes (Chakrabarti *et al.*, 1987; Kestler *et al.*, 1988; Naidu *et al.*, 1988). *In vitro*, at least, Nef function is not required for viral replication. There is, however, a strong selective pressure for Nef function in infected macaques and SIV Nef is essential for pathogenesis (Kestler *et al.*, 1991) (Section XIII.C).

SIV_{mac} Nef is a 29.5-kDa protein with only 38% amino acid homology to HIV-1 (Niederman *et al.*, 1991). SIV and HIV-2 *nef* genes code for larger Nef proteins than HIV-1. Like HIV-1 Nef, SIV_{mac} Nef may be myristoylated and retains a potential nucleotide binding domain, found in some of the membrane-associated G-proteins. Antibodies against Nef can be detected in approximately 70% of SIV_{mac}-infected macaques (Kirchhoff *et al.*, 1991). The antibody response to Nef does not correlate with disease or viremia (Kirchhoff *et al.*, 1991). Comparison of the amino acid sequences of HIV-1, HIV-2, and SIV_{mac} Nef proteins suggests a possible transcriptional role for Nef. Conserved within the *nef* sequences are leucine zipperlike motifs that lack a DNA binding domain and a carboxyl-terminal acidic region contained within two α-helices separated by a β-turn (Samuel *et al.*, 1991). Nef proteins apparently belong to a class of acidic, non-DNA-binding, leucine zipper transcription factors. This conclusion awaits further resolution since a role in transcription seems at variance with a cytoplasmic location identified for HIV-1 Nef (Franchini *et al.*, 1986; Hammes *et al.*, 1989).

H. Vif

The function of the *vif* gene product remains to be elucidated, particularly for SIV. Work with HIV-1 indicated that Vif is not required for viral assembly, but is essential for infectivity and transmission of HIV-1 (Fisher *et al.*, 1987; Strebel *et al.*, 1987). Proviral clones deficient in Vif are 1000-fold less infectious than wild-type virus. SIV_{mac} clones deleted of *vif* also appear to have the same phenotype. *vif*-deficient virus fails to infect rhesus PBMC and some human T-cell lines (Shibata *et al.*, 1990a,b). Yet, some established human cell lines can support viral replication of this mutant, suggesting that Vif is involved in the earliest stages of viral infection in a cell-type-dependent manner. In contrast, an SIV_{agm} clone deleted of Vif function was as fully infectious and cytopathogenic as the parental control virus (Shibata *et al.*, 1990a,b). This clone was assayed in only one established human T-cell line, and it remains to be determined whether full infectivity would be observed in other cell lines and in primary lymphoid cell cultures.

SIV_{mac} and SIV_{smm} Vif, whose amino coding regions overlap with the Pol coding regions, retain a significant degree of amino acid homology to each other and to HIV-2 (Myers *et al.*, 1992), particularly in the amino-terminus. There is less homology to SIV_{agm} and HIV-1 Vif.

I. Vpx

The *vpx* gene is unique to SIV and HIV-2, although the SIV_{agm} Vpx proteins may, in fact, be closer in sequence to HIV-1 Vpr protein than the Vpx protein of the other SIV and HIV-2 isolates (discussed below) (W. Hu *et al.*, 1989; Happer *et al.*, 1988, 1991; Tristem *et al.*, 1992). The SIV_{mac} *vpx* gene codes for a 12- to 16-kDa virion-associated protein (Yu *et al.*, 1991) which is immunogenic in experimentally infected rhesus macaques (Yu *et al.*, 1988). SIV_{mac} Vpx is a 14-kDa protein which is antigenic in experimentally infected macaques (Henderson *et al.*, 1988b). The Vpx protein is packaged in mature virions (Franchini *et al.*, 1988;

Henderson *et al.*, 1988b; Kappes *et al.*, 1993) and is localized primarily outside the virus core (Yu *et al.*, 1993). The molar ratio of Vpx in mature virions is equal to that of the Gag proteins (Henderson *et al.*, 1988b). The function of Vpx remains unknown. Vpx binds nucleic acids *in vitro* (Yu *et al.*, 1991; Henderson *et al.*, 1988b). Vpx function is dispensable for virus growth in some established human T-cell lines; replication kinetics of *vpx*-deficient virus is virtually indistinguishable from parental, wild-type virus (Yu *et al.*, 1991). This finding is similar to studies on HIV-2 Vpx (Yu *et al.*, 1991). Yet, in primary human macrophage cultures and in peripheral blood lymphocyte cultures freshly isolated from some donors, lack of Vpx can dramatically reduce viral replication, compared with the wild-type control (Yu *et al.*, 1991).

SIV_{mac} and SIV_{smm} Vpx, whose amino coding regions overlap with the Vif coding regions, retain a significant degree of amino acid homology to each other and to HIV-2 (Myers and Pavlakis, 1992). Less homology is observed to SIV_{agm} and HIV-1 Vpx.

J. Vpr

Like *vpx*, the SIV *vpr* gene product is a small 11-kDa protein, 97–101 amino acids, which is immunogenic in experimentally SIV-infected rhesus macaques (Yu *et al.*, 1990). There is a significant degree of homology between the *vpr* genes of SIV, HIV-2, and HIV-1. A recent analysis of Vpx and Vpr amino acid sequences from the SIV/HIV-2 isolates concluded that the two genes, which invariably are arrayed in tandem in the SIV/HIV-2 genomes, probably arose as a result of a duplication (Tristem *et al.*, 1992). The only exception is SIV_{agm} which has only one copy of the gene. Further, sequence alignments point to the hypothesis that all *vpx* and *vpr* genes arose from a single, ancestral gene; both genes retain sufficient, common amino acid residues to indicate an evolutionary relationship (Tristem *et al.*, 1990). The alignments also indicate that the SIV_{agm} *vpx* gene is, in fact, more similar to the *vpr* group of proteins than to the *vpx* group of proteins.

The function of Vpr remains to be elucidated. Like Vpx, Vpr is a virion-associated protein, the presence of which is not dependent on Vpx for packaging (Yu *et al.*, 1990). Vpr is localized within virions of Vpx virus. Similarly, Vpx is present in virions independent of the presence of Vpr. Mutated proviral clones deficient in *vpr* replicate in established human T-cell lines and in human PBL and enriched monocyte cultures with similar kinetics to the parental *vpr*$^+$ clone (Yu *et al.*, 1990). Vpr expression may be a requirement for *in vivo* pathogenesis (Lang *et al.*, 1993) (Section XIII.C).

X. SIV IN BODY FLUID AND MODES OF SIV TRANSMISSION

The nature of SIV in body secretions and the mode of transmission of SIV in the natural African monkey host, apart from doctors' needles, remain largely unknown. In experimentally infected macaques, SIV_{mac} has been cultured from vaginal secretions, milk, and saliva (C. J. Miller *et al.*, 1990; Mossman *et al.*, 1992). Studies comparing the amount of cell-free or cell-associated SIV in various

TABLE V. Body Fluids of Infected Macaques from Which SIV Has Been Isolated[a]

Fluid	References
Semen	Miller *et al.* (1989)
Vaginal secretions	Miller *et al.* (1989), Mossman *et al.* (1992)
Milk	McClure *et al.* (1991)
Saliva	Mossman *et al.* (1992)

[a] Virus isolations were generally only sporadically positive and insufficient data exist to compare the frequency or amount of infectious virus in cell-free or cell-associated form.

secretions from the natural or experimental simian host are lacking. Body fluids of infected macaques from which SIV has been isolated are listed in Table V.

In a breeding colony of sooty mangabeys at the Yerkes Primate Center, most of the animals seroconvert when they are older juveniles or young adults, suggesting, but by no means proving, that the primary mode of transmission may be sexual (Fultz *et al.*, 1990b). Among 138 adult animals tested, 57% were infected with SIV_{sm} based on virus isolation from PBMC. Female mangabeys were more commonly infected than males (66% versus 42%). About one-third of these animals were also infected with STLV-1. About 40% of the young mangabeys (<5 years of age) were seronegative for SIV_{sm} (McClure *et al.*, 1991). It is also possible that some young animals are latently infected and do not generate antibodies until later in life after virus activation. Indeed, more sensitive virologic assays (i.e., blood transfusion, PCR, secretion of SIV antibodies by PBMC stimulated *in vitro* with pokeweed mitogen) suggest that most, if not all, of the seronegative sooty mangabeys from the Yerkes colony may in fact be infected with SIV_{sm} and that the virus is transmitted to infants by other than sexual means (Villinger *et al.*, 1991; Jehuda-Cohen *et al.*, 1991a,b). However, these results remain to be confirmed. Evidence for sexual transmission of SIV was not observed in a semi-free-ranging group of mandrills (Cooper *et al.*, 1989).

SIV can be transmitted to female rhesus macaques by placing cell-free virus into the vaginal canal (C. J. Miller *et al.*, 1989, 1990) and SIV in both cell-free and cell-associated form can be isolated sporadically from the vaginal secretions of infected rhesus (C. J. Miller *et al.*, 1990). The source of SIV in the secretions is probably from SIV-infected cells located in the submucosa of the cervix and vagina and in the vaginal epithelium (C. J. Miller *et al.*, 1992a; Mossman *et al.*, 1992). In male macaques, SIV_{mac}-infected cells can be found at all levels of the reproductive tract and SIV can infect cells in the mucosal epithelium of the foreskin (C. J. Miller *et al.*, in press). The use of this SIV model for vaccine development aimed at genital mucosal immunity is covered in Section XIV.

Yet to be evaluated in the natural African simian host is the transmission of SIV from mother to infant transplacentally, via vaginal secretions and milk, or the horizontal, nonsexual transmission among adults. Human and rhesus trophoblast cells are susceptible to transient active infection *in vitro* with HIV-1 and SIV_{mac251} or SIV_{smB670}, respectively (Douglas *et al.*, 1991; Krugner-Higby *et al.*, 1993). A single episode of nonsexual SIV transmission, probably by physical contact, from an adult female African green monkey to an unrelated infant was observed at the Kenya primate center (Otsyula *et al.*, 1991). In summary, horizontal mother-to-infant and sexual transmission are thought to occur rather inefficiently in the natural African monkey hosts, but there is very little direct evidence for the natural routes of SIV transmission.

More data are available on SIV transmission in experimentally infected macaques. Recent evidence supports the transmission of SIV through the milk or vaginal secretions (McClure *et al.*, 1991; Mossman *et al.*, 1992) of experimentally infected rhesus macaques and suggests that SIV may be spread by the contamination of wounds or mucous membranes (including the conjunctiva) by infectious blood and saliva during fighting and biting (Jennings *et al.*, 1992; McClure *et al.*, 1991). Viremia is present soon after infection in most macaques, and a low level of virus is detectable in the saliva of some infected monkeys (Mossman *et al.*, 1992). It has been possible to document only an infrequent example of heterosexual transmission among SIV discordant mating pairs of macaques, e.g., transmission of SIV_{mne} from female to male *M. nemestrina* (Kuller *et al.*, 1992). Among SIV-infected macaques the most efficient route of natural virus transmission has been by biting and fighting among young adult males caged together (Jennings *et al.*, 1992). This most likely represents viral transmission by infected blood or saliva across oral mucosa. Insertive homosexual activity has not been observed among these animals.

Congenital transmission of SIV in experimentally infected macaques has been observed in about 25% of at risk pregnancies, similar to the prevalence of maternal–infant transmission of HIV-1 in humans. *In utero* transmission of SIV from an infected female macaque to her offspring has been reported (Daniel *et al.*, 1988a; Ochs *et al.*, 1991; Mossman *et al.*, 1992). Transmission *in utero* is thought to usually occur late in pregnancy associated with disruption of the placental barrier (Davison-Fairburn *et al.*, 1992). Prenatal infection of macaques and their mothers can be induced quite efficiently by direct inoculation of the fetuses or amniotic fluid during late gestation (Fazely *et al.*, 1993; Ochs *et al.*, 1992; Tarantal *et al.*, in press). Inoculation of fetuses early in gestation leads to a high fetal loss rate ($\approx$50%). After injecting SIV_{mac} into the amniotic fluid during late gestation, six of seven rhesus monkeys were born infected (Fazely *et al.*, 1993). All infected neonates showed signs of disease and, in contrast to their mothers, had absent CTL responses. Congenitally infected infants seroconvert at about 9–12 months of age, after waning of the passively transferred maternal antibody, and usually die shortly thereafter from fulminant simian AIDS. Maternally transmitted SIV apparently selects for a single variant found in the "swarm" of genetically distinct variants in the respective mothers (Martin-Amedee *et al.*, 1993).

The possibility of SIV transmission to two laboratory workers has recently been supported by the seroconversion to SIV of these individuals, neither of whom was at risk of infection with HIV-1 or HIV-2. One worker became infected from a needle stick about 2 years ago when collecting blood from an SIV-infected macaque (Khabbaz *et al.*, 1992a,b). Declining antibody titers suggest no persistent infection and virus could not be isolated from PBMC. The second worker became infected by handling SIV_{smB670}-infected samples without gloves while affected by active dermatitis. This person has increasing antibody titers, suggesting an ongoing SIV infection, and the virus (SIV_{smB670}) has recently been isolated from PBMC (Khabbaz *et al.*, 1994). A retrospective survey of laboratory workers and animal caretakers revealed SIV (or HIV-2) antibodies in 3 of 462 tested (Centers for Disease Control, 1992). Because this survey was anonymous, it is possible that one or two of the infected workers may be those previously reported. Further testing is planned for SIV seropositivity in laboratory workers involved in SIV research. The results prove that the cross-species transmission of SIV to humans is possible. In the absence, so far, of SIV isolation from the other two individuals, the possibility remains that they may have been only immunized with noninfectious SIV antigens or perhaps transiently infected.

The ability to infect macaques by injection of 200 μg of a lambda phage clone of SIV_{mac} (Letvin *et al.*, 1991) suggests that there is also a finite risk to laboratory workers handling plasmids containing SIV or other retroviral genomes (Portis *et al.*, 1992). The finding that SIV RNA could be pseudotyped with amphotropic MuLV or HIV-1 *in vitro* suggests that coinfection of cells with SIV and other retroviruses could lead to infection of new cell types in nature (Rizvi and Panganiban, 1992, 1993).

XI. HOST RANGE: VIRUS–CELL INTERACTIONS *IN VITRO*

The host range of SIV and HIV is limited to human and primate cells expressing the CD4 antigen as a receptor. None of the nonprimate cell lines tested have been permissive for SIV or HIV. This restriction in host range holds true for both biologic isolates of SIV and derived molecular clones. Nonprimate cells engineered to express the human CD4 molecule are able to bind recombinant gp120 of HIV-1, but are still resistant to infection with HIV-1 and SIV (Maddon *et al.*, 1988; Clapham *et al.*, 1991). CD4 expression on various human, rhesus, and AGM cell lines confers differential susceptibilities for HIV-1, HIV-2, and SIV strains, a property determined at the cell surface (Chapel *et al.*, 1990; Clapham *et al.*, 1991). The results suggest that events between virion attachment to CD4 and gp41-mediated membrane fusion are complex and may require surface components additional to CD4, a conclusion supported by SIV_{mac} membrane lipid mixing experiments with $CD4^+$ and $CD4^-$ human cell lines and susceptibility to infection (Statamatos and Duzgunes, 1993). *N*-linked oligosaccharides of SIV envelope gp140 are not required for the gp–CD4 interaction (Benjovad *et al.*, 1993). Recently, a serine protease molecule (CD26) present on activated T cells was reported to serve as a co-factor for entry of HIV in $CD4^+$ cells (Callebaut *et al.*, 1993).

A. SIV_{mac}

In vitro growth characteristics of the prototype SIV_{mac251} biologic clone in rhesus and human PBL (Kannagi *et al.*, 1985) and activation of MHC class II antigens in infected cells *in vitro* (Kannagi *et al.*, 1987) were mentioned in Section IV. The growth kinetics of SIV_{mac251} was later analyzed in long-term cultures of macaque PBMC infected *in vitro* (Montaner *et al.*, 1990). Viral replication was noted by *in situ* hybridization after 2 days, followed by detection of core protein (p27) by immunocytochemistry on day 4, reverse transcriptase activity on day 8, and budding particles by electron microscopy on day 14. Two bursts of viral replication occurred, the first at 3 weeks postinfection due to infection and subsequent depletion of $CD4^+$ lymphocytes, and the second 3 weeks later, due to infection of $CD8^+$ lymphocytes and remaining $CD4^+$ cells, resulting in the death of all cells by 39 days postinfection. In its infection of $CD8^+$ cells, SIV_{mac} differs from HIV-1 and HIV-2 strains (Dalgleish *et al.*, 1984; Klatzman *et al.*, 1984).

It was also shown that growth of SIV_{mac251} in human cells selected for mutant viral genomes encoding a truncated transmembrane protein (TMP) (gp32 instead of gp41) and possible loss of virulence (Kodama *et al.*, 1989b; Cranage *et al.*, 1989; Hirsch *et al.*, 1989a). This truncation reproducedly reverted to the full-length viral genome on passage in rhesus monkeys. Studies of HIV-1 particles pseudotyped with envelope glycoproteins of SIV_{mac} molecular clones (Section XIII.C) indicated that truncation of the cytoplasmic domain results in increased viral entry into human T cells (Zingler and Littman, 1993). By using site-directed mutagenesis, it was determined that truncation of the cytoplasmic domain of the TMP of SIV_{mac239} remains a block at the level of cell entry by altering the glycoprotein density of the viral envelope (Johnston *et al.*, 1993).

SIV_{mac251} had a restricted host range among human $CD4^+$ cells compared with HIV-1 or HIV-2 (Koenig *et al.*, 1989). This restricted tropism was associated with the inability of SIV_{mac} envelope to induce membrane fusion in the nonsusceptible cells. These findings suggested that additional steps in the viral entry pathway must occur after CD4 binding. Treatment of SIV_{mac}-infected T cells (HUT-78) with TNF-α, IFN-γ, IL-2, and GM-CSF increased virus production slightly (two- to threefold) based on increased RT activity (Lairmore *et al.*, 1991). Multiple SIV mRNA species were detected in an infected $CD4^+$ lymphoid cell line (Park *et al.*, 1991).

Infection of rhesus immature bone marrow cells with SIV_{mac251} *in vitro* led to a decreased clonogenic production that might help account for the hematological and immunological abnormalities associated with SIV infection (Beltz *et al.*, 1991). The bone marrow cell types permissive for SIV infection were not identified but were suspected to be in the monocyte/macrophage lineage. Another major *in vitro* finding from the same laboratory was the rapid lysis of $CD4^+$ T cells brought on by fusion with rhesus macrophages infected with SIV_{mac251} (McEntee *et al.*, 1991). Release of TNF-α and IL-6 by the $CD4^+$ cells during this fusion process could contribute to the degenerative and inflammatory lesions of SAIDS (McEntee *et al.*, 1992a). Neutralizing antibodies had the potential to limit the replication of SIV in cultured macrophages if given soon enough after infection (McEntee *et al.*, 1992b).

In related studies, SIV_{mac251} was readily isolated from bone marrow cells of infected monkeys and was shown to be harbored in macrophages rather than T lymphocytes (Watanabe *et al.*, 1990). A striking *in vivo* correlation was also shown between increased SIV_{mac} load in bone marrow cells and decreased hematopoietic progenitor cell colony growth. Inhibition of SIV_{mac} replication in bone marrow macrophages by treatment with anti-CD4 monoclonal antibody resulted in increased progenitor cell colony growth.

Deletion mutants of SIV_{mac251} lacking the integrase domains or the integrase and the *nef* gene were able to establish only a transient infection of permissive human T cells resulting in the expression of viral antigens from episomal viral DNA over a limited period of time (Prakash *et al.*, 1992; Vogel *et al.*, 1993). Such mutants might form the basis for an experimental live attenuated vaccine (see Section XIV.D).

Several groups have studied the effect of SIV_{mac} infection on macrophages *in vitro* using both biologic and molecular clones (see Section XIV.C). Productive infection of alveolar macrophages from SIV_{mac251}-infected animals was uncommon (Horvath *et al.*, 1991). Yet, alveolar macrophages from uninfected rhesus monkeys could be readily infected *in vitro* with uncloned SIV_{mac251} (Ringler *et al.*, 1989b). Using PCR the frequency of SIV DNA in alveolar and peripheral blood macrophages was highest early after infection and in the terminal stages of disease induced by uncloned SIV_{mac251} (Unger *et al.*, 1992). *In vitro* infection of alveolar macrophages with SIV_{mac251} did not cause a release of TNF-α or alter their response to lipopolysaccharide (Horvath *et al.*, 1991). However, TNF-α production was increased from cultured uninfected macrophages collected during the asymptomatic stage of SIV infection. Uninfected macrophages from animals with AIDS lost their capacity to produce TNF-α. Studies with macrophage-tropic and non-macrophage-tropic molecular clones of SIV_{mac} (Section XIV.C) have mapped the *in vitro* macrophage tropism to the *env* gene (Banapour *et al.*, 1991a,b; Mori *et al.*, 1992) and shown that macrophage tropism measured *in vitro* does not necessarily correlate with *in vivo* virulence (Banapour *et al.*, 1991a).

B. SIV_{mne}

Benveniste *et al.* (1990) described single-cell clones of SIV_{mne}-infected HUT-78 cells that produce large amounts of either the envelope glycoprotein gp120 or the nucleic acid binding Gag protein. These clones may be useful for purification of these retroviral proteins, in developing assays of protease inhibitors, and in preparing SIV envelope vaccines.

When added simultaneously with SIV_{mne} to uninfected macaque PBMC or thymocytes *in vitro*, superantigens induced increases in SIV DNA (Axberg *et al.*, 1993). In contrast, when added to T cells already infected with SIV, the superantigens reduced the level of viral DNA, suggesting deletion or anergy of SIV-infected lymphocytes. Thus, the effect of superantigen exposure on SIV expression may differ depending on whether the superantigen is present during initial infection of T cells or encounters T cells already infected with SIV.

C. SIV_{agm}

The initial $SIV_{agm\text{-}TYO\text{-}1}$ strain was shown to be tropic and cytopathic for certain human cell lines (Molt-4 clone 8, MT-2) that were $CD4^+$ (Ohta *et al.*, 1988). $CD4^-$ cell lines (Molt-3, Raji) could not be infected. Giant cell formation was observed in the susceptible cells as soon as 4 days after inoculation and reverse transcriptase (RT) activity in the supernatant was maximal 2 weeks after inoculation. The same monoclonal antibodies to CD4 (e.g., OKT_{4a}) that blocked infection of Molt-4 clone 8 T cells by HIV-1 were also able to block SIV_{agm} infection, suggesting that SIV_{agm} uses the same CD4 epitopes as HIV-1 for binding and that binding is required for SIV_{agm} infection. $SIV_{agm\text{-}TYO\text{-}1}$ could also replicate in fresh PBMC from uninfected AGM despite the inability of monoclonal antibodies against human CD4 antigen to react with most of these cells. This finding suggested that some molecules other than CD4 might act as surface receptors for SIV_{agm} in AGM lymphocytes. The SIV_{agm385} isolate also grew in some but not other $CD4^+$ cells (Daniel *et al.*, 1988b). Initial isolation of this SIV_{agm} strain depended on virus growth in the highly susceptible Molt-4 clone 8 cell line. Once isolated, SIV_{agm385} replicated efficiently in the SupT1, MT4, U937, and Jurkat T-cell lines but much less efficiently, if at all, in HUT-78, CEM, and uncloned Molt-4 cells. Negative results were obtained with normal PBMC from seronegative humans or AGM. The range of $CD4^+$ cells fully permissive for replication of SIV_{agm} isolates thus differed markedly from that of previous SIV isolates from macaques or HIV-1 and HIV-2. SIV_{agm} (Kraus *et al.*, 1989) was able to infect both $CD4^+$ and $CD4^-$ human cell lines indicating a second pathway of cell infection, possibly mediated by a cellular receptor distinct from the CD4 molecule (A. Werner *et al.*, 1990a). Cloning and sequencing of primate CD4 molecules revealed that they are about 90% identical to the human CD4 sequence except for the chimpanzee (98%) (Fomsgaard *et al.*, 1992). A low degree of sequence variation was found in individual animals within the same species.

Another remarkable difference between the cell tropism of SIV_{agm} versus SIV_{mac}, HIV-1, and HIV-2 was the effect of soluble CD4 on *in vitro* infection. Two independent studies with different isolates of SIV_{agm} showed that, whereas recombinant soluble human CD4 protein (rCD4) could block *in vitro* infection with HIV-1, HIV-2, and SIV_{mac}, the same rCD4 molecules markedly enhanced SIV_{agm} infection and subsequent cell fusion of human Molt-4 clone 8 cells (Allan *et al.*, 1990; A. Werner *et al.*, 1990b). This enhancing effect could be inhibited, i.e., neutralized, by CD4 monoclonal antibodies, by a monoclonal antibody mapping to the V3-like region of SIV_{agm} and by nonneutralizing antibodies from naturally infected AGM. It was suggested that this enhancement of SIV_{agm} infection by rCD4 resulted from the modulation, i.e., conformational change, of the viral membrane induced through gp120–CD4 binding leading to virus fusion and secondary activation of other viral binding sites or cell receptor molecules. A strong association of SIV_{agm} envelope glycoprotein heterodimers might contribute to this receptor-mediated activation (Allan *et al.*, 1992b). The restricted ability of SIV_{agm} to infect only some $CD4^+$ cell types and the evidence that HIV-1 can infect some $CD4^-$ cells are consistent with the involvement of other putative cell surface receptor molecules. The finding that nonneutralizing and V3 region antibodies were able to neutralize SIV_{agm} after exposure to rCD4 suggests that the CD4-induced conformational changes expose a cryptic neutralizing epitope in the V3 region that might be exploited for novel vaccine and therapeutic strategies. CD4-induced conformational changes showed similarities and differences between HIV-1, HIV-2, and SIV (Sattenau *et al.*, 1993).

$SIV_{smPBj14}$-induced lymphocyte proliferation (Fultz *et al.*, 1989) is described in Section V. Selective replication of SIV_{mne} in certain subsets of $CD4^+$ cells has been reported (Gallatin *et al.*, 1989).

XII. DETECTION OF SIV

Many of the same assay systems used to detect HIV have also been successfully employed for SIV. These include tests for the presence of viral antigens (ELISA, immunoblot, RIPA, immunofluorescence, and immunochemistry), viral nucleic acids (*in situ* hybridization and PCR), viral enzymatic functions (RT assay), and observable biological phenotypes such as cytopathic effect, syncytium formation, or the presence of viral particles (coculture experiments, plaque assays, and electron microscopy).

A. Assays for Viral Antigen

The common feature of these assays is the use of antiviral antibodies coupled with a means of visualizing either the antigen (e.g., RIPA) or the antigen–antibody complex (e.g., ELISA, immunoblot, immunofluorescence, and immunohistochemistry) (see also Chapter 8).

1. ELISA

The antigen-capture ELISA provides a convenient means of determining the amount of antigenemia in infected animals as well as quantifying the amount of cell-free virus released by infected cells *in vitro*. Although commercially available HIV-1 antigen-capture assays are capable of detecting SIV (Goldstein *et al.*, 1990), recently developed ELISAs, which employ monoclonal antibodies generated against SIV antigens, demonstrate greater sensitivities (Lairmore *et al.*, 1990, 1993; Lohman *et al.*, 1991; Thorstensson *et al.*, 1991). Capable of detecting as little as 100 μg/ml of viral p27 protein, these assays are more sensitive than RT assays (Lairmore *et al.*, 1990; Thorstensson *et al.*, 1991). The choice of antibody employed in the antigen-capture system determines the specificity of the assay. Utilizing antibodies with broad specificities allows for the detection of HIV-1, HIV-2, and SIV (Otteken *et al.*, 1992), while type-specific antibodies can be used to distinguish different SIV strains (Kodama *et al.*, 1991; Minassian *et al.*, 1988; Norrby *et al.*, 1989; Sutjipto *et al.*, 1990a; Kent *et al.*, 1991, 1992; Higgins *et al.*, 1992). Sera from monkeys infected with SIV may contain anti-gag antibodies that interfere with antigen-detection ELISA and thus give false-positive neutralization results (Burns and Desrosiers, 1992). However, these antibodies can almost always be removed by proper washing procedures (Albert *et al.*, 1993).

ELISAs can also be adapted for serosurveys by providing SIV antigen and screening for the presence of reactive antibodies in sera (Yee *et al.*, 1991). By using peptides corresponding to segments of SIV envelope as a source of antigen, immunogenetic epitopes within SIV_{mac} (Johnson *et al.*, 1988; Norrby *et al.*, 1991), SIV_{agm} (M. A. Miller *et al.*, 1992; Siegel *et al.*, 1992), SIV_{sm} (Johnson *et al.*, 1988; Norrby *et al.*, 1991; Samuelson *et al.*, 1993), and SIV_{mne} (Shafferman *et al.*, 1989a) have been identified. This approach to fine-mapping antibody responses has been termed site-directed serology.

2. Immunoblot

Immunoblot (or Western blot) assays, like ELISAs, have been used to identify viral antigens as well as antiviral antibodies. One major advantage of the immunoblot assay, however, is that it separates viral antigens according to their molecular weights by SDS-PAGE. Peeters and co-workers have recently used immunoblot analysis to identify a new SIV_{cpz} strain ($SIV_{cpz\text{-}ant}$) based on the unique molecular weight pattern of its viral proteins (Peeters *et al.*, 1992). The immunoblot assay has also been extensively used to determine the specificity of antiviral monoclonal antibodies because the molecular weight of the antigen recognized by a particular antibody can be easily calculated.

Immunoblots are commonly used in serosurveys in conjunction with other tests such as ELISAs or RIPAs. In one such study, a significant percentage of rhesus (5%) and human (38%) sera gave false-positive reactivity to SIV p27 by immunoblot (Lairmore *et al.*, 1990). Therefore, immunoblots should be interpreted with caution when screening human sera for anti-SIV antibodies.

3. RIPA

RIPA shares some features with the immunoblot assay in that it allows for viral antigens to be identified with respect to molecular weight and it can be adapted to detect antibodies which show reactivity to viral antigens. One unique feature of RIPA is that it allows interactions between viral antigens and other proteins to be analyzed since antibody reacts with viral antigen in a more "native" configuration. For example, RIPAs have been used to establish the multimeric nature of viral envelope proteins (Rey *et al.*, 1989). The cumbersome format of RIPA, however, has prevented its use in routine diagnostic procedures.

B. Assays for Viral RNA/DNA

1. *In Situ* Hybridization

In situ hybridization enables specific viral RNA or DNA sequences to be detected within individual cells. This technique coupled with immunohistochemical labeling using cell-type-specific markers allows for the identification of (1) the relative number of infected cells producing viral message, (2) the location of these cells, and (3) their cell type. Understandably, this approach has been used extensively in studies of SIV pathogenesis. Numerous infected tissues have been examined including brain (Chakrabarti *et al.*, 1991; Lackner *et al.*, 1991b; Hurtrel *et al.*, 1991, 1993; Beilke *et al.*, 1991), bone marrow (Kitagawa *et al.*, 1991; Sharma *et al.*, 1992, 1993b), lungs (Baskin *et al.*, 1991a; Chalifoux *et al.*, 1992), genital tract (C. J. Miller *et al.*, 1992c), liver (Gerber *et al.*, 1991), thymus (Baskin *et al.*, 1991b), lymph nodes (Montaner *et al.*, 1990; Sharma *et al.*, 1992a; Hartung *et al.*, 1992a), and gastrointestinal tract (Heise *et al.*, 1993). In addition, *in situ* hybridization has been used to examine at what level viral replication is restricted either *in vitro* (McEntee *et al.*, 1992b) or *in vivo* (Hurtrel *et al.*, 1993). While it has been demonstrated that *in situ* hybridization is more sensitive in detecting virus than are assays based on the recognition of viral antigen (Kitagawa *et al.*, 1991; Montaner *et al.*, 1990), this technique is not suitable for routine diagnostic work.

2. PCR

PCR has proven to be one of the most sensitive means available for detecting SIV (Baskin *et al.*, 1991b; Burns and Desrosiers, 1990; Kitchin *et al.*, 1990; Conway *et al.*, 1991; Jehuda-Cohen *et al.*, 1991a) and with the use of nested primer pairs, problems with high background and false positives can be minimized (Grankvist *et al.*, 1992; Unger *et al.*, 1992a). In addition to detecting SIV, PCR has proven extremely useful in (1) delineating the step(s) at which viral replication is blocked in restricted cells (Agy *et al.*, 1991), (2) assessing viral sequence diversity (Almond *et al.*, 1992b,c), (3) cloning SIV sequences (Almond *et al.*, 1990; Khan *et al.*, 1991b; Dewhurst *et al.*, 1990; Saksena *et al.*, 1992), and (4) the generation of SIV mutants (Kestler *et al.*, 1991; Mori *et al.*, 1992; Morrison and Desrosiers, 1993) and (5) identifying uncharacterized retroviruses (Donehower *et al.*, 1990). PCR assay has been adapted to detect viral RNA by including a reverse transcription step, which converts viral RNA into DNA—the template recognized in PCR. RT-PCR has been demonstrated to be a rapid, cost-effective, and sensitive means for detecting immunodeficiency viruses from both PBMC (Bagnarelli *et al.*, 1991; Blackbourn *et al.*, 1992) and plasma (Aoki *et al.*, 1992; Bruisten *et al.*, 1993).

With recent refinements in PCR methodologies, this technique has emerged as a highly sensitive quantitative assay for HIV (Delord *et al.*, 1992; Menzo *et al.*, 1992; Seshamma *et al.*, 1992; Piatak *et al.*, 1993a,b; Arens, 1993) and will undoubtedly be used in a similar fashion for evaluating the extent of SIV infections.

C. Enzymatic and Biological Assays

1. RT Assay

The RT assay is another standard method for detecting and quantifying retroviruses, including SIV (Johnson *et al.*, 1990b; Potts, 1990). In addition, RT assays provide a means of screening potentially therapeutic RT inhibitors *in vitro* (Baba *et al.*, 1991). Over the past several years, refinements have improved the format of the RT assay, making it suitable for screening larger numbers of samples (Spira *et al.*, 1986; Somogyi *et al.*, 1990; Gronowitz *et al.*, 1991) and use without radioisotopic materials (Eberle and Seibl, 1992; Urabe *et al.*, 1992). However, certain limitations of the RT assay still remain. Vasudevachari and co-workers have demonstrated that mycoplasma contamination of cell cultures inhibits HIV-RT activity to the extent that virus detection was obscured (Vasudevachari *et al.*, 1990). Unlike other antigen-based assays discussed above, the RT assay cannot be adapted to distinguish HIV-1, HIV-2, and SIV in a type-specific manner. Finally, since this assay detects an enzymatic function (i.e., RT activity), it is not necessarily a good indicator of the amount of infectious virus present in a sample (Kiernan *et al.*, 1988).

2. Infectivity Assays

The most commonly used method for quantifying SIV infectivity is the $TCID_{50}$ assay (50% tissue culture infectious dose). In this assay, samples are subjected to serial dilution and cocultured with susceptible indicator cells. Cells are then monitored for signs of infection by either an antigen assay, an RT assay, or direct visualization of viral cytopathic effects. In $TCID_{50}$ assays for SIV, the most commonly used target cells are CEMX174, a hybrid human T-cell/B-cell line (Salter *et al.*, 1985).

Plaque assays are another means of titering infectious virus, and while a plaque assay for SIV has been reported (McKeating *et al.*, 1989), it is not commonly used.

XIII. SIMIAN AIDS IN SIV-INFECTED MACAQUES

A. Virus Strains, Macaque Species, and Experimental Transmission

In contrast to the natural African simian hosts in which SIV is nonpathogenic, the Asian macaques, which are not infected with SIV in the wild, are remarkably susceptible to persistent infection with certain strains of SIV, resulting in the fatal immunosuppressive disease called simian AIDS (SAIDS) (for review, see Desrosiers and Letvin, 1987; Gardner and Luciw, 1989; Letvin and King, 1990; Hirsch and Johnson, 1992). As discussed earlier in this chapter, SIV-induced SAIDS is an experimental disease resulting from the inadvertent or purposeful infection of macaques in captivity with closely related strains of SIV, i.e., SIV_{mac} (Daniel *et al.*, 1985), SIV_{mne} (Benveniste *et al.*, 1986), SIV_{stm} (Lowenstine *et al.*, 1992), SIV_{sm} (Fultz *et al.*, 1986; Lowenstine *et al.*, 1986; Murphey-Corb *et al.*, 1986), all thought to derive originally in primate centers from naturally infected sooty mangabey or closely related African monkey species (Hirsch *et al.*, 1989b).

Most SIV isolates from African green monkeys (Miura *et al.*, 1989; Johnson and Hirsch, 1992) or a West African sooty mangabey (Marx *et al.*, 1991), although capable of inducing persistent infection, have not as yet induced disease in macaques. However, a pathogenic strain of SIV_{agm} from a vervet monkey has recently been reported (Dapolito *et al.*, 1993). Results of *in vivo* transmission in macaques of SIV isolates from mandrills, Sykes monkeys, or chimpanzees have not as yet been reported. All species of macaques appear susceptible to infection and disease and age or sex are not critical factors. However, similar to pediatric HIV infection fetal or newborn macaques are more susceptible to a rapid disease course because of their inability to mount a strong antiviral immune response (Bohm *et al.*, 1993; Klumpp *et al.*, 1993; Van Rompay, 1992b). The most commonly used macaque species in primate centers are the rhesus (*M. mulatta*), cynomolgus or longtailed (*M. fascicularis*), stumptailed (*M. arctoides*), and pigtailed (*M. nemestrina*).

Virtually 100% of macaques infected with the pathogenic SIV strains (i.e., SIV_{mac}, SIV_{mne}, SIV_{stm}, SIV_{sm}) develop fatal SAIDS within 2 months to 3 years depending, in part, on the vigor of the antiviral immune response including neutralizing antibodies (Letvin *et al.*, 1985; Kannagi *et al.*, 1986; Zhang *et al.*, 1988; Putkonen *et al.*, 1989; Makoschey *et al.*, 1993). Although capable of inducing fatal AIDS in pigtailed and rhesus macaques (Benveniste *et al.*, 1988; Morton *et al.*, 1989), SIV_{mne} infection of pigtailed macaques appears to be somewhat less virulent, i.e., exhibits a longer incubation period, than SIV_{mac}, SIV_{stm}, or SIV_{sm} infection of rhesus macaques. Ten of eleven cynomolgus monkeys infected with SIV_{mne} died of AIDS 43 to 155 weeks postinfection (Kuller *et al.*, 1990). More recently, however, a titered dose (10 ID_{50}) of SIV_{mne} caused fatal AIDS in three of six cynomolgus macaques after only 29–42 weeks postchallenge (Tsai *et al.*, 1994). Rhesus macaques infected with $SIV_{mneE11s}$ could be superinfected with SIV_{mac239} but the level of superinfection was reduced according to the level of preexisting immunity from the infection with SIV_{mne} (Lewis *et al.*, 1993a).

The most widely used model is the infection of juvenile or young adult (2–5 years) rhesus macaques with the prototype SIV_{mac251} biologic isolate, after which the median life span is about 11–12 months. A similar mean survival period is seen in cynomolgus monkeys inoculated with SIV_{sm} (Putkonen *et al.*, 1989) or rhesus monkeys infected with SIV_{stm} (Lowenstine *et al.*, 1992). In addition to the host immune response, genetic variability in resistance to SIV at the intracellular level, yet to be defined, probably also accounts for individual variation in the relative susceptibility to disease. Selection of SIV variants *in vitro* and *in vivo* may alter the relative pathogenicity. Long-term culture of SIV_{mac} in human T-cell lines appears to attenuate its virulence, possibly because of the selection of less pathogenic variants or truncation of the transmembrane envelope protein that occurs under these conditions (Kodama *et al.*, 1989b; Hirsch *et al.*, 1989a). For maximum virulence, SIV_{mac} stocks are, therefore, grown for only a few passages in fresh human or rhesus PBMC.

Systemic infection and eventual SAIDS can be readily induced by inoculating macaques systemically (i.e., IV or IM) or mucosally, i.e., across the intact genital mucosa (C. J. Miller *et al.*, 1989, 1992a), rectal mucosa (Cranage *et al.*, 1992b; Le Grand *et al.*, 1993; Pauza *et al.*, 1993), or conjunctiva (Conway *et al.*, 1991), with uncloned stocks of virulent cell-free SIV. Standardized biological stocks of SIV_{mac} (Daniel *et al.*, 1985; Kitchin *et al.*, 1990; Carlson *et al.*, 1990), SIV_{mne} (Benveniste *et al.*, 1986), and SIV_{sm} (Fultz *et al.*, 1986; Hirsch *et al.*, 1991; Murphey-Corb *et al.*, 1990b) have been titered *in vitro* and *in vivo* and specific assay systems and reagents are now available at several U.S. and European primate centers (see Section X). However, neutralization assays are not yet standardized and variable results can be attributed to use of different virus stocks, cell lines, assay procedures, and end points.

Once the threshold of infection is reached, the incubation period before induction of disease and the timeframe or strength of humoral antibody responses are essentially dose-independent (Daniel *et al.*, 1987). In general, the susceptibility of macaques to infection by IV inoculation with most of the SIV_{mac} and SIV_{sm} uncloned stocks is about two logs more sensitive than the *in vitro* susceptibility of macaque or human indicator cells, i.e., one animal minimal infectious dose (one ID) is about two logs less than the minimal tissue culture infectious dose (TCID). However, with the SIV_{mne} E11S biologic clone grown in the highly susceptible AA-2 human indicator cell line, the 50% monkey infectious dose is 100× the cell culture infectious dose (Tsai *et al.*, 1994). The reduction in infectivity thus could reflect growth in the established cell line. The dose of SIV_{mac} or SIV_{mne} required to establish infection by the rectal or conjunctival mucosal route is one to two logs greater than that required for intravenous infection, but the establishment of systemic infection occurs as quickly by either route. However, virological and clinical signs may evolve more slowly in intrarectally inoculated monkeys in comparison with IV infection controls and molecular sequencing studies suggest that less pathogenic variants with unique envelope and LTR sequences may be selected during rectal mucosal transmission (Meyer *et al.*, 1993; Pauza *et al.*,

1993; Trivedi *et al.*, 1993). Following low dose intrarectal challenge, rhesus monkeys remained seronegative but developed T-cell proliferative responses, and were relatively resistant to high dose SIV_{mac} challenge infection (Salvato *et al.*, 1993). Similarly, macaques previously exposed to subinfectious doses of SIV_{mne} given IV developed strong T-cell responses and were protected from subsequent IV virus challenge (Benveniste *et al.*, 1993). By contrast, experimental infection by the intact genital mucosa requires three to four logs more SIV_{mac} than does IV infection and establishment of systemic infection may be slightly delayed (Miller *et al.*, 1989). Not unexpectedly, the genital squamous cell vaginal surface constitutes more of a physical barrier to bloodstream SIV infection than does the thinner rectal or conjunctival mucosa.

Other routes of experimental SIV infection in macaques include maternal–fetal transmission from infected mothers or by direct inoculation of fetuses or amniotic fluid with SIV (see Section X). Cage transmission of SIV_{stm} and SIV_{mac} by biting among male macaques has also been observed (Jennings *et al.*, 1992). This probably represents transmission via infected blood across the oral mucosa. Thus, it is possible to use the SIV macaque model to experimentally re-create not only bloodstream and maternal–fetal transmission but also rectal, oral, and genital mucosal transmission of cell-free virus. Bloodstream infection with titered stocks of allogeneic PBMC-associated SIV has also been accomplished (Gardner *et al.*, 1992; Cranage *et al.*, 1992a; Johnson *et al.*, 1992a; Heeney *et al.*, 1992), but mucosal transmission of cell-associated SIV has not been successful to date (C. J. Miller, personal communication).

B. Description of Simian AIDS

The clinical, virologic, pathologic, and immunologic features of SAIDS, induced by certain strains of SIV, are very similar to those seen in humans infected with HIV-1 (for review, see Desrosiers and Letvin, 1987; Gardner and Luciw, 1989; Letvin and King, 1990; Desrosiers, 1990a; Hirsch and Johnson, 1992). SIV, as does HIV, uses the same CD4 as its major cellular receptor. As with HIV-1 infection in humans, SIV infection of macaques persists despite a strong humoral and cellular immune response in most animals and fatal immunosuppression inevitably ensues. In both systems, neutralization escape mutants arise and become the dominant phenotype (Joag *et al.*, 1993). The kinetics of viral infection and alteration of $CD4^+$ cells parallel those observed in human HIV-1 infection (Connor *et al.*, 1993) and provide reliable markers for disease progression. However, the incubation period is shorter (several months to several years) in the SIV macaque model. An initial peak of viremia and antigenemia, accompanied by a rash (Ringler *et al.*, 1987), occurs within the first several weeks after experimental infection, followed by a decline in these measurements as the immune system responds (Kannagi *et al.*, 1986; Putkonen *et al.*, 1989, 1992a). Depending on the vigor of the humoral and cellular immune response, the virus is held relatively in check and the animal remains relatively healthy for several months up to about 3 years. About 30–40% of rhesus macaques infected with SIV_{mac251} do not generate a strong humoral immune response to the virus, remain p27 antigenemic, and die within 1 year of challenge. The remaining 60–70% of the infected animals develop varying levels of SIV antibody and die within 1–4 years (Daniel *et al.*, 1987; Letvin and King, 1990).

Like HIV-1 infection of humans, the main clinical features of SIV infection of macaques are wasting, with a loss of 15–60% of body weight, and persistent diarrhea. The pathogenesis of wasting is not simply a result of malabsorption or chronic diarrhea, but probably involves a systemic process including release of cytokines such as TNF. SIV-infected macaques may develop a transient skin rash soon after infection, possibly the result of injury to SIV-infected or antigen-presenting Langerhans cells by cytotoxic T cells (Ringler *et al.*, 1987). In those animals capable of making a strong immune response to SIV infection, lymph nodes and spleen may be initially enlarged, showing follicular hyperplasia associated with a polyclonal B-cell hyperreactivity and a paracortical expansion of $CD8^+$ lymphocytes (Chalifoux *et al.*, 1987), similar to the pattern described in HIV-1 infections in man (Racz *et al.*, 1986). Lymph nodes of SIV immune-incompetent animals, by contrast, show little or no follicular hyperplasia and exhibit marked depletion of T-cell zones. Levels of viral replication and associated pathology were generally higher in visceral than peripheral lymph nodes after either intravenous or intrarectal infection (Emau *et al.*, 1993). The framework of follicles, consisting of follicular dendritic cells (FDC), becomes disrupted, fragmented, and indented and the lymph nodes eventually atrophy and show hyalinized, "burnt out" germinal centers and a paucity of follicles (Fig. 9A).

In early stages, immunohistochemical staining of lymph nodes may show a decrease in $CD4^+$ T cells and an increase in $CD8^+$ T cells and B cells (Ward *et al.*, 1987; Ringler *et al.*, 1989a). SIV antigen appears on dendritic reticulum cells within germinal centers as well as in scattered $CD4^+$ T cells, macrophages, and giant cells. A high viral load in lymph node follicles early in infection appears to indicate a more rapid disease course (Joag *et al.*, 1993a; Chakrabarti *et al.*, 1993; Rosenberg *et al.*, 1993). Similar findings of high viral load in follicular centers of HIV-infected humans during the incubation period have been reported recently (Pantaleo *et al.*, 1993a; Embretson *et al.*, 1993). During both acute and chronic stages of SIV infection, the reduction in blood $CD4^+$ cell levels is not reflected in lymph nodes, where the $CD4^+$ pool remains within normal limits (Rosenberg *et al.*, 1993). In the depleted nodes, $CD4^+$ T cells are profoundly lacking and SIV-infected cells are less numerous than in the asymptomatic stage. In the thymus, cortical thymocyte depletion accompanied by a compaction of the epithelium has been noted in SIV_{sm}-infected rhesus monkeys (Baskin *et al.*, 1991b).

Correlated with lymph node collapse in advanced SIV disease is a substantial increase in unintegrated viral DNA levels, possibly reflecting a transition in viral synthesis from primarily T cells which are now depleted to predominantly macrophages (Rosenberg *et al.*, 1993). By EM, SIV particles are present in the germinal centers and several cell types, including FDC, show tubuloreticular structures, indicative of IFN-α synthesis during an antiviral response (Joling *et al.*, 1992). By immunogold EM, SIV core antigen is observed on the surface of FDC. The observation that $CD8^+$ CTL (serine esterase granzyme B positive) are not located in germinal centers near SIV-infected FDC suggests that CTL do not mediate the follicle destruction. In summary, lymph nodes and spleen can serve as reservoirs for the virus and PBMC may not be a reliable indicator of virus burden.

SIV leptomeningitis and encephalopathy often occur early after infection and reveal infected monocyte/macrophages, glial nodules (Fig. 9H,I), and giant cells (Fig. 9G), indistinguishable from those described in HIV-1 encephalopathy (Wiley *et al.*, 1986; Letvin and King, 1990; Lackner *et al.*, 1991b; Sharer *et al.*, 1991; Hurtrel *et al.*, 1991; Chakrabarti *et al.*, 1991). Intrathecal synthesis of SIV env-specific antibodies is detectable in immune-competent infected monkeys (Sopper *et al.*, 1993) in whom the cerebrospinal fluid level of IL-6 greatly exceeds the blood level (Hemm *et al.*, 1993). Infection of the brain appears to be the result of transvascular spread of infected monocyte/macrophages. SIV isolation from cerebral spinal fluid (CSF) and elevation of quinolinic acid in CSF and serum were linked to symptomatic disease and poor survival (Jordan and Heyes, 1993). Microglia are also a prime target cell for SIV infection in the CNS (Brinkman *et al.*, 1993). Although only a small number of microglial cells are productively infected *in vivo*, the majority express MHC class II antigens, indicating a state of activation. In the brain, viral DNA is detected by *in situ* PCR in perivascular macrophages and endothelial cells (Mankowski *et al.*, 1993).

Recruitment of infected monocyte/macrophages to the CNS may be influenced by the elevated expression of vascular cell adhesion molecule-1 (VCAM-1) and HLA-DR on many arteriolar, venular, and capillary endothelial cells (Sasseville *et al.*, 1992; Beilke *et al.*, 1991). Blockade of the interaction of VCAM-1 and its leukocyte ligand, VLA-4, by VCAM antibodies represents a potential target for therapy of AIDS encephalitis. Cognitive and motor impairments are associated with SIV infection in rhesus monkeys (Murray *et al.*, 1992) along with elevated levels of quinalinic and kynurenic acid in the CSF, markers of macrophage activation (Heyes *et al.*, 1992). Diffuse astrogliosis in the cerebrocortical gray matter of SIV-infected rhesus monkeys has also been associated with motor and cognitive impairments (Rausch *et al.*, 1992; Bowden *et al.*, 1993). In contrast to HIV-infected humans, SIV-infected giant cells (syncytia) are associated with a giant cell pneumonitis (Fig. 9E) and, indeed, may be found in lymph node sinusoids (Fig. 9D) or any tissues of infected macaques, especially those animals that have a weak antibody response and a greater amount of virus in their tissues (Li *et al.*, 1991; Baskin *et al.*, 1992). Not surprisingly, the presence of SIV-infected macrophages in brain, lung, and other tissue can be correlated with the detection of macrophage-tropic viral strains (Desrosiers *et al.*, 1991; Sharma *et al.*, 1992b).

Infected CD4 T cells, macrophages, and follicular dendritic cells are seen by *in situ* hybridization and immunohistochemical staining in bone marrow (Kitagawa *et al.*, 1991) (Fig. 9K,L), lymph nodes and spleen (Wyand *et al.*, 1989a,b; Stahmer *et al.*, 1993), thymus (Ward *et al.*, 1987; Li *et al.*, 1993), GI tract (Heise *et al.*, 1993a,b) (Fig. 9J), joint synovial membranes (Roberts and Martin, 1991), CNS (for review, see Lackner *et al.*, 1991a), male and female reproductive tracts (Baskerville *et al.*, 1992; C. J. Miller *et al.*, 1992c, 1993b), and other tissues (Kaaya *et al.*, 1993) and granulomatous lesions (Horvath *et al.*, 1993) of SIV-infected macaques often early after infection. Hematopoietic dysfunction (Mandell *et al.*, 1993), hemolytic anemia (Hillyer *et al.*, 1991, 1993), enteropathy (Heise *et al.*, 1993; Mossman *et al.*, 1993), arthritis (Roberts and Martin, 1991), ocular disease (Conway *et al.*, 1990), and pneumonitis (Baskin *et al.*, 1991a) occur in association with these sites of SIV infection although the pathogenesis is unclear. Undoubtedly, the release of various cytokines by infected macrophages and $CD4^+$ T cells contributes to the pathology in each of these tissues (Baskerville *et al.*, 1990; Simon *et al.*, 1992). Other lesions of unknown pathogenesis rather often seen in SIV-infected macaques include membranous glomerulopathy and intimal thick-

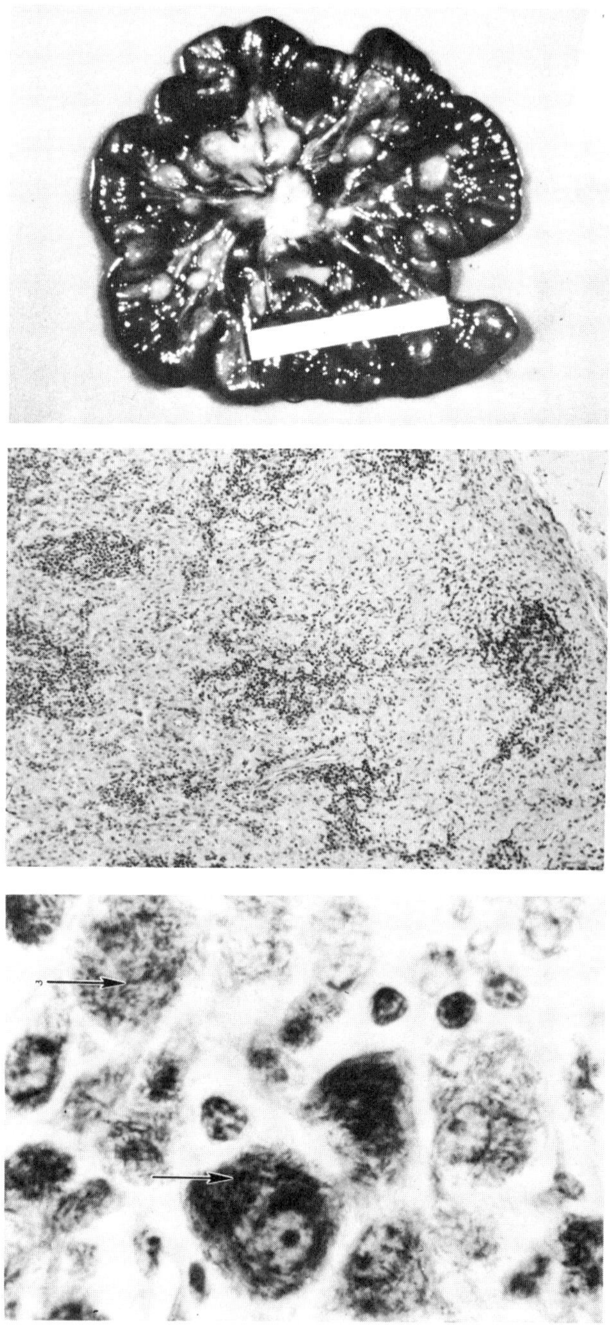

FIGURE 9. Histopathology of SIV-infected rhesus Macaques. (A) Enlarged mesenteric lymph nodes from infection with *Mycobacterium avium intracellulare* (MAI). (B) Lymph node from A showing depletion of lymphoid cells and proliferation of pale histiocytes (macrophages) in the sinusoids. H&E, ×10. (C) Same lymph node as in A and B. Note macrophages filled with dark-staining MAI rodlike organisms (arrows). PAS stain, ×100. (D) Syncytial giant cells in sinusoids of lymph node. H&E, ×63. (E) Syncytial giant cells in lung. H&E, ×63. (F) CMV inclusions (arrows) in lymph node with acute inflammation. H&E, ×100. (G) Syncytial giant cells in brain. H&E, ×100. (H) Glial nodule in brain

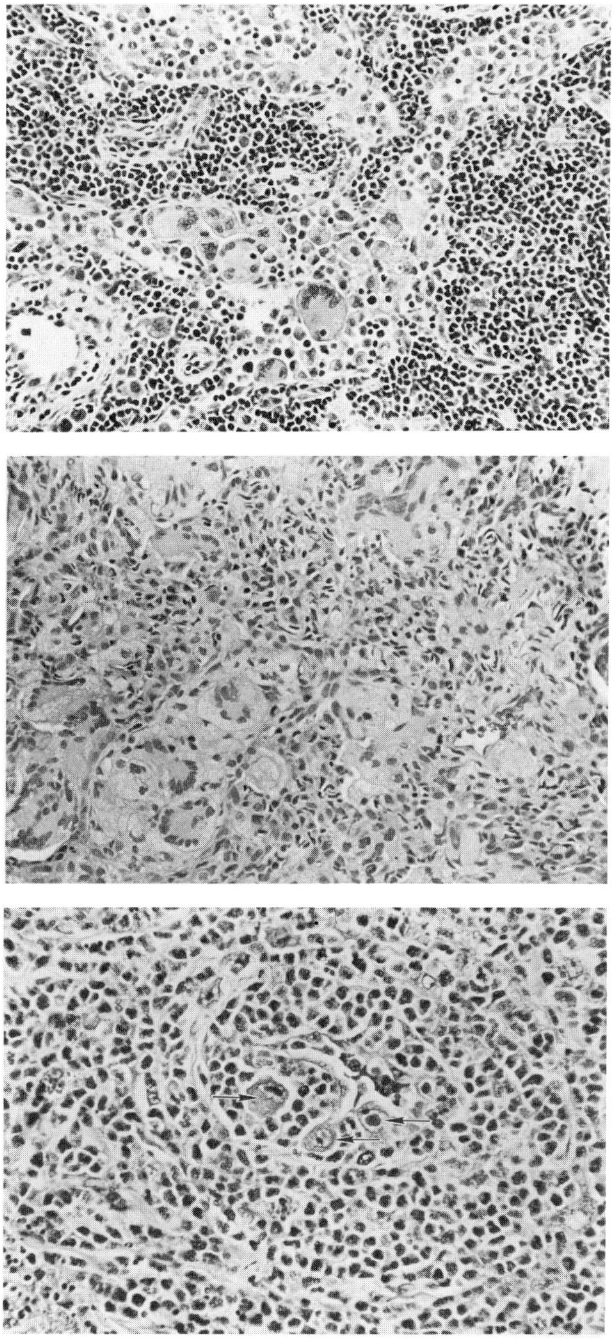

composed mainly of macrophages. Immunoperoxidase stain with CD68 monoclonal antibody to monocyte/macrophages. ×100. (I) SIV RNA in giant cells and glial nodule. *In situ* hybridization. ×100. (J) SIV RNA in Peyer's patch of small intestine. *In situ* hybridization. ×63. (K) SIV RNA in T cells of bone marrow. *In situ* hybridization for SIV RNA. Immunoperoxidase stain (not shown) indicated that these were CD3-positive T cells. ×100. (L) SIV RNA in bone marrow macrophages. *In situ* hybridization. ×100. (H–L courtesy of Satya Dandekar.)

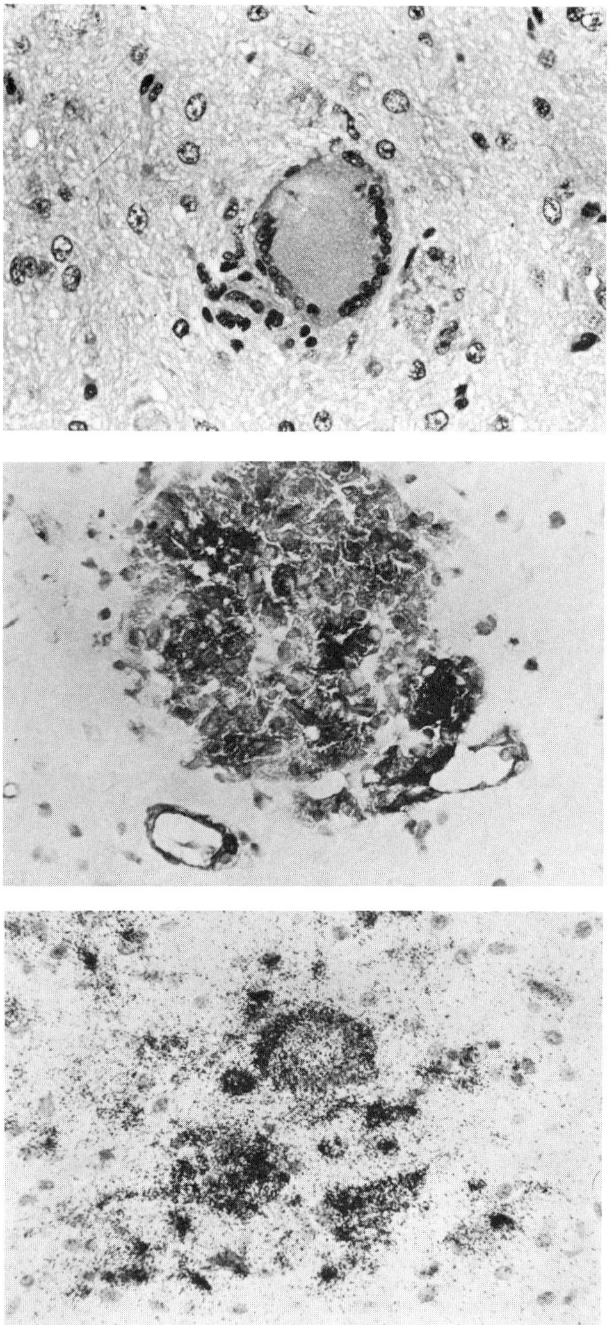

FIGURE 9. (*Continued*)

ening of the pulmonary artery (Chalifoux *et al.*, 1992). One SIV_{sm}-infected cynomolgus monkey with EBV-associated oral hairy leukoplakia has been reported (Feichtinger *et al.*, 1993).

Opportunistic infections observed with SAIDS and often associated with acute inflammation include cytomegalovirus in lungs, lymph nodes (Fig. 9F), GI tract, ganglia and nerves, adenovirus in gallbladder, common bile duct, and pan-

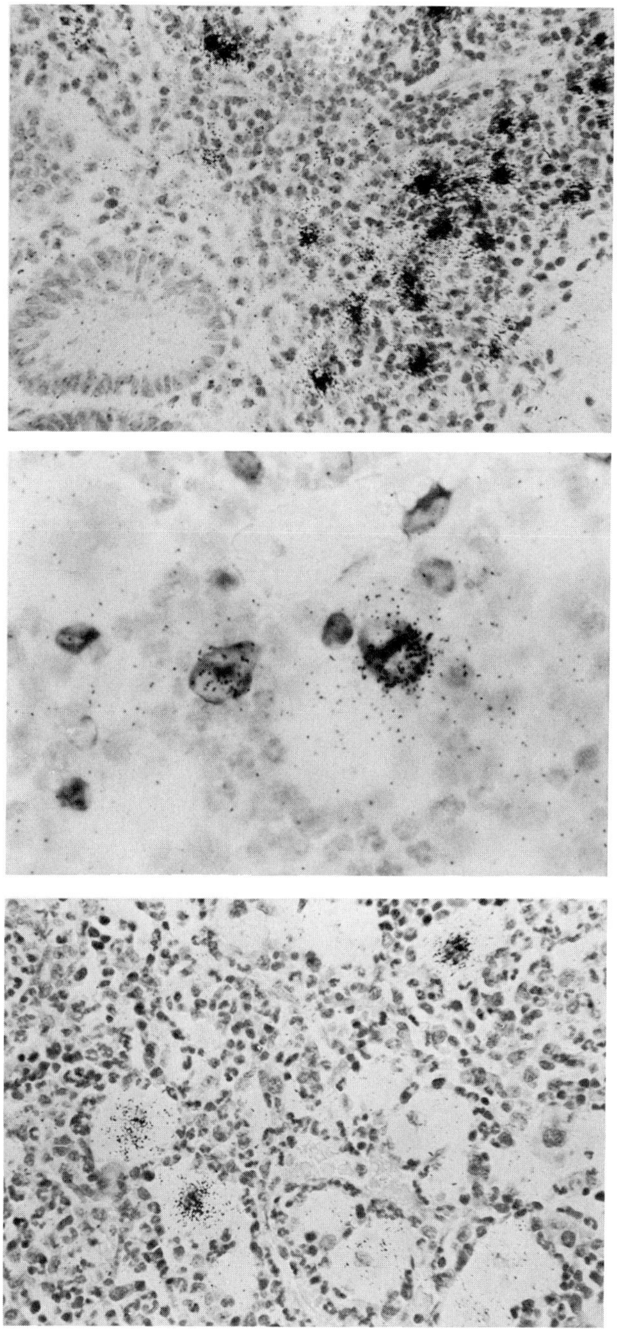

FIGURE 9. (*Continued*)

creas, SV40 in brain astroglial cells associated with progressive multifocal leukodystrophy, pneumocystis pneumonia, *Mycobacterium avium intracellulare* (Fig. 9A–C), candidiasis, cryptosporidiosis, and trichomoniasis (Lowenstine and Lerche, 1988; Baskin *et al.*, 1988; Baskin and Soike, 1989; McClure *et al.*, 1989; Baskerville *et al.*, 1991; Horvath *et al.*, 1992; Ilyinskii *et al.*, 1992). Variations in the incidence of these opportunistic infections recorded in macaques at different primate centers and in naturally occurring or experimentally induced cases of SAIDS (King *et al.*, 1990) are probably attributable to differences in the prevalence of the potential pathogens in the macaques at each location. SIV-infected rhesus macaques showed a delayed clearance of chancres caused by inoculated *Treponema pallidum*, apparently related to failure to produce antibody capable of opsonizing the organisms (Kuller *et al.*, 1992; Marra *et al.*, 1993).

Similar to human AIDS, lymphomas of B-cell origin occur sporadically as opportunistic tumors in monkeys with SIV-induced AIDS at different primate centers (Chalifoux *et al.*, 1987; Baskin *et al.*, 1986; Lowenstine *et al.*, 1992). Cynomolgus macaques infected with SIV_{sm} appear especially prone to such lymphomas; 9 of 24 cynomolgus monkeys developed B-cell lymphomas 5 to 15 months after inoculation with SIV_{sm} (Feichtinger *et al.*, 1990, 1992). The tumors are monoclonal, mostly extranodal including CNS, of high-grade malignancy and associated with an EBV-like herpesvirus. One can only speculate that a remarkable outbreak of lymphomas that occurred in more than 40 rhesus macaques at the CPRC in the late 1960s may also have been associated with SIV infection (Stowell *et al.*, 1971). The pathogenesis of the lymphomas is uncertain; the tumor cells are not infected with SIV but often express EBV nuclear antigens, suggesting a crucial role for cynomolgus-EBV in the onset of the lymphoma (Feichtinger *et al.*, 1992). Kaposi's sarcoma is not associated with SIV-induced AIDS, but similar lesions have been associated with infection of macaques by the immunosuppressive type D retrovirus (Giddens *et al.*, 1985) (SRV-2), as described later in this chapter.

In SIV_{mac}-infected macaques, as in HIV-1-infected humans, the balance of latent versus productive infection depends on the strength of the antiviral immune response and the presence of other immune activation factors (Kannagi *et al.*, 1986; Daniel *et al.*, 1987). Animals that die rapidly (within 2 to 7 months) have either no detectable antibody response or a low antibody response that recognizes only the viral envelope protein. These animals show a progressive decline both in total plasma immunoglobulin level and in number of $CD4^+$ T cells and have a high viral load including cell-free plasma viremia. From 20 to 50% of randomly selected rhesus macaques fall into this highly susceptible group. The other 50% or more that survive longer (1–4 years) remain persistently infected, but at a lower level, and have higher antibody titers to both core and envelope proteins, an increased plasma immunoglobulin level, and a slower reduction in $CD4^+$ T cells.

Generalized lymphadenopathy, polyclonal B-cell activation, increased serum neopterin levels (Popov *et al.*, 1992), and elevated levels of IL2 receptor-α expression (Ringler *et al.*, 1990) are further evidence of immune activation that persists until the terminal stages of illness. Activation of CD8 T cells is reflected by increased serum levels of soluble CD8 antigen and appearance of HLA-DR antigen. A correlation was observed between the development of disease in SIV-infected macaques, as in HIV-infected humans (Stricker *et al.*, 1987), and the presence of antibodies to an 18-kDa protein related to histone H_2B on the surface of activated lymphocytes, suggesting an autoimmune reaction (Fultz *et al.*, 1990a). In longer survivors the number of CD4 lymphocytes decreases soon after infection and then plateaus at $\leq 1000/mm^3$ until progressively decreasing to $\leq 400/mm^3$ shortly before death. Proliferative responses to T-cell-dependent B-cell mitogens, such as pokeweed mitogen, are markedly diminished before the

$CD4^+$ cell count becomes reduced. The combined assessment of cellular apoptosis, mitochondrial function and proliferative response to pokeweed mitogen has prognostic value in SIV infection (del Llano *et al.*, 1993). Loss of the $CD4^+$ cell pool in lymph nodes does not occur until CD4/CD8 ratios of <0.5 are detected in blood (Rosenberg *et al.*, 1993).

SIV can readily be isolated from PBMC, lymph nodes, spleen, salivary gland, brain (Bottiger *et al.*, 1991a), and other tissues throughout the course of the infection (Letvin *et al.*, 1985). Early in the course of disease, SIV infects a small number of actively proliferative $CD4^+$ T cells and resting memory T cells as well (Gallatin *et al.*, 1989; Willerford *et al.*, 1990). Infection of memory T cells early after infection may contribute to the immune dysfunction prior to significant lymphocytic depletion seen in SIV-infected macaques and HIV-infected humans (Clerici *et al.*, 1989). Stimulation of resting SIV-infected T cells by anti-CD3 and anti-CD28 monoclonal antibodies enhances viral replication (Polacino *et al.*, 1993).

PBMC obtained from monkeys prior to and $\frac{1}{2}$ to 3 years after infection with SIV_{mac} display no significant change in V_B or V_α gene family expression and no deletion of V_B-expressing cloned $CD4^+$ cell populations, suggesting that mechanisms other than superantigen-induced deletion must be responsible for $CD4^+$ lymphocyte loss in these animals (Chen *et al.*, 1993). Unstimulated $CD4^+$ T cells have little or no SIV DNA detectable in the genomic fraction but activation signals promote integration of proviral DNA in mature and immature T cells (Firpo *et al.*, 1992). In the terminal stages of SAIDS, numerous copies of unintegrated proviral DNA are present in nearly every tissue, lymphoid and nonlymphoid, examined (Hirsch *et al.*, 1991; Rosenberg *et al.*, 1992). Sequence analysis of molecular clones derived directly from tissue demonstrated a wide spectrum of related genotypes that are not represented in virus derived by cocultivation.

The SIV_{sm} PBj variant, isolated from a chronically infected pigtailed macaque, produces an acutely fatal disease in pigtailed macaques and less reliably in rhesus macaques (Fultz *et al.*, 1989; Lewis *et al.*, 1992b). Deaths occur within the first 2 weeks after inoculation and are associated with severe diarrhea and severe metabolic acidosis. The pathology reveals a diffuse nonulcerative lymphohistiocytic gastroenteritis with numerous SIV antigen-positive lymphocytes and macrophages in lymph nodes, spleen, GI-associated lymphoid tissues, and lamina propria of the GI tract. Animals dying acutely have detectable serum p27 antigenemia and no immune response. Surviving rhesus develop strong anti-SIV humoral immune responses but eventually die of SAIDS. The pathogenesis of the acute syndrome apparently involves generation of acute-phase reactants and lymphokines such as IL-6 (Lewis *et al.*, 1992b; Birx *et al.*, 1993) and TNF-α (Fultz, 1991).

C. Pathogenesis of Molecular Clones of SIV in Macaques

SAIDS has been readily induced by infection of macaques with at least five biologically active proviral (DNA) molecular clones of $SIV_{mac/sm}$ (Hirsch *et al.*, 1990; Johnson and Hirsch, 1992; Hirsch and Johnson, 1992; Luciw *et al.*, 1992), indicating that the virus is both necessary and sufficient for inducing fatal AIDS in these animals (Table IV). The prototype pathogenic molecular clone, SIV_{mac239}, is a T-cell-tropic virus that does not replicate in cultures of rhesus macrophages (Naidu *et al.*, 1988; Banapour *et al.*, 1991). SIV_{mac239} has been completely sequenced (Regier and Desrosiers, 1990). Rhesus macaques infected with SIV_{mac239} maintain a high viral load, produce antiviral antibodies, and die of SAIDS after 6 to 12 months (Kestler *et al.*, 1990a, 1991). Macaques infected with a *nef* deletion mutant of SIV_{mac239} have a lower viral load and do not become ill, indicating that the *nef* gene plays a major role in maintaining viral replication and pathogenesis (Kestler *et al.*, 1991).

Similar, more recent studies with a *vpr*$^-$ mutant of SIV_{mac239} suggest that this gene is also important for maximal SIV replication and persistence *in vivo* and for disease progression (Lang *et al.*, 1993). *vpr*$^-$ SIV_{mac239} mutants were not pathogenic *in vivo*. Pathogenic isolates of *vpr*$^-$ SIV_{mac} all had reversions to a *vpr*$^+$ phenotype, suggesting a requirement for *vpr* expression for disease progression. A macrophage-tropic virus variant (SIV_{mac316}), which evolved *in vivo* during the course of infection with SIV_{mac239}, was associated with the presence of SIV-related encephalitis and pneumonitis but was not essential for the development of SAIDS (Desrosiers *et al.*, 1991). This result was confirmed and expanded by Sharma *et al.* (1992a) who showed that SIV_{mac239} maintained its lymphocyte tropism in most of the inoculated monkeys and caused AIDS but not pneumonia or encephalitis. However, by serial intracerebral passage of SIV_{mac239}-infected bone marrow cells, a macrophage- and microglia-tropic, but not T-cell-tropic, SIV clone was derived that induced SIV encephalopathy and pneumonia but not immunodeficiency (Sharma *et al.*, 1992b). Sequence changes in the *env* gene of this macrophage-tropic neurovirulent SIV clone confer this cellular tropism (Anderson *et al.*, 1993). It is noteworthy that this neurovirulent strain showed poor neuroinvasiveness after systemic inoculation. Apparently, passage through the blood–brain barrier did not take place. These findings indicate the importance of tissue- or cell-specific selection of viral variants *in vivo* in relation to disease manifestations. Brain and lung appear to select for macrophage-tropic variants whereas spleen and lymph nodes select for lymphocytotropic variants. The different viral phenotypes may be selected by different cell types in bone marrow and other tissues and lead to the different clinical manifestations of disease (Kodawa *et al.*, 1993). Other closely related molecular clones of SIV_{mac} are either noninfectious (SIV_{mac142}) or weakly infectious (SIV_{mac251}, 1A11) *in vivo*, but have not induced disease (Naidu *et al.*, 1988; Marthas *et al.*, 1989; Luciw *et al.*, 1992).

Pigtailed macaques (*M. nemestrina*) infected with a molecular clone (Cl 8) of SIV_{mne} showed a decreased CD4 cell number after about 1 year, but had not yet developed SAIDS (Overbaugh and Rudensey, 1992). The SIV_{sm} H-3 and H-4 molecular clones caused fatal AIDS in rhesus macaques at 1–2 years postinfection; however, *nemestrina* macaques infected with SIV_{sm} H-4 remained healthy for over 2 years (Johnson *et al.*, 1990b). Virus from the cloned proviral halves of $SIV_{smPBj14}$ induced the same acute disease as the uncloned isolate (Fultz, 1991), but these clones were not stable (Dewhurst *et al.*, 1990). Molecular clones were also derived from nonacutely pathogenic derivatives of $SIV_{smPBj14}$ (Novembre *et al.*, 1991; Dewhurst *et al.*, 1992). Passage in pigtailed macaques of a molecular chimera made from a pathogenic molecular clone of $SIV_{smPBj14}$ and the nonlethal, related clone SIV_{smH4} indicated that the *env* gene in concert with another gene(s), not in the LTR, was required for acute disease induction (Novembre *et al.*, 1993b). Infectious molecular clones of SIV_{agm} ($SIV_{agm3,155}$), although infectious *in vivo*, have not induced disease in macaques (Johnson *et al.*, 1990b; Baier *et al.*, 1991), and virus has been difficult to recover from these animals. Biologically active molecular clones of SIV_{mdr} or SIV_{cpz} have not been reported.

In vitro testing of recombinants between molecular clones of the pathogenic, non-macrophage-tropic SIV_{mac239} and the 98% related nonpathogenic, macrophage-tropic $SIV_{mac1A11}$ indicated that gp130 Env and one or more genetic elements are essential for replication of SIV_{mac} in rhesus macrophages (Banapour *et al.*, 1991b). Using viral chimeras between SIV_{mac239} and SIV_{mac316} molecular clones, macrophage tropism was mapped to the *env* gene and included nine amino acid changes distributed over both the extracellular gp120 and the transmembrane gp41 portions of the envelope (Mori *et al.*, 1992). Surprisingly, the Env-dependent restricted replication of SIV_{mac239} in macrophages is not the result of limited virus entry (Mori *et al.*, 1993). Site-specific mutagenesis experiments identified single amino acids, 322 and 448, in the V3 and C4 domains of SIV_{mac239} gp120 that cooperate to effect virus entry into human CEMX174 cells and rhesus PBMC (Morrison and Desrosiers, 1993). These results show that the corresponding V3 sequences in HIV and SIV_{mac} both are involved in post-CD4 binding entry events and are able to determine, at least in part, differential cell tropism.

Site-specific mutagenesis of the SIV_{mac239} molecular clone also showed that the NF-κB binding site in the LTR is necessary for efficiency replication of the virus in primary alveolar macrophages but not in primary T cells *in vitro*. By comparing pseudotype particles of HIV-1 incorporating the envelope of SIV_{mac239}, which has a full-length cytoplasmic domain or the envelope of $SIV_{mac1A11}$, which has a truncated cytoplasmic domain, truncation of the cytoplasmic domain was shown to enhance substantially both cell fusion and infectivity for human T cells (Zingler and Littman, 1993).

In vivo testing of the SIV_{mac239} and $SIV_{mac1A11}$ recombinants indicated that at least two and perhaps more viral genes (e.g., *vpr*, *env*-TM, *nef*) are determinants for pathogenicity (Marthas *et al.*, 1993). As mentioned above, *in vivo* passage of SIV_{mac239} *nef* deletion and *vpr*$^-$ mutants showed that both of these viral genes are required for full pathogenic potential (Kestler *et al.*, 1991; Lang *et al.*, 1993). It is noteworthy that animals that are chronically infected at a low level with the SIV *nef* deletion mutant are resistant to superinfection with a high dose of virulent cloned or uncloned SIV_{mac} (Daniel *et al.*, 1992) and animals chronically infected at an even lower level with $SIV_{mac1A11}$, although superinfectible with a high dose of virulent uncloned SIV_{mac}, have a markedly delayed onset of disease (Marthas *et al.*, 1990) (see Section XV).

D. Immune Response of SIV-Infected Macaques

1. Humoral Immune Responses

The pattern of anti-SIV antibody response was shown to correlate with changes in total plasma immunoglobulin G (IgG), the absolute peripheral blood CD4$^+$ cell count, and the clinical disease course (Kannagi *et al.*, 1986; Daniel *et al.*, 1987). Two distinct patterns of antibody response were found in six rhesus monkeys infected with SIV_{mac}. Four animals that died within 5 months postinoculation developed low-titer anti-SIV antibody responses that recognized only the viral envelope protein and progressive declines in total plasma IgG levels and $CD4^+$ lymphocyte number. Two animals that lived more than 1 year before dying of AIDS developed high-titer anti-SIV antibody responses that recognized both viral envelope and core proteins, increases in total plasma IgG, and a later decrease in number of $CD4^+$ T cells. The longest living monkey was the only one to develop detectable SIV neutralizing antibodies. Broadly reactive neutralizing antibodies also correlated with longer survival in SIV infected cynomolgus macaques (Zhang *et al.*, 1988).

It is important to note that, in the SIV model, neutralizing antibodies generally do not correlate with immune protection, either in infected or vaccinated animals. This result does not simply reflect the lack of an accepted standard assay by which to measure SIV neutralization (McKeating *et al.*, 1989). Neutralizing antibodies from some SIV_{mac}-infected or -vaccinated animals bound to recombinant gp130 from homologous SIV isolate and recognized conformational determinants. The properties of these antibodies are quite similar to those that have been identified in sera from SIV-vaccinated or -infected macaques and from HIV-1-infected humans (Haigwood *et al.*, 1992).

In another study, six rhesus monkeys were inoculated i.v. with the SIV_{mac251} biologic isolate (Le Grand *et al.*, 1991). During primary infection, all six animals exhibited a general and persistent lymphadenopathy, a fever peak, a transient rash, and a transient thrombocytopenia. At 4–9 days postinoculation (pi), most animals exhibited a transient p27 antigenemia and by 1 week pi all PBL cultures were positive in their PBMC. By 21 days pi, all monkeys were PCR positive. Seroconversion as determined by ELISA occurred soon after infection in all six animals and reached high and persistent titers ranging between 10^5 and 10^6 by 3 months pi. Positive Western blot activity was detected in all six animals between 21 and 49 days pi. Three patterns of neutralization or antibody-dependent enhancement (ADE) of infection were observed. The latter process involves binding of virus to antibody and a subsequent increase (at least twofold) in SIV infection of cells via the Fc or C′ receptors (Robinson and Mitchell, 1990). In one pattern, two monkeys showed ADE and no neutralizing antibody between days 0 and 49 pi. In another pattern seen in one monkey, transient ADE at day 49 pi was followed by low neutralizing activity at day 100 pi. A third pattern seen in three monkeys showed successive peaks of ADE titers with no neutralizing activity before day 209 pi. At days 145 and 342 pi, two animals died of SAIDS with neurological symptoms and extensive vasculitis lesions; both animals had high ADE titers and no neutralizing antibodies. Thus, all sera from the SIV_{mac}-infected monkey exhibited ADE properties whereas no major neutralizing antibodies were detected in any of the animals.

In yet another study, plasma from four rhesus monkeys infected with SIV_{mac251} or SIV_{sm} enhanced SIV_{mac} infection of MT-2 cells (Montefiori *et al.*, 1990b). Enhancement required C′, CD4, and CD2. Titers of enhancing antibodies appeared to correlate with disease progression.

Although these last two studies suggest that enhancing antibodies may be present in SIV-infected macaques and contribute to disease, others have not been able to confirm this observation. It thus remains uncertain whether enhancement of endocytosis of SIV_{mac} via Fc or C′ receptors occurs on macaque macrophages or other cell types by either neutralizing or nonneutralizing plasma from SIV-infected monkeys.

Narayan and co-workers (Joag *et al.*, 1993) found that among eight rhesus macaques infected with molecularly cloned SIV_{mac239}, only one animal developed neutralizing antibodies to the virus, although all eight animals developed binding antibodies to the virus. The neutralizing antibodies collected late in infection neutralized virus isolated early in infection but failed to neutralize plasma virus isolated after the appearance of neutralizing antibodies. Only antigenic variants were isolated from blood, spleen, and lymph nodes. By contrast, viruses isolated from the other macaques that did not develop neutralizing antibodies were neutralized by the antibodies from the single monkey with such antibodies and were all of the parental phenotype. The findings suggest that development of neutralizing antibodies after infection is ineffective, since they select for neutralization-escape variants.

In summary, most macaques experimentally infected with SIV_{mac} mount a humoral immune response to viral proteins as detected by ELISA and Western blot. The PBMC of infected monkeys spontaneously produced anti-SIV antibodies *in vitro* (Zamorcli *et al.*, 1993). Only a few of these immune responders mount a significant neutralizing antibody response and these antibodies select for neutralization-escape variants. Mapping of B- and T-cell epitopes using sera from SIV-infected or -vaccinated monkeys is covered below (Section XIV.G). In those animals capable of generating a vigorous humoral antibody response, regardless of the level of neutralizing antibody, the initial viremia is quickly brought under control and the virus is held relatively in check for a period of time ($\geq$7 months) before waning of immunity, increase in virus level, and onset of SAIDS.

The mechanism of this temporary virus suppression is unknown, but it may involve antibody-dependent complement lysis of cell-free SIV (Spear *et al.*, 1992), antibody-dependent cellular cytotoxicity of SIV-infected cells (ADCC) (Ohkawa *et al.*, 1993), or induction of CTL activity (Yasutomi *et al.*, 1993) (see below). A similar pattern, but in a much more prolonged incubation period (mean 8 years), occurs with HIV-1 infection of humans. A minority of monkeys infected with SIV_{mac} generate only a weak or nondetectable humoral immune response. In these animals the viral load remains high and death from SAIDS usually occurs sooner, i.e., 2–6 months pi. The possible role of enhancing antibodies demonstrable *in vitro* in disease progression remains to be determined. However, the apparent enhancement of challenge infection and acceleration of disease seen after passive transfer of plasma from SIV-seropositive infected macaques suggests that such antibodies may indeed play an important role in pathogenesis *in vivo* (Gardner *et al.*, in press) (see Section XIV.F).

SIV-infected macaques generate antibodies to monkey MHC class I antigen, suggesting that SIV antigens may mimic monkey MHC or that self-MHC antigens acquired by the virus *in vivo* are presented in an altered form (Polyanskaya *et al.*, 1993). However, this effect is distinct from that seen in animals vaccinated with human cell-grown virus preparations (see Section XIV.A).

2. Cellular Immune Responses

a. T-Cell Proliferation

Most SIV-infected macaques, like most HIV-infected humans, mount a substantial cellular, as well as humoral, immune response to SIV antigens. Macaques infected with SIV_{mac} are able, until the terminal stage of SAIDS, to generate SIV-specific T-cell proliferative responses (McGraw *et al.*, 1990a,b; Ahmed-Ansari *et al.*, 1990). T cells from SIV-infected monkeys fail to proliferate in response to heat-inactivated SIV added directly to the culture. However, when SIV is processed by autologous antigen-presenting cells prior to culture with purified T cells, proliferative responses are uniformly demonstrated in SIV-infected monkeys, but not in uninfected controls. Proliferation in response to heat-inactivated SIV is mediated by $CD4^+$ T cells and is MHC class II-restricted. In contrast, the proliferative response to infectious SIV is mediated by both $CD4^+$ and $CD8^+$ T cells and is MHC class-restricted. This higher level of response may be caused by infectious virus presenting a more intense antigenic stimulus than the heat-inactivated virus. A decline in T-cell proliferative response is observed as disease progresses but it is unclear why this occurs.

A strong SIV Env-specific T-lymphocyte proliferative response was found in the absence of seroconversion in most macaques exposed intrarectally to subinfectious doses of SIV_{mne} whereas infectious doses of SIV_{mne} given by this route generated antibodies and weak or undetectable T-cell proliferative responses (Salvato *et al.*, 1993). Seronegative animals with T-cell proliferative responses to SIV antigens remain clinically well and are persistently positive for SIV DNA in PBMC (Emau *et al.*, 1993). MHC-restricted CTL specific for SIV Gag peptide has been observed in seropositive animals but not yet in seronegative animals. The observation of virus-specific T-cell proliferative responses in seronegative animals indicates a role for cellular immunity that may not include MHC-restricted CTL.

b. *Cytotoxic T Lymphocytes*

CTL directed against several viral proteins have been detected in SIV- and HIV-2-infected macaques by several groups (Vowels *et al.*, 1989; M. D. Miller *et al.*, 1990; Yamamoto *et al.*, 1990a; Venet *et al.*, 1992; Bourgault *et al.*, 1992a; Voss *et al.*, 1992). Although virus-specific CTL are able to suppress SIV replication *in vitro* (Kannagi *et al.*, 1990; Tsubota *et al.*, 1989), the effects of virus-specific CTL *in vivo* remain poorly understood. It also remains to be shown whether CTL escape mutants arise in SIV-infected macaques as reported in HIV-1-infected humans (Phillips *et al.*, 1991).

Letvin and colleagues first showed that $CD8^+$ lymphocytes cloned from PBL of SIV-infected macaques are able to reduce SIV replication *in vitro* (Kannagi *et al.*, 1990) and that this effect is primarily related to an MHC class I-restricted CTL killing of SIV-infected $CD4^+$ lymphocytes. Autologous target cells were B cells transformed by baboon herpesvirus (*H. papio*) and infected with recombinant vaccinia virus expressing the inserted SIV genes. They found that Gag-specific $CD8^+$, MHC class I-restricted CTL were detected in SIV-infected rhesus monkeys and the Gag epitope was defined as a nine-amino-acid peptide (M. D. Miller *et al.*, 1990; Yamamoto *et al.*, 1990a). Also characterized was the MHC class I molecule ("Mamu-A-01") that restricted this CTL Gag-specific response (Yamamoto *et al.*, 1990a). The Gag-specific CTL exhibited a restricted usage of V_B and J_B genes (Chen *et al.*, 1992). Gotch *et al.* (1993) described a CTL epitope shared between the gag protein of HIV-1, HIV-2 and SIV_{mac} and presented by a cynomolgus macaque MHC molecule designated allele III.

An Env-specific cell-mediated response was predominant and two distinct effector lymphocyte populations, $CD16^+$ MHC-unrestricted and $CD16^-$ MHC class I-restricted, mediated the SIV_{mac} Env-specific target cell lysis (Yamamoto *et al.*, 1990b). The precise nature of the $CD16^+$ Env-specific response is uncertain. It might be ADCC rather than natural killer (NK) cell mediated because it was detected in infected but not uninfected animals. Letvin and co-workers (Kannagi *et al.*, 1990) concluded that the functional role that the Gag- or Env-specific effector mechanisms play in the immune protection or immunopathogenesis of SIV infection remains unknown. However, the ability of $CD8^+$ lymphocytes to inhibit SIV_{mac} replication in MHC class I-restricted lymphocytes *in vitro* (Kannagi *et al.*, 1988) suggests that these CTL may inhibit SIV replication *in vivo*. Most recently these investigators have shown the presence of SIV_{mac}-specific CTL within a few days of infection temporarily associated with a decrease in viral replication (Yasutomi *et al.*, 1993a).

Vowels *et al.* (1989) also demonstrated MHC class I-restricted $CD8^+$ virus-specific CTL activity in SIV-infected rhesus but the viral epitopes were not defined. Target cells consisted of MHC haploidentical parental PBMC stimulated with mitogen and then pulsed with heat-killed SIV_{mac}. Inactivation of target cells with leucopeptin prior to CTL assay inhibited target cell generation, suggesting that viral antigens were processed by an endocytic pathway. This CTL activity was lost just before death.

J. P. Levy and colleagues (Venet *et al.*, 1992) also studied the CTL response of SIV-infected macaques against autologous B-cell lines (transformed by *H. papio*) presenting SIV_{mac} proteins after infection with recombinant vaccinia viruses. The SIV proteins studied were Env, Nef, Vif, Tat, and Vpx. Three different patterns of responses were defined in 12 animals; response against at least six of the eight target proteins in 6 animals; response against only two proteins in 3 animals; and response against none of the proteins in 3 animals. Both structural proteins (Env, Pol, Gag) and regulatory proteins (Nef, Rev, Vpx, Tat, Vif) represented efficient CTL targets. The animals responding against multiple proteins survived longer (>1 year) than the low-responder or nonresponder animals and a decrease in circulating CD4 cells occurred concurrently with the disappearance of the CTL response, suggesting that the pattern of response has prognostic value. No particular protein appeared to be recognized earlier than other ones and the loss of CTL response did not affect any one single protein more than another. Three epitopic peptides recognized by the macaque CTL were located in the central region of the Nef protein, indicating a high immunogenicity of this region (Bourgault *et al.*, 1992a).

Voss and Hunsmann (1993) also recently reported that the CTL activity might delay the development of disease in SIV-infected macaques. Further studies at very early time points after infection are needed to evaluate differences in the kinetics of recognition of different SIV proteins. The disappearance of the CTL response concomitant to clinical deterioration also occurs in HIV-infected patients, but the cause or effect relationship to pathogenesis remains unclear (Hoffenbach *et al.*, 1989; Tsubota *et al.*, 1989).

3. Non-MHC-Restricted $CD8^+$ T-Cell Suppression of SIV Replication

In addition to elimination of cells infected by SIV by MHC-restricted CTL activity, suppression of SIV replication in $CD4^+$ T cells *in vitro* can also occur by an MHC-unrestricted $CD8^+$ T-cell activity (Kannagi *et al.*, 1988). This suppression does not necessarily involve cell contact or lysis and does not eliminate infected CD4+ cells from the culture. A similar phenomenon has been described in HIV-1-infected humans (Walker *et al.*, 1986, 1989; Wiviott *et al.*, 1990; Kannagi *et al.*, 1990; Brinchmann *et al.*, 1990) and HIV-1-infected chimpanzees (Castro *et al.*, 1991). While CD8+ cells from uninfected humans do not show this activity, CD8+ cells from uninfected chimpanzees are capable of inhibiting HIV replication (Castro *et al.*, 1991). Removal of CD8+ cells from mixed PBMC cultures has facilitated the recovery of SIV from some seronegative high-risk sooty mangabeys (Villinger *et al.*, 1991) and humans (Walker *et al.*, 1986, 1989), respectively. These observations suggest that such CD8+ lymphocyte-mediated interference with SIV (or HIV) replication could play an important role in the maintenance of the asymptomatic carrier state. This suppression is presumably related to a soluble factor released by the CD8+ cells and yet to be characterized (Walker *et al.*, 1989). This factor may function, in part, by inhibiting viral transcription (Powell *et al.*, 1993).

4. Gamma–Delta T Cells

Gamma–delta T cells are another subset of T cells found in association with various epithelia, distinct from CTL and NK cells, that may contribute to antiviral immunosurveillance in an MHC-unrestricted fashion (Brenner *et al.*, 1986; Bank *et al.*, 1986). In rhesus monkeys, such gamma–delta T cells have been shown to effectively lyse cells infected with SIV (Malkovsky *et al.*, 1992). High levels of gamma–delta T cells have been observed in the peripheral blood of HIV-seropositive subjects (DePaoli *et al.*, 1991) and SIV-infected rhesus monkeys (Wallace *et al.*, 1993). It seems important to better understand the specificity and function of this T-cell subset in relation to the pathogenesis of AIDS (Janeway *et al.*, 1988).

5. Natural Killer Cells

NK activity against sensitive target cells (K562, Raji) and SIV pulsed target cells was significantly decreased in SIV_{mac}-infected rhesus macaques as compared with uninfected controls (Vowels *et al.*, 1990; O'Reilly and Hoover, 1991). Treatment of PBMC with IL-2 partially restored the NK activity. Early inactivity of NK cells might allow infected cells to escape early surveillance and facilitate establishment of persistent SIV infection. ADCC has not as yet been characterized and distinguished from NK activity in the SIV macaque model.

E. SIV Strain Variation *in Vivo*

The genetic evolution of molecularly cloned SIV in experimentally infected rhesus macaques has been examined mainly by analyzing clones of gp120 derived by PCR amplification directly from PBMC or lymph nodes. Burns and Desrosiers (1991) found that nucleotide substitutions (mostly G → A and A → G transitions) in the SIV_{mac239} gp120 accumulate at an average rate of 8.5 per 1000 per year, with the majority (81%) of nucleotide substitutions resulting in amino acid changes. In addition to point mutations, in-frame deletions were noted in cloned *env* genes. Variation occurred predominantly in five discrete regions, four of which correspond to variable regions V1, V2, V4, and V5 of HIV-1. However, the region corresponding to the variable V3 loop of HIV-1 was quite conserved among the SIV_{mac239} clones. These findings raise doubt as to whether the putative V3 loop in SIV_{mac} is immunodominant and subject to the same immune selection pressures as is the V3 loop in HIV-1. It is also unknown whether the V3 loop of SIV Env has an important role in determining cell tropism as it does with HIV-1 (Hwang *et al.*, 1991; Cheng-Mayer *et al.*, 1988).

Variation also occurred in the transmembrane gp41 subunit of the SIV_{mac} Env protein and the V6 domain in gp41 was identified as a strain-specific neutralizing determinant (Kodama *et al.*, 1991). By comparison, there was very little variation in the CD4 binding region. Wain-Hobson and colleagues found a broadly similar evolution of SIV quasispecies in SIV_{mac239}-infected macaques (Pelletier *et al.*, 1993). Almond *et al.* (1992a) also found similar results in the genetic evolution of the envelope gene in cynomolgus macaques infected with the uncloned 32H challenge stock of SIV_{mac251}. However, they found very little genetic variation in and around the cluster of potential glycosylation sites of the external portion of gp41 (Slade *et al.*, 1993a). At 1 month, the recovered viral *env* sequences were all closely related to the challenge virus, whereas *env* genes showed more diversity at 6 and 15 months. They also confirmed the lack of sequence variation in the HIV-1 *env* region equivalent to V3. The same investigators showed that sequence variation in the *env* gene of SIV_{mac} (32H) recovered from recombinant Env-vaccinated but challenge-infected macaques is predominantly in the V1 region (Almond *et al.*, 1993a). Genetic variation in the V1 and other regions, especially V4, distinguished between virus variants from blood and those from tissues (Slade *et al.*, 1993a).

SIV variable regions known to serve as targets for humoral or cell-mediated immunity are summarized above. The *env*, *rev*, and *nef* genes varied independently with all changes in the predominant protein sequences brought about by single nucleotide substitutions only (Slade *et al.*, 1993b). The results demonstrate that extensive sequence variability accumulates *in vivo* after infection with molecularly cloned SIV leading to a mixture of genotypes (i.e., quasispecies) and that selection occurs *in vivo* for changes in distinct variable regions of Env that lead to escape variants. The host immune response is considered the most likely source of this selection, a conclusion later confirmed by characterizing SIV mutants resistant to serum neutralization that arose during the course of persistent infection of rhesus monkeys (Burns *et al.*, 1993).

In the same variable regions, *env* gene mutations were detected in the SIV_{sm} H3 and H4 molecular clones in infected rhesus and pigtailed macaques (Johnson and Hirsch, 1991). In addition, in-frame stop codons in the *env* gene were frequent in viral DNA. Variations in only one or two residues in the relatively conserved V3 loop altered cellular tropism. The degree of viral heterogeneity in the spleen and lymph nodes was more extensive than in the brain or PBMC (Campbell and Hirsch, 1993).

To analyze *env* gene variation and determine whether a common genotype of SIV evolves in the course of SAIDS, a total of 160 *env* genes were PCR amplified, cloned, and sequenced from the tissue DNA of four pigtailed macaques infected with molecularly cloned SIV_{mne} (SIV_{mne} cl 8) (Overbaugh *et al.*, 1991). Common amino acid sequence changes were identified within V1, V4, and, in later stages of the disease, near V3. The predominant amino acid changes created new sites for *N*- and *O*-linked glycosylation that could contribute to escape from immunologic surveillance or to changes in cell tropism or cytopathology (e.g., syncytium formation) that could accelerate the onset of SAIDS (Overbaugh and Rudensey, 1992). These investigators also showed that the SIV variants that persist in cultures are similar to the variants that predominate in the host (Rudensey *et al.*, 1993).

The outstanding feature of the acutely pathogenic biologic and molecular clones of $SIV_{smPBj14}$ versus the parental virus ($SIV_{sm\text{-}9}$) or closely related nonacutely pathogenic molecular clones is the presence of an additional NF-κB binding site in the LTR, when compared with closely related SIV molecular clones (Dewhurst *et al.*, 1990; Courgnaud *et al.*, 1992). This duplication of the NF-κB enhancer element seemed a possible explanation for the increased virulence of this SIV variant (Novembre *et al.*, 1991). Analysis of the molecular diversity of $SIV_{smPBj14}$ proviral DNA in the intestinal tissues of an infected macaque revealed LTRs with one or two NF-κB binding sites with the single site variant most common (Novembre *et al.*, 1991). To further assess the importance of this duplication, virus chimeras were generated which had one or two NF-κB binding sites. None of these recombinants induced the acute disease syndrome. Viral recombinants made between acutely pathogenic and nonacutely pathogenic molecular clones of $SIV_{smPBj14}$ also showed no consistent difference in *in vitro* properties and demonstrated that the *env* genes, but not the LTR, was a determinant for acute disease induction although other viral genes were also involved (Novembre *et al.*, 1993b; Dewhurst *et al.*, 1992). The differing virulence of PBj14 molecular clones was reflected in their relative capacity for replication within lymphocytes and macrophages of the GI tract (O'Neil *et al.*, 1993).

F. Genetic Variation of SIV_{agm} during Natural and Experimental Infection

A high prevalence of SIV infection occurs naturally in healthy AGM with considerable genetic heterogeneity among the various SIV_{agm} isolates (Li *et al.*, 1989; Baier *et al.*, 1989; Johnson and Hirsch, 1992; Allan *et al.*, 1991). On experimental transmission to macaques, most of the SIV_{agm} strains tested have caused a low level of infection but, with perhaps two exceptions (Gravell *et al.*, 1989; Dopolito *et al.*, 1993), no disease. Thus, SIV_{agm} is usually nonpathogenic in both the natural and experimental hosts.

To determine the extent of genetic variation in SIV in the course of infection of the natural and artificial hosts, samples of PBMC DNA were evaluated by PCR amplification and sequencing in experimentally infected AGM and pigtailed macaques. In two AGM infected for 15–20 months with the SIV_{agm3} molecular clone, extensive variation in PBL was observed (~eight mutations per site per year) in the V1 and V2 regions of gp120 whereas the HIV-1 V3 loop homologue was highly conserved (Baier *et al.*, 1991). These results are similar to those obtained in SIV_{mac}- or SIV_{sm}-infected macaques and indicate that presence or lack of pathogenicity could not be correlated with any obvious difference in the extent of gp120 variation occurring during infection. In long-term naturally infected AGM, about 45% of proviruses analyzed were defective, comparable with the defectiveness of HIV-1 proviruses in seropositive individuals. No correlation was observed between antibody response and genetic variability. The results suggest that accumulation of HIV-1 defective variants is not the main explanation for HIV-1 pathogenicity.

Johnson and Hirsch (1992) amplified SIV sequences from PBL of four macaques at 20–52 weeks after infection with the molecularly cloned SIV_{agm155}. They found an extensive divergence of the *env* gene with the substitution rate averaging 10^{-2} to 10^{-3} substitution/site per year and as high as 10^{-1} nucleotide substitution/site per year in occasional hypermutant clones. An unusual propensity for G → A transitions leading to in-frame stop codons in the *env* gene was observed. The average rate of substitutions in the integrase domain of the *pol* gene was about an order of magnitude lower than the frequency of substitutions in the *env* gene. An equal degree of variation in the *env* gene occurred in animals that did and did not develop AIDS during the time observed. However, in this study (Johnson and Hirsch, 1992), AIDS appeared to be associated with the presence of *env* genes lacking the in-frame stop codons.

Studies of gp40 variation of SIV_{agm155} in experimentally infected AGM and pigtailed macaques showed that the natural form of this protein was full length and that, in contrast to the SIV_{mac} system (Hirsch *et al.*, 1989a; Kodama *et al.*, 1989b), a premature stop codon in the cytoplasmic tail was not rapidly removed after *in vivo* infection (Fomsgaard *et al.*, 1991, 1993). The frequency of nucleotide substitutions was 2- to 10-fold lower than reported for the $SIV_{mac/sm}$ model probably because of the lower level of viral replication in the infected animal (Johnson and Hirsch 1992). These findings could reflect the fidelity of the SIV_{agm3} reverse transcriptase enzyme which appears to be 2- to 20-fold more accurate than the homologous enzyme from HIV-1 (Manns *et al.*, 1991).

In contrast to macaques which select for full-length SIV *nef in vivo* (Kestler *et al.*, 1991), mangabeys do not appear to exert such rigorous selection nor is a full-length *nef* sufficient for SIV_{sm} disease in naturally infected mangabeys (Villinger *et al.*, 1993).

The results from the above studies taken together suggest that genetic variation *per se* does not account for the development of AIDS. The problem is obviously more complex. The findings of extensive genetic variation during experimental infection of macaques with SIV are similar to those that characterize all animal lentivirus infections (Haase 1986) and are major features of HIV infection. The *env* gene product is the most highly variable protein, and because it is a key determinant of viral host range and infectivity, persistence of infection may involve phenotypic selection for variant viruses that avoid elimination by the host immune response. HIV infectious viruses isolated late in the disease course tend to have an altered host range and to replicate more efficiently and be more cytopathic in tissue culture, suggesting an increased pathogenic potential *in vivo* (Cheng-Mayer *et al.*, 1988). Similar findings have yet to be reported in the SIV macaque model. However, it is uncertain whether these phenotypic changes in HIV directly contribute to the immune depletion or are a consequence of the increased viral replication secondary to the immune suppression. In mice and cats, many previous studies (Paquette *et al.*, 1989; Overbaugh *et al.*, 1988) with molecularly cloned oncoviruses have shown that a very subtle change in the Env protein can dramatically alter the pathogenicity and indicate that the genotype of the virus certainly does influence the pathogenesis.

G. Lack of Pathogenicity of SIV in the Natural Host

In their natural African primate hosts, the SIVs are relatively nonpathogenic. However, many more SIV-infected African monkeys must be observed into old age to confirm this lack of pathogenicity. The occurrence of immunodeficiency and lymphoproliferative disease in a 10-year-old AGM dually infected with SIV and STLV (Traina-Dorge *et al.*, 1992) suggests that SIV-related disease may occur under certain conditions in the natural host and that dual infection with SIV and STLV may contribute to the pathogenesis of both viruses. An understanding of the mechanism underlying the general lack of SIV pathogenicity in the natural host might help explain how closely related SIV strains cause disease when experimentally inoculated into macaque species; the results could direct attention to novel strategies aimed at preventing AIDS in HIV-infected humans. Similar information might be obtained from HIV-1-infected chimpanzees in which the infection is also nonpathogenic (for review, see Gardner and Luciw, 1989). The lack of HIV-1-induced disease in chimpanzees may possibly be related to the absence of chronic cellular activation and lack of a strong cellular immune response (Ferrari *et al.*, 1993). HIV-1 infected chimpanzees show an absence of T-cell dysfunction (e.g., anergy) *in vitro* as well as *in vivo* (Henley *et al.*, 1993). Although HIV-2 infection of rhesus macaques has been essentially nonpathogenic, it was recently found that an AIDS-like disease could be induced by serial passage of the HIV-2 EHO strain in *M. nemestrina* (McClure *et al.*, 1992).

Various virologic and immunologic parameters have been compared in SIV naturally infected African monkeys, SIV experimentally infected macaques, HIV-1 experimentally infected chimpanzees, and HIV-1 naturally infected humans. Such comparisons indicate that the lack of disease in the SIV-infected natural host cannot be attributed to any obvious differences in virus expression or immune response, because roughly similar viral loads, viral variation, cell tropisms, and antiviral humoral and cellular immune responses are found in the natural and experimental hosts (Fultz *et al.*, 1990a; Norley *et al.*, 1990; Hartung *et al.*, 1992a). Differences in cytokine levels produced by PBMC or of Th1/Th2 subsets in naturally infected sooty mangabeys and experimentally infected macaques might contribute to the presence or absence of disease (Mayne *et al.*, 1993; Ansari *et al.*, 1993).

In general, the level of infectious virus in PBMC or $CD4^+$ T cells in SIV-infected healthy AGM or sooty mangabeys (Fultz *et al.*, 1990a; Jehuda-Cohen *et al.*, 1990; Hartung *et al.*, 1992a) is comparable to that of asymptomatic SIV-infected macaques and HIV-infected individuals. Among seropositive AGM the number of cells carrying a viral genome, as measured by PCR, ranged from 1 to 15 per 10^5 PBMC, and the mean titer of infectious virus, as measured by endpoint dilution, was 10 TCID per 10^6 PBMC (Hartung *et al.*, 1992a). Plasma viremia was detected in 4 of 17 monkeys along with detection of SIV_{agm} expression in PBMC by *in situ* hybridization. Genetic variation of SIV_{agm} in the natural and experimental hosts is covered above (Section XIII.F). These findings indicate that SIV-infected AGM do not totally suppress viral replication *in vivo* and express a viral load and genetic variability equivalent to those seen for HIV-1 in asymptomatic humans (Hartung *et al.*, 1992a).

Major differences in the tropism of SIV_{agm} and HIV-1 have not been noted in that both infect predominantly $CD4^+$ T lymphocytes, monocytes, and macrophages. However, the monkeys do not experience the high viral load characteristic of progression to AIDS in HIV-1-infected humans. Like HIV, the major target molecule for SIV_{agm} is the simian analogue of the CD4 molecule (Kraus *et al.*, 1989). Both viruses also infect $CD4^-$ cells by unknown mechanisms (Werner *et al.*, 1990a). However, SIV_{agm} differs from HIV-1 in that soluble CD4 inhibits HIV-1 infection but enhances SIV_{agm} infection of $CD4^+$ cells (Allan *et al.*, 1990b; Werner *et al.*, 1990b). In addition, SIV_{agm} becomes more sensitive to neutralization after exposure to soluble CD4. Binding of CD4 to gp130 of SIV_{agm} may induce a conformational change and expose a previously hidden domain, thus allowing recognition of a second cellular receptor necessary for virus entry. Also, like HIV in humans, AGM monocytes are readily susceptible to *in vitro* infection and syncytium induction by SIV_{agm}. In contrast, SIV_{agm} is not cytopathic for AGM lymphocytes although it is cytopathic for certain human T-cell lines.

The functional immune response of AGM to SIV infection is quantitatively and qualitatively similar but by no means identical to macaques infected with $SIV_{mac/sm}$ or humans infected with HIV (Norley *et al.*, 1990). For example, in the naturally infected AGM, neutralizing antibody levels are low or absent and there is only a low antibody response to the core protein of SIV_{agm}. B-cell epitope mapping of the entire SIV_{agm3} envelope glycoprotein revealed that antibodies from naturally infected AGM recognized immunodominant regions corresponding to the HIV-1 V3 region, the transmembrane protein (Gann peptide), and the COOH-terminal area of gp130. The most immunodominant region was the transmembrane protein (Siegel *et al.*, 1992). Also, compared with the human HIV-1 system there is a substantially lower inhibitory effect of SIV_{agm} proteins on the proliferation of AGM lymphocytes. Similar to the situation with humans, infected AGM develop no detectable complement-activating antibodies but they do develop ADCC-activating antibodies as shown by lysis of SIV_{agm}-infected target cells exposed to antisera from infected animals using human PBL effector cells. NK lysis (target cells in the presence of effector cells alone) of SIV_{agm}-infected cells is not observed. SIV-infected AGM, like HIV-infected humans and SIV-infected macaques, also have detectable MHC-restricted envelope protein-specific CTL activity (Norley *et al.*, 1990). Furthermore, similar to HIV-1-infected humans (Brinchmann *et al.*, 1992) and SIV_{mac}-infected rhesus monkeys (Kannagi *et al.*, 1988), autologous $CD8^+$ cells from SIV_{agm}-infected AGM can inhibit SIV_{agm} replication *in vitro* by the production of a soluble factor (Cichutek and Norley, 1993). Perhaps the lack of pathogenesis of SIV infection in the natural hosts can be related, in part, to a lower antibody response and stronger cell-mediated response (i.e., Th1 versus Th2 response) (Clerici and Shearer, 1992).

In SIV-infected AGM and sooty mangabeys, as compared with infected macaques, an important difference at the cellular level is clearly the differential cytopathic effects of SIV_{sm} for $CD4^+$ lymphocytes (Ohta *et al.*, 1988; Kraus *et al.*, 1989; Fultz *et al.*, 1990a). As in HIV-infected humans, cytopathogenicity is reflected by both single-cell lysis and fusion of infected cells to form multinucleated end-stage syncytia. Failure of AGM and mangabeys to develop disease may be related, at least in part, to the fact that, in the natural host, SIV replicates in but is not cytopathic for $CD4^+$ cells. The PBj14 variant of $SIV_{smm\text{-}9}$, which induces acute lethal disease in both macaques and mangabeys, also shows extensive cytopathogenicity for both mangabey and macaque $CD4^+$ cells (Fultz *et al.*, 1989). Moreover, the PBj14-infected cultures undergo a $CD4^+$ and $CD8^+$ lymphocyte proliferative response *in vitro* consistent with the extensive lymphoid hyperplasia seen *in vivo*. These observations suggest at first glance that the *in vitro* cytopathic effects of SIV for CD4$^+$ cells from a given species of nonhuman primates are an *in vitro* correlate for disease induction *in vivo*. However, one cannot make a simple correlation between cytopathic effect on host CD4$^+$ cells *in vitro* and pathogenesis because only certain SIV strains (i.e., SIV_{sm}) and not others (i.e., SIV_{agm}, SIV_{mdr}) induce disease in macaques despite being cytopathic for macaque or human CD4$^+$ T cells *in vitro*. Moreover, syncytium induction in T cells or monocytes cannot alone account for disease because SIV_{agm} induces syncytia in AGM monocytes (Norley *et al.*, 1990). In summary, there is also no obvious explanation on the basis of viral cytopathogenicity for the lack of pathogenesis of SIV infection in the natural host.

In SIV_{mac}-infected macaques and HIV-1-infected humans, T-cell anergy and programmed cell death (apoptosis) are thought to result from inappropriate signaling by infected antigen-presenting cells (APC), i.e., monocyte/macrophages (Ameisen and Capron, 1991). The lack of apoptosis or disease in HIV-1-infected chimpanzees or SIV-infected AGM may be attributed to the inability of these viruses to interfere in the normal APC function of monocytes in the respective hosts (Gougeon *et al.*, 1993). Whether preservation of a normal APC mechanism accounts for the lack of T-cell dysfunction and depletion in SIV-infected AGM is unknown. However, since AGM monocytes are readily infected with SIV_{agm}, the presence or absence of infection of APC cannot account for the presence or absence of disease. In the natural host the findings suggest that intracellular genetic factors, yet to be determined, and not immunologic mechanisms, govern the control of SIV replication and account for the lack of T-cell dysfunction and immunosuppressive disease. However, the inability of SIV_{agm} to induce AIDS in experimentally infected macaques (Johnson and Hirsch, 1992), with rare exception (Gravell *et al.*, 1989; Dapolito *et al.*, 1993), suggests that the apathogenicity in AGM may in part also be a property of the virus.

The immunodepressive property of certain SIV strains might be related to the immunosuppressive and fusogenic region of the transmembrane protein (gp41) which corresponds to a conserved domain in the transmembrane protein (p15E) of mouse and cat leukemia viruses (Mathes *et al.*, 1978; Cianciolo *et al.*, 1985). A synthetic peptide representing this region of HIV-1 inhibits proliferation of T and B lymphocytes after mitogen stimulation (Ruegg *et al.*, 1989). The immunosuppressive peptide of HIV-1 exerted a stronger inhibitory effect on lymphocyte stimulation compared with the related peptide from SIV_{agm} (Norley *et al.*, 1990). That this viral protein (amino-terminus of gp41) may cause immunosuppression *in vivo* is supported by the recent observation that a synthetic peptide representing this region induces lymphopenia and immune tolerance to a linked peptide in both macaques and chimpanzees (Haynes *et al.*, 1993).

H. Studies on SIV/HIV-1 Chimeras *in Vitro* and in Macaques

Recombinants between molecular clones of SIV_{mac} and HIV-1 were produced in order to study viral determinants of pathogenesis and as a possible vaccine model. Such recombinants allow the HIV-1 envelope to be expressed in macaques in the context of the SIV core genes. Two groups independently produced chimeric viruses by substituting the *env* (plus *tat* and *rev*) gene of the pathogenic SIV_{mac239} clone with the *env* (plus *tat* and *rev*) gene of very similar HIV-1 clones and evaluated the viruses in the same monkey species (i.e., cynomolgus macaques). Shibata *et al.* (1991) produced a chimeric virus from molecular clones of HIV-1-NL432 and SIV_{mac239}. The HIV-1-NL432 clone was produced by ligating the 5′ and 3′ halves of the HIV-1-NY5 and the HIV-1-LAV proviral clones (Adachi *et al.*, 1986). Presumably, HIV-1-NL432 is not macrophage-tropic because the parental virus that contributed the *env* gene (i.e., HIV-1-LAV) does not replicate in macrophages. In addition, HIV-1-NL432 does not replicate in cultures of cynomolgus macaque PBMC (Shibata *et al.*, 1991). SIV_{mac239} encodes full-length open reading frames (ORFs) for Gag, Pol, Vif, Vpx, Vpr, Tat, Rev, and Env; however, the *nef* gene of this virus has a premature stop codon (Regier and Desrosiers, 1991). SIV_{mac239} replicates in cynomolgus (and rhesus) macaque PBMC and several human T-cell tumor lines but does not replicate efficiently in rhesus macrophages (Banapour *et al.*, 1991a; Mori *et al.*, 1992). The chimeric virus, designated NM, was constructed by substituting a restriction enzyme DNA fragment encoding the *tat*, *rev*, and *env* genes of SIV_{mac239} with a restriction enzyme DNA fragment from HIV-1-NL432 (Shibata *et al.*, 1991). Because of the construction strategy, NM virus is defective for both *vpr* and *nef*. NM virus replicates in human T-lymphoid cell lines and in cultures of cynomolgus macaque PBMC; information is not available on infection of human PBMC and either human or macaque macrophage cultures by this recombinant virus.

S. Sakuragi *et al.* (1992) infected two adult cynomolgus macaques (8 and 11 years old) with 10^4 $TCID_{50}$ of NM virus grown in a human T-cell line. At 2 and 12 weeks postinfection, viral DNA was detected in PBMC samples by PCR. Only one cynomolgus macaque showed virus by cocultivation of PBMC collected at 2 and 6 weeks postinfection. Southern blotting and Western blotting of virus recovered from this animal revealed that the chimeric structure was maintained *in vivo*. Western blot analysis of sera from both cynomolgus macaques showed seroconversion to anti-viral antibodies at 6 to 8 weeks postinfection, and these antibodies have persisted for 28 weeks. Both animals have remained healthy for 1 year after inoculation of the chimeric virus (S. Sakuragi *et al.*, 1992).

Li *et al.* (1992) produced SIV–HIV-1 chimeric virus (SHIV) from the HXBc2 isolate of HIV-1 and SIV_{mac239} (*nef* open) variant. HXBc2 virus is T-cell-tropic but does not replicate in macrophages (Fisher *et al.*, 1986). The LAV strain used by Shibata, Sakuragi, and co-workers (see above) and the HXBc2 strain are highly related with respect to sequence and probably represent the same HIV-1 isolate (Wain-Hobson *et al.*, 1991). Four chimeric viruses were cloned by replacing the *env*, *tat*, *rev* coding region of SIV_{mac239} (*nef* open) with the counterpart region of HXBc2 virus. Although these four chimeras differed with respect to splice acceptor sites for Tat and Rev, all produced infectious virus after transfection of CEMX174 cells and all replicated in PBMC cultures from cynomolgus macaques. Information on the replication potential of the HXBc2 virus in cynomolgus macaque PBMC has not been presented. One chimera, SHIV-4, was selected for *in vivo* studies. This chimera was later found to have a mutation in the *vpu* gene (J. Sodroski, personal communication). Four cynomolgus macaques were inoculated i.v. with 7×10^3 $TCID_{50}$ of SHIV-4 grown in cynomolgus macaque PBMC and titered on CEMX174 cells. At 2, 4, and 6 weeks postinfection, virus was recovered from PBMC of all four infected animals; radioimmunoprecipitation studies revealed that these viruses retained the chimeric structure (Li *et al.*, 1992). Two of these cynomolgus macaques became virus negative after 41 days whereas virus has been recovered from the remaining two cynomolgus macaques beyond 1 year after inoculation (J. Sodroski, personal communication). All animals have high anti-SHIV antibody titers that match the levels in SIV_{mac239} infected macaques and they have remained healthy. No significant changes in CD4 cell counts have occurred. Thus, further time is required to determine if these SIV–HIV-1 chimeras will be pathogenic in macaques.

XIV. VACCINES AGAINST SIV EXPERIMENTAL INFECTION OF MACAQUES

The similarities in viral genome structure and biologic properties between SIV and HIV and the ability to consistently induce AIDS in macaques with certain biologic and molecular clones of SIV make this an excellent model for study of HIV-1 vaccine strategies. Table VI summarizes the published vaccine efficacy

TABLE VI. Active Vaccination against SIV Infection of Macaques

Type of vaccine	Virus strain	Cell type in which challenge virus was grown	Efficacy	References
Inactivated whole virus	SIV_{mac}/ SIV_{sm}	Human T cells	Yes	Desrosiers *et al.* (1989b), Murphey-Corb *et al.* (1989), Carlson *et al.* (1990), Cranage *et al.* (1992), Johnson *et al.* (1992a), Putkonen *et al.* (1992), Hartung *et al.* (1992), LeGrand *et al.* (1992a), Stahl-Hennig *et al.* (1992), Marthes *et al.* (1992)
			No	Sutjipto *et al.* (1990b)
		Rhesus T cells	No	Gardner *et al.* (1992), Marthas *et al.* (1992b), Johnson *et al.* (1992), LeGrand *et al.* (1992), Cranage *et al.* (1993)
Inactivated cell-virus	SIV_{mac}	Human T cells	Yes	Stott *et al.* (1990)
Modified live virus	$SIV_{mac1A11}$	Human T cells	Yes	Marthas *et al.* (1990, 1992b)
		Rhesus T cells	No	Marthas *et al.* (1992a)
	SIV_{mac239} Δ nef	Rhesus T cells	Yes	Daniel *et al.* (1992)
Recombinant env	SIV_{mne}	NA	No	Hu *et al.* (1992a)
	SIV_{mac}/ SIV_{sm}	NA	Yes	Planelles *et al.* (1991), Giavedoni *et al.* (1993), Mills *et al.* (1992), Cranage *et al.* (1992a), Hoover *et al.* (1992), Javaherian *et al.* (1992), Daniel *et al.* (1992), Ahmad *et al.* (1994)
Recombinant env peptides	SIV_{mne}	NA	Yes	Shafferman *et al.* (1991)

results in this model. These results have been reviewed elsewhere (Daniel and Desrosiers, 1989; Lardner and Stott, 1990; Gardner and Hu, 1991; Warren and Dolatshahi, 1992, 1993). In the following subsections we summarize the current status of various active and passive vaccine approaches in the SIV macaque model.

A. Inactivated Whole Virus Vaccines

1. Observations on Protection

Many groups (Desrosiers *et al.*, 1989a; Murphey-Corb *et al.*, 1989; Carlson *et al.*, 1990; Stott *et al.*, 1990; Cranage *et al.*, 1993; Johnson *et al.*, 1992; Putkonen *et al.*, 1992b; Hartung *et al.*, 1992; Le Grand *et al.*, 1992a; Stahl-Hennig *et al.*, 1992) have reported successful protection of macaques against i.v. infection with SIV grown in human T cells by inactivated whole cell-associated or cell-free SIV vaccines. Efficacy is ≥90% in these trials. Conditions of the inactivated whole virus vaccine protection in general include challenge with a low dose, 10–200 animal infectious doses (ID), of the homologous virus grown in human T cells and given as cell-free inoculations either i.m., i.v., or intrarectally (i.r.) (Cranage *et al.*, 1992b) at the time of maximum immune response. The same inactivated whole virus and modified live virus vaccines that protected against low-dose i.v. challenge did not protect against vaginal challenge (Marthas *et al.*, 1992b). Almost 100% of the nonvaccinated control monkeys in each trial have become persistently infected starting about 2 weeks after inoculation. These vaccines did not protect against higher challenge doses ($>10^3$ ID). Importantly, using the same inactivated whole virus vaccines, only about 50% protection has been achieved against challenge with allogeneic rhesus PBMC infected with the homologous SIV strain (Gardner *et al.*, 1992; Cranage *et al.*, 1993; Johnson *et al.*, 1992a; Osterhaus *et al.*, 1992; Murphey-Corb, personal communication) and no protection has been achieved against i.v. challenge with a low dose of cell-free SIV grown in allogeneic rhesus PBMC. The apparent explanation for these discordant results between human and rhesus T-cell-grown challenge virus is that the SIV grown in human T-cell lines (e.g., 8166, HUT-78) picked up human MHC class I and II cellular antigens that were major targets of the immune protection achieved (see below). Because macaque T-cell lines are not available and have been very difficult to establish, an inactivated whole SIV vaccine grown in macaque cells has not yet been tested against SIV also grown in macaque cells.

Successful vaccines, grown in human T-cell lines, were made with whole inactivated SIV derived from biologic or molecular clones. Vaccine virus was purified by sucrose gradient or column chromatography and inactivated with formalin, β-propiolactone, or detergent. Native envelope glycoprotein purified by lectin affinity chromatography from an SIV_{sm} vaccine stock protected two of four monkeys (Murphey-Corb *et al.*, 1990b). Tween ether-disrupted SIV_{mac} also protected monkeys (Stahl-Hennig *et al.*, 1992). The total amount of immunogen used to confer protective immunity ranged from 500 μg to 2.8 mg and the schedule of immunizations consisted of a maximum of five inoculations over a 13-month period to a minimum of three inoculations over a 4-month interval. Higher doses of antigen were more effective than lower doses (Hartung *et al.*, 1992b). With each vaccine the antibody levels declined rapidly after each boost and challenge was done at the height of the immune response, 2 to 4 weeks after the final boost. In vaccine-protected monkeys, antiviral antibodies declined after challenge, reaching minimal levels by 4 to 6 months postchallenge. Vaccinated monkeys remained protected for only a relatively short time after the last boost. In one study (Murphey-Corb *et al.*, 1992c), two of four SIV_{sm}-vaccinated monkeys were susceptible to challenge infection with 10 ID of the same virus 8 months after the last boost, and in another study (Stott *et al.*, 1990), one of four SIV_{mac}-vaccinated monkeys was susceptible to infection with the same dose at 4 months after the last boost. Vaccine cross protection was obtained against two strains of SIV (SIV_{mac} and $SIV_{smDelta}$) that differ by about 17% in outer envelope nucleotide sequences and 10% in amino acid sequences (Murphey-Corb *et al.*, 1992c; Johnson *et al.*, 1992b; Putkonen *et al.*, 1992b). Cross protection of SIV_{mac}-vaccinated monkeys against challenge infection with HIV-2 (Cranage *et al.*, 1993), or of HIV-2-infected monkeys against challenge infection with SIV_{sm} (Putkonen *et al.*, 1990) was not possible probably because these strains are too divergent (~33% difference in *env* nucleotide sequences). The adjuvant used in most of the studies was muramyl dipeptide (MDP, SAF-A Syntex), although other successful vaccines were formulated with Freund's incomplete adjuvant (IFA), Quil A, or alum. There was no systemic or local adverse reactions to any of the inactivated whole virus vaccines and none of these vaccine preparations contained residual live virus based on thorough *in vitro* and *in vivo* assays done prior to challenge.

The type of immunity induced in the protected animals appeared to be "sterilizing," meaning that no evidence of transient infection was detected after challenge. The probable absence of any covert infection occurring in most of the vaccine-protected monkeys is supported by the absence after challenge of an anamnestic antibody response, by the failure to detect viral nucleic acid in blood or tissues (bone marrow, lymph nodes) by PCR, and by the failure to transmit infectious virus to naive monkeys by blood transfusion or lymph node extracts. However, in the first two SIV vaccine studies (Desrosiers *et al.*, 1989a; Murphey-Corb *et al.*, 1989), breakthroughs of infectious virus occurred in two apparently protected monkeys, one at 5 months and the other at 7 months after challenge, both accompanied by an anamnestic antibody response at that time. These two monkeys went on to develop SAIDS. Also, with an HIV-1 vaccine, one presumed protected chimpanzee broke through with infectious virus after 8 months (Girard *et al.*, 1991). Therefore, only time will tell whether the other vaccine-protected monkeys are completely free of latent virus or, if present, whether this latent virus will eventually be activated and cause disease.

These findings contrast with previously successful inactivated whole and recombinant envelope viral vaccines, including murine type C (Gardner *et al.*, 1985) and macaque type D retrovirus vaccines (Marx *et al.*, 1986; Hu *et al.*, 1989; Brody *et al.*, 1992) in which protection could be achieved against much higher challenge doses ($>10^6$ ID) and in which a transient infection and an anamnestic antibody response after challenge could usually be demonstrated in the protected animals.

2. Mechanism of Protection

The mechanism of SIV vaccine protection in these studies was initially puzzling because neither high levels of viral neutralizing antibody nor strong antiviral activity were induced (for review, see Gardner and Hu, 1991). ADCC-mediating antibodies were not detected in sera of these vaccine-protected monkeys (Ohkawa *et al.*, 1993). However, the inactivated whole SIV vaccines did induce antibodies that reacted with conformation-dependent, group-specific determinants on the SIV envelope gp130, similar in this respect to SIV antibodies found in SIV-infected macaques and HIV-1 antibodies found in HIV-1-infected humans (Haigwood *et al.*, 1992; Steimer *et al.*, 1991). Then, a critical observation (Stott *et al.*, 1991) that macaques immunized with uninfected human T cells (8166) used to grow the SIV were also protected against challenge with SIV grown in the same human cells, but not against SIV grown in rhesus cells (Cranage *et al.*, 1993), raised the possibility that immune responses to human T-cell antigens might be involved in protection. This idea was supported by the finding that vaccine protection was correlated with a higher level of anti-human cell antibody than antiviral antibody (Stott *et al.*, 1991; Langlois *et al.*, 1991). The specific cellular antigens contributing to this protection were then identified as mainly belonging to or associated with MHC class I and II molecules (Arthur *et al.*, 1992; Chan *et al.*, 1992; Cranage *et al.*, 1993). With the SIV_{mac} vaccine prepared in the United Kingdom, antibody response to the HLA class I (Chan *et al.*, 1992) and class II (Cranage *et al.*, 1993) molecules correlated with vaccine protection. With the SIV_{mac} vaccine made in the United States, the major cellular antigens contributing to protection were β_2-microglobulin, a molecule that noncovalently associates with the HLA class I α chain, and the HLA class II DR protein (Arthur *et al.*, 1992). Remarkably, there were more molecules of these cellular antigens per purified virion than of envelope gp120 (Arthur *et al.*, 1992). Moreover, antisera to these cellular antigens (but not to CD4) precipitated intact SIV and HIV grown in these same human cells and inhibited infection with these viruses *in vitro*. In addition, monkeys immunized with inactivated whole HIV-1 grown in human T cells (8166) were protected against SIV_{mac} grown in the same cells, but not against the same virus grown in rhesus cells (Cranage *et al.*, 1993). The strongest correlate of protection was anticellular rather than antiviral antibody levels (Cranage *et al.*, 1993).

The failure of inactivated whole SIV vaccines, previously found protective against cell-free SIV grown in human T cells, to protect against cell-free SIV grown in rhesus T cells was confirmed by all other groups working with these inactivated whole virus vaccines (Gardner *et al.*, 1992; Johnson *et al.*, 1992a; Le Grand *et al.*, 1992a; Cranage *et al.*, 1993; Titti *et al.*, 1993; Stahl-Hennig *et al.*, 1993; Putkonen *et al.*, 1993). It is therefore very likely that antibody responses to HLA class I and II antigens, perhaps acting synergistically with antibodies to gp120, are involved with the protection against SIV infection after immunization with inactivated whole SIV or HIV-1 grown in human cells. In line with this conclusion, two of four monkeys immunized with murine L cells expressing transfected human MHC class II DR antigen (Stott *et al.*, 1993), two of four monkeys immunized with purified HLA A1, B8 (W. Chan, personal communication) and two of two macaques immunized with purified HLA class II DR antigen (Arthur, personal communication) were protected against IV challenge with 10 ID of SIV_{mac251} grown in human cells. These studies show that protection can be mediated by immunity against HLA Class I and II xenoantigens and raises the possibility that similar allogeneic responses might be utilized for protection against HIV infection. It is also possible that different biologic properties, i.e., rapid high phenotypes, of the challenge virus grown in monkey as compared with human cells might also contribute to this lack of protection against rhesus cell-grown SIV (Stahl-Hennig *et al.*, 1993; Lubre and Hunsmann, 1993). Passive immunization results (see below) also supported the importance of anti-human cellular antibodies in the vaccine protection.

To encourage the development of an analogous inactivated whole HIV-1 vaccine for humans, it will be necessary to determine if such an SIV vaccine prepared in allogeneic rhesus cells can protect against rhesus cell-grown SIV. These studies are in progress (Murphey-Corb, personal communication; Cranage *et al.*, 1993). However, the specific and selective association of cellular proteins with SIV and HIV could have major implications for the infectious process, for pathogenesis, and for design of vaccines (Arthur *et al.*, 1992). In this regard, monkeys that expressed high levels of simian MHC class II on the surface of their PBMC were more resistant to disease induced by rhesus PBMC-grown SIV_{mac251} than were monkeys that expressed low levels of MHC II on their PBMC (Hall *et al.*, 1993).

3. Protection against SIV Disease

A few rhesus macaques, which were vaccinated with inactivated whole virus and became infected soon after challenge with a high dose of SIV_{mac}, had a lower level of initial viral infection than controls, and lived significantly longer ($\leq$3-fold) than controls before dying of SAIDS (Desrosiers *et al.*, 1989a; Sutjipto *et al.*, 1990b; Goldstein *et al.*, 1993). A similar finding was noted in HIV-2-infected cynomolgus macaques superinfected with SIV_{sm} (Putkonen *et al.*, 1990). The longer survival of these animals resembles that seen in some naive monkeys that are capable of mounting a vigorous immune response after experimental infection (Daniel *et al.*, 1987). Protection against disease was also seen in macaques immunized by a whole inactivated SIV vaccine but infected with rhesus cell-associated SIV challenge (Hirsch *et al.*, 1993b). Vaccine-induced immunity, therefore, may confer a degree of protection against early death in the monkeys that do become infected after systemic challenge. To date, SIV enhancing antibodies, which can be detected *in vitro* in some of the vaccinates before and after infection (Montefiori *et al.*, 1990a), have not been associated with accelerated disease after active vaccination. Therefore, the biologic importance *in vivo* of enhancing antibody detected *in vitro* in SIV infection (Montefiori *et al.*, 1990b), or after SIV vaccination (Montefiori *et al.*, 1990a) remains unknown. However, recent passive immunization experiments have suggested the possible enhancement of SIV infection and acceleration of disease (see Section XIV.F).

B. Macaques/SIV Env Peptides

Protection of pigtailed macaques (*M. nemestrina*) against i.v. challenge with 100 ID of homologous virus (SIV_{mne}/E11S) was also achieved with a vaccine of conserved SIV_{mne} envelope peptides expressed in *E. coli* as β-galactosidase fusion proteins (Shafferman *et al.*, 1991). The peptides were 10- to 21-residue segments of conserved, hydrophilic regions of SIV gp120 and transmembrane protein. They were different from the V3 loop peptides and were based on analogous peptides from regions of the HIV envelope previously found to be immunoreactive with HIV-infected patient sera (Shafferman *et al.*, 1989a,b). Animals received four inoculations over 1 year and were challenged i.v. 3 weeks after the last boost with 100 ID of homologous cell-free SIV_{mne}. Transient viral infection occurred in each vaccinated monkey as evidenced by a transient antibody rise and virus isolation or PCR detection of viral DNA in PBMC during the initial weeks after infection. However, the interval of detectable virus in the immunized monkeys was relatively brief and one monkey was never culture positive. At 60 weeks after infection, all immunized monkeys were healthy and have remained virus negative by culture, but latently infected based on PCR-positive PBMC, whereas all controls (immunized with β-Gal only) were persistently virus positive. Lymph node cells and PBMC were transferred from three immunized macaques to naive macaques, none of which became infected (Shafferman *et al.*, 1993). The degree of virus suppression in the vaccinated monkeys correlated directly with the level of neutralizing antibody on the day of challenge and with ELISA titers for antibody to the amino-terminus transmembrane peptide. Antibodies to human cellular antigens were not present in these animals. When the actively immunized animals were later boosted with the same Env peptides and challenged with the closely related but more virulent SIV_{mac}, they were readily superinfected (A. Shafferman, personal communication).

This study represents the first successful SIV recombinant subunit vaccine and adds evidence regarding the importance of envelope neutralizing antibody as a major determinant of immune protection against cell-free virus. Perhaps, because SIV_{mne} is less virulent than SIV_{mac} (Morton *et al.*, 1989), the monkeys were able to tolerate a transient infection under vaccine-induced immunity. This approach could represent an ideal vaccine model for humans if SIV_{mne}, like SIV_{mac}, were to be a relevant model for humans.

C. Macaques/SIV Recombinant Env Expressed in Vaccinia Virus, Baculovirus, or Mammalian Cells

Using the same SIV_{mne} model as above, complete protection against i.v. challenge infection was achieved with a combination recombinant vaccine strategy consisting of vaccinia Env priming and subunit Env boosts (Hu *et al.*, 1992a). Four macaques were immunized with a recombinant vaccinia virus expressing SIV_{mne} gp160 under the control of the vaccinia virus promoter. Cells infected with this vector produced the precursor gp160 as well as the cleavage products gp120 and gp32. Although all four animals showed envelope antibodies detectable by whole virus ELISA and by immunoblot and strong T-cell proliferative responses to SIV_{mne}, no neutralizing antibodies were detected. At weeks 62 and 70, all monkeys were therefore boosted with 0.5 mg of partially purified gp160 produced in recombinant baculovirus-infected insect cells. In all four monkeys a significant (>30-fold) increase in SIV antibody titers ensued and SIV neutralizing antibodies and a virus-specific lymphoproliferative response became detectable. This result follows the example of live vaccinia HIV-1 gp160 (HIVAC)-primed humans in whom boosting a year later with recombinant subunit HIV gp160 greatly enhanced neutralizing antibodies (Graham *et al.*, 1992). Antibodies to human cell antigens were not detected in the protected monkeys. These monkeys were able to completely resist i.v. challenge with 100 ID of the cloned homologous virus and are now 1 year postinoculation. All four unimmunized controls were readily infected but persistent infection was difficult to detect. This is the first example of vaccinia SIV Env protection in the macaque model. Macaques with preexisting immunity to vaccinia virus could also be protected against SIV_{mne}/Ells infected by the same recombinant vaccine (Hu *et al.*, 1993).

In contrast, all experiments using recombinant SIV_{mac} Env (gp120, gp160) immunogens expressed in vaccinia virus, baculovirus, or mammalian CHO cells, alone or in combination, have failed to protect macaques against the more virulent, uncloned SIV_{mac251} or 32H reisolate, whether or not neutralizing antibody was induced (Planelles *et al.*, 1991; Giavedoni *et al.*, 1992, 1993; Mills *et al.*, 1992; Cranage *et al.*, 1992a; Israel *et al.*, 1994; Javaherian *et al.*, 1992; Daniel *et al.*, 1992; Stott *et al.*, 1993; Mehtali *et al.*, 1993; Ahmad *et al.*, 1994). However, suppression of infection, e.g., lower levels of initial antigenemia, was often observed in the vaccinates compared with controls. One reason for the differences in these results may be that the SIV_{mne} challenge stock, in addition to being less virulent, is genetically more homogeneous (i.e., molecular clone) than the SIV_{mac} stock. Indeed, when the SIV_{mne}-protected monkeys described above (Hu *et al.*, 1992a) were later boosted with the same vaccine and then challenged with 10–100 ID of the more heterogeneous (uncloned) SIV_{mne} biologic isolate, they were all susceptible to superinfection (Hu *et al.*, 1992b). However, the vaccinates were able to suppress the infection with the uncloned virus. The findings suggest that the current recombinant HIV Env vaccines may not provide complete protective immunity but they may suppress viral infection. Whether this lower initial viral load will delay disease, as seems likely, remains to be determined.

Other novel SIV vaccine strategies, e.g., use of recombinant human adenovirus (Cheng *et al.*, 1992), recombinant fowlpox virus (Jenkins *et al.*, 1991) chimeric hepatitis B core antigen particles (Yon *et al.*, 1992), or recombinant BCG vaccines (Aldovini and Young, 1990), or virus-like particles (Giavedoni *et al.*, 1992b; Gonzales *et al.*, 1993) are in progress.

D. Attenuated Live SIV

In the history of viral vaccines, attenuated live virus has often been the most successful approach to immunize against conventional acute infectious disease viruses (Hilleman, 1992). Because of the possible activation of cellular proto-oncogenes by proviral integration and possible reversion to virulence, this type of vaccine for retroviruses is considered too dangerous for humans at this time. However, we may eventually have to resort to this approach if other strategies continue to be relatively ineffective. Moreover, much can be learned about protective immunity from this strategy in the experimental animal models. Modified live SIV_{mac} vaccines have been developed at the California and New England Primate Centers from mutants of the SIV_{mac251} biologic isolate. Perhaps not surprisingly, the results with this approach are by far the most optimistic in the SIV vaccine field.

The SIV mutant used at the California center ($SIV_{mac1A11}$) was derived from uncloned SIV_{mac251} and has an open *nef*, point mutations across the genome, and about 98% nucleotide sequence homology with the pathogenic SIV_{mac239} molecular clone (Marthas *et al.*, 1989; Unger *et al.*, 1992b). After a transient infection with $SIV_{mac1A11}$ lasting several weeks, rhesus monkeys seroconvert, become latently infected, remain weakly seropositive and healthy for ±3 years. By this time, all rhesus monkeys inoculated with the wild-type isolate (SIV_{mac251}) have died of SAIDS. Infectious virus can generally not be recovered from the PBMC or bone marrow from these animals after the first few weeks and PCR detection of latent viral DNA in these tissues is only infrequently positive. These animals latently infected with $SIV_{mac1A11}$ were protected against two i.v. challenges, given 11 months apart, with a low dose (1–10 ID) of cell-free SIV_{mac251}, grown in human PBMC (Marthas *et al.*, 1992a). However, these monkeys were not protected against vaginal challenge with the same SIV (Marthas *et al.*, 1992b) (see below). When the 1A11 latently infected monkeys were challenged with a high dose (10^2–10^3 ID) of the parental virus (SIV_{mac251}), also grown in human cells, they became superinfected, but they have survived much longer (>3 years) than the infected control monkeys (Marthas *et al.*, 1990). Similarly, after 1A11 latently infected animals were challenged with a low dose of SIV_{mac} grown in rhesus PBMC, they became superinfected, but the viral load in plasma and PBMC was lower than in unvaccinated controls (Marthas *et al.*, 1992a). These animals have not been observed long enough to determine whether disease onset will be delayed. These results suggest that latent infection with a modified live SIV molecular clone (1A11) can confer a degree of protection against i.v. challenge infection equal to or better than that obtained with inactivated whole virus vaccine and, perhaps more importantly, can also markedly delay the onset of disease after superinfection with a high dose of virulent SIV. The mechanism of protection, however, is not obvious. Antiviral antibody levels are low and antiviral CTL activity is not apparent after the initial period of active infection.

More dramatic results were obtained at the New England center using a *nef*-deleted mutant of SIV_{mac}. Initial pathogenesis studies showed that the *nef* gene of SIV_{mac239}, the pathogenic molecular clone, was required for maintaining high viral loads and inducing disease (Kestler *et al.*, 1991). Six of six monkeys infected with a *nef* deletion mutant remained healthy for 27 months despite the persistence of a low level of infectious virus that maintained this mutation. After this time, two of these animals were found resistant to superinfection with a low dose (10 ID) of SIV_{mac251} parental isolate and two other animals were resistant to the SIV_{mac239} pathogenic molecular clone (both SIV challenge viruses were grown in rhesus PBMC) (Daniel *et al.*, 1992). All controls were readily infected and died of SAIDS. Two of these protected monkeys were then rechallenged with 1000 $TCID_{50}$ of SIV_{mac251} 4 weeks after the first challenge. Both animals also resisted this high challenge dose, at least for the period observed (6 months).

Of all vaccines tried so far in the virulent SIV_{mac} system, the attenuated live virus approach has elicited the strongest (versus 1000 ID i.v.) and most durable (>2 years) immunity and has had the most dramatic effect on prolonging life in animals that are superinfected with SIV. The *nef* deletion mutant has also been the only vaccine able to completely protect monkeys against infection by SIV grown in rhesus PBMC. The higher level of viral replication *in vivo* with the *nef* mutant of SIV_{mac239} elicits a stronger and more durable humoral immune response compared with the 1A11 mutant of the same parental virus. This immune response may account for the stronger protection obtained with the *nef* mutant. However, the mechanism of protection, especially the role of cell-mediated immunity, needs to be further evaluated. Moreover, it is possible that protection occurs because of some kind of virus–virus interference at the intracellular level rather than via a conventional immune response. It remains to be seen how well this vaccine approach will protect against heterologous SIV strains, genital mucosal infection, and challenge with infected cells. The findings suggest the possibility of developing other attenuated live SIV and HIV vaccines by deleting *nef* and other genes (e.g., *vpu, vpx, vif*) (Desrosiers, 1992). Planned are immunogenicity and efficacy tests of similar HIV-1 deletion mutants in chimpanzees (Desrosiers, personal communication). Confirmation that monkeys infected with *nef* deletion mutants of SIV_{mac} are protected against challenge infection with SIV has recently been reported (Almond *et al.*, 1993b; Mehtali *et al.*, 1993).

E. Mucosal Immunity

Because, on a global scale, sexual transmission of HIV is the major route of virus spread, genital mucosal immunity must be a goal of AIDS vaccines. Vaccine protection must also be effective against cell-associated virus spread via this route and this implies the need for strong CMI, especially CTL activity. This vaccine strategy may take advantage of induction of immunity at one mucosal site which may then seed other mucosal sites with immune activated B and T cells via the "common mucosal immune system" (Mestecky, 1987). Most studies of mucosal immunization have been done in rodents. For example, vaginal immunization of rats has been done with a synthetic peptide from the HIV-1 envelope (O'Hagan *et al.*, 1992). Secretory immune responses have been induced in the mouse vagina by parenteral or intravaginal immunization with an immunostimulating complex (Thapar *et al.*, 1991). Also in mice, oral immunization with HIV gp120 in combination with liposome and/or cholera toxin elicited antigen-specific mucosal (IgA) and systemic (IgM, IgG) humoral immune responses (Merrill *et al.*, 1992). The SIV macaque model now appears quite suitable and timely for evaluating mucosal immunity with respect to candidate AIDS vaccines (for review, see C. J. Miller *et al.*, 1993).

The feasibility of experimental SIV_{mac} infection of macaques via the intact genital mucosa has been demonstrated, although a large cell-free virus inoculum is required (Miller *et al.*, 1989). Vaginal mucosal Langerhans cells appear infected during chronic SIV infection (C. J. Miller *et al.*, 1992), but it is not yet known if they are a target for acute infection in this model. SIV infection apparently damages the mucosal IgA immune system, impairing local SIV-specific IgA responses. Initial efforts to protect against vaginal mucosal infection by inactivated whole and modified live SIV vaccines that did protect against i.v. infection were not successful (Marthas *et al.*, 1992b). However, five of six macaques were recently shown to be protected against vaginal infection with cell-free SIV_{mac} by immunization with an inactivated whole SIV microcapsule vaccine given i.m. × 3 followed by boosts × 4 via the oral route or intratracheally (Marx *et al.*, 1993). Seven of eight controls were infected. After a second vaginal challenge, three of four i.m.-primed and mucosally boosted macaques remained protected. Although vaginal fluid contained SIV antibodies, the mechanism of protection remains unclear. Anti-human cellular antibodies were apparently not involved. Oral immunization alone gave no protection. These results provide the first evidence that protection against vaginal transmission of HIV is possible through vaccination.

An SIV Gag p27 vaccine made by yeast (Ty-VLP) covalently linked to cholera B subunit, was given to macaques vaginally or rectally followed by an oral boost (Lehner *et al.*, 1992). The vaccine elicited secretory IgA and IgG anti-p27 antibodies in vaginal or rectal fluid and in serum. Virus-specific T and B cells were stimulated in the regional lymph nodes. Challenge experiments with live virus are awaited.

Macaques immunized with SIV gp160 enterically coated liposomes by oral priming followed by i.m. boosts have made strong IgA antibody responses in the saliva and GI tract, but the monkeys have not yet been challenged (Thibodeau *et al.*, 1992). The possibility of oral tolerance must also be considered.

Very recently, two of two chimpanzees were shown to be susceptible to infection with about 300 HIV-1-infected allogeneic PBMC placed in the cervical os (Girard *et al.*, 1992). HIV-1-vaccinated chimpanzees will be challenged by this route next year.

Studies at the Wisconsin Primate Center suggest that the rectal mucosa represents a selective barrier to virus and that an effective mucosal vaccine may therefore need to target only a limited portion of the complex virus population (Meyer *et al.*, 1993).

Thus, it remains to be determined whether mucosal immunity is required to protect against genital or rectal infection with SIV or HIV and whether mucosal immunity can be induced by immunization.

F. Passive Immunotherapy

Passive immunization experiments in the SIV macaque model have given conflicting results (Table VII). In one report (Putkonen *et al.*, 1991), sera (9 ml/kg) from a healthy SIV-infected cynomolgus macaque protected 3 of 4 monkeys against infection from i.v. challenge 6 hr later with 10 ID of live SIV_{sm}. Prechallenge viral antibody titers in individual animals did not correlate with protection. In another study (Lewis *et al.*, 1993b), passive transfer of plasma (5 ml/kg) from pigtailed macaques that had been protected with a recombinant SIV_{mne} Env with SIV_{mne} peptide vaccine (Shafferman *et al.*, 1991) protected 3 of 6 recipients from challenge infection given 24 hr later. Moreover, 9 of 11 monkeys receiving

TABLE VII. Passive Vaccination against SIV Infection of Macaques

Source of antibodies	Virus strain	Efficacy	Reference
Infected macaque	SIV_{sm}	Yes	Putkonen *et al.* (1991)
	SIV_{mne}	Yes	Lewis *et al.* (1993b)
Vaccinated macaque	SIV_{mne}	Yes	Lewis *et al.* (1992)
	SIV_{mac}	Yes	Gardner *et al.* (in press)
Infected macaque	SIV_{mac}	No	Gardner *et al.* (in press)
	SIV_{mac}	No	Stott *et al.* (1993)
Vaccinated macaque	SIV_{mac}	Yes	Coulibaly *et al.* (1993)

plasma from a healthy SIV_{mne}-infected monkey were passively protected against this virus strain.

In contrast, recent results from UC Davis (Gardner *et al.*, in press) indicate that plasma (9 or 25 ml/kg) or purified Ig (400 μg/kg) with a high neutralizing antibody titer from healthy SIV_{mac}-infected rhesus monkeys failed to protect any of six recipients challenged 4–18 hr later or 24 hr before with 10 ID of homologous virus. Moreover, persistent antigenemia occurred promptly after challenge in five of the six animals. On the other hand, plasma from a vaccine protected monkey with little neutralizing antibody but high anticellular antibody levels protected three of eight recipients against the same challenge and these animals had only transient antigenemia. These passive antibody transfer findings are consistent with the important role of anti-human cellular antibodies as the mechanism of immune protection by the inactivated whole SIV vaccines and suggest the possibility of enhancing antibodies in the plasma of the SIV-infected donor animal. Failure of passive immunization using pools of macaque serum, purified Ig, or mouse monoclonal antibodies, to protect against SIV_{mac} challenge, despite high titers of neutralizing antibodies, has been reported recently by others (Stott *et al.*, 1993). The importance of anti-human cellular antibodies in immune protection from inactivated whole SIV vaccines was confirmed by Coulibaly *et al.* (1993), who showed that the passive protection was lost after absorption of the antibodies by the uninfected 8166 human cells.

Two recent studies on SIV vaccines are noteworthy and may have relevance to HIV vaccines. One study indicates that, rather than inducing sterilizing immunity, certain SIV whole virus and recombinant env vaccines lower the level of challenge viremia, a result that likely may lead to a longer survival (Giavedoni *et al.*, 1993; Ahmad *et al.*, 1994; Hoover *et al.*, 1993; Shafferman *et al.*, 1991; Hirsch *et al.*, 1993b; Hu *et al.*, 1992b; Daniel *et al.*, 1992). The second set of data indicate that macaques previously exposed to subinfectious or minimally infectious doses of SIV may subsequently be protected against higher dose infection; this result correlates with a lack of SIV-specific humoral immunity in the presence of strong SIV-specific T-cell immunity and a favorable balance of TH1 versus TH2 cell subsets (Benveniste *et al.*, 1993b; Pauza *et al.*, 1993a,b). In certain high-risk humans, the lack of acquiring HIV infection also correlates with strong anti-HIV T-cell immunity and TH1 > TH2 cells in the absence of humoral immunity (Clerici and Shearer, 1993). Thus, the growing sentiment is that AIDS vaccines should aim to induce a relative predominance of anti-HIV cellular versus humoral immunity (Salk *et al.*, 1993).

G. B- and T-Cell Epitopes of SIV Envelope and Core in SIV-Vaccinated or -Infected Monkeys

Reviews of the cellular and humoral antigenic epitopes in HIV and SIV have recently been published (Nixon *et al.*, 1992; Voss and Hunsmann, 1993). The functionally important regions within the envelope glycoprotein of SIV_{mac} and HIV-1 are shown in Fig. 10. Antibody responses to four SIV Env epitopes, expressed as β-galactosidase fusion proteins, were analyzed in three macaque species experimentally infected with SIV_{mne} (Shafferman *et al.*, 1989a). The four epitopes selected—one in the external envelope (amino- and carboxy-termini)

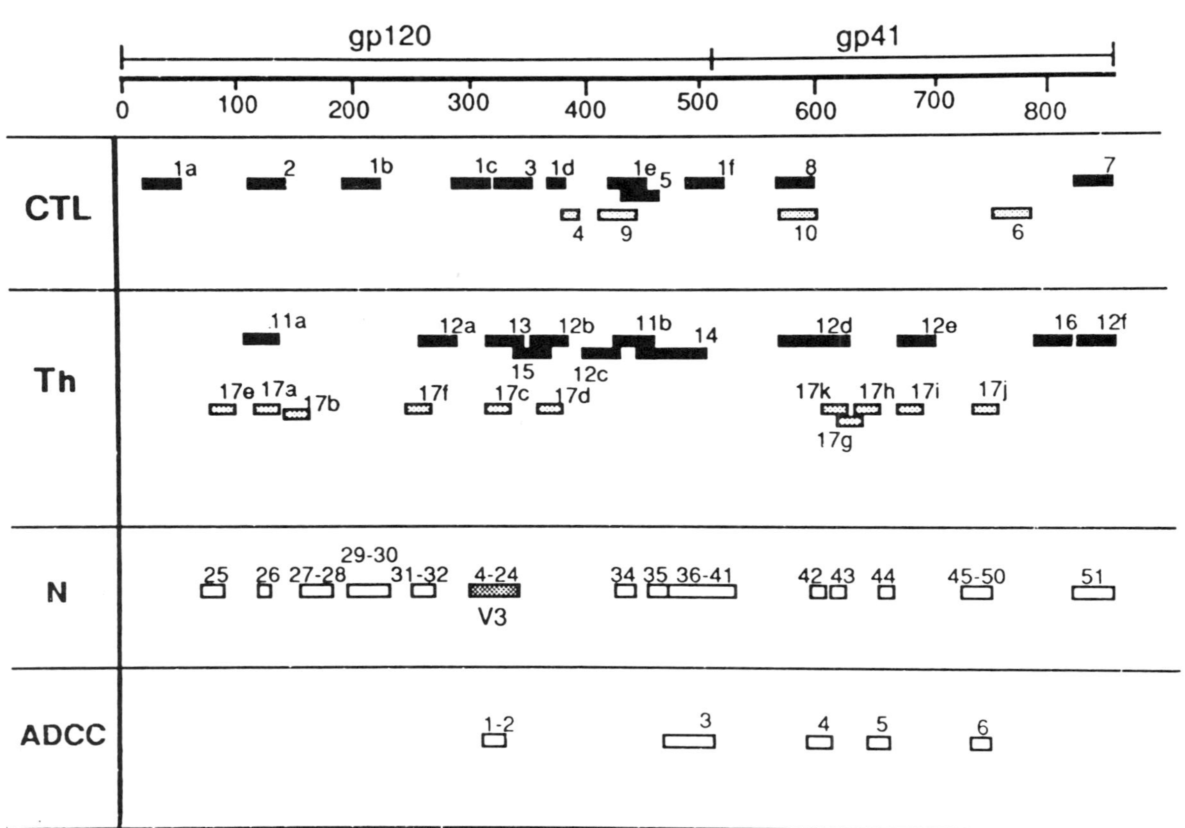

FIGURE 10. Humoral and cellular epitopes in HIV and SIV: functionally important regions within the envelope glycoprotein. For the CTL epitopes, black symbols represent regions with identified HLA restrictions, and which have been confirmed by more than one group. Gray symbols represent regions identified by one group only but where HLA restriction is known. (From Nixon *et al.*, 1992.)

and three in the transmembrane envelope (amino- and carboxy-termini)—corresponded to immunodominant HIV-1 Env epitopes recognized by HIV-1-infected humans (Shafferman *et al.*, 1989). The SIV_{mne} epitopes were hydrophilic and about 50% homologous in sequence to the corresponding HIV-1 Env peptides. The pattern of antibody response to the SIV Env epitopes in SIV_{mne}-infected macaques resembled that previously observed for the corresponding HIV-1 epitopes. The epitope (#582) located at the amino-terminus of the transmembrane envelope protein (gp41) was particularly immunodominant. This highly antigenic region of SIV and HIV-1 gp41 has previously been identified in other studies (Gnann *et al.*, 1987; Johnson *et al.*, 1988; Norrby *et al.*, 1989). These SIV_{mne} Env–Gal chimeric proteins, used as a vaccine, were capable of inducing neutralizing antibodies and conferring partial protection to SIV_{mne} challenge infection (Shafferman *et al.*, 1991) (see Section XIV).

Efforts to map neutralization epitopes of SIV_{mac}, SIV_{sm}, and HIV-2 Env have been made using synthetic peptides and monoclonal antibodies. Researchers in Sweden found that sera from monkeys or humans immunized with whole killed SIV_{sm} or HIV-2 or infected with SIV_{sm} or HIV-2 reacted mainly to synthetic peptides from the V3 region of HIV-2 (Mannervik *et al.*, 1992). Using sera of SIV_{sm}-infected cynomolgus macaques, four antibody regions were identified in the Env proteins corresponding to the second variable region (V2 aa 170–196), the region homologous to V3 in HIV-1 (aa 313–346), the carboxy-terminus of gp120 (aa 514–537), and the amino-terminus of the transmembrane proteins (aa 608–688) (Samuelsson *et al.*, 1993).

A synthetic peptide representing part of the putative V3 loop of SIV_{mac} induced a moderate titer of neutralizing antibody in goats and bound to antibodies in the sera of macaques protected by an inactivated whole SIV vaccine (Gardner *et al.*, 1991). However, there was, in general, a lack of specificity of this reaction with SIV vaccine protection centers and with neutralizing antibody titers and an inability to completely absorb SIV neutralizing antibody with this peptide. Yet another study (Javaherian *et al.*, 1992) using linear peptides corresponding to the V3 loop of SIV_{mac} envelope did not elicit neutralizing antibodies. These results indicated that the principal neutralizing determinant (PND) of SIV_{mac}, in contrast to HIV-1, is not represented by a linear peptide corresponding to the V3 loop of SIV envelope. More recently, the V3 loop of SIV_{mac} was shown to be a component of a conformation-dependent neutralizing domain (Javaherian *et al.*, in press). Furthermore, on serial analysis of SIV_{mac} molecular clones *in vivo* the V3 region of SIV_{mac} appeared much more conserved than the HIV-1 V3 loop (Burns and Desrosiers, 1991; Johnson *et al.*, 1990b; Overbaugh *et al.*, 1991; Almond *et al.*, 1992a).

Binding of sera from infected vaccine-protected monkeys to overlapping synthetic peptides representing the SIV_{mac} envelope glycoprotein revealed broadly reactive antigenic determinants (M. A. Miller *et al.*, 1992) and a strong B-cell neutralizing and T-helper cell epitope in the COOH-terminal (V4) region (amino acids 414–434) of SIV_{mac} (Torres *et al.*, 1993a). Two other B-cell epitopes, one corresponding to the V2 region (aa 166–181), and another which may be conformational (aa 254–313), were also identified (Benichou *et al.*, 1992, 1993; Torres *et al.*, 1993b). Sequence variation in the V1, V2, V4, and V5 regions of SIV molecular clones occurs during infection and may contribute to avoidance of immune recognition and to persistence of SIV infection (Burns and Desrosiers, 1991;

Johnson and Hirsch, 1992; Almond *et al.*, 1992a; Overbaugh and Rudensey, 1992). However, whether any of these B-cell epitopes are critical to immune protection remains to be proven.

A monoclonal antibody (MAb) that recognizes a type-specific neutralizing determinant in the transmembrane protein (gp41) (amino acids 106–110) of SIV_{mac} has been derived (Kodama *et al.*, 1991). Variation occurred in this epitope during persistent infection *in vivo*. Neutralizing MAbs directed at linear epitopes in the NH_2-terminal moiety (V2 region) of SIV_{mac} gp120 were also derived (Benichou *et al.*, 1992; Kent *et al.*, 1992). Neutralizing MAbs were also mapped to conformational epitopes in SIV gp120 (Kent *et al.*, 1992). We can conclude from these findings that the NH_2- and COOH-terminal portions of SIV gp120 and the NH_2-terminal portion of gp41 contain both conformational and linear neutralizing determinants.

Serum antibody reactivities to SIV_{mac} Env peptides in SIV_{mac} (32H)-infected macaques showed that the majority of infected animals recognized three linear antigenic determinants within the V1, V3, and C5 regions and the above-mentioned conformational epitope (McBride *et al.*, 1993). Animals immunized with a formalin-inactivated whole SIV_{mac} (32H) vaccine produced antibodies to the same peptides but fewer animals recognized the C5 region.

A putative complement-dependent enhancing antibody epitope has also been identified in SIV_{mac} based on homology with HIV-1 and HIV-2 enhancing epitopes (Robinson *et al.*, 1988; Robinson and Mitchell, 1990). This epitope (amino acids 601–635) (Gnann peptide) corresponds to an immunodominant domain of the HIV-1 transmembrane protein gp41 (amino acids 579–613). A second immunodominant domain in gp41 (amino acids 644–663) is also an enhancing epitope in HIV-1. Whether or not the *in vitro* demonstration of SIV enhancing antibodies in SIV-infected (Montefiori *et al.*, 1990b; Le Grand *et al.*, 1991) or vaccine-protected monkeys (Montefiori *et al.*, 1990a) has any *in vivo* biologic significance remains to be determined.

Monoclonal antibodies to the SIV_{mac} major core protein (p27) recognize nine different epitopes within the major core protein of SIV and HIV-1 and have allowed segregation of SIV isolates into six epitope groups, one of which is unique to SIV_{mac} (Higgins *et al.*, 1992). In another study (Kent *et al.*, 1992), 16 MAbs to SIV_{mac} p27 were made but only 3 of these showed any reactivity with a series of 22 overlapping peptides covering p27. These p27 MAbs are useful for the development of antigen capture assays that distinguish SIV isolates and immunohistochemical reagents for detection of SIV infection (Toedter *et al.*, 1990; Sutjipto *et al.*, 1990a) (Section XII).

In a recent collaborative study, 13 laboratories characterized a coded panel of ten murine MAbs to SIV_{mac251} envelope protein by several different binding and neutralization assays (D'Souza *et al.*, 1993). Despite the diverse neutralization assays employed, agreement between the different laboratories was sufficient to assign the MAbs to four neutralization groups based on ability to cross-neutralize and bind different SIV isolates. Moreover, there was a surprising degree of agreement between the laboratories on the relative potency of the MAbs. One group of antibodies neutralized and bound to both SIV_{mac} and SIV_{sm} isolates. A single antibody neutralized only SIV_{mac} but bound to SIV_{mac} and SIV_{sm} isolates. A third group neutralized only SIV_{mac} isolates and did not bind or neutralize SIV_{mac} or SIV_{sm} isolates. The fourth group of antibodies, the largest number, were nonneutralizing but did bind to both rhesus macaque and sooty mangabey isolates. This information will help to further define the neutralizing and binding epitopes of the SIV envelope that are critical to immune protection in comparison to HIV-1 and will facilitate development of improved diagnostic assays and candidate vaccines.

In a study of T-cell epitopes in the SIV Env protein, $CD4^+$ T-cell clones from macaques immunized with inactivated SIV or recombinant SIV Env proteins recognized one to seven peptides in four distinct regions of the gp130 including both variable and conserved domains (Jones *et al.*, 1992) (Fig. 11). These T-cell epitopes within conserved regions of the Env protein together with the B-cell epitopes recognized by neutralizing antibodies may be the basis of a synthetic vaccine design.

A CTL epitope in the SIV Gag protein and its MHC class I restriction molecule ("Mamu-A-01") have been described above (Section XIII.D.2.b) in SIV-infected macaques (Yamamoto *et al.*, 1990a). When monkeys of this particular MHC class I genotype were immunized with a live recombinant vaccinia virus expressing SIV_{mac} Gag, they generated a $CD8^+$ CTL response specific for the defined Gag epitope (Shen *et al.*, 1991, 1993). Similar results were obtained by Gotch *et al.* (1991). Immunization of monkeys with a synthetic peptide represent-

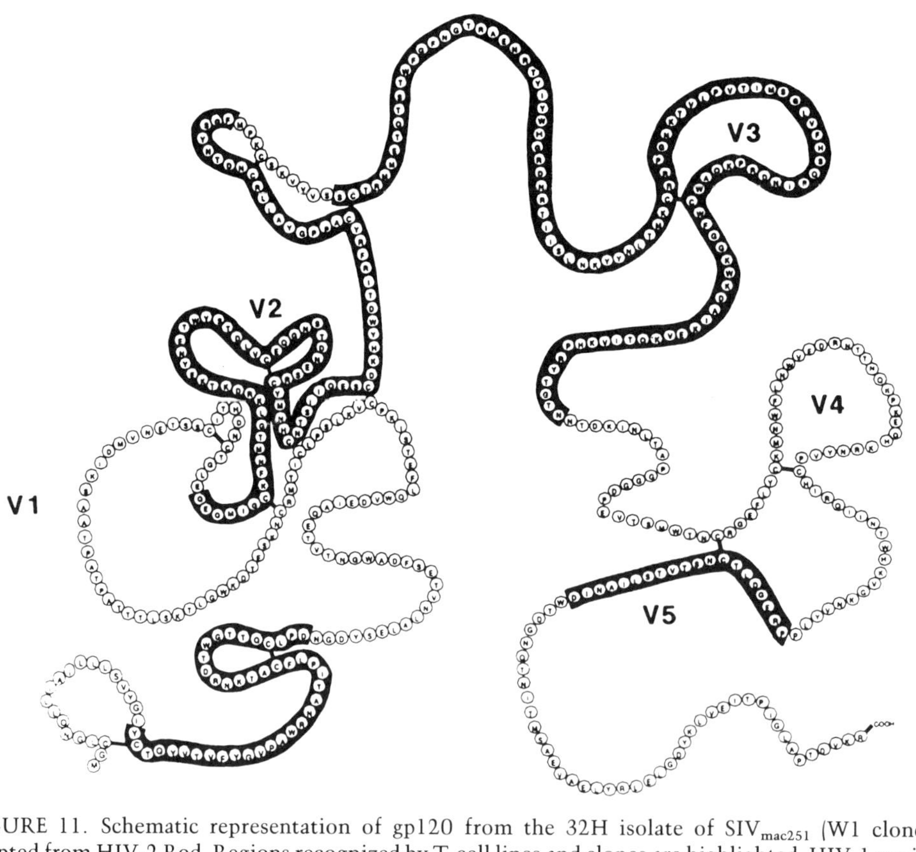

FIGURE 11. Schematic representation of gp120 from the 32H isolate of SIV_{mac251} (W1 clone), adapted from HIV-2 Rod. Regions recognized by T-cell lines and clones are highlighted. HIV-1 equivalent variable regions are indicated as V1, V2, V3, V4, and V5. (From Jones *et al.*, 1992.)

ing the Gag epitope (p11C) in mineral oil adjuvant also elicited MHC class I-restricted SIV-specific CD8$^+$ CTL in rhesus monkeys (Yasutomi *et al.*, 1993b).

XV. ANTIVIRAL AGENTS AGAINST SIV

The same attributes that make SIV infection of macaques an excellent model for AIDS pathogenesis and vaccine studies also make it an excellent model for antiviral therapy and immunotherapy. The model is now well defined and, for drug testing, the biologic endpoints for acute and chronic infection are well characterized. For acute infection, drug effects on viral replication and infection-susceptible cells can be measured easily by plasma antigen, seroconversion, virus titration, and lymphocyte subset analysis (Ljungdahl-Stahle *et al.*, 1992; Wyand, 1992). The same endpoints are assayed in the chronic infection model except that onset of disease is added. An advantage of the chronic model compared with human trials is that early intervention with highly experimental therapies is possible in the asymptomatic stage of infection (Wyand, 1992).

The principal target of HIV antiviral therapy in humans is the viral reverse transcriptase (RT) enzyme. However, virtually every step in the replication of HIV (or SIV) can serve as a target for new therapeutic intervention (Mitsuya *et al.*, 1990). Such targets include the CD4 receptor and the viral *tat, rev,* and protease genes. 3′-Azido deoxythymidine (AZT) and five of five nucleoside analogues were found to completely inhibit SIV RT and viral replication *in vitro* in highly permissive B- and T-cell lines (Tsai *et al.*, 1990) and the RT of SIV expressed in bacteria was shown to have the same *in vitro* kinetics of AZT inhibition as HIV-1 (Wu *et al.*, 1988; Prasad *et al.*, 1990). Nonnucleoside inhibitors of RT (i.e., TIBO derivatives) are also effective against SIV replication *in vitro* (Debyser *et al.*, 1991, 1992). AZT-resistant mutants of SIV were derived by *in vitro* selection (Van Rompay *et al.*, 1992a). Uninfected macaques have been used for pharmacokinetic and toxicity studies with several of the dideoxynucleosides and the responses were similar in both macaques and humans (Good *et al.*, 1990; Kaul *et al.*, 1989; Russell *et al.*, 1990; Boudinot *et al.*, 1990). When administered to near-term pregnant macaques, AZT was shown to readily cross the placenta but not to accumulate in the fetus (Lopez-Anaya *et al.*, 1990).

A number of studies have tested the *in vivo* prophylactic or therapeutic effect of anti-RT drugs on SIV expression or immune function and disease course in SIV-infected macaques (Table VIII). AZT, dideoxycytidine (ddC), 9-(2-phosphonylmethoxyethyl)adenine (PMEA), and foscarnet have all been shown to have a beneficial prophylactic effect on SIV by reducing SIV infectivity levels, or decline of CD4 cells, and delaying disease if given to SIV-infected monkeys before or within a few hours after infection, but little or no effect if given therapeutically after infection was established (Lundgren *et al.*, 1989, 1991; Martin *et al.*, 1990; McClure *et al.*, 1990; Fazely *et al.*, 1991; Balzarini *et al.*, 1991). The prophylactic or therapeutic effect of treatment during acute infection depends on the challenge dose and the time of starting the treatment. Most reports have used a fairly large infectious dose of SIV (>10 ID). More specifically, AZT given in a dose of 100 mg/kg per day × 28 days to rhesus monkeys starting 1 to 8 hr prior to i.v. challenge with 10–50 ID_{50} of SIV_{sm} did not prevent infection but did slightly suppress viremia and delay the fall in CD4 cells (Martin *et al.*, 1990). AZT treat-

TABLE VIII. Antiviral Agents Tested *in Vivo* in the SIV Macaque Model

Agent	Dose and schedule	Results	Reference
AZT	100 mg/kg/day × 28 days starting 1–72 hr after infection	Virus suppression Bone marrow function improved	Watanbe *et al.* (1989, 1991a)
AZT, ddC, or foscarnet	20 mg/kg every 8 hr × 7–9 days starting 8 hr before infection	Virus suppression Clinical improvement Delayed disease	Martin *et al.* (1990, in press)
PMEA	5–10 mg/kg/day × 28 days starting 1 day before infection	Virus suppression, dose dependent	Balzarini *et al.* (1991)
AZT	50–100 mg/kg/day starting 3 hr after infection	No effect on virus or disease	Fazely *et al.* (1991)
AZT + IFN-α	As a base—400,000 U/kg/day IFN-α	Virus supression	Fazely *et al.* (1991)
AZT	20 mg/kg × 10 days starting 8 hr before infection	Virus suppression	Lundgren *et al.* (1991)
3-Fluorothymidine (3FT)	20 mg/kg × 10 days starting 8 hr before infection	Virus suppression	Lundgren *et al.* (1991)
AZT	100 mg/kg × 14 days starting 1, 24, or 72 hr after infection with rapidly fatal $SIV_{smPBj14}$ strain	Virus suppression and delayed death in monkeys treated very early	McClure *et al.* (1991)
AZT, AZddT, D4T, or FLT	100 mg/kg × 14 days starting 24 hr before infection with SIV_{sm}	No effect on virus or disease	McClure *et al.* (1991)
AZT	50 mg/kg twice daily × 28 days starting 24 hr before infection	Virus suppression	Tsai *et al.* (1992)
AZT	50 mg/kg every 8 hr for 6–10 weeks starting 2 hr before infection with a low dose of SIV	Infection prevented	Van Rompay *et al.* (1992b)
AZT	50 mg/kg every 8 hr for 10 weeks starting 6 weeks after infection with a low dose of SIV	No effect on course of disease	Van Rompay *et al.* (1992b)
AZT	200 mg/kg/day × 208 days after infection established	Decreased antigenemia Delayed decrease in $CD4^+$ cells	Wyand (1992)
3FT	15 mg/kg/day × 10 days starting 8 hr before infection	No effect at high dose Prevented infection with low dose	Bottiger *et al.* (1992b)
CD4	2 mg i.m. every day × 50 days	Virus suppression Bone marrow function improved	Watanabe *et al.* (1989, 1991a)
α-Glucosidase inhibitor (SC49483)	250 and 750 mg/kg q.d. × 208 days	Decreased late antigenemia	Bryant *et al.* (1993)

ment started 24 or 72 hr after infection had less effect on these parameters. Most importantly, animals treated within 1 to 8 hr of infection tended to live slightly longer than those treated after 24 or 72 hr which resembled untreated controls (Martin *et al.*, 1993). AZT and IFN-α given for 21 days (Fazely *et al.*, 1991) and AZT or 3-fluorothymidine (3FT) given for 10 days (Lundgren *et al.*, 1991) decreased the amount of plasma p27 antigen during the initial burst of viral replication. The effect of 3FT treatment depended on the dose of virus given, a reduced dose of virus resulting in a later peak of antigenemia. At an inoculation of 2–10 ID, 3FT treatment prevented infection (Bottiger *et al.*, 1992b). A similar suppressive effect of AZT on plasma antigenemia was seen when AZT was given for 14 days starting within 1 hr after infection with the lethal SIV_{sm} variant PBj14 (McClure *et al.*, 1991). Infrequent but higher dosage of AZT (50 mg/kg twice daily) appeared more effective than the same total dose given three or four times daily (Tsai *et al.*, 1993a). AZT, 50 mg/kg given every 8 hr, prevented infection of infant rhesus monkeys with a low dose (4–10 ID) of SIV_{mac} but had no effect on the course of disease when treatment was started after infection had been established (Van Rompay *et al.*, 1992b).

These findings suggest that recent reports indicating that prophylaxis with AZT soon after human exposure to HIV-1 fails to prevent infection may reflect exposure to too high an infectious dose (Lange *et al.*, 1990; Looke and Grove, 1990). Long-term treatment (208 days) of chronically infected macaques with AZT (200 mg/kg per day) resulted in decreased levels of antigenemia and a delayed decrease in CD4 cells (Wyand *et al.*, 1992). Further observation is required to determine whether disease risk will be delayed. Similar results and slightly prolonged survival may occur with AZT treatment of infected humans (Fischl *et al.*, 1987; Chaisson *et al.*, 1988; Volberding *et al.*, 1990). The most prominent toxic effect seen in the AZT- and FLT-treated monkeys at the high doses used was anemia of the same kind as found in AIDS patients treated with AZT. Treatment of longtailed macaques (*M. fascicularis*), chronically infected with SIV_{mne}, with PMEA was much more effective in reducing the level of virus than was treatment with AZT (Tsai *et al.*, 1993c; Follis *et al.*, 1993). Prophylaxis with PMEA also had a much more dramatic anti-SIV effect as compared with AZT (Tsai *et al.*, 1993c). However, drug-resistant mutants do arise in PMEA-treated macaques (Sabo *et al.*, 1993).

Passive administration of recombinant soluble human CD4 to SIV-infected rhesus monkeys had a transient beneficial effect on virus expression accompanied by an improvement in bone marrow function but only as long as administered (50 days) (Watanabe *et al.*, 1989). It was then reported that soluble human CD4 elicited antibodies in rhesus monkeys that inhibited SIV replication *in vitro* (Watanabe *et al.*, 1991a). SIV-infected rhesus monkeys actively immunized with recombinant soluble human CD4 in Freund's complete adjuvant developed cross-reacting antibodies to their own CD4 and showed a dramatic decrease in SIV infectivity and concomitant increase in bone marrow colony-forming activity (Watanabe *et al.*, 1991b). Similarly, recombinant soluble CD4-immunized chimpanzees developed CD4 antibody responses with anti-HIV activity (Watanabe *et al.*, 1992). These findings suggest that a modified human CD4 immunogen might elicit a beneficial antibody response in HIV-infected individuals.

The hazards of bypassing animal models and using only the conventional laboratory-adapted virus strains and cell lines for measuring antiviral drug testing were recently emphasized by the finding that much higher concentrations of recombinant soluble CD4 are required to neutralize primary HIV-1 isolates as compared with laboratory strains of HIV-1 (Daar *et al.*, 1990). CD4 immunotoxins such as the CD4-pseudomonas exotoxin hybrid protein inhibit SIV and HIV-2 as well as HIV-1 in cell culture (Berger *et al.*, 1989; Ashorn *et al.*, 1992) but have not yet been tested *in vivo* in the SIV macaque model. Passive transfer to rhesus macaques of a CD4-specific mouse MAb capable of blocking *in vitro* replication of HIV and SIV resulted in extensive CD4 lymphocyte coating without inducing cell loss or immunosuppression (Reimann *et al.*, 1993). *In vivo* testing of this treatment is now under way in SIV-infected macaques.

The SIV macaque model should be particularly useful for measuring antiviral drug action against acute infection via different challenge routes and for testing the pathogenicity of drug-resistant mutants of SIV RT (Richman, 1990) or other viral targets. The genital mucosal route for experimental transmission of SIV is suitable for antiviral drug testing as illustrated by the partial inhibition of SIV vaginal infection by the spermicide nonoxynol-9 (C. J. Miller *et al.*, 1990). These kinds of studies are now timely because of the ability to induce SAIDS with molecular clones of SIV (Kestler *et al.*, 1990a). This model will also aid in the selection of candidate drugs or combinations of drugs when the number of HIV-infected individuals available for clinical trials are limited or for efforts to block vertical transmission of the virus where questions of possible fetal toxicity impede such studies being done on pregnant women. At present, the SIV model has not been used for therapy directed at the opportunistic infections (e.g., CMV) or B-cell lymphomas that develop in infected macaques (Feichtinger *et al.*, 1992) similar to HIV-infected humans. However, treatment of cynomolgus monkeys with four different compounds which inhibit herpesviruses did not affect the course of acute SIV infection (Bottiger and Oberg, 1993). These results indicate that transactivation by a replicating herpesvirus is not an essential factor in acute SIV infection *in vivo.*

Inhibition of SIV_{mac} replication in a human T-cell line (C8166) was demonstrated with peptide derivatives that block HIV-1 and SIV protease activity (Meek *et al.*, 1990; Ashorn *et al.*, 1990; Block *et al.*, 1993). Treated cultures displayed a lack of cleavage of the Gag precursor protein (p55) and accumulated immature lentiviral particles. Protease inhibitors have recently been tested against SIV infection *in vivo* (Martin *et al.*, 1994). Other novel antiviral agents such as antisense RNA, ribozymes, or transdominant mutants remain to be tested in this *in vivo* model.

Treatment of SIV-infected macaques with a MAb directed against the extracellular domain of the human IFN-α receptor blocked the binding of IFN-α to PBMC, inhibited the biologic action of IFN, and markedly reduced the loss of $CD4^+$ cells in SIV-infected animals (Tovey *et al.*, 1993). Total-body irradiation of SIV-infected macaques was well tolerated and may have therapeutic benefit (Fultz *et al.*, 1993).

Undoubtedly, the expanding AIDS epidemic will continue to foster early testing of experimental drugs (and vaccines) in human clinical trials. However, the need to evaluate novel therapies in this model system will also increase unless more effective AIDS therapies are found in the near future (Wyand *et al.*, 1992).

XVI. POSTINFECTIOUS IMMUNOTHERAPY

Based on Salk's hypothesis (Salk, 1987), several studies have been done on this model to test the efficacy of postinfectious immunotherapy with inactivated SIV vaccines. The first study used a γ-irradiated whole SIV vaccine with IFA adjuvant given once 4 months after infection (Gardner *et al.*, 1989). Another study used a glutaraldehyde-fixed whole virus-infected cell vaccine with Quil-A as adjuvant given 4, 8, 12, and 36 weeks after infection (Stott *et al.*, 1990). The third study used a formalin-inactivated whole SIV vaccine given with MDP administered every month for 11 months starting 1 month after infection (Murphey-Corb *et al.*, 1990a). None of these experiments showed any detectable change in virus titer, immune status, or clinical course, even though the inactivated whole virus vaccines were successful in protecting naive monkeys against challenge infection. Further efforts to show a beneficial immunotherapeutic effect of modified live virus or recombinant Env vaccines in this animal model are warranted, particularly in view of the apparent clinical benefit obtained by treating HIV-1-infected humans with recombinant HIV-1 Env vaccines (Redfield and Birx, 1992).

XVII. TYPE D RETROVIRUSES

The type D retrovirus genus comprises an exogenous group of simian retroviruses (SRV) that are indigenous to Asian macaques. Related endogenous type D viruses occur in the Old World spectacled langur (*Presbytis obscuris*) and New World squirrel monkey (Section II). SRV of macaques may have had its evolutionary origin from the endogenous langur viruses (PO-I-Lu) (Benveniste and Todaro, 1977). Only the exogenous type D viruses of macaques are pathogenic. In macaques, the type D viruses are horizontally transmitted, nononcogenic, but potentially immunosuppressive (Gardner *et al.*, 1988). Based on distinct neutralization patterns with sera from naturally infected monkeys, several serotypes (SRV 1–5) have been described at different primate centers. These viruses may cause a profound depletion of T and B cells and a fatal immunodeficiency syndrome in macaques resembling human AIDS. The immunosuppressive disease induced by type D retrovirus was initially called simian AIDS but now, to avoid confusion, this term is generally restricted to the fatal immunosuppressive disease caused by SIV. SRV is not closely related to HIV or SIV (Bryant *et al.*, 1985). The type D viruses are characterized by the formation of cytoplasmic type A precursor core particles, and extracellular mature, enveloped particles morphologically distinct from those formed by type B retroviruses or HIV and SIV (Fig. 12) (see below). The type D viruses share Mg^{2+} preference for RT activity, $tRNA^{Lys}$ primer, a similar genomic organization, and extensive genomic homology (see below).

A. Epidemiology

Since 1985, type D retroviruses (SRV) have been identified as the causative agents of a naturally occurring infectious immunodeficiency disease in eight spe-

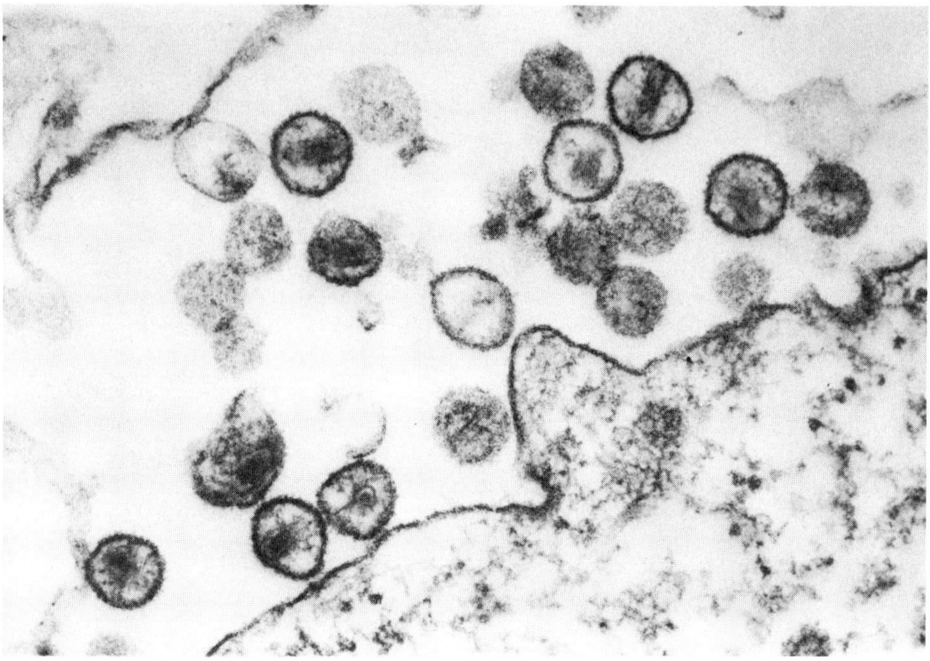

FIGURE 12. Electron micrograph of type D (SRV-1) virus particles (×125,000). [Thanks go to R. Munn (UC Davis) for his help with the electron microscopy.]

cies of macaques at five of the seven primate centers in the United States (Gardner *et al.*, 1988). Outbreaks of SRV-induced diseases continue to occur in macaques in primate facilities throughout the world. The NIH-sponsored primate centers that have been affected with this disease are New England, California, Oregon, Washington, and Wisconsin. SRV infection is highly prevalent in Asian macaques in captivity and feral macaque species in India and Indonesia are the natural hosts of the type D retroviruses. The prevalence of infection with the different SRV serotypes and associated diseases in these feral animals remains unknown. In certain breeding colonies such as that at the California Primate Center in 1983–1984, the prevalence of infection with SRV-1 or SRV-2 in adult macaques can be nearly 100% and the mortality may approach 50% before 2 years of age (Henrickson *et al.*, 1983, 1984; Lerche *et al.*, 1984). These viruses are not present in African nonhuman primates except possibly in talapoin monkeys which appear seropositive (Ilyinskii *et al.*, 1991).

The original type D retrovirus was isolated from a rhesus monkey mammary tumor in 1970 and was called Mason–Pfizer monkey virus (MPMV) (Chopra and Mason, 1970). A serologic survey of U.S. primate centers in the mid-1970s indicated that about 25% of all macaques had antibody reacting to MPMV core antigen (Fine and Arthur, 1981); this observation revealed the widespread distribution of SRV infection in captive macaques before the 1980s. Extensive serologic surveys in the last two decades of humans including primate center animal handlers have shown no proof of type D retrovirus infection despite numerous claims which were not supported by further analysis (Gardner *et al.*, 1977). Many putative human type D retrovirus isolates, particularly in Russia and Eastern Europe, were found to be contaminants of SRV growing in HeLa cells (Krause *et al.*, 1989). Recently, a type D virus closely related to MPMV was isolated from an AIDS lymphoma patient, but it remains to be confirmed that this is a bona fide human isolate (Bohannon *et al.*, 1991). All of the contemporary type D isolates associated with SAIDS are related to MPMV, have related core but distinct envelope antigens, and fall into five neutralization serotypes (SRV-1–SRV-5) (Marx *et al.*, 1985). SRV-1 is the serotype in macaques at the California (Marx *et al.*, 1984) and New England primate centers (Daniel *et al.*, 1984) and the SRV-2 serotype is present in macaques at the Oregon and Washington primate centers (Stromberg *et al.*, 1984; Marx *et al.*, 1985). The original MPMV is the third distinct serotype (SRV-3) (Arthur *et al.*, 1986), now known to be present in macaques at the Wisconsin Primate Center. Two further type D serotypes, SRV-4 and SRV-5, have been found in macaques at the University of California, Berkeley, and at the primate center in Beijing, China (P. Marx, personal communication).

Experimental transmission of the original MPMV in the early 1970s led to death in many infant rhesus monkeys from a wasting syndrome with thymic atrophy and profound neutropenia, anemia, lymphoid depletion, and opportunistic infections (Fine *et al.*, 1975). The features of this immunosuppressive syndrome in macaques were the same as those caused by SRV-1 and SRV-2 in the early 1980s. Reisolation and experimental transmission of MPMV from its initial source, a frozen sample of the spontaneous rhesus mammary carcinoma, confirmed the earlier observation that this virus, like SRV-1 and SRV-2, was immunosuppressive and apparently nononcogenic (Bryant *et al.*, 1985). Serial epidemiologic and virologic surveys have shown that type D retroviruses are the primary cause of almost all cases of spontaneous SAIDS in all of the primate facilities in which endemic infection with these viruses occurs (Gardner *et al.*, 1988; Daniel *et al.*, 1988a). However, variations in severity of this disease and its manifestations with the different SRV serotypes and the different species of macaques at each center are observed under conditions of both natural and experimental exposure. The SRV-2 serotype is associated not only with immunosuppression, but also with a proliferative disorder called retroperitoneal fibromatosis (RF) (see Fig. 14F) which has some features in common with Kaposi's sarcoma (Giddens *et al.*, 1985; Tsai *et al.*, 1986; Bryant *et al.*, 1986b). RF tissue DNA was able to transform NIH 3T3 cells, suggesting the presence of activated oncogenes (Grant *et al.*, 1993).

Direct physical contact is required for spread of SRV and the most likely route of natural transmission is by percutaneous inoculation of virus-containing saliva via biting and scratching (Lerche *et al.*, 1986, 1987). The following patterns of infection were observed among rhesus macaques at the California Primate Center: (1) SRV-1 persistent viremia with clinical SAIDS; (2) transient viremia with clinical recovery; (3) intermittent viremia suggesting reactivation of latent infection; (4) persistent viremia and virus shedding in several healthy animals (Kwang *et al.*, 1987). The persistently SRV-1 viremic monkeys were congenitally infected and immune tolerant to the virus. In a retrospective epidemiologic analysis of an SRV-1-induced epidemic of SAIDS at the California Primate Center, one healthy carrier was linked by direct physical contact to 34 cases of SAIDS over a 3-year period (Lerche *et al.*, 1986). SAIDS was experimentally transmitted to two juvenile rhesus by inoculation of SRV-1-containing saliva from this adult female monkey. Although SRV-1 could be isolated from PBMC and most bodily secretions of infected animals, the most plentiful source of virus was saliva which was the major natural route of virus transmission. The transmission of SRV in semen has not been evaluated. Although SRV-1 is present in vaginal secretions, the female-to-male sexual transmission of this virus also remains undetermined. Perinatal transmission of SRV-1 transplacentally or via milk occurs infrequently.

At the University of Washington, the persistence of SRV-2 infection was noted in pigtailed macaques (*M. nemestrina*) in which this infection is widespread (Moazed and Thouless, in press). Persistent viremia was evident in seronegative monkeys infected by vertical transmission. In these animals, proviral DNA was found in a wide range of tissues. Such immunotolerant animals would not be detected by serologic assays and would be a ready source for spread of the virus among contact monkeys.

B. Virus Morphology

The morphology of virus assembly and structure differs between the type D viruses, e.g., SRV, the type C oncoviruses, e.g., MuLV, and the lentiviruses, e.g., HIV/SIV/FIV. In the oncoviruses and lentiviruses, the capsid is assembled from individual precursor polyproteins at the cytoplasmic side of the plasma membrane. This assembly event occurs simultaneously with the process of virion budding from the membrane. In contrast, in type D and type B (mammary tumor virus) viruses, an immature capsid is preassembled within the cytoplasm as an intracytoplasmic A particle prior to transport to the plasma membrane. Myristylation of the Gag polyproteins is necessary to confer transport competency on the precursor type D capsids (i.e., intracytoplasmic A particles) (Rhee and Hunter, 1987). Mutational analysis showed that the matrix protein (p10) plays an important role in guiding the correct folding of the Gag polyprotein that occurs during assembly of immature capsids (Rhee and Hunter, 1990a). A single amino acid substitution within the matrix protein converts the type D morphology to that of a type C retrovirus, i.e., capsids are assembled at the plasma membrane instead of within the cytoplasm (Rhee and Hunter, 1990b). Thus, capsid assembly processes for different retroviruses differ only in the intracellular site to which capsid precursors are directed and this property is conferred by the matrix core protein.

The SAIDS type D viruses typically bud from the cell as complete ring-shaped nucleoids about 70 nm in diameter that represent the preassembled capsids or intracytoplasmic A particles. Extracellular mature type D particles are about 125 nm in diameter and have short irregularly arranged surface knobs of 6–8 nm (Fig. 12). The nucleoid and the core shell appear as central cylindrical or slightly conical structures. The nucleoid is clearly separated from the core shell by an electron-lucent space of about 8–10 nm on each side. The nucleoid of the type D viruses is similar to the lentiviruses in its central tubular configuration, but it differs from the lentiviruses in being less conical, not having one end contain the nucleoid, and having a visible space between the nucleoid and the shell. A detailed ultrastructural comparison of HIV/SIV and SRV is provided by Munn *et al.* (1985).

C. SRV Genetic Structure

SRV-1, SRV-2, and SRV-3 (MPMV) have been molecularly cloned and completely sequenced (Power *et al.*, 1986; Thayer *et al.*, 1987; Sonigo *et al.*, 1986).

SAIDS was induced with an infectious molecular clone of SRV-1 (Heidecker-Fanning *et al.*, 1987). All three viruses have a similar genetic organization with four separate translation frames encoding the group-specific antigen (Gag), protease (Prt), RNA-dependent DNA polymerase (Pol), and envelope glycoprotein (Env) (Fig. 13). The *prt* genes of SRV-1, SRV-2, SRV-3, HTLV-II, bovine leukosis virus (BLV), murine mammary tumor virus (MMTV), and hamster intracisternal A particles are in separate translational frames from the *gag* and *pol* genes.

Computer sequence analysis revealed that the type D retroviruses (SRV-1, MPMV) contain a gene segment in the polymerase gene that is lacking in the primate lentiviruses (McClure *et al.*, 1987, 1988). This gene represents a proteaselike element related to the aspartate proteinases and possessing dUTPase activity (Elder *et al.*, 1992). The role of this enzymatic activity in retroviral replication and pathogenesis awaits completion of *in vivo* studies with mutants. Each type D retrovirus has Mg^{2+} preference for RT activity and utilizes $tRNA^{Lys}$ as a primer for minus-strand DNA synthesis. Visna virus, MMTV, HIV-1, and SIV also utilize $tRNA^{Lys}$ as primer. However, the simian type D viruses have a markedly different genetic organization from HIV and SIV and lack extensive homology with these agents. Transactivation of LTR sequence-mediated gene expression is not a property of type D retrovirus replication. Each type D retrovirus has a Gag precursor polypeptide cleaved by the Prt enzyme into six proteins identi-

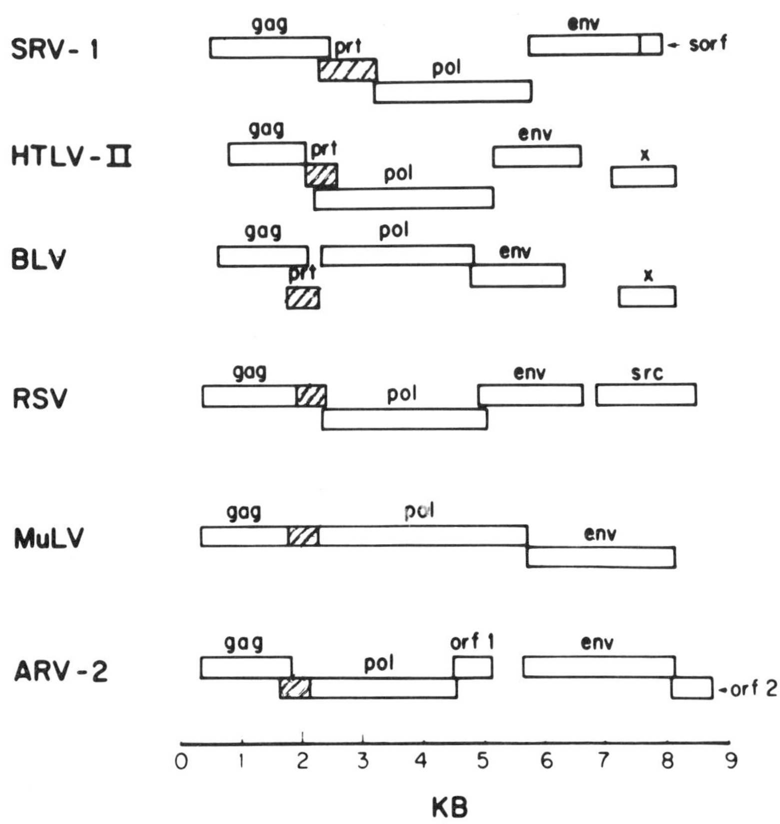

FIGURE 13. Genetic organization of SRV-1 versus that of other retroviruses. SRV-1, simian type D retrovirus, strain 1; HTLV-II, human T-cell leukemia virus, strain II; BLV, bovine leukemia virus; RSV, Rous sarcoma virus; MuLV, murine leukemia virus; ARV-2, SF-2 strain of HIV-1.

fied as p10, pp18, p12, p14, p27, and p4 (Henderson *et al.*, 1985). A comparison of amino acid mismatches based on nucleotide sequences shows that SRV-1 is more closely related to MPMV than is SRV-2. Whereas the Gag, Prt, Pol, and COOH-terminal Env domain of the three viruses differ only by 5–15%, the externally located NH_2-terminal domains differ by 17% in comparisons between SRV-1 and MPMV and by 42% between SRV-1 and SRV-2. The LTR of SRV-1 and MPMV are 88% homologous; SRV-2 shows 70% LTR homology with either virus. In keeping with their distinct neutralization serotypes, SRV-1 and SRV-2 show more envelope amino acid variation (≥40%) than seen so far between different HIV-1 isolates (≤25%).

The similarity between the gp70 glycoprotein of SRV and the endogenous type C viruses of baboons (BaEV) and cats (RD114) has suggested that the type D viruses are recombinants between type C and type B viruses of fairly recent evolutionary origin (Stephenson *et al.*, 1976).

D. Identification of SRV Receptors and Neutralization Epitopes

As mentioned above, the exogenous type D viruses of macaques are classified into five distinct neutralization serotypes, SRV 1–5. Despite the different neutralization specificities, all type D viruses exhibit interference in syncytial and pseudotype blocking assays, indicating the use of a common cell surface receptor. This receptor is also utilized by the endogenous type D retroviruses of squirrel monkey (SMRV) and langur (Po-1-Lu), by the endogenous type C retroviruses of the cat (RD114) and baboon (BaEV), and by the exogenous avian reticuloendotheliosis viruses (Sommerfelt and Weiss, 1990; Koo *et al.*, 1992; Weiss, 1993). Vesicular stomatitis pseudotype particles bearing envelope glycoproteins of each SRV strain, RD114 or BaEV were used to detect receptors expressed in human–rodent somatic cell hybrids segregating human chromosomes. The receptor gene for all of these viruses was localized to chromosome 19q13.1–13.2 (Sommerfelt *et al.*, 1990). Antibodies to human surface markers determined by genes mapped to chromosome 19 did not block virus–receptor interaction. The identity of the receptor thus remains to be determined.

Recently, a putative SRV-1 receptor of 58 kDa was identified in susceptible human Raji cells by immunoprecipitation with disrupted SRV-1 and purified gp70 (Torres *et al.*, 1991b; Benjamini *et al.*, 1991). Immunoprecipitation of this protein was inhibited by antibodies to the neutralization epitope (peptides 142–167) of SRV-1 gp70 (see below), suggesting that neutralization is the result of the prevention of SRV-1 binding to its cell receptor. Molecular cloning of this receptor, yet to be accomplished, will be required for its further characterization.

By using synthetic peptides, it has been possible to identify neutralization epitopes in the gp70 envelope glycoproteins of both SRV-1 and SRV-2 viruses (L. L. Werner *et al.*, 1990, 1991). In SRV-1 gp70, this neutralization epitope represents amino acids 147–162. In SRV-2 gp70, the neutralization epitope is amino acids 92–106. Mouse antibodies to these peptides are capable of neutralizing the *in vitro* infectivity of SRV-1 and SRV-2, respectively, with no cross neutralization (Torres *et al.*, 1991a; Malley *et al.*, 1991). Reciprocally, SRV-1 and SRV-2 peptides inhibit the respective SRV-1 and SRV-2 neutralizing antibody activity in immune monkey sera. Preliminary evidence indicates that antibodies to the SRV-1 peptide (147–162) prevent viral attachment to its cell receptor, as described above. T-cell epitopes in the SRV-1 and SRV-2 envelope gp70 have recently been characterized in mice (Malley *et al.*, 1992). It has yet to be shown, however, whether vaccines made of these B- or T-cell epitopes can induce neutralizing antibodies capable of protecting macaques against challenge infection. Two immunoregulatory T-cell populations that suppress T-cell proliferative responses to SRV-2 env peptides have been described in SRV-2 exposed celebes macaques (Malley *et al.*, 1993).

E. Clinical Features

The clinical features of SRV-induced SAIDS include an initial generalized lymphadenopathy and splenomegaly with reactive hyperplasia (Fig. 14A) accompanied by fever, weight loss, diarrhea, anemia, lymphopenia, granulocytopenia, and thrombocytopenia (Osborn *et al.*, 1984). Neurologic signs and symptoms are conspicuously absent. Necrotizing gingivitis (Noma) is occasionally observed. Despite the striking depletion of peripheral blood cellular elements, the bone marrow is frequently hypercellular (Fig. 14B) (MacKenzie *et al.*, 1986). Electrophoresis of sera of ill animals reveals hypoproteinemia, hypoalbuminemia, and hypogammaglobulinemia. SRV-1-infected macaques had increased levels of neopterin and quinolinic acid in cerebrospinal fluid and serum, probably as a result of secretion from immune stimulated macrophages (Heyes *et al.*, 1990, 1991). Numerous bacterial, protozoan, and viral agents have been identified including CMV and leukocyte-associated herpesvirus. Disseminated CMV has been the most frequent opportunistic infection (Fig. 14E). Concentrations of IgG, IgA, and IgM are decreased but complement component C3 is not changed and C4 is increased (Maul *et al.*, 1984). The absolute lymphocyte count decreases but the OKT4/OKT8 ratio remains unchanged compared with controls, reflecting an absolute decrease in both helper and suppressor T cells. Thus, in comparison with human AIDS and SIV-induced SAIDS, an inverted T-cell helper/suppressor ratio is not found in disease induced by the type D retrovirus. A decreased response to mitogens (Con A, PHA) and to allogeneic lymphoid cells occurs early and becomes more severe near death. Response to PWM is variable. IL-2 causes a complete or partial restoration of the response to the mitogens Con A and PHA. The virus-induced mechanism underlying these profound hematological disturbances remains unknown.

A major difference between SRV, SIV, and HIV is the broader immunosuppressive effect of the type D retrovirus on both T and B cells with a consequent depression of both T- and B-cell function. This difference is correlated with the broader tropism of the type D retrovirus for both T and B cells *in vitro* compared with the more restricted CD4 tropism of HIV and SIV. Unlike SIV and HIV infection of macaques and humans, respectively, SRV does not cause B-cell activation and hypergammaglobulinemia which is consistent with the histologic absence of plasma cells and the early impairment of B-cell function. Other differences between the type D retrovirus-induced disease and HIV-induced AIDS include an absence of *Pneumocystis carinii* as a common opportunistic infection and the non-occurrence of Kaposi's sarcoma (unless retroperitoneal fibrosis is the simian counterpart). The major similarities and differences between SRV and SIV are listed in Tables IX and X.

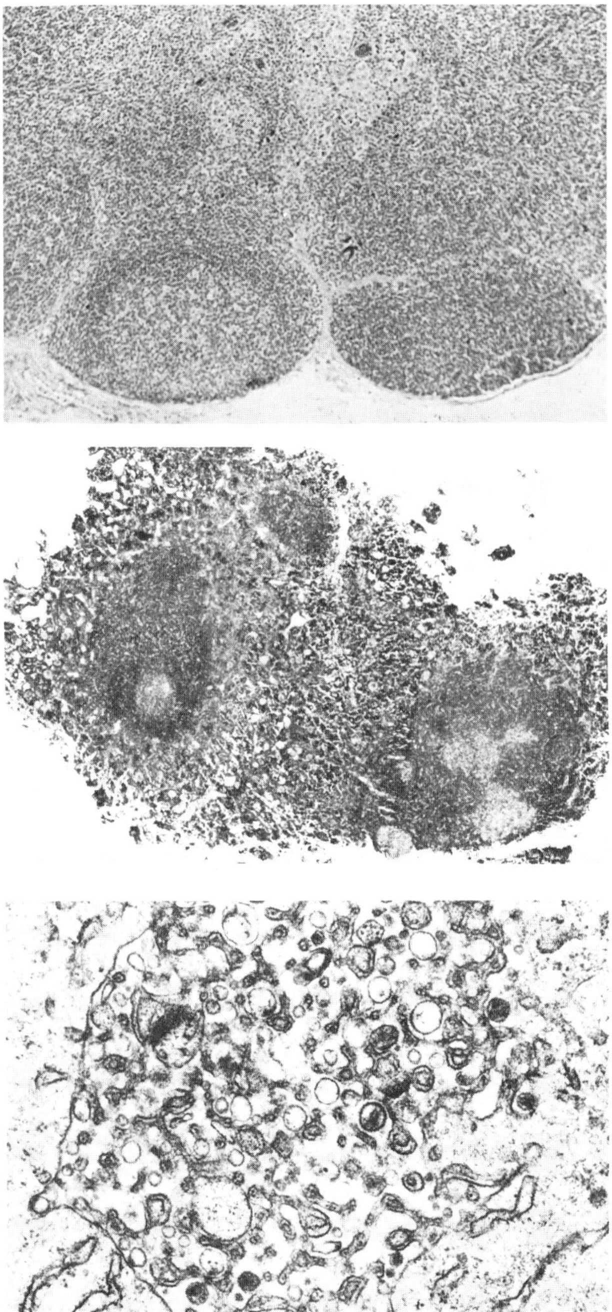

FIGURE 14. Representative histopathology of SRV infection of rhesus macaques. (A) Lymph node with reactive follicles seen during acute SRV infection. H&E, ×25. (B) Bone marrow with hypercellularity of hematopoietic elements and reactive lymphoid nodules. H&E, ×25. (C) Trapping of type D particles in germinal center of reactive lymph node. EM, ×16,500. (D) Depleted lymph node typical of SRV-induced SAIDS. H&E, ×63. (E) CMV inclusion with acute inflammation in liver. H&E, ×100. (F) Retroperitoneal fibromatosis (arrow) typical of SRV-2 infection. H&E, ×100. (G) SRV-1 gp20 antigen (arrow) in apocrine sweat gland of skin. Immunoperoxidase stain, ×25. (H) SRV-1 gp20 antigen in

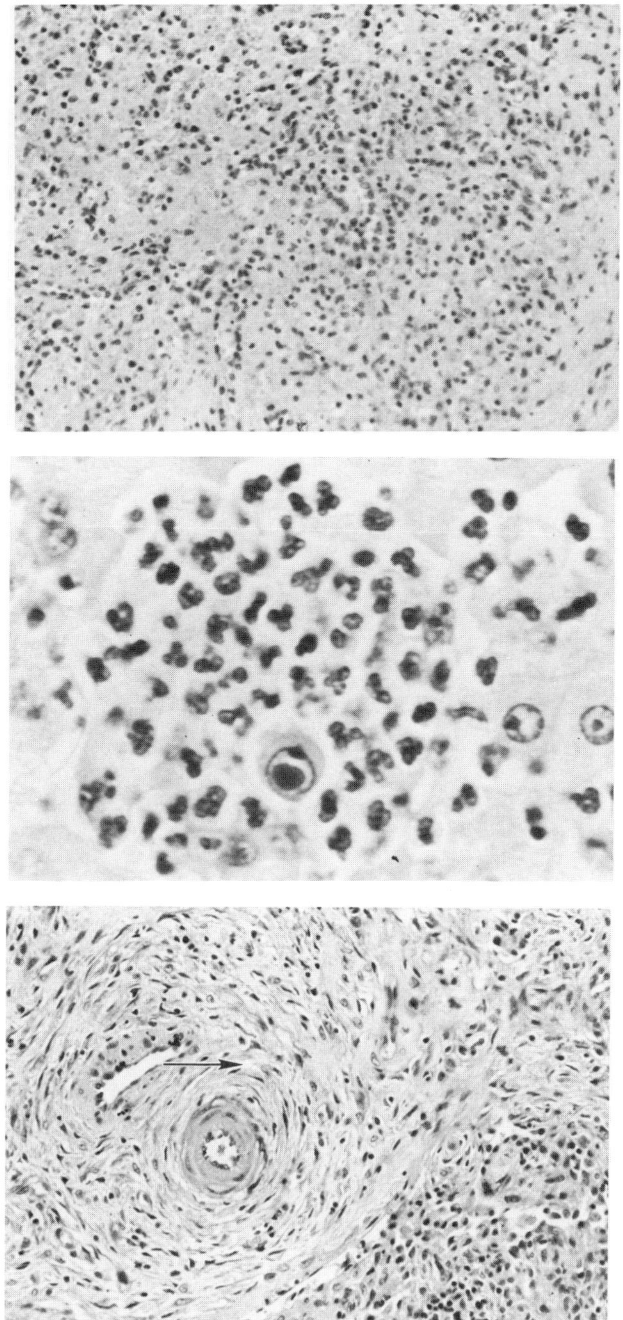

follicular center of a Peyer's patch (arrow) and in mucosal epithelial cells of the small intestine. Immunoperoxidase stain, ×25. (I) SRV-1 gp20 antigen in mucosal epithelial cells (arrow) of small intestine. Immunoperoxidase stain, ×75. (J) SRV-1 gp20 antigen in epithelial cells and lumina (arrow) of salivary gland. Immunoperoxidase stain, ×25. (K) SRV-1 gp20 RNA in epithelial cells of salivary gland. *In situ* hybridization. ×50. (L) Type D particles (arrow) of SRV-1 in acinar lumen of salivary gland. EM, ×16,000. (G–K courtesy of Andrew Lackner; C and L courtesy of Robert Munn.)

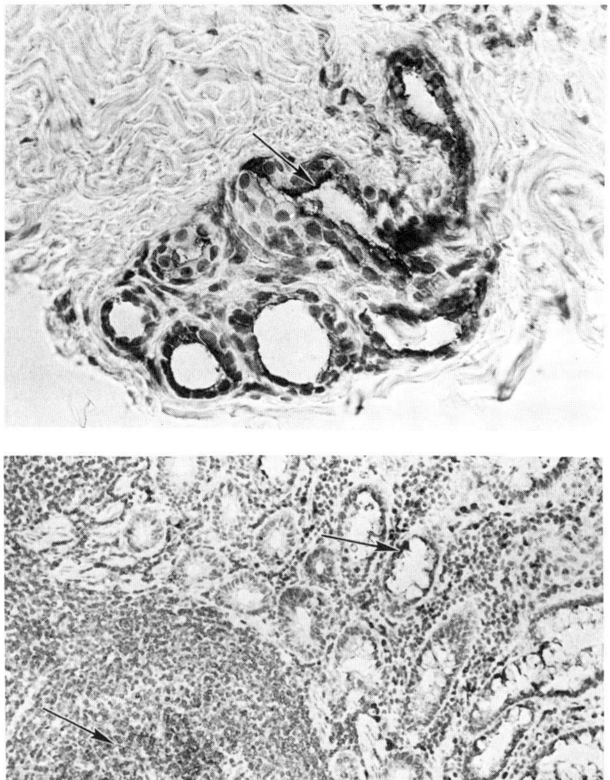

FIGURE 14. (*Continued*)

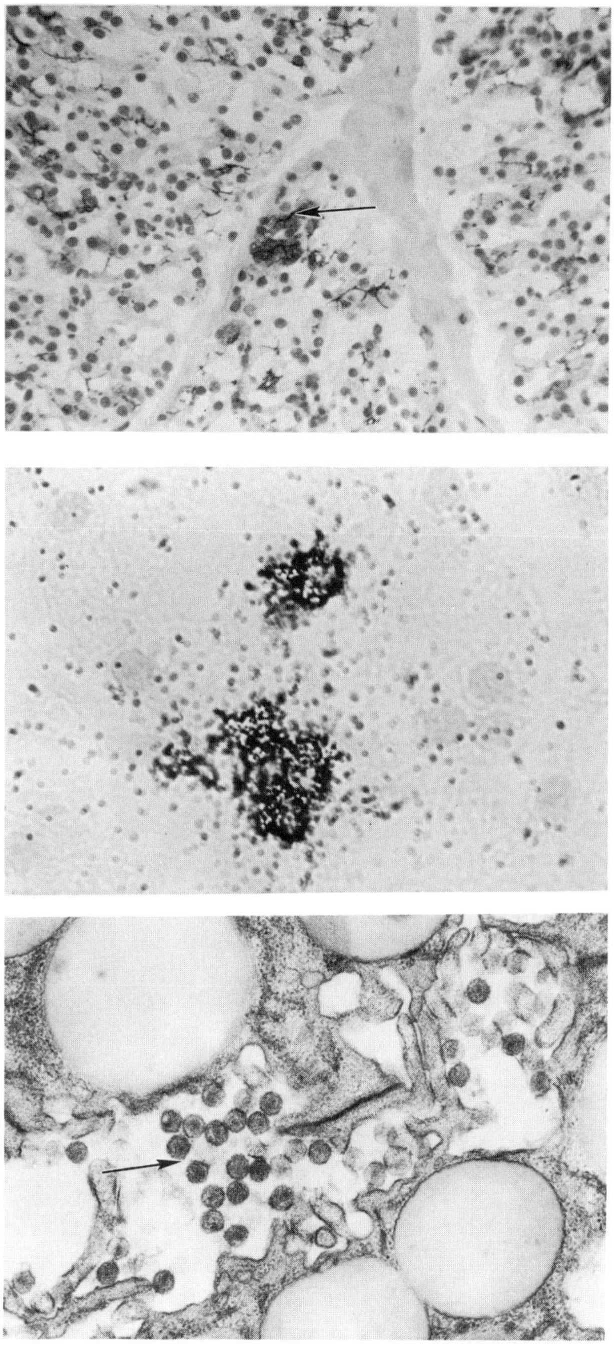

FIGURE 14. *(Continued)*

F. Experimental Transmission

The disease was first shown to be experimentally transmissible by blood and tissue homogenates in 1983 (London *et al.*, 1983; Gravell *et al.*, 1984b) and the causative type D viruses (SRV 1, 2) were isolated at the New England, California, and Washington primate centers and NIH in 1984 (Daniel *et al.*, 1984; Marx *et al.*, 1984; Stromberg *et al.*, 1984; Gravell *et al.*, 1984a). Intravenous inoculation of SRV-1 or SRV-2 harvested from tissue culture media causes the same spectrum of clinical disease as seen naturally in captive macaques (Maul *et al.*, 1986). All animals become infected, some die (≤40%) acutely 7 to 20 weeks after inoculation, some (≤30%) remain persistently infected, and some (≤30%) develop neutralizing antibody, became nonviremic, and remain healthy. Monkeys dying acutely have a high level of persistent viremia and no serum antibody response whereas monkeys with a more indolent clinical course have a low-grade viremia and only transient initial antibody response to the major core antigen (p27) (Kwang *et al.*, 1987). Monkeys that never become ill and are either nonviremic or transiently viremic develop high levels of serum antibody including neutralizing antibody to the virus envelope. Thus, in the SRV model system, one can correlate disease resistance with humoral antibody levels and neutralizing activity. These observations further establish the etiologic role of SRV in this fatal immunosuppressive disease. Conclusive proof of this etiology came later with the induction of an identical, fatal disease spectrum using molecularly cloned infectious SRV-1 (Heidecker-Fanning *et al.*, 1987) and the prevention of this disease with SRV-1 and SRV-2 vaccine preparations (see below).

G. Pathology

At necropsy, SRV-affected animals show severe depletion of lymphocytes in both germinal centers and paracortical regions of lymph nodes as well as an absence of plasma cells (Fig. 14D) (Daniel *et al.*, 1984; Osborn *et al.*, 1984). The histopathology of the lymph nodes is virtually indistinguishable from that of lymph nodes in the terminal stage of human AIDS (Meyer *et al.*, 1985) (Fig. 14). The distribution of SRV-1 in tissues of infected macaques was studied by virus isolation, electron microscopy, immunohistochemistry, and molecular hybridization (Lackner *et al.*, 1988). Virus could be isolated from PBMC, plasma, serum, urine, saliva, lymph nodes, tears, breast milk, cerebrospinal fluid, and vaginal secretions of sick monkeys as well as some healthy carriers. Separation of peripheral blood T and B cells by panning and fluorescent cell sorting indicated that animals with SAIDS harbor infectious type D retrovirus in both T and B cells but more in CD4 cells than in CD8 cells (Maul *et al.*, 1988). Virus was also detected in macrophages but not in mature neutrophils or platelets. Macrophage and neutrophil function is not impaired early in the course of disease (Le Grand *et al.*, 1985). Serial titrations suggested that only 0.1 to 1% of lymphocytes are infected in peripheral blood.

Testing of the susceptibility of normal macaque lymphocytes to infection with SRV *in vitro* was possible but proved difficult because of the inability to establish long-term cultures of macaque lymphocytes. SRV-1 will grow in established human T-cell lines such as HUT-78 as well as in certain EBV-transformed human B cells such as Raji cells. The ability of simian type D retroviruses to

TABLE IX. Similarities between SRV and SIV

- Exogenous lymphotropic retroviruses
- Induce fatal immunodeficiency in macaques
- Transmission by body secretions and close physical contact
- Experimentally transmissible with purified virus including molecular clones
- Persistent infection may occur despite vigorous immune response
- Survival correlates with vigor of humoral immune response
- Low levels of virus expression *in vivo*, mostly cell associated
- Syncytium induction *in vivo*
- Similar opportunistic infections

induce syncytia in Raji cells is the basis for an infectious virus assay (Daniel *et al.*, 1984) and serum inhibition of such syncytia is the basis for a neutralizing antibody assay. SRV-1 will also replicate in EBV transformed rhesus B cells (Van Kuyk *et al.*, 1991). Apart from syncytia, the virus has no direct cytolytic effect and is nontransforming. OKT4 antibodies do not block infection of Raji cells. The type D viruses also grow in human and rhesus fibroblasts and certain rhesus epithelial cells such as in the salivary glands. The virus replicates well in human osteosarcoma (HOS) and rhabdomyosarcoma (RD) cells (Marx *et al.*, 1984). As mentioned above, the receptor for the type D simian retroviruses in human cells has been mapped to chromosome 19 (Sommerfelt *et al.*, 1990). An antigen capture assay will detect SRV in cell culture and plasma (Tsai *et al.*, 1989).

Retroperitoneal fibromatosis (RF) or subcutaneous fibromatosis are highly vascular mesenchymal proliferative lesions, uniquely associated with SRV-2 infection of macaques (Giddens *et al.*, 1985). The RF lesions resemble Kaposi's sarcoma (KS) in humans. In RF cells, SRV-2 functions to increase IL-6 production (Roodman *et al.*, 1991) which may serve as an autocrine growth factor similar to the role shown for IL-6 in human KS cells *in vitro* (Miles *et al.*, 1990).

By immunohistochemistry with a monoclonal antibody to the SRV-1 transmembrane glycoprotein (gp20), viral protein was identified in cells of salivary gland (Fig. 14K), lymph node (Fig. 14H), spleen, thymus, and choroid plexus but not in the brain parenchyma of SAIDS monkeys (Lackner *et al.*, 1988). Viral

TABLE X. SRV Properties Differing from SIV

- Not closely related to HIV
- Cores form intracytoplasmic A particles
- Indigenous in macaques
- Recovery from infection possible
- Recovery correlates with neutralizing antibody
- Vaccine protection easier to achieve
- Virus tropism broader
- Virus variation less
- Viral genome less complex
- Viral genome encodes dUTPase
- B-cell response less marked
- B-cell lymphomas less common
- Neuropathy not present
- Retroperitoneal fibromatosis (SRV-2)
- Strong immune protection with inactivated whole virus and recombinant envelope vaccines
- Not transmissible to humans

antigen was also detected in mucosal epithelial cells of the mouth, GI tract (Fig. 14H,I), breast, and sweat glands (Fig. 14G) (Lackner *et al.*, 1989). Viral antigen was commonly detected in germinal centers of lymph node and spleen and virus particles were seen by EM at the same location (Fig. 14C), apparently in association with antigen-processing dendritic reticulum cells. The amount of viral antigen increased as the disease progressed and it appeared predominantly in the perifollicular capillary endothelial cells of the spleen. The only other *in vivo* site where abundant cell-free virus particles are detected by EM is the lumina of the salivary glands and ducts (Fig. 14L). Southern blot analysis revealed SRV-1 DNA in lymph nodes, salivary gland (Fig. 14K), and brain and SRV-2 DNA in lymph nodes, spleen, PBMC, and RF tissues but not in the skeletal muscle or liver of affected macaques (Bryant *et al.*, 1991). By *in situ* hybridization, SRV-1 RNA was detected in salivary gland and brain parenchyma of rhesus monkeys with no overt neurological symptoms. A finding of SRV-1 nucleic acid in the absence of detectable core antigen or neuropathology in the brain of monkeys suggested viral latency in the CNS. A partial transcriptional block to SRV expression in the brain parenchyma was suggested by the finding that very few cells were positive by *in situ* hybridization for viral RNA, seemingly too few to account for the signal seen by Southern blot for viral DNA. The cell types infected have not been identified and macrophages or giant cells characteristic of HIV-infected human brains or SIV-infected macaque brains have not been seen. No evidence of reactivation of type D virus from this latent state in the CNS has been observed over a number of years of observation. Despite the evidence for latency, cell-free SRV-1 could be isolated from the cerebrospinal fluid of over 50% of neurologically normal monkeys with SRV SAIDS. A few scattered epithelial cells in the choroid plexus appear to be the source of cell-free virus in the cerebrospinal fluid (Lackner *et al.*, 1988).

H. Control of SRV by Animal Husbandry and Vaccination

SRV infection in group-housed monkeys can be eliminated by a serial test (antiviral antibodies and virus isolation) and removal program (Lerche *et al.*, 1991). Because virus-positive monkeys may be seronegative (see Section XVII.A), it is necessary to include virus culture as part of the screening process. An inactivated whole SRV-1 vaccine and recombinant vaccines consisting of SRV-1, SRV-2, or SRV-3 envelope glycoproteins (gp70) expressed in live vaccinia virus have all been able to protect macaques against persistent infection and disease following challenge (Marx *et al.*, 1986; Hu *et al.*, 1989; Brody *et al.*, 1992). Protection correlated with the induction of neutralizing antibody and was long lasting (Benveniste *et al.*, 1993). A transient infection after challenge was detected in some of the SRV-vaccinated monkeys. SRV-1 and SRV-3 separately immunized monkeys showed cross protection against each of these virus strains, in keeping with their close genetic relationships. The SRV-1 and SRV-3 cross-reactive sera failed to neutralize SRV-2 *in vitro* because the latter is genetically more distinct (Brody *et al.*, 1992). SRV-2-vaccinated monkeys have yet to be challenged with SRV-1 or SRV-3. A polyvalent vaccine consisting of SRV-1 (or SRV-3) and SRV-2 Env antigens would probably be required to protect against the three most common SRV serotypes present in captive macaques. The possible role of human cellular antigens in the immune protection obtained with the inactivated whole SRV vaccines was not studied, but such human antigens would not be present in the highly successful recombinant envelope vaccines. It is important now to test the efficacy of these SRV vaccines under cage exposure conditions. It would be very useful to protect against the horizontal transmission of SRV among macaques because the resultant morbidity and mortality continue to be a major practical problem for primate facilities and the infection presents an important scientific concern for those using these animals for research.

XVIII. CONCLUSIONS

This extensive review has portrayed the exciting and rapidly developing research on SIV and SRV, two retrovirus genera that cause fatal immunodeficiency disease in captive macaques and serve as excellent animal models of AIDS. We have emphasized the natural history, molecular biology, and pathogenesis of these agents with special attention given to SIV. In part, we have covered the efforts at vaccine control and antiviral therapy using these models. Similarities and differences between SIV and SRV have been noted (Tables IX and X). Most significantly, monkeys can recover spontaneously from SRV infection and recombinant envelope vaccines can solidly protect against experimental SRV infection. Clearly, if humans were infected with a type D retrovirus, the likelihood of a successful vaccine would be much greater than the likelihood of an equally efficacious HIV vaccine. Because SIV is a lentivirus much more closely related to HIV than is SRV, SIV has essentially replaced SRV as a model for human AIDS. Nevertheless, it is important to use the SRV vaccines to protect captive macaques against infection by the virus and its disease that continues to plague primate facilities supplying these animals for biomedical research.

Much remains to be learned about mechanisms of pathogenesis of SIV- and SRV-induced SAIDS. Suppression of the immune system by SRV is associated with a more general ablative effect on many lymphoid and other hematopoietic cell types, similar in this respect and in its natural history features to infection of cats with certain strains of FeLV (Hardy, 1993). Since neither SRV nor FeLV are directly cytolytic, the generalized cellular depletions that they cause must have indirect mechanisms, most likely involving cytokines such as TNF, IL-1, IL-6, and interferons released from infected or activated lymphocytes. Alternatively, an effect of the "immunosuppressive peptide" present in the amino end of the transmembrane protein could be involved (Section XIII.G). An autoimmune component such as antilymphocytic antibodies may also contribute to the disease course. A pathogenic effect related to viral load is supported by the finding that SRV infection is fatal primarily only in those monkeys that cannot mount an adequate immune response and in which the viral load is therefore high. A major difference between SRV and SIV is that SRV is much more genetically stable and therefore escape mutants that can avoid immune surveillance seldom if ever arise. Consequently, in contrast to SIV infection, many SRV-infected monkeys can generate a sufficiently strong immune response to contain the virus in a latent state or completely eliminate it.

The mechanisms of immunosuppression induced by lentiviruses such as SIV, HIV, and FIV involve a more central role for CD4 T-helper lymphocyte dysfunction and depletion, via direct virus-mediated cytopathic effects and virus-specific immune responses. In addition, indirect mechanisms mediated by the gp120–

CD4 interaction can be involved, such as T-cell anergy, inappropriate cell signaling, superantigen-induced stimulation, clonal depletion, and programmed cell death (apoptosis) (Gougeon *et al.*, 1993; Levy, 1993; Pantaleo *et al.*, 1993b). High rates of virus mutation allow for the formation of neutralizing escape mutants, constant replication of virus particles, and eventual exhaustion of the immune system. A switch to a more cytopathic viral phenotype occurs in disease progression. Dysfunction and loss of $CD4^+$ T cells, the hallmark of lentiviral AIDS, accounts in large part for the failure of the immune system to eliminate or hold the virus in check and accounts for the inevitable fatal outcome of the infection. Suppression of T-helper cell function before marked depletion of these cells could be related to products of SIV, e.g., gp130 and Tat or SIV-induced immunoregulatory cytokines such as TGF-β and IL-10 (Shearer and Clerici, 1993). B-cell hyperresponsiveness is much more evident with SIV than with SRV, which puts more onus on the inappropriate activation of the immune system in the pathogenesis of SIV than of SRV. The molecular mimicry between MHC and HIV could explain some of these mechanisms and lead to a graft-versus-host-like disease (Hoffmann *et al.*, 1991; Habeshaw *et al.*, 1992). Neuropathology is unique to SIV and other lentiviruses because of active infection of CNS microglial cells and migration of infected monocytes from the bloodstream into the CNS. Probably more than one of these mechanisms are involved in $CD4^+$ cell depletion during different phases of the disease.

It seems very likely that common denominators or final common pathways exist in the pathogenesis of both SRV and SIV. With both viruses the lymph nodes, thymus, and spleen are the major reservoir of virus trapping. Immune stimulation and secretion of cytokines along with release of other yet to be defined immunosuppressive factors from activated lymphocytes and macrophages can be involved. A deeper understanding of the early events in the course of infection with either SRV or SIV should clarify the role of the lymphoid organs in pathogenesis and give us more insight into the indirect mechanisms that underlie the fatal immunosuppressive retrovirus infections of macaques and man.

Finally, these precious nonhuman primate resources must be carefully maintained because they will undoubtedly continue to be a vital tool in understanding viral pathogenesis (Gardner, 1993). Moreover, they are immensely valuable for research into the treatment and prevention of AIDS and other chronic diseases.

ACKNOWLEDGMENTS. We would like to thank Kate Maney for preparing and editing the manuscript and Robert Munn for assistance in photography.

XIX. REFERENCES

Adachi, A., Gendelman, H. E., Koenig, S., Folks, T., Willey, R., Rabson, A., and Martin, M. A., 1986, Production of acquired immunodeficiency syndrome-associated retrovirus in human and nonhuman cells transfected with an infectious molecular clone, *J. Virol.* **59**:284.

Agy, M. B., Foy, K., Gale, M. J., Benveniste, R. E., Clark, E. A., and Katze, M. G., 1991, Viral and cellular gene expression in $CD4^+$ human lymphoid cell lines infected by the simian immunodeficiency virus, SIV/Mne, *Virology* **183**:170.

Agy, M. B., Frumkin, L. R., Corey, L., Coombs, R. W., Wolinsky, S. M., Koehler, J., Morton, W. R., and Katze, M. G., 1992, Infection of Macaca nemestrina by human immunodeficiency virus type-1, *Science* **257**:103.

Ahmad, S., Lohman, B., Marthas, M., Giavedoni, L., El-Amad, Z., Haigwood, N. L., Scandella, C. J., Gardner, M. B., Luciw, P. A., and Yilma, T., 1994, Reduced virus load in rhesus macaques immunized with recombinant gp160 and challenged with simian immunodeficiency virus, *AIDS Res Hum Retroviruses* **10**:195.

Ahmed-Ansari, A., Powell, J. D., Jensen, P. E., Yehuda-Cohen, T., McClure, H. M., Anderson, D., Fultz, P. N., and Sell, K. W., 1990, Requirements for simian immunodeficiency virus antigen-specific in vitro proliferation of T cells from infected rhesus macaques and sooty mangabeys, *AIDS* **4**:399.

Albert, J., Bjorling, E., Von Gegerfelt, A., Scarlatti, G., Zhang, Y.-J., Fenyo, E.-M., and Thorstensson, R., 1993, Antigen detection is a reliable method for evaluating HIV/SIV neutralization assays [letter], *AIDS Res. Hum. Retroviruses* **9**:501.

Aldovini, A., and Young, R. A., 1990, Development of a BCG recombinant vehicle for candidate AIDS vaccines, *Int. Rev. Immunol.* **7**:79.

Allan, J. S., 1992, Viral evolution and AIDS, *J. NIH Res.* **4**:51.

Allan, J. S., Kanda, P., Kennedy, R. C., Cobb, E. K., Anthony, M., and Eichberg, J. W., 1990a, Isolation and characterization of simian immunodeficiency viruses from two subspecies of African green monkeys, *AIDS Res. Hum. Retroviruses* **6**:275.

Allan, J. S., Strauss, J., and Buck, D. W., 1990b, Enhancement of SIV infection with soluble receptor molecules, *Science* **247**:1084.

Allan, J. S., Short, M., Taylor, M. E., Su, S., Hirsch, V. M., Johnson, P. R., Shaw, G. M., and Hahn, B. H., 1991a, Species-specific diversity among simian immunodeficiency viruses from African green monkeys, *J. Virol.* **65**:2816.

Allan, J. S., Whitehead, E. M., Strout, K., Short, M., Kanda, P., Hart, T. K., and Bugelski, P. J., 1992b, Strong association of simian immunodeficiency virus (SIVagm) envelope glycoprotein heterodimers: Possible role in receptor-mediated activation, *AIDS Res. Hum. Retroviruses* **8**:2011.

Almond, N., Page, M., Mills, K., Jenkins, A., Ling, C., Thorpe, R., Kitchin, P., and Williams, M., 1990, The production and purification of PCR-derived recombinant simian immunodeficiency virus p27 gag protein; its use in detecting serological and T-cell responses in macaques [published erratum appears in *J. Virol. Methods* **30**:229], *J. Virol. Methods* **28**:305.

Almond, N., Jenkins, A., Heath, A. B., Taffs, L. F., and Kitchin, P., 1992a, The genetic evolution of the envelope gene of simian immunodeficiency virus in cynomolgus macaques infected with a complex virus pool, *Virology* **191**:996.

Almond, N., Jones, S., Heath, A. B., and Kitchin, P. A., 1992c, The assessment of nucleotide sequence diversity by the polymerase chain reaction is highly reproducible, *J. Virol. Methods* **40**:37.

Almond, N., Jenkins, A., Slade, A., Heath, A., Cranage, M., and Kitchin, P., 1992b, Population sequence analysis of a simian immunodeficiency virus (32H reisolate of SIVmac251): A virus stock used for international vaccine studies, *AIDS Res. Hum. Retroviruses* **8**:77.

Almond, N., Jenkins, A., Heath, A. B., and Kitchin, P., 1993a, Sequence variation in the env gene of simian immunodeficiency virus recovered from immunized macaques is predominantly in the V1 region, *J. Gen. Virol.* **74**:865.

Almond, N., Rud, E., Kent, K., Chan, L., Page, M., Jones, W., Mills, K., Clarke, B., Kitchin, P., and Stott, J., 1993b, The molecular characterisation of vaccine trials using recombinant envelope immunogens and live SIV containing distinct nef genes, *Nonhuman Primate Models for AIDS Research Symposium* (Abstract).

Ameisen, J. C., and Capron, A., 1991, Cell dysfunction and depletion in AIDS: The programmed cell death hypothesis [see comments], *Immunol. Today* **12**:102.

Anderson, M. G., and Clements, J. E., 1991, Comparison of the transcriptional activity of the long terminal repeats of simian immunodeficiency viruses SIVmac251 and SIVmac239 in T-cell lines and macrophage cell lines, *J. Virol.* **65**:51.

Anderson, M. G., and Clements, J. E., 1992, Two strains of SIVmac show differential transactivation mediated by sequences in the promoter, *Virology* **191**:559.

Anderson, M. G., Hauer, D., Sharma, D. P., Joag, S. V., Narayan, O., Zink, M. C., and Clements, J. E., 1993, Analysis of envelope changes acquired by SIVmac239 during neuroadaption in rhesus macaques, *Virology* **195**:616.

Ansari, A. A., Mayne, A., Hunt, D., and Villinger, F., 1993, TH1/TH2 subset analysis in PBMC samples of sooty mangabeys and rhesus macaques naturally and experimentally infected with SIV, *Puerto Rico: Tenth Annual Symposium on Nonhuman Primate Models for AIDS Research Symposium* (Abstract).

Aoki, S. S., Yarchoan, R., Kageyama, S., Hoekzema, D. T., Pluda, J. M., Wyvill, K. M., Broder, S., and

Mitsuya, H., 1992, Plasma HIV-1 viremia in HIV-1 infected individuals assessed by polymerase chain reaction, *AIDS Res. Hum. Retroviruses* **8**:1263.

Arens, M., 1993, Use of probes and amplification techniques for the diagnosis and prognosis of human immunodeficiency virus (HIV-1) infections, *Diagn. Microbiol. Infect. Dis.* **16**:165.

Arthur, L. O., Gilden, R. V., Marx, P. A., and Gardner, M. B., 1986, Simian acquired immunodeficiency syndrome, *Prog. Allergy* **1986**:332.

Arthur, L. O., Bess, J. W., Jr., Sowder, R. C., II, Benveniste, R. E., Mann, D. L., Chermann, J.-C., and Henderson, L. E., 1992b, Cellular proteins bound to immunodeficiency viruses: Implications for pathogenesis and vaccines, *Science* **258**:1935.

Arya, S. K., 1988, Human and simian immunodeficiency retroviruses: Activation and differential transactivation of gene expression, *AIDS Res. Hum. Retroviruses* **4**:175.

Ashorn, P., McQuade, T. J., Thaisrivongs, S., Tomasselli, A. G., and Tarpley, W. G., and Moss, B., 1990, An inhibitor of the protease blocks maturation of human and simian immunodeficiency viruses and spread of infection, *Proc. Natl. Acad. Sci. USA* **87**:7472.

Ashorn, P., Moss, B., and Berger, E. A., 1992, Activity of CD4-Pseudomonas exotoxin against cells expressing diverse forms of the HIV and SIV envelope glycoproteins, *J. AIDS* **5**:70.

Axberg, I., Polacino-Firpo, P., and Clark, E. A., 1993, Staphylococcal enterotoxins increase simian immunodeficiency virus (SIV) levels during initial infection of macaque lymphocytes, but decrease SIV levels in cells already infected with SIV, *Berlin: IXth International Conference on AIDS* 190 (Abstract).

Baba, M., De, C. E., Tanaka, H., Ubasawa, M., Takashima, H., Sekiya, K., Nitta, I., Umezu, K., Nakashima, H., and Mori, S., 1991, Potent and selective inhibition of human immunodeficiency virus type 1 (HIV-1) by 5-ethyl-6-phenylthiouracil derivatives through their interaction with the HIV-1 reverse transcriptase, *Proc. Natl. Acad. Sci. USA* **88**:2356.

Bagnarelli, P., Menzo, S., Manzin, A., Giacca, M., Varaldo, P. E., and Clementi, M., 1991, Detection of human immunodeficiency virus type 1 genomic RNA in plasma samples by reverse-transcription polymerase chain reaction, *J. Med. Virol.* **34**:89.

Baier, M., Werner, A., Cichutek, K., Garber, C., Muller, C., Kraus, G., Ferdinand, F. J., Hartung, S., Papas, T. S., and Kurth, R., 1989, Molecularly cloned simian immunodeficiency virus SIVagm3 is highly divergent from other SIVagm isolates and is biologically active in vitro and in vivo, *J. Virol.* **63**:5119.

Baier, M., Behr, E., Kurth, R., and Konig, H., 1990a, Cloning and expression of the complete SIVagm pol region in E. coli. Purification and partial characterization of the reverse transcriptase, *Arzneim. Forsch.* **40**:1284.

Baier, M., Garber, C., Muller, C., Cichutek, K., and Kurth, R., 1990b, Complete nucleotide sequence of a simian immunodeficiency virus from African green monkeys: A novel type of intragroup divergence, *Virology* **176**:216.

Baier, M., Dittmar, M. T., Cichutek, K., and Kurth, R., 1991, Development in vivo of genetic variability of simian immunodeficiency virus, *Proc. Natl. Acad. Sci. USA* **88**:8126.

Balzarini, J., Naesens, L., Slachmuylders, J., Niphuis, H., Rosenberg, I., Holy, A., Schellekens, H., and De Clercq, E., 1991, 9-(2-Phosphonylmethoxyethyl)adenine (PMEA) effectively inhibits retrovirus replication in vitro and simian immunodeficiency virus infection in rhesus monkeys, *AIDS* **5**:21.

Banapour, B., Marthas, M. L., Munn, R. J., and Luciw, P. A., 1991a, In vitro macrophage tropism of pathogenic and nonpathogenic molecular clones of simian immunodeficiency virus (SIVmac), *Virology* **183**:12.

Banapour, B., Marthas, M. L., Ramos, R. A., Lohman, B. L., Unger, R. E., Gardner, M. B., Pedersen, N. C., and Luciw, P. A., 1991c, Identification of viral determinants of macrophage tropism for simian immunodeficiency virus SIVmac, *J. Virol.* **65**:5798.

Bank, I., DePinho, R. A., Brenner, M. B., Cassimeris, J., Alt, F. W., and Chess, L., 1986, A functional T3 molecule associated with a novel heterodimer on the surface of immature human thymocytes, *Nature* **322**:179.

Barry, P. A., Pratt-Lowe, E., Alcendor, D. J., Unger, R. E., and Luciw, P. A., 1990, Molecular interactions between human immunodeficiency virus type 1 and human cytomegalovirus, *Ann. N.Y. Acad. Sci.* **616**:54.

Barry, P. A., Pratt-Lowe, E., Unger, R. E., and Luciw, P. A., 1991, Cellular factors regulate transactivation of human immunodeficiency virus type 1, *J. Virol.* **65**:1392.

Baskerville, A., Ramsay, A., Cranage, M. P., Cook, N., Cook, R. W., Dennis, M. J., Greenaway, P. J.,

Kitchin, P. A., and Stott, E. J., 1990, Histopathological changes in simian immunodeficiency virus infection, *J. Pathol.* **162**:67.

Baskerville, A., Dowsett, A. B., Cook, R. W., Dennis, M. J., Cranage, M. P., and Greenaway, P. J., 1991, Pneumocystis carinii pneumonia in simian immunodeficiency virus infection: Immunohistological and scanning and transmission electron microscopical studies, *J. Pathol.* **164**:175.

Baskerville, A., Ramsay, A. D., Addis, B. J., Dennis, M. J., Cook, R. W., Cranage, M. P., and Greenaway, P. J., 1992a, Interstitial pneumonia in simian immunodeficiency virus infection, *J. Pathol.* **167**:241.

Baskerville, A., Cook, R. W., Dennis, M. J., Cranage, M. P., and Greenaway, P. J., 1992b, Pathological changes in the reproductive tract of male rhesus monkeys associated with age and simian AIDS, *J. Comp. Pathol.* **107**:49.

Baskin, G. B., and Soike, K. F., 1989, Adenovirus enteritis in SIV-infected rhesus monkeys [letter], *J. Infect. Dis.* **160**:905.

Baskin, G. B., Martin, L. N., Rangan, S. R. S., Gormus, B. J., Murphey-Corb, M., Wolf, R. H., and Soike, K. F., 1986, Transmissible lymphoma and simian acquired immunodeficiency syndrome in rhesus monkeys, *J. Natl. Cancer Inst. Monogr.* **77**:127.

Baskin, G. B., Murphey-Corb, M., Watson, E. A., and Martin, L. N., 1988, Necropsy findings in rhesus monkeys experimentally infected with cultured simian immunodeficiency virus (SIV)/Delta, *Vet. Pathol.* **25**:456.

Baskin, G. B., Murphey-Corb, M., Martin, L. N., Soike, K. F., Hu, F. S., and Kuebler, D., 1991a, Lentivirus-induced pulmonary lesions in rhesus monkeys (Macaca mulatta) infected with simian immunodeficiency virus, *Vet. Pathol.* **28**:506.

Baskin, G. B., Murphey-Corb, M., Martin, L. N., Davison, F. B., Hu, F. S., and Kuebler, D., 1991b, Thymus in simian immunodeficiency virus-infected rhesus monkeys, *Lab. Invest.* **65**:400.

Baskin, G. B., Murphey-Corb, M., Roberts, E. D., and Didier, P. J., 1991c, Correlates of SIV-induced lesions in the central nervous system of experimentally infected rhesus monkeys, *Seattle: 9th Annual Symposium on Nonhuman Primate Models for AIDS* (Abstract).

Baskin, G. B., Murphey-Corb, M., Roberts, E. D., Didier, P. J., and Martin, L. N., 1992, Correlates of SIV encephalitis in rhesus monkeys, *J. Med. Primatol.* **21**:59.

Beilke, M. A., Riding, In, D., Hamilton, R., Stone, G. A., Jordan, E. K., Brashears, G., Nusbaum, W., Huddleston, D., Gibbs, C. J., Jr., and Gravell, M., 1991, HLA-DR expression in macaque neuroendothelial cells in vitro and during SIV encephalitis, *J. Neuroimmunol.* **33**:129.

Bellas, R. E., Hopkins, N., and Li, Y., 1993, The NF-κB binding site is necessary for efficient replication of simian immunodeficiency virus of macaques in primary macrophages but not in T cells in vitro, *J. Virol.* **67**:2908.

Beltz, L., Narayan, O., Adams, R. J., Noga, S. J., and Donnenberg, A. D., 1991a, Recovery of the simian immunodeficiency virus (SIV) and depression of colony formation in in vitro infected progenitor cell-enriched rhesus bone marrow (BM), *J. Med. Primatol.* **20**:144.

Benichou, S., Legrand, R., Nakagawa, N., Faure, T., Traincard, F., Vogt, G., Dormont, D., Tiollais, P., Kieny, M. P., and Madaule, P., 1992, Identification of a neutralizing domain in the external envelope glycoprotein of simian immunodeficiency virus, *AIDS Res. Hum. Retroviruses* **8**:1165.

Benichou, S., Venet, A., Beyer, C., Tiollais, P., and Maduale, P., 1993, Characterization of B-cell epitopes in the envelope glycoproteins of simian immunodeficiency virus, *Virology* **194**:870.

Benjamini, E., Torres, J. V., Werner, L. L., and Malley, A., 1991, Isolation and characterization of the neutralizable epitope of simian retrovirus-1 (SRV-1) and of the cell receptor for the virus, *Adv. Exp. Med. Biol.* **303**:71.

Benjouad, A., Babas, T., Montagnier, L., and Bahraoui, E., 1993, N-linked oligosaccharides of simian immunodeficiency virus envelope glycoproteins are dispensable for the interaction with the CD4 receptor, *Biochem. Biophys. Res. Commun.* **190**:311.

Benveniste, R. E., and Todaro, G. J., 1977, Evolution of primate oncornaviruses: An endogenous virus from langur (Presbytis spp.) with related virogene sequences in other Old World monkeys, *Proc. Natl. Acad. Sci. USA* **74**:4557.

Benveniste, R. E., Lieber, M. M., Livingston, D. M., Sherr, C. J., Todaro, G. J., and Kalter, S. S., 1974, Infectious C-type virus isolated from a baboon placenta, *Nature* **248**:17.

Benveniste, R. E., Arthur, L. O., Tsai, C. C., Sowder, R., Copeland, T. D., Henderson, L. E., and Oroszlan, S., 1986, Isolation of a lentivirus from a macaque with lymphoma: Comparison with HTLV-III/LAV and other lentiviruses, *J. Virol.* **60**:483.

Benveniste, R. C., Morton, W. R., Clark, E. A., Tsai, C. C., Ochs, H. D., Ward, J. M., Kuller, L., Knott, W. B., Hill, R. W., Gale, M. J., and Thouless, M. E., 1988, Inoculation of baboons and macaques with simian immunodeficiency virus/Mne, a primate lentivirus closely related to human immunodeficiency virus type 2, *J. Virol.* **62**:2091.

Benveniste, R. E., Raben, D., Hill, R. W., Knott, W. B., Drummond, J. E., Arthur, L. O., Jahrling, P. B., Morton, W. R., Henderson, L. E., and Heidecker, G., 1989, Molecular characterization and comparison of simian immunodeficiency virus isolates from macaques, mangabeys, and African green monkeys, *J. Med. Primatol.* **18**:287.

Benveniste, R. E., Hill, R. W., Eron, L. J., Csaikl, U. M., Knott, W. B., Henderson, L. E., Sowder, R. C., Nagashima, K., and Gonda, M. A., 1990, Characterization of clones of HIV-1 infected HuT 78 cells defective in gag gene processing and of SIV clones producing large amounts of envelope glycoprotein, *J. Med. Primatol.* **19**:351.

Benveniste, R. E., Hill, R. W., Knott, W. B., Tsai, C.-C., Kuller, L., and Morton, W. R., 1993a, Detection of serum antibodies in Ethiopian baboons that cross-react with SIV, HTLV-I, and type D retroviral antigens, *J. Med. Primatol.* **22**:124.

Benveniste, R., Kuller, L., Hu, S.-L., Clark, E., Clerici, M., Shearer, G., and W. Morton, 1993b, Prior exposure to subinfectious doses of SIV protects macaques from subsequent virus challenge, *Madison, WI:11th Annual Nonhuman Primate Models for AIDS* (Abstract #8), 8.

Benveniste, R. E., Kuller, L., Roodman, S. T., Hu, S.-L., and Morton, W. R., 1993c, Long-term protection of macaques against high-dose type D retrovirus challenge after immunization with recombinant vaccinia virus expressing envelope glycoproteins, *J. Med. Primatol.* **22**:74.

Berger, E. A., Clouse, K. A., Chaudhary, V. K., Chakrabarti, S., FitzGerald, D. J., Pastan, I., and Moss, B., 1989, CD4-Pseudomonas exotoxin hybrid protein blocks the spread of human immunodeficiency virus infection in vitro and is active against cells expressing the envelope glycoproteins from diverse primate immunodeficiency retroviruses, *Proc. Natl. Acad. Sci. USA* **86**:9539.

Berkhout, B. A., Gatignol, A., Rabson, A. B., and Jeang, K.-T., 1990, TAR-independent activation of the HIV-1 LTR; evidence that TAT requires specific regions of the promoter, *Cell* **62**:757.

Binninger, D., Ennen, J., Bonn, D., Norley, S. G., and Kurth, R., 1991, Mutational analysis of the simian immunodeficiency virus SIVmac nef gene, *J. Virol.* **65**:5237.

Birx, D. L., Lewis, M. G., Vahey, M., Tencer, K., Zack, P. M., Brown, C. R., Jahrling, P. B., Tosato, G., Burke, D., and Redfield, R., 1993, Association of interleukin-6 in the pathogenesis of acutely fatal $SIV_{smm/PBj\text{-}14}$ in pigtailed macaques, *AIDS Res. and Hum. Retroviruses* **9**:1123.

Black, P. L., Downs, M. B., Lewis, M. G., Ussery, M. A., Dreyer, G. B., Petteway, S. J., and Lambert, D. M., 1993, Antiretroviral activities of protease inhibitors against murine leukemia virus and simian immunodeficiency virus in tissue culture, *Antimicrob. Agents Chemother.* **37**:71.

Blackbourn, D. J., Chuang, L. F., Sutjipto, S., Killam, K. J., McCready, P. M., Doi, R. H., Li, Y., and Chuang, R. Y., 1992, Detection of simian immunodeficiency virus RNA from infected rhesus macaques by the polymerase chain reaction, *J. Virol. Methods* **37**:109.

Blakeslee, J. R., Jr., Sowder, W. G., and Baulu, J., 1985, Wild African green monkeys of Barbados are HTLV negative [letter], *Lancet* **1**:525.

Bohannon, R. C., Donehower, L. A., and Ford, R. J., 1991, Isolation of a type D retrovirus from B-cell lymphomas of a patient with AIDS, *J. Virol.* **65**:5663.

Bohm, Jr., Martin, L. N., Davison-Fairburn, B., Baskin, G. B., and Murphey-Corb, M., 1993, Neonatal disease induced by SIV infection of the rhesus monkey (*Macaca mulatta*), *AIDS Res. and Hum. Retroviruses* **9**:1131.

Bosch, M. L., Earl, P. L., Fargnoli, K., Picciafuoco, S., Giombini, F., Wong-Staal, F., and Franchini, G., 1989, Identification of the fusion peptide of primate immunodeficiency viruses, *Science* **244**:694.

Bottiger, D., and B. Oberg, Lack of involvement of transactivating herpesviruses in acute SIV infection in monkeys, 1993, *Berlin: IXth International Conference on AIDS* (Abstract WS-A11-4), 27.

Bottinger, D., Keys, B., Putkonen, P., Walther, L., Fenyo, E. M., Norrby, E., Oberg, B., and Chiodi, F., 1991a, Simian immunodeficiency virus (SIVsm) isolation from blood and brain of experimentally infected macaques, *AIDS* **5**:445.

Bottiger, D., Stahle-Ljungdahl, E., and Oberg, B., 1991b, Treatment of acute SIV infection in cynomolgus monkeys with 2′,3′-dideoxyinosine (ddI) and 2′,3′-dideoxythymidine (d4T), *Antiviral Chem. Chemother.* **2**:357.

Bottiger, D., Vrang, L., and Oberg, B., 1992a, Influence of the infectious dose of simian immunodeficiency virus on the acute infection in cynomologus monkeys and on the effect of treatment with 3′-fluorothymidine, *Antiviral Chem. Chemother.* **3**:267.

Bottiger, D., Putkonen, P., and Oberg, B., 1992b, Prevention of HIV-2 and SIV infections in cynomolgus macaques by prophylactic treatment with 3'-fluorothymidine, *AIDS Res. Hum. Retroviruses* **8**:1235.

Boudinot, F. D., Schinazi, R. F., Gallo, J. M., McClure, H. M., Anderson, D. C., Doshi, K. J., Kambhampathi, P. C., and Chu, C. K., 1990, 3'-Azido-2',3'-dideoxyuridine (AzddU): Comparative pharmacokinetics with 3'-azido-3'-deoxythymidine (AZT) in monkeys, *AIDS Res. Hum. Retroviruses* **6**:219.

Bourgault, I., Venet, A., and Levy, J. P., 1992a, Three epitopic peptides of the simian immunodeficiency virus Nef protein recognized by macaque cytolytic T lymphocytes, *J. Virol.* **66**:750.

Bourgault, I., Venet, A., and Levy, J. P., 1992b, Immunization of rhesus macaques by lipopeptides induces cytolytic T lymphocytes (CTL), *Puerto Rico: Tenth Annual Symposium on Nonhuman Primate Models for AIDS* (Abstract).

Bowden, D., Agy, M., Anderson, D., Kimpo, C., Ha, J., and Sackett, G., 1993, Neurobehavioral AIDS induced by SIVmne or HIV-1 in infant pig-tailed macaques (Macaca nemestrina), *Berlin: IXth International Conference on AIDS* 193 (Abstract, p. 193, PO-A17·0349).

Braddock, M., Chambers, A., Wilson, W., Esnouf, M. P., Adams, S. E., Kingsman, A. J., and Kingsman, S. M., 1989, HIV-1 TAT "activates" presynthesized RNA in the nucleus, *Cell* **58**:269.

Braddock, M., Thorburn, A. M., Chambers, A., Elliott, G. D., Anderson, G. J., Kingsman, A. J., and Kingsman, S. M., 1990, A nuclear translational block imposed by the HIV-1 U3 region is relieved by the Tat-TAR interaction, *Cell* **62**:1123.

Brenner, M. B., McLean, J., Dialynas, D. P., Strominger, J. L., Smith, J. A., Owen, F. L., Seidman, J. G., Ip, S., Rosen, F., and Krangel, M. S., 1986, Identification of a putative second T-cell receptor, *Nature* **322**:145.

Brinchmann, J. E., Gaudernack, G., and Vartdal, F., 1990, $CD8^+$ T cells inhibit HIV replication in naturally infected $CD4^+$ T cells, *J. Immunol.* **144**:2961.

Brinchmann, J. E., Gaudernack, G., Thorsby, E., and Vartdal, F., 1992, Staphylococcal exotoxin superantigens induce human immunodeficiency virus type 1 expression in naturally infected $CD4^+$ T cells, *J. Virol.* **66**:5924.

Brinkmann, R., Schwinn, A., Muller, J., Stahl-Hennig, C., Coulibaly, C., Hunsmann, G., Czub, S., Rethwilm, A., Dorries, R., and Ter Meulen, V., 1993, In vitro and in vivo infection of rhesus monkey microglial cells by simian immunodeficiency virus, *Virology* **195**:561.

Brody, B. A., Hunter, E., Kluge, J. D., Lasarow, R., Gardner, M., and Marx, P. A., 1992, Protection of macaques against infection with simian type D retrovirus (SRV-1) by immunization with recombinant vaccinia virus expressing the envelope glycoproteins of either SRV-1 or Mason-Pfizer monkey virus (SRV-3), *J. Virol.* **66**:3950.

Bruisten, S., van, G. B., Koppelman, M., Rasch, M., van, S. D., Schukkink, R., Beyer, R., Weigel, H., Lens, P., and Huisman, H., 1993, Detection of HIV-1 distribution in different blood fractions by two nucleic acid amplification assays, *AIDS Res. Hum. Retroviruses* **9**:259.

Bryant, M. L., Yamamoto, J., Luciw, P., Munn, R., Marx, P., Higgins, J., Pedersen, N., Levine, A., and Gardner, M. B., 1985, Molecular comparison of retroviruses associated with human and simian AIDS, *Hematol. Oncol.* **3**:187.

Bryant, M. L., Gardner, M. B., Marx, P. A., Maul, D. H., Lerche, N. W., Osborn, K. G., Lowenstine, L. J., Bodgen, A., Arthur, L. O., and Hunter, E., 1986a, Immunodeficiency in rhesus monkeys associated with the original Mason-Pfizer monkey virus, *J. Natl. Cancer Inst.* **77**:957.

Bryant, M. L., Marx, P. A., Shiigi, S. M., Wilson, B. J., McNulty, W. P., and Gardner, M. B., 1986b, Distribution of type D retrovirus sequences in tissues of macaques with simian acquired immune deficiency and retroperitoneal fibromatosis, *Virology* **150**:149.

Bryant, M. L., Ratner, L., Duronio, R. J., Kishore, N. S., Devadas, B., Adams, S. P., and Gordon, J. I., 1991, Incorporation of 12-methoxydodecanoate into the human immunodeficiency virus 1 gag polyprotein precursor inhibits its proteolytic processing and virus production in a chronically infected human lymphoid cell line, *Proc. Natl. Acad. Sci. USA* **88**:2055.

Bryant, M., Mueller, R., Partis, R., Stolzenbach, J., Wyand, M., Tiemeier, D., and Marr, J., 1993, Efficacy and safety of α-glucosidase inhibitor prodrug SC-49483 in rhesus monkeys infected with SIV, *Berlin: IXth International Conference on AIDS* (Abstract, p27, WS-A11-2).

Burns, D. P., and Desrosiers, R. C., 1990, Sequence variability of simian immunodeficiency virus in a persistently infected rhesus monkey, *J. Med. Primatol.* **19**:317.

Burns, D. P., and Desrosiers, R. C., 1991, Selection of genetic variants of simian immunodeficiency virus in persistently infected rhesus monkeys, *J. Virol.* **65**:1843.

Burns, D. P., and Desrosiers, R. C., 1992, A caution on the use of SIV/HIV gag antigen detection systems in neutralization assays, *AIDS Res. Hum. Retroviruses* **8**:1189.

Burns, D. P. W., Collignon, C., and Desrosiers, R. C., 1993, Simian immunodeficiency virus mutants resistant to serum neutralization arise during persistent infection of rhesus monkeys, *J. Virol.* **67**:4104.

Callebaut, C., Krust, B., Jacotot, E., and Hovanessian, A. G., 1993, T cell activation antigen, CD26, as a cofactor for entry of HIV in $CD4^+$ cells, *Science* **262**:2045.

Campbell, B. J., and Hirsch, V. M., 1993, Extensive env gene heterogeneity in tissues from SIV infected macaques, *Nonhuman Primate Models for AIDS Research Symposium* (Abstract).

Carlson, J. R., McGraw, T. P., Keddie, E., Yee, J. L., Rosenthal, A., Langlois, A. J., Dickover, R., Donovan, R., Luciw, P. A., and Jennings, M. B., 1990, Vaccine protection of rhesus macaques against simian immunodeficiency virus infection, *AIDS Res. Hum. Retroviruses* **6**:1239.

Castro, B. A., Nepomuceno, M., Lerche, N. W., Eichberg, J. W., and Levy, J. A., 1991, Persistent infection of baboons and rhesus monkeys with different strains of HIV-2, *Virology* **184**:219.

Centers for Disease Control, 1992, Anonymous survey for simian immunodeficiency virus (SIV) seropositivity in SIV-laboratory researchers—United States, *Morb. Mortal. Wkly. Rep.* **41**:814.

Chaisson, R., Leuther, M. D., Allain, J.-P., Nusinoff-Lehrman, S., Boone, G. S., Feigal, D., and Volberding, P., 1988, Effect of zidovudine on serum human immunodeficiency virus core antigen levels, *Arch. Intern. Med.* **148**:2151.

Chakrabarti, L., Guyader, M., Alizon, M., Daniel, M. D., Desrosiers, R. C., Tiollais, P., and Sonigo, P., 1987, Sequence of simian immunodeficiency virus from macaque and its relationship to other human and simian retroviruses, *Nature* **328**:543.

Chakrabarti, L., Hurtrel, M., Maire, M. A., Vazeux, R., Dormont, D., Montagnier, L., and Hurtrel, B., 1991, Early viral replication in the brain of SIV-infected rhesus monkeys, *Am. J. Pathol.* **139**:1273.

Chakrabarti, L., Cumont, M.-C., Montagnier, L., and Hurtrel, B., 1993, Sequential analysis of viral load in lymph nodes of SIV infected rhesus macaques: Correlates with disease evolution, *Berlin: IXth International Conference on AIDS* 26 (Abstract).

Chalifoux, L. V., Ringler, D. J., King, N. W., Sehgal, P. K., Desrosiers, R. C., Daniel, M. D., and Letvin, N. L., 1987, Lymphadenopathy in macaques experimentally infected with the simian immunodeficiency virus (SIV), *Am. J. Pathol.* **128**:104.

Chalifoux, L. V., Simon, M. A., Pauley, D. R., MacKey, J. J., Wyand, M. S., and Ringler, D. J., 1992, Arteriopathy in macaques infected with simian immunodeficiency virus, *Lab. Invest.* **67**:338.

Chan, W. L., Rodgers, A., Hancock, R. D., Taffs, F., Kitchin, P., Farrar, G., and Liew, F. Y., 1992, Protection in simian immunodeficiency virus-vaccinated monkeys correlates with anti-HLA class I antibody response, *J. Exp. Med.* **176**:1203.

Chapel, A., Bourges, J. F., Bensussan, A., d'Auriol, L., Vilmer, E., and Dormont, D., 1990, Lack of permissivity of human monoclonal $CD4^+$ lymphocytes to HIV, *Cancer Detect. Prev.* **14**:445.

Chen, Z. W., Yamamoto, H., Watkins, D. I., Levinson, G., and Letvin, N. L., 1992, Predominant use of a T-cell receptor V beta gene family in simian immunodeficiency virus Gag-specific cytotoxic T lymphocytes in a rhesus monkey, *J. Virol.* **66**:3913.

Chen, Z. W., Kou, Z.-C., Shen, L., Reimann, R. A., and Letvin, N. L., 1993, Conserved T cell receptor repertoire in SIVmac-infected rhesus monkeys, *Nonhuman Primate Models for AIDS Research Symposium* (Abstract).

Cheng, S. M., Blume, M., Lee, S. G., Hung, P. P., Hirsch, V. M., and Johnson, P. R., 1990a, Coexpression of biologically active simian immunodeficiency virus (SIV) Rev and Env in an SV40 system: The SIV rev gene regulates env expression, *Virology* **177**:816.

Cheng, S. M., Blume, M., Lee, S. G., Hung, P. P., Hirsch, V. M., and Johnson, P. R., 1990b, The simian immunodeficiency virus (SIV) rev gene regulates env expression, *J. Med. Primatol.* **19**:167.

Cheng, S. M., Lee, S. G., Ronchetti-Blume, M., Virk, K. P., Mizutani, S., Eichberg, J. W., Davis, A., Hung, P. P., Hirsch, V. M., Chanock, R. M., Purcell, R. H., and Johnson, P. R., 1992, Coexpression of the simian immunodeficiency virus env and rev proteins by a recombinant human adenovirus host range mutant, *J. Virol.* **66**:6721.

Cheng-Mayer, C., Seto, D., Tateno, M., and Levy, J. A., 1988, Biologic features of HIV-1 that correlate with virulence in the host, *Science* **240**:80.

Cheng-Mayer, C., Iannello, P., Shaw, K., Luciw, P. A., and Levy, J. A., 1989, Differential effects of nef on HIV replication: Implications for viral pathogenesis in the host, *Science* **246**:1629.

Chopra, H. C., and Mason, M. M., 1970, A new virus in a spontaneous mammary tumor of a rhesus monkey, *Cancer Res.* **30**:2081.

Cianciolo, G. J., Copeland, T. D., Oroszlan, S., and Snyderman, R., 1985, Inhibition of lymphocyte proliferation by a synthetic peptide homologous to retroviral envelope proteins, *Science* **230**:453.

Cichutek, K., and Norley, S., 1993, Lack of immune suppression in SIV-infected natural hosts, *AIDS* **7**(Suppl 1):S25.

Ciochon, R. L., and Chiarelli, A. B. (eds.), 1980, *Evolutionary Biology of the New World Monkeys and Continental Drift*, Plenum Press, New York.

Clapham, P. R., Blanc, D., and Weiss, R. A., 1991, Specific cell surface requirements for the infection of CD4-positive cells by human immunodeficiency virus types 1 and 2 and by simian immunodeficiency virus, *Virology* **181**:703.

Clerici, M., and Shearer, G. M., 1993, A TH1 → TH2 switch is a critical step in the etiology of HIV infection, *Immunol. Today* **14**:107.

Clerici, M., Stocks, N. I., Zajac, R. A., Boswell, R. N., Bernstein, D. C., Mann, D. L., Shearer, G. M., and Berzofsky, J. A., 1989, Interleukin-2 production used to detect antigenic peptide recognition by T-helper lymphocytes from asymptomatic HIV-seropositive individuals, *Nature* **339**:383.

Colombini, S., Arya, S. K., Reitz, M. S., Jagodzinski, L., Beaver, B., and Wong, S. F., 1989, Structure of simian immunodeficiency virus regulatory genes, *Proc. Natl. Acad. Sci. USA* **86**:4813.

Connor, R. I., Mohri, H., Yunzhen, C., and Ho, D. D., 1993, Increased viral burden and cytopathicity correlate temporally with $CD4^+$ T-lymphocyte decline and clinical progression in human immunodeficiency virus type 1-infected individuals, *J. Virol.* **67**:1772.

Conway, M. D., Didier, P., Fairburn, B., Soike, K. F., Martin, L., Murphey-Corb, M., Meiners, N., and Insler, M. S., 1990, Ocular manifestation of simian immunodeficiency syndrome (SAIDS), *Curr. Eye Res.* **9**:759.

Conway, M. D., Davison-Fairburn, B., Martin, L. N., Insler, M. S., and Murphey-Corb, M., 1991, Infection of rhesus monkeys with topical instillation of simian immunodeficiency virus (SIV)B670 into the conjunctival sac, *J. Med. Primatol.* **20**:152.

Cooper, R., Feistner, A., Evans, S., Tsujimoto, H., and Hayami, M., 1989, A lack of evidence of sexual transmission of a simian immunodeficiency agent in a semifree-ranging group of mandrills [letter], *AIDS* **3**:764.

Coulibaly, C., Makoschey, B., Stahl-Hennig, C., Petry, H., Luke, W., and Hunsmann, B., 1993, Passive immunization against SIVmac with purified hyperimmune IgG, *Berlin: IXth International Conference on AIDS* 252 (Abstract).

Courgnaud, V., Laure, F., Fultz, P. N., Montagnier, L., Brechot, C., and Sonigo, P., 1992, Genetic differences accounting for evolution and pathogenicity of simian immunodeficiency virus from a sooty mangabey monkey after cross-species transmission to a pig-tailed macaque, *J. Virol.* **66**:414.

Cranage, M. P., Almond, N., Jenkins, A., and Kitchin, P. A., 1989, Transmembrane protein of SIV [letter], *Nature* **342**:349.

Cranage, M., Stott, J., Mills, K., Ashworth, T., Taffs, F., Farrar, G., Chan, L., Dennis, M., Putkonen, P., Biberfeld, G., Murphey-Corb, M., Page, M., Baskerville, A., Kitchin, P., and Greenaway, P., 1992a, Vaccine studies with 32H reisolate of SIVmac251: An overview, *AIDS Res. Hum. Retroviruses* **8**:1479 (Abstract).

Cranage, M. P., Baskerville, A., Ashworth, L. A., Dennis, M., Cook, N., Sharpe, S., Farrar, G., Rose, J., Kitchin, P. A., and Greenaway, P. J., 1992b, Intrarectal challenge of macaques vaccinated with formalin-inactivated simian immunodeficiency virus, *Lancet* **339**:273.

Cranage, M. P., Polyanskaya, N., McBride, B., Cook, N., Ashworth, L. A. E., Dennis, M., Baskerville, A., Greenaway, P. J., Corcoran, T., Kitchin, P., Rose, J., Murphey-Corb, M., Desrosiers, R. C., Stott, E. J., and Farrar, G. H., 1993, Studies on the specificity of the vaccine effect elicited by inactivated simian immunodeficiency virus, *AIDS Res. Hum. Retroviruses* **9**:13.

Cullen, B. R., 1986, Trans-activation of human immunodeficiency virus occurs via a bimodal mechanism, *Cell* **46**:973.

Daar, E. S., Li, X. L., Moudgil, T., and Hod, D. D., 1990, High concentrations of recombinant soluble CD4 are required to neutralize primary human immunodeficiency virus type 1 isolates, *Proc. Natl. Acad. Sci. USA* **87**:6574.

Dalgleish, A. G., Beverley, P. C., Clapham, P. R., Crawford, D. H., Greaves, M. F., and Weiss, R. A., 1984, The CD4 (T4) antigen is an essential component of the receptor for the AIDS retrovirus, *Nature* **312**:763.

Daniel, M. D., and Desrosiers, R. C., 1989, Use of simian immunodeficiency virus for evaluation of AIDS vaccine strategies, *AIDS* **3**(Suppl. 1):S131.

Daniel, M. D., King, N. W., Letvin, N. L., Hunt, R. D., Sehgal, P. K., and Desrosiers, R. C., 1984, A new type D retrovirus isolated from macaques with an immunodeficiency syndrome, *Science* **223**:602.

Daniel, M. D., Letvin, N. L., King, N. W., Kannagi, M., Sehgal, P. K., Hunt, R. D., Kanki, P. J., Essex, M., and Desrosiers, R. C., 1985, Isolation of T-cell tropic HTLV-III-like retrovirus from macaques, *Science* **228**:1201.

Daniel, M. D., Letvin, N. L., Sehgal, P. K., Hunsmann, G., Schmidt, D. K., King, N. W., and Desrosiers, R. C., 1987, Long-term persistent infection of macaque monkeys with the simian immunodeficiency virus, *J. Gen. Virol.* **68**:3183.

Daniel, M. D., Letvin, N. L., Sehgal, P. K., Schmidt, D. K., Silva, D. P., Solomon, K. R., Hodi, F. S., Ringler, D. J., Hunt, R. D., King, N. W., and Desrosiers, R. C., 1988a, Prevalence of antibodies to 3 retroviruses in a captive colony of macaque monkeys, *Int. J. Cancer* **41**:601.

Daniel, M. D., Li, Y., Naidu, Y. M., Durda, P. J., Schmidt, D. K., Troup, C. D., Silva, D. P., MacKey, J. J., Kestler, H. W., III, Sehgal, P. K., King, N. W., Ohta, Y., Hayami, M., and Desrosiers, R. C., 1988b, Simian immunodeficiency virus from African green monkeys, *J. Virol.* **62**:4123.

Daniel, M. D., Kirchhoff, F., Czajak, S. C., Sehgal, P. K., and Desrosiers, R. C., 1992, Protective effects of a live attenuated SIV vaccine with a deletion in the nef gene, *Science* **258**:1938.

Dapolito, G. A., Elkins, W. R., London, W. T., Goldstein, S., Johnson, P. R., and Hirsch, V. M., 1993, Characterization of a highly pathogenic SIVagm isolate, *Nonhuman Primate Models for AIDS Research Symposium* (Abstract).

Davison-Fairburn, B., Baskin, G., and Murphey-Corb, M., 1992, Maternal–fetal transmission of SIV in 2 species of macaques, *Puerto Rico: Tenth Annual Symposium on Nonhuman Primate Models for AIDS* (Abstract).

Dayton, A. I., Sodroski, J. G., Rosen, C. A., Goh, W. C., and Haseltine, W. A., 1986, The trans-activator gene of the human T cell lymphotropic virus type III is required for replication, *Cell* **44**:941.

Debouck, C., 1991, Substrate specificity of the human (type 1) and simian immunodeficiency virus proteases, *Adv. Exp. Med. Biol.* **306**:407.

Debyser, Z., Pauwels, R., Andries, K., Desmyter, J., Kukla, M., Janssen, P. A., and De, C. E., 1991, An antiviral target on reverse transcriptase of human immunodeficiency virus type 1 revealed by tetrahydroimidazo-[4,5,1-jk] [1,4]benzodiazepin-2 (1H)-one and -thione derivatives, *Proc. Natl. Acad. Sci. USA* **88**:1451.

Debyser, Z., De Vreese, K., Pauwels, R., Yamamoto, N., Anne, J., De Clerca, E., and Desmyter, J., 1992, Differential inhibitory effects of TIBO derivatives on different strains of simian immunodeficiency virus, *J. Gen. Virol.* **73**:1799.

Delchambre, M., Gheysen, D., Thines, D., Thiriart, C., Jacobs, E., Verdin, E., Horth, M., Burny, A., and Bex, F., 1989, The GAG precursor of simian immunodeficiency virus assembles into virus-like particles, *EMBO J.* **8**:2653.

Del Llano, A. M., Amieiro-Puig, J. P., Kraiselburd, E. N., Kessler, M. J., Malaga, C. A., and Lavergne, J. A., 1993, The combined assessment of cellular apoptosis, mitochondrial function and proliferative response to pokeweed mitogen has prognostic value in SIV infection, *J. Med. Primatol.* **22**:147.

Delord, B., Ottmann, M., Schrive, M. H., Ragnaud, J. M., Seigneurin, J. M., and Fleury, H. J., 1992, Analysis of HIV-1 expression in vivo with in situ hybridization and the polymerase chain reaction, *Mol. Cell. Probes* **6**:215.

Denham, W. W., 1987, *West Indian Green Monkeys: Problems in Historical Biogeography*, Karger, Basel.

DePaoli, P., Gennari, D., Martelli, P., Basaglia, G., Crovatto, M., Battistin, S., and Santini, G., 1991, A subset of gamma delta lymphocytes is increased during HIV-1 infection, *Clin. Exp. Immunol.* **83**:187.

Desrosiers, R. C., 1986, Origin of the human AIDS virus, *Nature* **319**:728.

Desrosiers, R. C., 1990a, The simian immunodeficiency viruses, *Annu. Rev. Immunol.* **8**:557.

Desrosiers, R. C., 1990b, A finger on the missing link, *Nature* **345**:288.

Desrosiers, R. C., 1992, HIV with multiple gene deletions as a live attenuated vaccine for AIDS, *AIDS Res. Hum. Retroviruses* **8**:1457.

Desrosiers, R. C., and Letvin, N. L., 1987, Animal models for acquired immunodeficiency syndrome, *Rev. Infect. Dis.* **9**:438.

Desrosiers, R. C., and Ringler, D. J., 1989, Use of simian immunodeficiency viruses for AIDS research, *Intervirology* **30**:301.

Desrosiers, R. C., Wyand, M. S., Kodama, T., Ringler, D. J., Arthur, L. O., Sehgal, P. K., Letvin, N. L., King, N. W., and Daniel, M. D., 1989a, Vaccine protection against simian immunodeficiency virus infection, *Proc. Natl. Acad. Sci. USA* **86**:6353.

Desrosiers, R. C., Daniel, M. D., and Li, Y., 1989b, HIV-related lentiviruses of nonhuman primates, *AIDS Res. Hum. Retroviruses* **5**:465.

Desrosiers, R. C., Hansen-Moosa, A., Mori, K., Bouvier, D. P., King, N. W., Daniel, M. D., and Ringler, D. J., 1991, Macrophage-tropic variants of SIV are associated with specific AIDS-related lesions but are not essential for the development of AIDS, *Am. J. Pathol.* **139**:29.

Dewhurst, S., Embreton, J. E., Anderson, D. C., Mullins, J. I., and Fultz, P. N., 1990, Sequence analysis and acute pathogenicity of molecularly cloned smm-PBj14, *Nature* **345**:636.

Dewhurst, S., Embretson, J. E., Fultz, P. N., and Mullins, J. I., 1992, Molecular clones from a non-acutely pathogenic derivative of SIVsmmPBj14: Characterization and comparison to acutely pathogenic clones, *AIDS Res. Hum. Retroviruses* **8**:1179.

Dietrich, U., Adamski, M., Kreutz, R., Seipp, A., Kuhnel, H., and Rubsamen-Waigmann, H., 1989, A highly divergent HIV-2 related isolate, *Nature* **342**:948.

Dillon, P. J., Nelbock, P., Perkins, A., and Rosen, C. A., 1990, Function of the human immunodeficiency virus types 1 and 2 Rev proteins is dependent on their ability to interact with a structured region present in env gene mRNA, *J. Virol.* **64**:4428.

Doms, R. W., Earl, P. L., Chakrabarti, S., and Moss, B., 1990, Human immunodeficiency virus types 1 and 2 and simian immunodeficiency virus env proteins possess a functionally conserved assembly domain, *J. Virol.* **64**:3537.

Donehower, L. A., Bohannon, R. C., Ford, R. J., and Gibbs, R. A., 1990, The use of primers from highly conserved pol regions to identify uncharacterized retroviruses by the polymerase chain reaction, *J. Virol. Methods* **28**:33.

Douglas, G. C., Fry, G. N., Thirkill, T., Holmes, E., Hakim, H., Jennings, M., and King, B. F., 1991, Cell-mediated infection of human placental trophoblast with HIV in vitro, *AIDS Res. Hum. Retroviruses* **7**:735.

D'Souza, M. P., Kent, K. A., Thiriart, C., Milman, G., 1993, International collaboration comparing neutralization and binding assays for monoclonal antibodies to simian immunodeficiency virus, *AIDS Res. Hum. Retroviruses*, **9**:415.

Earl, P. L., Doms, R. W., and Moss, B., 1990, Oligomeric structure of the human immunodeficiency virus type 1 envelope glycoprotein, *Proc. Natl. Acad. Sci. USA* **87**:648.

Eberle, J., and Seibl, R., 1992, A new method for measuring reverse transcriptase activity by ELISA, *J. Virol. Methods* **40**:347.

Elder, J. H., Lerner, D. L., Hasselkus-Light, C. S., Fontenot, D. J., Hunter, E., Luciw, P. A., Montelaro, R. C., and Phillips, T. R., 1992, Distinct subsets of retroviruses encode dUTPase, *J. Virol.* **66**:1791.

Emau, P., McClure, H. M., Isahakia, M., Else, J. G., and Fultz, P. N., 1991, Isolation from African Sykes' monkeys (Cercopithecus mitis) of a lentivirus related to human and simian immunodeficiency viruses, *J. Virol.* **65**:2135.

Emau, P., Uno, H., Zimbric, M., Bro, S., and Pauza, C. D., 1993, Differential localization of SIV replication in lymphoid tissues of rhesus monkeys after intravenous or intrarectal infection, *Madison, WI: 11th Annual Symposium on Nonhuman Primate Models for AIDS Research Symposium* (Abstract, p. 70).

Embretson, J., Zapancic, M., Ribas, J. L., Burke, A., Racz, P., Tenner-Racz, K., and Haase, A. T., 1993, Massive covert infection of helper T lymphocytes and macrophages by HIV during the incubation period of AIDS, *Nature* **362**:359.

Essex, M., 1994, Simian immunodeficiency virus in people, *N. Engl. J. Med.* **330**:209.

Essex, M., and Kanki, P. J., 1988, The origins of the AIDS virus, *Sci. Am.* **259**:64.

Fazely, F., Sharma, P. L., Fratazzi, C., Greene, M. F., Wyand, M. S., Memon, M. A., Penninck, D., and Ruprecht, R. M., 1993, Simian immunodeficiency virus infection via amniotic fluid: A model to study fetal immunopathogenesis and prophylaxis, *J. AIDS* **6**:107.

Feichtinger, H., Putkonen, P., Parravicini, C., Li, S. L., Kaaya, E. E., Bottiger, D., Biberfeld, G., and Biberfeld, P., 1990, Malignant lymphomas in cynomolgus monkeys infected with simian immunodeficiency virus, *Am. J. Pathol.* **137**:1311.

Feichtinger, H., Kaaya, E., Putkonen, P., Li, S. L., Ekman, M., Gendelman, R., Biberfeld, G., and Biberfeld, P., 1992, Malignant lymphoma associated with human AIDS and with SIV-induced immunodeficiency in macaques, *AIDS Res. Hum. Retroviruses* **8**:339.

Feichtinger, H., Kaaya, E., Ekman, M., Fend, F., Biberfeld, G., and Biberfeld, P., 1993, Oral hairy leukoplakia in SIV-infected cynomolgus macaques, *Berlin: IXth International Conference on AIDS* 193 (Abstract, p. 193).

Ferrari, G., Ottinger, J., Place, C., Nigida, S. M., Arthur, L. O., and Weinhold, K. J., 1993, The impact of HIV-1 infection on phenotypic and functional parameters of cellular immunity in chimpanzees, *AIDS Res. Hum. Retroviruses* **9**:647.

Fine, D. L., and Arthur, L. O., 1981, Expression of natural antibodies against endogenous and horizontally transmitted macaque retroviruses in captive primates, *Virology* **112**:49.

Fine, D. L., Landon, J. C., Pienta, R. J., Kubicek, M. T., Valerio, M. G., Loeb, W. F., and Chopra, H. C., 1975, Responses of infant rhesus monkeys to inoculation with Mason-Pfizer monkey virus materials, *J. Natl. Cancer Inst.* **54**:651.

Firpo, P. P., Axberg, I., Scheibel, M., and Clark, E. A., 1992, Macaque CD4⁺ T-cell subsets: Influence of activation on infection by simian immunodeficiency viruses (SIV), *AIDS Res. Hum. Retroviruses* **8**:357.

Fischl, M. A., Richman, D. D., Grieco, M. H., Gottlieb, M. S., Volberding, P. A., Laskin, O. L., Leedom, J. M., Groopman, J. E., Mildvan, D., and Schooley, R. T., 1987, The efficacy of azidothymidine (AZT) in the treatment of patients with AIDS and AIDS-related complex. A double-blind, placebo-controlled trial, *N. Engl. J. Med.* **317**:185.

Fisher, A. G., Feinberg, M. B., Josephs, S. F., Harper, M. E., Marselle, L. M., Reyes, G., Gonda, M. A., Aldovini, A., Debouk, C., Gallo, R. C., and Wong-Staal, F., 1986, The trans-activator gene of HTLV-III is essential for virus replication, *Nature* **320**:367.

Fisher, A. G., Ensoli, B., Ivanoff, L., Chamberlain, M., Petteway, S., Ratner, L., Gallo, R. C., and Wong-Staal, F., 1987, The sor gene of HIV-1 is required for efficient virus transmission in vitro, *Science* **237**:888.

Flugel, R. M., Zentgraf, H., Munk, K., and Darai, G., 1978, Activation of an endogenous retrovirus from Tupaia (tree shrew), *Nature* **271**:543.

Follis, K. E., Tsai, C.-C., Grant, R., Sabo, A., and Bischofberger, N., 1993, Evaluation of PMEA efficacy in macaques chronically infected with simian immunodeficiency virus (SIV), *Madison, WI: 11th Annual Symposium on Nonhuman Primate Models for AIDS* (Abstract, p. 101).

Fomsgaard, A., Hirsch, V. M., Allan, J. S., and Johnson, P. R., 1991, A highly divergent proviral DNA clone of SIV from a distinct species of African green monkey, *Virology* **182**:397.

Fomsgaard, A., Hirsch, V. M., and Johnson, P. R., 1992, Cloning and sequences of primate CD4 molecules: Diversity of the cellular receptor for simian immunodeficiency virus/human immunodeficiency virus, *Eur. J. Immunol.* **22**:2973.

Fomsgaard, A., Johnson, P. R., London, W. T., and Hirsch, V. M., 1993, Genetic variation of the SIVagm transmembrane glycoprotein in naturally and experimentally infected primates, *AIDS* **7**:1041.

Franchini, G., Robert-Guroff, M., Ghrayeb, J., Chang, N. T., and Wong-Staal, F., 1986, Cytoplasmic localization of the HTLV-III 3′ orf protein in cultured T cells, *Virology* **155**:593.

Franchini, G., Collalti, E., Arya, S. K., Fenyo, E. M., Biberfeld, G., Zagury, J. F., Kanki, P. J., Wong-Staal, F., and Gallo, R. C., 1987a, Genetic analysis of a new subgroup of human and simian T-lymphotropic retroviruses: HTLV-IV, LAV-2, SBL-6669, and STLV-IIIagm, *AIDS Res. Human Retro.* **3**:11.

Franchini, G., Gurgo, C., Guo, H-G., Gallo, R. C., Collalti, E., Fargnoli, K. A., Hall, L. F., Wong-Staal, F., and Reitz, M. S., Jr., 1987b, Sequence of simian immunodeficiency virus and its relationship to the human immunodeficiency viruses, *Nature* **328**:539.

Franchini, G., Rusche, J. R., O'Keefe, T. J., and Wong-Staal, F., 1988, The human immunodeficiency virus type 2 (HIV-2) contains a novel gene encoding a 16 kD protein associated with mature virions, *AIDS Res. Hum. Retroviruses* **4**:243.

Fukasawa, M., Miura, T., Hasegawa, A., Morikawa, S., Tsujimoto, H., Miki, K., Kitamura, T., and Hayami, M., 1988, Sequence of simian immunodeficiency virus from African green monkey, a new member of the HIV/SIV group, *Nature* **333**:457.

Fultz, P. N., 1991, Replication of an acutely lethal simian immunodeficiency virus activates and induces proliferation of lymphocytes, *J. Virol.* **65**:4902.

Fultz, P. N., and Tao, B., 1993, Molecular and biologic evolution of acutely lethal SIVsmmPBj14, *Berlin: IXth International Conference on AIDS* 190 (Abstract, p. 190).

Fultz, P. N., McClure, H. M., Anderson, D. C., Swenson, R. B., Anand, R., and Srinivasan, A., 1986, Isolation of a T-lymphotropic retrovirus from naturally infected sooty mangabey monkeys (Cercocebus atys), *Proc. Natl. Acad. Sci. USA* **83**:5286.

Fultz, P. N., McClure, H. M., Anderson, D. C., and Switzer, W. M., 1989, Identification and biologic characterization of an acutely lethal variant of simian immunodeficiency virus from sooty mangabeys (SIV/SMM), *AIDS Res. Hum. Retroviruses* **5**:397.

Fultz, P. N., Anderson, D. C., McClure, H. M., Dewhurst, S., and Mullins, J. I., 1990a, SIVsmm infection of macaque and mangabey monkeys: Correlation between in vivo and in vitro properties of different isolates, *Dev. Biol. Stand.* **72**:253.

Fultz, P. N., Gordon, T. P., Anderson, D. C., and McClure, H. M., 1990b, Prevalence of natural infection with simian immunodeficiency virus and simian T-cell leukemia virus type 1 in a breeding colony of sooty mangabey monkeys, *AIDS* **4**:619.

Fultz, P. N., Schwiebert, R., Su, L., and Salter, M., 1993, Total lymphoid irradiation of SIV-infected macaques as a potential therapy for HIV infection, *Nonhuman Primate Models for AIDS Research Symposium* (Abstract).

Gajdusek, D. C., 1990, Subacute spongiform encephalopathies: Transmissible cerebral amyloidoses caused by unconventional viruses, in: *Virology* (B. N. Fields and D. M. Knipe, eds.), pp. 2289–2324, Raven Press, New York.

Gallatin, W. M., Gale, M. J., Hoffman, P. A., Willerford, D. M., Draves, K. E., Benveniste, R. E., Morton, W. R., and Clark, E. A., 1989, Selective replication of simian immunodeficiency virus in a subset of $CD4^+$ lymphocytes, *Proc. Natl. Acad. Sci. USA* **86**:3301.

Gao, F., Yue, L., White, A. T., Pappas, P. G., Barchue, J., Hanson, A. P., Greene, B. M., Sharp, P. M., Shaw, G. M., and Hahn, B. H., 1992, Human infection by genetically diverse SIVSM-related HIV-2 in west Africa, *Nature* **358**:495.

Gardner, M. B., 1993, The importance of nonhuman primate research in the battle against AIDS: A historical perspective, *J. Med. Primatol.* **22**:86.

Gardner, M. B., and Hu, S. L., 1991, SIV vaccines, 1991—A year in review, *AIDS* **5**(Suppl. 2):S115.

Gardner, M. B., and Luciw, P. A., 1988, Simian immunodeficiency viruses and their relationship to the human immunodeficiency virus, *AIDS* **2**:S3.

Gardner, M. B., and Luciw, P. A., 1989, Animal models of AIDS, *FASEB J.* **3**:2593.

Gardner, M. B., and Luciw, P. A., 1992, Simian retroviruses, in: *AIDS and Other Manifestations of HIV Infection* (G. P. Wormser, ed.), pp. 127–144, Raven Press, New York.

Gardner, M. B., and Stott, J., 1990, Progress in development of SIV vaccines: A review, *AIDS* **4**(Suppl. 1):S137.

Gardner, M. B., Rahseed, S., Shimizu, S., Rongey, R. W., Henderson, B. E., McAllister, R. M., Klement, V., Charman, H. P., Gilden, R. V., Heberling, R. L., and Huebner, R. J., 1977, Search for RNA Tumor Virus in Humans. In: *Cold Spring Harbor Conferences on Cell Proliferation, vol. 4, Origins of Human Cancer,* edited by H. H. Hiatt, J. D. Watson, and J. A. Winsten. Cold Spring Harbor Laboratory, pp. 1235–1251.

Gardner, M. B., Pedersen, N., Marx, P., Henrickson, R., Luciw, P., and Gilden, R., Vaccination against virally induced animal tumors, 1985, In: *Immunity to Cancer,* edited by A. E. Reif and M. S. Mitchell. Academic Press, Inc.: New York, p. 605–617.

Gardner, M. B., Luciw, P., Lerche, N., and Marx, P., 1988, Non-human primate retrovirus isolates and AIDS, in: *Immunodeficiency Disorders and Retroviruses* (K. Perk, ed.), pp. 171–226, Academic Press, New York.

Gardner, M. B., Jennings, M., Carlson, J., Lerche, N., McGraw, T., Luciw, P., and Marx, P., 1989, Postexposure immunotherapy of SIV infected rhesus with an SIV immunogen, *J. Med. Primatol.* **18**:321.

Gardner, M. B., Carlson, J. R., Jennings, M., Rosenthal, A., Langlois, A., Haynes, B., Bolognesi, D., and Palker, T. J., 1991, SIV protection of rhesus monkeys, in: *Biotechnology Therapeutics,* Repligen, pp. 9–19.

Gardner, M., Yamamoto, J., Marthas, M., Miller, C., Jennings, M., Rosenthal, A., Luciw, P., Planelles, V., Yilma, T., and Giavedoni, L., 1992, SIV and FIV vaccine studies at UC Davis: 1991 update, *AIDS Res. Hum. Retroviruses* **8**:1495.

Gardner, M., Rosenthal, A., Jennings, M., Antipa, L., and Robinson, E., 1994, Passive immunization of rhesus macaques against SIV. *J. Med. Primatol.*, in press.

Gelderblom, H. R., 1991, Assembly and morphology of HIV: Potential effect of structure on viral function, *AIDS* **5**:617.

Genetic organization of a chimpanzee lentivirus related to HIV-1 [editorial], 1991, *Dis. Markers* **9**:53.

Gerber, M. A., Chen, M. L., Hu, F. S., Baskin, G. B., and Petrovich, L., 1991, Liver disease in rhesus monkeys infected with simian immunodeficiency virus, *Am. J. Pathol.* **139**:1081.

Giavedoni, L. D., Jones, L., Gardner, M. B., Gibson, H. L., Ng, C. T., Barr, P. J., and Yilma, T., 1992a, Vaccinia virus recombinants expressing chimeric proteins of human immunodeficiency virus and gamma interferon are attenuated for nude mice, *Proc. Natl. Acad. Sci. USA* **89**:3409.

Giavedoni, L. D., Parodi, L., Ahmad, S., Bassiri, M., and Yilma, T., 1992b, Immune enhancement and attenuation of vaccinia virus recombinants expressing empty viral particles of SIV and lymphokines, *Puerto Rico: Tenth Annual Symposium on Nonhuman Primate Models for AIDS* (Abstract).

Giavedoni, L. D., Planelles, V., Haigwood, N. L., Ahmad, S., Kluge, J. D., Marthas, M. L., Gardner, M. B., Luciw, P. A., and Yilma, T. D., 1993, Immune response of rhesus macaques to recombinant SIVgp130 does not protect from challenge infection, *J. Virol.* **67**:577.

Giddens, W. E., Jr., Tsai, C.-C., Morton, W. R., Ochs, H. D., Knitter, G. H., and Blakley, G. A., 1985, Retroperitoneal fibromatosis and acquired immunodeficiency syndrome in macaques, *Am. J. Pathol.* **119**:253.

Girard, M., Kieny, M. P., Pinter, A., Barre-Sinoussi, F., Nara, P., Kolbe, H., Kusumi, K., Chaput, A., Reinhart, T., and Muchmore, E., 1991, Immunization of chimpanzees confers protection against challenge with human immunodeficiency virus, *Proc. Natl. Acad. Sci. USA* **88**:542.

Girard, M., Mahoney, J., Rimsky, L., Barre-Sinoussi, F., Muchmore, E., Weinhold, K., and Fultz, P., 1992, HIV-1 genital infection: A chimpanzee model, *Paris: 7th Cent Gardes Meeting* (Abstract).

Gnann, J. W. J., Nelson, J. A., and Oldstone, M. B., 1987, Fine mapping of an immunodominant domain in the transmembrane glycoprotein of human immunodeficiency virus, *J. Virol.* **61**:2639.

Goldstein, S., Engle, R., Olmsted, R. A., Hirsch, V. M., and Johnson, P. R., 1990, Detection of SIV antigens by HIV-1 antigen capture immunoassays, *J. AIDS* **3**:98.

Goldstein, S., London, W. T., Elkins, W. R., and Hirsch, V. M., 1993, Whole inactivated SIV vaccine protects macaques from infection with cell-free SIV generated in human but not macaque cells, *Madison, WI: 11th Annual Symposium on Nonhuman Primate Models for AIDS* (Abstract, p. 7).

Gonzalez, S. A., Affranchino, J. L., Gelderblom, H. R., and Burny, A., 1993, Assembly of the matrix protein of simian immunodeficiency virus into virus-like particles, *Virology* **194**:548.

Good, R. C., 1968, Biology of the mycobacterioses. Simian tuberculosis: Immunologic aspects, *Ann. N.Y. Acad. Sci.* **154**:200.

Good, S. S., Koble, C. S., Crouch, R., Johnson, R. L., Rideout, J. L., and de Miranda, P., 1990, Isolation and characterization of an ether glucuronide of zidovudine, a major metabolite in monkeys and humans, *Drug Metab. Dispos.* **18**:321.

Gormus, B. J., Ohashi, D. K., Ohkawa, S., Walsh, G. P., Meyers, W. M., Brennan, P. J., and Trygg, C., 1988a, Serologic responses to Mycobacterium leprae-specific phenolic glycolipid-I antigen in sooty mangabey monkeys with experimental leprosy, *Int. J. Lepr. Other Mycobact Dis.* **56**:537.

Gotch, F. M., Hovell, R., Delchambre, M., Silvera, P., and McMichael, A. J., 1991, Cytotoxic T-cell response to simian immunodeficiency virus by cynomolgus macaque monkeys immunized with recombinant vaccinia virus, *AIDS* **5**:317.

Gotch, F., Nixon, D., Gallimore, A., McAdam, S., and McMichael, A., 1993, Cytotoxic T lymphocyte epitopes shared between HIV-1, HIV-2 and SIV, *J. Med. Primatol.* **22**:119.

Gougeon, M. L., Colizzi, V., Dalgleish, A., and Montagnier, L., 1993, New concepts in AIDS pathogenesis, *AIDS Res. Hum. Retroviruses* **9**:287.

Graham, B. S., Belshe, R. B., Clements, M. L., Dolin, R., Corey, L., Wright, P. F., Gorse, G. J., Midthun, K., Keefer, M. C., and Roberts, N. J., Jr., 1992, Vaccination of vaccinia-naive adults with human immunodeficiency virus type 1 gp160 recombinant vaccinia virus in a blinded, controlled, randomized clinical trial. The AIDS Vaccine Clinical Trials Network, *J. Infect. Dis.* **166**:244.

Grankvist, O., Bredberg, R. U., Gustafsson, A., Albert, J., Albino, P., Andreasson, P. A., Naucler, A., Biberfeld, G., and Wadell, G., 1992, Improved detection of HIV-2 DNA in clinical samples using a nested primer-based polymerase chain reaction, *J. AIDS* **5**:286.

Grant, R. F., Malinak, C. J., Wu, H., and Tsai, C.-C., 1993, Studies on tumor associated genes and the transforming potential of DNA from retroperitoneal fibromatosis, *Nonhuman Primate Models for AIDS Research Symposium* (Abstract).

Grant, S. K., Deckman, I. C., Minnich, M. D., Culp, J., Franklin, S., Dreyer, G. B., Tomaszek, T. J., Debouck, C., and Meek, T. D., 1991, Purification and biochemical characterization of recombinant simian immunodeficiency virus protease and comparison to human immunodeficiency virus type 1 protease, *Biochemistry* **30**:8424.

Grant, S. K., Deckman, I. C., Culp, J. S., Minnich, M. D., Brooks, I. S., Hensley, P., Debouck, C., and Meek, T. D., 1992, Use of protein unfolding studies to determine the conformational and dimeric stabilities of HIV-1 and SIV proteases, *Biochemistry* **31**:9491.

Gravell, M., London, W. T., Hamilton, R. S., Sever, J. L., Kapikian, A. Z., Murti, G., Arthur, L. O., Gilden, R. V., Osborn, K. G., Marx, P. A., Henrickson, R. V., and Gardner, M. B., 1984a, Transmission of simian AIDS with type D retrovirus isolate, *Lancet* **1**:334.

Gravell, M., London, W. T., Sever, J. L., Houff, S. A., Madden, D. L., Dalakas, M. C., Osborn, K. G., Maul, D. H., Henrickson, R. V., Marx, P. A., Lerche, N., Prahalada, S., and Gardner, M. B., 1984b, Transmission of simian acquired immunodeficiency syndromes (SAIDS) with blood or filtered plasma, *Science* **223**:74.

Gravell, M., London, W. T., Hamilton, R. S., Stone, G., and Monzon, M., 1989, Infection of macaque monkeys with simian immunodeficiency virus from African green monkeys: Virulence and activation of latent infection, *J. Med. Primatol.* **18**:247.

Grief, C., Hockley, D. J., Fromholc, C. E., and Kitchin, P. A., 1989, The morphology of simian immunodeficiency virus as shown by negative staining electron microscopy, *J. Gen. Virol.* **70**:2215.

Gronowitz, J. S., Neumuller, M., Lennerstrand, J., Bhikhabhai, R., Unge, T., Weltman, H., and Kallander, C. F., 1991, Carrier bound templates for single tube reverse transcriptase assays and for combined purification and activity analyses, with special reference to HIV, *Biotechnol. Appl. Biochem.* **13**:127.

Haase, A. T., 1986, Pathogenesis of lentivirus infections, *Nature* **322**:130.

Habeshaw, J., Hounsell, E., and Dalgleish, A., 1992, Does the HIV envelope induce a chronic graft-versus-host-like disease? *Immunol. Today* **13**:207.

Hahn, B. H., Kong, L. I., Lee, S. W., Kumar, P., Taylor, M. E., Arya, S. K., and Shaw, G. M., 1987, Relation of HTLV-4 to simian and human immunodeficiency-associated viruses, *Nature* **330**:184.

Hahn, B. H., Jin, M. J., Hui, H., Robertson, D. L., Allan, J. S., Shaw, G. M., and Sharp, P. M., 1993, A mosaic SIVagm genome provides evidence for simian/simian cross-species transmission in the wild, *Nonhuman Primate Models for AIDS Research Symposium* (Abstract).

Haigwood, N. L., Misher, L., Chin, S. M., Blair, M., Planelles, V., Scandella, C. J., Steimer, K. S., Gardner, M. B., and Yilma, T., 1992, Characterization of group specific antibodies in primates: Studies with SIV envelope in macaques, *J. Med. Primatol.* **21**:82.

Hall, G., Ashworth, T., Sharpe, S., Cook, R., Dennis, M., Andrews, C., and Cranage, M., 1993, Does the amount of rhesus MHC class II expressed in blood monocytes affect the pathology of macaques infected with SIV? *Nonhuman Primate Models for AIDS Research Symposium* (Abstract).

Hammes, S. R., Dixon, E. P., Malim, M. H., Cullen, B. R., and Greene, W. C., 1989, Nef protein of human immunodeficiency virus type 1: Evidence against its role as a transcriptional inhibitor, *Proc. Natl. Acad. Sci. USA* **86**:9549.

Hardy, W. D., Jr., 1993, Feline oncoretroviruses, in: *The Retroviridae*, Vol. 2 (J. A. Levy, ed.), pp. 109–180, Plenum Press, New York.

Hartung, S., Boller, K., Cichutek, K., Norley, S. G., and Kurth, R., 1992a, Quantitation of a lentivirus in its natural host: Simian immunodeficiency virus in African green monkeys, *J. Virol.* **66**:2143.

Hartung, S., Norley, S. G., Ennen, J., Cichutek, K., Plesker, R., and Kurth, R., 1992b, Vaccine protection against SIVmac infection by high- but not low-dose whole inactivated virus immunogen, *J. AIDS* **5**:461.

Haynes, B. F., Arthur, L. O., Frost, P., Matthews, T. J., Langlois, A. J., Palker, T. J., Hart, M. K., Scearce, R. M., Jones, D. M., McDanel, C., Ottinger, J., Bolognesi, D., and Weinhold, K. J., 1993,

Conversion of an immunogenic human immunodeficiency virus (HIV) envelope synthetic peptide to a tolerogen in chimpanzees by the fusogenic domain of HIV gp41 envelope protein, *J. Exp. Med.* **177**:717.

Heberling, R. L., Barker, S. T., Kalter, S. S., Smith, G. C., and Helmke, R. J., 1977, Oncornavirus: Isolation from a squirrel monkey (Saimiri sciureus) lung culture, *Science* **195**:289.

Heeney, J., Jonker, R., Koornstra, W., Dubbes, R., Niphuis, H., Di Rienzo, A. M., Gourgeon, M. L., and Montagnier, L., 1993, The resistance of HIV-infected chimpanzees to progression to AIDS correlates with absence of HIV-related T-cell dysfunction, *J. Med. Primatol.* **22**:194.

Heeney, J. L., de Vries, P., Dubbes, R., Koornstra, W., Niphuis, H., ten Haaft, P., Boes, J., Dings, M. E. M., Morein, B., and Osterhaus, A. D. M. E., 1992, Comparison of protection from homologous cell-free vs cell-associated SIV challenge afforded by inactivated whole SIV vaccines, *J. Med. Primatol.* **21**:126.

Heidecker-Fanning, G., Lerche, N. W., Lowenstine, L. J., Lackner, A. A., Osborn, K. G., Gardner, M. B., and Marx, P. A., 1987, Induction of simian acquired immune deficiency syndrome (SAIDS) with a molecular clone of a type D SAIDS retrovirus, *J. Virol.* **61**:3066.

Heise, C., Vogel, P., Miller, C. J., Halsted, C. H., and Dandekar, S., 1993a, Simian immunodeficiency virus infection of the gastrointestinal tract of rhesus macaques: Functional, pathological, and morphological changes, *Am. J. Pathol.* **142**:1759.

Heise, C., Vogel, P., Miller, C. J., Lackner, A., and Dandekar, S., 1993, Distribution of SIV infection in the gastrointestinal tract of rhesus macaques at early and terminal stages of AIDS, *J. Med. Primatol.* **22**:187.

Hemm, S., Sopper, S., Meixensberger, J., Coulibaly, C., Stahl-Hennig, C., Hunsmann, G., ter Meulen, V., and Dorries, R., 1993, Kinetics of humoral and cellular immunological parameters in cerebrospinal fluid and blood of SIVmac infected rhesus monkeys, *Berlin: IXth International Conference on AIDS* 190 (Abstract).

Henderson, L. E., Sowder, R., Smythers, G., Benveniste, R. E., and Oroszlan, S., 1985, Purification and N-terminal amino acid sequence comparisons of structural proteins from retrovirus-D/Washington and Mason-Pfizer monkey virus, *J. Virol.* **55**:778.

Henderson, L. E., Benveniste, R. E., Sowder, R., Copeland, T. D., Schultz, A. M., and Oroszlan, S., 1988a, Molecular characterization of gag proteins from simian immunodeficiency virus (SIVMne), *J. Virol.* **62**:2587.

Henderson, L. E., Sowder, R. C., Copeland, T. D., Benveniste, R. E., and Oroszlan, S., 1988b, Isolation and characterization of a novel protein (x-orf product) from SIV and HIV-2, *Science* **241**:199.

Henderson, L. E., Sowder, R. C., Copeland, T. D., Oroszlan, S., and Benveniste, R. E., 1990, Gag precursors of HIV and SIV are cleaved into six proteins found in the mature virions, *J. Med. Primatol.* **19**:411.

Hendry, R. M., Well, M. A., Phelan, M. A., Schneider, A. L., Epstein, J. S., and Quinnan, G. V., 1986, Antibodies to simian immunodeficiency virus in African green monkeys in Africa in 1957–62, *Lancet* **2**:455.

Henrickson, R. V., Osborn, K. G., Madden, D. L., Anderson, J. H., Maul, D. H., Sever, J. L., Ellingsworth, L. R., Lowenstine, L. J., and Gardner, M. B., 1983, Epidemic of acquired immunodeficiency in rhesus monkeys, *Lancet* **1**:388.

Henrickson, R. V., Maul, D. H., Lerche, N. W., Osborn, K. G., Lowenstine, L. J., Prahalada, S., Sever, J. L., Madden, D. L., and Gardner, M. B., 1984, Clinical features of simian acquired immunodeficiency syndrome (SAIDS) in rhesus monkeys, *Lab. Anim. Sci.* **34**:146.

Heyes, M. P., Gravell, M., London, W. T., Eckhaus, M., Vickers, J. H., Yergey, J. A., April, M., Blackmore, D., and Markey, S. P., 1990, Sustained increases in cerebrospinal fluid quinolinic acid concentrations in rhesus macaques (Macaca mulatta) naturally infected with simian retrovirus type-D, *Brain Res.* **531**:148.

Heyes, M. P., Lackner, A., Kaufman, S., and Milstein, S., 1991, Cerebrospinal fluid and serum neopterin and biopterin in D-retrovirus-infected rhesus macaques (Macaca mulatta): Relationship to clinical and viral status, *AIDS* **5**:555.

Heyes, M. P., Jordan, E. K., Lee, K., Saito, K., Frank, J. A., Snoy, P. J., Markey, S. P., and Gravell, M., 1992, Relationship of neurologic status in macaques infected with the simian immunodeficiency virus to cerebrospinal fluid quinolinic acid and kynurenic acid, *Brain Res.* **570**:237.

Higgins, J. R., Sutjipto, S., Marx, P. A., and Pedersen, N. C., 1992, Shared antigenic epitopes of the major core proteins of human and simian immunodeficiency virus isolates, *J. Med. Primatol.* **21**:265.

Hilleman, M. R., 1992, Historical and contemporary perspectives in vaccine developments: From the vantage of cancer, *Prog. Med. Virol.* **39**:1.

Hillyer, C. D., Brodie, A. R., Ansari, A. A., Anderson, D. C., and McClure, H. M., 1991, Severe autoimmune hemolytic anemia in SIVsmm9-infected Macaca mulatta, *J. Med. Primatol.* **20**:156.

Hillyer, C. D., Klumpp, S. A., Hall, J. M., Lackey, D. A., Ansari, A. A., and McClure, H. M., 1993, Multifactorial etiology of anemia in SIV-infected rhesus macaques: decreased BFU-E formation, serologic evidence of autoimmune hemolysis and an exuberent erythropoietin response, *J. Med. Primatol.* **22**:253.

Hirsch, V. M., and Johnson, P. R., 1992, Pathogenesis of experimental SIV infection of macaques, *Semin. Virol.* **3**:175.

Hirsch, V. M., Edmondson, P., Murphey-Corb, M., Arbeille, B., Johnson, P. R., and Mullins, J. I., 1989a, SIV adaption to human cells, *Nature* **341**:573.

Hirsch, V. M., Olmstedt, R. A., Murphey-Corb, M., Purcell, R. H., and Johnson, P. R., 1989b, An African primate lentivirus (SIVsm) closely related to HIV-1, *Nature* **339**:389.

Hirsch, V. M., Zack, P. M., and Johnson, P. R., 1990, Molecular characterization of SIV in tissues from experimentally infected macaques, *J. Med. Primatol.* **19**:287.

Hirsch, V. M., Zack, P. M., Vogel, A. P., and Johnson, P. R., 1991, Simian immunodeficiency virus infection of macaques: End-stage disease is characterized by widespread distribution of proviral DNA in tissues, *J. Infect. Dis.* **163**:976.

Hirsch, V. M., Dapolito, G. A., Goldstein, S., McClure, H., Emau, P., Fultz, P. N., Isahakia, M., Lenroot, R., Myers, G., and Johnson, P. R., 1993a, A distinct African lentivirus from Sykes' monkeys, *J. Virol.* **1517**:1528.

Hirsch, V. M., Goldstein, S., Hynes, N., London, W. T., Elkins, W. R., Zack, P. M., Byrum, R., and Johnson, P. R., 1993b, Whole inactivated SIV vaccine protects from disease but not infection after cell-associated SIV challenge, *Madison, WI: 11th Annual Symposium on Nonhuman Primate Models for AIDS* (Abstract, p. 12).

Hirsch, V. M., McGann, C., Dapolito, G., Goldstein, S., Ogen-Odoi, A., Biryawaho, B., Lakwo, T., and Johnson, P. R., 1993c, Identification of a new subgroup of SIVagm in tantalus monkeys, *Virology* **197**:426.

Hirsch, V. M., Lewis, M. G., Johnson, P. R., Klumpp, S., Anderson, D. C., and McClure, H. M., 1993c, Viral genetic determinants in SIVsmmPBj pathogenesis, *Nonhuman Primate Models for AIDS Research Symposium* (Abstract).

Hoffenbach, A., Langlade-Demoyen, P., Dadaglio, G., Vilmer, E., Michel, F., Mayaud, C., Autran, B., and Plata, F., 1989, Unusually high frequencies of HIV-specific cytotoxic T lymphocytes in humans, *J. Immunol.* **142**:452.

Hoffmann, G. W., Kion, T. A., and Grant, M. D., 1991, An idiotypic network model of AIDS immunopathogenesis, *Proc. Natl. Acad. Sci. USA* **88**:3060.

Holmberg, C. A., Henrickson, R., Anderson, J., and Osborn, B. I., 1985, Malignant lymphoma in a colony of Macaca arctoides, *Vet. Pathol.* **22**:42.

Honjo, S., Narita, T., Kobayashi, R., Hiyaoka, A., Fujimoto, K., Takasaka, M., Sakakibara, I., Mukai, R., Ishikawa, K., and Ohta, Y., 1990, Experimental infection of African green monkeys and cynomolgus monkeys with a SIVAGM strain isolated from a healthy African green monkey, *J. Med. Primatol.* **19**:9.

Horvath, C. J., Desrosiers, R. C., Sehgal, P. K., King, N. W., and Ringler, D. J., 1991, Effect of simian immunodeficiency virus infection on tumor necrosis factor-alpha production by alveolar macrophages, *Lab. Invest.* **65**:280.

Horvath, C. J., Simon, M. A., Bergsagel, D. J., Pauley, D. R., King, N. W., Garcea, R. L., and Ringler, D. J., 1992, Simian virus 40-induced disease in rhesus monkeys with simian acquired immunodeficiency syndrome, *Am. J. Pathol.* **140**:1431.

Horvath, C. J., Hunt, R. D., Simon, M. A., Sehgal, P. K., and Ringler, D. J., 1993, An immunohistologic study of granulomatous inflammation in SIV-infected rhesus monkeys, *J. Leukocyte Biol.* **53**:532.

Hu, S.-L., Zarling, J. M., Chinn, J., Travis, B. M., Moran, P. A., Sias, J., Kuller, L., Morton, W. R., Heidecker, G., and Benveniste, R. E., 1989, Protection of macaques against simian AIDS by immunization with a recombinant vaccinia virus expressing the envelope glycoproteins of simian type D retrovirus, *Proc. Natl. Acad. Sci. USA* **86**:7213.

Hu, S.-L., Stallard, V., Misher, L., Langlois, A. J., Kuller, L., Morton, W. R., and Benveniste, R. E., 1992a, Immunization with recombinant SIVmnegp160 vaccines partially protects macaques against uncloned homologous virus challenge, *Puerto Rico: Tenth Annual Symposium on Non-human Primate Models for AIDS* (Abstract).

Hu, S.-L., Abrams, K., Barber, G. N., Moran, P., Zarling, J. M., Langlois, A. J., Kuller, L., Morton, W. R., and Benveniste, R. E., 1992b, Protection of macaques against SIV infection by subunit vaccines of SIV envelope glycoprotein gp160, *Science* **255**:456.

Hu, S.-L., Stallard, V., Abrams, K., Barber, G. N., Kuller, L., Langlois, A. J., Morton, W. R., and Benveniste, R. E., 1993, Protection of vaccinia-primed macaques against SIVmne infection by combination immunization with recombinant vaccinia virus and SIVmne gp160, *J. Med. Primatol.* **22**:92.

Hu, W., Vander, H. N., and Ratner, L., 1989, Analysis of the function of viral protein X (VPX) of HIV-2, *Virology* **173**:624.

Huet, T., Cheynier, R., Meyerhans, A., Roelants, G., and Wain-Hobson, S., 1990, Genetic organization of a chimpanzee lentivirus related to HIV-1, *Nature* **345**:356.

Hurtrel, B., Chakrabarti, L., Hurtrel, M., Maire, M. A., Dormont, D., and Montagnier, L., 1991, Early SIV encephalopathy, *J. Med. Primatol.* **20**:159.

Hurtrel, B., Chakrabarti, L., Hurtrel, M., and Montagnier, L., 1993, Target cells during early SIV encephalopathy, *Res. Virol.* **144**:41.

Hwang, S. S., Boyle, T. J., Lyerly, H. K., and Cullen, B. R., 1991, Identification of the envelope V3 loop as the primary determinant of cell tropism in HIV-1, *Science* **253**:71.

Ilyinskii, P., Daniel, M., Lerche, N., and Desrosiers, R., 1991, Antibodies to type D retrovirus in talapoin monkeys, *J. Gen. Virol.* **72**:453.

Ilyinskii, P. O., Daniel, M. D., Horvath, C. J., and Desrosiers, R. C., 1992, Genetic analysis of simian virus 40 from brains and kidneys of macaque monkeys, *J. Virol.* **66**:6353.

Israel, Z. R., Edmonson, P. F., Maul, D. H., O'Neil, S. P., Mossman, S. P., Thriart, Cl., Fabry, L., Van Opstal, O., Bruck, C., Bex, F., Burny, A., Fultz, P. N., Mullins, J. I., and Hoover, E. A., 1994, Incomplete protection, but suppression of virus burden, elicited by subunit simian immunodeficiency virus vaccines, *J. Virol.* **68**:1843–1853.

Jacks, T., Power, M. D., Masiarz, F. R., Luciw, P. A., Barr, P. J., and Varmus, H. E., 1988, Characterization of ribosomal frameshifting in HIV-1 gag–pol expression, *Nature* **331**:280.

Jakobovits, A., Smith, D. H., Jakobovits, E. B., and Capon, D. J., 1988, A discrete element 3′ of human immunodeficiency virus 1 (HIV-1) and HIV-2 mRNA initiation sites mediates transcriptional activation by an HIV trans activator, *Mol. Cell. Biol.* **8**:2555.

Janeway, C. A., Jr., Carding, S., Jones, B., Murray, J., Portoles, P., Rasmussen, R., Rojo, J., Saizawa, K., West, J., and Bottomly, K., 1988, $CD4^+$ T cells: Specificity and function, *Immunol. Rev.* **101**:39.

Javaherian, K., Langlois, A. J., Schmidt, S., Kaufmann, M., Cates, N., Langedijk, J. P. M., Meloen, R. H., Desrosiers, R. C., Burns, D. P. W., Bolognesi, D. P., LaRosa, G., and Putney, S. D., 1992, The principal neutralization determinant of simian immunodeficiency virus differs from that of human immunodeficiency virus type 1, *Proc. Natl. Acad. Sci. USA* **89**:1418.

Javaherian, K., Langlois, A. J., Montefiori, D. C., Kent, K. A., Ryan, K. A., Wyman, P. D., Stott, J., Bolognesi, D. P., Murphey-Corb, M., and LaRosa, G. J., 1994, Studies on the confirmation-dependent neutralizing epitopes of simian immunodeficiency virus envelope protein, *J. Virol.* **68**:2624–2631.

Jehuda-Cohen, T., Slade, B. A., Powell, J. D., Villinger, F., De, B., Folks, T. M., McClure, H. M., Sell, K. W., and Ahmed-Ansari, A., 1990, Polyclonal B-cell activation reveals antibodies against human immunodeficiency virus type 1 (HIV-1) in HIV-1-seronegative individuals, *Proc. Natl. Acad. Sci. USA* **87**:2973.

Jehuda-Cohen, T., Powell, J. D., Sell, K. W., Villinger, F., Lockwood, E., McClure, H. M., and Ahmed-Ansari, A., 1991a, Transmission of retroviral infection by transfusion of seronegative blood in nonhuman primates, *J. Infect. Dis.* **163**:1223.

Jehuda-Cohen, T., Villinger, F., Powell, J. D., McClure, H. M., Sell, K. W., and Ahmed-Ansari, A., 1991b, Presence of SIV antibodies in the sera of infants born to SIV-seronegative monkeys [letter], *J. AIDS* **4**:204.

Jenkins, S., Gritz, L., Fedor, C. H., O'Neill, E. M., Cohen, L. K., and Panicali, D. L., 1991, Formation of lentivirus particles by mammalian cells infected with recombinant fowlpox virus, *AIDS Res. Hum. Retroviruses* **7**:991.

Jennings, M., Lowenstine, L., Gardner, M., Rosenthal, A., Antipa, L., Cotterman, R., and Yee, J., 1992, Horizontal transmission of SIV in juvenile male rhesus without protection by inactivated, whole-virus vaccine, *Fifth Annual Meeting of the National Cooperative Vaccine Development Groups for AIDS* (Abstract).

Joag, S. V., Anderson, M. G., Clements, J. E., McEntee, M. F., Sharma, D. P., Adams, R. J., and Narayan, O., 1993a, Antigenic variation of molecularly cloned SIVmac239 during persistent infection in a rhesus macaque, *Virology* **195**:406.

Joag, S. V., Adams, R. J., Foresman, L., and Narayan, O., 1993b, Phases of SIV infection in rhesus macaques: Role of lymphoid organs in pathogenesis, *Nonhuman Primate Models for AIDS Research Symposium* (Abstract).

Johnson, P. R., Parks, D. E., Norrby, E., Lerner, R. A., Purcell, R. H., and Chanock, R. M., 1988, Site-directed ELISA identifies a highly antigenic region of the simian immunodeficiency virus transmembrane glycoprotein, *AIDS Res. Hum. Retroviruses* **4**:159.

Johnson, P. R., and Hirsch, V. M., 1992, Genetic variation of simian immunodeficiency viruses in nonhuman primates, *AIDS Res. Hum. Retroviruses* **8**:367.

Johnson, P. R., Fomsgaard, A., Allan, J., Gravell, M., London, W. T., Olmsted, R. A., and Hirsch, V. M., 1990a, Simian immunodeficiency viruses from African green monkeys display unusual genetic diversity, *J. Virol.* **64**:1086.

Johnson, P. R., Goldstein, S., London, W. T., Fomsgaard, A., and Hirsch, V. M., 1990b, Molecular clones of SIVsm and SIVagm: Experimental infection of macaques and African green monkeys, *J. Med. Primatol.* **19**:279.

Johnson, P. R., Montefiori, D. C., Goldstein, S., Hamm, T. E., Zhou, J., Kitov, S., Haigwood, N. L., Misher, L., London, W. T., Gerin, J. L., Allison, A., Purcell, R. H., Chanock, R. M., and Hirsch, V. M., 1992a, Inactivated whole SIV vaccine in macaques: Evaluation of protective efficacy against challenge with cell-free virus or infected cells, *AIDS Res. Hum. Retroviruses* **8**:1501.

Johnson, P. R., Montefiori, D. C., Goldstein, S., Hamm, T. E., Zhou, J., Kitov, S., Haigwood, N. L., Misher, L., London, W. T., and Gerin, J. L., 1992b, Inactivated whole-virus vaccine derived from a proviral DNA clone of simian immunodeficiency virus induces high levels of neutralizing antibodies and confers protection against heterologous challenge, *Proc. Natl. Acad. Sci. USA* **89**:2175.

Johnston, P. B., Dubay, D. W., and Hunter, E., 1993, Truncations of the simian immunodeficiency virus transmembrane protein confer expanded virus host range by removing a block to virus entry into cells, *J. Virol.* **67**:3077.

Joling, P., Van Wichen, D. F., Parmentier, H. K., Biberfeld, P., Bottiger, D., Tschopp, J., Rademakers, L. H. P. M., and Schuurman, H.-J., 1992, Simian immunodeficiency virus (SIVsm) infection of cynomolgus monkeys: Effects on follicular dendritic cells in lymphoid tissue, *AIDS Res. Hum. Retroviruses* **8**:2021.

Jones, W. C., Barnard, A. L., Slade, A., and Mills, K. H. G., 1992, Heterogeneity in the recognition of the simian immunodeficiency virus envelope glycoprotein by $CD4^+$ T cell clones from immunized macaques, *J. Immunol.* **149**:3120.

Jordan, E. K., and Heyes, M. P., 1993, Virus isolation and quinolinic acid in primary and chronic simian immunodeficiency virus infection, *AIDS* **7**:1173.

Jurkiewicz, E., Nakamura, H., Schneider, J., Yamamoto, N., Hayami, M., and Hunsmann, G., 1986, Structural analysis of p19 and p24 core polypeptides of primate lymphotropic retroviruses (PLRV), *Virology* **150**:291.

Kaaya, E., Li, S. L., Feichtinger, H., Stahmer, I., Putkonen, P., Mandache, E., Mgaya, E., Biberfeld, G., and Biberfeld, P., 1993, Accessory cells and macrophages in the histopathology of SIVsm-infected cynomolgus monkeys, *Res. Virol.* **144**:81.

Kalter, S. S., and Heberling, R. L., 1988, B virus (Herpesvirus simiae) infection, *AM News* **54**:71.

Kanki, P. J., Alroy, J., and Essex, M., 1985a, Isolation of T-lymphotropic retrovirus related to HTLV-III/LAV from wild-caught African green monkeys, *Science* **230**:951.

Kanki, P. J., Kurth, R., Becker, W., Dreesman, G., McLane, M. F., and Essex, M., 1985b, Antibodies to simian T-lymphotropic retrovirus type III in African green monkeys and recognition of STLV-III viral proteins by AIDS and related sera, *Lancet* **1**:1330.

Kanki, P. J., McLane, M. F., King, N. W., Letvin, N. L., Hunt, R. D., Sehgal, P., Daniel, M. D., Desrosiers, R. C., and Essex, M., 1985c, Serologic identification and characterization of a macaque T-lymphotropic retrovirus closely related to HTLV-III, *Science* **230**:1199.

Kanki, P. J., Barin, F., M'Boup, S., Allan, J. S., Romet-Lemonne, J., Marlink, R., McLane, M. F., Lee, T. H., Arbeille, B., Denis, F., and Essex, M., 1986, New human T-lymphotropic retrovirus related to simian T-lymphotropic virus type III (STLV-IIIagm), *Science* **232**:238.

Kannagi, M., Yetz, J. M., and Letvin, N. L., 1985, In vitro growth characteristics of simian T-lymphotropic virus type III, *Proc. Natl. Acad. Sci. USA* **82**:7053.

Kannagi, M., Kiyotaki, M., Desrosiers, R. C., Reimann, K. A., King, N. W., Waldron, L. M., and Letvin, N. L., 1986, Humoral immune responses to T cell tropic retrovirus simian T lymphotropic virus type III in monkeys with experimentally induced acquired immune deficiency-like syndrome, *J. Clin. Invest.* **78**:1229.

Kannagi, M., Kiyotaki, M., King, N. W., Lord, C. I., and Letvin, N. L., 1987, Simian immunodeficiency virus induces expression of class II major histocompatibility complex structures on infected target cells in vitro, *J. Virol.* **61**:1421.

Kannagi, M., Chalifoux, L. V., Lord, C. I., and Letvin, N. L., 1988, Suppression of simian immunodeficiency virus replication in vitro by $CD8^+$ lymphocytes, *J. Immunol.* **140**:2237.

Kannagi, M., Masuda, T., Hattori, T., Kanoh, T., Nasu, K., Yamamoto, N., and Harada, S., 1990, Interference with human immunodeficiency virus (HIV) replication by $CD8^+$ T cells in peripheral blood leukocytes of asymptomatic HIV carriers in vitro, *J. Virol.* **64**:3399.

Kao, S. Y., Calman, A. F., Luciw, P. A., and Peterlin, B. M., 1987, Anti-termination of transcription within the long terminal repeat of HIV-1 by tat gene product, *Nature* **330**:489.

Kappes, J. C., Morrow, C. D., Lee, S. W., Jameson, B. A., Kent, S. B. H., Hood, L. E., Shaw, G. M., and Hahn, B. H., 1988, Identification of a novel retroviral gene unique to human immunodeficiency virus type 2 and simian immunodeficiency virus SIVmac, *J. Virol.* **62**:3501.

Kappes, J. C., Conway, J. A., Lee, S. W., Shaw, G. M., and Hahn, B. H., 1991, Human immunodeficiency virus type 2 vpx protein augments viral infectivity, *Virology* **184**:197.

Kappes, J. C., Parkin, J. S., Conway, J. A., Kim, J., Brouillette, C. G., Shaw, G. M., and Hahn, B. H., 1993, Intracellular transport and virion incorporation of vpx requires interaction with other virus type-specific components, *Virology* **193**:222.

Kaul, S., Dandekar, K. A., and Pittman, K. A., 1989, Analytical method for the quantification of 2′,3′-didehydro-3′-deoxythymidine, a new anti-human immunodeficiency virus (HIV) agent, by high-performance liquid chromatography (HPLC) and ultraviolet (UV) detection in rat and monkey plasma, *Pharm. Res.* **6**:895.

Kawakami, T. G., Huff, S. D., Buckley, P. M., Dungworth, D. L., and Snyder, S. P., 1972, C-type virus associated with gibbon lymphosarcoma, *Nature* **235**:171.

Kent, K. A., Gritz, L., Stallard, G., Cranage, M. P., Collignon, C., Thiriart, C., Corcoran, T., Silvera, P., and Stott, E. J., 1991, Production of monoclonal antibodies to simian immunodeficiency virus envelope glycoproteins, *AIDS* **5**:829.

Kent, K. A., Rud, E., Corcoran, T., Powell, C., Thiriart, C., Collignon, C., and Stott, E. J., 1992, Identification of two neutralizing and 8 non-neutralizing epitopes on simian immunodeficiency virus envelope using monoclonal antibodies, *AIDS Res. Hum. Retroviruses* **8**:1147.

Kestler, H. W., III, Li, Y., Naidu, Y. M., Butler, C. V., Ochs, M. F., Jaenel, G., King, N. W., Daniel, M. D., and Desrosiers, R. D., 1988, Comparison of simian immunodeficiency virus isolates, *Nature* **331**:619.

Kestler, H., Naidu, Y. M., Kodama, T., King, N. W., Daniel, M. D., Li, Y., and Desrosiers, R. C., 1989, Use of infectious molecular clones of simian immunodeficiency virus for pathogenesis studies, *J. Med. Primatol.* **18**:305.

Kestler, H., Kodama, T., Ringler, D., Marthas, M., Pedersen, N., Lackner, A., Regier, D., Sehgal, P., Daniel, M., King, N., and Desrosiers, R., 1990, Induction of AIDS in rhesus monkeys by molecularly cloned simian immunodeficiency virus, *Science* **248**:1109.

Kestler, H. W., Ringler, D. J., Mori, K., Panicali, D. L., Sehgal, P. K., Daniel, M. D., and Desrosiers, R. C., 1991, Importance of the nef gene for maintenance of high virus loads and for development of AIDS, *Cell* **65**:651.

Khabbaz, R., Rowe, T., Murphey-Corb, M., Heneine, W., Schable, C., George, J. R., Pau, C., Parekh, B., and Folks, T., 1992a, Seroconversion to simian immunodeficiency virus (SIV) in two laboratory workers, *Puerto Rico: Tenth Annual Symposium on Nonhuman Primate Models for AIDS* (Abstract 9).

Khabbaz, R. F., Rowe, T., Murphey-Corb, M., Heneine, W. M., Schable, C. A., George, J. R., Pau, C. P., Parekh, B. S., Lairmore, M. D., and Curran, J. W., 1992b, Simian immunodeficiency virus needlestick accident in a laboratory worker, *Lancet* **340**:271.

Khabbaz, R. F., Heneine, W., George, J. R., Parekh, B., Rowe, T., Woods, T., Switzer, W. M., McClure, H. M., Murphey-Corb, M., and Folks, T. M., 1994, Brief Report: Infection of a laboratory worker with simian immunodeficiency virus, *N. Engl. J. Med.* **330**:172.

Khan, A. S., Galvin, T. A., Jennings, M. B., Gardner, M. B., and Lowenstine, L. J., 1991a, SIV of stump-tailed macaque (SIVstm) is a divergent Asian isolate, *J. Med. Primatol.* **20**:167.

Khan, A. S., Galvin, T. A., Lowenstine, L. J., Jennings, M. B., Gardner, M. B., and Buckler, C. E., 1991b, A highly divergent simian immunodeficiency virus (SIVstm) recovered from stored stump-tailed macaque tissues, *J. Virol.* **65**:7061.

Kiernan, R. E., McPhee, D. A., and Doherty, R. R., 1988, Quantitation of infectious human immunodeficiency virus using a modified plaque forming assay, *J. Virol. Methods* **22**:303.

Kikukawa, R., Koyanagi, Y., Harada, S., Kobayashi, N., Hatanaka, M., and Yamamoto, N., 1986, Differential susceptibility to the acquired immunodeficiency syndrome retrovirus in cloned cells of human leukemic T-cell line Molt-4, *J. Virol.* **57**:1159.

King, N. W., Chalifoux, L. V., Ringler, D. J., Wyand, M. S., Sehgal, P. K., Daniel, M. D., Letvin, N. L., Desrosiers, R. C., Blake, B. J., and Hunt, R. D., 1990, Comparative biology of natural and experimental SIVmac infection in macaque monkeys: A review, *J. Med. Primatol.* **19**:109.

Kirchhoff, F., Voss, G., Nick, S., Stahl, H. C., Coulibaly, C., Frank, R., Jentsch, K. D., and Hunsmann, G., 1991, Antibody response to the negative regulatory factor (nef) in experimentally infected macaques: Correlation with viremia, disease progression, and seroconversion to structural viral proteins, *Virology* **183**:267.

Kitagawa, M., Lackner, A. A., Martfeld, D. J., Gardner, M. B., and Dandekar, S., 1991, Simian immunodeficiency virus infection of macaque bone marrow macrophages correlates with disease progression in vivo, *Am. J. Pathol.* **138**:921.

Kitchin, P. A., Almond, N., Szotyori, Z., Fromholc, C. E., McAlpine, L., Silvera, P., Stott, E. J., Cranage, M., Baskerville, A., and Schild, G., 1990a, The use of the polymerase chain reaction for the detection of simian immunodeficiency virus in experimentally infected macaques, *J. Virol. Methods* **28**:85.

Klatzmann, D., Champagne, E., Chamaret, S., Gruest, J., Guetard, F., Hercend, T., Gluckman, J., and Montagnier, L., 1984, T-lymphocyte T4 molecule behaves as the receptor for human retrovirus LAV, *Nature* **312**:767.

Klumpp, S. A., Novembre, F. J., Anderson, D. C., Simon, M. A., Ringler, D. J., and McClure, H. M., 1993, Clinical and pathologic findings in infant rhesus macaques infected with SIVsmm by maternal transmission, *J. Med. Primatol.* **22**:169.

Kodama, T., Silva, D. P., Daniel, M. D., Phillips-Conroy, J. E., Jolly, C. J., Rogers, J., and Desrosiers, R. C., 1989a, Prevalence of antibodies to SIV in baboons in their native habitat, *AIDS Res. Hum. Retroviruses* **5**:337.

Kodama, T., Wooley, D. P., Naidu, Y. M., Kestler, H. W., Daniel, M. D., Li, Y., and Desrosiers, R. C., 1989b, Significance of premature stop codons in env of simian immunodeficiency virus, *J. Virol.* **63**:4709.

Kodama, T., Burns, D. P., Silva, D. P., Veronese, F. D., and Desrosiers, R. C., 1991, Strain-specific neutralizing determinant in the transmembrane protein of simian immunodeficiency virus, *J. Virol.* **65**:2010.

Kodama, T., Mori, K., Kawahara, T., Ringler, D. J., and Desrosiers, R. C., 1993, Analysis of simian immunodeficiency virus sequence variation in tissues of rhesus macaques with simian AIDS, *J. Virol.* **67**:6522.

Koenig, S., Hirsch, V. M., Olmsted, R. A., Powell, D., Maury, W., Rabson, A., Fauci, A. S., Purcell, R. H., and Johnson, P. R., 1989, Selective infection of human $CD4^+$ cells by simian immunodeficiency virus: Productive infection associated with envelope glycoprotein-induced fusion, *Proc. Natl. Acad. Sci. USA* **86**:2443.

Koo, H. M., Gu, J., Varela-Echavarria, A., Ron, Y., and Dougherty, J. P., 1992, Reticuloendotheliosis type C and primate type D oncoretroviruses are members of the same receptor interference group, *J. Virol.* **66**:3448.

Kornfeld, H., Riedel, N., Viglianti, G., Hirsch, V., Mullins, J. I., 1987, Cloning of HTLV-4 and its relation to simian and human immunodeficiency viruses, *Nature* **326**:610.

Kraus, G., Werner, A., Baier, M., Binniger, D., Ferdinand, F. J., Norley, S., and Kurth, R., 1989, Isolation of human immunodeficiency virus-related immunodeficiency viruses from African green monkeys, *Proc. Natl. Acad. Sci. USA* **868**:2892.

Krause, H., Wunderlich, V., and Uckert, W., 1989, Molecular cloning of a type D retrovirus from human cells (PMFV) and its homology to simian acquired immunodeficiency type D retroviruses, *Virology* **173**:214.

Krohn, K. J. E., Hakkarainen, K., Aavik, E., Dewhurst, S., Sadaie, R., and Mullins, J. I., 1993, Transcomplementation of simian immunodeficiency virus rev with human T-cell leukemia virus type 1 rex, *J. Virol.* **67**:5681.

Krugner-Higby, L., Kucera, L., Lerche, N., Sever, J., Fucillo, W., Allan, J., and Benveniste, R., 1990, Serological survey for two simian retroviruses in macaques and African green monkeys, *Lab. Anim. Sci.* **40**:24.

Krugner-Higby, L. A., Schultz, K. T., Johnson, K. J., and Golos, T. G., 1993, Replication of SIV-mac251 and SIVB670 in vitro in placental syncytiotrophoblasts, *Madison, WI: 11th Annual Symposium on Nonhuman Primate Models for AIDS* (Abstract p 6).

Kuller, L., Morton, W. R., Benveniste, R. E., Tsai, C. C., Clark, E. A., Gale, M. J., Hu, S. L., Thouless, M. E., and Katze, M. G., 1990, Inoculation of Macaca fascicularis with simian immunodeficiency virus, SIVmne immunologic, serologic, and pathologic changes, *J. Med. Primatol.* **19**:367.

Kuller, L., Benveniste, R. E., Watanabe, R., Tsai, C. C., and Morton, W. R., 1992, Transmission of SIVMne from female to male Macaca nemestrina, *J. Med. Primatol.* **21**:299.

Kwang, H. S., Pedersen, N. C., Lerche, N. W., Osborn, K. G., Marx, P. A., and Gardner, M. B., 1987, Viremia, antigenemia, and serum antibodies in rhesus macaques infected with simian retrovirus type 1 and their relationship to disease course, *Lab. Invest.* **56**:591.

Lackner, A. A., Rodriguez, M. H., Bush, C. E., Munn, R. J., Kwang, H.-S., Moore, P. F., Osborn, K. G., Marx, P. A., Gardner, M. B., and Lowenstine, L. J., 1988, Distribution of a macaque immunosuppressive type D retrovirus in neural, lymphoid, and salivary tissues, *J. Virol.* **62**:2134.

Lackner, A. A., Schiodt, M., Armitage, G. C., Moore, P. F., Munn, R. J., Marx, P. A., Gardner, M. B., and Lowenstine, L. J., 1989, Mucosal epithelial cells and Langerhans cells are targets for infection by the immunosuppressive type D retrovirus simian AIDS retrovirus serotype 1, *J. Med. Primatol.* **18**:195.

Lackner, A. A., Dandekar, S., and Gardner, M. B., 1991a, Neurobiology of simian and feline immunodeficiency virus infections, *Brain Pathol.* **1**:12.

Lackner, A. A., Smith, M. O., Munn, R. J., Martfeld, D. J., Gardner, M. B., Marx, P. A., and Dandekar, S., 1991b, Localization of simian immunodeficiency virus in the central nervous system of rhesus monkeys, *Am. J. Pathol.* **139**:609.

Lackner, A. A., Vogel, P., Hoogenboom, E., Sandekar, S., Kluge, J. D., and Marthas, M., 1991c, Pathogenic (SIVmac-239) and nonpathogenic (SIVmac-1A11) molecular clones of SIV have distinct tissue distributions that vary with length of infection, *Seattle: 9th Annual Symposium on Nonhuman Primate Models for AIDS* (Abstract).

Lairmore, M. D., Lerche, N. W., Schultz, K. T., Stone, C. M., Brown, B. G., Hermann, L. M., Yee, J. A., and Jennings, M., 1990, SIV, STLV-I and type D retrovirus antibodies in captive rhesus macaques and immunoblot reactivity to SIV p27 in human and rhesus monkey sera, *AIDS Res. Hum. Retroviruses* **6**:1233.

Lairmore, M. D., Post, A. A., Goldsmith, C. S., and Folks, T. M., 1991, Cytokine enhancement of simian immunodeficiency virus (SIV/mac) from a chronically infected cloned T-cell line (HuT-78), *Arch. Virol.* **121**:43.

Lairmore, M. D., Hofheinz, D. E., Letvin, N. L., Stoner, C. S., Pearlman, S., and Toedter, G. P., 1993, Detection of simian immunodeficiency virus and human immunodeficiency virus type 2 capsid antigens by a monoclonal antibody-based antigen capture assay, *AIDS Res. Hum. Retroviruses* **9**:565.

Lang, S. M., Weeger, M., Stahl-Hennig, C., Coulibaly, C., Hunsmann, G., Muller, J., Muller-Hermelink, H., Fuchs, D., Wachter, H., and Daniel, M. M., 1993, Importance of vpr for infection of rhesus monkeys with simian immunodeficiency virus, *J. Virol.* **67**:902.

Lange, J. M., Boucher, C. A., Hollak, C. E., Wiltink, E. H., Reiss, P., van Royen, E. A., Roos, M., Danner, S. A., and Goudsmit, J., 1990, Failure of zidovudine prophylaxis after accidental exposure to HIV-1 [see comments], *N. Engl. J. Med.* **322**:1375.

Langlois, A. J., Weinhold, K. J., Matthews, T. J., and Bolognesi, D. P., 1991, In vitro assays for detecting neutralizing and fusion-inhibiting antibodies to SIVMAC251, *AIDS Res. Hum. Retroviruses* **7**:713.

Laspia, M. F., Rice, A. P., and Mathews, M. B., 1989, HIV-1 Tat protein increases transcriptional initiation and stabilizes elongation, *Cell* **59**:283.

Le, S. Y., Malim, M. H., Cullen, B. R., and Maizel, J. V., 1990, A highly conserved RNA folding region coincident with the Rev response element of primate immunodeficiency viruses, *Nucleic Acids Res.* **18**:1613.

Le Grand, E. K., Donovan, R. M., Marx, P. A., Moulton, J. E., Cheung, A. T., Lewis, A. E., and Gardner, M. B., 1985, Monocyte function in rhesus monkeys with simian acquired immune deficiency syndrome, *Vet. Immunol. Immunopathol.* **10**:131.

Le Grand, R., Vogt, G., Chapel, A., and Dormont, D., 1991, Antibody-dependent enhancement and neutralization pattern of sera from SIV-infected or HIV-2-vaccinated rhesus monkeys, *J. Med. Primatol.* **20**:172.

Le Grand, R., Vogt, G., Vaslin, B., Roques, P., and Dormont, D., 1992a, Specific and non-specific immunity and protection of macaques against SIV infection, *Vaccine* **10**:873.

Le Grand, R., Kress, J. P., Vogt, G., Vaslin, B., Roques, P., Theodoro, F., Aubertin, A. M., and Dormont, D., 1992b, Modification of viral load after immune stimulation of SIV-infected rhesus monkeys, *Puerto Rico: Tenth Annual Symposium on Nonhuman Primate Models for AIDS* (Abstract).

Le Grand, R., Vaslin, B., Vogt, G., Theodoro, F., Noack, O., Gras, G., Aubertin, A. M., and Dormont, D., 1993, Early stage kinetics of cellular and seric viremia after intra-rectal SIV infection of rhesus macaques, *Berlin: IXth International Conference on AIDS* (Abstract, p. 192).

Lehner, T., Bergmeier, L. A., Panagiotidi, C., Tao, L., Brookes, R., Klavinskis, L. S., Walker, P., Walker, J., Ward, R. G., and Hussain, L., 1992, Induction of mucosal and systemic immunity to a recombinant simian immunodeficiency viral protein, *Science* **258**:1365.

Lerche, N. W., Henrickson, R. V., Maul, D. H., and Gardner, M. B., 1984, Epidemiologic aspects of an outbreak of acquired immunodeficiency in rhesus monkeys (Macaca mulatta), *Lab. Anim. Sci.* **34**:146.

Lerche, N. W., Osborn, K. G., Marx, P. A., Prahalada, S., Maul, D. H., Lowenstine, L. J., Munn, R. J., Bryant, M. L., Henrickson, R. V., and Arthur, L. O., 1986, Inapparent carriers of simian acquired immune deficiency syndrome type D retrovirus and disease transmission with saliva, *J. Natl. Cancer Inst.* **77**:489.

Lerche, N. W., Marx, P. A., Osborn, K. G., Maul, D. H., Lowenstine, L. J., Bleviss, M. L., Moody, P., Henrickson, R. V., and Gardner, M. B., 1987, Natural history of endemic type D retrovirus infection and acquired immune deficiency syndrome in group-housed rhesus monkeys, *J. Natl. Cancer Inst.* **79**:847.

Lerche, N. W., Marx, P. A., and Gardner, M. B., 1991, Elimination of type D retrovirus infection from group-housed rhesus monkeys using serial testing and removal, *Lab. Anim. Care* **41**:123.

Letvin, N. L., and King, N. W., 1990, Immunologic and pathologic manifestations of the infection of rhesus monkeys with simian immunodeficiency virus of macaques, *J. AIDS* **3**:1023.

Letvin, N. L., Daniel, M. D., Sehgal, P. K., Desrosiers, R. C., Hunt, R. D., Waldron, L. M., MacKey, J. J., Schmidt, D. K., Chalifoux, L. V., and King, N. W., 1985, Induction of AIDS-like disease in macaque monkeys with T-cell tropic retrovirus STLV-III, *Science* **230**:71.

Letvin, N. L., Lord, C. I., King, N. W., Wyand, M. S., Myrick, K. V., and Haseltine, W. A., 1991, Risks of handling HIV [letter], *Nature* **349**:573.

Levy, J. A., 1993, Pathogenesis of human immunodeficiency virus infection, *Microbiol. Rev.* **57**:183.

Lewis, M. G., Zack, P. M., Elkins, W. R., and Jahrling, P. B., 1992b, Infection of rhesus and cynomolgus macaques with a rapidly fatal SIV (SIVsmm/PBj) isolate from sooty mangabeys, *AIDS Res. Hum. Retroviruses* **8**:1631.

Lewis, M. G., Bellah, S., McKinnon, K., Yalley-Ogunro, J., Greenhouse, J., Elkins, W. R., Rosenberg, Y., Gartner, S., and Eddy, G., 1993a, Dual infection of rhesus macaques with SIV-E11s and SIV239, *Nonhuman Primate Models for AIDS Research Symposium* (Abstract).

Lewis, M. G., Elkins, W. R., McCutchan, F. E., Benveniste, R. E., Lai, C. Y., Montefiori, D. C., Burke, D. S., Eddy, G. A., and Shafferman, A., 1993b, Passively transferred antibodies directed against conserved regions of SIV envelope protect macaques from SIV infection, *Vaccine* **11**:1347.

Li, J., Lord, C. I., Haseltine, W., Letvin, N. L., and Sodroski, J., 1992, Infection of cynomolgus monkeys with a chimeric HIV-1/SIVmac virus that expresses the HIV-1 envelope glycoproteins, *J. AIDS* **5**:639.

Li, S. L., Kaaya, E. E., Feichtinger, H., Putkonen, P., Parravicini, C., Bottiger, D., Biberfeld, G., and Biberfeld, P., 1991, Monocyte/macrophage giant cell disease in SIV-infected cynomolgus monkeys, *Res. Virol.* **142**:173.

Li, S. L., Kaaya, E. E., Feichtinger, H., Ekman, M., Biberfeld, G., and Biberfeld, P., 1993, Thymic changes in SIVsm infected cynomolgus monkeys, *Berlin: IXth International Conference on AIDS* 193. (Abstract)

Li, Y., Naidu, Y., Fultz, P., Daniel, M. D., and Desrosiers, R. C., 1989a, Genetic diversity of simian immunodeficiency virus, *J. Med. Primatol.* **18**:261.

Li, Y., Naidu, Y. M., Daniel, M. D., and Desrosiers, R. C., 1989b, Extensive genetic variability of simian immunodeficiency virus from African green monkeys, *J. Virol.* **63**:1800.

Ljungdahl-Stahle, E., Trojnar, J., Oberg, B., and Wahren, B., 1992, Early appearance of simian immunodeficiency virus (SIV) antigen and antibodies as variables in evaluating antiviral drugs in macaques, *J. Virol. Methods* **37**:43.

Loh, P. C., 1993, Spumaviruses, in: *The Retroviridae*, Vol. 2 (J. A. Levy, ed.), p. 361, Plenum Press, New York.

Lohman, B. L., Higgins, J., Marthas, M. L., Marx, P. A., and Pedersen, N. C., 1991, Development of simian immunodeficiency virus isolation, titration, and neutralization assays which use whole blood from rhesus monkeys and an antigen capture enzyme-linked immunosorbent assay, *J. Clin. Microbiol.* **29**:2187.

London, W. T., Madden, D. L., Gravell, M., Dalakas, M. C., Hauff, S. A., Sever, J. L., Henrickson, R. V., Maul, D. H., Osborn, K. G., and Gardner, M. B., 1983, Experimental transmission of simian acquired immunodeficiency syndrome (SAIDS) and Kaposi's-like skin lesions, *Lancet* **2**:869.

Looke, D. F., and Grove, D. I., 1990, Failed prophylactic zidovudine after needlestick injury [letter], *Lancet* **335**:1280.

Lopez-Anaya, A., Unadkat, J. D., Schumann, L. A., and Smith, A. L., 1990, Pharmacokinetics of zidovudine (azidothymidine). I. Transplacental transfer, *J. AIDS* **3**:959.

Lowenstine, L. J., and Lerche, N. W., 1988, Retrovirus infections of nonhuman primates: a review, *J. Zoo Anim. Med.* **19**:168.

Lowenstine, L. J., Pedersen, N. C., Higgins, J., Pallis, K. C., Uyeda, A., Marx, P., Lerche, N. W., Munn, R. J., and Gardner, M. B., 1986, Seroepidemiologic survey of captive Old-World primates for antibodies to human and simian retroviruses, and isolation of a lentivirus from sooty mangabeys (Cercocebus atys), *Int. J. Cancer* **38**:563.

Lowenstine, L. J., Lerche, N. W., Yee, J. L., Uyeda, A., Jennings, M. B., Munn, R. J., McClure, H. M., Anderson, D. C., Fultz, P. N., and Gardner, M. B., 1992, Evidence for a lentiviral etiology in an epizootic of immune deficiency and lymphoma in stump-tailed macaques (Macaca arctoides), *J. Med. Primatol.* **21**:1.

Luciw, P. A., and Leung, N. J., 1992, Mechanisms of retrovirus replication, in: *The Retroviridae*, Vol. 1 (J. A. Levy, ed.), p. 159, Plenum Press, New York.

Luciw, P. A., Cheng-Mayer, C., and Levy, J. A., 1987, Mutational analysis of the human immunodeficiency virus: The orf-B region down-regulates virus replication, *Proc. Natl. Acad. Sci. USA* **84**:1434.

Luciw, P. A., Shaw, K. E., Unger, R. E., Planelles, V., Stout, M. W., Lackner, J. E., Pratt-Lowe, E., Leung, N. J., Banapour, B., and Marthas, M. L., 1992, Genetic and biological comparisons of pathogenic and nonpathogenic molecular clones of simian immunodeficiency virus (SIVmac), *AIDS Res. Hum. Retroviruses* **8**:395.

Lüke, W., and Hunsmann, G., 1993, Protection of monkeys by a split vaccine against SIVmac depends upon biological properties of the challenge virus, *AIDS* **7**:787.

Lüke, W., Hoefer, K., Moosmayer, D., Nickel, P., Hunsmann, G., and Jentsch, K. D., 1990, Partial purification and characterization of the reverse transcriptase of the simian immunodeficiency virus TYO-7 isolated from an African green monkey, *Biochemistry* **29**:1764.

Lundgren, B., Ljundahl-Stahle, E., Wahren, B., Norrby, E., and Oberg, B., 1989, Antiviral effects of AZT, foscarnet and DDC in SIV-infected macaques (Macaca fascicularis), *Maastricht, Netherlands: International TNO Meeting on Animal Models in AIDS* (Abstract).

Lundgren, B., Bottiger, D., Ljungdahl-Stahle, E., Norrby, E., Stahle, L., Wahren, B., and Oberg, B., 1991, Antiviral effects of 3′-fluorothymidine and 3′-azidothymidine in cynomolgus monkeys infected with simian immunodeficiency virus, *J. AIDS* **4**:489.

McBride, B., Corthals, G., Rud, E., Kent, K., Webster, S., Cook, N., and Cranage, M. P., 1993, Comparison of serum antibody reactivities to a conformational and to linear antigenic sites in the external envelope glycoprotein of simian immunodeficiency virus (SIVmac) induced by infection and vaccination, *J. Gen. Virol.* **74**:1033.

McClure, H. M., Anderson, D. C., Fultz, P. N., Ansari, A. A., Lockwood, E., and Brodie, A., 1989, Spectrum of disease in macaque monkeys chronically infected with SIV/SMM, *Vet. Immunol. Immunopathol.* **21**:13.

McClure, H. M., Anderson, D. C., Ansari, A. A., Fultz, P. N., Klumpp, S. A., and Schinazi, R. F., 1990, Nonhuman primate models for evaluation of AIDS therapy, *Ann. N.Y. Acad. Sci.* **616**:287.

McClure, H. M., Anderson, D. C., Fultz, P. N., Ansari, A. A., Jehuda-Cohen, T., Villinger, F., Klumpp, S. A., Switzer, W., Lockwood, E., and Brodie, A., 1991, Maternal transmission of SIVsmm in rhesus macaques, *J. Med. Primatol.* **20**:182.

McClure, J., Misher, L., Schmidt, A., Steele, J., Morton, W., and Hu, S.-L., 1992, In vivo passage of HIV-2 EHO in Macaca nemestrina results in persistent infection, disease progression, and altered cellular tropism, *Puerto Rico: Tenth Annual Symposium on Nonhuman Primate Models for AIDS* (Abstract).

McClure, M. A., Johnson, M. S., and Doolittle, R. F., 1987, Relocation of a protease-like gene segment between two retroviruses, *Proc. Natl. Acad. Sci. USA* **84**:2693.

McClure, M. A., Johnson, M. S., Feng, D.-F., and Doolittle, R. F., 1988, Sequence comparisons of retroviral proteins: Relative rates of change and general phylogeny, *Proc. Natl. Acad. Sci. USA* **85**:2469.

McClure, M. O., 1991, The simian immunodeficiency viruses, *Mol. Aspects Med.* **12**:247.

McClure, M. O., Sattentau, Q. J., Beverley, P. C. L., Hearn, J. P., Fitzgerald, A. K., Zuckerman, A. J., and Weiss, R. A., 1987, HIV infection of primate lymphocytes and conservation of the CD4 receptor, *Nature* **330**:487.

McEntee, M. F., Sharma, D. P., Zink, M. C., Adams, R. J., Flexner, C., Clements, J. E., and Narayan, O., 1991, Rhesus monkey macrophages infected with simian immunodeficiency virus cause rapid lysis of CD4-bearing lymphocytes, *J. Gen. Virol.* **72**:317.

McEntee, M. F., Gorrell, M. D., Adams, R. J., Narayan, O., and Pitha, P., 1992a, Tumour necrosis factor and interleukin 6 production during interaction between activated $CD4^+$ lymphocytes and simian immunodeficiency virus-infected macrophages, *J. Gen. Virol.* **73**:1107.

McEntee, M. F., Zinc, M. C., Anderson, M. G., Farzadegan, H., Adams, R. J., Kent, K. A., Stott, E. J., Clements, J. E., and Narayan, O., 1992b, Neutralizing antibodies modulate replication of simian immunodeficiency virus SIVmac in primary macaque macrophages, *J. Virol.* **66**:6200.

McGraw, T. P., Vowels, B. R., Gardner, M. B., Ahmed-Ansari, A., and Gershwin, M. E., 1990a, Simian immunodeficiency virus-specific T-cell-mediated proliferative response of infected rhesus macaques, *AIDS* **4**:191.

McGraw, T. P., Vowels, B. R., Gershwin, M. E., and Gardner, M. B., 1990b, Cellular immune response of SIV-infected rhesus macaques, *J. Med. Primatol.* **19**:177.

McKeating, J. A., McKnight, A., McIntosh, K., Clapham, P. R., Mulder, C., and Weiss, R. A., 1989, Evaluation of human and simian immunodeficiency virus plaque and neutralization assays, *J. Gen. Virol.* **70**:3327.

MacKenzie, M., Lowenstine, L., Lalchandani, R., Lerche, N., Osborn, K., Spinner, A., Bleviss, M., Henrickson, R., and Gardner, M., 1986, The hematologic abnormalities in simian acquired immune deficiency syndrome (SAIDS): Peripheral blood and bone marrow findings, *Lab. Anim. Sci.* **36**:14.

Maddon, P. J., McDougal, J. S., Clapham, P. R., Dalgleish, A. G., Jamal, S., Weiss, R. A., and Axel, R., 1988, HIV infection does not require endocytosis of its receptor, CD4, *Cell* **54**:865.

Makoschey, B., Stahl-Hennig, C., Coulibaly, C., and Hunsmann, G., 1993, Neutralizing antibodies in SIVmac infected macaques in relation to the course of disease, *Berlin: IXth International Conference on AIDS* (Abstract, p. 190).

Malim, M. H., Bohnlein, S., Fenrick, R., Le, S. Y., Maizel, J. V., and Cullen, B. R., 1989a, Functional comparison of the Rev trans-activators encoded by different primate immunodeficiency virus species, *Proc. Natl. Acad. Sci. USA* **86**:8222.

Malim, M. H., Hauber, J., Le, S. Y., Maizel, J. V., and Cullen, B. R., 1989b, The HIV-1 rev trans-activator acts through a structured target sequence to activate nuclear export of unspliced viral mRNA, *Nature* **338**:254.

Malkovsky, M., Bartz, S. R., MacKenzie, D., Radtke, B. E., Wallace, M., Manning, J., Pauza, C. D., and Fisch, P., 1992, Are gamma delta T cells important for the elimination of virus-infected cells? *J. Med. Primatol.* **21**:113.

Malley, A., Torres, J., Benjamini, E., Pangares, N., and Axthelm, K., 1991a, T cell epitopes on the envelope glycoproteins of simian retrovirus 1 and 2, *Seattle: 9th Annual Symposium on Nonhuman Primate Models for AIDS* (Abstract).

Malley, A., Werner, L., Benjamini, E., Leung, C. Y., Torres, J., Pangares, N., Shiigi, S., and Axthelm, M., 1991b, Characterization of T- and B-cell epitopes of a simian retrovirus (SRV-2) envelope protein, *J. Med. Primatol.* **20**:177.

Malley, A., Pangares, N., Mayo, S. K., Zeleny-Pooley, M., Torres, J. V., Benjamini, E., and Axthelm, M. C., 1992, A type D SRV-2 virus-specific CD8 T cell that regulates virus-induced T cell proliferation in Celebes macaques, *Puerto Rico: Tenth Annual Symposium on Nonhuman Primate Models for AIDS* (Abstract).

Malley, A., Pangares, N., Mayo, S. K., Zeleny-Pooley, M., Torres, J. V., Benjamin, E., and Axthelm, M. K., 1993, Type D SRV-2 virus-specific $CD8^+$ and $CD4^-$ $CD8^-$ T cells that regulate virus-induced T-cell proliferation in Celebes macaques, *J. Med. Primatol.* **22**:80.

Mandell, C. P., Jain, N. C., Miller, C., Marthas, M., and Dandekar, S., 1993, Early hematologic changes in rhesus macaques infected with pathogenic and nonpathogenic isolates of SIVmac, *J. Med. Primatol.* **22**:177.

Mankowski, J. L., Spelman, J. P., Ressetar, H. G., Clements, J. E., and Zink, M. C., 1993, Pathogenesis of encephalitis caused by SIVmac, *Madison, WI: 11th Annual Symposium on Nonhuman Primate Models for AIDS Research Symposium* (Abstract, p. 19).

Mannervik, M., Putkonen, P., Ruden, U., Kent, K. A., Norrby, E., Wahren, B., and Broliden, P. A., 1992, Identification of B-cell antigenic sites on HIV-2 gp125, *J. AIDS* **5**:177.

Manns, A., Konig, H., Baier, M., Kurth, R., and Grosse, F., 1991, Fidelity of reverse transcriptase of the simian immunodeficiency virus from African green monkey, *Nucleic Acids Res.* **19**:533.

Marra, C. M., Kuller, L., Shaffer, J. M., Morton, W. R., and Lukehart, S. A., 1993, The impact of SIV on the immunological response to syphilis, *Berlin: IXth International Conference on AIDS* (Abstract, p. 42).

Marthas, M. L., Banapour, B., Sutjipto, S., Siegel, M. E., Marx, P. A., Gardner, M. B., Pedersen, N. C., and Luciw, P. A., 1989, Rhesus macaques inoculated with molecularly cloned simian immunodeficiency virus, *J. Med. Primatol.* **18**:311.

Marthas, M. L., Sutjipto, S., Higgins, J., Lohman, B., Torten, J., Luciw, P. A., Marx, P. A., and Pedersen, N. C., 1990, Immunization with a live, attenuated simian immunodeficiency virus (SIV) prevents early disease but not infection in rhesus macaques challenged with pathogenic SIV, *J. Virol.* **64**:3694.

Marthas, M. L., Sutjipto, S., Miller, C., Higgins, J., Torten, J., Unger, R., Marx, P. A., and Pedersen, N. C., 1991, Whole-inactivated and live, virulence-attenuated SIV vaccines protect rhesus macaques against infection by intravenous but not vaginal routes of challenge, *Seattle: 9th Annual Symposium on Nonhuman Primate Models for AIDS* (Abstract).

Marthas, M., McChesney, M., Lohman, B., Miller, C., Higgins, J., Torten, J., Pedersen, N. C., Gardner, M., and Yilma, T., 1992a, A live, virulence-attenuated SIV vaccine protects rhesus macaques against infection by intravenous challenge with virulent, cell-free SIV grown in human, but not rhesus cells, *Fifth Annual Meeting of the National Cooperative Vaccine Development Groups for AIDS* (Abstract).

Marthas, M. L., Miller, C. J., Sutjipto, S., Higgins, J., Torten, J., Lohman, B. L., Unger, R. E., Ramos, R. A., Kiyono, H., McGhee, J. R., Marx, P. A., and Pedersen, N. C., 1992b, Efficacy of live-attenuated and whole-inactivated simian immunodeficiency virus vaccines against vaginal challenge with virulent SIV, *J. Med. Primatol.* **21**:99.

Marthas, M., Ramos, R. A., Lohman, B. L., Van Rompay, K. K. A., Unger, R. E., Miller, C. J., Banapour, B., Pedersen, N. C., and Luciw, P., 1993, Viral determinants of simian immunodeficiency virus (SIV) virulence in rhesus macaques assessed using attenuated and pathogenic molecular clones of SIVmac, *J. Virol.* **67**:6047.

Martin, I., Defrise-Quertain, F., Mandieau, V., Nielsen, N. M., Saermark, T., Burny, A., Brasseur, R., Ruysschaert, J. M., and Vandenbranden, M., 1991, Fusogenic activity of SIV (simian immunodeficiency virus) peptides located in the GP32 NH_2 terminal domain, *Biochem. Biophys. Res. Commun.* **175**:872.

Martin, L. N., Murphey-Corb, M., Soike, K. F., and Davison-Fairburn, B., 1990, Effects of 28 day treatment with azidothymidine (AZT) initiated after infection with simian immunodeficiency virus (SIV), *Antiviral Res.* **1**(Suppl.):114.

Martin, L. N., Murphey-Corb, M., Soike, K. F., Davison-Fairburn, B., and Baskin, G. B., 1993, Effects of initiation of 3′-azido,3′-deoxythymidine treatment at different times after infection of rhesus monkeys with simian immunodeficiency virus, *J. Infect. Dis.* **168**:825.

Martin, L. N., Soike, K. F., Murphey-Corb, M., Bohm, R. P., Roberts, E. D., Kakuk, T. J., Thaisrivongs, S., Vidmar, T. J., Ruwart, M. J., Davio, S. R., and Tarpley, W. G., 1994, Effects of a peptidomimetic inhibitor of retroviral proteases (U-75875) on simian immunodeficiency virus infection in rhesus monkeys, in press.

Martin-Amedee, A., Davison-Fairburn, B., Trichel, A., and Murphey-Corb, M., 1993, Analysis of genetic diversity in maternally-transmitted SIV-infected rhesus macaque neonates, *Nonhuman Primate Models for AIDS Research Symposium* (Abstract, p. 3).

Marx, P. A., Maul, D. H., Osborn, K. G., Lerche, N. W., Moody, P., Lowenstine, L. J., Henrickson, R. V., Arthur, L. O., Gilden, R. V., Gravell, M., London, W. T., Sever, J. L., Levy, J. A., Munn, R. J., and Gardner, M. B., 1984, Simian AIDS: Isolation of a type D retrovirus and transmission of the disease, *Science* **223**:1083.

Marx, P. A., Osborn, K. G., Maul, D. H., Lerche, N. W., Lowenstine, L. J., Kluge, J. D., Zaiss, C. P., Henrickson, R. V., Shiigi, S. M., Wilson, B. J., Malley, A., Olson, L. C., McNulty, W. P., Arthur, L. O., Gilden, R. V., Barker, C. S., Hunter, E., Munn, R. J., Heidecker, G., Bryant, M. L., and Gardner, M. B., 1985, Isolation of a new serotype of simian acquired immune deficiency syndrome type D retrovirus from Celebes black macaques (Macaca nigra) with immune deficiency and retroperitoneal fibromatosis, *J. Virol.* **56**:571.

Marx, P. A., Pedersen, N., Lerche, N., Osborn, K., Lowenstine, L., Lackner, A., Maul, D., Kluge, J. D., Zaiss, C., Sharpe, V., Spinner, A., and Gardner, M., 1986, Prevention of simian acquired immune deficiency syndrome with a formalin-inactivated type D retrovirus vaccine, *J. Virol.* **60**:431.

Marx, P. A., Munn, R. J., and Joy, K. I., 1988, Computer emulation of thin section electron microscopy predicts an envelope-associated icosadeltahedral capsid for human immunodeficiency virus, *Lab. Invest.* **58**:112.

Marx, P. A., Li, Y., Lerche, N. W., Sutjipto, S., Gettie, A., Yee, J. A., Brotman, B. H., Prince, A. M., Hanson, A., and Webster, R. G., 1991, Isolation of a simian immunodeficiency virus related to human immunodeficiency virus type 2 from a west African pet sooty mangabey, *J. Virol.* **65**:4480.

Marx, P. A., Compans, R. W., Gettie, A., Staas, J. K., Gilley, R. M., Mulligan, M. J., Vamshchikov, G. V., Chen, D., and Eldridge, J. H., 1993, Protection against vaginal SIV transmission with microencapsulated vaccine, *Science,* **260**:1323.

Mathes, L. E., Olsen, R. G., Hebebrand, L. C., Hoover, E. A., and Schaller, J. P., 1978, Abrogation of lymphocyte blastogenesis by a feline leukaemia virus protein, *Nature* **274**:687.

Maul, D. H., Miller, C. H., Marx, P., Bleviss, M. L., Madden, D. L., Henrickson, R., and Gardner, M. B., 1984, Immune defects in simian acquired immunodeficiency syndrome, *Vet. Immunol. Immunopathol.* **8**:201.

Maul, D. H., Lerche, N. W., Osborn, K. G., Marx, P. A., Zaiss, C., Spinner, A., Kluge, J. D., MacKenzie, M. R., Lowenstine, L. J., Bryant, M. L., Blakeslee, J. R., Henrickson, R. V., and Gardner, M. B., 1986, Pathogenesis of simian AIDS in rhesus macaques inoculated with type D retroviruses, *Am. J. Vet. Res.* **47**:863.

Maul, D. H., Zaiss, C. P., MacKenzie, M. R., Shiigi, S. M., Marx, P. A., and Gardner, M. B., 1988, Simian retrovirus D serogroup 1 has a broad cellular tropism for lymphoid and nonlymphoid cells, *J. Virol.* **62**:1768.

Mayne, A., Villinger, F., Findley, H., and Ansari, A. A., 1993, Quantitative differences in the levels of IL-2 and IFN-gamma produced by PBMC of naturally SIV infected sooty mangabeys and rhesus macaques experimentally infected with SIV, *Madison, WI: 11th Annual Symposium on Nonhuman Primate Models for AIDS* (Abstract, p. 76.).

Meek, T. D., Lambert, D. M., Dreyer, G. B., Carr, T. J., Tomaszek, T. A., Jr., Moore, M. L., Strickler, J. E., Debouck, C., Hyland, L. J., and Matthews, T. J., 1990, Inhibition of HIV-1 protease in infected T-lymphocytes by synthetic peptide analogues, *Nature* **343**:90.

Mehtali, M., Aubertin, A. M., Bayer, C., Venet, A., Riviere, Y., and Kieny, M. P., 1993, Different approaches towards an HIV vaccine using SIV as a model, *Berlin: IXth International Conference on AIDS* 39 (Abstract).

Menzo, S., Bagnarelli, P., Giacca, M., Manzin, A., Varaldo, P. E., and Clementi, M., 1992, Absolute quantitation of viremia in human immunodeficiency virus infection by competitive reverse transcription and polymerase chain reaction, *J. Clin. Microbiol.* **30**:1752.

Merrill, K. W., Pietrobon, P. J. F., Fujihaski, K., McGhee, J. R., and Kiyono, H., 1992, Mucosal vaccine for AIDS: Oral immunization with HIV gp120 in combination with liposome and cholera toxin induces antigen-specific IgA producing cells, *Fifth Annual Meeting of the National Cooperative Vaccine Development Groups for AIDS* (Abstract).

Mestecky, J., 1987, The common mucosal immune system and current strategies for induction of immune responses in external secretions, *J. Clin. Immunol.* **7**:265.

Meyer, K. K., Trivedi, P., Streblow, D. N., Emau, P., and Pauza, C. D., 1993, Selection of an SIVmac LTR sequence upon mucosal inoculation of rhesus macaques, *Madison, WI: 11th Annual Symposium on Nonhuman Primate Models for AIDS Research Symposium* (Abstract, p. 13.).

Meyer, P. R., Ormerod, L. D., Osborn, K. G., Lowenstine, L. J., Henrickson, R. V., Modlin, R. L., Smith, R. E., Gardner, M. B., and Taylor, C. R., 1985, An immunopathologic evaluation of lymph nodes from monkey and man with acquired immune deficiency syndrome and related conditions, *Hematol. Oncol.* **3**:199.

Miles, S. A., Rezai, A. R., Salazar-Gonzalez, J. F., Vander Meyden, M., Stevens, R. H., Logan, D. M., Mitsuyasu, R. T., Taga, T., Hirano, T., and Kishimoto, T., 1990, AIDS Kaposi sarcoma-derived cells produce and respond to interleukin 6, *Proc. Natl. Acad. Sci. USA* **87**:4068.

Miller, C. J., Alexander, N. J., Sutjipto, S., Lackner, A. A., Gettie, A., Hendrickx, A. G., Lowenstine, L. J., Jennings, M., and Marx, P. A., 1989, Genital mucosal transmission of simian immunodeficiency virus: Animal model for heterosexual transmission of human immunodeficiency virus, *J. Virol.* **63**:4277.

Miller, C. J., Alexander, N. J., Sutjipto, S., Joye, S. M., Hendrickx, A. G., Jennings, M., and Marx, P. A., 1990, Effect of virus dose and nonoxynol-9 on the genital transmission of SIV in rhesus macaques, *J. Med. Primatol.* **19**:401.

Miller, C. J., Alexander, N. J., Vogel, P., Anderson, J., and Marx, P. A., 1992a, Mechanism of genital transmission of SIV: A hypothesis based on transmission studies and the location of SIV in the genital tract of chronically infected female rhesus macaques, *J. Med. Primatol.* **21**:64.

Miller, C. J., McChesney, M., and Moore, P. F., 1992b, Langerhans cells, macrophages and lymphocyte subsets in the cervix and vagina of rhesus macaques, *Lab. Invest.* **67**:628.

Miller, C. J., Vogel, P., Alexander, N. J., Sutjipto, S., Hendrickx, A. G., and Marx, P. A., 1992c, Localization of SIV in the genital tract of chronically infected female rhesus macaques, *Am. J. Pathol.* **141**:655.

Miller, C. J., Ramos, R. A., Vogel, P., Isahakia, M., Marthas, M. L., and Marx, P. A., 1992d, SIVagm in lymphoid tissues of experimentally-infected African green monkeys, *Puerto Rico: Tenth Annual Symposium on Nonhuman Primate Models for AIDS* (Abstract).

Miller, C. J., McGhee, J. R., and Gardner, M. B., 1993a, Biology of disease: Mucosal immunity, HIV transmission, and AIDS, *Lab. Invest.* **68**:129.

Miller, C. J., Vogel, P., Alexander, N. J., Dandekar, S., Hendrickx, A. G., and Marx, P. A., 1993b, Pathology and localization of SIV in the reproductive tract of chronically infected male rhesus macaques, *Lab. Invest.*, in press.

Miller, M. A., Murphey-Corb, M., and Montelaro, R. C., 1992, Identification of broadly reactive continuous antigenic determinants of simian immunodeficiency virus glycoproteins, *AIDS Res. Hum. Retroviruses* **8**:1153.

Miller, M. D., Lord, C. I., Stallard, V., Mazzara, G. P., and Letvin, N. L., 1990, The gag-specific cytotoxic T lymphocytes in rhesus monkeys infected with the simian immunodeficiency virus of macaques, *J. Immunol.* **144**:122.

Miller, M. D., Gould, F. S., Shen, L., Woods, R. M., Koenig, S., Mannino, R. J., and Letvin, N. L., 1992, Vaccination of rhesus monkeys with synthetic peptide in a fusogenic proteoliposome elicits simian immunodeficiency virus-specific $CD8^+$ cytotoxic T lymphocytes, *J. Exp. Med.* **176**:1739.

Mills, K. H. G., Jones, W. C., Kitchin, P. A., Page, M., Chan, W. L., Taffs, F., Bruck, C., Thiriart, C., deWilde, M., Rud, E., and Stott, E. J., 1992, Role of SIV-specific cellular and humoral immunity in protection against SIV infection in macaques, *Puerto Rico: Tenth Annual Symposium on Nonhuman Primate Models for AIDS* (Abstract 41).

Mitsuya, H., Yarchoan, R., and Broder, S., 1990, Molecular targets for AIDS therapy, *Science* **249**:1533.

Miura, T., Tsujimoto, H., Fukasawa, M., Ohta, Y., Honjo, S., and Hayami, M., 1989, Genetic analysis and infection of SIVAGM and SIVMND, *J. Med. Primatol.* **18**:255.

Moazed, T. C., and Thouless, M. E., 1993, Viral persistence of simian type D retrovirus (SRV-2/W) in naturally infected pig-tailed macaques (M. nemestrina), *J. Med. Primatol.*, in press.

Montaner, L. J., Ringler, D. J., Wyand, M. S., Mattmuller, M. R., MacKey, J. J., Schmidt, D. K., Daniel, M. D., and King, N. W., 1990, Study of long-term cultures of simian immunodeficiency virus (SIVmac 251)-infected peripheral blood lymphocytes, *Lab. Invest.* **63**:242.

Montefiori, D. C., Murphey-Corb, M., Desrosiers, R. C., and Daniel, M. D., 1990a, Complement-mediated, infection-enhancing antibodies in plasma from vaccinated macaques before and after inoculation with live simian immunodeficiency virus, *J. Virol.* **64**:5223.

Montefiori, D. C., Robinson, W. E., Jr., Hirsch, V. M., Modliszewski, A., Mitchell, W. M., and Johnson, P. R., 1990b, Antibody-dependent enhancement of simian immunodeficiency virus (SIV) infection in vitro by plasma from SIV-infected rhesus macaques, *J. Virol.* **64**:113.

Mori, K., Ringler, D. J., Kodama, T., and Desrosiers, R. C., 1992, Complex determinants of macrophage tropism in env of simian immunodeficiency virus, *J. Virol.* **66**:2067.

Mori, K., Ringler, D. J., and Desrosiers, R. C., 1993, Restricted replication of simian immunodeficiency virus strain 239 in macrophages is determined by env but is not due to restricted entry, *J. Virol.* **67**:2807.

Morrison, H. G., and Desrosiers, R. C., 1993, A PCR-based strategy for extensive mutagenesis of a target DNA sequence, *Biotechniques* **14**(3):454.

Morton, W. R., Kuller, L., Benveniste, R. E., Clark, E. A., Tsai, C. C., Gale, M. J., Thouless, M. E., Overbaugh, J., and Katze, M. G., 1989, Transmission of the simian immunodeficiency virus SIVmne in macaques and baboons, *J. Med. Primatol.* **18**:237.

Mossman, S. P., O'Neil, S. P., Hoover, E. A., and Maul, D. H., 1992, Vertical transmission of SIV-mac239 in rhesus macaques, *Puerto Rico: Tenth Annual Symposium on Nonhuman Primate Models for AIDS* (Abstract, p. 31.).

Mossman, S. P., O'Neil, S. P., Maul, D. H., Fultz, P. N., and Hoover, E. A., 1993, Enteropathy in pigtail macaques inoculated with SIVsmmPBj14-bcl.3, *Madison, WI: 11th Annual Symposium on Nonhuman Primate Models for AIDS* (Abstract, p. 77.).

Muller, J. G., Krenn, V., Czub, S., Stahl, H. C., Coulibaly, C., Hunsmann, G., and Muller, H. H., 1993, The thymic epithelial reticulum and interdigitating cells in SIV-induced thymus atrophy and its comparison with other forms of thymus involution, *Res. Virol.* **144**:93.

Muller, M. C., Saksena, N. K., Nerrienet, E., Chappey, C., Herve, V. M. A., Durand, J.-P., Legal-Campodonico, P., Lang, M.-C., Digoutte, J.-P., Georges, A. J., Georges-Courbot, M.-C., Sonigo, P., and Barre-Sinoussi, F., 1993, Simian immunodeficiency viruses from central and western Africa: Evidence for a new species-specific lentivirus in tantalus monkeys, *J. Virol.* **67**:1227.

Munn, R. J., Marx, P. A., Yamamoto, J. K., and Gardner, M. B., 1985, Ultrastructural comparison of the retroviruses associated with human and simian acquired immunodeficiency syndrome, *Lab. Invest.* **53**:194.

Murphey-Corb, M., Martin, L. N., Rangan, S. R. S., Baskin, G. B., Gormus, B. J., Wolf, R. H., Andes, W. A., West, M., and Montelaro, R. C., 1986, Isolation of an HTLV-III-related retrovirus from macaques with simian AIDS and its possible origin in asymptomatic mangabeys, *Nature* **321**:435.

Murphey-Corb, M., Martin, L. N., Davison-Fairburn, B., Montelaro, R. C., Miller, M., West, M., Ohkawa, S., Baskin, G. B., Zhang, J. Y., and Putney, S. D., 1989, A formalin-inactivated whole SIV vaccine confers protection in macaques, *Science* **246**:1293.

Murphey-Corb, M., Davison-Fairburn, B., Ohkawa, S., Martin, L., and Baskin, G., 1990a, Immunization of healthy SIV-infected macaques with a formalin inactivated whole virus vaccine has no apparent effect on disease progression and survival, *New Orleans: 8th Annual Symposium on Nonhuman Primate Models for AIDS* (Abstract 39, p. 55.).

Murphey-Corb, M., Martin, L. N., Davison-Fairburn, B., Montelaro, R. C., Miller, M., West, M., Ohkawa, S., Baskin, G. B., Zhang, J. Y., and Putney, S., 1990b, A formalin inactivated whole SIV vaccine and a glycoprotein-enriched subunit vaccine confers protection against experimental challenge with pathogenic live SIV in rhesus monkeys, *Dev. Biol. Stand.* **72**:273.

Murphey-Corb, M., Davison-Fairburn, B., and Baskin, G., 1992a, Maternal–fetal transmission of SIV in the macaque, *Paris: 7th Cent Gardes Meeting* (Abstract, p. 38.).

Murphey-Corb, M., Davison-Fairburn, B., Ohkawa, S., and Martin, L., 1992b, Passive protection of neonatal rhesus monkeys born of mothers immunized with whole inactivated SIV, *Puerto Rico: Tenth Annual Symposium on Nonhuman Primate Models for AIDS* (Abstract, p. 125.).

Murphey-Corb, M., Ohkawa, S., Davison-Fairburn, B., Martin, L. N., Baskin, G. B., Langlois, A. J., McIntee, M., Narayan, O., and Gardner, M. B., 1992c, A formalin-fixed whole SIV vaccine induces protective responses that are cross-protective and durable, *AIDS Res. Hum. Retroviruses* **8**:1475.

Murray, E. A., Rausch, D. M., Lendvay, J., Sharer, L. R., and Eiden, L. E., 1992, Cognitive and motor impairments associated with SIV infection in rhesus monkeys, *Science* **255**:1246.

Myers, G., and Pavlakis, G. N., 1992, Evolutionary potential of complex retroviruses, in: *The Retroviridae*, Vol. 1 (J. A. Levy, ed.), pp. 51–105, Plenum Press, New York.

Myers, G., MacInnes, K., and Korber, B., 1992, The emergence of simian/human immunodeficiency viruses, *AIDS Res. Hum. Retroviruses* **8**:373.

Naidu, Y. M., Kestler, H. W., III, Li, Y., Butler, C. V., Silva, D. P., Schmidt, D. K., Troup, C. D., Sehgal, P. K., Sonigo, P., Daniel, M. D., and Desrosiers, R. C., 1988, Characterization of infectious molecular clones of simian immunodeficiency virus (SIVmac) and human immunodeficiency virus type 2: Persistent infection of rhesus monkeys with molecularly cloned SIVmac, *J. Virol.* **62**:4691.

Napier, J. R., and Napier, P. H., 1986, *The Natural History of the Primates,* MIT Press, Cambridge, Mass.

Nara, P. L., Smit, L., Dunlop, N., Hatch, W., Merges, M., Waters, D., Kelliher, J., Gallo, R. C., Fischinger, P. J., and Goudsmit, J., 1990, Emergence of viruses resistant to neutralization by V3-specific antibodies in experimental human immunodeficiency virus type 1 IIIB infection of chimpanzees, *J. Virol.* **64**:3779.

Narayan, O., and Clements, J. E., 1989, Biology and pathogenesis of lentiviruses, *J. Gen. Virol.* **70**:1617.

Nicol, I., Messinger, D., Dubouch, P., Bernard, J., Desportes, I., Jouffre, R., Snart, R., Nara, P., Gallo, R. C., and Zagury, D., 1989, Use of Old World monkeys for acquired immunodeficiency syndrome research, *J. Med. Primatol.* **18**:227.

Niederman, T. M., and Ratner, L., 1992, Functional analysis of HIV1 and SIV Nef proteins, *Res. Virol.* **143**:43.

Niederman, T. M., Thielan, B. J., and Ratner, L., 1989, Human immunodeficiency virus type 1 negative factor is a transcriptional silencer, *Proc. Natl. Acad. Sci. USA* **86**:1128.

Niederman, T. M., Hu, W., and Ratner, L., 1991, Simian immunodeficiency virus negative factor suppresses the level of viral mRNA in COS cells, *J. Virol.* **65**:3538.

Nishino, Y., Ohki, K., Kimura, T., Morikawa, S., Mikami, T., and Ikuta, K., 1992, Major core proteins, p24s, of human, simian, and feline immunodeficiency viruses are partly expressed on the surface of the virus-infected cells, *Vaccine* **10**:677.

Nixon, D. F., Broliden, K., Ogg, G., and Broliden, P. A., 1992, Cellular and humoral antigenic epitopes in HIV and SIV [editorial], *J. Immunol.* **76**:515.

Norley, S. G., Kraus, G., Ennen, J., Bonilla, J., Konig, H., and Kurth, R., 1990, Immunological studies of the basis for the apathogenicity of simian immunodeficiency virus from African green monkeys, *Proc. Natl. Acad. Sci. USA* **87**:9067.

Norrby, E., Biberfeld, G., Johnson, P. R., Parks, D. E., Houghten, R. A., and Lerner, R. A., 1989, The chemistry of site-directed serology for HIV infections, *AIDS Res. Hum. Retroviruses* **5**:487.

Norrby, E., Parks, D. E., Utter, G., Houghten, R. A., and Lerner, R. A., 1989b, Immunochemistry of the dominating antigenic region Ala582 to Cys604 in the transmembranous protein of simian and human immunodeficiency virus, *J. Immunol.* **143**:3602.

Norrby, E., Putkonen, P., Bottiger, B., Utter, G., and Biberfeld, G., 1991, Comparison of linear antigenic sites in the envelope proteins of human immunodeficiency virus (HIV) type 2 and type 1, *AIDS Res. Hum. Retroviruses* **7**:279.

Novembre, F. J., Hirsch, V. M., McClure, H. M., and Johnson, P. R., 1991, Molecular diversity of SIVsmm/PBj and a cognate variant, SIVsmm/PGg, *J. Med. Primatol.* **20**:188.

Novembre, F. J., Hirsch, V. M., McClure, H. M., Fultz, P. N., and Johnson, P. R., 1992, SIV from stump-tailed macaques: Molecular characterization of a highly transmissible primate lentivirus, *Virology* **186**:783.

Novembre, F. J., Chege, G., Ishahakia, M., and McClure, H. M., 1993a, Serosurvey of nonhuman primates in Kenya: Evidence for SIVsmm-related infection in a baboon, *Madison, WI: 11th Annual Symposium on Nonhuman Primate Models for AIDS* (Abstract, p. 48.).

Novembre, F. J., Johnson, P. R., Lewis, M. G., Anderson, D. C., Klumpp, S., McClure, H. M., and Hirsch, V. M., 1993b, Multiple viral determinants contribute to pathogenicity of the acutely lethal simian immunodeficiency virus SIVsmmPBj variant, *J. Virol* **67**:2466.

Ochs, H. D., Morton, W. R., Kuller, L. D., Zhu, Q., Tsai, C.-C., Agy, M. B., and Benveniste, R. E., 1993, Intra-amniotic inoculation of pigtailed macaque (Macaca nemestrina) fetuses with SIV and HIV-1, *J. Med. Primatol.* **22**:162.

Ochs, H. D., Morton, W. R., Tsai, C. C., Thouless, M. E., Zhu, Q., Kuller, L. D., Wu, Y. P., and Benveniste, R. E., 1991, Maternal–fetal transmission of SIV in macaques: Disseminated adenovirus infection in an offspring with congenital SIV infection, *J. Med. Primatol.* **20**:193.

O'Hagan, D. T., Rafferty, D., McKeating, J. A., and Illum, L., 1992, Vaginal immunization of rats with a synthetic peptide from human immunodeficiency virus envelope glycoprotein, *J. Gen. Virol.* **73**:2141.

Ohkawa, S., Wilson, L. A., Miller, M., Montelaro, R., Martin, L., and Murphey-Corb, M., 1993, ADCC function in sera from infected but not from immunized rhesus monkeys, *Madison, WI: 11th Annual Symposium on Nonhuman Primate Models for AIDS Research Symposium* (Abstract, p. 61.).

Ohta, Y., Masuda, T., Tsujimoto, H., Ishikawa, K., Kodama, T., Morikawa, S., Nakai, M., Honjo, S., and Hayami, M., 1988, Isolation of simian immunodeficiency virus from African green monkeys and seroepidemiologic survey of the virus in various non-human primates, *Int. J. Cancer* **41**:115.

Ohta, Y., Tsujimoto, H., Ishikawa, K., Yamamoto, H., Doi, Y., Honjo, S., Arita, M., and Hayami, M., 1989, No evidence for the contamination of live oral poliomyelitis vaccines with simian immunodeficiency virus [letter], *AIDS* **3**:183.

O'Neil, S. P., Mossman, S. P., Maul, D. H., Israel, Z. R., Dean, G. A., Fultz, P. N., and Hoover, E. A., 1993, Comparative pathogenicity of molecular and biologic clones of SIVsmmPBj14, *Nonhuman Primate Models for AIDS Research Symposium* (Abstract).

O'Reilly, K. L., and Hoover, E. A., 1991, Natural killer activity in early and chronic SIV infection, *Seattle: 9th Annual Symposium on Nonhuman Primate Models for AIDS* (Abstract).

Osborn, K. G., Prahalada, S., Lowenstine, L. J., Gardner, M. B., Maul, D. H., and Henrickson, R. V., 1984, The pathology of an epizootic of acquired immunodeficiency in rhesus macaques, *Am. J. Pathol.* **114**:94.

Osterhaus, A., de Vries, P., Morein, B., Akerblom, L., and Heeney, J., 1992b, Comparison of protection afforded by whole virus ISCOM versus MDP adjuvanted formalin-inactivated SIV vaccines from IV cell-free or cell-associated homologous challenge, *AIDS Res. Hum. Retroviruses* **8**:1507.

Otsyula, M., Suleman, M., Gettie, A., Isahakia, M., and Marx, P. A., 1991, Transmission of simian immunodeficiency virus (SIV) in naturally infected African green monkeys (AGMs), *Seattle: 9th Annual Symposium on Nonhuman Primate Models for AIDS* (Abstract).

Otteken, A., Nick, S., Bergter, W., Voss, G., Faisst, A. C., Stahl, H. C., and Hunsmann, G., 1992, Identification of a gag protein epitope conserved among all four groups of primate immunodeficiency viruses by using monoclonal antibodies, *J. Gen. Virol.* **73**:2721.

Overbaugh, J., and Rudensey, L. M., 1992, Alterations in potential sites for glycosylation predominate during evolution of the simian immunodeficiency virus envelope gene in macaques, *J. Virol.* **66**:5937.

Overbaugh, J., Rudensey, L. M., Papenhausen, M. D., Benveniste, R. E., and Morton, W. R., 1991, Variation in simian immunodeficiency virus env is confined to V1 and V4 during progression to simian AIDS, *J. Virol.* **65**:7025.

Pantaleo, G., Graziosi, C., Demarest, J. F., Butini, L., Montroni, M., Fox, C. H., Orenstein, J. M., Kotler, D. P., and Fauci, A. S., 1993a, HIV infection is active and progressive in lymphoid tissue during the clinically latent stage of disease, *Nature* **362**:355.

Pantaleo, G., Graziosi, C., and Fauci, A. S., 1993b, New concepts in the immunopathogenesis of human immunodeficiency virus infection, *N. Engl. J. Med.* **328**:327.

Paquette, Y., Hanna, Z., Savard, P., Brousseau, R., Robitaille, Y., and Jolicoeur, P., 1989, Retrovirus-induced murine motor neuron disease: Mapping the determinant of spongiform degeneration within the envelope gene, *Proc. Natl. Acad. Sci. USA* **86**:3896.

Park, I. W., Steen, R., and Li, Y., 1991, Characterization of multiple mRNA species of simian immunodeficiency virus from macaques in a $CD4^+$ lymphoid cell line, *J. Virol.* **65**:2987.

Parkin, N. T., Cohen, E. A., Darveau, A., Rosen, C., Haseltine, W., and Sonenberg, N., 1988, Mutational analysis of the 5′ non-coding region of human immunodeficiency virus type 1: Effects of secondary structure on translation, *EMBO J.* **7**:2831.

Paul, J. R., 1971, *A History of Poliomyelitis,* Yale University Press, New Haven.

Pauza, C. D., Emau, P., Salvato, M. S., Trivedi, P., MacKenzie, D., Malkovsky, M., Uno, H., and Schultz, K. T., 1993a, Pathogenesis of SIVmac251 after atraumatic inoculation of the rectal mucosa in rhesus monkeys, *J. Med. Primatol.* **22**:154.

Pauza, C. D., Trivedi, P., Meyer, K. K., Streblow, D. N., Malkovsky, M., Salvato, M. S., Schultz, K. T., and Emau, P., 1993b, Intramucosal transmission of primate retroviruses, *Madison, WI: 11th Annual Nonhuman Primate Models for AIDS* (Abstract 2), p. 2.

Peeters, M., Honore, C., Huet, T., Bedjabaga, L., Ossari, S., Bussi, P., Cooper, R. W., and Delaporte, E., 1989, Isolation and partial characterization of an HIV-related virus occurring naturally in chimpanzees in Gabon, *AIDS* **3**:625.

Peeters, M., Fransen, K., Delaporte, E., Van den Haesevelde, M., Gershy-Damet, G. M., Kestens, L., van der Groen, G., and Piot, P., 1992, Isolation and characterization of a new chimpanzee lentivirus (simian immunodeficiency virus isolate cpz-ant) from a wild-captured chimpanzee, *AIDS* **6**:447.

Peeters, M., Janssens, W., Heyndrickx, L., Williams, B., Fransen, K., Van den Haesevelde, M., and Heeny, J., 1993, Role of V3 loop in neutralization of sequential SIVcpz-ant isolates in a naturally SIVcpz-ant infected chimpanzee, *Berlin: IXth International Conference on AIDS* (Abstract, p. 41.).

Pelletier, E., Cheynier, R., Letvin, N., and Wain-Hobson, S., 1993, Evolution of SIV quasispecies in animals infected with a molecularly cloned provirus, *Berlin: IXth International Conference on AIDS* (Abstract, p. 159.).

Phillips, R. E., Rowland-Jones, S., Nixon, D. F., Gotch, F. M., Edwards, J. P., Ogunlesi, A. O., Elvin, J. G., Rothbard, J. A., Bangham, C. R. M., Rizza, C. R., and McMichael, A. J., 1991, Human immunodeficiency virus genetic variation that can escape cytotoxic T cell recognition, *Nature* **354**:453.

Piatak, M., Jr., Luk, K. C., Williams, B., and Lifson, J. D., 1993a, Quantitative competitive polymerase chain reaction for accurate quantitation of HIV DNA and RNA species, *BioTechniques* **14**:70.

Piatak, M., Jr., Saag, M. S., Yang, L. C., Clark, S. J., Kappes, J. C., Luk, K. C., Hahn, B. H., Shaw, G. M., and Lifson, J. D., 1993b, High levels of HIV-1 in plasma during all stages of infection determined by competitive PCR, *Science* **259**:1749.

Planelles, V., Haigwood, N. L., Marthas, M. L., Mann, K. A., Scandella, C., Lidster, W. D., Shuster, J. R., VanKuyk, R., Marx, P. A., Gardner, M. B., and Luciw, P. A., 1991, Functional and immunological characterization of SIV envelope glycoprotein produced in genetically engineered mammalian cells, *AIDS Res. Hum. Retroviruses* **7**:889.

Polacino, P. S., Liang, H. A., Firpo, E. J., and Clark, E. A., 1993, T-cell activation influences initial DNA synthesis of simian immunodeficiency virus in resting T lymphocytes from macaques, *J. Virol.* **67**:7008.

Polyanskaya, N., Cook, N., Webster, S., and Cranage, M. P., 1993, Species specific anti-cell antibody responses in infection and vaccination with SIV, *Berlin: IXth International Conference on AIDS* (Abstract, p. 254.).

Polzien, F., Scharf, J. G., Luke, W., and Hunsmann, G., 1992, CD4-binding of gp130 micelles isolated from SIVagmTYO-7, *AIDS Res. Hum. Retroviruses* **8**:1171.

Popov, J., McGraw, T., Hofmann, B., Vowels, B., Shum, A., Nishanian, P., and Fahey, J. L., 1992, Acute lymphoid changes and ongoing immune activation in SIV infection, *J. AIDS* **5**:391.

Portis, J. L., McAtee, F. J., and Kayman, S. C., 1992, Infectivity of retroviral DNA in vivo [letter], *J. AIDS* **5**:1272.

Potts, B. J., 1990, "Mini" reverse transcriptase (RT) assay, in: *Techniques in HIV Research* (A. Aldovini and B. D. Walker, eds.), Stockton Press, New York.

Powell, J. D., Bednarik, D. P., Folks, T. M., Jehuda-Cohen, T., Villinger, F., Sell, K. W., and Ansari, A. A., 1993, Inhibition of cellular activation of retroviral replication by $CD8^+$ T cells derived from non-human primates, *Clin. Exp. Immunol.* **91**:473.

Power, M. D., Marx, P. A., Bryant, M. L., Gardner, M. B., Barr, P. J., and Luciw, P. A., 1986, The nucleotide sequence of a type D retrovirus, SRV-1 etiologically-linked with the simian acquired immunodeficiency syndrome, *Science* **231**:1567.

Prakash, K., Ranganathan, P. N., Mettus, R., Reddy, P., Srinivasan, A., and Plotkin, S., 1992, Generation of deletion mutants of simian immunodeficiency virus incapable of proviral integration, *J. Virol.* **66**:167.

Prasad, V. R., Myrick, K., Haseltine, W., and Goff, S. P., 1990, Expression of enzymatically active reverse transcriptase of simian immunodeficiency virus in bacteria: Sensitivity to nucleotide analogue inhibitors, *Virology* **179**:896.

Putkonen, P., Bottiger, B., Warstedt, K., Thorstensson, R., Albert, J., and Biberfeld, G., 1989, Experimental infection of cynomolgus monkeys (Macaca fascicularis) with HIV-2, *J. AIDS* **2**:359.

Putkonen, P., Thorstensson, R., Ghavamzadeh, L., Albert, J., Hild, K., Biberfeld, G., and Norrby, E., 1991b, Prevention of HIV-2 and SIVsm infection by passive immunization in cynomolgus monkeys, *Nature* **352**:436.

Putkonen, P., Kaaya, E. E., Bottiger, D., Li, S. L., Nilsson, C., Biberfeld, P., and Biberfeld, G., 1992a, Clinical features and predictive markers of disease progression in cynomolgus monkeys experimentally infected with simian immunodeficiency virus, *AIDS* **6**:257.

Putkonen, P., Thorstensson, R., Cranage, M., Nilsson, C., Ghavamzadeh, L., Albert, J., Greenaway, P., and Biberfeld, G., 1992c, A formalin inactivated whole SIVmac vaccine in Ribi adjuvant protects against homologous and heterologous SIV challenge, *J. Med. Primatol.* **21**:108.

Putkonen, P., Nilsson, C., Hild, K., Benthin, R., Cranage, M., Aubertin, A. M., and Biberfeld, G., 1993, Whole inactivated SIV vaccine grown on human cells fails to protect against homologous SIV grown on simian cells, *J. Med. Primatol.* **22**:100.

Racz, P., Tenner-Racz, K., Kahl, C., Feller, A. C., Kern, P., and Dietrich, M., 1986, Spectrum of morphologic changes of lymph nodes from patients with AIDS or AIDS-related complexes, *Prog. Allergy* **37**:81.

Rangan, S. R. S., Martin, L. N., Bozelka, B. E., Wang, N., and Gormus, B. J., 1986, Epstein–Barr virus-related herpesvirus from a rhesus monkey (Macaca mulatta) with malignant lymphoma, *Int. J. Cancer* **38**:425.

Rausch, D. M., Murray, E. A., Lendvay, J., Sharer, L. R., Heyes, M., Nohr, D., Weihe, E., and Eidden, L. E., 1992, Correlation of immunological, clinical, neuropathological, and neurochemical parameters with motor and cognitive impairments observed in SIV-infected rhesus macaques, *Puerto Rico: Tenth Annual Symposium on Nonhuman Primate Models for AIDS* (Abstract).

Redfield, R. R., and Birx, D. L., 1992, HIV-specific vaccine therapy: Concepts, status, and future directions, *AIDS Res. Hum. Retroviruses* **8**:1051.

Regier, D. A., and Desrosiers, R. C., 1990, The complete nucleotide sequence of a pathogenic molecular clone of simian immunodeficiency virus, *AIDS Res. Hum. Retroviruses* **6**:1221.

Regier, D. A., and Desrosiers, R. C., 1991, The complete nucleotide sequence of a pathogenic molecular clone of simian immunodeficiency virus, *Virology* **184**:513.

Reimann, K. A., Burkly, L. C., Burrus, B., Waite, B. C. D., Lord, C. I., and Letvin, N. L., 1993, In vivo administration to rhesus monkeys of a CD4-specific monoclonal antibody capable of blocking AIDS virus replication, *AIDS Res. Hum. Retroviruses* **9**:199.

Rey, M. A., Laurent, A. G., McClure, J., Krust, B., Montagnier, L., and Hovanessian, A. G., 1990, Transmembrane envelope glycoproteins of human immunodeficiency virus type 2 and simian immunodeficiency virus SIV-mac exist as homodimers, *J. Virol.* **64**:922.

Rhee, S. S., and Hunter, E., 1987, Myristylation is required for intracellular transport but not for assembly of D-type retrovirus capsids, *J. Virol.* **61**:1045.

Rhee, S. S., and Hunter, E., 1990a, Structural role of the matrix protein of type D retroviruses in gag polyprotein stability and capsid assembly, *J. Virol.* **64**:4383.

Rhee, S. S., and Hunter, E., 1990b, A single amino acid substitution within the matrix protein of a type D retrovirus converts its morphogenesis to that of a type C retrovirus, *Cell* **63**:77.

Rice, A. P., and Mathews, M. B., 1988, Transcriptional but not translational regulation of HIV-1 by the tat gene product, *Nature* **332**:551.

Richman, D. D., 1990, Zidovudine resistance of human immunodeficiency virus [see comments], *Rev. Infect. Dis.* **12**(Suppl. 5):S507.

Ringler, D. J., Hancock, W. W., King, N. W., Letvin, N. L., Daniel, M. D., Desrosiers, R. C., and Murphy, G. F., 1987, Immunophenotypic characterization of the cutaneous exanthem of SIV-infected rhesus monkeys, *Am. J. Pathol.* **126**:199.

Ringler, D. J., Wyand, M. S., Walsh, D. G., MacKey, J. J., Chalifoux, L. V., Popovic, M., Minassian, A. A., Sehgal, P. K., Daniel, M. D., Desrosiers, R. C., and King, N. W., 1989a, Cellular localization of simian immunodeficiency virus in lymphoid tissues. I. Immunohistochemistry and electron microscopy, *Am. J. Pathol.* **134**:373.

Ringler, D. J., Wyand, M. S., Walsh, D. G., MacKey, J. J., Sehgal, P. K., Daniel, M. D., Desrosiers, R. C., and King, N. W., 1989b, The productive infection of alveolar macrophages by simian immunodeficiency virus (SIV), *J. Med. Primatol.* **18**:217.

Ringler, D. J., Walsh, D. G., Chalifoux, L. V., MacKey, J. J., Daniel, M. D., Desrosiers, R. C., King, N. W., and Hancock, W. W., 1990, Soluble and membrane-associated interleukin 2 receptor-alpha expression in rhesus monkeys infected with simian immunodeficiency virus, *Lab. Invest.* **62**:435.

Ritter, G. D. J., Mulligan, M. J., Lydy, S. L., and Compans, R. W., 1993, Cell fusion activity of the simian immunodeficiency virus envelope protein is modulated by the intracytoplasmic domain, *Virology* **197**:255.

Rizvi, T. A., and Panganiban, A. T., 1992, Propagation of SIV vectors by genetic complementation with a heterologous env gene, *AIDS Res. Hum. Retroviruses* **8**:89.

Rizvi, T. A., and Panganiban, A. T., 1993, Simian immunodeficiency virus RNA is efficiently encapsidated by human immunodeficiency virus type 1 particles, *J. Virol.* **67**:2681.

Robert-Guroff, M., Popovic, M., Gartner, S., Markham, P., Gallo, R. C., and Reitz, M. S., 1990, Structure and expression of tat-, rev-, and nef-specific transcripts of human immunodeficiency virus type 1 in infected lymphocytes and macrophages, *J. Virol.* **64**:3391.

Roberts, E. D., and Martin, L. N., 1991, Arthritis in rhesus monkeys experimentally infected with simian immunodeficiency virus (SIV/DELTA), *Lab. Invest.* **65**:637.

Robinson, W. E., Jr., and Mitchell, W. M., 1990, Neutralization and enhancement of in vitro and in vivo HIV and simian immunodeficiency virus infections, *AIDS* **4**(Suppl. 1):S151.

Robinson, W. E., Jr., Montefiori, D. C., and Mitchell, W. M., 1988, Will antibody-dependent enhancement of HIV-1 infection be a problem with AIDS vaccine? *Lancet* **1**:830.

Roodman, S. T., Woon, M. D., Hoffmann, J. W., Theodorakis, P., Tsai, C. C., and Wu, H. N., 1991, Interleukin-6 and retroperitoneal fibromatosis from SRV-2-infected macaques with simian AIDS, *J. Med. Primatol.* **20**:201.

Rosen, C. A., Sodroski, J. G., Goh, W. C., Dayton, A. I., Lippke, J., and Haseltine, W. A., 1986, Post-transcriptional regulation accounts for the trans-activation of the human T-lymphotropic virus type III, *Nature* **319**:555.

Rosenberg, Y. J., White, B. D., Papermaster, S. F., Zack, P., Jarling, P. B., Eddy, G. A., Burke, D. S., and Lewis, M. G., 1991, Variation in T-lymphocyte activation and susceptibility to SIVPBj-14-induced acute death in macaques, *J. Med. Primatol.* **20**:206.

Rosenberg, Y. J., Zack, P. M., Greenhouse, J. J., Leon, E. C., Papermaster, S. F., White, B. D., Eddy, G. A., Burke, D. S., and Lewis, M. G., 1992, Immunological deterioration and proviral levels in LN of SIV-infected macaques predicts progression to AIDS, *Puerto Rico: Tenth Annual Symposium on Nonhuman Primate Models for AIDS* (Abstract).

Rosenberg, Y. J., Zack, P. M., White, B. D., Papermaster, S. F., Elkins, W. R., Eddy, G. A., and Lewis, M. G., 1993, Decline in the $CD4^+$ lymphocyte population in the blood of SIV-infected macaques is not reflected in lymph nodes, *AIDS Res. Hum. Retroviruses* **9**:639.

Rudensey, L. M., Papenhausen, M. D., and Overbaugh, J., 1993, Replication and persistence of simian immunodeficiency virus variants after passage in macaque lymphocytes and established human cell lines, *J. Virol.* **67**:1727.

Ruegg, C. L., Monell, C. R., and Strand, M., 1989, Inhibition of lymphoproliferation by a synthetic peptide with sequence identity to gp41 of human immunodeficiency virus type 1, *J. Virol.* **63**:3257.

Russell, J. W., Whiterock, V. J., Marrero, D., and Klunk, L. J., 1990, Disposition in animals of a new anti-HIV agent: 2',3'-didehydro-3'-deoxythymidine, *Drug Metab. Dispos.* **18**:153.

Sabo, A., Grant, R., Follis, K., Bischofberger, N., and Tsai, C.-C., 1993, Characterization of SIV isolates from macaques treated with the nucleoside analog, 9-(2-phosphonylmethoxyethyl)-adenine (PMEA), *Nonhuman Primate Models for AIDS Research Symposium* (Abstract).

Sakai, H., Shibata, R., Miura, T., Hayami, M., Ogawa, K., Kiyomasu, T., Ishimoto, A., and Adachi, A., 1990, Complementation of the rev gene mutation among human and simian lentiviruses, *J. Virol.* **64**:2202.

Sakai, H., Shibata, R., Sakuragi, J., Kiyomasu, T., Kawamura, M., Hayami, M., Ishimoto, A., and Adachi, A., 1991, Compatibility of rev gene activity in the four groups of primate lentiviruses, *Virology* **184**:513.

Sakai, H., Sakuragi, J., Sakuragi, S., Shibata, R., and Adachi, A., 1992a, Functional analysis of biologically distinct genetic variants of simian immunodeficiency virus isolated from a mandrill, *Virology* **189**:161.

Sakai, H., Sakuragi, J., Sakuragi, S., Shibata, R., Hayami, M., Ishimoto, A., and Adachi, A., 1992b, Genetic characterization of simian immunodeficiency virus isolated from an African mandrill, *Arch. Virol.* **125**:1.

Sakai, H., Sakuragi, S., Sakuragi, J.-I., Kawamura, M., Shibata, R., and Adachi, A., 1992c, Sequences responsible for efficient replication of simian immunodeficiency virus SIVmnd in cells of the monocyte/macrophage lineage, *J. Gen. Virol.* **73**:2989.

Saksena, N. K., Dwyer, D. E., Muller, M., Durand, J. P., Herve, V., Lang, M. C., and Barre, S. F., 1992, PCR amplification of large genomic fragments from human and simian immunodeficiency virus infected cell lines, *Life Sci.* **50**:1973.

Sakuragi, J., Fukasawa, M., Shibata, R., Sakai, H., Kawamura, M., Akari, H., Kiyomasu, T., Ishimoto, A., Hayami, M., and Adachi, A., 1991, Functional analysis of long terminal repeats derived from four strains of simian immunodeficiency virus SIVAGM in relation to other primate lentiviruses, *Virology* **185**:455.

Sakuragi, J., Sakai, H., Sakuragi, S., Shibata, R., Wain-Hobson, S., Hayami, M., and Adachi, A., 1992b, Functional classification of simian immunodeficiency virus isolated from a chimpanzee by transactivators, *Virology* **189**:354.

Sakuragi, S., Shibata, R., Mukai, R., Komatsu, T., Fukasawa, M., Sakai, H., Sakuragi, J.-I., Kawamura, M., Ibuki, K., Hayami, M., and Adachi, A., 1992, Infection of macaque monkeys with a chimeric human and simian immunodeficiency virus, *J. Gen. Virol.* **73**:2983.

Salk, J., 1987, Prospects for the control of AIDS by immunizing seropositive individuals, *Nature* **327**:473.

Salk, J., Bretscher, P. A., Salk, P. L., Clerici, M., and Shearer, G. M., 1993, A strategy for prophylactic vaccination against HIV, *Science* **260**:1270.

Salter, R. D., Howell, D. N., and Cresswell, P., 1985, Genes regulating HLA class I antigen expression in T–B lymphoblast hybrids, *Immunogenetics* **21**:235.

Salvato, M. S., Emau, P., Malkovsky, M., Schultz, K., McKinzie, D., Johnson, E., and Pauza, C. D., 1993, Cellular immune responses in rhesus macaques with inapparent SIV infection, *Madison, WI: 11th Annual Symposium on Nonhuman Primate Models for AIDS Research Symposium* (Abstract, p. 24.).

Samuel, K. P., Hodge, D. R., Chen, Y. M., and Papas, T. S., 1991, Nef proteins of the human immunodeficiency viruses (HIV-1 and HIV-2) and simian immunodeficiency virus (SIV) are structurally similar to leucine zipper transcriptional activation factors, *AIDS Res. Hum. Retroviruses* **7**:697.

Samuelsson, A., Bjorling, E., Putkonen, P., Utter, G., Chiodi, F., and Norrby, E., 1993, Identification of four antibody-binding sites in the envelope proteins of simian immunodeficiency virus, SIVsm, *AIDS* **7**:159.

Sasseville, V. G., Newman, W. A., Lackner, A. A., Smith, M. O., Lausen, N. C. G., Deall, D., and Ringler, D. J., 1992, Elevated vascular cell adhesion molecule-1 in AIDS encephalitis induced by simian immunodeficiency virus [Short Communication], *Am. J. Pathol.* **141**:1021.

Sattentau, Q. J., Moore, J. P., Vignaux, F., Traincard, F., and Poignard, P., 1993, Conformational changes induced in the envelope glycoproteins of the human and simian immunodeficiency viruses by soluble receptor binding, *J. Virol.* **67**:7383.

Schneider, J., and Hunsmann, G., 1988, Editorial review. Simian lentiviruses—the SIV group, *AIDS* **2**:1.

Schneider, J., Jurkiewicz, E., Hayami, M., Desrosiers, R., Marx, P., and Hunsmann, G., 1987, Serological and structural comparison of HIV, SIVmac, SIVagm and SIVsm, four primate lentiviruses, *Ann. Inst. Pasteur/Virol.* **138**:93.

Schrier, J. E., Harper, J. S., and Povar, M. L., 1992, Update: Nonhuman primate importation, *Lab. Primate Newsl.* **31**:1.

Scinicariello, F., English, W. J., and Hilliard, J., 1993, Identification by PCR of meningitis caused by herpes B virus [letter], *Lancet* **341**:1660.

Selby, M. J., and Peterlin, B. M., 1990, Trans-activation by HIV-1 tat via a heterologous RNA binding protein, *Cell* **62**:769.

Selby, M. J., Bain, E. S., Luciw, P. A., and Peterlin, B. M., 1989, Structure, sequence, and position of the stem-loop in tar determine transcriptional elongation by tat through the HIV-1 long terminal repeat, *Genes Dev.* **3**:547.

Seshamma, T., Bagasra, O., Oakes, J. W., and Pomerantz, R. J., 1992, A quantitative reverse transcriptase-polymerase chain reaction for HIV-1-specific RNA species, *J. Virol. Methods* **40**:331.

Shafferman, A., Layne, A., Sadoff, J., Burke, D. S., Morton, W. R., and Benveniste, R. E., 1989a, Antibody recognition of SIVmac envelope peptides in plasma from macaques experimentally infected with SIV/Mne, *AIDS Res. Hum. Retroviruses* **5**:327.

Shafferman, A., Lennox, J., Grosfeld, H., Sadoff, J., Redfield, R. R., and Burke, D. S., 1989b, Patterns of antibody recognition of selected conserved amino acid sequences from the HIV envelope in sera from different stages of HIV infection, *AIDS Res. Hum. Retroviruses* **5**:33.

Shafferman, A., Jahrling, P. B., Benveniste, R. E., Lewis, M. G., Phipps, T. J., Eden-McCutchan, F., Sadoff, J., Eddy, G. A., and Burke, D. S., 1991, Protection of macaques with a simian immunodeficiency virus envelope peptide vaccine based on conserved human immunodeficiency virus type 1 sequences, *Proc. Natl. Acad. Sci. USA* **88**:7126.

Shafferman, A., Lewis, M. G., McCutchan, F. E., Benveniste, R. E., Jahrling, P. B., Burke, D. S., and Eddy, G. A., 1993, Prevention of transmission of simian immunodeficiency virus from vaccinated macaques that developed transient virus infection following challenge, *Vaccine* **11**:848.

Sharer, L. R., Michaels, J., Murphey-Corb, M., Hu, F. S., Kuebler, D. J., Martin, L. N., and Baskin, G. B., 1991, Serial pathogenesis study of SIV brain infection, *J. Med. Primatol.* **20**:211.

Sharma, D. P., Anderson, M., Zink, M. C., Adams, R. J., Donnenberg, A. D., Clements, J. E., and Narayan, O., 1992a, Pathogenesis of acute infection in rhesus macaques with a lymphocytotropic strain of simian immunodeficiency virus, *J. Infect. Dis.* **166**:738.

Sharma, D. P., Zink, M. C., Anderson, M., Adams, R., Clements, J. E., Joag, S. V., and Narayan, O., 1992b, Derivation of neurotropic simian immunodeficiency virus from exclusively lymphocytotropic parental virus: Pathogenesis of infection in macaques, *J. Virol.* **66**:3550.

Shen, L., Chen, Z. W., Miller, M. D., Stallard, V., Mazzara, G. P., Panicali, D. L., and Letvin, N. L., 1991, Recombinant virus vaccine-induced SIV-specific $CD8^+$ cytotoxic T lymphocytes, *Science* **252**:440.

Shen, L., Mazzara, G. P., DiSciullo, S. O., Panicali, D. L., and Letvin, N. L., 1993, Immunization with lentivirus-like particles elicits a potent SIV-specific recall cytotoxic T-lymphocyte response in rhesus monkeys, *AIDS Res. Hum. Retroviruses* **9**:129.

Shibata, R., Adachi, A., Sakai, H., Ishimoto, A., Miura, T., and Hayami, M., 1990a, Mutational analysis of simian immunodeficiency virus from African green monkeys and human immunodeficiency virus type 2, *J. Med. Primatol.* **19**:217.

Shibata, R., Miura, T., Hayami, M., Sakai, H., Ogawa, K., Kiyomasu, T., Ishimoto, A., and Adachi, A., 1990b, Construction and characterization of an infectious DNA clone and of mutants of simian immunodeficiency virus isolated from the African green monkey, *J. Virol.* **64**:307.

Shibata, R., Sakai, H., Ogawa, K., Ishimoto, A., and Adachi, A., 1990c, Comparative studies on tat mutants of three primate lentiviruses, *Arch. Virol.* **114**:243.

Shibata, R., Sakai, H., Kiyomasu, T., Ishimoto, A., Hayami, M., and Adachi, A., 1990d, Generation and characterization of infectious chimeric clones between human immunodeficiency virus type 1 and simian immunodeficiency virus from an African green monkey, *J. Virol.* **64**:5861.

Siegel, F., Norley, S., Hartung, S., and Kurth, R., 1992, B-cell epitope mapping of the entire SIVagm envelope glycoprotein including fine mapping of immunogenic regions, *J. AIDS* **5**:583.

Simon, M. A., Chalifoux, L. V., and Ringler, D. J., 1992, Pathologic features of SIV-induced disease and the association of macrophage infection with disease evolution, *AIDS Res. Hum. Retroviruses* **8**:327.

Slade, A., Jones, S., Almond, N., and Kitchin, P., 1993a, Simian immunodeficiency virus (mac 251-32H) transmembrane protein sequence remains conserved throughout the course of infection in macaques, *AIDS Res. Hum. Retroviruses* **9**:133.

Slade, A., Jones, S., Jenkins, A., Bootman, J., Almond, N., and Kitchin, P., 1993b, The quantitation and sequence characterization of proviral DNA in the blood and tissues of two SIV infected macaques, *Berlin: IXth International Conference on AIDS* 192 (Abstract).

Sommerfelt, M. A., and Weiss, R. A., 1990, Receptor interference groups of 20 retroviruses plating on human cells, *Virology* **176**:58.

Sommerfelt, M. A., Williams, B. P., McKnight, A., Goodfellow, P. N., and Weiss, R. A., 1990, Localization of the receptor gene for type D simian retroviruses on human chromosome 19, *J. Virol.* **64**:6214.

Somogyi, P. A., Gyuris, A., and Foldes, I., 1990, A solid phase reverse transcriptase micro-assay for the detection of human immunodeficiency virus and other retroviruses in cell culture supernatants, *J. Virol. Methods* **27**:269.

Sonigo, P., Barker, C., Hunter, E., and Wain-Hobson, S., 1986, Nucleotide sequence of Mason-Pfizer monkey virus: An immunosuppressive D-type retrovirus, *Cell* **45**:375.

Sopper, S., Hemm, S., Meixensberger, J., Coulibaly, C., Stahl-Hennig, C., Hunsmann, G., Fleckenstein, B., ter Meulen, V., and Dorries, R., 1993, Dynamics of the immune system response in cerebrospinal fluid and blood of SIVmac-infected rhesus monkeys, *J. Med. Primatol.* **22**:138.

Spear, G. R., Takefman, Sullivan, B. L., Landay, A. L., Jennings, M. B., and Carlson, J. R., 1992, Evidence for complement as an effector mechanism in challenge resistance of macaques vaccinated with human-cell-derived SIV, in press.

Spear, G. R., Takefman, D. M., Sullivan, B. L., Landay, A. L., Jennings, M. B., and Carlson, J. R., 1993, Anti-cellular antibodies in sera from vaccinated macaques can induce complement-mediated virolysis of human immunodeficiency virus and simian immunodeficiency virus, *Virology* **195:**475.

Spies, C. P. and Compans, R. W., 1993, Alternate pathways of secretion of simian immunodeficiency virus envelope glycoproteins, *J. Virol.* **67:**6535.

Spies, C. P., Ritter, D., Mulligan, M. J., and Compans, R. W., 1994, Truncation of the cytoplasmic domain of the simian immunodeficiency virus envelope glycoprotein alters the conformation of the external domain, *J. Virol.* **68:**585.

Spira, T. J., Bozeman, L. H., Holman, R. C., Warfield, D. T., Phillips, S. K., and Feorino, P. M., 1987, Micromethod for assaying reverse transcriptase of human T-cell lymphotropic virus type III/lymphadenopathy-associated virus, *J. Clin. Microbiol.* **25:**97.

Stahl-Hennig, C., Voss, G., Nick, S., Petry, H., Fuchs, D., Wachter, H., Coulibaly, C., Luke, W., and Hunsmann, G., 1992, Immunization with Tween-ether-treated SIV adsorbed to aluminum hydroxide protects monkeys against experimental SIV infection, *Virology* **186:**588.

Stahl-Hennig, C., Voss, G., Dittmer, U., Coulibaly, C., Petry, H., Makoschey, B., and Aubertin, A. M., 1993, Protection of monkeys by a split vaccine against SIVmac depends on biological properties of the challenge virus, *Berlin: IXth International Conference on AIDS* 253 (Abstract).

Stahmer, I., de Mesquita, R., Ordonez, C., Ekman, M., Biberfeld, G., and Biberfeld, P., 1993, Interaction of follicular dendritic cells (FDC) with SIV in the AIDS-model of experimentally SIV-infected cynomolgous monkeys, *Berlin: IXth International Conference on AIDS* 194 (Abstract).

Stamatatos, L., and Duzgunes, N., 1993, Simian immunodeficiency virus (SIVmac251) membrane lipid mixing with human $CD4^+$ and $CD4^-$ cell lines in vitro does not necessarily result in internalization of the viral core proteins and produce infection, *J. Gen. Virol.* **74:**1043.

Steimer, K. S., Scandella, C. J., Skiles, P. V., and Haigwood, N. L., 1991, Neutralization of divergent HIV-1 isolates by conformation-dependent human antibodies to Gp120, *Science* **254:**105.

Stephenson, J. R., Hino, S., Garrett, E. W., and Aaronson, S. A., 1976, Immunological cross reactivity of Mason-Pfizer monkey virus with type C RNA viruses endogenous to primates, *Nature* **261:**609.

Stott, E. J., Kitchin, P. A., Page, M., Flanagan, B., Taffs, L. F., Chan, W. L., Mills, K. H. G., Silvera, P., and Rodgers, A., 1991, Anti-cell antibody in macaques [letter], *Nature* **353:**393.

Stott, E. J., Chan, W. L., Mills, K. H., Page, M., Taffs, F., Cranage, M., Greenaway, P., and Kitchin, P., 1990, Preliminary report: Protection of cynomolgus macaques against simian immunodeficiency virus by fixed infected-cell vaccine, *Lancet* **336:**1538.

Stott, E. J., Kitchin, P., Mills, K., Page, M., Taffs, F., and Chan, W. L., 1993, The role of cellular and viral antigens in protection against SIV infection of macaques, *Berlin: IXth International Conference on AIDS* 39 (Abstract, p. 39.).

Stowell, R. E., Smith, E. K., Espana, C., and Nelson, V. G., 1971, Outbreak of malignant lymphoma in rhesus monkeys, *Lab. Invest.* **25:**476.

Strebel, K., Daugherty, D., Clouse, K., Cohen, D., Folks, T., and Martin, M. A., 1987, The HIV "A" (sor) gene product is essential for virus infectivity, *Nature* **328:**728.

Stricker, R. B., McHugh, T. M., Moody, D. J., Morrow, W. J. W., Stites, D. P., Shuman, M. A., and Levy, J. A., 1987, An AIDS-related cytotoxic autoantibody reacts with a specific antigen on stimulated $CD4^+$ T cells, *Nature* **327:**710.

Stromberg, K., Benveniste, R. E., Arthur, L. O., Rabin, H., Giddens, W. E., Jr., Ochs, H. D., Morton, W. R., and Tsai, C.-C., 1984, Characterization of exogenous type D retrovirus from a fibroma of a macaque with simian AIDS and fibromatosis, *Science* **224:**289.

Sugamura, K., and Hinuma, Y., 1993, Human retroviruses: HTLV-I and HTLV-II, in: *The Retroviridae,* Vol. 2 (J. A. Levy, ed.), pp. 399–435, Plenum Press, New York.

Sutjipto, S., Kodama, T., Yee, J., Gettie, A., Jennings, M., Desrosiers, R. C., and Marx, P. A., 1990a, Characterization of monoclonal antibodies that distinguish simian immunodeficiency virus isolates from each other and from human immunodeficiency virus types 1 and 2, *J. Gen. Virol.* **71:**247.

Sutjipto, S., Pedersen, N. C., Miller, C. J., Gardner, M. B., Hanson, C. V., Gettie, A., Jennings, M., Higgins, J., and Marx, P. A., 1990b, Inactivated simian immunodeficiency virus vaccine failed to protect rhesus macaques from intravenous or genital mucosal infection but delayed disease in intravenously exposed animals, *J. Virol.* **64:**2290.

Tao, J., and Frankel, A. D., 1993, Electrostatic interactions modulate the RNA-binding and transactivation specificities of the human immunodeficiency virus and simian immunodeficiency virus Tat proteins, *Proc. Natl. Acad. Sci. USA* **90**:1571.

Tarantal, A. F., Marthas, M. L., McChesney, M. B., Miller, C. J., Lerche, N. W., Pedersen, N. C., and Hendrickx, A. G., 1994, Effects of SIV infection on fetal development and pregnancy outcome in the rhesus macaque (Macaca mulatta), *Pediatric AIDS and HIV Infection: Fetus to Adolescent*, in press.

Thapar, M. A., Parr, E. L., Bozzola, J. J., and Parr, M. B., 1991, Secretory immune responses in the mouse vagina after parenteral or intravaginal immunization with an immunostimulating complex (ISCOM), *Vaccine* **9**:129.

Thayer, R. M., Power, M. D., Bryant, M. L., Gardner, M. B., Barr, P. J., and Luciw, P. A., 1987, Sequence relationships of type D retroviruses which cause simian acquired immunodeficiency syndrome, *Virology* **157**:317.

Thibodeau, L., Tremblay, C., and Lachapelle, L., 1992, Oral priming followed by parenteral immunization with HIV-immunosomes induces HIV-1-specific salivary and circulatory IgA in mice and rabbits, *AIDS Res. Hum. Retroviruses* **8**:1379.

Thorstensson, R., Walther, L., Putkonen, P., Albert, J., and Biberfeld, G., 1991, A capture enzyme immunoassay for detection of HIV-2/SIV antigen, *J. AIDS* **4**:374.

Titti, F., Koanga Mogtomo, M. L., Borsetti, A., Geraci, A., Sernicola, L., Panzini, G., Turillazzi, G. P., Baroncelli, S., Giovannetti, A., and Zamarchi, R., 1993, Immunization of Macaca fascicularis with inactivated SIV preparations: challenge with human- or monkey-derived SIV and the effects of a longer immunization schedule, *J. Med. Primatol.* **22**:110.

Todaro, G. J., Benveniste, R. E., Sherr, C. J., Schlom, J., Schidlovsky, G., and Stephenson, J. R., 1978a, Isolation and characterization of a new type D retrovirus from the Asian primate Presbytis obscurus (spectacled langur), *Virology* **84**:189.

Todaro, G. J., Benveniste, R. E., Sherwin, S. A., and Sherr, C. J., 1978b, MAC-1, a new genetically transmitted type C virus of primates: "Low frequency" activation from stumptail monkey cell cultures, *Cell* **13**:775.

Todaro, G. J., Sherr, C. J., Sen, A., King, N., Daniel, M. D., and Fleckenstein, B., 1978c, Endogenous New World primate type C viruses isolated from owl monkey (Aotus trivirgatus) kidney cell line, *Proc. Natl. Acad. Sci. USA* **75**:1004.

Toedter, G., Stoner, C., Pearlman, S., Brewer, K., Charie, L., Cueto, C., Marquez, M., and Hofheinz, D., 1990, Improved detection of SIV p27 by antigen capture ELISA, *New Orleans: 8th Annual Symposium on Nonhuman Primate Models for AIDS* (Abstract, p. 113.).

Tomonaga, K., Katahira, J., Fukasawa, M., Hassan, M. A., Kawamura, M., Akari, H., Miura, T., Goto, T., Nakai, M., and Suleman, M., 1993, Isolation and characterization of simian immunodeficiency virus from African white-crowned mangabey monkeys (Cercocebus torquatus lunulatus), *Arch. Virol.* **129**:77.

Tong-Starksen, S. E., Welsh, T. M., and Peterlin, B. M., 1990, Differences in transcriptional enhancers of HIV-1 and HIV-2. Response to T cell activation signals, *J. Immunol.* **145**:4348.

Torres, J. V., Werner, L. L., Malley, A., and Benjamini, E., 1991b, The induction of neutralizing antibodies by synthetic peptides of the envelope protein of type D simian retrovirus-1 (SRV-1), *Mol. Immunol.* **28**:907.

Torres, J. V., Werner, L. L., Malley, A., and Benjamini, E., 1991c, Neutralization epitope in the envelope glycoprotein of simian retrovirus-1 (SRV-1) and identification of the virus receptor, *J. Med. Primatol.* **20**:218.

Torres, J. V., Malley, A., Banapour, B., Anderson, D. E., Axthelm, M. K., Gardner, M. B., and Benjamini, E., 1993a, An epitope on the surface envelope glycoprotein (gp130) of simian immunodeficiency virus (SIVmac) involved in viral neutralization and T cell activation, *AIDS Res. Hum. Retroviruses* **9**:423.

Torres, J. V., Anderson, D. E., Malley, A., Banapour, B., Axthelm, M. K., Benjamini, E., and Gardner, M. B., 1993b, SIV envelope glycoprotein epitopes recognized by antibodies from infected or vaccinated rhesus macaques, *J. Med. Primatol.* **22**:129.

Tovey, M. G., Lebon, P., Meyer, F., Hurtrel, B., Gresser, I., Venet, A., Girard, M., and Aubertin, A.-M., 1993, Antibody to the human IFN α receptor reduces the loss of $CD4^+$ T cells in macaques infected with the simian immunodeficiency virus (SIV), *Nonhuman Primate Models for AIDS Research Symposium* (Abstract).

Traina-Dorge, V., Blanchard, J., Martin, L., and Murphey-Corb, M., 1992, Immunodeficiency and lymphoproliferative disease in an African green monkey dually infected with SIV and STLV-I, *AIDS Res. Hum. Retroviruses* **8**:97.

Tristem, M., Marshall, C., Karpas, A., Petrik, J., and Hill, F., 1990, Origin of vpx in lentiviruses [letter], *Nature* **347**:341.

Tristem, M., Marshall, C., Karpas, A., and Hill, F., 1992, Evolution of the primate lentiviruses: Evidence from vpx and vpr, *EMBO J.* **11**:3405.

Trivedi, P., Emau, P., and Pauza, C. D., 1993, The mucosal barrier selects for SIV with unique envelope sequences, *Madison, WI: 11th Annual Symposium on Nonhuman Primate Models for AIDS Research Symposium* (Abstract, p. 50.).

Tsai, C. C., Giddens, W. E., Jr., Ochs, H. D., Morton, W. R., Knitter, G. H., Blakley, G. A., and Benveniste, R. E., 1986, Retroperitoneal fibromatosis and acquired immunodeficiency syndrome in macaques: Clinical and immunologic studies, *Lab. Anim. Sci.* **36**:119.

Tsai, C. C., Yarnall, M., Follis, K. E., and Benveniste, R. E., 1989, Antigen capture assay for detection of simian type D retroviruses in cell cultures and plasma samples, *Lab. Anim. Sci.* **39**:554.

Tsai, C. C., Follis, K. E., Yarnall, M., Deaver, L. E., Benveniste, R. E., and Sager, P. R., 1990, In vitro screening for antiretroviral agents against simian immunodeficiency virus (SIV), *Antiviral Res.* **14**:87.

Tsai, C.-C., Follis, K. E., Grant, R. F., Nolte, R. E., Bartz, C. R., Benveniste, R. E., and Sager, P. R., 1993, Effect of dosing frequency on ZDV prophylaxis in macaques infected with simian immunodeficiency virus, *J. Acq. Immune Syndr.* **6**:1086.

Tsai, C.-C., Follis, K. E., Grant, R. F., Nolte, R. E., Wu, H., and Benveniste, R. E., 1993, Infectivity and pathogenesis of titered dosages of simian immunodeficiency virus (SIV) experimentally inoculated into longtailed macaques (*Macaca fascicularis*), *Lab. Anim. Sci.* **43**:411–416.

Tsai, C.-C., Follis, K., Sabo, A., and Grant, R., 1993b, Distribution of SIVmne in tissues and organs of Macaca fascicularis treated with AZT or PMEA, *Nonhuman Primate Models for AIDS Research Symposium* (Abstract).

Tsai, C.-C., Follis, K. E., Grant, R., Sabo, A., Nolte, R., Bartz, C., Bischofberger, N., and Benveniste, R., 1993c, Comparison of the efficacy of AZT and PMEA treatment on acute SIVmne infection in macaques, *Nonhuman Primate Models for AIDS Research Symposium* (Abstract).

Tsubota, H., Ringler, D. J., Kannagi, M., King, N. W., Solomon, K. R., MacKey, J. J., Walsh, D. G., and Letvin, N. L., 1989, $CD8^+$ $CD4^-$ lymphocyte lines can harbor the AIDS virus in vitro [see comments], *J. Immunol.* **143**:858.

Tsujimoto, H., Seiki, M., Nakamura, H., Watanabe, T., Sakakibara, I., Sasagawa, A., Honjo, S., Hayami, M., and Yoshida, M., 1985, Adult T-cell leukemia-like disease in monkey naturally infected with simian retrovirus related to human T-cell leukemia virus type I, *Jpn. J. Cancer Res.* **76**:911.

Tsujimoto, H., Cooper, R. W., Kodama, T., Fukasawa, M., Miura, T., Ohta, Y., Ishikawa, K. I., Nakai, M., Frost, E., Roelants, G. E., Roffi, J., and Hayami, M., 1988, Isolation and characterization of simian immunodeficiency virus from mandrills in Africa and its relationship to other human and simian immunodeficiency viruses, *J. Virol.* **62**:4044.

Unger, R. E., Stout, M. W., and Luciw, P. A., 1991, Simian immunodeficiency virus (SIVmac) exhibits complex splicing for tat, rev, and env mRNA, *Virology* **182**:177.

Unger, R. E., Marthas, M. L., Lackner, A. A., Pratt-Lowe, E., Lohman, B. L., Van Rompay, K., and Luciw, P. A., 1992a, Detection of simian immunodeficiency virus DNA in macrophages from infected rhesus macaques, *J. Med. Primatol.* **21**:74.

Unger, R. E., Marthas, M. L., Pratt-Lowe, E., Padrid, P. A., and Luciw, P. A., 1992b, The nef gene of simian immunodeficiency virus SIVmac1A11, *J. Virol.* **66**:5432.

Urabe, T., Sano, K., Tanno, M., Mizoguchi, J., Otani, M., Lee, M. H., Takasaki, T., Kusakabe, H., Imagawa, D. T., and Nakai, M., 1992, A non-radioisotopic reverse transcriptase assay using biotin-11-deoxyuridinetriphosphate on primer-immobilized microtiter plates, *J. Virol. Methods* **40**:145.

Vaishnav, Y. N., and Wong, S. F., 1991, The biochemistry of AIDS, *Annu. Rev. Biochem.* **60**:577.

VanKuyk, R. W., Acevedo, R. A., Torres, J., Levy, N. B., Planelles, V., Munn, R. J., Unger, R. E., Gardner, M. B., and Luciw, P. A., 1991, Characterization of rhesus B lymphoblastoid cell lines infected with simian type D retrovirus, *AIDS Res. Hum. Retroviruses* **7**:899.

Van Rompay, K. K. A., Marthas, M. L., Lohman, B. L., and Pedersen, N. C., 1992a, Characterization of AZT susceptibility of biological SIVmac and molecular clones, and emergence of AZT resistant mutants in vitro, *Puerto Rico: Tenth Annual Symposium on Nonhuman Primate Models for AIDS* (Abstract).

Van Rompay, K. K. A., Marthas, M. L., Ramos, R. A., Mandell, C. P., McGowan, E. K., Joye, S. M., and Pedersen, N. C., 1992c, Simian immunodeficiency virus (SIV) infection of infant rhesus macaques as a model to test antiretroviral drug prophylaxis and therapy: Oral 3'-azido-3'deoxythymidine prevents SIV infection, *Antimicrob. Agents* **36**:2381.

Vasudevachari, M. B., Mast, T. C., and Salzman, N. P., 1990, Suppression of HIV-1 reverse transcriptase activity by mycoplasma contamination of cell cultures, *AIDS Res. Hum. Retroviruses* **6**:411.

Venet, A., Bourgault, I., Aubertin, A. M., Kieny, M. P., and Levy, J. P., 1992, Cytotoxic T lymphocyte response against multiple simian immunodeficiency virus (SIV) proteins in SIV-infected macaques, *J. Immunol.* **148**:2899.

Veronese, F. D., Joseph, B., Copeland, T. D., Oroszlan, S., Gallo, R. C., and Sarngadharan, M. G., 1989, Identification of simian immunodeficiency virus SIVmac env gene products, *J. Virol.* **63**:1416.

Viglianti, G. A., and Mullins, J. I., 1988, Functional comparison of transactivation by simian immunodeficiency virus from rhesus macaques and human immunodeficiency virus type I, *J. Virol.* **62**:4523.

Viglianti, G. A., Sharma, P. L., and Mullins, J. I., 1990, Simian immunodeficiency virus displays complex patterns of RNA splicing, *J. Virol.* **64**:4207.

Villinger, F., Powell, J. D., Jehuda-Cohen, T., Neckelmann, N., Vuchetich, M., De, B., Folks, T. M., McClure, H. M., and Ansari, A. A., 1991, Detection of occult simian immunodeficiency virus SIVsmm infection in asymptomatic seronegative nonhuman primates and evidence for variation in SIV gag sequence between in vivo- and in vitro-propagated virus, *J. Virol.* **65**:1855.

Villinger, F., Lauro, S., Mayne, A., Folks, T. M., and Ansari, A. A., 1993, Sequence analysis of the SIVsmm nef gene in naturally and experimentally infected sooty mangabey monkeys and macaques, *Nonhuman Primate Models for AIDS Research Symposium* (Abstract).

Vogel, M., Cichutek, K., Norley, S., and Kurth, R., 1993, Self-limiting infection by int/nef-double mutants of simian immunodeficiency virus, *Virology* **193**:115.

Volberding, P. A., Lagakos, S. W., Koch, M. A., Pettinelli, C., Myers, M. W., Booth, D. K., Balfour, H. H., Jr., Reichman, R. C., Bartlett, J. A., and Hirsch, M. S., 1990, Zidovudine in asymptomatic human immunodeficiency virus infection. A controlled trial in persons with fewer than 500 CD4-positive cells per cubic millimeter. The AIDS Clinical Trials Group of the National Institute of Allergy and Infectious Diseases [see comments], *N. Engl. J. Med.* **322**:941.

Voss, G., and Hunsmann, G., 1993, Cellular immune response to SIV_{mac} and HIV-2 in macaques: Model for the human HIV-1 infection, *J. AIDS* **6**:969.

Voss, G., Nick, S., Stahl-Hennig, C., Coulibaly, C., Petry, H., Luke, W., and Hunsmann, G., 1992, Potential significant of the cellular immune response against the macaque strain of simian immunodeficiency virus (SIVMAC) in immunized and infected rhesus macaques, *J. Gen. Virol.* **73**:2273.

Vowels, B. R., Gershwin, M. E., Gardner, M. B., Ahmed-Ansari, A., and McGraw, T. P., 1989, Characterization of simian immunodeficiency virus-specific T-cell-mediated cytotoxic response of infected rhesus macaques, *AIDS* **3**:785.

Vowels, B. R., Gershwin, M. E., Gardner, M. B., and McGraw, T. P., 1990, Natural killer cell activity of rhesus macaques against retrovirus-pulsed $CD4^+$ target cells, *AIDS Res. Hum. Retroviruses* **6**:905.

Wain-Hobson, S., Vartanian, J.-P., Henry, M., Chenciner, N., Cheynier, R., Delassus, S., Martins, L. P., Sala, M., Nugeyre, M.-T., Guetard, D., Klatzmann, D., Gluckman, J.-C., Rozenbaum, W., Barre-Sinoussi, F., and Montagnier, L., 1991, LAV revisited: Origins of the early HIV-1 isolates from Institut Pasteur, *Science* **252**:961.

Walker, C. M., Moody, D. J., Stites, D. P., and Levy, J. A., 1986, $CD8^+$ lymphocytes can control HIV infection in vitro by suppressing virus replication, *Science* **234**:1563.

Walker, C. M., Moody, D. J., Stites, D. P., and Levy, J. A., 1989, $CD8^+$ T lymphocyte control of HIV replication in cultured $CD4^+$ cells varies among infected individuals, *Cell. Immunol.* **119**:470.

Wallace, M., Pauza, C. D., and Malkovsky, M., 1993, Antiviral cytotoxic responses of primate gamma-delta T cells isolated by magnetic cell separation, *Nonhuman Primate Models for AIDS Research Symposium* (Abstract).

Ward, J. M., O'Leary, T. J., Baskin, G. B., Benveniste, R., Harris, C. A., Nara, P. L., and Rhodes, R. H., 1987, Immunohistochemical localization of human and simian immunodeficiency viral antigens in fixed tissue sections, *Am. J. Pathol.* **18**:217.

Warren, J. T., and Dolatshahi, M., 1992, Worldwide survey of AIDS vaccine challenge studies in nonhuman primates: Vaccines associated with active and passive immune protection from live virus challenge, *J. Med. Primatol.* **21**:139.

Warren, J. T., and Dolatshahi, M., 1993, First updated and revised survey of worldwide HIV and SIV vaccine challenge studies in nonhuman primates: progress in first and second order studies, *J. Med. Primatol.* **22**:203.

Watanabe, M., Reimann, K. A., DeLong, P. A., Liu, T., Fisher, R. A., and Letvin, N. L., 1989, Effect of recombinant soluble CD4 in rhesus monkeys infected with simian immunodeficiency virus of macaques, *Nature* **337**:267.

Watanabe, M., Ringler, D. J., Nakamura, M., DeLong, P. A., and Letvin, N. L., 1990, Simian immunodeficiency virus inhibits bone marrow hematopoietic progenitor cell growth, *J. Virol.* **64**:656.

Watanabe, M., Chen, Z. W., Tsubota, H., Lord, C. I., Levine, C. G., and Letvin, N. L., 1991a, Soluble human CD4 elicits an antibody response in rhesus monkeys that inhibits simian immunodeficiency virus replication, *Proc. Natl. Acad. Sci. USA* **88**:120.

Watanabe, M., Levine, C. G., Shen, L., Fisher, R. A., and Letvin, N. L., 1991b, Immunization of simian immunodeficiency virus-infected rhesus monkeys with soluble human CD4 elicits an antiviral response, *Proc. Natl. Acad. Sci. USA* **88**:4616.

Watanabe, M., Boyson, J. E., Lord, C. I., and Letvin, N. L., 1992, Chimpanzees immunized with recombinant soluble CD4 develop anti-self CD4 antibody responses with anti-human immunodeficiency virus activity, *Proc. Natl. Acad. Sci. USA* **89**:5103.

Weigler, B. J., Roberts, J. A., Hird, D. W., Lerche, N. W., and Hilliard, J. K., 1990, A cross sectional survey for B virus antibody in a colony of group housed rhesus macaques, *Lab. Anim. Sci.* **40**:257.

Weiss, R. A., 1993, Cellular receptors and viral glycoproteins involved in retrovirus entry, in: *The Retroviridae*, Vol. 2 (J. A. Levy, ed.), pp. 1–108, Plenum Press, New York.

Werner, A., Winskowsky, G., Cichutek, K., Norley, S. G., and Kurth, R., 1990b, Productive infection of both $CD4^+$ and $CD4^-$ human cell lines with HIV-1, HIV-2 and SIVagm, *AIDS* **4**:537.

Werner, A., Winskowsky, G., and Kurth, R., 1990c, Soluble CD4 enhances simian immunodeficiency virus SIVagm infection, *J. Virol.* **64**:6252.

Werner, L. L., Malley, A., Torres, J. V., Leung, C. Y., Kwang, H. S., and Benjamini, E., 1990, Synthetic peptides of envelope proteins of two different strains of simian AIDS retrovirus (SRV-1 and SRV-2) represent unique antigenic determinants for serum neutralizing antibodies, *Mol. Immunol.* **27**:1103.

Werner, L. L., Torres, J. V., Leung, C. Y., Kwang, H. S., Malley, A., and Benjamini, E., 1991, Immunobiological properties of a recombinant simian retrovirus-1 envelope protein and a neutralizing monoclonal antibody directed against it, *Mol. Immunol.* **28**:819.

Wiley, C. A., Schrier, R. D., Nelson, J. A., Lampert, P. W., and Oldstone, M. B. A., 1986, Cellular localization of human immunodeficiency virus infection within the brains of acquired immune deficiency syndrome patients, *Proc. Natl. Acad. Sci. USA* **83**:7089.

Willerford, D. M., Gale, M. J., Benveniste, R. E., Clark, E. A., and Gallatin, W. M., 1990, Simian immunodeficiency virus is restricted to a subset of blood $CD4^+$ lymphocytes which includes memory cells, *J. Immunol.* **144**:3779.

Wiviott, L. D., Walker, C. M., and Levy, J. A., 1990, $CD8^+$ lymphocytes suppress HIV production by autologous $CD4^+$ cells without eliminating the infected cells from culture, *Cell. Immunol.* **128**:628.

Wright, C. M., Felber, B. K., Paskalis, H., and Pavlakis, G. N., 1986, Expression and characterization of the trans-activator of HTLV-III/LAV virus, *Science* **234**:988.

Wu, J. C., Chernow, M., Boehme, R. E., Suttmann, R. T., McRoberts, M. J., Prisbe, E. J., Matthews, T. R., Marx, P. A., Chuang, R. Y., and Chen, M. S., 1988, Kinetics and inhibition of reverse transcriptase from human and simian immunodeficiency viruses, *Antimicrob. Agents Chemother.* **32**:1887.

Wyand, M. S., 1992, The use of SIV-infected rhesus monkeys for the preclinical evaluation of AIDS drugs and vaccines, *AIDS Res. Hum. Retroviruses* **8**:349.

Wyand, M. S., Ringler, D. J., Mattmuller, M., Daniel, M. D., Desrosiers, R. C., and King, N. W., 1989a, Detection of simian immunodeficiency virus in macaque lymph nodes with a SIVmac envelope probe, *J. Med. Primatol.* **18**:209.

Wyand, M. S., Ringler, D. J., Naidu, Y. M., Mattmuller, M., Chalifoux, L. V., Sehgal, P. K., Daniel, M. D., Desrosiers, R. C., and King, N. W., 1989b, Cellular localization of simian immunodeficiency virus in lymphoid tissues, *Am. J. Pathol.* **134**:385.

Yamamoto, H., Miller, M. D., Tsubota, H., Watkins, D. I., Mazzara, G. P., Stallard, V., Panicali, D. L., Aldovini, A., Young, R. A., and Letvin, N. L., 1990a, Studies of cloned simian immunodeficiency virus-specific T lymphocytes. gag-specific cytotoxic T lymphocytes exhibit a restricted epitope specificity, *J. Immunol.* **144**:3385.

Yamamoto, H., Miller, M. D., Watkins, D. I., Snyder, G. B., Chase, N. E., Mazzara, G. P., Gritz, L., Panicali, D. L., and Letvin, N. L., 1990b, Two distinct lymphocyte populations mediate simian immunodeficiency virus envelope-specific target cell lysis, *J. Immunol.* **145**:3740.

Yasutomi, Y., Reimann, K. A., Lord, C. I., Miller, M. D., and Letvin, N. L., 1993a, Simian immunodeficiency virus-specific $CD8^+$ lymphocyte response in acutely infected rhesus monkeys, *J. Virol.* **67**:1707.

Yasutomi, Y., Palker, T. J., Gardner, M. B., Haynes, B. F., and Letvin, N. L., 1993b, Synthetic peptide in mineral oil adjuvant elicits simian immunodeficiency virus-specific $CD8^+$ cytotoxic lymphocytes in rhesus monkeys, *J. Immunol.* **151**:5096.

Yee, J. L., Jennings, M. B., Carlson, J. R., and Lerche, N. W., 1991, A simple, rapid immunoassay for the detection of simian immunodeficiency virus antibodies, *Lab. Anim. Care* **41**:119.

Yon, J., Rud, E., Corcoran, T., Kent, K., Rowlands, D., and Clarke, B., 1992, Stimulation of specific immune responses to simian immunodeficiency virus using chimeric hepatitis B core antigen particles, *J. Gen. Virol.* **73**:2569.

Yu, X. F., Ito, S., Essex, M., and Lee, T. H., 1988, A naturally immunogenic virion-associated protein specific for HIV-2 and SIV, *Nature* **335**:262.

Yu, X. F., Matsuda, M., Essex, M., and Lee, T. H., 1990, Open reading frame vpr of simian immunodeficiency virus encodes a virion-associated protein, *J. Virol.* **64**:5688.

Yu, X. F., Yu, Q. C., Essex, M., and Lee, T. H., 1991, The vpx gene of simian immunodeficiency virus facilitates efficient viral replication in fresh lymphocytes and macrophage, *J. Virol.* **65**:5088.

Yu, X., Matsuda, Z., Yu, Q. C., Lee, T. H., and Essex, M., 1993, Vpx of simian immunodeficiency virus is localized primarily outside the virus core in mature virions, *J. Virol.* **67**:4386.

Zamarchi, R., Veronese, M. L., Titti, F., Geraci, A., Verani, P., Rossi, G. B., Amadori, A., and Chieco-Bianchi, L., 1993, *In vitro* spontaneous production of anti-SIV antibodies is a reliable tool in the follow-up of protection of SIV-vaccinated monkeys, *AIDS Res and Hum Retroviruses* **9**:1139.

Zhang, J. Y., Martin, L. N., Watson, E. A., Montelaro, R. C., West, M., Epstein, L., and Murphey-Corb, M., 1988, Simian immunodeficiency virus/delta-induced immunodeficiency disease in rhesus monkeys: Relation of antibody response and antigenemia, *J. Infect. Dis.* **158**:1277.

Zhang, Y.-J., Ohman, P., Putkonen, P., Albert, J., Walther, L., Stalhandske, P., Biberfeld, G., and Fenyo, E. M., 1993, Autologous neutralizing antibodies to SIV_{sm} in cynomolgus monkeys correlate to prognosis, *Virology* **197**:609.

Zingler, K., and Littman, D. R., 1993, Truncation of the cytoplasmic domain of the simian immunodeficiency virus envelope glycoprotein increases env incorporation into particles and fusogenicity and infectivity, *J. Virol.* **67**:2824.

CHAPTER 6

Molecular Features of the Human T-Cell Leukemia Virus
Mechanisms of Transformation and Leukemogenicity

PATRICK L. GREEN AND IRVIN S. Y. CHEN

I. DISEASES ASSOCIATED WITH HUMAN T-CELL LEUKEMIA VIRUS INFECTION

A. Human T-Cell Leukemia Virus Type I

The first oncogenic human retrovirus was identified in a T-lymphoblastoid cell line that had been established from a patient diagnosed with cutaneous T-cell lymphoma (Poiesz *et al.*, 1980). In retrospect, it is likely that this patient had adult T-cell leukemia (ATL). Shortly thereafter, another cell line derived from a patient with ATL was shown to harbor a type C retrovirus, and to produce antigens reactive with sera obtained from ATL patients (Hinuma *et al.*, 1981; Yoshida *et al.*, 1982). The viruses produced from these two distinct cell lines were determined to be identical and were designated human T-cell leukemia virus type I (HTLV-I) (Popovic *et al.*, 1982). It has subsequently been established by seroepidemiology and molecular studies that HTLV-I is the etiologic agent of ATL. ATL is endemic to regions of southern Japan, central Africa, northeastern South America, the Caribbean basin, and the southeastern United States (Blattner *et al.*, 1982; Catovsky *et al.*, 1982; Blayney *et al.*, 1983; Bunn *et al.*, 1983; Merino *et al.*, 1984; Saxinger *et al.*, 1984; Su *et al.*, 1985; Biggar *et al.*, 1985). In addition, a

PATRICK L. GREEN • Department of Microbiology and Immunology, Vanderbilt University School of Medicine, Nashville, Tennessee 37232-2363. IRVIN S. Y. CHEN • Departments of Medicine and Microbiology and Immunology, University of California School of Medicine and Jonsson Comprehensive Cancer Center, Los Angeles, California 90024-1736.

The Retroviridae, Volume 3, edited by Jay A. Levy. Plenum Press, New York, 1994.

significant number of intravenous drug abusers in the United States and Europe have been shown to be infected with HTLV (Tedder *et al.*, 1984; Jason *et al.*, 1985; Sandler, 1986; Robert-Guroff *et al.*, 1986; Lee *et al.*, 1989) (see Sugamura and Hinuma, 1993).

1. Adult T-Cell Leukemia

The majority of HTLV-I-infected individuals are asymptomatic carriers displaying no apparent clinical symptoms. HTLV-I proviral DNA in infected asymptomatic individuals is only detectable by the polymerase chain reaction (PCR), reflecting the relatively low number of infected lymphocytes. Molecular analysis of HTLV sequences in patient lymphocytes reveals a polyclonal integration pattern.

Preleukemic-ATL patients are generally diagnosed with leukocytosis and/or morphologically abnormal lymphocytes, but display few clinical symptoms (Gallart *et al.*, 1983; Kinoshita *et al.*, 1985). At the pre-ATL stage, clonal populations of cells can be detected indicating a selective outgrowth of infected cells, some of which may progress to the malignancies found in ATL. Approximately 30% of pre-ATL patients progress to chronic/smoldering ATL, characterized by skin lesions and low levels of circulating leukemic cells (Shimoyama *et al.*, 1983; Yamaguchi *et al.*, 1983; Kawano *et al.*, 1985). Progression to acute ATL can occur in a matter of months and is characterized by an aggressive clonal malignancy of $CD4^+$ T lymphocytes (Yoshida *et al.*, 1982; Yamada, 1983; Morimoto *et al.*, 1985). It is important to note that less than 1% of HTLV-I-infected individuals develop ATL (Kondo *et al.*, 1987; Murphy *et al.*, 1989); ATL generally appears in early adulthood after a long latency period, approximately 20 to 30 years following infection (Kawano *et al.*, 1985).

ATL cells are monoclonal or oligoclonal by the pattern of HTLV-I integration (Yoshida *et al.*, 1982; Seiki *et al.*, 1984), and also by the clonal rearrangement pattern of the T-cell antigen receptor β gene (Matsuoka *et al.*, 1988). Although a single provirus is present in the leukemic cell, there is no detectable viral gene expression. Long-term *in vitro* culture of the leukemic cells usually results in expression of the virus and virion release (Seiki *et al.*, 1982; Yoshida *et al.*, 1982). This pattern of viral pathogenesis suggests that HTLV-I is a weak carcinogen and that the viral gene products may be required for initiating but not maintaining the leukemic phenotype. Presumably other genetic alterations within the cell are necessary to maintain this phenotype. Chromosomal abnormalities are commonly found in ATL cells and often reflect the degree of severity of the disease (Fukuhara *et al.*, 1983; Sanada *et al.*, 1985), again suggesting that somatic mutations are critical progression events in the development of the tumorigenic phenotype (see Sugamura and Hinuma, 1993).

2. HTLV-Associated Myelopathy/Tropical Spastic Paraparesis

One of the more interesting aspects of HTLV-I infection is its association with a disease completely different from ATL. HTLV-I has been implicated as a causative agent in the human neurological disorder known as HTLV-associated myelopathy (HAM) or tropical spastic paraparesis (TSP) (Gessain *et al.*, 1985; Osame *et al.*, 1986, 1987a; Bhagavati *et al.*, 1988; Jacobson *et al.*, 1988), a chronic demyelinating disease. The disease is characterized by weakness and spasticity of the extremities, hyperflexia, and mild peripheral sensory loss (Osame *et al.*, 1987a; Vernant *et al.*, 1987). Earlier HTLV-I disease associations were based on the presence of anti-HTLV-I antibodies in the serum and cerebrospinal fluid (CSF) of HAM/TSP patients (Osame *et al.*, 1987b; Nishimura *et al.*, 1988). These findings were further supported by the detection of HTLV-I sequences in the blood of individuals diagnosed with HAM/TSP using the PCR method (Kwok *et al.*, 1988).

It is not yet clear whether the HTLV-I that is responsible for neurological disorders such as HAM/TSP is identical to that responsible for ATL. One group who isolated and characterized a virus from the CSF of a patient with HAM/TSP found a related but not identical HTLV-I, as determined by restriction fragment mapping (Sarin *et al.*, 1989). However, another group reported that T-cell lines established from the CSF of patients with HAM produced virus indistinguishable from HTLV-I; also determined by restriction fragment mapping analysis (Nishimura *et al.*, 1988). In addition, some biological differences were noted between an HTLV-I ATL isolate and viruses produced by T-cell lines derived from the peripheral blood and CSF of TSP patients (Jacobson *et al.*, 1988).

Although the pathogenesis of HAM/TSP is poorly understood, it appears to be distinct from ATL. Development of disease can occur within a few years following infection, whereas ATL usually requires 20–30 years. Also, analysis of HTLV-I sequences in the peripheral blood from HAM/TSP patients generally demonstrates a polyclonal pattern of integration, in contrast to the clonal pattern detected in cells obtained from ATL patients (Yoshida *et al.*, 1987; Greenberg *et al.*, 1989a). It has been reported that HAM/TSP patients have 10- to 100-fold higher levels of proviral DNA than in HTLV-I-infected asymptomatic carriers (Gessain *et al.*, 1990; Kira *et al.*, 1991). In addition, the percentage of polyclonally infected cells in HAM/TSP patients is approximately 3-fold higher than in asymptomatic carriers (Yoshida *et al.*, 1989). One study has revealed an ongoing state of lymphocyte activation in HAM/TSP patients, reflected by a large number of circulating mononuclear cells able to spontaneously proliferate *in vitro* (Jacobson *et al.*, 1988). Also, an increase in the level of antibodies to HTLV-I in the serum and the CSF of patients with HAM/TSP has been observed (Gessain *et al.*, 1988; Dalgleish *et al.*, 1988). It has been proposed that immune activation of T lymphocytes in HAM/TSP may represent an autocrine phase induced by viral *trans*-activation and coordinate expression of interleukin-2 (IL-2) and the viral *tax/rex* gene products in circulating T cells (Tendler *et al.*, 1990). Thus, an increase in virus load and viral antibody levels may exacerbate these symptoms by disrupting immune regulation and triggering the onset of autoimmunity.

3. Other Disease Associations

HTLV-I has also been linked to and may be involved in the etiology of mycosis fungoides (Hall *et al.*, 1991; Whittaker and Luzzatto, 1993) and Sezary syndrome, another group of cutaneous T-cell lymphomas (CTCL). Clinical features of mycosis fungoides and Sezary syndrome are similar to ATL in that cutaneous involvement may be extensive, but in contrast to ATL, proliferating leukemic cells from CTCL patients are predominantly $CD4^+$ and $CD25^-$ (Manzari *et al.*,

1987). In addition, the disease progresses more slowly than ATL, whereas visceral involvement may not occur for many years following diagnosis. The etiology of mycosis fungoides remains elusive, but serological studies suggest involvement of HTLV-I infection with a small subset of cases (Gallo *et al.*, 1983; Wantzin *et al.*, 1986). Two independent molecular studies revealed the presence of defective HTLV-I proviral DNA in the blood and in cutaneous lesions of 10% of mycosis fungoides patients analyzed (Hall *et al.*, 1991; Whittaker and Luzzatto, 1993). However, a third study was unable to confirm these results (Bazarbachi *et al.*, 1993). One explanation for this apparent contradiction is that mycosis fungoides may have different etiologies that depend on the geographical origin of the patients.

Reports have implicated HTLV-I involvement in multiple sclerosis (MS), a neurologic disorder which bears similarities to HAM/TSP. However, this possibility remains an area of considerable controversy. Two groups have reported the detection of HTLV-I sequences in MS patients using PCR (Greenberg *et al.*, 1989b; Reddy *et al.*, 1989), whereas several other groups have failed to PCR-amplify HTLV-I sequences from MS patient DNA (Bangham *et al.*, 1989; Richardson *et al.*, 1989; Chen *et al.*, 1990).

B. Human T-Cell Leukemia Virus Type II

Human T-cell leukemia virus type II (HTLV-II) is related to HTLV-I, but its association with human malignancy is less substantiated. Only two isolates have been associated with an atypical form of hairy cell leukemia involving T cells (Saxon *et al.*, 1978; Rosenblatt *et al.*, 1986). In one case there was evidence of both T- and B-cell lymphoproliferative disorders (Rosenblatt *et al.*, 1988b). In this patient, HTLV-II was also found to be integrated in a clonal population of cells and the leukemic cells did not express viral RNA. This finding is consistent with the properties of HTLV-I in infected ATL patients.

Additional cases of HTLV-II in association with disease will be required to definitively link HTLV-II infection to a human disorder. If the characteristics of HTLV-II infection and pathogenicity are similar to HTLV-I, such as long latency and the low percentage of infected individuals who develop disease, then the search for disease association with HTLV-II infection will be difficult. HTLV-I and -II are approximately 70% homologous at the nucleotide sequence level. Thus, it has been difficult distinguishing between the two viruses by serological methods because of cross-reactivity. By PCR analysis, several groups have reported a relatively high ratio of HTLV-II infection in a population of intravenous drug abusers in New Orleans (Lee *et al.*, 1989), as well as Italy (Zella *et al.*, 1990) and Argentina (Bouzas *et al.*, 1991). Also, populations of native Americans in New Mexico and Florida are endemic for HTLV-II infection (Hjelle *et al.*, 1990; Hjelle *et al.*, 1993; Levine *et al.*, 1993). To date, two HTLV-II-infected patients have developed chronic neurodegenerative syndrome similar to HAM (Hjelle *et al.*, 1992). These newly identified HTLV-II-infected populations may provide an opportunity to study the natural history and pathogenicity of this virus in humans.

II. GENETIC STRUCTURE OF HTLV

A. Structural and Enzymatic Proteins

As with other retroviruses, HTLV has an RNA genome present in the virion, which upon entry into the cell is transcribed into a DNA copy by the virally encoded reverse transcriptase enzyme. The viral DNA then integrates into the host cell chromosome, where in this proviral form, it can replicate by directing the synthesis of the viral proteins using the host machinery. The HTLV provirus is similar to proviruses of other type C retroviruses, containing the long terminal repeats (LTR) and the three essential genes *gag, pol,* and *env* (Fig. 1). The LTR contains the *cis*-acting sequences important in the regulation of viral gene expression. The full-length viral mRNA encodes the *gag, pol,* and protease gene products and is also utilized as the genomic RNA packaged into the virion (Table I). The *gag* gene encodes the virion core proteins initially synthesized as a precursor, which is subsequently cleaved to form the mature matrix (MA, p19), capsid (CA, p24), and nucleocapsid (NC, p15) (Copeland *et al.*, 1983; Hattori *et al.*, 1984; Oroszlan *et al.*, 1984). The viral protease open reading frame overlaps the 3′ end of the *gag* gene and the 5′ portion of the *pol* gene. The viral protease cleaves the Gag precursor to generate the mature viral core proteins. The *pol* gene encodes several enzymatic activities, which include the reverse transcriptase, integrase, and RNase H. A 4.3-kb subgenomic mRNA, in which one intron is removed by splicing (singly spliced), encodes the *env* gene products. Env is also synthesized as a precursor, which is cleaved to form the surface glycoprotein (SU, gp46) and the transmembrane protein (TM, p21) (Table I).

B. Regulatory Proteins: Tax and Rex

Both HTLV-I and -II contain novel sequences at the 3′ end of the genome which distinguish HTLV from most other replication-competent viruses (Fig. 1). This region, initially referred to as X or pX, encodes the two *trans*-regulatory proteins Tax and Rex (Seiki *et al.*, 1982; Shimotohno *et al.*, 1985) made from a 2.1-kb subgenomic doubly spliced mRNA via translation in separate but overlapping reading frames. The Tax and Rex proteins are essential for viral replication (Chen *et al.*, 1985; Seiki *et al.*, 1988) (Table I). Similar open reading frames that encode regulatory genes analogous to *tax* and *rex* have also been described for bovine leukemia virus (Sagata *et al.*, 1984) (see Chapter 2) and simian T-cell leukemia virus type I (Watanabe *et al.*, 1986) (see Chapter 4). In addition, the human immunodeficiency viruses types I and II (HIV-1 and HIV-2), which cause acquired immunodeficiency syndrome (AIDS) in humans, encode functionally analogous genes, *tat* (Sodroski *et al.*, 1985b; Dayton *et al.*, 1986) and *rev* (Feinberg *et al.*, 1986; Sodroski *et al.*, 1986).

1. Tax

HTLV-I and -II tax genes encode nuclear proteins of 40 and 37 kD, respectively. Tax functionally *trans*-activates expression from the HTLV LTR to in-

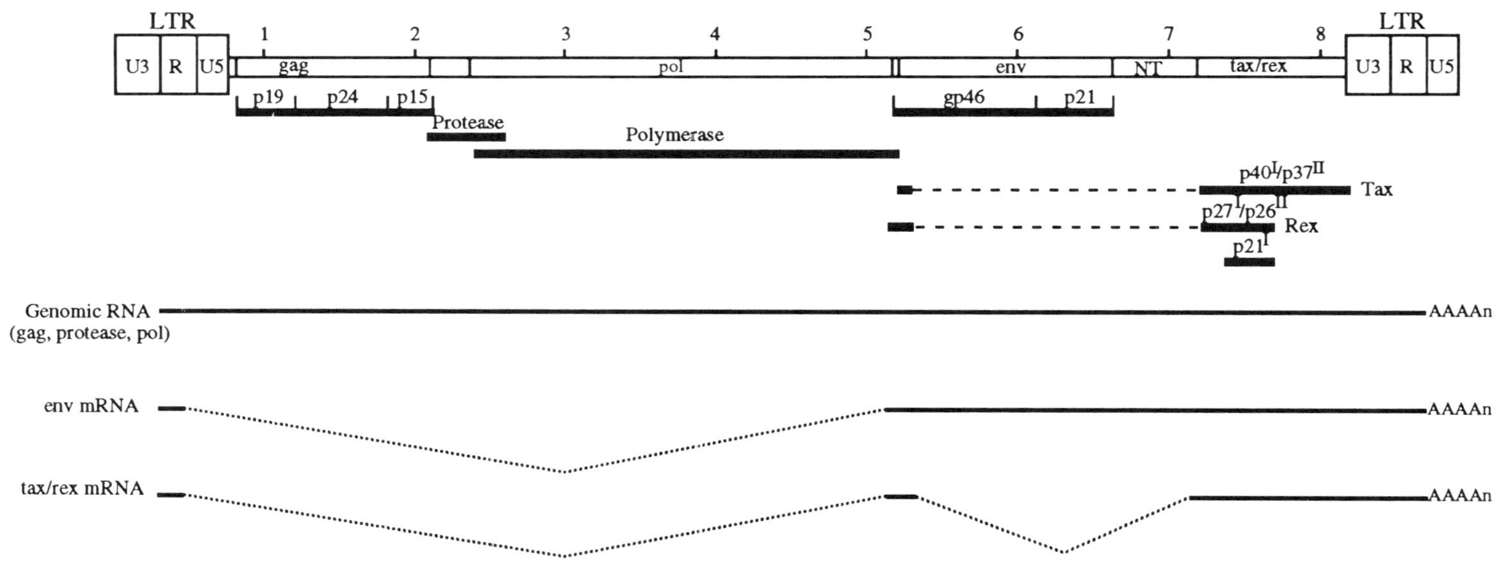

FIGURE 1. Genome organization of the HTLV-I and HTLV-II (Seiki *et al.*, 1983; Shimotohno *et al.*, 1985). The structure and organization of the HTLV provirus genome, including the long terminal repeats (LTRs) and genes encoding the known proteins are shown at the top of the figure. Numbers above the genome depict length in kilobases. Sizes and positions of the individual proteins encoded by the provirus are shown below the genome. Sizes of HTLV-I and HTLV-II Tax and Rex protein products are indicated by superscript I and II, respectively. The structure of the three species of mRNA produced are shown at the bottom of the figure.

TABLE I. Gene Products of HTLV-I and -II[a]

Gene	Proteins (kDa)	Definition	Function
gag			
HTLV-I & -II	p19	Matrix protein (MA)	Structural-virion assembly
	p24	Capsid protein (CA)	Structural-virion assembly
	p15	Nucleocapsid protein (NC)	Structural-virion assembly
pro			
HTLV-I & -II	p14?	Protease (PR)	Processing mature Gag
pol			
HTLV-I & -II	p95?	Polymerase and other enzymatic proteins (RT)	Reverse transcription, integration, and RNase H activity
env			
HTLV-I & -II	gp46	Surface glycoprotein (SU)	Receptor recognition and binding
	p21	Transmembrane protein (TM)	Envelope anchor and membrane fusion
tax			
HTLV-I	p40	*trans*-activator protein	Transcriptional activation of the viral promoter (LTR)
HTLV-II	p37		
rex			
HTLV-I	p27	Regulator of expression	Posttranscriptional; positive regulator of *gag/pol* and *env* mRNA
HTLV-II	p26		
HTLV-I	p21	—	—

[a] Nomenclature is from Leis *et al.* (1988). Data for HTLV-I from Seiki *et al.* (1983); data for HTLV-II from Shimotohno *et al.* (1985). ?, protease and polymerase protein size is based on amino acid sequence derived from nucleotide sequence data.

crease viral RNA transcription (Lee *et al.*, 1984; Slamon *et al.*, 1984, 1985; Cann *et al.*, 1985; Felber *et al.*, 1985; Fujisawa *et al.*, 1985; Sodroski *et al.*, 1985a). The *cis*-acting sequences that Tax responds to are contained within the unique 3′ (U3) region of the LTR (Fig. 2), about 75–225 nucleotides upstream of the transcriptional start site (Brady *et al.*, 1987). Tax is a nuclear protein that does not directly bind DNA as determined by gel shift and footprinting assays. Thus, its mechanism of action is indirect, presumably through interaction and/or modification of cellular transcription factors.

Viral transcriptional regulatory proteins have historically been linked to perturbation of cell processes. Many viral and cellular oncogenes have been described to act *in trans* and aberrantly alter transcription at cellular loci. Proteins with functions similar to Tax have been described in several DNA tumor viruses such as papovaviruses (SV40 large T antigen), herpesviruses (immediate early proteins), adenoviruses (E1A protein), and papillomaviruses. These proteins modulate viral transcription and direct their replication cycle into early and late stages of protein synthesis. In addition, several of these *trans*-activating proteins play a crucial role in cellular transformation. For example, the E1A proteins of adenovirus are necessary and, under certain conditions, sufficient to induce cellular transformation (Ruley, 1983). Interestingly, Tax also *trans*-activates expression

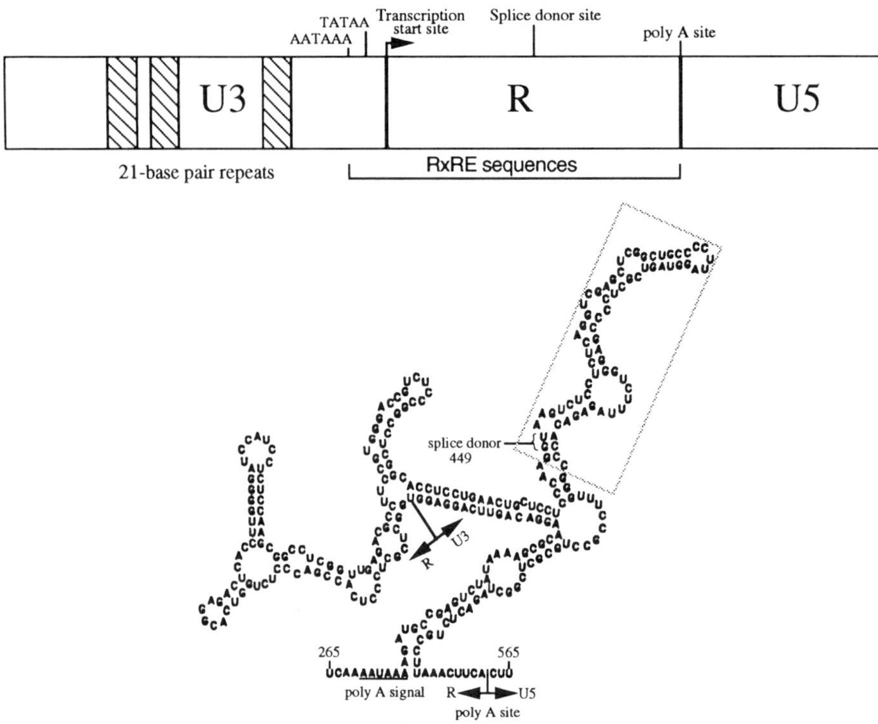

FIGURE 2. Structure of the HTLV long terminal repeat (LTR). The size of the proviral LTR is 755 and 763 nucleotides (nt) for HTLV-I and HTLV-II, respectively; the sizes of U3, R, and U5 are 352 and 314 nt, 227 and 249 nt, and 176 and 200 nt for HTLV-I and HTLV-II, respectively (Seiki *et al.*, 1983; Shimotohno *et al.*, 1985). Sequences in the LTR unique to HTLV include the 21-base pair repeats or Tax response element (TRE) and the Rex-response element (RxRE). The RNA folding pattern of the subregion of the HTLV-II LTR shown to confer Rex responsiveness is shown below. This stem loop structure was identified using VAX computer sequence analysis (Black *et al.*, 1991a; Kim *et al.*, 1991). A similar secondary structure has been predicted for this same region of HTLV-I (Hanly *et al.*, 1989; Toyoshima *et al.*, 1990). This secondary structure brings the poly A signal and the poly A site into close proximity. The HTLV-II RxRE has been recently mapped to a smaller subregion depicted by the boxed-in region; this region has been shown to be critical for HTLV-II Rex binding *in vitro* and transient Rex function *in vivo* (Black *et al.*, 1991b).

of cellular gene products which may account for one mechanism of transformation by HTLV. Tax *trans*-activation will be discussed in more detail below.

2. Rex

Unlike Tax, the Rex regulatory protein does not directly regulate HTLV transcription, but acts at a posttranscriptional level by selectively augmenting the cytoplasmic expression of genome-length and singly spliced *env* mRNA (Inoue *et al.*, 1986a, 1987; Hikada *et al.*, 1988; Ohta *et al.*, 1988; Rosenblatt *et al.*, 1988a; Seiki *et al.*, 1988; Hanly *et al.*, 1989; Ahmed *et al.*, 1990; Kim *et al.*, 1991). Translation of these transcripts into structural and enzymatic proteins is required for virion assembly and formation of infectious particles. As a result of the positive regulation of the *gag/pol* and *env* mRNAs, Rex indirectly inhibits the expression of doubly spliced mRNA which encodes both Tax and Rex. It has been hypothesized that Rex plays an important role in the transition from the early to the late or productive phase of HTLV replication. Furthermore, the modulation of Rex production or function may also determine whether a productively infected cell becomes latent.

HTLV-I Rex is a 27-kDa nuclear phosphoprotein which is concentrated in the nucleoli (Siomi *et al.*, 1988). A stretch of basic amino acid residues at the NH_2-terminus is responsible for this subcellular localization (Siomi *et al.*, 1988; Nosaka *et al.*, 1989). The Rex nuclear localization signal is also able to direct heterologous proteins to the nucleolus. A truncated HTLV-I Rex protein of 21 kDa is translated from the same open reading frame and this protein is thought to initiate at an internal start codon of the *tax/rex* doubly spliced mRNA (Kiyokawa *et al.*, 1985). Several groups have recently identified, using RNA PCR, a distinct RNA species with the potential to encode the p21X protein (Berneman *et al.*, 1992; Orita *et al.*, 1991; Ciminale *et al.*, 1992). Further analysis has suggested that this mRNA is expressed primarily from deleted proviruses (Bhat *et al.*, 1993; Orita *et al.*, 1993a). The function of p21 has not been determined; one study has indicated that it is not required for the regulation of the incompletely spliced mRNAs (Zazopoulos *et al.*, 1990).

Two Rex proteins of 26 and 24 kDa have been detected and characterized in HTLV-II-infected cells (Shima *et al.*, 1986; Rosenblatt *et al.*, 1988a; Green *et al.*, 1990). Structural analysis has indicated that p26 and p24 Rex have indistinguishable amino acid sequences and that the apparent size difference is the result of posttranslational modification by phosphorylation (Green *et al.*, 1991). A truncated form of Rex similar to HTLV-I p21 has not been described for HTLV-II. However, a mRNA capable of encoding an analogous protein has been detected in HTLV-II-infected cells (Orita *et al.*, 1993b).

The function of Rex depends on a *cis*-acting sequence termed the Rex response element (RxRE). The HTLV-I RxRE has been mapped to the U3/R region of the 3′ LTR (Seiki *et al.*, 1988; Hanly *et al.*, 1989; Ahmed *et al.*, 1990). Deletion mutational analysis indicated that the RxRE is located between nucleotides 302 and 560 of the 3′ LTR, and must be maintained in the RNA molecule in the sense orientation (Seiki *et al.*, 1988). The location of the RxRE in the 3′ LTR does not explain how Rex differentially regulates viral mRNAs since all viral messages contain this region. The HTLV-II RxRE has been mapped to the 5′ (Ohta *et al.*, 1988; Rosenblatt *et al.*, 1988a; Black *et al.*, 1991a) or 3′ LTR (Kim *et al.*, 1991) (Fig. 2). One group has specifically mapped the RxRE to the R region of the 5′ LTR containing sequences downstream of the splice donor between nucleotides 465 and 520 (Fig. 2) (Rosenblatt *et al.*, 1988a; Black *et al.*, 1991a). Following transcription and RNA processing, this sequence would be found only in the *gag/pol* mRNA and not in the spliced *env* mRNA, and thus would not account for the Rex-dependent increase in *env* mRNA in the cytoplasm. Thus, similar to HTLV-I, other RxRE elements appear to be required for HTLV-II Rex to differentially regulate the incompletely spliced mRNAs.

It has been proposed that the RxRE of both HTLV-I and -II can form a stable stem loop RNA secondary structure (Hanly *et al.*, 1989; Ahmed *et al.*, 1990; Toyoshima *et al.*, 1990; Ballaun *et al.*, 1991; Bogerd *et al.*, 1991) (Fig. 2). In addition, the RxRE has been shown to play a critical role in the polyadenylation of HTLV-I transcripts (Ahmed *et al.*, 1991). The folding of the RxRE may serve to juxtapose the AAUAAA hexamer and the GU-rich elements essential for polyadenylation. It remains to be determined whether the location of RxRE elements of HTLV-I and -II are truly distinct or if the difference resides in the experimental approaches used to investigate the location of the elements.

Biochemical characterization of both HTLV-I and -II Rex activity has indicated that purified Rex directly binds RxRE sequences *in vitro* in both gel mobility shift and immunobinding assays (Ballaun *et al.*, 1991; Black *et al.*, 1991b; Bogerd *et al.*, 1991; Grassmann *et al.*, 1991; Yip *et al.*, 1991). Rex binding to the RxRE was shown to directly correlate with function *in vivo*. Recent evidence has suggested that phosphorylation of Rex positively regulates the binding to viral mRNA molecules containing the RxRE *in vitro*, and may play a role in regulating viral mRNA expression (Green *et al.*, 1992). There have been several studies describing HTLV-I *rex* gene mutations and their phenotypes (Rimsky *et al.*, 1989; Bohnlein *et al.*, 1991; Hofer *et al.*, 1991). Mutational analysis of HTLV-I Rex has identified two classes of mutations lacking biological activity (Rimsky *et al.*, 1989). One class has been described as a recessive-negative and mutant proteins of this class are not appropriately localized to the nucleus. With one exception, all of the Rex recessive-negative mutants are within the nuclear localization signal. The second class of mutants which are dispersed throughout the coding sequence correspond to *trans*-dominant repressors which inhibit the function of the wild-type Rex protein. The Rex *trans*-dominant repressor mutants are capable of binding the RxRE with high affinity *in vitro*, providing supportive evidence of a binding domain in addition to an activation or effector domain (Bogerd *et al.*, 1991; Hope *et al.*, 1991). One report has provided evidence that a cluster of these Rex *trans*-dominant mutants between amino acids 54 and 69 may encompass a domain important for Rex oligomerization (Weichselbraun *et al.*, 1992; Bogerd and Green, 1993).

There is evidence that HTLV-II Rex may have a second function, in addition to regulating the expression of the incompletely spliced mRNAs. At high levels of Rex, a downregulation or inhibition of Tax-mediated *trans*-activation is observed (Rosenblatt *et al.*, 1988a). Analysis of cap site-initiated mRNA reveals a reduction in the level of total mRNA, indicating that this effect is at the level of transcription or RNA stability. This finding suggests that a threshold level of Rex is required for optimal induction of transcription. The Rex transcriptional effect is Tax-dependent and is mediated via *cis*-acting sequences in U3 of the 5′LTR and most likely involves interaction of Rex with Tax and/or other cellular transcription factors.

3. Regulation of Gene Expression

It is apparent that Tax and Rex play a critical role in regulating the infection cycle of HTLV. Moreover, this regulation may be crucial in the ability of HTLV to establish chronic and latent infection in man. Based on our current knowledge of the regulation of the HTLV infection cycle, the following scenario of HTLV gene regulation could be envisioned (Fig. 3). Immediately after infection of the target cell and formation of the provirus, initial transcription is controlled by cellular factors. Initially, all of the viral mRNA is spliced into the doubly spliced

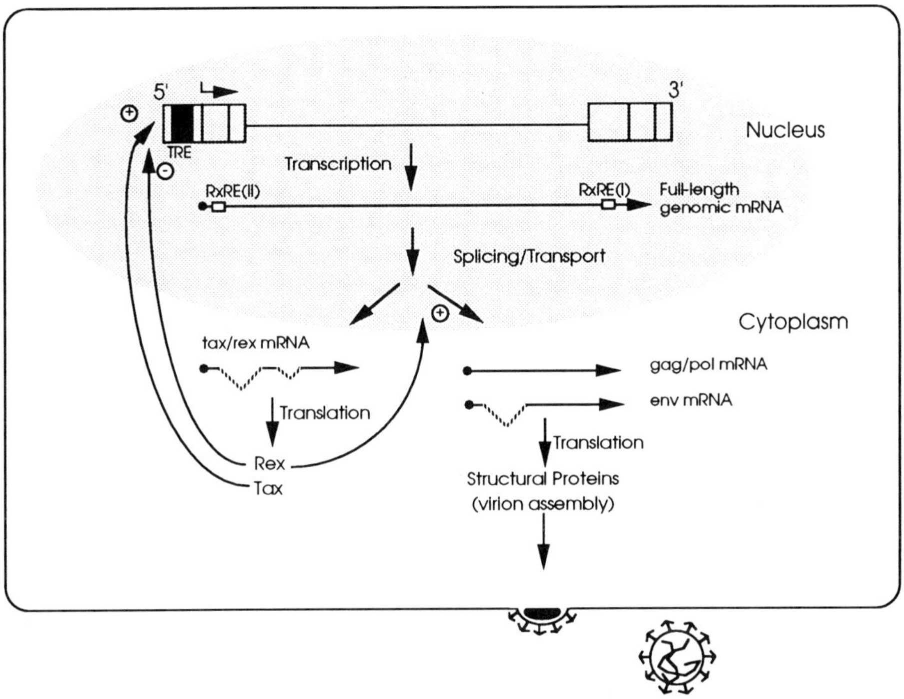

FIGURE 3. Regulation of HTLV gene expression. Initial transcription of the provirus is mediated by cellular transcription factors in which genomic length mRNA is transcribed. This RNA is differentially spliced to generate mRNAs for the *trans*-regulatory proteins (Tax and Rex), and the structural and enzymatic proteins (Gag, Env, and Pol). The Tax protein acts via the Tax-response element (TRE) to increase transcription from the viral LTR. The Rex protein increases the level of cytoplasmic *gag/pol* mRNA and *env* mRNA (+) at the expense of *tax/rex* mRNA. This regulation is post-transcriptional (splicing/transport) and requires the Rex-response element (RxRE) in the RNA. Rex also negatively regulates Tax-mediated *trans*-activation in a dose dependent manner (−) which could be important in viral latency. TRE, 21-base pair repeats or Tax response element; RxRE(I) and RxRE(II), Rex response element for HTLV-I and HTLV-II, respectively.

tax/rex mRNA. *tax* translation is favored over *rex* because of sequences surrounding their respective methionine initiation codons. Thus, initially Tax and negligible or low levels of Rex are produced which potentiate Tax-mediated *trans*-activation. Expression of Rex appears to reach a critical threshold, which results in the production of high levels of *gag/pol* and *env* mRNA at the expense of *tax/rex* mRNA. Thus, the virus is entering the late phase of replication in which Gag, Pol, and Env proteins are produced, resulting in virion assembly and release. Overabundance of Rex ultimately could result in an autocrine repression of viral gene transcription. This may occur mechanistically, either by transport of incompletely spliced RNAs to the cytoplasm and/or by inhibition of Tax-mediated *trans*-activation of virus gene expression. Ultimately, the intracellular levels of Tax and Rex could be reduced, resulting in little or no viral gene transcription. This feedback control of viral gene product expression may be critical to the establishment of and release from a latent state of infection and may allow evasion from immunosurveillance in the host. The role of this complex regulation in the development of disease is not clear.

III. HTLV TRANSFORMATION

A. *In Vitro* Transformation

The growth potential of normal human peripheral blood T lymphocytes plated in culture is limited. Primary T-cell survival time in culture is approximately 3 to 4 weeks, although the life span may be extended by the introduction of exogenous lymphokines into the culture. An increased understanding of the pathogenesis of HTLV has been gained by studying the properties of the virus *in vitro*. Both HTLV-I and -II will transform normal human T cells *in vitro* (Chosa *et al.*, 1982; Yamamoto *et al.*, 1982; Chen *et al.*, 1983; Miyoshi *et al.*, 1983a; Popovic *et al.*, 1983). Transformation is defined as the ability of T cells infected with HTLV to proliferate indefinitely in the absence of exogenous interleukin-2 (IL-2). HTLV is poorly infectious, and efficient *in vitro* infection is successful only by cocultivation of target cells with X-irradiated virus producer cells; cell-free HTLV is generally noninfectious. The precise mechanism for T-lymphocyte transformation has not been determined.

The HTLV cell surface receptor has not been identified, although it has been localized to human chromosome 17 (Summerfelt *et al.*, 1988) (see Weiss, 1993). Interestingly, the HTLV cellular receptor is present on a wide range of human and animal cells, as determined by the ability of HTLV to infect and/or transform a variety of animal T lymphocytes. In addition to T cells, HTLV infection has been demonstrated in human cells of various origins including B cells (Yamamoto *et al.*, 1982; Koyanagi *et al.*, 1984), macrophages (Koyangi *et al.*, 1993), immature bone marrow cells (Markham *et al.*, 1984), osteosarcoma cells (Clapham *et al.*, 1983), endothelial cells (Hoshino *et al.*, 1983; Weiss *et al.*, 1985a), and fibroblasts (Yoshikura *et al.*, 1984). HTLV-I and -II will infect and immortalize T lymphocytes from monkeys (Miyoshi *et al.*, 1982), rabbits (Miyoshi *et al.*, 1983b), rats (Tateno *et al.*, 1984), and cats (Hoshino *et al.*, 1984). Mouse cells appear to lack the HTLV receptor and are not susceptible to infection (Summerfelt *et al.*, 1988).

Coculture of HTLV producer cells with peripheral blood or cord blood T cells initially results in a polyclonal expansion of infected T-cell subpopulations approximately 2 to 4 weeks after coculture. HTLV-transformed cells have increased endogenous expression of IL-2 as well as the IL-2 receptor (IL-2R). This increase in IL-2 and IL-2R expression results in autocrine stimulation of infected cells and has been hypothesized as an important event in T-cell transformation by HTLV. Further passage of infected cell cultures results in a selective outgrowth of one or a few dominant cell clones. The majority of transformed T cells are of the $CD4^+$ phenotype, although $CD8^+$ cells can also be transformed (Harper *et al.*, 1986).

Although *in vitro* transformation by HTLV provides a valuable model system for studying the mechanism of leukemogenesis by HTLV, cells transformed *in vitro* are distinct from those isolated from ATL patients. These differences are exemplified by analyzing dominant clones arising from sustaining HTLV-transformed cells in culture. Dominant clones that arise from the polyclonal population of infected cells *in vitro* clearly are the result of selection pressures in culture. These selection pressures may not be representative of selection *in vivo*, which results in the monoclonal population of leukemic cells found in ATL patients. This point is most clearly demonstrated by comparing *in vitro* transformed cells and ATL tumor cells. Virus expression is detectable in cells transformed *in vitro* and tumor cells that have been adapted to grow in culture, whereas extremely low viral RNA expression is detected in circulating leukemic cells of ATL patients (Yoshida *et al.*, 1982; Seiki *et al.*, 1984). In addition, culturing ATL cells *in vitro* results in the expansion of clonal populations of cells which are generally distinct from the clones of cells that predominate in patients. Comparison of the properties of *in vitro* transformed cells and tumor cells illustrates that the process of *in vitro* transformation by HTLV does not directly parallel tumor induction *in vivo.*

It has been proposed that virus-induced cellular transformation is simply another means of HTLV replication (Chen *et al.*, 1989). As stated above, the primary mode of HTLV infection is through cell-to-cell contact and not through cell-free infection. Transformation and abnormal proliferation of T cells carrying an HTLV proviral genome would clearly facilitate virus spread and would be advantageous for survival. Under these circumstances, virus replication (provirus replication) would be achieved during host cell division rather than by productive infection followed by the recruitment of newly infected cell clones.

B. *In Vivo* Transformation

Very little is known about the cellular progression events *in vivo* which result in HTLV-induced disease. Infection is characterized by long clinical latency during which virus expression appears to be silent in most infected cells. The genome of HTLV lacks sequences homologous to a known oncogene and does not integrate at specific sites within the host genome (Seiki *et al.*, 1982; Weiss *et al.*, 1984, 1985b; Shimotohno *et al.*, 1985). Thus, HTLV appears to transform cells, and ultimately induce disease by an unknown novel mechanism that is distinct from other oncogenic retroviruses of its class. Although direct evidence is lacking for the events leading to the manifestation of HTLV-associated disease, it most likely involves an initial polyclonal phase followed by progression to monoclonal malignancy. The molecular mechanisms by which HTLV transforms T lymphocytes *in vitro* may reflect successful infection and the early polyclonal proliferation phase that occurs in humans. Evidence supporting this possibility has been provided by a recent report on the investigation of infection of the related bovine leukemia virus (BLV) in sheep. It was demonstrated that viral gene expression proceeds through early and late stages in lymphocytes taken from infected animals (Powers and Radke, 1992). Karyotypic abnormalities have been detected in ATL cells and often have been indicative of the degree of severity of disease (Fukuhara *et al.*, 1983; Sanada *et al.*, 1985), suggesting that somatic mutations may play a critical role in the development of the tumorigenic phenotype. The lack of a good animal model system has prevented investigation of events required for HTLV-induced leukemogenesis. HTLV infection has been transmitted to both rabbits and rats; however, no lymphoproliferative disorders have yet been described (Tateno *et al.*, 1984; Uemura *et al.*, 1987). Successful studies investigating HIV infection of mice with severe combined immunodeficiency (SCID) reconstituted with human lymphoid tissues suggest that this animal model may also be an appropriate *in vivo* system to investigate HTLV pathogenesis (Namikawa *et al.*, 1988; Feuer *et al.*, 1993).

IV. THE ROLE OF TAX IN HTLV TRANSFORMATION

A. Tax *Trans*-Activation of Cellular Genes

Much investigation is focused on the mechanism of HTLV transformation. It has been proposed that Tax, the *trans*-activator protein, may be responsible for initiating events leading to cellular transformation. As described above, the sequences responsive to Tax reside in the U3 region of the LTR. This region contains a transcriptional enhancer consisting of three copies of an imperfectly conserved 21-base-pair (bp) repeat (Josephs *et al.*, 1984; Fujisawa *et al.*, 1986; Shimotohno *et al.*, 1986; Kitado *et al.*, 1987; Rosen *et al.*, 1987) (Fig. 2). Deletion analysis of the HTLV-I or -II LTRs linked to a reporter gene demonstrated that two or more copies of the 21-bp repeat in either orientation are sufficient for *trans*-activation mediated by Tax (Paskalis *et al.*, 1986; Shimotohno *et al.*, 1986; Brady *et al.*, 1987). Also, *trans*-activation of the viral LTR by Tax is not restricted to cells of the T-cell lineage, and is observed in many cell types (Sodroski *et al.*, 1985a; Fujisawa *et al.*, 1986; Shah *et al.*, 1986).

The three 21-bp repeat sequences contain a TGACG core sequence that is also found in the E1A-activated early genes of adenovirus (Lee *et al.*, 1987) and in the cyclic AMP (cAMP) response element (Montminy *et al.*, 1986). The 21-bp repeats of HTLV can direct cAMP- and TPA (12-*O*-tetradecanoyl-phorbol-13-acetate)-dependent transcription (Jeang *et al.*, 1988a; Fujisawa *et al.*, 1989; Giam and Xu, 1989; Poteat *et al.*, 1989; Tan *et al.*, 1989a). Interestingly, the core cAMP responsive consensus sequence in itself is not Tax responsive (Fujisawa *et al.*, 1989). Moreover, HTLV LTR-linked gene expression is superinduced by cAMP in the presence of Tax (Poteat *et al.*, 1989). Thus, it appears that a similar sequence is recognized by transcription factors playing a role in Tax *trans*-activation and cAMP response, but the two pathways of induction are independent. The significance of HTLV LTR activation in infected T cells *in vivo* by cAMP remains to be established.

The mechanism of Tax-mediated *trans*-activation is not well understood. Tax is a nuclear protein which does not appear to directly bind DNA nor does its activity require the synthesis of new host cell proteins (Jeang *et al.*, 1988b; Giam and Xu, 1989). Thus, Tax may interact with and/or modify preexisting host cell proteins in a manner that alters its ability to bind DNA or to be recognized by the transcriptional machinery, which then associates with the appropriate regulatory sequences. Alternatively, Tax may indirectly bind DNA via host cell proteins and serve as an adaptor or mediator of the interaction between transcription factors bound at different sites. One experiment has suggested that the localization of Tax at the promoter region is important for its activity. A GAL4–Tax fusion protein but not Tax alone, *trans*-activated a construct containing GAL4 binding sites (Fujisawa *et al.*, 1991). Also, Tax has been shown to bind indirectly via host cell factors to sequences between the second and third 21-bp repeat (Marriott *et al.*, 1989). However, a comparison of nuclear extracts from HTLV-I-transformed T cells which do or do not express functional Tax indicates that none of the observed features of the DNase I footprint pattern correlate directly with the presence of Tax (Altman *et al.*, 1988; Nyborg *et al.*, 1988).

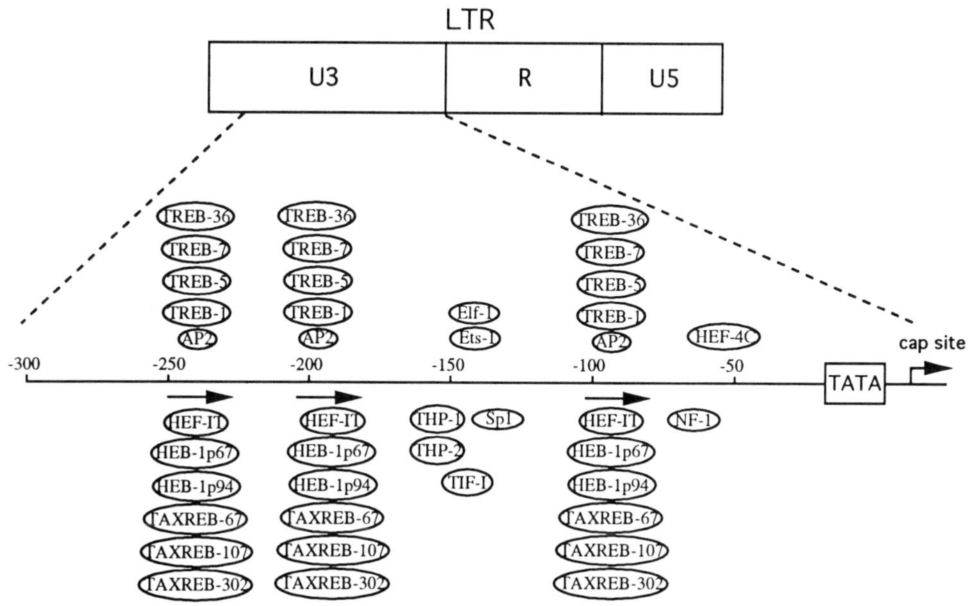

FIGURE 4. Summary of host cell protein binding sites within the HTLV-I LTR. The factors which have been identified and their approximate binding sites are schematically indicated. HEF-IT, Sp1, NF-1, and HEF-4C were purified from the HTLV-I transformed T-cell line, SLB-1. TREB-1, TIF-1, AP2, Sp1, NF-1, HEF-4C, HEB-1p67, and HEB-1p94 were purified or cloned from the transformed human fibroblast cell line, HeLA. TAXREB67, TAXREB107, and TAXREB302 were cloned from the human T-cell line, Jurkat. TREB-5, TREB-7, and TREB-36 THP-1, and THP-2 were cloned from the HTLV-I transformed T-cell line, HUT-102. The position of the 21-bp repeats, the TATA box, and the cap site (designated +1) are indicated.

1. Interaction with Cellular Proteins

Numerous cellular proteins that bind to the 21-bp repeat Tax-response region have been identified using various techniques (Fig. 4). Proteins identified using DNA affinity chromatography include a group of polypeptides ranging from 35 to 43 kDa purified from HeLa cell extracts, termed Tax response element binding protein 1 (TREB-1) (Tan *et al.*, 1989b), and a group of polypeptides of 41, 43, and 59 kDa [host expression factor IT (HEF-IT)] purified from lymphocyte extracts (Nyborg and Dynan, 1990). A combination of techniques which include DNA affinity chromatography, glycerol gradient sedimentation, and UV cross-linking have identified lymphocyte polypeptides of 32, 36, 42, 50, and 110 kDa which bind the 21-bp repeats (Beimling and Moelling, 1990). Two cellular proteins, HTLV enhancer binding (HEB)-1p67 and HEB-1p94, have been purified by DNA affinity chromatography (Lombar-Platet and Jalinot, 1993). Also, six proteins with molecular sizes of 29 (TREB5), 55 (TREB7), 29 (TREB36), 22 (TaxREB302), 34 (TaxREB107), and 52 (TAXREB67) kDa were identified by screening cDNA libraries for expression of proteins that bind the 21-bp repeat oligonucleotide (Yoshimura *et al.*, 1990; Tsujimoto *et al.*, 1991; Morita *et al.*, 1993; Nyunoya *et al.*, 1993). It is possible that several of these polypeptides are identical. Other proteins, including AP-2, SP1, nuclear factor 1 (NF-1), HEF-4C,

Tax interaction factor-1 (TIF-1), TaxRE-2 holding protein (THP)-1 and THP-2, and products of the proto-oncogene *ets* family (Ets-1 and Elf-1) bind to the HTLV-I promoter at sequences outside the 21-bp repeats (Clark *et al.*, 1993; Gitlin *et al.*, 1991; Marriott *et al.*, 1990; Nyborg *et al.*, 1990; Tanimura *et al.*, 1993) (Figure 4). Tax-dependent *in vitro* transcription systems have been developed for dissecting the biochemical events orchestrated by Tax (Zhao and Giam, 1991; Matthews *et al.*, 1992). These studies have shown that *trans*-activation by Tax requires direct interaction with cellular factors which specifically bind the 21-bp repeats. More recently, evidence indicates that Tax transactivation is mediated through enhanced binding of basic-region leucine zipper (b-ZIP) DNA binding proteins (activating transcription factor [ATF]-1, ATF-2, and cAMP response element binding protein [CREB] to the 21-bp repeats (Franklin *et al.*, 1993; Wagner and Green, 1993). This increased DNA binding may result from Tax enhancing dimerization of these b-ZIP proteins, and thus facilitates the DNA binding reaction.

After the discovery that Tax positively regulated HTLV gene expression, emphasis has turned to studying Tax regulation of T-cell genes, particularly those genes involved in cellular proliferation. Early indications that Tax might alter cellular gene expression resulted from transient and stable transfections of a Tax-expression plasmid in Jurkat T cells (Inoue *et al.*, 1986b; Wano *et al.*, 1988). Tax was found to *trans*-activate the expression of a particular set of endogenous genes, which are normally only transiently induced by antigenic or mitogenic stimuli in T lymphocytes. These include IL-2 (Maruyama *et al.*, 1987; Siekevitz *et al.*, 1987a; Ruben *et al.*, 1988), IL-2R (Inoue *et al.*, 1986b; Cross *et al.*, 1987; Siekevitz *et al.*, 1987a; Leung and Nabel, 1988), and granulocyte–macrophage colony-stimulating factor (GM-CSF) (Wano *et al.*, 1988; Nimer *et al.*, 1989). The induction of IL-2, IL-2R, and GM-CSF occurs only in certain T-cell lines, suggesting that a T-cell-specific factor(s) or a specific stage of differentiation is required. The expression of these genes may have significant implications since ATL tumor cells and established HTLV-infected cell lines constitutively express IL-2R on their surface (Hattori *et al.*, 1981; Depper *et al.*, 1984), and some cell lines also produce IL-2. Also, HTLV-I and -II-infected and transformed cell lines constitutively produce GM-CSF (Chen *et al.*, 1983; Chan *et al.*, 1986). Sincc ATL cells do not express substantial amounts of viral proteins, including Tax, Tax could be responsible for initiating molecular changes that lead to these manifestations of HTLV infection, but may not be required for maintaining the leukemic phenotype.

2. Relation to Transformation

The majority of T lymphocytes within a host are quiescent and only undergo cell division in response to antigenic stimulation. Also, the majority of stimulated lymphocytes have a limited life span. In contrast, T cells infected with HTLV can be induced to proliferate indefinitely. Of particular importance is the activation of the IL-2 and IL-2R gene which could result in autocrine stimulation of proliferation of T cells. One study investigating HTLV transformation has suggested that induction of IL-2 and IL-2R expression is involved in the initial phase of polyclonal expansion of transformed lymphocytes (Kimata and Ratner, 1991). *Trans*-activation of the IL-2R appears to be mediated by the transcription factor NF-κB or NF-κB-like factors (Ballard *et al.*, 1988; Leung and Nabel, 1988; Ruben *et al.*, 1988; Lindholm *et al.*, 1991). The IL-2R promoter region contains a 12-bp element homologous to the NF-κB binding site, previously identified as the κ immunoglobulin gene enhancer (Pierce *et al.*, 1988). This motif is also required for transcriptional activation of IL-2R by mitogens such as phytohemagglutinin (PHA) and phorbol 12-myristate 13-acetate (PMA) (Bohnlein *et al.*, 1988; Lowenthal *et al.*, 1988; Radonovich and Jeang, 1989). Interestingly, the HIV-1 LTR which is also *trans*-activated by Tax contains two NF-κB binding sites.

One polypeptide, first described as HIV enhancer 86-kDa A (HIVEN86A) and now known to be the proto-oncogene c-*rel*, has been characterized and is apparently directly responsible for this activation (Siekevitz *et al.*, 1987b; Ballard *et al.*, 1988; Pierce *et al.*, 1988; Bohnlein *et al.*, 1989). Thus, Tax-mediated *trans*-activation of the IL-2R gene or the HIV-1 LTR may be the result of Tax inducing synthesis of or modifying HIVEN86A and possibly other factors that normally participate in mitogen activation. The identification of mutant Tax proteins selectively defective for *trans*-activation for the HTLV LTR or NF-κB-responsive promoters suggests that the mechanism by which Tax *trans*-activates the HTLV LTR and NF-κB-responsive promoters is different (Ruben *et al.*, 1989; Smith and Greene, 1991).

In addition to IL-2 and IL-2R, Tax can *trans*-activate expression of GM-CSF which is clearly apparent in HTLV-transformed T-cell lines which show constitutive high-level production of GM-CSF (Wano *et al.*, 1988; Nimer *et al.*, 1989). This would be consistent with the frequent clinical observation of granulocytosis and eosinophilia in ATL patients (Wachsman *et al.*, 1986). The sequences responsive to Tax are apparently unique to the GM-CSF promoter element and are not found in IL-2 and IL-2R promoter or the HTLV LTR (Nimer *et al.*, 1988; Nimer, 1991). Tax is unable to exert its positive effects on GM-CSF expression in non-T-cell lineages or in immature T cells such as CEM. Thus, T-cell-specific factors present only in mature T lymphocytes are required in concert with Tax for activating GM-CSF expression.

HTLV Tax also *trans*-activates the proto-oncogenes c-*sis* (Ratner, 1989) and c-*fos* (Fujii *et al.*, 1988). c-*sis* encodes platelet-derived growth factor β and c-*fos* has been proposed to be involved in cell growth and differentiation. Tax has also been shown to *trans*-activate the major histocompatibility class I antigen (Maruyama *et al.*, 1987; Sawada *et al.*, 1990). Interestingly, Tax has been shown to downregulate expression of the β-polymerase gene, a cellular enzyme involved in host cell DNA repair (Jeang *et al.*, 1990). It has been proposed that Tax may indirectly contribute to the increasing chromosomal damage and the number of oncogenic mutations that accumulate in HTLV-I-infected cells by reducing the intracellular levels of β-polymerase. Tax has also been shown to increase the expression of parathyroid hormone related protein, IL-6, leukemia inhibitory factor, and lymphotoxin (Ejima *et al.*, 1993; Lal *et al.*, 1993; Tschachler *et al.*, 1993). Therefore, the effect of Tax on the expression of the cellular genes listed above may initiate a series of aberrant transcriptional regulation events, ultimately resulting in cellular transformation.

B. Cellular Transformation by Exogenous Tax

Attempts to transform cells *in vitro* with cloned Tax-expression vectors have had limited success, and have provided evidence that Tax can be oncogenic, but not necessarily for T cells. Introduction of Tax into NIH3T3 cells and Rat-1 cells induced growth beyond saturation density, as well as growth in soft agar (Tanaka *et al.*, 1990). A clonal transformed Rat-1 cell line contained a single copy of *tax*, formed large colonies in soft agar, and induced tumors rapidly in nude mice. Revertants of the Rat-1 transformed cell line were isolated that had lost the transformed phenotype; Southern blot analysis indicated loss of the *tax* gene as well (Yamaoka *et al.*, 1992). Reintroduction of the *tax* gene restored the transformed phenotype. Thus, these results suggest that the transformed phenotype is a direct result of Tax and not the consequence of secondary mutations of cellular genes.

To date, the direct transformation of primary blood T lymphocytes by introduction of Tax-expression vectors by transfection or infection has been unsuccessful. Some success has been obtained by infecting human cord blood lymphocytes *in vitro* with a transformation-defective *Herpesvirus saimiri* vector that expressed Tax. *H. saimiri* normally infects New World primate and human T lymphocytes but does not result in cellular transformation (Fleckenstein and Desrosiers, 1982). Infection of mitogen-activated cord blood lymphocytes with *H. saimiri/tax* recombinants resulted in the appearance of immortalized T-cell lines with cell surface phenotypes similar to ATL tumor cells, but these cells remained truly different from HTLV-transformed cells in that they remained dependent on IL-2 for growth (Grassmann *et al.*, 1989, 1992). It should be noted that the presence of large amounts of *Herpesvirus* sequence in the vector complicates the identification of sequences required for transformation. For example, Tax may be *trans*-activating a *Herpesvirus* transformation gene. Recent reports also indicate that soluble extracellular Tax may modulate the expression of immunoregulatory cytokine genes in both HTLV-infected and uninfected primary blood lymphocytes (Marriott *et al.*, 1989; Lindholm *et al.*, 1992). However, this Tax stimulatory effect requires prior stimulation of cells by PHA. Overall, these results suggest that Tax may initiate the transformation process and immortalize lymphocytes, but that additional genetic events are required for the complete transformed phenotype.

C. Tax Transformation *in Vivo*

To date, the demonstration of Tax oncogenicity *in vivo* has been limited. The best evidence stems from the development of transgenic mice expressing Tax from the HTLV LTR. Tumors developed in mice at multiple sites between 13 and 17 weeks of age, but the histological type of these tumors was of mesenchymal origin (neurofibromas), not of lymphoid origin (Nerenberg *et al.*, 1987), the natural target for HTLV infection in humans *in vivo*. Characterization of tumors, as well as established tumor cell lines indicated expression of high levels of Tax, IL-2, IL-2R, and GM-CSF, a reflection of events seen with HTLV-induced transformation of lymphoid cells *in vitro* (Green *et al.*, 1989a). Clinical manifestations in these mice include splenomegaly and peripheral granulocytosis most likely the result of overexpression of GM-CSF, as well as morphological abnormalities of the thymus (Hinrichs *et al.*, 1987; Nerenberg *et al.*, 1987). Additionally, transgenic mice display exocrinopathy involving the salivary and lachrymal glands, with the degree of pathology corresponding to the levels of Tax expression (Green *et al.*, 1989b). The pathology in *tax* transgenic mice resembles Sjorgren's syndrome, a disease reported to be associated with HTLV infection in humans (Gessain *et al.*, 1985; Vernant *et al.*, 1988). It is important to note that some transgenic lines were tumor-free despite efficient expression of *tax*, suggesting once again that *tax* expression may initiate the transformation process and induces additional cellular alterations that are required for the fully transformed phenotype. Evidence supporting this theory has been provided by a study in which transplanted HTLV-I-Tax-transformed tumors in mice can be ablated by antisense oligonucleotide inhibition of NFκB but not *tax* gene expression (Kitajima *et al.*, 1992).

PCR amplification of leukemic cells from patients with ATL indicated that a high percentage of ATL patients carry deleted HTLV-I genomes in their tumor cells. Further analysis has indicated that the pX region containing the *tax/rex* gene and the 3′ LTR are preferentially retained in leukemic cells (Korber *et al.*, 1991). This maintenance of the *tax* gene *in vivo*, although under many circumstances in a latent transcriptional state, is consistent with a role for Tax in contributing to oncogenesis in humans.

V. MITOGENIC STIMULATION OF CELLS BY HTLV

There is evidence that HTLV can affect T-cell proliferation via cell surface stimulation. It has been demonstrated that HTLV-I purified particles are mitogenic for resting T cells obtained from peripheral blood or thymocytes (Gazzolo and Duc Dodon, 1987; Dumontet *et al.*, 1988). This activation was independent of accessory cells and was equally efficient on both $CD4^+$ and $CD8^+$ T cells, and proliferation parallels induction of IL-2 and IL-2R gene expression. Virus-induced proliferation was reported to be mediated by the external envelope glycoprotein gp46, since human sera containing antibodies to HTLV-I, as well as a monoclonal antibody to HTLV-I gp46 neutralize the mitogenic effect (Gazzolo and Duc Dodon, 1987). Furthermore, the fact that virus inactivated by UV light treatment was still mitogenic for T cells suggests that the binding of virions to cell surface receptors is sufficient to induce T-cell activation. A second study providing additional evidence indicated that mitogenic stimulation by noninfectious HTLV-I virions induced the production of large quantities of HIV-1 from human peripheral blood lymphocytes infected with HIV-1 (Zack *et al.*, 1988). These results suggest that the HTLV-I cell surface receptor delivers a mitogenic signal after contact with viral particles or, alternatively, that viral receptors could be associated with membrane antigens involved in T-cell activation. Further investigation of T-cell activation by HTLV-I in relationship to the major known pathways of normal T-cell activation has indicated that HTLV-I interferes with the activation pathway mediated by the CD2 molecule, but not with the pathway initiated by the CD3/T-cell receptor complex (Duc Dodon *et al.*, 1989; Maguer *et al.*, 1993). Another study found that the mitogenic activity was mediated through the CD2/lymphocyte associated antigen 3 activation pathway. However, the mitogenic activity was restricted to HTLV producing T-cells suggesting that HTLV particles alone are not intrinsically mitogenic (Kimata *et al.*, 1993).

The mitogenic activity of HTLV may play a role in initially maintaining viral replication. Retrovirus production is generally blocked in quiescent cells; cells must be induced to traverse through S phase for productive infection to occur (Varmus *et al.*, 1977; Chen and Temin, 1982; Zack *et al.*, 1990). Thus, the HTLV-I cell surface mitogenic effect can be important in establishing infection as well as increasing, by polyclonal proliferation, the number of appropriate target cells. It is conceivable that this mitogenic stimulation is an important step in the immortalization process. There have been several reports demonstrating that envelope glycoproteins from the murine leukemia viruses serve as mitogenic ligands by binding to cell surface receptors, and are major determinants in the induction of leukemia in mice (Oliff *et al.*, 1984; Holland *et al.*, 1985; Li *et al.*, 1990).

VI. MODEL FOR HTLV TRANSFORMATION

A complex series of events follows infection of the cell by HTLV. These events are mediated by the viral particle and the *trans*-activator protein Tax and are incorporated in the following model of HTLV-induced leukemogenesis (Fig. 5). The interaction of HTLV virions with the cell surface receptor via Env directly stimulates T-cell proliferation and facilitates establishment of infection. Thus, viral particles stimulate T-cell division before infection is established. Activated T lymphocytes are more susceptible to HTLV transformation (Merl *et al.*, 1984), and retroviruses require cell proliferation to establish a productive infection (Temin and Kassner, 1974; Varmus *et al.*, 1977; Chen and Temin, 1982). Once productive infection is established, Tax destabilizes the transcriptional regulation of cellular genes including IL-2, IL-2R, GM-CSF, c-*fos*, c-*sis*, and likely many others not yet described. Ultimately, cellular transformation can result via autocrine and/or paracrine mechanisms.

HTLV can be envisioned to lead to continued cell proliferation by an autocrine mechanism in which factors produced by cells can act on themselves to stimulate proliferation. For example, Tax induces expression of IL-2R and weak production of IL-2 resulting in an infected T-cell population constitutively expressing IL-2R on its surface. IL-2 production could be dramatically increased following further cell stimulation by antigen. Thus, IL-2 expression would be synergistically activated, and the IL-2 produced would act back on the entire population of IL-2R-expressing T cells predisposing them to autonomous growth. The induction of cell division may increase the number of secondary mutations that accumulate in the cell genome. Presumably additional genetic changes, such as the development of chromosomal abnormalities characteristic of ATL tumor cells, are required for complete transformation.

Tax-mediated *trans*-activation of cellular genes may lead to aberrant growth by a paracrine mechanism. As discussed above, Tax induces GM-CSF expression. GM-CSF secreted by HTLV-infected T cells can stimulate macrophages and myeloid precursor cells to proliferate and differentiate. Activated macrophages can subsequently stimulate T lymphocytes via release of effector molecules such as IL-1 (Unanue and Allen, 1987). IL-1 increases the production of IL-2 and IL-2R in T cells. The overexpression of lymphokine and the inappropriate expression of

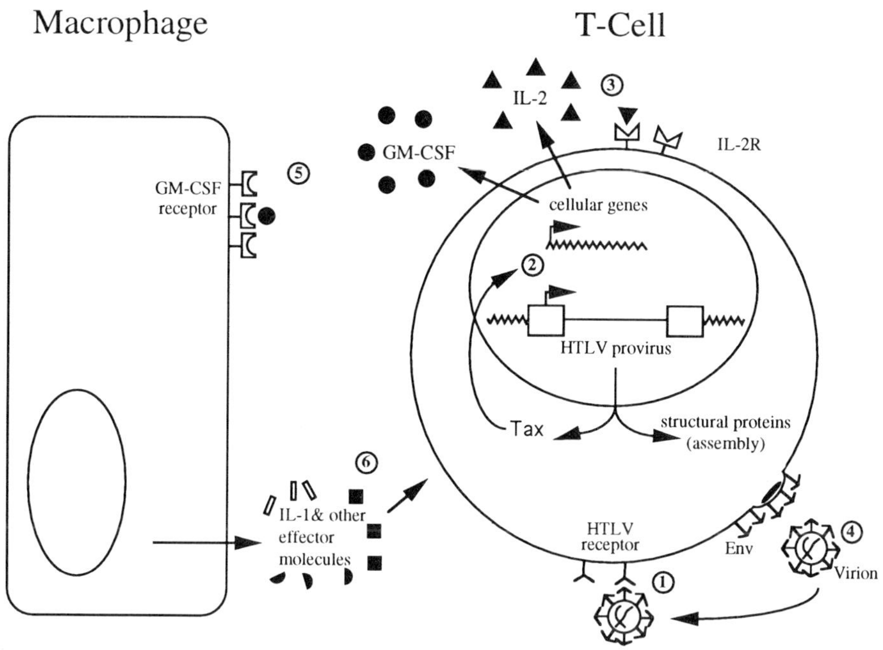

FIGURE 5. A model for HTLV-mediated T-cell transformation: Events that induce mitogenesis or may be involved in transformation are represented. (1) T-cell activation is initially mediated by HTLV virion interaction with an unidentified cellular receptor resulting in mitogenic stimulation and productive infection. (2) Proviral transcription and initially Tax production results in *trans*-activation of cellular genes, including interleukin 2 (IL-2) and IL-2 receptor (IL-2R). (3) IL-2 may have an autocrine effect on T-cell growth via interaction with IL-2R. (4) Production of new virions can induce additional mitogenic effects by acting back on the infected cell. Aberrant modulation by Tax of other cellular genes such as GM-CSF, *c-fos*, and *c-sis* may also be involved in deregulation of T-cell growth. (5) GM-CSF secreted by T-cells may have an effect on T-cell growth by first activating or increasing the number of macrophages that subsequently (6) activate T-cells via release of IL-1 and other effector molecules.

proto-oncogenes in HTLV-infected T cells, which normally occurs only upon antigenic stimulation, may have a direct role in the cascade of events resulting in HTLV tumorigenesis.

This hypothetical model implicates at least two viral gene products in cellular transformation *in vitro*, and the early stages of leukemogenesis *in vivo:* Env interacting with its cellular receptor at the cell surface and Tax acting in the nucleus. It is conceivable that HTLV transformation is a by-product of the provirus trying to replicate itself. HTLV particles mitogenically stimulate T-cell division by interacting with a receptor on the cell surface that may be normally involved in T-cell activation responses. The binding event triggers deregulated expression of IL-2 and IL-2R, gene products which play a crucial role in the orchestrated activation of T cells and the normal immune response. Sustained proliferation can be maintained by Tax *trans*-activation of cellular genes and may not require antigenic stimulation at the cell surface as would normal uninfected T lymphocytes.

Tumorigenesis *in vivo* is more complex and is only partially represented by this model since additional unknown secondary events and/or factors are presumably required to induce the complete malignant phenotype. It is interesting that HTLV transformation is reminiscent of the requirements for transformation of primary rodent fibroblasts *in vitro*. In those experiments, transformation required the coexpression of two oncogenes, one which acts at the cell membrane and one which acts in the nucleus (Land *et al.*, 1983; Ruley, 1983). This observation leads us to the conclusion that multiple hits or events are required for the complete transformed phenotype.

VII. CONCLUSIONS

HTLV-I and perhaps in some cases HTLV-II are associated with various forms of leukemia and neurodegenerative disorders in humans. Attempts to understand the pathogenic mechanism of HTLV in humans have been complicated by the extremely long latency and the low percentage of infected individuals who develop disease. HTLV can transform T lymphocytes *in vitro*, but is important to stress that the transformed cells display different characteristics than leukemic cells *in vivo*, most likely because of different selection pressures. It appears that the initiation of the transformation process *in vivo* can be mimicked *in vitro* and that these events are probably more characteristic of the early stages of infection in humans.

The overall pathogenic mechanisms of HTLV and its role in human disease remain to be elucidated. Progression events that occur *in vivo* and result in the development of leukemia must be more fully characterized. The development of a relevant animal model system will be of utmost importance to study HTLV infection, transformation, and tumorigenesis *in vivo*. It will be important to explore the determinants of HTLV that cause neurological disease versus leukemia which may lead to insights in the causes of autoimmune diseases. To date, it is unclear how or why HTLV exists in a transcriptionally repressed or latent state in leukemic cells. Viral latency presumably allows the virus to evade the host immune response; however, the host factors and/or possibly viral factors that regulate latency remain to be determined. Future studies in these areas should bring a greater understanding of the role HTLV plays in the oncogenic process and possibly lead to the development of novel therapeutic strategies.

VIII. REFERENCES

Ahmed, Y. F., Hanly, S. M., Malim, M. H., Cullen, B. R., and Greene, W. C., 1990, Structure–function analyses of the HTLV-I Rex and HIV-1 Rev RNA response elements: Insights into the mechanism of Rex and Rev action, *Genes Dev.* **4**:1014–1022.

Ahmed, Y. F., Gilmartin, G. M., Hanly, S. M., Nevins, J. R., and Greene, W. C., 1991, The HTLV-I Rex response element mediates a novel form of mRNA polyadenylation, *Cell* **64**:727–737.

Altman, R., Harooch, D., Garcia, J. A., and Gaynor, R. B., 1988, Human T-cell leukemia virus types I and II exhibit different DNase I protection patterns, *J. Virol.* **62**:1339–1346.

Ballard, D. W., Bohnlein, E., Lowenthal, J. W., Wano, Y., Franza, B. R., and Greene, W. C., 1988, HTLV-I *tax* induces cellular proteins that activate the κB element in the IL-2 receptor α gene, *Science* **241**:1652–1655.

Ballaun, C., Farrington, G. R., Dobrovnik, M., Rusche, J., Hauber, J., and Bohnlein, E., 1991, Functional analysis of human T-cell leukemia virus type I Rex-response element: Direct RNA binding of Rex protein correlates with *in vivo* activity, *J. Virol.* **65**:4408–4413.

Bangham, C. R. M., Nightingale, S., Cruickshank, J. K., and Daenke, S., 1989, PCR analysis of DNA from multiple sclerosis patients for the presence of HTLV-I, *Science* **246**:821–823.

Bazarbachi, A., Saal, F., Laroche, L., Flageul, B., Peries, J., and de The, H., 1993, HTLV-I provirus and mycosis fungoides, *Science* **259**:1470–1471.

Beimling, P., and Moelling, K., 1990, Tax-independent binding of multiple cellular factors to Tax-response element DNA of HTLV-I, *Oncogene* **5**:361–368.

Berneman, Z. N., Gartenhaus, R. B., Reitz, M. S., Blattner, W. A., Manns, A., Hanchard, B., Ikehara, O., Gallo, R. C., and Klotman, M. E., 1992, Expression of alternately spliced human T-lymphotropic virus type I pX mRNA in infected cell lines and in primary uncultured cells from patients with adult T-cell leukemia/lymphoma and healthy carriers, *Proc. Natl. Acad. Sci. USA* **89**:3005–3009.

Bhagavati, S., Ehrlich, G., Kula, R. W., Kwok, S., Sninsky, J., Udani, V., and Poiesz, B. J., 1988, Detection of human T-cell lymphoma/leukemia virus type I DNA and antigen in spinal fluid and blood of patients with chronic progressive myelopathy, *N. Engl. J. Med.* **318**:1141–1147.

Bhat, N. K., Adachi, Y., Samuel, K. P., and Derse, D., 1993, HTLV-1 gene expression by defective proviruses in an infected T-cell line, *Virology* **196**:15–24.

Biggar, R. J., Johnson, B. K., Oster, C., Sarin, P. S., Ocheng, D., Tukei, P., Nsanze, H., Alexander, S., Bodner, A. J., Siongok, T., Gallo, R. C., and Blattner, W. A., 1985, Regional variation in prevalence of antibody against human T-lymphotropic virus types I and III in Kenya, East Africa, *Int. J. Cancer* **35**:763–767.

Black, A. C., Chen, I. S. Y., Arrigo, S. J., Ruland, C. T., Chin, E., Allogiamento, T., and Rosenblatt, J. D., 1991a, Different *cis*-acting regions of the HTLV-II 5′ LTR are involved in regulation of gene expression by Rex, *Virology* **181**:433–444.

Black, A. C., Ruland, C. T., Yip, M. T., Luo, J., Tran, B., Kalsi, A., Quan, E., Chen, I. S. Y., and Rosenblatt, J. D., 1991b, Human T-cell leukemia virus type II Rex binding and activity requires an intact splice donor site and a specific RNA secondary structure, *J. Virol.* **65**:6645–6653.

Blattner, W. A., Kalyanaraman, V. S., Robert-Guroff, M., Lister, T. A., Galton, D. A. G., Sarin, P. S., Crawford, M. H., Catovsky, D., Greaves, M., and Gallo, R. C., 1982, The human type-C retrovirus, HTLV, in blacks from the Caribbean region, and relationship to adult T-cell leukemia/lymphoma, *Int. J. Cancer* **30**:257–264.

Blayney, D. W., Blattner, W. A., Robert-Guroff, M., Jaffe, E. S., Fischer, R. I., Bunn, P. A., Patton, M. G., Rarick, H. R., and Gallo, R. C., 1983, The human T-cell leukemia/lymphoma virus (HTLV) in southeastern United States, *J. Am. Med. Assoc.* **250**:1048–1052.

Bogerd, H. P., Huckaby, G. L., Ahmed, Y. F., Hanly, S. M., and Greene, W. C., 1991, The type 1 human T-cell leukemia virus (HTLV-I) Rex trans-activator binds directly to the HTLV-I Rex and the type 1 human immunodeficiency virus Rev RNA response elements, *Proc. Natl. Acad. Sci. USA* **88**:5704–5708.

Bogerd, H., and Greene, W. C., 1993, Dominant negative mutants of human T-cell leukemia virus type I Rex and human immunodeficiency virus type 1 Rev fail to multimerize in vivo, *J. Virol.* **67**:2496–2502.

Bohnlein, E., Lowenthal, J., Siekevitz, M., Ballard, D., Franza, B. R., and Greene, W. C., 1988, The same inducible nuclear proteins regulate mitogen activation of both the interleukin-2 receptor-alpha gene and type 1 HIV, *Cell* **53**:827–836.

Bohnlein, E., Siekevitz, M., Ballard, D. W., Lowenthal, J. W., Rimsky, L., Bogerd, H., Hoffman, J., Wano, Y., Franza, B. R., and Greene, W. C., 1989, Stimulation of the human immunodeficiency virus type 1 enhancer by the human T-cell leukemia virus type I *tax* gene product involves the action of inducible cellular proteins, *J. Virol.* **63**:1578–1586.

Bohnlein, S. F., Pirker, F. P., Hofer, L., Bachmayer, H., Bohnlein, E., and Hauber, J., 1991, Trans-dominant repressors for human T-cell leukemia virus type I Rex and human immunodeficiency virus type 1 Rev function, *J. Virol.* **65**:81–88.

Bouzas, M. B., Muchinik, G., Zapiola, I., De Rosa, M. F., Perez-Bianco, R., Cahn, P., Gallo, D., and Hansen, C., 1991, Human T-cell lymphotropic virus type II infection in Argentina, *J. Infect. Dis.* **164**:1026–1027.

Brady, J., Jeang, K. T., Durall, J., and Khoury, G., 1987, Identification of the p40x-responsive regulatory sequences within the human T-cell leukemia virus type I long terminal repeat, *J. Virol.* **61**:2175–2181.

Bunn, P. A., Jr., Schechter, G. P., Jaffe, E., Blayney, D. W., Young, R. C., Matthews, M. J., Blattner,

W. A., Broder, S., Robert-Guroff, M., and Gallo, R. C., 1983, Clinical course of retrovirus-associated adult T-cell lymphoma in the United States, *N. Engl. J. Med.* **309**:257–264.

Cann, A. J., Rosenblatt, J. D., Wachsman, W., Shah, N. P., and Chen, I. S. Y., 1985, Identification of the gene responsible for human T-cell leukemia virus transcriptional regulation, *Nature* **318**:571–574.

Catovsky, D., Rose, M., Goolden, A. W. G., White, J. M., Bourikas, G., Brownell, A. I., Blattner, W. A., Greaves, M. F., Galton, D. A. G., McCluskey, D. R., Lampert, I., Ireland, R., Bridges, J. M., and Gallo, R. C., 1982, Adult T-cell lymphoma-leukaemia in blacks from the West Indies, *Lancet* **1**:639–647.

Chan, J. Y., Slamon, D. J., Nimer, S. D., Golde, D. W., and Gasson, J. C., 1986, Regulation of expression of human granulocyte macrophage colony stimulating factor (GM-CSF), *Proc. Natl. Acad. Sci. USA* **83**:8669–8673.

Chen, I. S. Y., and Temin, H. M., 1982, Establishment of infection by spleen necrosis virus: Inhibition in stationary cells and the role of secondary infection, *J. Virol.* **41**:183–191.

Chen, I. S. Y., Quan, S. G., and Golde, D. W., 1983, Human T-cell leukemia virus type II transforms normal human lymphocytes, *Proc. Natl. Acad. Sci. USA* **80**:7006–7009.

Chen, I. S. Y., Slamon, D. J., Rosenblatt, J. D., Shah, N. P., Quan, S. G., and Wachsman, W., 1985, The x gene is essential for HTLV replication, *Science* **229**:54–58.

Chen, I. S. Y., Cann, A. J., Lugo, J. P., and Rosenblatt, J. D., 1989, Models for mechanisms for transformation of the human T-cell leukemia viruses, in: *Retroviruses and Disease* (M. E. Pullman, H. Hanafusa, and A. Printer, eds.), p. 113, Academic Press, New York.

Chen, I. S. Y., Haislip, A. M., Myers, L. W., Ellison, G. W., and Merrill, J. E., 1990, Failure to detect human T-cell leukemia virus-related sequences in multiple sclerosis blood, *Arch. Neurol.* **47**:1064–1065.

Chosa, T. N., Yamamoto, Y., Tanaka, Y., Koyanagi, Y., and Hinuma, Y., 1982, Infectivity dissociated from transforming activity in a human retrovirus adult T-cell leukemia virus, *Gann* **73**:844–847.

Ciminale, V., Pavlakis, G. N., Derse, D., Cunningham, C. P., and Felber, B. K., 1992, Complex splicing in the human T-cell leukemia virus (HTLV) family of retroviruses: Novel mRNAs and proteins produced by HTLV type I, *J. Virol.* **66**:1737–1745.

Clapham, P., Nagy, K., Cheingsong-Popov, R., Exley, M., and Weiss, R. A., 1983, Productive infection and cell-free transmission of human T-cell leukemia virus in a nonlymphoid cell line, *Science* **222**:1125–1127.

Clark, N. M., Smith, M. J., Hilfinger, J. M., and Markovitz, D. M., 1993, Activation of the human T-cell leukemia virus type I enhancer is mediated by binding sites for Elf-1 and pets factor, *J. Virol.* **67**:5522–5528.

Copeland, T. D., Oroszlan, S., Kalyanaraman, V. S., Sarngadharan, M. G., and Gallo, R. C., 1983, Complete amino acid sequence of human T-cell leukemia virus structural protein p15, *FEBS Lett.* **162**:390–395.

Cross, S. L., Feinberg, M. B., Wolf, J. B., Holbrook, N. J., Wong-Staal, F., and Leonard, W. J., 1987, Regulation of the human interleukin-2 receptor alpha promoter: Activation of a nonfunctional promoter by the transactivator gene of HTLV-I, *Cell* **49**:47–56.

Dalgleish, A., Richardson, J., Matutes, E., Cruickshank, K., Newell, A., Sinclair, A., Thorpe, R., Brasher, M., Weber, J., Catovsky, D., and Rudge, P., 1988, HTLV-I infection in tropical spastic paraparesis: Lymphocytes culture and serologic response, *AIDS Res. Hum. Retroviruses* **6**:475–485.

Dayton, A. I., Sodroski, J. G., Rosen, C. A., Goh, W. C., and Haseltine, W. A., 1986, The *trans*-activator gene of human T-cell lymphotropic virus type III is required for replication, *Cell* **44**:941–947.

Depper, J. M., Leonard, W. J., Krönke, M., Waldmann, T. A., and Greene, W. C., 1984, Augmented T cell growth factor receptor expression in HTLV-I infected human leukemic T cells, *J. Immunol.* **133**:1691–1695.

Duc Dodon, M., Bernard, A., and Gazzolo, L., 1989, Peripheral T-lymphocyte activation by human T-cell leukemia virus type 1 interferes with the CD2 but not with the CD3/TCR pathway, *J. Virol.* **63**:5413–5419.

Dumontet, C., Duc Dudon, M., Gazzalo, L., and Gerlier, D., 1988, Human T-cell leukemia virus type I-induced proliferation of human thymocytes requires the presence of a comitogen, *Cell. Immunol.* **112**:391–401.

Ejima, E., Rosenblatt, J. D., Massari, M., Quan, E., Stephens, D., Rosen, C. A., and Prager, D., 1993, Cell-type-specific transactivation of the parathyroid hormone-related protein gene promoter by the human T-cell leukemia virus type I (HTLV-I) tax and HTLV-II tax proteins, *Blood* **81**:1017–1024.

Feinberg, M. B., Jarrett, R. F., Aldovini, A., Gallo, R. C., and Wong-Staal, F., 1986, HTLV-III expression and production involve complex regulation at the levels of splicing and translation of viral RNA, *Cell* **46**:807–817.

Felber, B. K., Paskalis, H., Kleinman-Ewing, C., Wong-Staal, F., and Pavlakis, G. N., 1985, The pX protein of HTLV-I is a transcriptional activator of its long terminal repeats, *Science* **229**:675–679.

Feuer, G., Zack, J. A., Harrington, W. J., Jr., Valderama, R., Rosenblatt, J. D., Wachsman, W., Baird, S. M., and Chen, I. S. Y., 1993, Establishment of human T-cell leukemia virus type I T-cell lymphomas in severe combined immunodeficient mice, *Blood* **82**:722–31.

Fleckenstein, B., and Desrosiers, R. C., 1982, Herpesvirus saimiri and Herpesvirus ateus, in: *The Herpesviruses* (B. Roizman, ed.), pp. 253–332, Plenum Press, New York.

Franklin, A. A., Kubik, M. F., Uittenbogaard, M. N., Brauweiler, A., Utaisincharoen, P., Matthews, M. H., Dynan, W. S., Hoeffler, J. P., and Nyborg, J. K., 1993, Transactivation by the human T-cell leukemia virus Tax protein is mediated through enhanced binding of activating transcription factor-2 (ATF-2) ATF-2 response and cAMP element-binding protein (CREB), *J. Biol. Chem.* **268**:21225–21231.

Fujii, M., Sassone-Corsi, P., and Verma, I. M., 1988, c-*fos* promoter trans-activation by the tax_1 protein of human T-cell leukemia virus type I, *Proc. Natl. Acad. Sci. USA* **85**:8526–8530.

Fujisawa, J., Seiki, M., Kiyokawa, T., and Yoshida, M., 1985, Functional activation of the long terminal repeat of human T-cell leukemia virus type I by a trans-acting factor, *Proc. Natl. Acad. Sci. USA* **82**:2277–2281.

Fujisawa, J., Seiki, M., Sata, M., and Yoshida, M., 1986, A transcriptional enhancer sequence of HTLV-I is responsible for trans-activation mediated by $p40^{xI}$ of HTLV-I, *EMBO J.* **5**:713–718.

Fujisawa, J.-I., Toita, M., and Yoshida, M., 1989, A unique enhancer element for the transactivator ($p40^{tax}$) of human T-C leukemia virus type I that is distinct from cyclic AMP- and 12-O-tetradecanoylphorbol-13-acetate-response elements, *J. Virol.* **63**:3234–3239.

Fujisawa, J., Toita, M., Yoshimura, T., and Yoshida, M., 1991, The indirect association of human T-cell leukemia virus *tax* protein with DNA results in transcriptional activation, *J. Virol.* **65**:4525–4528.

Fukuhara, S., Hinuma, Y., Gotoh, Y.-I., and Uchino, H., 1983, Chromosome aberrations in T lymphocytes carrying adult T-cell leukemia-associated antigens (ATLA) from healthy adults, *Blood* **61**:205–207.

Gallart, T., Anegon, I., Woessner, S., de la Fuente, R., Florensa, L., Sans-Sabrafren, C., and Vives, J., 1983, Clinically benign adult T-cell chronic lymphocytosis with unusual phenotype, *Lancet* **1**:769–770.

Gallo, R. C., Kalyanaraman, V. S., Sarngadharan, M. G., Sliski, A., Vonderheid, E. C., Maeda, M., Nakao, Y., Yamada, K., Ito, Y., Gutensohn, N., Murphy, S., Bunn, P. A., Jr., Catovsky, D., Greaves, M. F., Blayney, D. W., Blattner, W., Jarrett, W. F. H., zur Hausen, H., Seligmann, M., Brouet, J. C., Haynes, B. F., Jegasothy, B. V., Jaffe, E., Cossman, J., Broder, S., Fisher, R. I., Golde, D. W., and Robert-Guroff, M., 1983, Association of the human type C retrovirus with a subset of adult T-cell cancers, *Cancer Res.* **43**:3892–3899.

Gazzolo, L., and Duc Dodon, M., 1987, Direct activation of resting T lymphocytes by human T-lymphotropic virus type I, *Nature* **326**:714–717.

Gessain, A., Barin, F., Vernant, J. C., Gout, O., Maurs, L., Calender, A., and De The, G., 1985, Antibodies to human T-lymphotropic virus type I in patients with tropical spastic paraparesis, *Lancet* **2**:407–409.

Gessain, A., Caudie, C., Gout, O., Vernant, J. C., Maurs, L., Giordano, C., Malone, G., Tournier-Lasserve, E., Essex, M., and De The, G., 1988, Intrathecal synthesis of antibodies to human T lymphotropic virus type I patients with endemic tropical spastic paraparesis, *J. Infect. Dis.* **157**:1226–1234.

Gessain, A., Saal, F., Gout, O., Daniel, M. T., Flandrin, G., De The, G., Peries, J., and Sigaux, F., 1990, High human lymphotropic virus type I proviral DNA load with polyclonal integration in peripheral blood mononuclear cells of French West Indian, Guianese, and African patients with tropical spastic paraparesis, *Blood* **75**:428–433.

Giam, C. Z., and Xu, Y. L., 1989, HTLV-I Tax gene product activates transcription via preexisting cellular factors and cAMP responsive element, *J. Biol. Chem.* **264**:15236–15241.

Gitlin, S. D., Bosselut, R., Gegonne, A., Ghysdael, J., and Brady, J. H., 1991, Sequence specific interaction of the Ets1 protein with the long terminal repeat of the human T-lymphotropic virus type I, *J. Virol.* **65**:5513–5523.

Grassmann, R., Dengler, C., Muller-Fleckenstein, I., Fleckenstein, B., McGuire, K., Dokhelar, M.-C., Sodroski, J. G., and Haseltine, W. A., 1989, Transformation to continuous growth of primary human T lymphocytes by human T-cell leukemia virus type I X-region genes transduced by a *Herpesvirus saimiri* vector, *Proc. Natl. Acad. Sci. USA* **86**:3351–3355.

Grassmann, R., Berchtolds, S., Aepinus, C., Ballaun, C., Bohnlein, E., and Fleckenstein, B., 1991, In vitro binding of human T-cell leukemia virus Rex proteins to the Rex-response element of viral transcripts, *J. Virol.* **65**:3721–3727.

Grassmann, R., Berchtolds, S., Radant, I., Alt, M., Fleckenstein, B., Sodroski, J. G., Haseltine, W. A., and Ramstedt, U., 1992, Role of the human T-cell leukemia virus type 1 X region proteins in immortalization of primary human lymphocytes in culture, *J. Virol.* **66**:4570–4575.

Green, J. E., Begley, G., Wagner, D. K., Waldmann, T. A., and Jay, G., 1989a, Transactivation of GM-CSF and the interleukin 2 receptor in transgenic mice carrying the HTLV-I tax gene, *Mol. Cell. Biol.* **9**:4731–4737.

Green, J. E., Hinrichs, S. H., Vogel, J., and Jay, G., 1989b, Exocrinopathy resembling Sjogren's syndrome in HTLV-I tax transgenic mice, *Nature* **341**:72–74.

Green, P. L., Xie, Y., and Chen, I. S. Y., 1990, The internal methionine codons of the HTLV-II *rex* gene are not required for $p24^{Rex}$ production or virus replication and transformation, *J. Virol.* **64**:4914–4921.

Green, P. L., Xie, Y., and Chen, I. S. Y., 1991, The Rex proteins of HTLV-II differ by serine phosphorylation, *J. Virol.* **65**:546–550.

Green, P. L., Yip, M. T., Xie, Y., and Chen, I. S. Y., 1992, Phosphorylation regulates RNA binding by the human T-cell leukemia virus Rex protein, *J. Virol.* **66**:4325–4330.

Greenberg, S. J., Jacobson, S., Waldmann, T. A., and McFarlin, D. E., 1989a, Molecular analysis of HTLV-I proviral integration and T cell receptor rearrangement indicates that T cells in tropical spastic paraparesis are polyclonal, *J. Infect. Dis.* **159**:741–744.

Greenberg, S. J., Ehrlich, G. D., Abbott, M. A., Hurwitz, B. J., Waldmann, T. A., and Poiesz, B. J., 1989b, Detection of sequences homologous to human retroviral DNA in multiple sclerosis by gene amplification, *Proc. Natl. Acad. Sci. USA* **786**:2878–2882.

Hall, W. W., Liu, C. R., Schneewind, O., Takahashi, H., Kaplan, M., Roupe, G., and Vahlne, A., 1991, Deleted HTLV-I provirus in blood and cutaneous lesions of patients with mycosis fungoides, *Science* **253**:317–319.

Hanly, S. M., Rimsky, L. T., Malim, M. H., Kim, J. H., Hauber, J., Duc Dodon, M., Le, S.-Y., Maizel, J. V., Cullen, B. R., and Greene, W. C., 1989, Comparative analysis of the HTLV-I Rex and HIV-1 Rev *trans*-regulatory proteins and their RNA response elements, *Genes Dev.* **3**:1534–1544.

Harper, M. E., Kaplan, M. H., Marselle, L. M., Pahwa, S. G., Chayt, K. J., Sarngadharan, M. G., Wong Staal, F., and Gallo, R. C., 1986, Concomitant infection with HTLV-I and HTLV-III in a patient with T8 lymphoproliferative disease, *N. Engl. J. Med.* **315**:1073–1078.

Hattori, S., Kiyokawa, T., Imagawa, K.-I., Shimizu, F., Hashimura, E., Seiki, M., and Yoshida, M., 1984, Identification of *gag* and *env* gene products of human T-cell leukemia virus (HTLV), *Virology* **136**:338–347.

Hattori, T., Uchiyama, T., Toibana, T., Takatsuki, K., and Uchino, H., 1981, Surface phenotype of Japanese adult T-cell leukemia cells characterized by monoclonal antibodies, *Blood* **58**:645–647.

Hikada, M., Inoue, J., Yoshida, M., and Seiki, M., 1988, Post-transcriptional regulator (*rex*) of HTLV-I initiates expression of viral structural proteins but suppresses expression of regulatory proteins, *EMBO J.* **7**:519–523.

Hinrichs, S. H., Nerenberg, M., Reynolds, K., Khoury, G., and Jay, G., 1987, A transgenic mouse model for human neurofibromatosis, *Science* **237**:1340–1343.

Hinuma, Y., Nagata, K., Hanaoka, M., Nakai, M., Matsumoto, T., Kinoshita, K.-I., Shirakawa, S., and Miyoshi, I., 1981, Adult T-cell leukemia: Antigen in an ATL cell line and detection of antibodies to the antigen in human sera, *Proc. Natl. Acad. Sci. USA* **78**:6476–6480.

Hjelle, B., Scalf, R., and Swenson, S., 1990, High frequency of human T-cell leukemia/lymphoma virus type II infection in New Mexico blood donors: Determination by sequence specific oligonucleotide hybridization, *Blood* **76**:450–454.

Hjelle, B., Appenzeller, O., Mills, R., Alexander, S., Torrez-Martinez, N., Jahnke, R., and Ross, G., 1992, Chronic neurodegenerative disease associated with HTLV-II infection, *Lancet* **339:** 645–646.

Hjelle, B., Zhu, S. W., Takahashi, H., Ijichi, S., and Hall, W. W., 1993, Endemic human T-cell leukemia virus type II infection in southwestern US Indians involves two prototype variants of virus, *J. Infect. Dis.* **168:**737–40.

Hofer, L., Weichselbraun, I., Quick, S., Farrington, G. K., Bohnlein, E., and Hauber, J., 1991, Mutational analysis of the human T-cell leukemia virus type I *trans*-acting *rex* gene product, *J. Virol.* **65:**3379–3383.

Holland, C. A., Hartley, J. W., Rowe, W. P., and Hopkins, N., 1985, At least four viral genes contribute to the leukemogenicity of murine retrovirus MCF 247 in AKR mice, *J. Virol.* **53:**158–165.

Hope, T. J., Bond, B. L., McDonald, B., Klein, N. P., and Parslow, T. G., 1991, Effector domains of human immunodeficiency virus type 1 Rev and human T-cell leukemia virus type I Rex are functionally interchangeable and share an essential peptide motif, *J. Virol.* **65:**6001–6007.

Hoshino, H., Esumi, H., Miwa, M., Shimoyama, M., Minato, K., Tobinai, K., Hirose, M., Watanabe, S., Inada, N., Kinoshita, K., Kamihira, S., Ichimaru, M., and Sugimura, T., 1983, Establishment and characterization of 10 cell lines derived from patients with adult T-cell leukemia, *Proc. Natl. Acad. Sci. USA* **80:**6061–6065.

Hoshino, H., Tanaka, H., Shimotohno, K., Miwa, M., Nagai, M., Shimoyama, M., and Sugimura, T., 1984, Immortalization of peripheral blood lymphocytes of cats by human T-cell leukemia virus, *Int. J. Cancer* **34:**513–517.

Inoue, J., Seiki, M., and Yoshida, M., 1986a, The second pX product p27xIII of HTLV-I is required for gag gene expression, *FEBS Lett.* **209:**187–190.

Inoue, J., Seiki, M., Taniguchi, T., Tsuru, S., and Yoshida, M., 1986b, Induction of interleukin-2 receptor gene by p40xI encoded by human T-cell leukemia virus type I, *EMBO J.* **5:**2883–2888.

Inoue, J. I., Yoshida, M., and Seiki, M., 1987, Transcriptional ($p40^{x}$) and post-transcriptional ($p27^{xIII}$) regulators are required for the expression and replication of human T-cell leukemia virus type I genes, *Proc. Natl. Acad. Sci. USA* **84:**3653–3657.

Jacobson, S., Raine, C. S., Mingioli, E. S., and McFarlin, D. E., 1988, Isolation of an HTLV-1-like retrovirus from patients with tropical spastic paraparesis, *Nature* **331:**540–543.

Jason, J. M., McDougal, J. S., Cabradilla, C., Kalyanaraman, V. S., and Evatt, B. L., 1985, Human T-cell leukemia virus (HTLV-I) p24 antibody in New York City blood product recipients, *Am. J. Hematol.* **20:**129–137.

Jeang, K. T., Boros, I., Brady, J., Radonovich, M., and Khoury, G., 1988a, Characterization of cellular factors that interact with the human T-cell leukemia virus type I p40x-responsive 21-base-pair sequence, *J. Virol.* **62:**4499–4509.

Jeang, K. T., Shank, P. R., and Kumar, A., 1988b, Transcriptional activation of homologous viral long terminal repeats by the human immunodeficiency virus type I or the human T-cell leukemia virus type I *tat* proteins occurs in the absence of *de novo* protein synthesis, *Proc. Natl. Acad. Sci. USA* **85:**8291–8295.

Jeang, K. T., Widen, S. G., Semmes, O. J., and Wilson, S. H., 1990, HTLV-I transactivator protein, Tax is a *trans*-repressor of human β-polymerase gene, *Science* **247:**1082–1084.

Josephs, S. F., Wong-Staal, F., Mazari, V., Gallo, R. C., Sodroski, J., Trus, M. D., Perkins, D., Patarca, R., and Haseltine, W. A., 1984, Long terminal repeat structure of an American isolate of type I human T-cell leukemia virus, *Virology* **139:**340–345.

Kawano, F., Yamaguchi, K., Nishimura, H., Tsuda, H., and Takatsuki, K., 1985, Variation in the clinical courses of adult T-cell leukemia, *Cancer* **55:**851–856.

Kim, J. H., Kaufman, P. A., Hanly, S. M., Rimsky, L. T., and Greene, W. C., 1991, Rex transregulation of human T-cell leukemia virus type II gene expression, *J. Virol.* **65:**405–414.

Kimata, J. T., and Ratner, L., 1991, Temporal regulation of viral and cellular gene expression during human T-lymphotropic virus type I-mediated lymphocyte immortalization, *J. Virol.* **65:**4398–4407.

Kimata, J. T., Palker, T. J., and Ratner, L., 1993, The mitogenic activity of human T-cell leukemia virus type I is T-cell associated and requires the CD2/LFA-3 activation pathway, *J. Virol.* **67:**3134–3141.

Kinoshita, K., Amagasaki, T., Ikeda, S., Suzuyama, J., Toriya, K., Nishino, K., Tagawa, M., Ichimaru, M., Kamihira, S., Yamada, Y., Momita, S., Kusano, M., Morikawa, T., Fujita, S., Ueda, Y., Ito, N., and Yoshida, M., 1985, Preleukemic state of adult T cell leukemia: Abnormal T lymphocytosis induced by human adult T cell leukemia–lymphoma virus, *Blood* **66**:120–127.

Kira, J., Koyanagi, Y., Yamada, T., Itoyama, Y., Goto, I., Yamamoto, N., Sasaki, H., and Sakaki, Y., 1991, Increased HTLV-I proviral DNA in HTLV-I-associated myelopathy; a quantitative polymerase reaction study, *Ann. Neurol.* **29**:194–201.

Kitado, H., Chen, I. S. Y., Shah, N. P., Cann, A. J., Shimotohno, K., and Fan, H., 1987, U3 sequences from the HTLV-I and -II LTRs confer pX protein responsiveness to a murine leukemia virus LTR, *Science* **235**:901–904.

Kitajima, I., Shinohara, T., Bilakovics, J., Brown, D. A., Xu, X., and Nerenberg, M., 1992, Ablation of transplanted HTLV-I Tax transformed tumors in mice by antisense inhibition of NF-κB, *Science* **258**:1792–1795.

Kiyokawa, T., Seiki, M., Iwashita, S., Imagawa, K., Shimizu, F., and Yoshida, M., 1985, $p27^{xIII}$ and $p21^{xIII}$, proteins encoded by the pX sequence of human T-cell leukemia virus type I, *Proc. Natl. Acad. Sci. USA* **82**:8359–8363.

Kondo, T., Kono, H., Nonaka, H., Miyamoto, N., Yoshida, R., Bando, F., Inoue, H., Miyoshi, I., Hinuma, Y., and Hanaoka, M., 1987, Risk of adult T-cell leukaemia/lymphoma in HTLV-I carriers, *Lancet* **2**:159.

Korber, B., Okayama, A., Donnelly, R., Tachibana, N., and Essex, M., 1991, Polymerase chain reaction analysis of defective human T-cell leukemia virus type I proviral genomes in leukemic cells of patients with adult T-cell leukemia, *J. Virol.* **65**:5471–5476.

Koyanagi, Y., Yamamoto, N., Kobayashi, N., Hirai, K., Konishi, H., Takeuchi, K., Tanaka, Y., Hatanaka, M., and Hinuma, Y., 1984, Characterization of human B-cell lines harboring both adult T-cell leukemia (ATL) virus and Epstein–Barr virus derived from ATL patients, *J. Gen. Virol.* **65**:1781–1789.

Koyanagi, Y., Itoyama, Y., Nakamura, N., Takamatsu, K., Kira, J., Iwamasa, T., Goto, I., and Yamamoto, N., 1993, In vivo infection of human T-cell leukemia virus type I in non-T cells, *Virology* **196**:25–33.

Kwok, S., Kellogg, D., Ehrlich, G., Poiesz, B., Bhagavati, S., and Sninsky, J. J., 1988, Characterization of a sequence of human T cell leukemia virus type I from a patient with chronic progressive myelopathy, *J. Infect. Dis.* **158**:1193–1197.

Lal, R. B., Rudolph, D., Buckner, C., Pardi, D., and Hooper, W. C., 1993, Infection with human T-lymphotropic viruses leads to constitutive expression of leukemia inhibitory factor and interleukin-6, *Blood* **81**:1827–1832.

Land, H., Parada, L. F., and Weinberg, R. A., 1983, Tumorigenic conversion of primary embryo fibroblasts requires at least two cooperating oncogenes, *Nature* **304**:596–602.

Lee, H., Swanson, P., Shorty, V. S., Zack, J. A., Rosenblatt, J. D., and Chen, I. S. Y., 1989, High rate of HTLV-II infection in seropositive IV drug abusers from New Orleans, *Science* **244**:471–475.

Lee, K. A. W., Hai, T. Y., SivaRaman, L., Thimmappaya, B., Hurst, H. C., Jones, N. C., and Green, M. R., 1987, A cellular protein, activating transcription factor, activates transcription of multiple E1a-inducible adenovirus early promoters, *Proc. Natl. Acad. Sci. USA* **84**:8355–8359.

Lee, T. H., Coligan, J. E., Sodroski, J. G., Haseltine, W. A., Salahuddin, S. Z., Wong-Staal, F., Gallo, R. C., and Essex, M., 1984, Antigens encoded by the 3′-terminal region of human T-cell leukemia virus: Evidence for a functional gene, *Science* **226**:57–61.

Leis, J., Baltimore, D., Bishop, J. M., Coffin, J., Fleissuer, E., Goff, S. T., Oroszlan, S., Robinson, H., Skalka, A. M., Tewin, H. M., and Vogt, V., 1988, Standardised and simplified nomenclature for proteins common to all retroviruses, *J. Virol.* **62**:1808–1809.

Leung, K., and Nabel, G. J., 1988, HTLV-I *trans*-activator induces interleukin-2 receptor expression through an NFκB-like factor, *Nature* **333**:776–778.

Levine, P. H., Jacobson, S., Elliott, R., Cavallero, A., Colclough, G., Dorry, C., Stephenson, C., Knigge, R. M., Drummond, J., and Nishimura, M., 1993, HTLV-II infection in Florida Indians, *AIDS Res. Hum. Retroviruses* **9**:123–127.

Li, J. P., D'Andrea, A. D., Lodish, H. F., and Baltimore, D., 1990, Activation of cell growth by binding of Friend spleen focus-forming virus gp55 glycoprotein to the erythropoietin receptor, *Nature* **343**:762–764.

Lindholm, P., Marriott, S., Gitlin, S., Bohan, C., and Brady, J., 1991, Induction of nuclear factor NF-κB DNA binding activity after exposure of lymphoid cells to soluble Tax_1 protein, *New Biol.* **2**:1034–1043.

Lindholm, P. F., Reid, R. L., and Brady, J. N., 1992, Extracellular Tax_1 protein stimulates tumor necrosis factor-β and immunoglobulin kappa light chain expression in lymphoid cells, *J. Virol.* **66**:1294–1302.

Lombard-Platet, G., and Jalinot, P., 1993, Purification by DNA affinity precipitation of the cellular factors HEB1-p67 and HEB1-p94 which bind specifically to the human T-cell leukemia virus type-I 21 bp enhancer, *Nucleic Acids Res.* **21**:3935–3942.

Lowenthal, J. W., Bohnlein, E., Ballard, D. W., and Greene, W. C., 1988, Regulation of interleukin 2 receptor alpha subunit (Tac or CD25 antigen) gene expression: Binding of inducible nuclear proteins to discrete promoter sequences correlates with transcriptional activation, *Proc. Natl. Acad. Sci. USA* **88**:4468–4472.

Maguer, V., Casse-Ripoll, H., Gazzolo, L., and Duc Dodon, M., 1993, Human T-cell leukemia virus type I-induced proliferation of human immature $CD2^+CD3^+$ thymocytes, *J. Virol.* **69**:5529–5537.

Manzari, V., Gismond, A., Barillari, G., Morrone, S., Modesta, A., Albonici, L., De Marchis, L., Fazio, V., Gradilone, A., Zani, M., Frati, L., and Santoni, A., 1987, HTLV-V: A new human retrovirus isolated in a Tac-negative T-cell lymphoma/leukemia, *Science* **238**:1581–1583.

Markham, P. D., Salahuddin, S. Z., Macchi, B., Robert-Guroff, M., and Gallo, R. C., 1984, Transformation of different phenotypic types of human bone marrow T-lymphocytes by HTLV-I, *Int. J. Cancer* **33**:13–17.

Marriott, S. J., Boros, I., Duvall, J. F., and Brady, J. N., 1989, Indirect binding of human T-cell leukemia virus type I tax to a response element in the viral long terminal repeat, *Mol. Cell. Biol.* **9**:4152–4160.

Marriott, S. J., Lindholm, P. F., Brown, K. M., Gitlin, S. D., Duvall, J. F., Radonovich, M. F., and Brady, J. N., 1990, A 36 kilodalton cellular transcription factor mediates an indirect interaction of human T-cell leukemia/lymphoma virus type I TAX_1 with a response element in the viral long terminal repeat, *Mol. Cell. Biol.* **10**:4192–4201.

Maruyama, M., Shibuya, H., Harada, H., Hatakeyama, M., Seiki, M., Fujita, T., Inoue, J., Yoshida, M., and Taniguchi, T., 1987, Evidence for aberrant activation of the interleukin-2 autocrine loop by HTLV-I-encoded $p40^{xI}$ and T3/Ti complex triggering, *Cell* **48**:343–350.

Matsuoka, M., Hagiya, M., Hattori, T., Asou, N., Maeda, S., Shimada, K., Tsai, S. C., Sakano, H., and Takatsuki, K., 1988, Gene rearrangements of T cell receptor β and γ chains in HTLV-I infected primary neoplastic T cells, *Leukemia* **2**:84–90.

Matthews, M.-A. H., Markowitz, R.-B., and Dynan, W. S., 1992, In vitro activation of transcription by the human T-cell leukemia virus type I Tax protein, *Mol. Cell. Biol.* **12**:1986–1996.

Merino, F., Robert-Guroff, M., Clark, J., Biondo-Bracho, M., Blattner, W. A., and Gallo, R. C., 1984, Natural antibodies to human T-cell leukemia/lymphoma virus in healthy Venezuelan populations, *Int. J. Cancer* **34**:501–506.

Merl, S., Kloster, B., Moore, J., Hubbell, C., Tomar, R., Davey, F., Kalinowski, D., Planas, A., Ehrlich, G., Clark, D., Comis, R., and Poiesz, B., 1984, Efficient transformation of previously activated and dividing T lymphocytes by human T cell leukemia–lymphoma virus, *Blood* **64**:967–974.

Miyoshi, I., Taguchi, H., Fujishita, M., Yoshimoto, S., Kubonishi, I., Ohtsuki, Y., Shiraishi, Y., and Akagi, T., 1982, Transformation of monkey lymphocytes with adult T-cell leukemia virus, *Lancet* **1**:1016.

Miyoshi, I., Kubonishi, I., Yoshimoto, S., Akagi, T., Ohtsuki, Y., Shiraishi, Y., Nagata, K., and Hinuma, Y., 1983a, Type C virus particles in a cord T-cell line derived by co-cultivating normal human cord leukocytes and human leukaemic T cells, *Nature* **294**:770–771.

Miyoshi, I., Yoshimoto, S., Taguchi, H., Kubonishi, I., Fujishita, M., Ohtsuki, Y., and Shiraishi, Y., 1983b, Transformation of rabbit lymphocytes with T-cell leukemia virus, *Gann* **74**:1–4.

Montminy, M. R., Sevarino, K. A., Wagner, J. A., Mandel, G., and Goodman, R. H., 1986, Identification of the cyclic-AMP responsive element in the rat somatostatin gene, *Proc. Natl. Acad. Sci. USA* **83**:6682–6686.

Miromoto, C., Matsuyama, T., Oshige, C., Tanaka, H., Hercend, T., Reinherz, E. L., and Schlossman, S. F., 1985, Functional and phenotypic studies of Japanese adult T-cell leukemia cells, *J. Clin. Invest.* **75**:836–843.

Morita, T., Sato, T., Nyunoya, H., Tsujimoto, A., Takahara, J., Irino, S., and Shimotohno, K., 1993, Isolation of a cDNA clone encoding DNA-binding protein (TAXREB107) that binds specifically to domain C of the tax-responsive enhancer element in the long terminal repeat of human T-cell leukemia virus type I, *AIDS Res. Hum. Retroviruses* **9**:115–121.

Murphy, E. L., Hanchard, B., Figueroa, J. P., Gibbs, W. N., Lofters, W. S., Campbell, M., Goedert, J. J., and Blattner, W. A., 1989, Modeling the risk of adult T-cell leukemia/lymphoma in persons infected with human T-lymphotropic virus type I, *Int. J. Cancer* **43**:250–253.

Namikawa, R., Kaneshima, H., Lieberman, M., Weissman, I. L., McCune, J. M., 1988, Infection of SCID-hu mouse by HIV-1, *Science* **242**:1684–1686.

Nerenberg, M., Hinrichs, S. H., Reynolds, R. K., Khoury, G., and Jay, G., 1987, The *tat* gene of human T-lymphotropic virus type I induces mesenchymal tumors in transgenic mice, *Science* **237**:1324–1329.

Nimer, S. D., 1991, Tax responsiveness of the GM-CSF promoter is mediated by mitogen-inducible sequences other that kappa B, *New Biol.* **3**:997–1004.

Nimer, S. D., Morita, E. A., Martis, M. J., Wachsmann, W., and Gasson, J. C., 1988, Characterization of the human granulocyte macrophage colony stimulating factor promoter region by genetic analysis: Correlation with DNase I footprinting, *Mol. Cell. Biol.* **8**:1979–1984.

Nimer, S. D., Gasson, J. C., Hu, K., Smalberg, I., Chen, I. S. Y., and Rosenblatt, J. D., 1989, Activation of the GM-CSF promoter by HTLV-I and HTLV-II *tax* proteins, *Oncogene* **4**:671–676.

Nishimura, M., Adachi, A., Maeda, M., Akiguchi, I., Fujita, M., Kameyama, M., and Ishimoto, A., 1988, Analysis of the provirus genome integrated in T cell lines established from the cerebrospinal fluid of patients with human T lymphotropic virus type-I-associated myelopathy (HAM), *J. Neuroimmunol.* **20**:33–37.

Nosaka, T., Siomi, H., Adachi, Y., Ishibashi, M., Kubota, S., Maki, M., and Hatanaka, M., 1989, Nucleolar targeting signal of human T-cell leukemia virus type I Rex-encoded protein is essential for cytoplasmic accumulation of unspliced viral mRNA, *Proc. Natl. Acad. Sci. USA* **86**:9798–9802.

Nyborg, J. K., and Dynan, W. S., 1990, Interaction of cellular proteins with the human T-cell leukemia virus type I transcriptional control region: Purification of cellular proteins that bind the 21-base pair repeat elements, *J. Biol. Chem.* **265**:8230–8236.

Nyborg, J. K., Dynan, W. S., Chen, I. S. Y., and Wachsman, W., 1988, Binding of host-cell factors to DNA sequences in the long terminal repeat of human T-cell leukemia virus type I: Implications for viral gene expression, *Proc. Natl. Acad. Sci. USA* **85**:1457–1461.

Nyunoya, H., Morita, T., Sato, T., Honma, S., Tsujimoto, A., and Shimotohno, K., 1993, Cloning of a cDNA encoding a DNA-binding protein TAXREB302 that is specific for the tax-responsive enhancer of HTLV-I, *Gene* **126**:251–255.

Ohta, M., Nyunoya, H., Tanako, H., Okamoto, T., Akagi, T., and Shimotohno, K., 1988, Identification of a *cis*-regulatory element involved in accumulation of human T-cell leukemia virus type II genomic mRNA, *J. Virol.* **62**:4445–4449.

Oliff, A., Signorelli, K., and Collins, L., 1984, The envelop gene and long terminal repeat sequences contribute to the pathogenic phenotype of helper-independent Friend viruses, *J. Virol.* **51**:788–794.

Orita, S., Saiga, A., Takagi, S., Tanaka, T., Okumura, K., Aono, Y., Hinuma, Y., and Igarashi, H., 1991, A novel alternately spliced viral mRNA transcribed in cells infected with human T-cell leukemia virus type I is mainly responsible for expressing p21X protein, *FEBS Lett.* **295**:127–134.

Orita, S., Kobayashi, H., Aono, Y., Saiga, A., Maeda, M., and Igarashi, H., 1993a, p21X mRNA is expressed as a singly spliced pX transcript from defective provirus genomes having a partial deletion of the pol-env region in human T-cell leukemia virus type 1-infected cells, *Nucleic Acids Res.* **21**:3799–3807.

Orita, S., Sato, S., Aono, Y., Minoura, N., Yamashita, T., Hinuma, Y., and Igarashi, H., 1993b, Identification of novel singly spliced pX mRNA transcripts common to all human T-cell leukemia virus type 1-related retroviruses, *Virus Genes* **7**:197–204.

Oroszlan, S., Copeland, T. D., Kalyanaraman, V. S., Sarngadharan, M. G., Schultz, A. M., and Gallo, R. C., 1984, Chemical analyses of human T-cell leukemia virus structural proteins, in: *Human T-cell Leukemia/Lymphoma Viruses* (R. C. Gallo, M. E. Essex, and L. Gross, eds.), pp. 101–110, Cold Spring Harbor Laboratory, Cold Spring Harbor, N.Y.

Osame, M., Izumo, S., Igata, A., Matsumoto, M., Matsumoto, T., Sonoda, S., Tara, M., and Shibata, Y., 1986, Blood transfusion with HTLV-I associated myelopathy, *Lancet* **2**:104–105.

Osame, M., Igata, A., Matsumoto, M., Usuku, K., Izumo, S., and Kosaka, K., 1987a, HTLV-I associated myelopathy: A report of 85 cases, *Ann. Neurol.* **22**:116.

Osame, M., Matsumoto, M., Usuku, K., Izumo, S., Ijichi, N., Amitani, H., Tara, M., and Igata, A., 1987b, Chronic progressive myelopathy associated with elevated antibodies to human T-lymphotropic virus type I and adult T-cell leukemia-like cells, *Ann. Neurol.* **21**:117–122.

Paskalis, H., Felber, B. K., and Pavlakis, G. N., 1986, Cis-acting sequences responsible for the transcriptional activation of human T-cell leukemia virus type I constitute a conditional enhancer, *Proc. Natl. Acad. Sci. USA* **83**:6558–6562.

Pierce, J. W., Lenardo, M., and Baltimore, D., 1988, Oligonucleotide that binds nuclear factor NF-κB as a lymphoid-specific and inducible enhancer element, *Proc. Natl. Acad. Sci. USA* **85**:1482–1486.

Poiesz, B. J., Ruscetti, F. W., Gazdar, A. F., Bunn, P. A., Minna, J. D., and Gallo, R. C., 1980, Detection and isolation of type C retrovirus particles from fresh and cultured lymphocytes of a patient with cutaneous T-cell lymphoma, *Proc. Natl. Acad. Sci. USA* **77**:7415–7419.

Popovic, M., Reitz, M. S., Jr., Sarngadharan, M. G., Robert-Guroff, M., Kalyanaraman, V. S., Nakao, Y., Miyoshi, I., Minowada, J., Yoshida, M., Ito, Y., and Gallo, R. C., 1982, The virus of Japanese adult T-cell leukaemia is a member of the human T-cell leukaemia virus group, *Nature* **300**:63–66.

Popovic, M., Lange-Wantzin, G., Sarin, P. S., Mann, D., and Gallo, R. C., 1983, Transformation of human umbilical cord blood T cells by human T-cell leukemia/lymphoma virus, *Proc. Natl. Acad. Sci. USA* **80**:5402–5406.

Poteat, H. T., Kaidson, P., McGuire, K., Park, L., Sodroski, J., and Haseltine, W., 1989, Response of the human T-cell leukemia virus type I long terminal repeat to cyclic AMP, *J. Virol.* **63**:1604–1611.

Powers, M. A., and Radke, K., 1992, Activation of bovine leukemia virus transcription in lymphocytes from infected sheep: Rapid transition through early to late gene expression, *J. Virol.* **66**:4769–4777.

Radonovich, M., and Jeang, K.-T., Activation of the human T-cell leukemia virus type I long terminal repeat by 12-O-tetradecanoylphorbol-13-acetate and by Tax (p40x) occurs through similar but functionally distinct target sequences, *J. Virol.* **63**:2987–2994.

Ratner, L., 1989, Regulation of expression of the c-sis proto-oncogene, *Nucleic Acids Res.* **17**:4101–4115.

Reddy, E. P., Sandberg-Wollheim, M., Mettus, R. V., Ray, P. E., DeFreitas, E., and Koprowski, H., 1989, Amplification and molecular cloning of HTLV-I sequences from DNA of multiple sclerosis patients, *Science* **243**:529–533.

Richardson, J. H., Wucherpfenning, K. W., Endo, N., Rudge, P., Dalgleish, A. G., and Hafler, D. A., 1989, PCR analysis of DNA from multiple sclerosis patients for the presence of HTLV-I, *Science* **246**:821.

Rimsky, L., Duc Dodon, M., Dixon, E. P., and Greene, W. C., 1989, *Trans*-dominant inactivation of HTLV-I and HIV-1 gene expression by mutation of the HTLV-I Rex transactivator, *Nature* **341**:453–456.

Robert-Guroff, M., Weiss, S. H., Giron, J. A., Jennings, A. M., Ginzburg, H. M., Margolis, I. B., Blattner, W. A., and Gallo, R. C., 1986, Prevalence of antibodies to HTLV-I, -II, and -III in intravenous drug abusers from an AIDS endemic region, *J. Am. Med. Assoc.* **255**:3133–3137.

Rosen, C. A., Park, R., Sodroski, J. G., and Haseltine, W. A., 1987, Multiple sequence elements are required for regulation of human T-cell leukemia virus gene expression, *Proc. Natl. Acad. Sci. USA* **84**:4919–4923.

Rosenblatt, J. D., Golde, D. W., Wachsman, W., Jacobs, A., Schmidt, G., Quan, S., Gasson, J. C., and Chen, I. S. Y., 1986, A second HTLV-II isolate associated with atypical hairy-cell leukemia, *N. Engl. J. Med.* **315**:372–375.

Rosenblatt, J. D., Cann, A. J., Slamon, D. J., Smalberg, I. S., Shah, N. P., Fujii, J., Wachsman, W., and Chen, I. S. Y., 1988a, HTLV-II trans-activation is regulated by two overlapping nonstructural genes, *Science* **240**:916–919.

Rosenblatt, J. D., Giorgi, J. V., Golde, D. W., Ben Eraza, J., Wu, A., Winberg, C. D., Glaspy, J., Wachsman, W., and Chen, I. S. Y., 1988b, Integrated HTLV-II genome in CD8+ T-cells from a patient with atypical hairy cell leukemia: Evidence for distinct T- and B-cell lymphoproliferative disorders, *Blood* **71**:363–366.

Ruben, S., Poteat, H., Tan, T. H., Kawakami, K., Roeder, R., Haseltine, W., and Rosen, C. A., 1988, Cellular transcription factors and regulation of IL-2 receptor gene expression by HTLV-I *tax* gene product, *Science* **241**:89–92.

Ruben, S. M., Perkins, A., and Rosen, C. A., 1989, Activation of NF-κB by HTLV-I transactivator protein Tax requires an additional factor present in lymphoid cells, *New Biol.* **1**:275–284.

Ruley, H. E., 1983, Adenovirus early region 1A enables viral and cellular transforming genes to transform primary cells in culture, *Nature* **304**:602–606.

Sagata, N., Yasunaga, T., Ogawa, Y., Tsuzuku-Kawamura, J., and Ikawa, Y., 1984, Bovine leukemia virus: Unique structural features of its long terminal repeats and its evolutionary relationship to human T-cell leukemia virus, *Proc. Natl. Acad. Sci. USA* **81**:4741–4745.

Sanada, I., Tanaka, R., Kumagai, E., Tsuda, H., Nishimura, H., Yamaguchi, K., Kawano, F., Fujiwara, H., and Takatsuki, K., 1985, Chromosomal aberrations in adult T cell leukemia: Relationship to the clinical severity, *Blood* **65**:649–654.

Sandler, S. G., 1986, HTLV-I and -II: New risks for recipients of blood transfusions? *J. Am. Med. Assoc.* **256**:2245–2246.

Sarin, P. S., Rodgers-Johnson, P., Sun, D. K., Thornton, A. H., Morgan, O. S. C., Gibbs, W. N., Mora, C., McKhann, G., II, Gajduske, D. C., and Gibbs, C. J., Jr., 1989, Comparison of a human T-cell lymphotropic virus type I strain from cerebrospinal fluid of a Jamaican patient with tropical spastic paraparesis with a prototype human T-cell lymphotropic virus type I, *Proc. Natl. Acad. Sci. USA* **86**:2021–2025.

Sawada, M., Suzumura, A., Yoshida, M., and Marunouchi, T., 1990, Human T-cell leukemia virus type I *trans* activator induces class I major histocompatibility complex antigen expression in glial cells, *J. Virol.* **64**:4002–4006.

Saxinger, W., Blattner, W. A., Levine, P. H., Clark, J., Biggar, R., Hoh, M., Moghissi, J., Jacobs, P., Wilson, L., Jacobson, R., Crookes, R., Strong, M., Ansari, A. A., Dean, A. G., Nkrumah, F. K., Mourali, N., and Gallo, R. C., 1984, Human T-cell leukemia virus (HTLV-I) antibodies in Africa, *Science* **225**:1473–1476.

Saxon, A., Stevens, R. H., Quan, S. G., and Golde, D. W., 1978, Immunologic characterization of hairy cell leukemias in continuous culture, *J. Immunol.* **120**:777–782.

Seiki, M., Hattori, S., and Yoshida, M., 1982, Human adult T-cell leukemia virus: Molecular cloning of the provirus DNA and the unique terminal structure, *Proc. Natl. Acad. Sci. USA* **79**:6899–6902.

Seiki, M., Hattori, S., Yoshida, M., 1983, Human adult T-cell leukemia virus: complete nucleotide sequence of the provirus genome integrated in leukemia cell DNA. *Proc. Natl. Acad. Sci. USA* **80**:3618–3622.

Seiki, M., Eddy, R., Shows, T. B., and Yoshida, M., 1984, Nonspecific integration of the HTLV provirus genome into adult T-cell leukaemia cells, *Nature* **309**:640–642.

Seiki, M., Inoue, J.-I., Hikada, M., and Yoshida, M., 1988, Two *cis*-acting elements responsible for posttranscriptional *trans*-regulation of gene expression of human T-cell leukemia virus type I, *Proc. Natl. Acad. Sci. USA* **85**:7124–7128.

Shah, N. P., Wachsman, W., Souza, L., Cann, A. J., Slamon, D. J., and Chen, I. S. Y., 1986, Comparison of the transactivation properties of HTLV-I and HTLV-II x proteins, *Mol. Cell. Biol.* **6**:3626–3631.

Shima, H., Takano, M., Shimotohno, K., and Miwa, M., 1986, Identification of $p26^{xb}$ and $p24^{xb}$ of human T-cell leukemia virus type II, *FEBS Lett.* **209**:289–294.

Shimotohno, K., Takahashi, Y., Shimizu, N., Gojobori, T., Chen, I. S. Y., Golde, D. W., Miwa, M., and Sugimura, T., 1985, Complete nucleotide sequence of an infectious clone of human T-cell leukemia virus type II: A new open reading frame for the protease gene, *Proc. Natl. Acad. Sci. USA* **82**:3101–3105.

Shimotohno, K., Takano, M., Terunchi, T., and Miwa, M., 1986, Requirement of multiple copies of a 21-nucleotide sequence in the U3 regions of human T-cell leukemia virus type I and type II long terminal repeats for trans-activation of transcription, *Proc. Natl. Acad. Sci. USA* **83**:8112–8116.

Shimoyama, M., Minato, K., Tobinai, K., Nagai, M., Setoya, T., Takenaka, T., Ishihara, K., Watanabe, S., Hoshina, H., Miwa, M., Kinoshita, M., Okabe, S., Fukushima, N., and Inada, N., 1983, Atypical adult T-cell leukemia–lymphoma: Diverse clinical manifestations of adult T-cell leukemia–lymphoma, *Jpn. J. Clin. Oncol.* **13**:165–188.

Siekevitz, M., Feinberg, M. B., Holbrook, N., Wong-Staal, F., and Greene, W. C., 1987a, Activation of interleukin 2 and interleukin 2 receptor (Tac) promoter expression by the trans-activator (tat) gene product of human T-cell leukemia virus, type I, *Proc. Natl. Acad. Sci. USA* **84**:5389–5393.

Siekevitz, M., Josephs, S. F., Dukovich, M., Peffer, N., Wong-Staal, F., and Greene, W. C., 1987b, Activation of the HIV-1 LTR by T cell mitogens and the trans-activator protein of HTLV-I, *Science* **238**:1575–1578.

Siomi, H., Shida, H., Nam, S. H., Nosaka, T., Maki, M., and Hatanaka, M., 1988, Sequence requirements for nucleolar localization of human T-cell leukemia virus type I pX protein which regulates viral RNA processing, *Cell* **55**:197–209.

Slamon, D. J., Shimotohno, K., Cline, M. J., Golde, D. W., and Chen, I. S. Y., 1984, Identification of the putative transforming protein of the human T-cell leukemia viruses HTLV-I and HTLV-II, *Science* **226**:61–65.

Slamon, D. J., Press, M. F., Souza, L. M., Cline, M. J., Golde, D. W., Gasson, J. C., and Chen, I. S. Y., 1985, Studies of the putative transforming protein of the type I human T-cell leukemia virus, *Science* **228**:1427–1430.

Smith, M. R., and Greene, W. C., 1991, Identification of HTLV-I tax trans-activator mutants exhibiting novel transcriptional phenotypes, *Genes Dev.* **4**:1875–1885.

Sodroski, J., Rosen, C., Goh, W. C., and Haseltine, W., 1985a, A transcriptional activator protein encoded by the x-lor region of the human T-cell leukemia virus, *Science* **228**:1430–1434.

Sodroski, J., Rosen, C., Wong-Staal, F., Salahuddin, S. K., Popovic, M., Arya, S., Gallo, R. C., and Haseltine, W. A., 1985b, *Trans*-acting transcriptional regulation of human T-cell leukemia virus type III long terminal repeat, *Science* **227**:171–173.

Sodroski, J., Goh, W. C., Rosen, C., Dayton, A., Terwillinger, E., and Haseltine, W., 1986, A second posttranscriptional *trans*-activator gene required for HTLV-III replication, *Nature* **321**:412–417.

Su, I.-J., Chan, H.-L., Kuo, T.-T., Eimoto, T., Maeda, Y., Kikuchi, M., Kuan, Y.-Z., Shih, L.-Y., Chen, M.-J., and Takeshita, M., 1985, Adult T-cell leukemia/lymphoma in Taiwan, *Cancer* **56**:2217–2220.

Sugamura, K., and Hinuma, Y., 1993, Human retroviruses: HTLV-I and HTLV-II, in: *The Retroviridae,* Vol. 2 (J. A. Levy, ed.), pp. 399–435, Plenum Press, New York.

Summerfelt, M. A., Williams, B. P., Clapham, P. R., Solomon, E., Goodfellow, P. N., and Weiss, R. A., 1988, Human T-cell leukemia viruses use a receptor determined by human chromosome 17, *Science* **242**:1557–1559.

Tan, T. H., Jia, R., and Roder, R. G., 1989a, Utilization of signal transduction pathway by the human T-cell leukemia virus type-I transcriptional activator tax, *J. Virol.* **63**:3761–3768.

Tan, T. H., Horikoshi, M., and Roder, R. G., 1989b, Purification and characterization of multiple nuclear factors that bind to the TAX-inducible enhancer within the human T-cell leukemia virus type I long terminal repeat, *Mol. Cell. Biol.* **9**:1733–1745.

Tanaka, A., Takahashi, C., Yamaoka, S., Nosaka, T., Maki, M., and Hatanaka, M., 1990, Oncogenic transformation by the *tax* gene of human T-cell leukemia virus type I *in vitro, Proc. Natl. Acad. Sci. USA* **87**:1071–1075.

Tanimura, A., Teshima, H., Fujisawa, J., and Yoshida, M., 1993, A new regulatory element that augments the Tax-dependent enhancer of human T-cell leukemia virus type 1 and cloning of cDNAs encoding its binding proteins, *J. Virol.* **67**:5375–5382.

Tateno, M., Kondo, N., Itoh, T., Chubachi, T., Togashi, T., and Yoshiki, T., 1984, Rat lymphoid lines with human T-cell leukemia virus production. I. Biological and serological characterization, *J. Exp. Med.* **159**:1105–1116.

Tedder, R. S., Shanson, D. C., Jeffries, D. J., Cheingsong-Popov, R., Dalgleish, A., Clapham, P., Nagy, K., and Weiss, R. A., 1984, Low prevalence in the UK of HTLV-I and HTLV-II infection in subjects with AIDS, with extended lymphadenopathy, and at risk of AIDS, *Lancet* **2**:125–128.

Temin, H. M., and Kassner, V. K., 1974, Replication of reticuloendotheliosis viruses in cell culture: Acute infection, *J. Virol.* **13**:291–297.

Tendler, C. L., Greenberg, S. J., Blattner, W. A., Manns, A., Murphy, E., Fleisher, T., Hanchard, B., Morgan, O., Burton, J. D., Nelson, D. L., and Waldmann, T. A., 1990, Transactivation of interleukin 2 and its receptor induces immune activation in human T-cell lymphotropic virus type I-associated myelopathy, *Proc. Natl. Acad. Sci. USA* **87**:5218–5222.

Toyoshima, H., Itoh, M., Inoue, J.-I., Seiki, M., Takaku, F., and Yoshida, M., 1990, Secondary structure of the human T-cell leukemia virus type I Rex-responsive element is essential for Rex regulation of RNA processing and transport of unspliced RNAs, *J. Virol.* **64**:2825–2832.

Tschachler, E., Bohnlein, E., Felzmann, S., and Reitz, M. S., Jr., 1993, Human T-lymphotropic virus type I tax regulates the expression of the human lymphotoxin gene, *Blood* **81**:95–100.

Tsujimoto, A., Nyunoya, H., Morita, T., Sato, T., and Shimitohno, K., 1991, Isolation of cDNAs for DNA binding proteins which specifically bind to a *tax*-responsive enhancer element in the long terminal repeat of human T-cell leukemia virus type I, *J. Virol.* **65**:1420–1426.

Uemura, Y., Kotani, S., Yoshimoto, S., Fujishita, M., Yamashita, M., Ohtsuki, Y., Taguchi, H., and Miyoshi, I., 1987, Mother-to-offspring transmission of human T cell leukemia virus type I in rabbits, *Blood* **69**:1255–1258.

Unanue, E. R., and Allen, P. M., 1987, The basis for the immunoregulatory role of macrophages and other accessory cells, *Science* **236**:551–557.

Varmus, H. E., Padgett, T., Heasley, S., Simon, G., and Bishop, J. M., 1977, Cellular functions are required for the synthesis and integration of avian sarcoma virus-specific DNA, *Cell* **11**:307–319.

Vernant, J. C., Maurs, L., Gessain, A., Barin, F., Gout, O., Delaporte, J., Sanhadji, K., Buisson, G., and de The, G., 1987, Endemic tropical spastic paraparesis associated with human T-lymphotropic virus type I: A clinical and seroepidemiological study of 25 cases, *Ann. Neurol.* **21**:123–130.

Vernant, J. C., Buisson, G., Magdeleine, J., De-Thore, J., Jouannelle, A., Neisson-Vernant, C., and Monplaisir, N., 1988, T-lymphocyte alveolitis, tropical spastic paresis, and Sjogren syndrome, *Lancet* **1**:177.

Wachsman, W., Golde, D. W., and Chen, I. S. Y., 1986, HTLV and human leukemia: Perspectives 1986, *Semin. Hematol.* **23**:246–256.

Wagner, S., and Green, M. R., 1993, HTLV-I Tax protein stimulation of DNA binding of bZIP proteins by enhancing dimerization, *Science* **262**:395–399.

Wano, Y., Feinberg, M., Hosking, J. B., Bogerd, H., and Greene, W. C., 1988, Stable expression of the *tax* gene of type I human T-cell leukemia virus in human T cells activates specific cellular genes involved in growth, *Proc. Natl. Acad. Sci. USA* **85**:9733–9737.

Wantzin, G. L., Thomsen, K., Nissen, N. I., Saxinger, C., and Gallo, R. C., 1986, Occurrence of human T cell lymphotropic virus (type I) antibodies in cutaneous T cell lymphoma, *J. Am. Acad. Dermatol.* **15**:598–602.

Watanabe, T., Seiki, M., Hirayama, Y., and Yoshida, M., 1986, Human T-cell leukemia virus type I is a member of the African subtype of simian viruses (STLV), *Virology* **148**:383–388.

Weichselbraun, I., Berger, J., Dobrovnik, M., Bogerd, H., Grassmann, R., Greene, W. C., Hauber, J., and Bohnlein, E., 1992, Dominant-negative mutants are clustered in a domain of the human T-cell leukemia virus type I Rex protein: Implications for *trans* dominance, *J. Virol.* **66**:4540–4545.

Weiss, R. A., 1993, Cellular receptors and viral glycoproteins involved in retrovirus entry, in: *The Retroviridae*, Vol. 2 (J. A. Levy, ed.), pp. 1–108, Plenum Press, New York.

Weiss, R., Teich, N., Varmus, H., and Coffin, J., eds., 1984, *RNA Tumor Viruses*, Cold Spring Harbor Laboratory, Cold Spring Harbor, N.Y.

Weiss, R. A., Clapham, P., Nagy, K., and Hoshino, H., 1985a, Envelope properties of human T-cell leukemia viruses, *Curr. Top. Microbiol. Immunol.* **115**:235–246.

Weiss, R., Teich, N., Varmus, H., and Coffin, J., eds., 1985b, *RNA Tumor Viruses*, Supplements and Appendixes, Cold Spring Harbor Laboratory, Cold Spring Harbor, N.Y.

Whittaker, S. J., and Luzzatto, L., 1993, HTLV-I provirus and mycosis fungoides, *Science* **259**:1470.

Yamada, Y., 1983, Phenotypic and functional analysis of leukemic cells from 16 patients with adult T-cell leukemia/lymphoma, *Blood* **61**:192–199.

Yamaguchi, K., Nishimura, H., Kohrogi, H., Jono, M., Miyamoto, Y., and Takatsuki, K., 1983, A proposal for smoldering adult T-cell leukemia: A clinicopathologic study of five cases, *Blood* **62**:758–766.

Yamamoto, N., Okada, M., Koyanagi, Y., Kannagi, Y., Kannagi, M., and Hinuma, Y., 1982, Transformation of human leukocytes by cocultivation with an adult T cell leukemia virus producer cell line, *Science* **217**:737–739.

Yamaoka, S., Tobe, T., and Hatanaka, M., 1992, Tax protein of human T-cell leukemia virus type I is required for maintenance of the transformed phenotype, *Oncogene* **7**:433–437.

Yip, M. T., Dynan, W. S., Green, P. L., Black, A. C., Arrigo, S. J., Torbati, A., Heaphy, S., Ruland, C., Rosenblatt, J. D., and Chen, I. S. Y., 1991, HTLV-II Rex protein binds specifically to RNA sequences of the HTLV LTR but poorly to the HIV-1 RRE, *J. Virol.* **65**:2261–2272.

Yoshida, M., Miyoshi, I., and Hinuma, Y., 1982, Isolation and characterization of retrovirus from cell lines of human adult T-cell leukemia and its implication in the disease, *Proc. Natl. Acad. Sci. USA* **79**:2031–2035.

Yoshida, M., Osame, M., Usuku, K., Matsumoto, M., and Igata, A., 1987, Viruses detected in HTLV-I-associated myelopathy and adult T-cell leukaemia are identical on DNA blotting, *Lancet* **1**:1085–1086.

Yoshida, M., Osame, M., Misaomi, K., Toita, M., Kuwasaki, N., Nishida, Y., Niraki, Y., Takahashi, K., Nomura, K., Sonoda, S., Eiraku, N., Ijichi, S., and Usuku, K., 1989, Increased replication of HTLV-I in HTLV-I-associated myelopathy, *Ann. Neurol.* **26**:331–335.

Yoshikura, H., Nishida, J., Yoshida, M., Kitamura, Y., Takaku, F., and Ikeda, S., 1984, Isolation of HTLV derived from Japanese adult T cell leukemia patients in human diploid fibroblast strain IMR90 and the biological characters of the infected cells, *Int. J. Cancer* **33**:745–749.

Yoshimura, T., Fujisawa, J., and Yoshida, M., 1990, Multiple cDNA clones encoding nuclear proteins that bind to the tax-dependent enhancer of HTLV-I: All contain a leucine zipper structure and a basic amino acid domain, *EMBO J.* **9**:2537–2544.

Zack, J. A., Cann, A. J., Lugo, J. P., and Chen, I. S. Y., 1988, AIDS virus production from infected peripheral blood T cells following HTLV-I-induced mitogenic stimulation, *Science* **240**:1026–1029.

Zack, J. A., Arrigo, S. J., Weitsman, S. R., Go, A. S., Haislip, A., and Chen, I. S. Y., 1990, HIV-1 entry into quiescent primary lymphocytes: Molecular analysis reveals a labile, latent viral structure, *Cell* **61**:213–222.

Zazopoulos, E., Sodroski, J. G., and Haseltine, W. A., 1990, p21rex protein of HTLV-I, *J. AIDS* **3**:1135–1139.

Zella, D., Mori, L., Sala, M., Ferrante, P., Casoli, C., Magnani, G., Achilli, G., Cattaneo, E., Lori, F., and Bertazzoni, U., 1990, HTLV-II infection in Italian drug abusers, *Lancet* **336**:575–576.

Zhao, L.-J., and Giam, C.-Z., 1991, Interaction of the human T-cell lymphotropic virus type I (HTLV-I) transcriptional activator Tax with cellular factors that bind specifically to the 21-base-pair repeats in the HTLV-I enhancer, *Proc. Natl. Acad. Sci. USA* **88**:11445–11449.

CHAPTER 7

Retroviruses and Their Role in Cancer

HUNG FAN

I. INTRODUCTION

Historically, retroviruses were first isolated and studied because they efficiently cause tumors in experimental and domestic animals (Rous, 1911; Ellerman and Bang, 1908). Work was particularly intense during the 1970s, under the "War on Cancer" and the Special Virus Cancer Program. Important discoveries from retrovirology such as oncogenes and proto-oncogenes (Bishop, 1987) have had profound implications for understanding tumorigenesis, including for spontaneous human tumors. Retroviruses remain some of the best model systems for studying carcinogenesis in the whole organism, since they induce tumors with extremely high efficiency. The focus of this chapter will be retroviral carcinogenesis in small animal models. Major attention will be paid to nonacute retroviruses, since they induce disease through multiple steps, with analogies to non-virus-induced cancer as well.

In the whole animal or human, cancer is the culmination of a series of changes at the subcellular, cellular, and organismal levels. Alterations in the incipient tumor cell must occur, and in addition host immunological defenses against tumors must fail. Histopathologists have identified various premalignant states (e.g., hyperplasia, dysplasia, carcinoma *in situ*), and malignant tumors themselves often undergo progression. Recently, at least four molecular changes in tumor cells have been identified in the pathway to human colon carcinoma, including inactivation of tumor suppressor genes and activation of proto-oncogenes (Cho and Vogelstein, 1992). In the context of retroviral carcinogenesis, the retrovirus may induce one or several changes leading to tumor develop-

HUNG FAN • Department of Molecular Biology and Biochemistry and Cancer Research Institute, University of California, Irvine, California 92717.

The Retroviridae, Volume 3, edited by Jay A. Levy. Plenum Press, New York, 1994.

ment. However, it is now clear that even for the most potent retroviruses, carcinogenesis is still a multistep process.

In considering retroviral oncogenesis, the retrovirus may have a direct or indirect role. Direct roles result from action of a viral gene product or control sequence in the tumor cell, resulting in increased cell proliferation. While direct roles by retroviruses are well characterized, indirect roles may be important as well. In indirect mechanisms, viral infection alters cells other than the tumor cells, with indirect effects on the target cells. For example, viral modification of the immune system can lead to tumors, e.g., secondary malignancies in AIDS (Fauci, 1988). Other indirect mechanisms will be discussed below.

With regard to carcinogenesis, retroviruses can be divided into two classes: acute transforming retroviruses and nonacute retroviruses (Teich *et al.*, 1984). Nonacute retroviruses are standard replication-competent retroviruses. They typically induce neoplasms with relatively long latencies, and they do not morphologically transform cells in culture. Acute transforming retroviruses are derived from nonacute retroviruses, and they carry oncogenes which endow them with the capacity to rapidly induce tumors and morphologically transform cells in culture (Duesberg and Vogt, 1970). Viral oncogenes result from retroviral capture of normal cell proto-oncogenes—genes encoding proteins that play roles in normal growth control (Stehelin *et al.*, 1976). The relationship between a typical acute transforming virus, its parental retrovirus, and the cellular proto-oncogene is shown in Fig. 1. During the capture process, alterations in the proto-oncogene occur (deletions, fusions, point mutations) that convert it into an oncogene (Bishop and Varmus, 1984). Most acute transforming retroviruses are replication-defective; they are typically propagated as mixtures of the acute transforming retrovirus and a replication-competent retrovirus. Thus, tumorigenesis by acute transforming retroviruses can be viewed as viral introduction of a potent transforming gene (the oncogene) into cells, which induces them to develop into tumors.

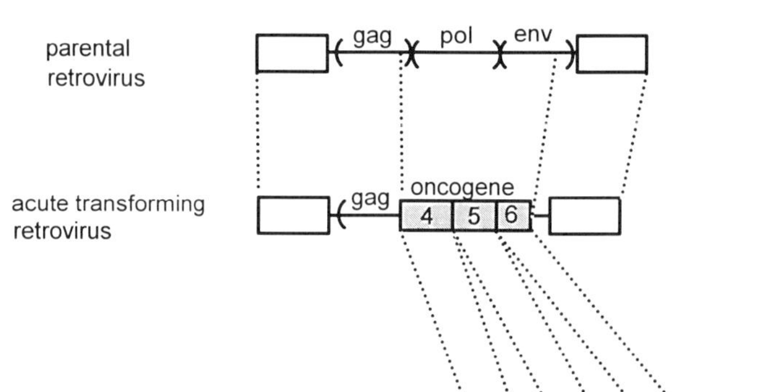

FIGURE 1. Organization of acute transforming retroviruses. The relationship of a typical acute transforming retrovirus to the parental retrovirus and its corresponding cellular proto-oncogene is shown. Most acute transforming retroviruses have lost viral sequences, leading them to be replication-defective by themselves. Their oncogenes often correspond to a subset of the exons of the proto-oncogene. Point mutations within the oncogene relative to the proto-oncogene are also often found. Viral oncogenes also lack the introns of the corresponding proto-oncogene.

II. TUMORIGENESIS BY ACUTE TRANSFORMING RETROVIRUSES

Numerous acute transforming retroviruses have been intensively studied. Most attention has been focused on identifying their oncogenes and defining the mechanisms of oncogene action. Some of the well-studied acute transforming retroviruses (and their oncogenes) are shown in Table I. Retroviral oncogenes will be considered in detail in other chapters in this series. However, two aspects of acute retroviral pathogenicity deserve consideration.

TABLE I. Acute Transforming Retroviruses

Virus	Host	Tumor type	Oncogene	
Rous sarcoma virus	Chicken	Sarcoma	*v-src:*	Tyrosine-specific protein kinase
Fujinami virus	Chicken	Sarcoma	*v-fps:*	Tyrosine-specific protein kinase
PRCII	Chicken	Sarcoma	*v-fps:*	
MC29	Chicken	Myelocytomas, carcinomas, sarcomas	*v-myc:*	Nuclear protein, interacts with DNA binding protein (*max*)
Avian myeloblastosis virus (AMV)	Chicken	Myeloblastosis	*v-myb:*	Nuclear DNA-binding protein
Avian erythroblastosis virus (AEV)	Chicken	Erythroblastosis	*v-erbB:*	Tyrosine-specific protein kinase; cellular homologue is EGF receptor
			v-erbA:	Nuclear DNA-binding protein; cellular homologue is thyroid hormone receptor
S-17	Chicken	Sarcoma	*v-jun:*	Nuclear DNA-binding protein; cellular homologue is part of Ap-1 transcription factor
Moloney murine sarcoma virus (M-MSV)	Mouse	Sarcoma	*v-mos:*	Serine-threonine protein kinase
Kirsten MSV	Mouse	Sarcoma	*v-ras*^{K}:	G-protein that signals in growth
Harvey MSV	Mouse	Sarcoma	*v-ras*^{H}:	G-protein that signals in growth
Abelson MuLV	Mouse	B lymphoma	*v-abl:*	Tyrosine-specific protein kinase
FBJ osteosarcoma virus	Mouse	Osteosarcoma	*v-fos:*	Nuclear DNA binding protein; cellular homologue binds *c-jun* protein
Friend spleen focus-forming virus	Mouse	Erythroleukemia	*gp52:*	Deleted MCF-type envelope glycoprotein
Feline sarcoma virus (McDonough strain)	Cat	Sarcoma	*v-fms:*	Tyrosine-specific protein kinase; cellular homologue is CSF-1 receptor
Snyder–Theilen FeSV	Cat	Sarcoma	*v-fes:*	Tyrosine-specific protein kinase; cellular homologue is equivalent to chicken *c-fps*
Simian sarcoma virus	Monkey	Sarcoma	*v-sis:*	Cellular homologue is PDGF beta chain

A. Determinants of Tumor Specificity

An interesting feature of tumorigenesis by acute transforming retroviruses is that each particular virus induces tumors of a specific (or a small number of) cell types when inoculated into susceptible hosts (see Table I). For instance, Rous sarcoma virus induces fibrosarcomas, Abelson murine leukemia virus induces B lymphomas, and avian erythroblastosis virus induces erythroleukemia. Several possible explanations for this specificity exist:

1. The oncogene may transform cells of a particular cell type, resulting in the observed tumor specificity.
2. The tissue specificity of the enhancers in the viral long terminal repeat (LTR) may determine in what cell types the oncogene is efficiently expressed (see Luciw and Leung, 1992).
3. Fusion of viral protein sequences to the oncogene protein or truncation of proto-oncogene domains may influence tissue specificity of transformation.

Evidence for or against each of these possibilities will be discussed.

1. Oncogene Effects

The strongest arguments for the transduced oncogene influencing specificity of the resulting tumor type come from the behaviors of independently isolated acute transforming retroviruses carrying the same oncogene. Such naturally occurring isolates often induce the same kinds of tumors. Two avian viruses that appear to induce early myeloid tumors both carry the *myb* oncogene—AMV and E26 (Radke *et al.*, 1982; Klempnauer *et al.*, 1982). The chicken Fujinami and PRCII sarcoma viruses carry the chicken *fps* oncogene (Shibuya and Hanafusa, 1982), which is equivalent to the feline *fes* oncogene present in Snyder–Theilen feline sarcoma virus (Groffen *et al.*, 1983). Kirsten murine sarcoma virus (MSV), Harvey MSV, and the Rasheed rat sarcoma virus all have transduced one of two closely related *c-ras* genes from the rat genome (Rasheed *et al.*, 1983; Gonda *et al.*, 1982; Chang *et al.*, 1982), and they all induce fibrosarcomas.

Nevertheless, exceptions to this general pattern also exist. For instance, HZ2 feline sarcoma virus (which induces fibrosarcomas) carries a transduced form of the feline *c-abl* oncogene (Bergold *et al.*, 1987), while Abelson MuLV (which induces B-lymphoid tumors) carries the same oncogene but transduced from the mouse genome (Goff *et al.*, 1980). Several avian retroviruses carrying the *myc* oncogene have been independently isolated. Depending on the virus, the resulting diseases range from myelocytomatosis (myeloid leukemia) alone (CMII, FH3) through myelocytomatosis, carcinomas, and sarcomas (MC29), to carcinomas and sarcomas (MH2 and OK10) (Teich *et al.*, 1984). In addition, several FeLV derivatives containing transduced feline *c-myc* have been isolated from FeLV-induced lymphoid tumors (Levy *et al.*, 1984; Neil *et al.*, 1984), and these *myc*-containing viruses induce lymphosarcomas in cats (Levy *et al.*, 1988).

The issue of tumor specificity has also been addressed by the generation of new acute transforming retroviruses by molecular cloning. Several *src*-containing murine retroviruses have been generated by inserting the chicken *v-src* gene from Rous sarcoma virus into murine leukemia virus backbones (Anderson and Scolnick, 1983; Feuerman *et al.*, 1985, 1988). In all of these cases, the only frank tumors resulting were fibrosarcomas, like the tumors induced by RSV in chickens. There was, however, some suggestion as well for induction of erythroid hyperplasia by one of the viruses (Anderson and Scolnick, 1983) but erythroleukemia was not observed. Yet, an MuLV containing the *v-myc* sequences from avian MC29 virus was also generated in an analogous fashion (Brightman *et al.*, 1988b). While this virus transforms mouse fibroblasts and myeloid cells in culture, much like MC29 in chicken cells, it induces lymphoid leukemia when injected into mice (Brightman *et al.*, 1986).

Thus, while comparative virology suggests that the nature of the oncogene may play an important role in tumor specificity, there are clearly exceptions indicating that other aspects of viral organization prevail in some cases.

2. LTR Effects

The enhancers in many retroviral LTRs are preferentially active in particular tissues. In the case of nonacute retroviruses, they are major determinants in the disease specificity (see below). It would be reasonable that the LTRs also govern disease specificity for acute transforming retroviruses, since high-level expression of an oncogene in a particular cell would clearly potentiate tumorigenesis of that cell. Evidence consistent with this possibility comes from myeloproliferative sarcoma virus (MPSV) and Moloney murine sarcoma virus (M-MSV) (Stocking *et al.*, 1985). Both of these mouse retroviruses carry the *v-mos* oncogene. However, MPSV induces both myeloid leukemias and fibrosarcomas, while M-MSV exclusively induces fibrosarcomas. The difference in pathogenic behavior of these two viruses is governed by the LTRs. Substitution of the MPSV LTR onto the M-MSV genome results in a virus that induces both myeloid leukemias and fibrosarcomas, and vice versa (Stocking *et al.*, 1985).

Experiments with certain other acute transforming viruses, however, have not found strong effects of the LTRs on disease specificity. Green and Risser (Green *et al.*, 1988) investigated this issue by substituting the F-MuLV LTR onto Abelson MuLV, but they found that B lymphomas were still induced. Jolicoeur *et al.* (1991) changed the LTRs on MuLVs carrying the *v-abl*, *v-ras*, and *v-fos* oncogenes, and found no effect on the resulting disease. In experiments on *v-src*- and *v-myc*-containing MuLVs, substitution of the MPSV and F-MuLV LTRs onto these viruses did not change the phenotypes of the resulting tumors [fibrosarcoma for M-MuLV (*src*) and B- and T-lymphomas for M-MuLV (*myc*)] (Feuer *et al.*, 1993). However, alterations in the tumor latency were observed. Thus, while LTR specificities may influence tumor specificity for some acute transforming retroviruses, they do not appear to be the primary factors for many others.

3. Viral–*onc* Fusion Proteins

The fusion of viral protein to oncogene protein in acute transforming retroviruses could potentially affect the tissue specificity of the transforming protein by altering stability or affinity for substrates in particular cells. Evidence for this possibility exists in the case of Abelson MuLV, which encodes a *gag–abl* fusion protein (Goff *et al.*, 1981). Deletions within the *gag* domain reduce the ability of Ab-MuLV to transform lymphoid cells in culture, primarily because of decreased stability of the oncogene protein (Prywes *et al.*, 1983). This finding indicates a role for *gag* protein in the lymphoid transformation. The converse has been observed in the *v-myc*-containing avian retrovirus FH3 (Chen *et al.*, 1989). FH3, which encodes a *gag–myc* fusion protein, induces cranial myelocytomas and transforms macrophages but not fibroblasts in culture. However, spontaneous FH3 variants appear during *in vitro* passage which have acquired fibroblast transformation capacity. These variants show deletions in the *gag* portions of the fusion protein (CA and NC) (Tikhonenko and Linial, 1992). Thus, for FH3, some *gag* sequences restrict transformation of fibroblasts. A comparison of other avian *myc*-containing viruses is also noteworthy. As indicated above, these viruses induce various combinations of disease, ranging from pure myeloid to pure carcinoma/sarcoma. All of the viruses make different forms of viral–*myc* fusion proteins, suggesting the importance of the nature of the fusion proteins (Bishop and Varmus, 1984). However, this possibility has not been systematically investigated from this vantage point.

Nevertheless, other situations exist where the nature of the viral–*onc* fusion protein appears secondary in disease specificity. As described above, different retroviruses carrying the same oncogene often induce the same tumor types. Also, when two murine retroviruses expressing the Rous sarcoma virus *src* gene were compared, a virus expressing an M-MuLV *gag–src* fusion protein induced fibrosarcomas just as does an M-MuLV expressing *v-src* alone (Feuerman *et al.*, 1988).

In summary, for most cases, the disease specificity for each acute transforming retrovirus is probably influenced by multiple factors, with stronger effects from the nature of the oncogene, modulated by the fusion of viral sequences to the oncogene protein. LTRs are probably less important in most cases. Other factors that can influence the types of tumors resulting from an acute transforming retrovirus are the age and route of inoculation, perhaps reflecting access to target cells for tumorigenic transformation.

B. Secondary Events in Tumorigenesis by Acute Transforming Retroviruses

The fact that acute transforming retroviruses require secondary events for tumor formation, even though they carry oncogenes, has often been overlooked. Several examples of this phenomenon will be described here. The first case is Rous sarcoma virus tumorigenesis in chickens. When RSV is injected into young birds, tumors form at the site of inoculation (e.g., the wing web), but not at distant sites. Nevertheless, these birds are viremic for RSV, and it is likely that other nontransformed cells are virus-infected. Indeed, it has been shown that RSV infection of developing embryos does not interfere with normal embryogenesis, even though substantial levels of $pp60^{src}$ transforming protein are expressed in many tissues (Dolberg and Bissell, 1984). However, if RSV-infected young birds are subsequently subjected to wounding (e.g., a wing clip), an additional tumor will form at the site of the wound (Dolberg *et al.*, 1985). This finding indicates that the physical injury (wing clip or initial injection) may induce local changes (mitoses, release of growth factors) that supply secondary events contributing to tumorigenesis. Subsequent studies have indicated that local release of transforming growth factor-β (TGF-β) supplies the crucial secondary event necessary for tumor formation (Sieweke *et al.*, 1989, 1990).

A second example involves tumorigenesis by Abelson MuLV. Early experiments with Ab-MuLV indicated that tumorigenicity was dependent on the helper virus: Moloney MuLV was effective, whereas an endogenous MuLV was not (Rosenberg and Baltimore, 1978). This dependence was subsequently mapped to the LTR of the helper M-MuLV, and a common insertion site for the M-MuLV helper (*ahi-1*) was identified (Poirier *et al.*, 1988). Presumably this process represents activation of a secondary cellular proto-oncogene analogous to LTR activations observed with nonacute retroviruses (see below); such an activation apparently contributes to tumorigenesis by Ab-MuLV.

A third example involves the Friend erythroleukemia complex. This virus mixture consists of an acute transforming virus (Friend SFFV) and a helper virus (F-MuLV) (Teich *et al.*, 1984). The SFFV component can induce erythroleukemia in the absence of the helper virus (Wolff *et al.*, 1986; Wolff and Ruscetti, 1985). Friend SFFV encodes a truncated form of recombinant (MCF-like) envelope glycoprotein (Dresler *et al.*, 1979; Ruscetti *et al.*, 1979) that is capable of rendering erythroid precursors independent of erythropoietin by binding to the erythropoietin receptor (Li *et al.*, 1990) (see Section III.C.3). In animals infected with polycythemia-inducing strains of the Friend complex, an early stage of polyclonal erythroid expansion is observed, driven by the SFFV component (Sitbon *et al.*, 1986). At later times, monoclonal erythroid tumors emerge, and they show two hallmarks of secondary tumorigenic events: (1) SFFV proviral insertion and activation of a common insertion site [*spi-1*, an *ets*-like transcription factor (Moreau-Gachelin *et al.*, 1988; Paul *et al.*, 1991)] and (2) deletion or inactivation of the p53 tumor suppressor gene (Dreyfus *et al.*, 1990; Mowat *et al.*, 1985). Thus, two secondary events in addition to the oncogene are required for leukemogenesis by Friend SFFV.

A final example involves a murine retrovirus generated by molecular cloning which expresses the mouse *c-myc* gene (Baumbach *et al.*, 1986). This virus is not pathogenic when injected into BALB/c mice by itself. However, if the mice are subsequently treated with pristane (which elicits a macrophage response), then myeloid tumors result (Baumbach *et al.*, 1987). Moreover, the tumors show rearrangements (activations) of the CSF-1 receptor gene; CSF-1 is an essential growth factor for myelomonocytic cells. Thus, in addition to overexpression of *c-myc* by the retrovirus, two other processes are required for tumorigenesis: myeloid cell growth stimulation and activation of the CSF-1 receptor.

III. TUMORIGENESIS BY NONACUTE RETROVIRUSES

Nonacute retroviruses induce long-latency neoplasms in a variety of animals (reviewed in Teich *et al.*, 1984). The avian viruses include avian leukosis virus (ALV) associated with B lymphomas or erythroid leukemia, depending on host genetic background (Hayward *et al.*, 1981), and myeloblastosis-associated virus (MAV) linked to osteopetrosis and nephroblastoma (Smith, 1982). The murine viruses include various strains of murine leukemia virus (see Table II) and murine mammary tumor virus (MMTV) associated with mammary carcinomas (Kozak and Ruscetti, 1992). In some cases, carcinogenic murine viruses are endogenous

TABLE II. Frequently Studied Murine Leukemia Viruses

Virus	Disease
Moloney MuLV (M-MuLV)	T lymphoma
AKR-derived MuLVs	
Gross passage A	T lymphoma
SL3-3	T lymphoma
Akv (AKR endogenous MuLV)	Not pathogenic
Radiation leukemia virus (RadLV)	T lymphoma
Friend MuLV (F-MuLV)	Erythroleukemia; myeloid leukemia
Rauscher MuLV (R-MuLV)	Erythroleukemia; myeloid leukemia
Graffi MuLV	Myeloid leukemia (chloroleukemia)
Soule MuLV	Myeloid leukemia

viruses activated from the germ line of certain inbred mouse strains (e.g., Akv MuLV from the AKR strain, and MMTV from the GR strain). Feline leukemia virus (FeLV) is another important model system because it efficiently induces lymphosarcomas in outbred cats (Jarrett *et al.*, 1964a,b; see Hardy, 1993). The principles involved in nonacute retroviral carcinogenesis are similar for different classes of virus, so these concepts will be discussed first. Detailed illustrations will then be presented for murine nonacute retroviruses.

A. LTR Activation of Proto-oncogenes

The best understood mechanism involved in nonacute retroviral carcinogenesis is LTR activation of proto-oncogenes. This process was first elucidated by Hayward and collaborators for B lymphomas induced by avian leukosis virus (Hayward *et al.*, 1981). During retroviral replication, proviral DNA is inserted into many (virtually random) sites in the host genome (see Luciw and Leung, 1992). However, when independent ALV-induced tumors were examined for proviral insertion, they were found at a specific chromosomal locus—in the vicinity of the *c-myc* proto-oncogene (Hayward *et al.*, 1981). Further analyses showed that the proviral insertion led to transcriptional activation of *c-myc* expression by readthrough from the cryptic promoter in the ALV downstream LTR. This process is diagrammed in Fig. 2. This mechanism of proto-oncogene activation was termed "promoter insertion." A variation of this process has been observed in ALV-induced erythroblastosis in chickens of a different genetic background (Fung *et al.*, 1983). The *erbB* proto-oncogene is activated; however, transcription is initiated in the upstream LTR (as for standard proviral transcription) followed by readthrough past the downstream LTR into *erbB* sequences. In this case, an RNA splicing event removes most of the viral sequences from the fusion transcript, yielding an mRNA that overexpresses *erbB*. A similar process has been observed in Moloney MuLV activation of the *c-myb* proto-oncogene in myeloid tumors (Shen-Ong and Wolff, 1987; Nason-Burchenal and Wolff, 1993). This alternative process is diagrammed in Fig. 3. In either of these cases, high-efficiency transcription from the retroviral LTR leads to overexpression of a proto-oncogene product, which provides an unregulated growth stimulus to the developing tumor cell.

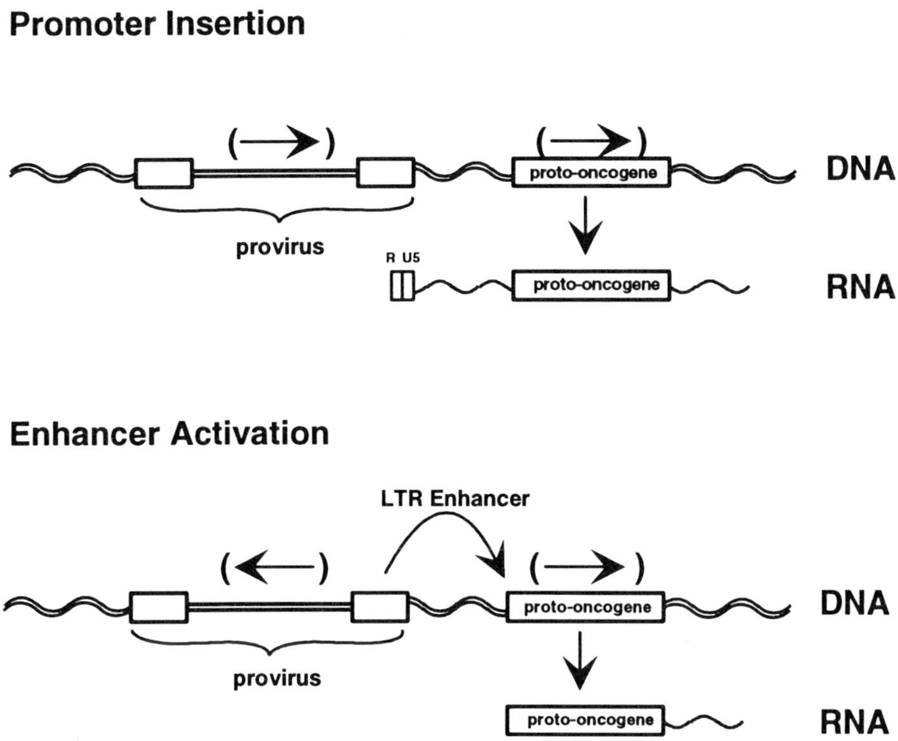

FIGURE 2. Proto-oncogene activation during non-acute retrovirus leukemogenesis. The promoter insertion mechanism is illustrated at the top of the diagram. In this case, transcription initiates in the downstream LTR of an inserted provirus and proceeds into the proto-oncogene coding sequences. The result is over-expression of the proto-oncogene (as a hybrid transcript) under control of the retroviral LTR. The enhancer activation mechanism is illustrated at the bottom of the diagram. In this case, the LTR enhancer activates transcription from the proto-oncogene via its own promoter. Processing of the primary transcript (e.g. mRNA splicing) may also occur, but it is not shown in the figure.

Subsequent to the initial discovery of promoter insertion, other virus–tumor systems were found in which classical promoter insertion does not occur, although site-specific insertion of proviruses in the vicinity of proto-oncogenes is found. This was observed by others in ALV-induced B lymphomas (Payne *et al.*, 1982), and for tumors induced by other nonacute retroviruses such as MMTV (Nusse and Varmus, 1982) and murine leukemia viruses. In these cases, the provirus was inserted next to the proto-oncogene in the wrong configuration for classical promoter insertion; e.g., in the opposite transcriptional orientation (Payne *et al.*, 1982; Nusse and Varmus, 1982; Cuypers *et al.*, 1984b) or downstream of the proto-oncogene (Payne *et al.*, 1982), or at sites spanning over 1 kb (Nusse and Varmus, 1982). Nevertheless, in all of these cases, overexpression of the proto-oncogenes occurred. The explanation was that the enhancer sequences in retroviral LTRs were activating transcription from the proto-oncogene's own promoter. This possibility was consistent with the fact that enhancers are *cis*-acting DNA elements that can activate adjacent promoters in an orientation-, position-, and distance-independent manner. Enhancer activation is also illustrated in Fig. 2.

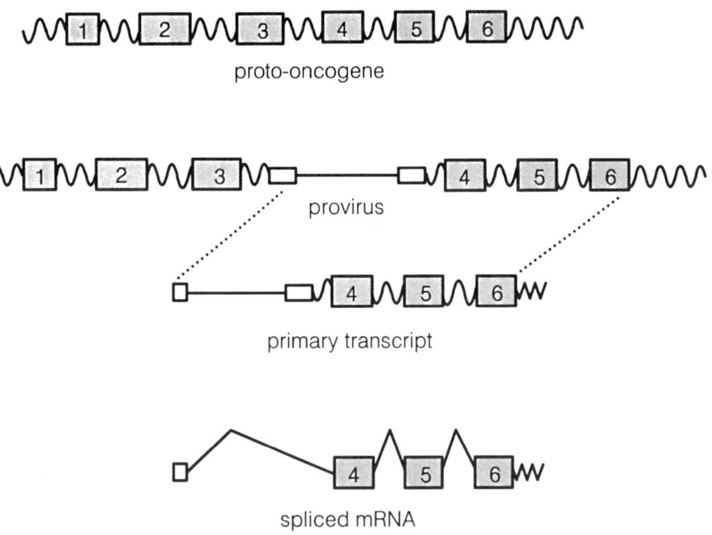

FIGURE 3. An alternative form of promoter-insertion. In this situation, transcription is under control of the virus, beginning in the upstream LTR of the provirus, with readthrough of the normal cleavage/polyadenylation site into downstream exons of the proto-oncogene. mRNA splicing then eliminates all but the extreme 5′ viral sequences from the mature mRNA.

Taken together, these mechanisms can be termed LTR activation of proto-oncogenes, with promoter insertion being a special case. For some well-studied systems (e.g., ALV-induced B lymphoma, Moloney MuLV-induced T lymphoma), the majority of tumors show LTR activation of one or a small number of proto-oncogenes. In fact, study of nonacute retrovirus-induced tumors has led to identification of new proto-oncogenes that have never been captured into acute transforming retroviruses: e.g., *pim-1* [M-MuLV–lymphoma (Cuypers *et al.*, 1984b)] and *int-1/wnt-1, int-2, int-3,* and *wnt-3* [MMTV-induced mammary tumors (Dickson *et al.*, 1984; Gallahan and Callahan, 1987; Roelink *et al.*, 1990)]. For some virus–tumor systems, common insertion sites have not yet been identified for a majority of the tumors (Fan *et al.*, 1986; Hanecak *et al.*, 1988). It is unclear if this finding represents activation of as yet unidentified proto-oncogenes, or if some nonacute retroviruses induce tumors without LTR activation of proto-oncogenes.

LTR activation of proto-oncogenes also provides a rationale for (at least part of) the long tumor latency for nonacute retroviruses. Since retroviral DNA integration occurs at multiple (random) sites in the host DNA, the likelihood of insertion near any given proto-oncogene is low on a probability basis. Thus, multiple rounds of infection (and time) may be required before proviral insertion in the vicinity of the crucial proto-oncogene for tumorigenesis to occur. This could possibly be amplified if multiple proto-oncogene activations are required (see below).

B. The LTR Is the Major Determinant in Pathogenicity

For nonacute retroviruses, the LTRs are the major determinants for pathogenicity (reviewed by Fan, 1990). This observation has largely been addressed by generation of chimeric viruses via molecular cloning, in which the parents differ in pathogenic properties. The earliest experiments compared related pathogenic and nonpathogenic avian viruses (Robinson *et al.*, 1982), and pathogenicity was found to correlate with the origin of the U3 region of the LTR. Subsequently, similar results were found for murine retroviruses (DesGroseillers *et al.*, 1983; DesGroseillers and Jolicoeur, 1984; Lenz and Haseltine, 1983; Lenz *et al.*, 1984). In general, nonacute retroviruses with LTRs that support high-efficiency expression are more pathogenic. In more detailed analyses, pathogenicity was found to be determined by the nature of core elements in the transcriptional enhancers (Speck *et al.*, 1990; Hallberg *et al.*, 1991). This finding can be readily rationalized in light of LTR activation of proto-oncogenes (Fig. 1): in order for efficient activation of a proto-oncogene to take place in a tumor cell, the enhancer sequences must be highly active.

Other experiments have shown that the tissue specificity of nonacute retroviral disease is also controlled by the nature of the LTR enhancers. The best documented case has been with chimeras between Moloney MuLV (T lymphoma) and Friend MuLV (erythroleukemia) (see Fig. 4). Replacement of the M-MuLV LTR with the F-MuLV LTR will convert M-MuLV into a virus that induces erythroleukemia, and the converse is also true (Chatis *et al.* 1983, 1984). Further experiments with refined M-MuLV–F-MuLV chimeras indicated that the enhancer sequences were the crucial elements responsible for disease specificity (Li *et al.* 1987). Transient expression assays with LTR–reporter gene constructs indicated that the M-MuLV LTR expresses more efficiently in T-lymphoid cell lines, while the F-MuLV LTR works poorly in those lines but is efficient in erythroid cell lines (Li *et al.*, 1987; Boral *et al.*, 1989). These results are again easily rationalized in terms of LTR activation of proto-oncogenes: in order for a nonacute retrovirus to induce a neoplasm in a particular differentiated cell type, the LTR enhancers must be highly active in that cell type to efficiently activate an adjacent proto-oncogene. Other similar studies with chimeric viruses have mapped determinants for non-T/non-B lymphoma in the LTR of amphotropic MuLV (DesGroseillers and Jolicoeur, 1984), and for T lymphoma in the LTR of ecotropic CasBr MuLV (Jolicoeur and DesGroseillers, 1985).

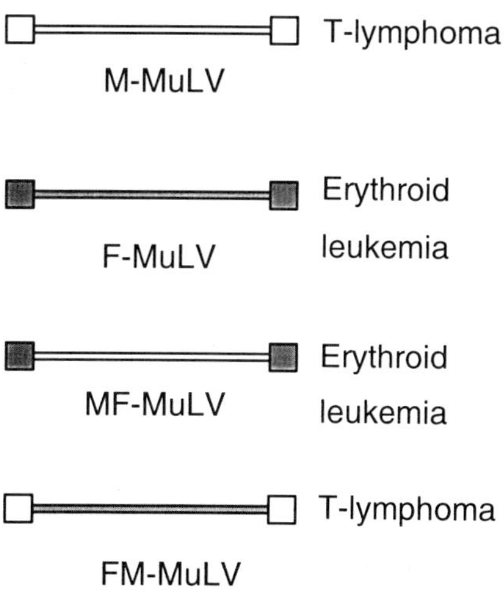

FIGURE 4. Pathogenic behavior of chimeric MuLVs. Chimeric MuLV's between Moloney MuLV and Friend MuLV were generated by molecular cloning (Chatis *et al.* 1983). The disease specificity of the chimera is dictated by the origin of the LTRs.

The role of the LTRs in nonacute retrovirus leukemogenicity has also been investigated by mutational analyses in the LTRs. Site-specific mutations in particular sequence binding motifs in the M-MuLV and SL3-3 MuLV LTRs have highlighted the importance of the "core" and LVb (or Ets-1) sites in overall pathogenicity (Speck *et al.*, 1990; Hallberg *et al.*, 1991). Interaction between different sequence motifs within the enhancers are also important in determining the overall specificity of the enhancers (and resulting disease). Exchanges of subregions of the enhancers between F-MuLV and M-MuLV indicated that interaction between different binding elements is important (Golemis *et al.*, 1989). Moreover, deletion of downstream GC-rich sequences from the enhancers appears to relax the specificity of other elements in the enhancers, resulting in induction of leukemias of multiple lineages (Speck *et al.*, 1990; Hanecak *et al.*, 1991).

C. Leukemogenesis by Murine Leukemia Viruses

A more detailed discussion of leukemogenesis by murine leukemia viruses will be presented here. These viruses have been well studied, and they illustrate most of the important principles about nonacute retroviral leukemogenesis. Some of the most-studied systems include M-MuLV [T lymphoma (Fan *et al.*, 1991)], spontaneous leukemia of AKR mice [induced by the genetically transmitted Akv virus, T lymphoma (Gross, 1983b)], the SL3-3 virus [derived from Akv, T lymphoma (Lenz and Haseltine, 1983)], and F-MuLV [erythroleukemia (Gross, 1983c)] (Table II).

Mice infected with M-MuLV develop T lymphoma with a mean latency of 3–4 months. Leukemogenesis is most effective when the virus is injected into newborns; M-MuLV is poorly leukemogenic in adult mice. Infected mice show two symptoms at early times prior to appearance of the frank leukemia: accelerated thymus atrophy and mild splenomegaly (Storch *et al.*, 1985; Davis *et al.*, 1986). The resulting leukemias generally appear as thymic and/or splenic enlargement along with tumors in the nodes. The tumors are mono- or oligoclonal, as judged by proviral integration patterns or T-cell receptor gene rearrangements (Cuypers *et al.*, 1984b; Steffen and Weinberg, 1978). AKR mice develop T lymphoma with a longer latency (ca. 9 months), with similar final symptoms. It is somewhat more difficult to define preleukemic physiological changes in AKR mice, since they spontaneously self-infect and there are no uninfected mice to use for comparison.

1. Proto-oncogene Activations in MuLV-Induced Tumors

For most MuLVs studied, tumor cells show proviral insertions at specific chromosomal loci. These loci have generally corresponded to known proto-oncogenes or new genes that fit the description of proto-oncogenes, i.e., normal cell genes that are overexpressed or otherwise activated in the tumors. Studies of different MuLVs have identified a multitude of activated proto-oncogenes, as summarized in Table III. In contrast to tumors induced by avian retroviruses, where only one proto-oncogene activation has been observed for each type of virus-induced leukemia (Hayward *et al.*, 1981; Fung *et al.*, 1983), leukemogenesis by MuLVs is generally characterized by activation of a wider collection of proto-

TABLE III. Proto-oncogenes/Common Insertion Sites in MuLV-Induced Tumors

Virus	Disease	Common insertion site or activated proto-oncogene
Moloney MuLV	T lymphoma	
	In mice	*c-myc, pim-1, pvt-1/mis-1/mlvi-1, lck, pim-2*[a]
	In rats	*c-myc, pvt-1/mis-1/mlvi-1, mlvi-2, mlvi-3, mlvi-4, dsi-1, lck, tpl-1/ets-1*[a], *tpl-2*[a]
	Myeloid leukemia	*c-myb*, IL-3
AKR/Gross MuLV	T lymphoma	*c-myc, gin-1, pim-1*
Friend MuLV	Erythroleukemia	*fli-1, fre-2*
	Myeloid leukemia	*fis-1, fim-1, evi-1/fim-3, c-fms/fim-2*
Endogenous MuLV (AKXD mice)	Myeloid leukemia	*evi-1/fim-3, evi-2*
RadLV	T lymphoma	*vin-1*

[a] Insertions associated with tumor progression.

oncogenes. For instance in the mouse, M-MuLV-induced T lymphomas are commonly associated with activation of at least four proto-oncogenes (*c-myc, pim-1, pim-2,* and *pvt-1*) (Steffen, 1984; Cuypers *et al.*, 1984b; Selten *et al.*, 1984; Graham *et al.*, 1985). Under conditions where M-MuLV induces myeloid leukemia (injection in adults followed by pristane priming), the *c-myb* proto-oncogene is activated (Shen-Ong and Wolff, 1987). When M-MuLV induces T lymphomas in rats, another collection of proto-oncogenes show activation, some of which are equivalent to those activated in mice [*c-myc, pvt-1/mlvi-1* (Koehne *et al.*, 1989; Villeneuve *et al.*, 1986; Steffen, 1984)], while others are different [*mlvi-2, mlvi-3, mlvi-4, dsi-1* (Vijaya *et al.*, 1987; Tsichlis *et al.*, 1985)]. The functions of proto-oncogenes activated in MuLV-induced tumors are summarized in Table IV.

In many cases, MuLV-induced tumors may show evidence of proviral activation of more than one proto-oncogene [e.g., *c-myc* and *pim-1* in M-MuLV-induced T lymphomas in the mouse (Cuypers *et al.*, 1986; Fan *et al.*, 1988)]. This finding might reflect activation of more than one proto-oncogene in the tumor cells, or an oligoclonal tumor in which different cells have activations of different proto-oncogenes. In at least some cases, the latter is true. This conclusion is evident from tumor transplantation experiments (Cuypers *et al.*, 1986), and from the fact that the extent of proto-oncogene rearrangement (evident from the intensity of rearranged bands) may be different for two proto-oncogenes in the same tumor mass. Yet, tumor transplantation has also clearly identified cases in which one tumor cell shows activation of more than one proto-oncogene (Cuypers *et al.*, 1986). Indirect evidence for the activation of multiple proto-oncogenes also comes from studies of transgenic mice. Mice carrying an overexpressed *pim-1* transgene do not show high rates of spontaneous leukemia development. However, if they are infected with M-MuLV, tumors rapidly appear; these tumors show proviral activation of either *c-myc* or *n-myc* (van Lohuizen *et al.*, 1989). Thus, overexpression of *myc* and *pim-1* in the same cell results in high-efficiency leukemogenesis. Similarly, mice transgenic for overexpressed *c-myc* show accelerated disease upon M-MuLV infection (van Lohuizen *et al.*, 1991). In this case, several common proviral insertion sites were identified, including three novel ones.

TABLE IV. Functions of Proto-oncogenes or Insertion Sites Activated by Murine Leukemia Viruses

Proto-oncogene or insertion site	Function	References
c-myb	Sequence-specific DNA-binding protein	Shen-Ong and Wolff (1987), Wolff *et al.* (1988), Biedenkapp *et al.* (1988)
c-myc	Site-specific DNA-binding protein; dimerizes with *max* protein	Steffen (1984), Selten *et al.* (1984), Blackwood and Eisenman (1991), Blackwell *et al.* (1990)
dsi-1	?	Vijaya *et al.* (1987)
evi-1	Zinc finger-containing protein	Mucenski *et al.* (1988), Morishita *et al.* (1988)
evi-2	Leucine zipper-containing protein	Buchberg *et al.* (1990)
fim-1	?	Sola *et al.* (1986)
fim-2/c-fms	Receptor for CSF-1; tyrosine protein kinase	Sola *et al.* (1986, 1988)
fim-3/CB-1	Maps near *evi-1*; insertions here activate *evi-1*	Bordereaux *et al.* (1987), Sola *et al.* (1988), Bartholomew *et al.* (1989)
fis-1	?	Silver and Buckler (1986)
fli-1	*ets*-family DNA-binding protein; maps near *ets-1*	Ben-David *et al.* (1990, 1991), Sels *et al.* (1992)
fli-2	Heterogeneous nuclear ribonucleoprotein A1; involved in mRNA splicing	Ben-David *et al.* (1992)
fre-2	Function unknown; maps near *Fv-2*	R. Friedrich (unpublished)
gin-1	?	Villemur *et al.* (1987)
lck	Tyrosine protein kinase; associates with CD4 protein in T lymphocytes	Veillette *et al.* (1988), Voronova and Sefton (1986), Marth *et al.* (1985)
mlvi-2	?	Tsichlis *et al.* (1985)
mlvi-3	?	Tsichlis *et al.* (1985)
mlvi-4	Maps near *c-myc*; function unknown	Tsichlis *et al.* (1989, 1990), Lazo *et al.* (1990a)
pim-1	Serine/threonine protein kinase	Cuypers *et al.* (1984b), Saris *et al.* (1991)
pim-2	?	Breuer *et al.* (1989)
pvt-1/mis-1/ mlvi-1	Transcript identified, protein not known; located near *c-myc*	Graham *et al.* (1985), Villeneuve *et al.* (1986), Koehne *et al.* (1989), Tsichlis *et al.* (1983, 1989)
p53	Tumor suppressor gene product	Mowat *et al.* (1985), Ben David *et al.* (1988)
tpl-1	*ets*-family DNA-binding protein; maps near *ets-1*	Bear *et al.* (1989)
tpl-2	Presumptive serine/threonine protein kinase	Patriotis *et al.* (1993)
vin-1	Cyclin D	Tremblay *et al.* (1992)

2. MuLV-Induced Preleukemic Events

While LTR activation of proto-oncogenes is clearly an important process for MuLV leukemogenesis, virus-induced preleukemic events have also been documented. Indeed, these preleukemic changes are also essential for tumor development, and this necessity contributes to the extended latency period of leukemogenesis (reviewed by Fan *et al.*, 1991). Another interesting feature is that the preleukemic changes may occur in hematopoietic compartments distinct from the organ where the tumor ultimately develops. This observation is well illustrated for spontaneous AKR leukemia and M-MuLV-induced leukemia (both T-lymphoid) (Haran-Ghera, 1980; Asjo *et al.*, 1985). Potentially leukemic cells were identified by transplantation of cells from preleukemic animals into histocompatible but genetically distinct recipients. It was found that tumors of donor origin could develop in the recipient animals. Moreover, at early (preleukemic) times, potentially leukemic cells were present in the bone marrow and spleens of donor animals, but *not* in the thymus (where the tumors ultimately develop) (Haran-Ghera, 1980; Asjo *et al.*, 1985). Further experiments indicated that the potentially leukemic cells from the bone marrow have the properties of thymocyte precursors (prothymocytes) (Haran-Ghera *et al.*, 1978; Reisner *et al.*, 1980). This finding suggests that preleukemic changes occur in prothymocytes in the bone marrow that then migrate to the thymus where they undergo maturation and other neoplastic changes, e.g., LTR activation of proto-oncogenes. Tumor transplantation experiments of thymocytes from preleukemic AKR mice have further subdivided potentially leukemic cells. In younger animals, thymuses contain cells which can develop into leukemias if injected back into the thymus of a recipient. In older animals, thymuses also contain cells that can establish tumors in recipients if inoculated either intravenously or intrathymically (Hays *et al.*, 1989a). Thus, these experiments define thymus-dependent and independent preleukemic cells in the thymus. Further experiments indicate that the thymus-dependent cells have properties of prothymocytes (Hays *et al.*, 1989b).

Different stages in preleukemic cell development in AKR mice have also been identified by flow cytometry. In these experiments, leukemia was accelerated by intrathymic injection of a pathogenic MCF recombinant. As discussed below, MCF recombinants are derivatives of MuLVs that arise during *in vivo* passage in the mouse; they contain distinct *env* regions. A series of preleukemic stages in the thymus could be observed, including polyclonal expansion of cells, followed by outgrowth of a mono- or oligoclonal population and the subsequent development of the frank tumor (O'Donnell *et al.*, 1984). The timing of proviral insertions next to *c-myc* and *pim-1* was also investigated, and the results indicated that proviral insertion is generally a late event in this leukemogenic process (O'Donnell *et al.*, 1985).

In M-MuLV-induced leukemia, another preleukemic state characterized by mild splenomegaly has been observed (Storch *et al.*, 1985). This splenomegaly is accompanied by increases in hematopoietic progenitors of multiple lineages (myeloid and erythroid, and presumably lymphoid), even though the resulting disease is T lymphoid (Davis *et al.*, 1987). The relevance of this preleukemic splenomegaly to leukemogenesis was emphasized by the fact that a poorly pathogenic enhancer variant of M-MuLV (Mo + PyF101 M-MuLV) does not induce splenic hyperplasia (Davis *et al.*, 1985, 1987). It should be noted that the splenic hyperplasia represents polyclonal expansion of cells, as measured by Southern blot hybridizations for proviral DNA insertion patterns (B. K. Brightman and H. Fan, unpublished). Thus, M-MuLV infects the animal at least twice during leukemogenesis: first to induce splenic hyperplasia, and second to insertionally activate proto-oncogenes. The mechanism of splenic hyperplasia induction has been investigated further. It is noteworthy that hyperplasia does not appear to result from direct M-MuLV infection of hematopoietic progenitors, but by infection of some other cell type (Brightman *et al.*, 1990). Moreover, splenic hyperplasia appears to result indirectly from hematopoietic defects in the bone marrow (Li and Fan, 1990). Indeed, induction of splenic hyperplasia by marrow destruction by ^{89}Sr can complement the leukemogenic defect of Mo + PyF101 M-MuLV (Li and Fan, 1991). A model of these events is shown in Fig. 5.

3. MCF Recombinant Viruses in MuLV Leukemogenesis

One of the characteristic virological events associated with MuLV leukemogenesis is the appearance of mink cell focus-inducing (MCF or polytropic) MuLV recombinants (Hartley *et al.*, 1977). They were initially defined by their ability to form foci in mink lung cells. These viruses are recombinants in the viral *env*

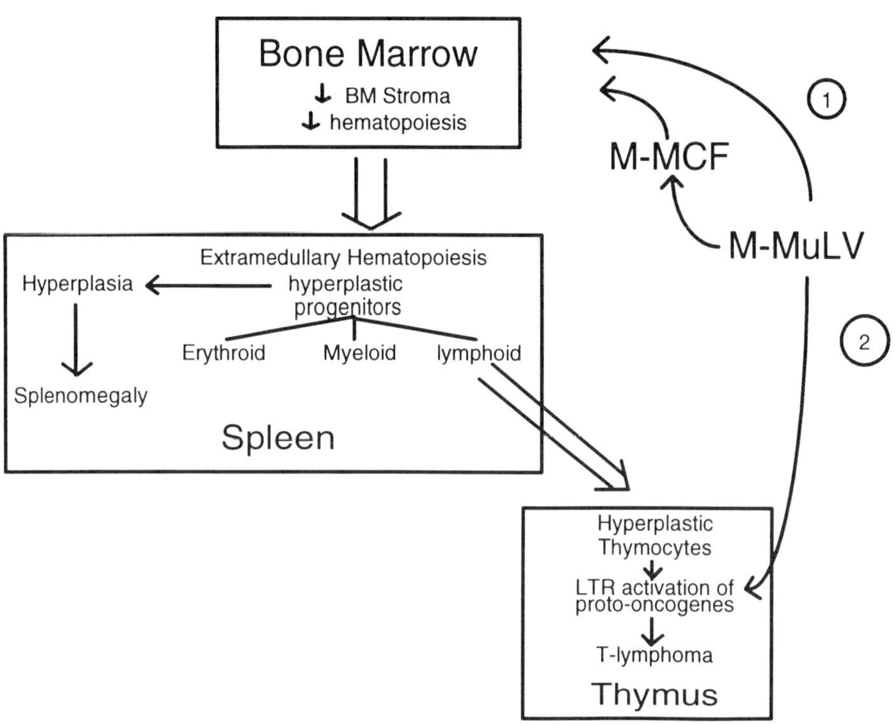

FIGURE 5. A model for M-MuLV leukemogenesis. In this model (Fan *et al.* 1991), M-MuLV infects the animal at both early (1) and late (2) times. The first infection, in combination with an MCF recombinant, leads to defects in bone marrow stroma and bone marrow hematopoiesis. Resulting extramedullary hematopoiesis leads to splenomegaly, and increased progenitors of multiple lineages. If a hyper-plastic lymphoid progenitor migrates to the thymus and undergoes thymic differentiation, LTR activation of proto-oncogenes may occur during a second infection. Subsequently a T-lymphoma emerges. Double lines indicate movement of hematopoietic stem cells between different organs.

region between the initially infecting ecotropic MuLV and endogenous MuLV proviral sequences present in all mice (Stoye *et al.*, 1991). The resulting MCF virus contains polytropic *env* sequences, giving the virus the ability to infect cells via a different cellular receptor. MCF receptors are present on both mouse and nonmouse cells (Kozak and Ruscetti, 1992). The fact that MCF recombinants have been detected in virtually all MuLV-induced leukemias in mice suggests that they play a role in disease induction, and the appearance of high levels of MCF viruses at or near the time of tumor appearance has led to the suggestion that they are the "proximal leukemogens" (Hartley *et al.*, 1977). In fact, MCF viruses may play multiple roles in MuLV leukemogenesis, depending on the virus–host system.

MCF recombinants were first observed in AKR mice, which develop leukemia spontaneously as a result of activation of an endogenous replication-competent MuLV, Akv MuLV (Hartley *et al.*, 1977). The crucial role of MCF viruses in AKR leukemogenesis was evident from the fact that generally all newly acquired proviruses in tumors were of the MCF type, even though the animals were viremic for Akv MuLV from birth (Quint *et al.*, 1981; Herr and Gilbert, 1983). When several independently isolated AKR MCF viruses were compared, they could be divided into two categories based on their pathogenicity. Pathogenic MCFs accelerate the leukemia in AKR mice when inoculated into young animals, whereas nonpathogenic MCFs do not (Lung *et al.*, 1983). Molecular comparisons of pathogenic and nonpathogenic MCFs indicate that they differ in their *env* and LTR regions. The nonpathogenic MCFs (Class II) result from a single recombination in the *env* region (Fig. 6). The pathogenic MCFs (Class I) result from two recombinations (Stoye *et al.*, 1991; Thomas and Coffin, 1982; Evans and Malik, 1987): they contain polytropic *env* sequences, but they also contain LTRs from endogenous xenotropic virus sequences as the result of a second recombination event (see Kozak and Ruscetti, 1992).

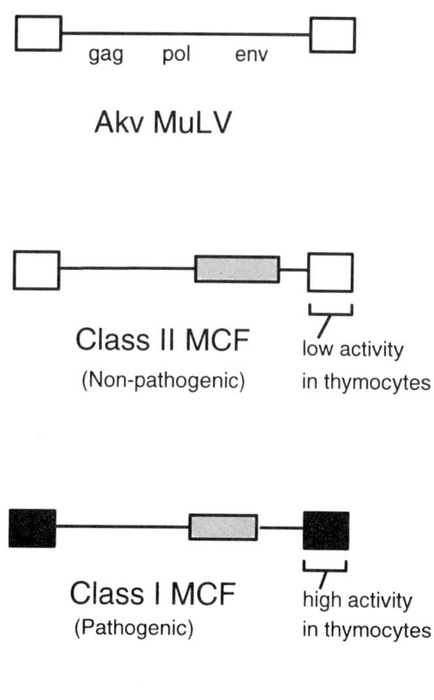

FIGURE 6. Class I and II MCF recombinant viruses. The organization of class I and II MCF viruses relative to parental endogenous Akv MuLV is shown. Class II MCFs arise by a single recombination event which exchanges endogenous polytropic envelope glycoprotein sequences (gray rectangle) into the Akv MuLV genome. Class I MCFs have a further alteration, in which the LTR from an endogenous xenotropic MuLV replaces the Akv MuLV LTR. This LTR has high transcriptional activity in thymocytes, while the Akv MuLV LTR has low activity. Class I MCFs are pathogenic in AKR leukemia acceleration tests, while Class II MCFs are not.

Other inbred strains of mice such as CWD also develop spontaneous leukemia from endogenously activated Akv MuLV, and in these mice it has been proposed that Class II MCFs are intermediates in Class I MCF formation (Thomas *et al.*, 1984). The importance of the LTR change for MCF pathogenicity can be attributed to the fact that the enhancers of Akv M-MuLV are relatively weak in T-lymphoid cells (Yoshimura *et al.*, 1985), while the new enhancers in Class I MCF LTRs are much more active. Thus, in order for efficient LTR activation of proto-oncogenes to occur (see above), an LTR change is required. Class I MCF LTRs have a tandemly duplicated region in their enhancers (Fig. 7). It has been shown that removal of one copy of the tandem repeat greatly reduces pathogenicity, in concert with a reduction in enhancer activity in T-lymphoid cells (Holland *et al.*, 1989).

Mechanistically, the polytropic *env* sequences in MCF recombinants could have several different roles in leukemogenesis. First, since MCF viruses infect cells via a different receptor than ecotropic MuLVs, they should be able to continue infection in animals that are already extensively infected with MuLV. This process could be particularly important in mice such as AKR which might express high levels of endogenous Akv MuLV *env* protein on thymocytes. Surface *env* protein would render them resistant to superinfection by the same virus; establishment of new proviruses is necessary for LTR activation of proto-oncogenes to occur, so MCF-mediated superinfection could facilitate this process. A second possibility is that MCF *env* protein could affect the growth properties of cells. Injection of MCF viruses into AKR mice is thymolytic, which might facilitate repopulation of the thymus with virus-induced potentially leukemic cells of the bone marrow or spleen (N. Haran-Ghera, personal communication). Recently, another role for MCF glycoprotein has been suggested. As described above, a deleted form of MCF glycoprotein (gp52) can bind to the erythropoietin receptor and stimulate epo-independent growth of epo-dependent cells (Li *et al.*, 1990) (Fig. 8). Similarly, infection of IL-2-dependent cells with MCF virus renders them independent of IL-2 for growth, presumably by binding of the MCF glycoprotein to the IL-2 receptor (Li and Baltimore, 1991). Since many thymocytes and T lymphocytes have IL-2 receptors on their cell surface, infection of these cells with MCF virus might provide them with an autocrine mitogenic stimulus and contrib-

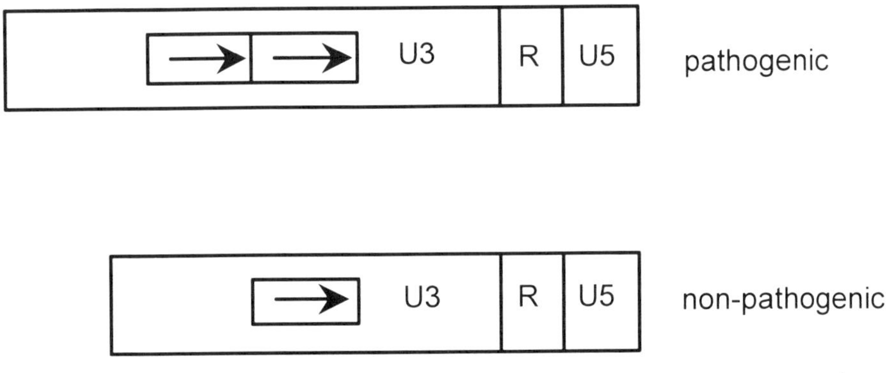

FIGURE 7. Variant non-pathogenic Class I MCF MuLVs. Pathogenic Class I MCF viruses from AKR mice generally contain two tandemly repeated copies of the enhancer sequences in the LTR (indicated by the boxes with arrows). The non-pathogenic variants usually contain only one copy of the tandem repeats.

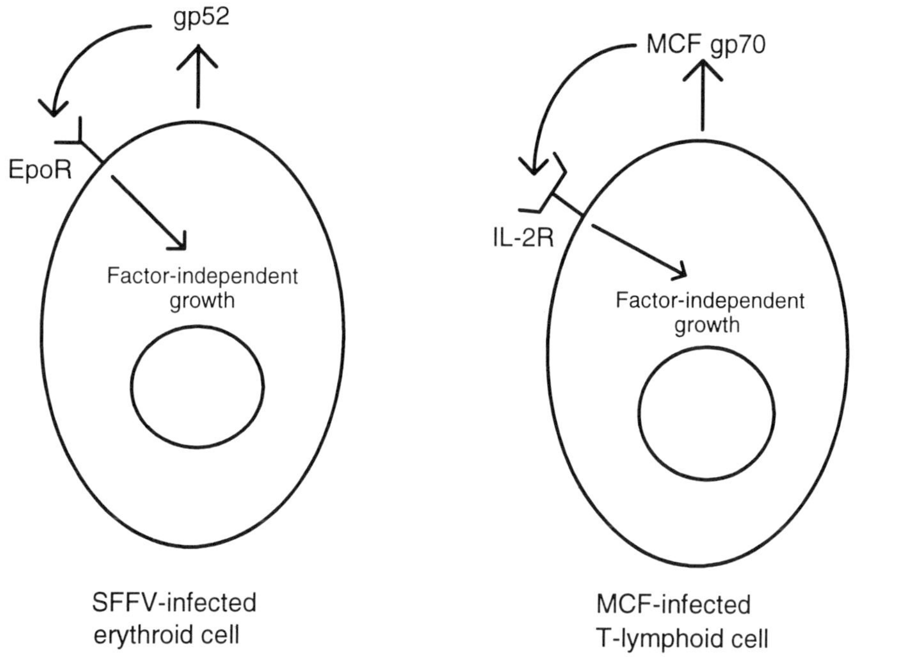

FIGURE 8. Autocrine growth loops stimulated by MuLVs. The mechanism of autocrine growth stimulation by Friend SFFV is shown on the left side of the figure. gp52 protein binds to the erythropoietin receptor, and can render SFFV-infected erythroid stem cells erythropoietin-independent for growth. A similar situation may exist for T-lymphoid cells infected with MCF recombinants. The MCF surface glycoprotein gp70 can bind to the interleukin-2 receptor and stimulate IL-2-independent growth in model systems. It is possible that this process occurs as well in MCF-infected T-lymphoid cells.

ute to neoplastic progression. The concept of autocrine mechanisms in MuLV-induced leukemogenesis was previously proposed by McGrath and Weissman (1979) (see below).

Studies of M-MuLV have also provided insights into the role of MCFs in pathogenesis. Unlike Akv MuLV, the M-MuLV LTR is highly active in T-lymphoid cells; thus, Moloney MCFs result from a single recombination in the *env* region without alterations in the LTR (Bosselman *et al.*, 1982). Tumors from M-MuLV-infected mice show insertions of either input M-MuLV or M-MCF proviruses adjacent to proto-oncogenes (Cuypers *et al.*, 1984a); nevertheless, in virtually all infected mice MCFs can be detected, beginning at quite early times (Evans and Morrey, 1987). In M-MuLV-infected mice, there is evidence that MCFs play a role at early (preleukemic) times. In particular, the preleukemic bone marrow defect and consequent splenic hyperplasia (see above) appear to involve MCF recombinants (Li and Fan, 1990). This conclusion results from the fact that the poorly pathogenic Mo + PyF101 M-MuLV (which does not induce splenic hyperplasia) does not generate MCF recombinants *in vivo* (Brightman *et al.*, 1991b). This implies that the preleukemic events that Mo + PyF101 M-MuLV is unable to induce involve generation of MCF recombinants. The current working hypothesis is that combined infection of bone marrow stroma by M-MuLV and M-MCF results in growth inhibition, which leads to defects in marrow hema-

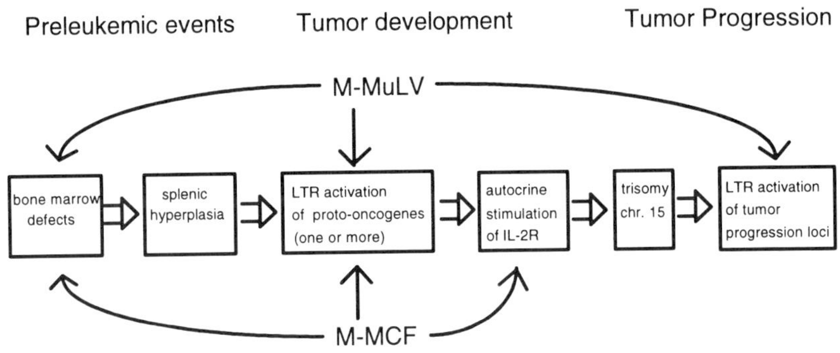

FIGURE 9. Steps in M-MuLV-induced leukemogenesis. A pathway involved in M-MuLV leukemogenesis is shown. Steps at which M-MuLV and M-MCF infections occur are indicated.

topoiesis and compensatory extramedullary hematopoiesis (splenic hyperplasia, Fig. 9) (Li and Fan, 1990; Fan *et al.*, 1991).

Despite the strong evidence for involvement of MCFs in MuLV leukemogenesis, it is also clear that they are not absolutely required for disease. First, M-MuLV can induce T lymphoma in rats (Steffen *et al.*, 1979; Tsichlis *et al.*, 1983). However, rats lack the endogenous retroviruses that give rise to MCF *env* protein in mice. As a result, when M-MuLV is inoculated into rats, no MCF recombinants are formed, yet T lymphoma develops efficiently. Second, when Mo + PyF101 M-MuLV is infected into mice, it does induce T lymphoma, although with markedly less efficiency than wild-type M-MuLV (Brightman *et al.*, 1991b). Thus, even though Mo + PyF101 M-MuLV does not generate MCFs in mice, it is able to induce T lymphoma. Third, a genetic locus *Rmcf* ^{R} has been identified that renders mice incapable of generating MCFs when inoculated with Friend MuLV (Hartley *et al.*, 1983). *Rmcf* ^{R} mice were originally identified by resistance to developing erythroleukemia when infected with F-MuLV. However, closer examination reveals that F-MuLV does induce leukemia in *Rmcf* ^{R} mice, although with an extended latency and shift from erythroid to myeloid tumors (Ruscetti *et al.*, 1985). The most reasonable explanation is that if MCFs can form, then they contribute to efficient MuLV leukemogenesis, acting at potentially several different steps. However, in the absence of MCFs, MuLVs can still induce disease, but with lower efficiency.

4. Tumor Progression

In MuLV-infected animals, the virus can also contribute to tumor progression. This fact has been demonstrated by *in vitro* passage of M-MuLV-induced rat T lymphomas (Bear *et al.*, 1989). Continued *in vitro* passage resulted in acquisition of new M-MuLV proviruses by reinfection. Molecular cloning and screening identified several common sites of proviral insertion in these "progressed" lines, and they were also found to represent common insertion sites in primary M-MuLV-induced rat lymphomas (particularly after *in vivo* passage of the tumor cells in nude mice). They have been designated tumor progression loci (*tpl*). So far, two such loci have been isolated and characterized (*tpl-1*, *tpl-2*). *tpl-1* was found to be related to the *ets-1* transcription factor/proto-oncogene (Bear *et al.*,

1989), while *tpl-2* is probably a serine–threonine protein kinase (Patriotis *et al.*, 1993).

5. Chromosomal Alterations in MuLV-Induced Tumors

The study of chromosomal abnormalities in tumors (naturally occurring as well as experimentally induced) has been very important in identifying genetic changes that take place during tumorigenesis. For instance, certain tumor types show specific chromosomal translocations. These translocations reflect activation of a normal cell proto-oncogene either by (1) movement of the proto-oncogene to a highly active chromosomal environment [e.g., translocation of the *c-myc* gene into an immunoglobulin locus in B lymphomas such as human Burkitt's lymphoma (Marcu *et al.*, 1983; Dalla-Favera *et al.*, 1982; Taub *et al.*, 1982)] or by (2) generation of a novel fusion protein between the proto-oncogene and another cell gene [e.g., the *bcr–abl* fusion protein characteristic of Philadelphia chromosome-positive chronic myelogenous leukemia and B lymphoma (Heisterkamp *et al.*, 1983; Konopka *et al.*, 1984)]. Overexpression of proto-oncogenes in naturally occurring cancers can also occur by gene amplification [e.g., *n-myc* in neuroblastomas (Brodeur *et al.*, 1984) and *her2/neu* in human mammary carcinoma (Slamon *et al.*, 1987, 1989)]. Gene amplification can be detected in some cases by homogeneously staining regions (HSRs) on chromosomes (Brodeur *et al.*, 1984), or as double minute chromosomes (Oliner *et al.*, 1992).

More recently, the contributions of tumor suppressor gene mutations during carcinogenesis have become of great interest. The normal function of these genes appears to be inhibition of abnormal cell growth. In tumors, loss of tumor suppressor gene function is frequently observed. At the chromosomal level, this loss is sometimes manifested by a characteristic chromosomal deletion in one chromosome (Weinberg, 1991). This event often represents physical deletion of the region containing the tumor suppressor gene from one chromosome, accompanied by specific mutation in the tumor suppressor gene on the homologous (undeleted) chromosome. With the advent of advanced chromosomal mapping techniques such as restriction fragment length polymorphism (RFLP) analysis, it is now possible to detect tumor suppressor gene deletions in tumors that are not apparent in chromosomal karyotype studies. Hallmarks of tumor suppressor genes are "loss of heterozygosity" for specific chromosomal regions.

One tumor suppressor gene that is frequently altered in tumor tissues is p53 (Cho and Vogelstein, 1992). In the case of p53, mutant p53 acts dominantly by forming protein oligomers with wild-type p53 protein and inactivating it (Finlay *et al.*, 1989). Monoclonal antibodies also exist that can detect mutant p53 by an altered staining pattern in tumor cells (Lu *et al.*, 1992; Martinez *et al.*, 1991).

Chromosomal studies of MuLV-induced tumors have revealed one characteristic change: trisomy for mouse chromosome 15 (Wiener *et al.*, 1978). The trisomy presumably reflects growth or tumorigenic advantage conferred by an additional copy of some gene from this chromosome. Identification of the critical gene on chromosome 15 has not been achieved. One interesting possibility is that the critical gene could be *c-myc*, which maps to this chromosome. Indeed, one study of M-MuLV-induced T lymphomas showed that in several tumors containing trisomy of chromosome 15 and also proviral activation of *c-myc*, the chromosome containing the activated *c-myc* locus was duplicated, while the unaffected chromosome was not (Cuypers *et al.*, 1986). However, in the same study, not all tumors with proviral insertions near *c-myc* showed trisomy of chromosome 15, so this duplication is not absolutely required for leukemogenesis, but it might promote the process.

Tumor suppressor gene mutations in MuLV-induced leukemias have not been extensively examined. However, as described above for the Friend SFFV acute transforming virus, one of the secondary changes observed in Friend SSFV erythroleukemia cells is deletion of p53 tumor suppressor genes (Mowat *et al.*, 1985; Dreyfus *et al.*, 1990). Thus, the possibility that tumor suppressor gene alterations occur in leukemias induced by nonacute MuLVs seems reasonable. Preliminary experiments on MuLV-induced tumors suggest that many of them contain p53 mutations (N. Haran-Ghera, personal communication).

6. Potential Immunological Mechanisms of MuLV Leukemogenesis

The role of host immune responses in modulating MuLV infection and leukemogenesis will be addressed in more detail below. However, at this point it will be useful to mention two models invoking host immune responses to viral infection in the leukemogenic process itself. The basis for one model was the finding that mice infected with M-MuLV showed an anti-gp70 (envelope) T-lymphocyte response (Lee and Ihle, 1981a; Lee *et al.*, 1981; but see below). Moreover, in a survey of several inbred mouse strains, there was a correlation between those that developed an anti-gp70 T-cell response (measured by antigen-stimulated proliferation) and the susceptibility to leukemogenesis by M-MuLV (Lee and Ihle, 1981b). In particular, one strain (CBA/N) that did not show an anti-gp70 T-cell response also was resistant to leukemogenesis by M-MuLV. This finding led to the model that an antiviral immune response expands T-lymphocyte division, thus creating an enlarged pool of rapidly dividing cells that could undergo leukemogenic transformation by secondary events (Ihle and Lee, 1982). In an extension of this model, the activated virus-specific T lymphocytes could be releasing cytokines that expand a pool of other hematopoietic targets for further transformation.

As noted above, a second immunological model for MuLV leukemogenesis has also been proposed (McGrath and Weissman, 1979; McGrath *et al.*, 1980). It was suggested that M-MuLV virus particles (or their MCF derivatives) bind to M-MuLV-specific T lymphocytes and mitogenically activate them. This activation would lead to rapid expansion of these M-MuLV-specific T lymphocytes. One of these mitogenically activated lymphocytes would ultimately undergo additional changes and develop into a T lymphoma. However, this model does not account for the fact that many of the T lymphomas induced by M-MuLV have phenotypes corresponding to immature thymocytes [e.g., $CD4^{+}CD8^{+}$ "double positives" (Brightman *et al.*, 1988a; Lazo *et al.*, 1990b)]. Such cells would not respond to their cognate antigens by division. Another difficulty with this model is that it is now understood that T lymphocytes recognize their cognate antigens as degraded peptides held in association with proteins of the major histocompatibility (MHC) complex (Bjorkman and Parham, 1990), and the specific binding between T lymphocyte and antigen is of rather low affinity (Jorgensen *et al.*, 1992). Thus, the original observation of specific binding of virus particles to lymphoma cells (McGrath and Weissman, 1979) would not likely correspond to a T-cell–antigen interaction.

7. A Summary of the Multistep Process of MuLV Leukemogenesis

The preceding sections have illustrated the multiple steps in MuLV leukemogenesis, and the fact that viral infection is involved in many of them. The progression of these steps is illustrated in Fig. 9. While this model has been developed on the basis of M-MuLV infection, it is likely that many (if not all) steps are applicable to other MuLVs. As indicated, leukemogenesis entails early (preleukemic) events, in which the virus induces splenic hyperplasia and appearance of potentially leukemic cells. The preleukemic phase is then followed by LTR activation of proto-oncogenes. Subsequent tumor progression may also involve activation of different proto-oncogenes. In mice, MCF recombinant viruses appear to play roles at all stages of the disease, although they are not absolutely required. Early in the process, they are involved in establishing splenic hyperplasia, perhaps by inhibiting bone marrow hematopoiesis. For viruses such as Akv-1, exchange of the LTRs during MCF formation is also critical for efficient LTR activation of proto-oncogenes in target tissues. Finally, MCF *env* glycoprotein may bind IL-2 receptor in tumor cells and further induce mitogenesis in an autocrine fashion. Sometime during this series of events, development of chromosome 15 trisomy may also occur, presumably reflecting increased gene dosage of some allele on this chromosome.

While the scheme presented in Fig. 9 suggests that the events in M-MuLV leukemogenesis follow a set order, this may not necessarily be the case. An interesting example has been shown for myeloid leukemias induced by intravenous inoculation of M-MuLV into BALB/c mice accompanied by pristane priming (Shen-Ong and Wolff, 1987). In this case, M-MuLV proviral insertion next to the *c-myb* locus occurs in 100% of myeloid tumors (Shen-Ong and Wolff, 1987). In particular, highly sensitive reverse transcriptase–polymerase chain reaction (RT-PCR) detection techniques (Mukhopadhyaya and Wolff, 1992) revealed that the M-MuLV proviral insertion at *c-myb* occurs quite soon after infection (within 12 days), even though tumors do not develop until several months later (Nason-Burchenal and Wolff, 1993). Thus, in adult mice infected with M-MuLV, LTR activation of a proto-oncogene can occur early after infection, but cells containing this activation apparently remain dormant until other events occur that cooperate with the *c-myb* activation in inducing leukemia development. More recent experiments on BALB/c or NIH Swiss mice infected with M-MuLV as newborns indicate that the same *c-myb* activations can be detected at early times, even though these animals ultimately develop T lymphomas (L. Wolff, B. Belli, and H. Fan, unpublished). It is noteworthy that the resulting T lymphomas do not contain *c-myb* activations.

IV. HOST EFFECTS ON MuLV INFECTION AND LEUKEMOGENESIS

The preceding sections have predominantly considered MuLV leukemogenesis from the perspective of the virus and the developing tumor cell. However, the host response to viral infection is equally important in determining whether viral infection will proceed and whether tumors will develop. This topic will be discussed in the following sections.

A. Host Genes Affecting MuLV Infection and Leukemogenesis

The availability of inbred mouse strains has greatly facilitated genetic analyses of MuLV susceptibility. Early on, it was observed that different strains had characteristic susceptibility or resistance to leukemogenesis by various MuLVs. Crosses between high- and low-susceptibility strains then allowed genetic mapping of the resistance/susceptibility loci. Initially these loci were functionally defined as conferring resistance or susceptibility to tumorigenicity by a particular virus. For instance, mapping of resistance alleles for leukemogenesis by Friend SFFV has led to characterization of loci called *Fv-1*, *Fv-2*, etc. (Lilly, 1967, 1970; Pincus *et al.*, 1971). Subsequently, the modes of action of these different loci have proved to be diverse. In some cases, viral resistance loci have subsequently been assigned to previously known genes, e.g., those of the MHC (*H-2* in the mouse). A summary of the host genes affecting MuLV leukemogenesis is given in Table V.

1. Host Genes That Affect Virus Replication

Of the genes that have the strongest effect on MuLV leukemogenesis, those that affect virus replication in the animal are most prominent. The best known gene in this class is *Fv-1* (Pincus *et al.*, 1971; Kozak and Ruscetti, 1992). There are two *Fv-1* alleles: *Fv-1*^{n} and *Fv-1*^{b}, named for the prototypic strains NIH Swiss and BALB/c. MuLVs can be divided into those that are N-tropic and those that are B-tropic. N-tropic viruses replicate much more efficiently in cells of *Fv-1*$^{n/n}$ genotype than in cells of *Fv-1*$^{b/b}$ genotype. Conversely, B-tropic viruses replicate much more efficiently in *Fv-1*^{b} cells than in *Fv-1*^{n} cells. In the *Fv-1* system, resistance is

TABLE V. Host Genes That Affect MuLV Infection and Leukemogenesis

Genes affecting virus replication in the animal
Fv-1: governs susceptibility of cells to N- and B-tropic viruses; not important for NB-tropic viruses
Rmcf^{r}: blockage of MCF virus propagation due to spontaneous expression of endogenous polytropic (MCF-type) glycoprotein
Fv-4: blockage of ecotropic MuLV infection due to spontaneous expression of endogenous ecotropic MuLV glycoprotein
Genes affecting hematopoietic stem cell growth
Fv-2: controls percentage of cycling erythroid precursors
S1: encodes "stem cell factor"—growth factor for hematopoietic stem cells
W: encodes stem cell factor receptor in hematopoietic cells; same as the *c-kit* proto-oncogene
Genes affecting the immune response
MHC genes
H-2D: important for F-MuLV; where the *rFv-1* resistance determinant maps
I-A: important for F-MuLV
Qa-1/TL: important for F-MuLV; where the *rFv-2* resistance determinant maps
H-2K: important for RadLV
rFv-3: controls the humoral response to F-MuLV infection

dominant: cells from *Fv-1*$^{n/b}$ mice, resulting from crosses between *Fv-1*$^{n/n}$ and *Fv-1*$^{b/b}$ mice, are relatively resistant to infection by either N- or B-tropic viruses. With respect to the virus, N–B restriction is governed by determinants in the *gag* region, and CA protein (the major capsid protein or p30) in particular (Gautsch *et al.*, 1978; DesGroseillers and Jolicoeur, 1983; Boone *et al.*, 1983). When viruses are inoculated into nonpermissive cells (e.g., N-tropic virus into a B-type cell), the restricted step is after viral penetration and uncoating, but before integration of proviral DNA (Jolicoeur and Baltimore, 1976; Sveda and Soeiro, 1976; Yang *et al.*, 1980). However, the exact mechanism remains to be determined.

N–B restriction governed by *Fv-1* is important in *in vivo* infection and leukemogenesis. For N-tropic viruses such as Akv MuLV (the leukemogenic endogenous virus of AKR mice), Gross, and SL3-3 MuLVs, leukemia efficiently occurs only in *Fv-1*$^{n/n}$ mice such as AKR, NFS, or NIH Swiss. This finding reflects the requirement of efficient viral infection and spread for disease. In contrast, in AKR mice that are congenic for *Fv-1*^{b} (AKR/*Fv-1*^{b}), the endogenous Akv MuLV does not propagate efficiently, and T lymphoma does not efficiently occur. However, with very long latency, B lymphomas appear (Haran-Ghera *et al.*, 1993).

Continued passage of N-tropic MuLVs in B-type cells (and vice versa) can result in development of NB-tropic variants. NB-tropic MuLVs can replicate with equal efficiency in either N- or B-type cells; they are also not restricted in heterozygous *Fv-1*$^{n/b}$ cells. Several of the highly leukemogenic exogenous MuLVs such as Moloney, Friend, and Rauscher MuLVs are NB-tropic, perhaps reflecting extensive *in vivo* passage of these viruses (or the tumor cells that gave rise to them) through mice prior to isolation. As would be expected, leukemogenesis by these viruses is not affected substantially by the nature of the *Fv-1* genotype of the infected mouse.

A second mouse locus affecting efficiency of MuLV leukemogenesis is the *Rmcf* locus (Hartley *et al.*, 1983). As discussed above, formation of MCF *env* gene recombinants is inhibited when mice carrying the *Rmcf*^{R} gene are infected with F-MuLV. Concordantly, the efficiency of leukemogenesis by F-MuLV is reduced in *Rmcf*^{R} mice. For this locus, resistance is dominant to sensitivity, and *Rmcf*$^{R/s}$ mice do not generate MCFs (Ruscetti *et al.*, 1981, 1985). The mechanism of *Rmcf*^{R} resistance has been shown to result from the expression of an endogenous MCF-type (polytropic) *env* glycoprotein (gp70 or SU) on various hematopoietic cells in mice of this genotype (Ruscetti *et al.*, 1981, 1985; Buller *et al.*, 1988, 1989). The resistance reflects the phenomenon called interference or resistance to superinfection. A cell infected with an MuLV cannot be superinfected by a second MuLV that uses the same cellular receptor (see Kozak and Ruscetti, 1992). This result is caused by blockage of the receptors on the infected cell by saturation binding with *env* protein encoded by the first infecting virus. Thus, in *Rmcf*^{R} mice, cells expressing endogenous MCF-type glycoprotein are resistant to infection by MCF recombinants, and MCFs do not propagate efficiently in these animals (Ruscetti *et al.*, 1985; Buller *et al.*, 1988). The *Rmcf*^{R} allele does not actually encode the expressed polytropic MuLV. Rather, this gene appears to govern expression of polytropic *env* protein in certain cells of the hematopoietic lineage (Frankel *et al.*, 1989).

One important feature of *Rmcf*^{R} restriction in animals *in vivo* is that not all MuLVs are subject to its effect. In particular, inoculation of M-MuLV into *Rmcf*^{R} mice leads to development of leukemia with the same time course as for

M-MuLV in *Rmcf* ^{s} mice (Brightman *et al.*, 1991a). The lack of *Rmcf* ^{R} restriction for M-MuLV versus F-MuLV has been mapped to the LTR of M-MuLV (Brightman *et al.*, 1991a). This difference may reflect the fact that *Rmcf* ^{R}-controlled gp70 expression occurs predominantly on erythroid and myeloid stem cells, and less efficiently on lymphoid stem cells (Buller *et al.*, 1989). Thus, for an MCF derived from F-MuLV, replication occurs efficiently only in cell types that support the tissue-specific enhancers of the F-MuLV LTR (erythroid and myeloid cells), and these cells are blocked from infection by *Rmcf* ^{R}-controlled MCF-type gp70. In contrast, M-MuLV enhancers are active in lymphoid and myeloid cells; infection of lymphoid cells by an M-MCF would not be subject to resistance even in *Rmcf* ^{R} mice, since those cells would be susceptible to infection (no expression of endogenous polytropic gp70) and they would also support transcription from the M-MuLV LTR.

A third mouse gene conferring resistance/susceptibility to MuLVs is *Fv-4* (Odaka *et al.*, 1981). Mice carrying the *Fv-4* ^{r} allele are resistant to infection by ecotropic MuLVs. Thus, leukemogenesis by MuLVs is inhibited in these mice. Molecular analysis has revealed that the *Fv-4* ^{r} allele corresponds to an endogenous MuLV provirus of the ecotropic type (Ikeda *et al.*, 1985). Thus, as for *Rmcf* ^{R}, the basic mechanism of viral resistance appears to be viral interference caused by expression of an endogenous MuLV gp70. However, while *Rmcf* ^{R} is apparently a regulatory gene that governs expression of endogenous MuLV polytropic gp70, *Fv-4* ^{r} is the structural gene itself for the interfering endogenous ecotropic gp70.

2. Host Genes That Affect Hematopoietic Stem Cell Growth

Another class of host genes that influence leukemogenesis by MuLVs are those that affect growth of hematopoietic stem cells. This finding is supported by the observation that continual cell division of infected cells is important for allowing accumulation of other changes that ultimately result in leukemic transformation. One gene in this class, *Fv-2*, was identified on the basis of relative sensitivity of mouse strains to leukemogenesis by the Friend SSFV/F-MuLV complex (Lilly, 1970). Subsequent studies indicated that *Fv-2* is involved in regulation of normal erythroid precursor cell division (Suzuki and Axelrad, 1980; Hangoc *et al.*, 1987; Bondurant *et al.*, 1985). In particular, a much higher percentage of erythroid precursors in *Fv-2* ^{ss} mice are cycling than for *Fv-2* ^{rr} mice. This effect is specific for erythroid cells, and not for other hematopoietic precursors such as those of the myeloid lineage.

Two other genes that influence MuLV leukemogenicity by effects on hematopoietic stem cell growth are the mouse *W* and *Sl* loci (MacDonald *et al.*, 1980; Bennett *et al.*, 1968; Steeves *et al.*, 1968). The mouse *W* gene codes for a growth factor receptor present in hematopoietic stem cells of various lineages and maturity, including relatively early (and multipotent) stem cells The *W* gene has been shown to be the same as the *c-kit* proto-oncogene (Tan *et al.*, 1990; Chabot *et al.*, 1988). *c-kit* is a tyrosine kinase-type growth factor receptor that responds to binding of its ligand by intracellular phosphorylation of proteins on tyrosine residues, which results in growth stimulation (Chabot *et al.*, 1988; Besmer *et al.*, 1986). Recent studies have shown that the product of the mouse *Sl* (*steel*) locus is the cognate growth factor for the *W* gene (*c-kit*) growth factor receptor (Huang *et al.*, 1990; Copeland *et al.*, 1990; Anderson *et al.*, 1990). Mutations in either *W* or *Sl* reduce the extent of hematopoiesis for multiple lineages. In parallel, animals carrying mutations at these two loci show reduced sensitivity to leukemia development after infection by various MuLVs (MacDonald *et al.*, 1980; Bennett *et al.*, 1968; Steeves *et al.*, 1968).

3. Host Genes That Affect the Immune Response

Depending on the virus and tumor system, the host immune response can play a major or minor (or no) role in determining the time course of disease in MuLV-infected mice. Early (neonatal) infection by an endogenous or exogenous MuLV is generally the most efficient means of leukemia induction (see below). In these cases, immunological tolerance to the infecting MuLV often results from the early MuLV expression. Thus, for situations such as endogenous leukemia of AKR mice and neonatal infections with M- or F-MuLV, relatively few antiviral immune responses occur, and host genes have relatively little effect in modulating leukemogenesis by these viruses. Yet, when MuLVs are inoculated into adult mice, antiviral immune responses can affect the efficiency of leukemogenesis (see Section IV.B).

The host immune response to MuLV infection is under the control of several genes. The effects of these genes are generally weaker than those such as *Fv-1;* they are often evident only in experimental systems where secondary mutations or methods of virus inoculation favor relatively slow time courses of disease. One popular system has been injection of the Friend virus complex (SSFV + F-MuLV helper) into adult mice (Chesebro *et al.*, 1990). Depending on the strain of mouse, the severity of disease ranges from mild (transient splenomegaly followed by recovery) to severe (prolonged splenomegaly followed by death). Genetic crosses between resistant and susceptible mice have identified several genes that affect severity of the disease—*rFv-1, rFv-2, rFv-3* (Chesebro and Wehrly, 1978, 1979). Further analysis indicated that they influence humoral or cellular immune responses to the virus.

Three separate regions of the MHC (*H-2* in the mouse) have been shown to affect recovery from Friend virus-induced disease (Chesebro *et al.*, 1990). The *H-2* region is a complex chromosomal region that includes structural genes for Class I (H-2K and H-2D) and Class II MHC (I-A) proteins, related proteins (Q and T1), as well as some cytokines (TNF-α and -β). The function of Class I and II MHC proteins is understood: these proteins bind antigenic peptides and display them to T lymphocytes, where specific binding and activation of the T lymphocytes occurs (Bjorkman and Parham, 1990). Class I MHC proteins are found on most cell types, and they present antigen to CD8 surface protein-containing T lymphocytes—cytotoxic T lymphocytes (CTLs) and suppressor T lymphocytes. Class II MHC proteins are found on B lymphocytes and macrophages ("antigen-presenting cells"), and they present antigen to CD4-positive T-helper lymphocytes. Antigen-activated T-helper lymphocytes secrete growth factors necessary for growth and maturation of both CTLs and B lymphocytes.

Genetic mapping indicated that different alleles at the *I-A, H-2D,* and *Qa-1/Tla* regions all affect resistance to Friend virus disease (Chesebro *et al.*, 1990). (The previously defined *rFv-1* and *rFv-2* loci correspond to the resistance alleles in *H-2D* and *Qa-1/Tla,* respectively.) In the case of *I-A,* the critical gene was shown to be the structural gene for Class II MHC (Miyazawa *et al.*, 1988, 1992), and it was suggested that resistance results from efficient presentation of viral antigen by macrophages and other antigen-presenting cells. In the case of *H-2D*, resistance was associated with efficient antiviral CTL responses, and resistance could be mapped to a Class I MHC structural gene (Miyazawa *et al.*, 1992; Polsky and Lilly, 1991). Moreover, the fact that resistance is dominant in genetic crosses suggests that the resistant form of the Class I MHC protein can interfere with presentation of viral proteins to virus-specific CTLs by the sensitive form, or it interferes with the generation of virus-specific CTLs. Less is known about the mechanistic details of *rFv-2* resistance, besides its mapping to the *Qa-1/Tla* region. Resistance to other MuLVs (RadLV and Gross MuLV) has also been mapped to the *H-2* region (Lonai and Haran-Ghera, 1977; Meruelo *et al.*, 1977; Chesebro *et al.*, 1990), although resistance alleles map to different genes in the region (e.g., *H-2K* for RadLV).

A third gene affecting the immune response to Friend virus disease has also been identified, *rFv-3*. This gene is not located in the *H-2* region. The resistance allele *rFv-3*^{r} confers the ability to raise antiviral antibodies (Chesebro and Wehrly, 1979). These antibodies apparently inhibit cell-to-cell spread of virus early during infection, and thus decrease the ability to induce disease.

B. The Immune Response to MuLV Infection

MuLVs have the capacity to invoke immune responses in infected animals. However, depending on the virus and method of infection, immunological responses play smaller or larger roles in modulating the course of infection and pathogenesis.

1. Infection of Newborn Animals

It has long been known that infection of neonatal animals with MuLVs is the most efficient means of inducing leukemia. This observation was first made by Gross (1951) for Gross passage A MuLV, and it is the standard method of infection for leukemogenesis by exogenous viruses such as M-MuLV. In addition, several high-leukemia-incidence inbred strains of mice (e.g., AKR) spontaneously activate endogenous MuLVs shortly after birth; this event leads to spread of the endogenous MuLV and induction of leukemia. Other mice that show similar behavior are strains that genetically transmit the normally exogenous M-MuLV genome [*mov* mice (Jaenisch, 1980; Jaenisch *et al.*, 1975)]; these were generated by Jaenisch via infection of undifferentiated mouse embryos *in vitro* followed by reimplantation into pseudopregnant foster mothers.

Mice infected with MuLVs as newborns generally show little antiviral immunity. In the case of cell-mediated immunity, virus-specific CTLs usually cannot be detected (Chieco-Bianchi *et al.*, 1974; Green, 1984; Korostoff *et al.*, 1990). The early report describing M-MuLV-specific CTLs in infected mice (see above; Lee *et al.*, 1981) remains controversial. Presumably, absence of virus-specific CTLs results from elimination of virus-specific thymocytes during T-lymphoid maturation by tolerance-inducing mechanisms. Yet, there is evidence for development of humoral immune responses in mice infected as newborns. While circulating levels of virus-specific antibodies cannot be detected in the blood (Chieco-Bianchi *et al.*, 1988; Hirsch *et al.*, 1969; Oldstone *et al.*, 1972), this observation results from sequestration of antibody–antigen complexes in the kidneys in at least some cases. Sera from M-MuLV-infected animals do not show virus-neutralizing activity *in vitro*. Moreover, M-MuLV-infected mice generally show high circulating levels of infectious virus. Thus, while neonatally infected animals may raise antibodies to MuLV infection, this humoral response does not appear to affect the *in vivo* pathogenesis. Overall, antiviral immune activities have little effect in mice infected with MuLVs as newborns.

2. MuLV Infection of Adults

In contrast, substantial immunological responses occur in mice infected with MuLVs as adults. As discussed above, cell-mediated immunity is the critical factor, and antiviral CTLs appear to control virus production and appearance of disease (Chieco-Bianchi *et al.*, 1988; Robertson *et al.*, 1992). In the case of M-MuLV, infection of adult C57B16 mice leads to the production of virus-specific CTLs (Biasi *et al.*, 1991). Moreover, these animals do not establish high levels of virus, and they do not develop lymphomas. Furthermore, interruption of cell-mediated immunity by treatment of animals with antibody to CD4 (which would eliminate T-helper lymphocytes) prior to M-MuLV infection inhibits formation of virus-specific CTLs (Biasi *et al.*, 1990, 1991). Concomitantly, these animals establish M-MuLV infection and develop T lymphomas. Thus, cell-mediated immunity appears to control viral infection in mice infected with M-MuLV as adults.

In contrast to M-MuLV, F-MuLV can establish infection when inoculated into adults. As described in the preceding section, the relative efficiency of F-MuLV infection and pathogenesis (erythroleukemia) is influenced by the immune response of the animals. In particular, various genes encoded by the MHC (*H-2*) region (*H-2D*, *I-A*, and *Qa-1/Tla*) affect antiviral immunity. The fact that the immune response is controlled by genes affecting Class I and II MHC indicates that both antiviral T helpers and CTLs are involved (Robertson *et al.*, 1992). For F-MuLV, the CTL responses are specific for the viral envelope protein (gp70 or SU), while the T-helper responses are directed against viral *gag* protein (Klarnet *et al.*, 1989). The CTL response has the larger effect in controlling viral replication and pathogenesis (Robertson *et al.*, 1992).

3. AKR.Fv-1^b Mice

Another system in which anti-MuLV CTL responses can be detected is that of AKR.Fv-1^b mice. Conventional AKR mice spontaneously activate their endogenous N-tropic Akv MuLV, which ultimately leads to leukemia. However, AKR mice congenic for the *Fv-1*^{b} allele will not efficiently propagate spontaneously activated Akv MuLV because they are restrictive for replication of N-tropic viruses. As a result, AKR.Fv-1^b mice have a low incidence of spontaneous leukemia. However, with long latency, the mice develop spontaneous B and T lymphomas; MCF recombinant viruses can be detected in the T-lymphomas (Haran-Ghera *et al.*, 1993). This observation presumably represents breakthrough of virus replication past the Fv-1 restriction barrier. Treatment of AKR.Fv-1^b mice with anti-

CD8 monoclonal antibodies significantly increases the frequency and rate of B-cell leukemia development. These findings suggest that antiviral immune responses are at least in part responsible for the disease latency. The presence of antiviral CTLs in AKR.Fv-1^b mice is plausible, since the lack of early virus expression might allow the later development of virus-specific CTLs. Extensive virus expression during the neonatal period might lead to deletion of virus-specific CTLs during neonatal thymic maturation.

Another system for the study of CTLs in AKR.Fv-1^b mice has employed doubly congenic AKR.H-2^b.Fv-1^b animals (Green, 1984). Mice of the *H*-2^b MHC haplotype are high CTL responders for Akr MuLV, while those of *H*-2^k haplotype (e.g., AKR) are low responders. As a result, while it is difficult to detect anti-AKR MuLV CTLs in AKR.Fv-1^b mice by standard assays, they can be detected in AKR.H-2^b.Fv-1^b mice. The CTLs were found to be directed against *env* glycoprotein (Azuma *et al.*, 1987). It is noteworthy that in these mice, the anti-viral CTL response declines with age, even though the concentration of virus-specific CTLs does not change (Green, 1987; Wegmann *et al.*, 1991). This finding has been attributed to the appearance of $CD8^+$ (suppressor) T lymphocytes that suppress the activity of the virus-specific CTLs (Rich *et al.*, 1992).

C. Timing of Viral Infection

As described above, MuLV infection of neonatal mice is the most efficient means for inducing disease. To a major extent this fact reflects the lack of substantial antiviral immune responses, which allows efficient viral infection to proceed. However, it is also possible that infection of specific hematopoietic or nonhematopoietic target cells can only occur in very young animals. Substantial hematopoietic development takes place during the first week after birth in mice (e.g., movement of progenitor cells from the bone marrow to the thymus). Possibly critical target cells for infection only exist for a short time after birth. Support for this hypothesis has been reported for AKR mice. Passive immunotherapy of newborn AKR mice with antibodies to MuLV substantially reduces the subsequent appearance of leukemia (Schwarz *et al.*, 1979; Huebner *et al.*, 1976). Subsequent studies indicated that the suppressive effect was dependent on administration of the antibody within a few days of birth (Weinhold *et al.*, 1984). Administration of antibody during the critical period inhibits the eventual appearance of productively infected cells (infectious centers or ICCs) in the thymus, although not in other organs such as the spleen and bone marrow. Furthermore, the earliest effects of antibody administration appear to be the inhibition of infection of a subpopulation of bone marrow cells (Buckheit *et al.*, 1987, 1988). This infection is apparently required for ultimate spread of virus to the thymus and development of disease. The nature of the critical infected bone marrow cells has yet to be determined: they could be hematopoietic progenitors, or stromal cells that secondarily confer infection to hematopoietic cells.

D. Influence of Sex on MuLV Leukemogenesis

For many MuLVs (M-MuLV, AKR endogenous MuLV, F-MuLV), leukemia occurs more rapidly in females than in males (Lilly *et al.*, 1975). The basis for this finding has not been determined. However, sex hormones have been implicated by the fact that ovariectomy decreases the incidence of leukemia in AKR mice, and orchidectomy increases the incidence (McEndy *et al.*, 1944). Glucocorticoid response elements (GREs) have been found in the enhancer regions of several MuLVs, including M-MuLV (DeFranco and Yamamoto, 1986; Speck and Baltimore, 1987) and the SL3-3 derivative of AKR MuLV (Celander and Haseltine, 1987). While the glucocorticoid responsiveness of these GREs has not been demonstrated in all cases (Overhauser and Fan, 1985), it is nevertheless possible that they differentially confer hormone-responsive transcriptional activity to these viruses in certain cells of female mice. As a result, proto-oncogene activation or other virus-driven events might be more efficient in females.

E. Requirements of Cell Division for Leukemogenesis

There is a general requirement for continual cell division of target tissues for MuLV leukemogenesis. This finding could reflect the fact that retroviral infection and continued viral expression is most efficient in dividing cells (Luciw and Leung, 1992); indeed, cell division is required for integration of proviral DNA (Brown, 1990). Surveys of viral replication in infected animals indicate that hematopoietic organs and the intestines are sites of high virus expression in animals infected as newborns (Simon *et al.*, 1982; Feuer and Fan, 1990). These tissues are the most prominent ones with high mitotic rates in adults.

There are several illustrations of the importance of cell division in leukemogenesis by MuLVs. BALB/c mice infected as adults with M-MuLV will develop myeloid leukemia if treated with pristane (Wolff *et al.*, 1988). The pristane treatment elicits granuloma formation and high-level myeloid hematopoiesis. Yet, if the pristane treatment is omitted, no myeloid tumors form. Thus, development of M-MuLV-induced myeloid tumors is dependent on secondary treatments that elicit cell division in the myeloid lineage. Another example is infection of germfree (gnotobiotic) mice (Isaak *et al.*, 1988). Germfree mice show a greatly increased latency for leukemogenesis by M-MuLV, which may result from the fact that the immune system is relatively inactive in the absence of antigenic stimulation from microbial infections.

The requirement of an intact thymus for development of T lymphomas in AKR mice or in M-MuLV-infected mice is pertinent as well (see, e.g., Haran-Ghera *et al.*, 1992). If mice are thymectomized, they do not develop T lymphoma, even if thymectomy occurs after viral infection or activation. The thymus is required for maturation of thymocytes into T lymphocytes, and considerable cell division takes place during this process. Thymocyte division and differentiation concomitant with viral infection is apparently necessary for development of T lymphomas.

The requirement for continual cell division in the target cells could reflect various processes. As described above, sequential infection of developing tumor cells may result in insertional activation of tumor progression loci, or infection with MCF recombinants. While incipient tumor cells are developing independence from growth factors, a continual supply of growth factors may be necessary for these cells until they achieve full autonomy. In addition, spontaneous secondary genetic mutations (inactivation of tumor suppressor genes, activation of proto-oncogenes) in the developing tumor cells may be facilitated by rapid cell division.

V. TUMORIGENESIS BY MOUSE MAMMARY TUMOR VIRUS

Another important tumorigenesis system for nonacute retroviruses is mouse mammary tumor virus (MMTV). Historically, MMTV was initially identified by the study of inbred mouse strains with high mammary tumor incidence [e.g., C3H, GR, RIII (Gross, 1983a)]. In a classic series of experiments, Bittner (1942) showed that a high incidence of mammary tumors could be transferred to low-incidence BALB/c mice by foster nursing them on C3H mothers. The transmissible agent was eventually shown to be MMTV, a virus that was congenitally transmitted in C3H mice (Gross, 1983a). Mammary tumors in these mice showed newly acquired copies of MMTV provirus, indicative of viral infection (Cohen and Varmus, 1980). For the GR and RIII strains, the tumorigenic MMTV was an endogenous MMTV provirus that spontaneously activated, analogous to the AKV MuLV provirus carried in AKR mice (Traina *et al.*, 1981). The number of endogenous MMTV-related proviruses in mice is lower than the number of MuLV-related proviruses: depending on the strain, most mice have less than 10 endogenous MMTV-related proviruses (Traina *et al.*, 1981), while they will contain 50–100 endogenous MuLV-related proviruses (Frankel *et al.*, 1989, 1990).

It should be noted that other viruses besides MMTV can also produce mammary tumors in inbred mice. For instance, polyoma virus can induce such tumors in C3H mice (Berebbi *et al.*, 1988; Freund *et al.*, 1992). Moreover, spontaneous mammary carcinomas occur in mice lacking MMTV. These observations suggest that besides carrying active endogenous MMTV, some mouse strains can have nonviral genetic factors that contribute to development of mammary tumors.

A. MMTV Structure and Organization

MMTV is classified as a type B retrovirus, on the basis of its morphology. It is thus distinct from MuLVs, which are type C retroviruses. As with all replication-competent retroviruses, MMTV has *gag*, *pol*, and *env* genes that are expressed as viral proteins by the same mechanisms used by all retroviruses (see Luciw and Leung, 1992). However, the MMTV LTR is unique among retroviruses in that it encodes a protein, and it is hormonally responsive.

The protein coding capacity of the MMTV LTR was apparent from initial DNA sequencing studies (Peters *et al.*, 1982). The open reading frame (3′ ORF) in the LTR is expressed as a spliced subgenomic mRNA (Robertson and Varmus, 1979), and it can encode a putative protein of approximately 36 kDa. The function of the protein encoded by the 3′ ORF was unknown for many years, and speculation centered on its being involved in cell transformation and/or regulation (van Klaveren and Bentvelzen, 1988; Salmons *et al.*, 1990). However, recently this protein has been shown to play a role in *in vivo* infection by functioning as a superantigen. Superantigens have the characteristics of simultaneously binding to MHC proteins and T-cell receptor protein variable regions (Coffin, 1992). As a result, they activate all T lymphocytes bearing a particular T-cell receptor gene class. In the short term, this event results in mitogenesis of these T lymphocytes, but in the long term, these T lymphocytes are clonally deleted from the T-cell repertoire.

The first indication that MMTV encodes a superantigen arose from the finding that there was a complete concordance in genetic mapping between genetic loci specifying the murine *mls* antigens and endogenous MMTV proviruses (Frankel *et al.*, 1991). The *mls* antigens had been previously shown to function as superantigens (Acha-Orbea *et al.*, 1992). Subsequent experiments have shown that it is the protein encoded in the 3′ ORF of the LTR that encodes the *mls* superantigen (Acha-Orbea *et al.*, 1991; Pullen *et al.*, 1992; Marrack *et al.*, 1991; Beutner *et al.*, 1992). Thus, this MMTV gene is now referred to as SAG for superantigen.

The role of the SAG gene in viral infection is to facilitate MMTV infection of T lymphocytes. The normal route of infection for MMTV is through the milk, and experiments by Buller *et al.* (1989) have suggested that initial MMTV infection occurs in T and B lymphocytes in the digestive tract; subsequently infection is found in other sites such as the mammary gland. Thus, expression of SAG would lead to extensive proliferation of T lymphocytes of a particular T-cell receptor class (e.g., Vbeta-14 for the C3H MMTV). This proliferation may facilitate productive viral infection. Recent experiments with transgenic mice provide strong confirmation of the role of MMTV SAG. Mice transgenic for the C3H MMTV SAG show clonal deletion of T cells of the Vbeta-14$^+$ class. Moreover, these mice are now resistant to *in vivo* infection with C3H MMTV, but not other MMTVs whose SAGs bind to different T-cell receptors of different Vbeta classes (Golovkina *et al.*, 1992). Thus, elimination of T lymphocytes responsive to an MMTV SAG renders the animal resistant to infection by that virus. This finding implies that SAG-mediated T-lymphocyte infection is necessary for *in vivo* propagation of virus.

The second unique feature of the MMTV LTR is that it is transcriptionally responsive to glucocorticoid hormones. MMTV transcription is strongly activated by glucocorticoids such as dexamethasone (Ringold *et al.*, 1975). This result is presumably related to the fact that mammary tissue is responsive to various corticosteroid hormones. Extensive analyses on the LTR sequences responsible for hormone responsiveness have been carried out (Chandler *et al.*, 1983). The hormone-responsive region consists of several hundred nucleotides (−450 to −150) in the U3 region of the LTR. This region contains several copies of glucocorticoid response elements (GREs)—DNA elements that bind the glucocorticoid receptor (Scheidereit and Beato, 1984). In addition to the positively acting GREs (which confer increased transcription in the presence of glucocorticoids), a negative regulatory element (NRE) is also present in this region (Hsu *et al.*, 1988). Deletion of the NRE increases basal transcription from the LTR in the absence of hormone.

B. Stages of Mammary Tumorigenesis

As with most tumors, MMTV-induced neoplasia results from several steps (Fig. 10). In the case of MMTV, some of these steps can be identified histologically (Knepper *et al.*, 1989). During MMTV tumorigenesis, mammary hyperpla-

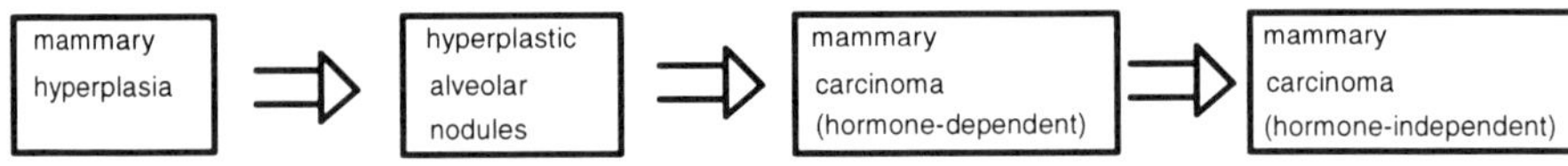

FIGURE 10. Stages in MMTV tumorigenesis. Stages that occur during development of MMTV-induced tumors are shown.

sia occurs first, followed by the development of hyperplastic alveolar nodules (HANs). These premalignant HANs can be isolated and propagated by transplantation into cleared mammary fat pads where they form hyperplastic outgrowths (HOGs). HOGs can be isolated, and molecular studies carried out. Some of the HOGs further develop into mammary tumors. Initially, most MMTV-induced mammary tumors are hormone-dependent, i.e., they grow only when glucocorticoid hormones are present. However, subsequently, hormone-independent mammary tumors develop. This result will be discussed further below.

C. MMTV Activation of Proto-oncogenes

Study of MMTV-induced mammary tumors has led to identification of several proto-oncogenes that are activated by the MMTV LTR: *int-1* (now called *wnt-1*) (Nusse and Varmus, 1982), *int-2* (Peters *et al.*, 1983), *int-3* (Gallahan and Callahan, 1987), and *wnt-3* (Roelink *et al.*, 1990). *wnt-1, int-2,* and *wnt-3* proto-oncogene products are growth factors. The *int-2* gene product is related to basic fibroblast growth factor (Yoshida *et al.*, 1987). These proto-oncogenes are normally not expressed in mammary tissue; ectopic expression under control of the MMTV LTR leads to abnormal cell growth. MMTV LTR activation occurs by enhancer activation rather than promoter insertion: in virtually all cases, the position and orientation of the MMTV LTR relative to the proto-oncogene are such that readthrough transcription from the LTR cannot occur (Nusse and Varmus, 1982). MMTV LTR activations in mammary tumors are rather unusual in that compared with other retroviral insertional activations, which are often found in close proximity to the affected proto-oncogene, MMTV proviral insertions may occur at considerable distance (more than 1 kb) from the affected *int* locus.

One noteworthy feature of *int* gene activations is that they may function in a paracrine fashion, i.e., the product may affect a neighboring cell (Fig. 11). When the *int-1* gene was expressed in mouse NIH-3T3 fibroblasts via a retroviral vector, the NIH-3T3 cells were not transformed. However, cocultivation of the *int-1*-expressing NIH-3T3 cells with mammary epithelial cells resulted in transformation of the mammary cells (Brown *et al.*, 1986). Another indication of the paracrine function of *int-1* came from tumor transplantation experiments (Mester *et al.*, 1987). MMTV tumors containing *int-1* activations were passaged in animals. Initially, the tumors were hormone-dependent, but with passage, hormone-independent tumors emerged. The emergence of hormone-independent tumors coincided with the appearance of cells with insertions at *int-2.* Most importantly, some of these cells lacked the *int-1* insertions of the original hormone-dependent tumors. These results have been interpreted as indicating that the original tumor was a mixed population of cells. Some contained the *int-1*

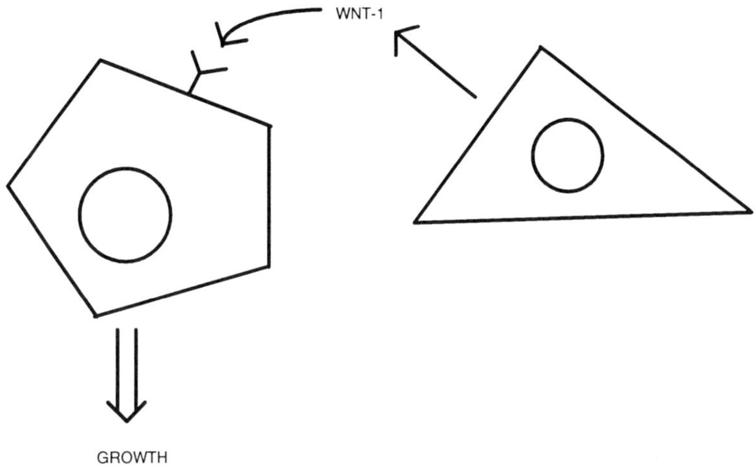

FIGURE 11. Paracrine stimulation of cell growth. Paracrine stimulation of cell growth is illustrated for *int-1/wnt-1*. If this protein is produced by one cell, it can stimulate an adjacent cell to grow if that cell has receptors for *wnt-1*. This process has been demonstrated in cell culture (Brown *et al.* 1986), and similarly persistent production of such a growth-stimulating protein *in vivo* might be involved in MMTV-induced tumors.

insertion and others lacked the insertion but stimulated cells in a paracrine fashion. Eventual emergence of the hormone-independent tumors presumably resulted from proviral insertions at *int-2* in those cells lacking the *int-1* insertion.

While the results in the preceding paragraph suggest that there may be a temporal order of proto-oncogene activations (*int-1* preceding *int-2*), this is clearly not always the case. The frequency of insertions at different *int* loci is influenced by the genetic background of the mouse (Marchetti *et al.*, 1991). In addition, preneoplastic lesions (HANs) have been compared with end-stage tumors for proviral insertions (Morris *et al.*, 1990). The results indicated that insertions at either *int-1* or *int-2* could be found in both preneoplastic HANs as well as tumors, and there was no correlation with disease progression.

Transgenic mouse lines expressing a number of proto-oncogenes under the control of the MMTV LTR have been generated. These include lines transgenic for *wnt-1* (Jorgensen *et al.*, 1992), *int-2* (Muller *et al.*, 1990; Pellegrini and Basilico, 1987; Jhappan *et al.*, 1992), *neu* (Tremblay *et al.*, 1989), and *c-myc* (Leder *et al.*, 1986). All of these lines show mammary gland hyperplasia, and development of mammary adenocarcinomas with longer latency. Presumably the tumors result from additional spontaneous changes that occur in the hyperplastic cells.

D. MMTV-Induced T Lymphomas

In addition to mammary tumors, MMTV can induce T lymphomas. These tumors are often observed in male mice of strains that spontaneously activate endogenous MMTVs. Male mice are much less prone to development of mammary tumors, so the appearance of T lymphomas with long latency is more readily observed. When these lymphomas are studied, they all show newly acquired copies of MMTV proviral DNA. Moreover, the LTRs of these MMTV proviruses show deletions in the U3 region (Michalides and Wagenaar, 1986; Lee *et al.*, 1987). These deletions generally result in three changes: (1) loss of the coding capacity for the SAG protein, (2) deletion of the negative regulatory element in the LTR (Hsu *et al.*, 1988), and (3) generation of T-lymphoid-specific enhancer sequences by juxtaposition of different MMTV LTR sequences (Michalides and Wagenaar, 1986; Lee *et al.*, 1987). A replication-competent variant MMTV carrying such deletions that is highly T lymphomagenic has been cloned (Ball *et al.*, 1988). Presumably the modified MMTV LTRs lead to high-level expression in T lymphocytes and insertional activation of proto-oncogenes, although these putative proto-oncogenes have not been identified. Given the realization that MMTV normally infects T lymphocytes *in vivo*, the induction of T lymphomas is not surprising.

VI. CONCLUDING REMARKS

The unifying theme in this chapter is that carcinogenesis is a multistep process. In the case of retrovirus-induced tumors, any given virus may carry out one or more steps in the process, while the others presumably occur by nonviral mechanisms (e.g., spontaneous mutations, expression of cellular growth factors).

For acute transforming retroviruses, the primary transforming factors are the oncogenes carried by these viruses. The exact organization of a transforming virus (including the identity and nature of the oncogene) governs the type of tumor induced and the rapidity of disease development. Nevertheless, multiple steps are still required for tumorigenesis. In some cases, these additional steps can be mediated by the transforming virus (e.g., Friend SFFV insertional activation of *spi-1/ets-1*) or the helper virus (e.g., helper virus activation of *ahi-1* in Ab-MuLV-induced B lymphomas). In other cases, they may occur spontaneously (e.g., some of the p53 deletions in Friend SFFV-induced erythroleukemia).

Nonacute retroviruses are a rich field for the study of multistep carcinogenic processes. LTR activation of proto-oncogenes is a unifying mechanism for tumorigenesis by these viruses. For most well-studied systems, proviral insertion near one or more proto-oncogenes (or the presence of common-insertion sites) has been found. However, it should be noted that for some viral systems, infrequent activation of known proto-oncogenes or insertion sites has been observed [e.g., long-latency B lymphomas induced by an SV40 enhancer-containing variant of M-MuLV (Hanecak *et al.*, 1988)]. One explanation is that the relevant insertion sites have not yet been identified. Alternatively, leukemogenesis for these viruses might not involve LTR activation of proto-oncogenes (see discussion on HTLV-I and BLV below).

As described in this chapter, MuLVs can carry out multiple steps in the tumorigenic process in addition to activation of proto-oncogenes. In at least some systems, they can induce preleukemic changes (splenomegaly, induction of potentially leukemic cells), and they can promote tumor progression by activation of tumor progression loci. In addition, MCF recombinant viruses can play roles at all stages of the disease process, including induction of preleukemic changes, insertional activation of proto-oncogenes, and establishment of autocrine loops by stimulation of growth factor receptors. While MCF recombinant viruses are not absolutely required for tumorigenesis, they clearly contribute to the efficiency of MuLV-induced disease. Presumably in their absence (e.g., *Rmcf* [r] mice or in rats), other nonviral mechanisms mediate the changes that cooperate with proviral activation of proto-oncogenes.

This review has not dealt with carcinogenesis by HTLV-I and BLV (see Sugamura and Hinuma, 1993). These related viruses are noteworthy in inducing disease with an extremely long latency (many years). Moreover, tumors induced by HTLV-I and BLV do not show common sites of proviral insertion. The leading hypothesis of how these viruses induce tumors relates to their *tax* transactivating protein. *tax* not only transactivates transcription from the HTLV or BLV LTR, but it can transactivate expression from several cellular genes, including IL-2, IL-2 receptor, and several other growth factors. Human T lymphocytes can be immortalized by infection with HTLV-I, perhaps by establishment in the infected cells of an autocrine loop involving IL-2 and IL-2R (Sugamura and Hinuma, 1993). Conceivably all other events involved in leukemogenesis besides establishment of hyperplastic lymphocytes occur by non-virus-mediated means. This possibility may explain the long delay between viral infection and appearance of tumors.

Studies on tumorigenesis by all of these retroviruses may provide insights into carcinogenesis by other processes as well. An example concerns the activation of cellular proto-oncogenes in various human tumors (Bishop, 1987). One of the secondary events that occurs in Friend SFFV-induced erythroleukemias is deletion or inactivation of the cellular p53 gene. This inactivation of the p53 gene (a tumor suppressor gene) has recently been recognized as a common genetic alteration in many kinds of human cancers (e.g., Cho and Vogelstein, 1992). Another example is the preleukemic state in M-MuLV-infected mice (see Section III.C.2). This state involves bone marrow defects and accompanying extramedullary hematopoiesis in the spleen. While it is virus-driven in the case of M-MuLV-infected mice, there are parallels to human (nonviral) leukemia. Conditions such as myelodysplasia, in which there is a defect in marrow hematopoiesis, frequently resolve as leukemia. Thus, the study of retrovirus-induced cancers will continue to provide important insights into general mechanisms of carcinogenesis.

VII. REFERENCES

Acha-Orbea, H., Shakhov, A. N., Scarpellino, L., *et al.*, 1991, Clonal deletion of V beta 14-bearing T cells in mice transgenic for mammary tumour virus, *Nature* **350**:207–211.

Acha-Orbea, H., Held, W., Scarpellino, L., and Shakhov, A. N., 1992, Mls: A link between immunology and retrovirology, *Int. Rev. Immunol.* **8**:327–336.

Anderson, D. M., Lyman, S. D., Baird, A., *et al.*, 1990, Molecular cloning of mast cell growth factor, a hematopoietin that is active in both membrane bound and soluble forms [published erratum appears in *Cell* 1990 **63**(5):following 1112], *Cell* **63**:235–243.

Anderson, S. M., and Scolnick, E. M., 1983, Construction and isolation of a transforming murine retrovirus containing the src gene of Rous sarcoma virus, *J. Virol.* **46**:594–605.

Asjo, B., Skoog, L., Palminger, I., *et al.*, 1985, Influence of genotype and the organ of origin on the subtype of T-cell in Moloney lymphomas induced by transfer of preleukemic cells from athymic and thymus-bearing mice, *Cancer Res.* **45**:1040–1045.

Azuma, H., Phillips, J. D., and Green, W. R., 1987, Clonal heterogeneity of anti-AKR/Gross leukemia virus cytotoxic T lymphocytes. Evidence for two distinct antigen systems, *J. Immunol.* **139**:2464–2473.

Ball, J. K., Diggelmann, H., Dekaban, G. A., *et al.*, 1988, Alterations in the U3 region of the long terminal repeat of an infectious thymotropic type B retrovirus, *J. Virol.* **62**:2985–2993.

Bartholomew, C., Morishita, K., Askew, D., *et al.*, 1989, Retroviral insertions in the CB-1/Fim-3 common site of integration activate expression of the Evi-1 gene, *Oncogene* **4**:529–534.

Baumbach, W. R., Keath, E. J., and Cole, M. D., 1986, A mouse c-myc retrovirus transforms established fibroblast lines in vitro and induces monocyte–macrophage tumors in vivo, *J. Virol.* **59**:276–283.

Baumbach, W. R., Stanley, E. R., and Cole, M. D., 1987, Induction of clonal monocyte–macrophage tumors in vivo by a mouse c-myc retrovirus: Rearrangement of the CSF-1 gene as a secondary transforming event, *Mol. Cell Biol.* **7**:664–671.

Bear, S. E., Bellacosa, A., Lazo, P. A., *et al.*, 1989, Provirus insertion in Tpl-1, an Ets-1-related oncogene, is associated with tumor progression in Moloney murine leukemia virus-induced rat thymic lymphomas, *Proc. Natl. Acad. Sci. USA* **86**:7495–7499.

Ben-David, Y., Prideaux, V. R., Chow, V., Benchimol, S., and Bernstein, A., 1988, Inactivation of the p53 oncogene by internal deletion or retroviral integration in erythroleukemic cell lines induced by Friend leukemia virus, *Oncogene* **3**:179–185.

Ben-David, Y., Giddens, E. B., and Bernstein, A., 1990, Identification and mapping of a common proviral integration site Fli-1 in erythroleukemia cells induced by Friend murine leukemia virus, *Proc. Natl. Acad. Sci. USA* **87**:1332–1336.

Ben-David, Y., Giddens, E. B., Letwin, K., and Bernstein, A., 1991, Erythroleukemia induction by Friend murine leukemia virus: Insertional activation of a new member of the ets gene family, Fli-1, closely linked to c-ets-1, *Genes Dev.* **5**:908–918.

Ben-David, Y., Bani, M. R., Chabot, B., De Koven, A., and Bernstein, A., 1992, Retroviral insertions downstream of the heterogeneous nuclear ribonucleoprotein A1 gene in erythroleukemia cells: Evidence that A1 is not essential for cell growth, *Mol. Cell Biol.* **12**:4449–4455.

Bennett, M., Steeves, R. A., Cudkowicz, G., Mirand, E. A., and Russell, L. B., 1968, Mutant Sl alleles of mice affect susceptibility to Friend spleen focus-forming virus, *Science* **162**:564–565.

Berebbi, M., Dandolo, L., Hassoun, J., Bernard, A. M., and Blangy, D., 1988, Specific tissue targeting of polyoma virus oncogenicity in athymic nude mice, *Oncogene* **2**:149–156.

Bergold, P. J., Wang, J. Y., Hardy, W. D., Jr., Littau, V., Johnson, E., and Besmer, P., 1987, Structure and origins of the HZ2-feline sarcoma virus, *Virology* **158**:320–329.

Besmer, P., Murphy, J. E., George, P. C., *et al.*, 1986, A new acute transforming feline retrovirus and relationship of its oncogene v-kit with the protein kinase gene family, *Nature* **320**:415–421.

Beutner, U., Frankel, W. N., Cote, M. S., Coffin, J. M., and Huber, B. T., 1992, Mls-1 is encoded by the long terminal repeat open reading frame of the mouse mammary tumor provirus Mtv-7, *Proc. Natl. Acad. Sci. USA* **89**:5432–5436.

Biasi, G., Mazzocchi, M., Facchinetti, A., *et al.*, 1990, Induction of Moloney murine sarcoma virus tolerance in adult mice by anti-CD4 monoclonal antibody treatment, *Cancer Res.* **50**: 5703S–5706S.

Biasi, G., Facchinetti, A., Panozzo, M., Zanovello, P., Chieco-Bianchi, L., and Collavo, D., 1991, Moloney murine leukemia virus tolerance in anti-CD4 monoclonal antibody-treated adult mice, *J. Immunol.* **147**:2284–2289.

Biedenkapp, H., Borgmeyer, U., Sippel, A. E., and Klempnauer, K. H., 1988, Viral myb oncogene encodes a sequence-specific DNA-binding activity, *Nature* **335**:835–837.

Bishop, J. M., 1987, The molecular genetics of cancer, *Science* **235**:305–311.

Bishop, J. M., and Varmus, H., 1984, Functions and origins of retroviral transforming genes, in: *RNA Tumor Viruses: Molecular Biology of Tumor Viruses,* 2nd ed. (R. Weiss, N. Teich, H. Varmus, and J. Coffin, eds.), pp. 999–1108, Cold Spring Harbor Laboratory, Cold Spring Harbor, N.Y.

Bittner, J. J., 1942, Milk-influence of breast tumors in mice, *Science* **95**:462–463.

Bjorkman, P. J., and Parham, P., 1990, Structure, function, and diversity of class I major histocompatibility complex molecules, *Annu. Rev. Biochem.* **59**:253–288.

Blackwell, T. K., Kretzner, L., Blackwood, E. M., Eisenman, R. N., and Weintraub, H., 1990, Sequence-specific DNA binding by the c-Myc protein, *Science* **250**:1149–1151.

Blackwood, E. M., and Eisenman, R. N., 1991, Max: A helix–loop–helix zipper protein that forms a sequence-specific DNA-binding complex with Myc, *Science* **251**:1211–1217.

Bondurant, M. C., Koury, M. J., and Krantz, S. B., 1985, The Fv-2 gene controls induction of erythroid burst formation by Friend virus infection in vitro: Studies of growth regulators and viral replication, *J. Gen. Virol.* **66**:83–96.

Boone, L. R., Myer, F. E., Yang, D. M., *et al.*, 1983, Reversal of Fv-1 host range by in vitro restriction endonuclease fragment exchange between molecular clones of N-tropic and B-tropic murine leukemia virus genomes, *J. Virol.* **48**:110–119.

Boral, A. L., Okenquist, S. A., and Lenz, J., 1989, Identification of the SL3-3 virus enhancer core as a T-lymphoma cell-specific element, *J. Virol.* **63**:76–84.

Bordereaux, D., Fichelson, S., Sola, B., Tambourin, P. E., and Gisselbrecht, S., 1987, Frequent involvement of the fim-3 region in Friend murine leukemia virus-induced mouse myeloblastic leukemias, *J. Virol.* **61**:4043–4045.

Bosselman, R. A., Van Straaten, F., Van Beveren, C., Verma, I. M., and Vogt, M., 1982, Analysis of the env gene of a molecularly cloned and biologically active Moloney mink cell focus-forming proviral DNA, *J. Virol.* **44**:19–31.

Breuer, M. L., Cuypers, H. T., and Berns, A., 1989, Evidence for the involvement of pim-2, a new common proviral insertion site, in progression of lymphomas, *EMBO J.* **8**:743–748.

Brightman, B. K., Pattengale, P. K., and Fan, H., 1986, Generation and characterization of a recombinant Moloney murine leukemia virus containing the v-myc oncogene of avian MC29 virus: In vitro transformation and in vivo pathogenesis, *J. Virol.* **60**:68–81.

Brightman, B. K., Chandy, K. G., Spencer, R. H., Gupta, S., Pattengale, P. K., and Fan, H., 1988a, Characterization of lymphoid tumors induced by a recombinant murine retrovirus carrying the avian v-myc oncogene. Identification of novel (B-lymphoid) tumors in the thymus, *J. Immunol.* **141**:2844–2854.

Brightman, B. K., Chandy, K. G., Spencer, R. H., Gupta, S., Pattengale, P. K., and Fan, H., 1988b, Characterization of lymphoid tumors induced by a recombinant murine retrovirus carrying the avian v-myc oncogene. Identification of novel (B-lymphoid) tumors in the thymus, *J. Immunol.* **141**:2844–2854.

Brightman, B. K., Davis, B. R., and Fan, H., 1990, Preleukemic hematopoietic hyperplasia induced by Moloney murine leukemia virus is an indirect consequence of viral infection, *J. Virol.* **64**:4582–4584.

Brightman, B. K., Li, Q. X., Trepp, D. J., and Fan, H., 1991a, Differential disease restriction of Moloney and Friend murine leukemia viruses by the mouse Rmcf gene is governed by the viral long terminal repeat, *J. Exp. Med.* **174**:389–396.

Brightman, B. K., Rein, A., Trepp, D. J., and Fan, H., 1991b, An enhancer variant of Moloney murine leukemia virus defective in leukemogenesis does not generate detectable mink cell focus-inducing virus in vivo [published erratum appears in *Proc. Natl. Acad. Sci. USA* 1991 **88**:5066], *Proc. Natl. Acad. Sci. USA* **88**:2264–2268.

Brodeur, G. M., Seeger, R. C., Schwab, M., Varmus, H. E., and Bishop, J. M., 1984, Amplification of N-myc in untreated human neuroblastomas correlates with advanced disease stage, *Science* **224**:1121–1124.

Brown, A. M., Wildin, R. S., Prendergast, T. J., and Varmus, H. E., 1986, A retrovirus vector expressing the putative mammary oncogene int-1 causes partial transformation of a mammary epithelial cell line, *Cell* **46**:1001–1009.

Brown, P. O., 1990, Integration of retroviral DNA, *Curr. Top. Microbiol. Immunol.* **157**:19–48.

Buchberg, A. M., Bedigian, H. G., Jenkins, N. A., and Copeland, N. G., 1990, Evi-2, a common integration site involved in murine myeloid leukemogenesis, *Mol. Cell Biol.* **10**:4658–4666.

Buckheit, R. W., Jr., Bolognesi, D. P., and Weinhold, K. J., 1987, The effects of leukemosuppressive immunotherapy on bone marrow infectious cell centers in AKR mice, *Virology* **157**:387–396.

Buckheit, R. W., Jr., Bolognesi, D. P., and Weinhold, K. J., 1988, The role of bone marrow and thymic elements in the initiation and spread of virus production in the AKR thymus, *Virology* **166**:533–541.

Buller, R. S., Sitbon, M., and Portis, J. L., 1988, The endogenous mink cell focus-forming (MCF) gp70 linked to the Rmcf gene restricts MCF virus replication in vivo and provides partial resistance to erythroleukemia induced by Friend murine leukemia virus, *J. Exp. Med.* **167**:1535–1546.

Buller, R. S., Van Zant, G., Eldridge, P. W., and Portis, J. L., 1989, A population of murine hematopoietic progenitors expresses an endogenous retroviral gp70 linked to the Rmcf gene and associated with resistance to erythroleukemia, *J. Exp. Med.* **169**:865–880.

Celander, D., and Haseltine, W. A., 1987, Glucocorticoid regulation of murine leukemia virus transcription elements is specified by determinants within the viral enhancer region, *J. Virol.* **61**:269–275.

Chabot, B., Stephenson, D. A., Chapman, V. M., Besmer, P., and Bernstein, A., 1988, The proto-oncogene c-kit encoding a transmembrane tyrosine kinase receptor maps to the mouse W locus, *Nature* **335**:88–89.

Chandler, V. L., Maler, B. A., and Yamamoto, K. R., 1983, DNA sequences bound specifically by glucocorticoid receptor in vitro render a heterologous promoter hormone responsive in vivo, *Cell* **33**:489–499.

Chang, E. H., Gonda, M. A., Ellis, R. W., Scolnick, E. M., and Lowy, D. R., 1982, Human genome contains four genes homologous to transforming genes of Harvey and Kirsten murine sarcoma viruses, *Proc. Natl. Acad. Sci. USA* **79**:4848–4852.

Chatis, P. A., Holland, C. A., Hartley, J. W., Rowe, W. P., and Hopkins, N., 1983, Role for the 3′ end of the genome in determining disease specificity of Friend and Moloney murine leukemia viruses, *Proc. Natl. Acad. Sci. USA* **80**:4408–4411.

Chatis, P. A., Holland, C. A., Silver, J. E., Frederickson, T. N., Hopkins, N., and Hartley, J. W., 1984, A 3′ end fragment encompassing the transcriptional enhancers of nondefective Friend virus confers erythroleukemogenicity on Moloney leukemia virus, *J. Virol.* **52**:248–254.

Chen, C., Biegalke, B. J., Eisenman, R. N., and Linial, M. L., 1989, FH3, a v-myc avian retrovirus with limited transforming ability, *J. Virol.* **63**:5092–5100.

Chesebro, B., and Wehrly, K., 1978, Rfv-1 and Rfv-2, two H-2-associated genes that influence recovery from Friend leukemia virus-induced splenomegaly, *J. Immunol.* **120**:1081–1085.

Chesebro, B., and Wehrly, K., 1979, Identification of a non-H-2 gene (Rfv-3) influencing recovery from viremia and leukemia induced by Friend virus complex, *Proc. Natl. Acad. Sci. USA* **76**:425–429.

Chesebro, B., Miyazawa, M., and Britt, W. J., 1990, Host genetic control of spontaneous and induced immunity to Friend murine retrovirus infection, *Annu. Rev. Immunol.* **8**:477–499.

Chieco-Bianchi, L., Sendo, F., Aoki, T., and Barrera, O. L., 1974, Immunologic tolerance to antigens associated with murine leukemia viruses: T-cell unresponsiveness, *J. Natl. Cancer Inst.* **52**:1345–1350.

Chieco-Bianchi, L., Collavo, D., and Biasi, G., 1988, Immunologic unresponsiveness to murine leukemia virus antigens: Mechanisms and role in tumor development, *Adv. Cancer Res.* **51**:277–306.

Cho, K. R., and Vogelstein, B., 1992, Genetic alterations in the adenoma–carcinoma sequence, *Cancer* **70**:1727–1731.

Coffin, J. M., 1992, Superantigens and endogenous retroviruses: A confluence of puzzles, *Science* **255**:411–413.

Cohen, J. C., and Varmus, H. E., 1980, Proviruses of mouse mammary tumor virus in normal and neoplastic tissues from GR and C3Hf mouse strains, *J. Virol.* **35**:298–305.

Copeland, N. G., Gilbert, D. J., Cho, B. C., *et al.*, 1990, Mast cell growth factor maps near the steel locus on mouse chromosome 10 and is deleted in a number of steel alleles, *Cell* **63**:175–183.

Cuypers, H. T., Selten, G., Quint, W., *et al.*, 1984a, Murine leukemia virus-induced T-cell lymphomagenesis: integration of proviruses in a distinct chromosomal region, *Cell* **37**:141–150.

Cuypers, H. T., Selten, G., Quint, W., *et al.*, 1984b, Murine leukemia virus-induced T-cell lymphomagenesis: Integration of proviruses in a distinct chromosomal region, *Cell* **37**:141–150.

Cuypers, H. T., Selten, G. C., Zijlstra, M., De Goede, R. E., Melief, C. J., and Berns, A. J., 1986, Tumor progression in murine leukemia virus-induced T-cell lymphomas: Monitoring clonal selections with viral and cellular probes, *J. Virol.* **60**:230–241.

Dalla-Favera, R., Bregni, M., Erikson, J., Patterson, D., Gallo, R. C., and Croce, C. M., 1982, Human c-myc onc gene is located on the region of chromosome 8 that is translocated in Burkitt lymphoma cells, *Proc. Natl. Acad. Sci. USA* **79**:7824–7827.

Davis, B., Linney, E., and Fan, H., 1985, Suppression of leukaemia virus pathogenicity by polyoma virus enhancers, *Nature* **314**:550–553.

Davis, B. R., Chandy, K. G., Brightman, B. K., Gupta, S., and Fan, H., 1986, Effects of nonleukemogenic and wild-type Moloney murine leukemia virus on lymphoid cells in vivo: Identification of a preleukemic shift in thymocyte subpopulations, *J. Virol.* **60**:423–430.

Davis, B. R., Brightman, B. K., Chandy, K. G., and Fan, H., 1987, Characterization of a preleukemic state induced by Moloney murine leukemia virus: Evidence for two infection events during leukemogenesis, *Proc. Natl. Acad. Sci. USA* **84**:4875–4879.

DeFranco, D., and Yamamoto, K. R., 1986, Two different factors act separately or together to specify functionally distinct activities at a single transcriptional enhancer, *Mol. Cell Biol.* **6**:993–1001.

DesGroseillers, L., and Jolicoeur, P., 1983, Physical mapping of the Fv-1 tropism host range determinant of BALB/c murine leukemia viruses, *J. Virol.* **48**:685–696.

DesGroseillers, L., and Jolicoeur, P., 1984, Mapping the viral sequences conferring leukemogenicity and disease specificity in Moloney and amphotropic murine leukemia viruses, *J. Virol.* **52**:448–456.

DesGroseillers, L., Villemur, R., and Jolicoeur, P., 1983, The high leukemogenic potential of Gross passage A murine leukemia virus maps in the region of the genome corresponding to the long terminal repeat and to the 3′ end of env, *J. Virol.* **47**:24–32.

Dickson, C., Smith, R., Brookes, S., and Peters, G., 1984, Tumorigenesis by mouse mammary tumor virus: Proviral activation of a cellular gene in the common integration region int-2, *Cell* **37**:529–536.

Dolberg, D. S., and Bissell, M. J., 1984, Inability of Rous sarcoma virus to cause sarcomas in the avian embryo, *Nature* **309**:552–556.

Dolberg, D. S., Hollingsworth, R., Hertle, M., and Bissell, M. J., 1985, Wounding and its role in RSV-mediated tumor formation, *Science* **230**:676–678.

Dresler, S., Ruta, M., Murray, M. J., and Kabat, D., 1979, Glycoprotein encoded by the Friend spleen focus-forming virus, *J. Virol.* **30**:564–575.

Dreyfus, F., Sola, B., Fichelson, S., *et al.*, 1990, Rearrangements of the Pim-1, c-myc, and p53 genes in Friend helper virus-induced mouse erythroleukemias, *Leukemia* **4**:590–594.

Duesberg, P. H., and Vogt, P. K., 1970, Differences between the ribonucleic acids of transforming and nontransforming avian tumor viruses, *Proc. Natl. Acad. Sci. USA* **67**:1673–1680.

Ellerman, V., and Bang, O., 1908, Experimentelle Leukamie bei Huhnern, *Zentralbl. Bakteriol.* **46**:595–609.

Evans, L. H., and Malik, F. G., 1987, Class II polytropic murine leukemia viruses (MuLVs) of AKR/J mice: Possible role in the generation of class I oncogenic polytropic MuLVs, *J. Virol.* **61**:1882–1892.

Evans, L. H., and Morrey, J. D., 1987, Tissue-specific replication of Friend and Moloney murine leukemia viruses in infected mice, *J. Virol.* **61**:1350–1357.

Fan, H., 1990, Influences of the long terminal repeats on retrovirus pathogenicity, *Semin. Virol.* **1**:165–174.

Fan, H., Mittal, S., Chute, H., Chao, E., and Pattengale, P. K., 1986, Rearrangements and insertions in the Moloney murine leukemia virus long terminal repeat alter biological properties in vivo and in vitro, *J. Virol.* **60**:204–214.

Fan, H., Chute, H., Chao, E., and Pattengale, P. K., 1988, Leukemogenicity of Moloney murine leukemia viruses carrying polyoma enhancer sequences in the long terminal repeat is dependent on the nature of the inserted polyoma sequences, *Virology* **166**:58–65.

Fan, H., Brightman, B. K., Davis, B. R., and Li, Q. X., 1991, Leukemogenesis by Moloney murine leukemia virus, in: *Viruses That Affect the Immune System* (H. Y. Fan, I. S. Y. Chen, N. Rosenberg, and W. Sugden, eds.), pp. 155–174, American Society for Microbiology, Washington, D.C.

Fauci, A. S., 1988, The human immunodeficiency virus: Infectivity and mechanisms of pathogenesis, *Science* **239**:617–622.

Feuer, G., and Fan, H., 1990, Substitution of murine transthyretin (prealbumin) regulatory sequences into the Moloney murine leukemia virus long terminal repeat yields infectious virus with altered biological properties, *J. Virol.* **64**:6130–6140.

Feuer, G., Lee, W. T., Pattengale, P. K., and Fan, H., 1993, Specificity of disease induced by M-MuLV chimeric retroviruses containing v-myc or v-src is not determined by the LTR, *Virology* **195**:286–291.

Feuerman, M. H., Davis, B. R., Pattengale, P. K., and Fan, H., 1985, Generation of a recombinant Moloney murine leukemia virus carrying the v-src gene of avian sarcoma virus: Transformation in vitro and pathogenesis in vivo, *J. Virol.* **54**:804–816.

Feuerman, M. H., Lee, W. T., Pattengale, P. K., and Fan, H., 1988, Comparison of three recombinant murine leukemia viruses carrying the v-src oncogene of avian sarcoma virus: Differences in in vitro transformation and in vivo pathogenicity, *Mol. Carcinog.* **1**:57–66.

Finlay, C. A., Hinds, P. W., and Levine, A. J., 1989, The p53 proto-oncogene can act as a suppressor of transformation, *Cell* **57**:1083–1093.

Frankel, W. N., Stoye, J. P., Taylor, B. A., and Coffin, J. M., 1989, Genetic identification of endogenous polytropic proviruses by using recombinant inbred mice, *J. Virol.* **63**:3810–3821.

Frankel, W. N., Stoye, J. P., Taylor, B. A., and Coffin, J. M., 1990, A linkage map of endogenous murine leukemia proviruses [published erratum appears in *Genetics* 1990 **125**:455], *Genetics* **124**:221–236.

Frankel, W. N., Rudy, C., Coffin, J. M., and Huber, B. T., 1991, Linkage of Mls genes to endogenous mammary tumour viruses of inbred mice [see comments], *Nature* **349**:526–528.

Freund, R., Dawe, C. J., Carroll, J. P., and Benjamin, T. L., 1992, Changes in frequency, morphology, and behavior of tumors induced in mice by a polyoma virus mutant with a specifically altered oncogene, *Am. J. Pathol.* **141**:1409–1425.

Fung, Y. K., Lewis, W. G., Crittenden, L. B., and Kung, H. J., 1983, Activation of the cellular oncogene c-erbB by LTR insertion: Molecular basis for induction of erythroblastosis by avian leukosis virus, *Cell* **33**:357–368.

Gallahan, D., and Callahan, R., 1987, Mammary tumorigenesis in feral mice: Identification of a new int locus in mouse mammary tumor virus (Czech II)-induced mammary tumors, *J. Virol.* **61**:66–74.

Gautsch, J. W., Elder, J. H., Schindler, J., Jensen, F. C., and Lerner, R. A., 1978, Structural markers on core protein p30 of murine leukemia virus: Functional correlation with Fv-1 tropism, *Proc. Natl. Acad. Sci. USA* **75**:4170–4174.

Goff, S. P., Gilboa, E., Witte, O. N., and Baltimore, D., 1980, Structure of the Abelson murine leukemia virus genome and the homologous cellular gene: Studies with cloned viral DNA, *Cell* **22**:777–785.

Goff, S. P., Witte, O. N., Gilboa, E., Rosenberg, N., and Baltimore, D., 1981, Genome structure of Abelson murine leukemia virus variants: Proviruses in fibroblasts and lymphoid cells, *J. Virol.* **38**:460–468.

Golemis, E., Li, Y., Fredrickson, T. N., Hartley, J. W., and Hopkins, N., 1989, Distinct segments within the enhancer region collaborate to specify the type of leukemia induced by nondefective Friend and Moloney viruses, *J. Virol.* **63**:328–337.

Golovkina, T. V., Chervonsky, A., Dudley, J. P., and Ross, S. R., 1992, Transgenic mouse mammary tumor virus superantigen expression prevents viral infection, *Cell* **69**:637–645.

Gonda, M. A., Young, H. A., Elser, J. E., *et al.*, 1982, Molecular cloning, genomic analysis, and biological properties of rat leukemia virus and the onc sequences of Rasheed rat sarcoma virus, *J. Virol.* **44**:520–529.

Graham, M., Adams, J. M., and Cory, S., 1985, Murine T lymphomas with retroviral inserts in the chromosomal 15 locus for plasmacytoma variant translocations, *Nature* **314**:740–743.

Green, P. L., Kaehler, D. A., McKearn, J., and Risser, R., 1988, Substitution of the LTR of Abelson murine leukemia virus does not alter the cell type of virally induced tumors, *Oncogene* **2**:585–592.

Green, W. R., 1984, Genetic control of the induction of cytolytic T lymphocyte responses to AKR/Gross viral leukemias. II. Negative control by the Fv-1 locus in AKR mice of responder H-2b haplotype, *J. Immunol.* **132**:2665–2671.

Green, W. R., 1987, Induction of anti-AKR/Gross virus cytolytic T lymphocytes in AKR.H-2b:Fv-1b congenic mice: Age-dependent conversion to a nonresponder phenotype, *J. Immunol.* **138**:1602–1606.

Groffen, J., Heisterkamp, N., Shibuya, M., Hanafusa, H., and Stephenson, J. R., 1983, Transforming genes of avian (v-fps) and mammalian (v-fes) retroviruses correspond to a common cellular locus, *Virology* **125**:480–486.

Gross, L., 1951, "Spontaneous" leukemia developing in C3H mice following inoculation, in infancy, with Ak-leukemia extracts, or Ak-embryos, *Proc. Soc. Exp. Biol. Med.* **76**:27–32.

Gross, L., 1983a, Mouse mammary carcinoma, in: *Oncogenic Viruses* (L. Gross, ed.), pp. 262–304, Pergamon Press, Elmsford, N.Y.

Gross, L., 1983b, Mouse leukemia, in: *Oncogenic Viruses* (L. Gross, ed.), pp. 305–458, Pergamon Press, Elmsford, N.Y.

Gross, L., 1983c, The Friend virus, in: *Oncogenic Viruses* (L. Gross, ed.), pp. 522–560, Pergamon Press, Elmsford, N.Y.

Hallberg, B., Schmidt, J., Luz, A., Pedersen, F. S., and Grundstrom, T., 1991, SL3-3 enhancer factor 1 transcriptional activators are required for tumor formation by SL3-3 murine leukemia virus, *J. Virol.* **65**:4177–4181.

Hanecak, R., Pattengale, P. K., and Fan, H., 1988, Addition or substitution of simian virus 40 enhancer sequences into the Moloney murine leukemia virus (M-MuLV) long terminal repeat yields infectious M-MuLV with altered biological properties, *J. Virol.* **62**:2427–2436.

Hanecak, R., Pattengale, P. K., and Fan, H., 1991, Deletion of a GC-rich region flanking the enhancer element within the long terminal repeat sequences alters the disease specificity of Moloney murine leukemia virus, *J. Virol.* **65**:5357–5363.

Hangoc, G., Lu, L., and Broxmeyer, H. E., 1987, The comparative enhancing effects of prostaglandin E1 on colony formation by erythroid progenitor (BFU-E) cells from bone marrow of mice differing in the Fv-2 locus, *Leuk. Res.* **11**:501–507.

Haran-Ghera, N., 1980, Potential leukemic cells among bone marrow cells of young AKR/J mice, *Proc. Natl. Acad. Sci. USA* **77**:2923–2926.

Haran-Ghera, N., Rubio, N., Leef, F., and Goldstein, G., 1978, Characteristics of preleukemia cells induced in mice, *Cell. Immunol.* **37**:308–314.

Haran-Ghera, N., Peled, A., Brightman, B. K., and Fan, H., 1992, Termination of the B cell lymphoma dormant state in thymectomized AKR mice, *J. Immunol.* **148**:2947–2952.

Haran-Ghera, N., Peled, A., Brightman, B. K., and Fan, H., 1993, Leukemogenesis in AKR.Fv-1b congenic mice, *Cancer Res.* **53**:3433–3488.

Hardy, W. D., Jr., 1993, Feline oncoretroviruses, in: *The Retroviridae,* Vol. 2 (J. A. Levy, ed.), pp. 109–180, Plenum Press, New York.

Hartley, J. W., Wolford, N. K., Old, L. J., and Rowe, W. P., 1977, A new class of murine leukemia virus associated with development of spontaneous lymphomas, *Proc. Natl. Acad. Sci. USA* **74**:789–792.

Hartley, J. W., Yetter, R. A., and Morse, H. C., 3D, 1983, A mouse gene on chromosome 5 that restricts infectivity of mink cell focus-forming recombinant murine leukemia viruses, *J. Exp. Med.* **158**:16–24.

Hays, E. F., Bristol, G. C., Lugo, J. P., and Wang, X. F., 1989a, Progression to development of lymphoma in the thymus of AKR mice treated neonatally with SL 3-3 virus, *Exp. Hematol.* **17**:1116–1121.

Hays, E. F., Bristol, G. C., McDougall, S., Klotz, J. L., and Kronenberg, M., 1989b, Development of lymphoma in the thymus of AKR mice treated with the lymphomagenic virus SL 3-3, *Cancer Res.* **49**:4225–4230.

Hayward, W. S., Neel, B. G., and Astrin, S. M., 1981, Activation of a cellular onc gene by promoter insertion in ALV-induced lymphoid leukosis, *Nature* **290**:475–480.

Heisterkamp, N., Stephenson, J. R., Groffen, J., *et al.*, 1983, Localization of the c-abl oncogene adjacent to a translocation break point in chronic myelocytic leukaemia, *Nature* **306**:239–242.

Herr, W., and Gilbert, W., 1983, Somatically acquired recombinant murine leukemia proviruses in thymic leukemias of AKR/J mice, *J. Virol.* **46**:70–82.

Hirsch, M. S., Allison, A. C., and Harvey, J. J., 1969, Immune complexes in mice infected neonatally with Moloney leukaemogenic and murine sarcoma viruses, *Nature* **223**:739–740.

Holland, C. A., Thomas, C. Y., Chattopadhyay, S. K., Koehne, C., and O'Donnell, P. V., 1989, Influence of enhancer sequences on thymotropism and leukemogenicity of mink cell focus-forming viruses, *J. Virol.* **63**:1284–1292.

Hsu, C. L., Fabritius, C., and Dudley, J., 1988, Mouse mammary tumor virus proviruses in T-cell lymphomas lack a negative regulatory element in the long terminal repeat, *J. Virol.* **62**:4644–4652.

Huang, E., Nocka, K., Beier, D. R., *et al.*, 1990, The hematopoietic growth factor KL is encoded by the Sl locus and is the ligand of the c-kit receptor, the gene product of the W locus, *Cell* **63**:225–233.

Huebner, R. J., Gilden, R. V., Toni, R., *et al.*, 1976, Prevention of spontaneous leukemia in AKR mice by type-specific immunosuppression of endogenous ecotropic virogenes, *Proc. Natl. Acad. Sci. USA* **73**:4633–4635.

Ihle, J. N., and Lee, J. C., 1982, Possible immunological mechanisms in C-type viral leukemogenesis in mice, *Curr. Top. Microbiol. Immunol.* **98**:85–101.

Ikeda, H., Laigret, F., Martin, M. A., and Repaske, R., 1985, Characterization of a molecularly cloned retroviral sequence associated with Fv-4 resistance, *J. Virol.* **55**:768–777.

Isaak, D. D., Bartizal, K. F., and Caulfield, M. J., 1988, Decreased pathogenicity of murine leukemia virus-Moloney in gnotobiotic mice, *Leukemia* **2**:540–544.

Jaenisch, R., 1980, Retroviruses and embryogenesis: Microinjection of Moloney leukemia virus into midgestation mouse embryos, *Cell* **19**:181–188.

Jaenisch, R., Fan, H., and Croker, B., 1975, Infection of preimplantation mouse embryos and of newborn mice with leukemia virus: Tissue distribution of viral DNA and RNA and leukemogenesis in the adult animal, *Proc. Natl. Acad. Sci. USA* **72**:4008–4012.

Jarrett, W. F. H., Crawford, E. M., Martin, W. B., and Davie, F., 1964a, Leukaemia in the cat. A virus-like particle associated with leukaemia (lymphosarcoma), *Nature* **202**:567–568.

Jarrett, W. H. F., Martin, W. B., Crighton, G. W., Dalton, R. G., and Stewart, M. F., 1964b, Leukaemia in the cat. Transmission experiments with leukaemia (lymphosarcoma), *Nature* **202**:566–567.

Jhappan, C., Gallahan, D., Stahle, C., *et al.*, 1992, Expression of an activated Notch-related int-3 transgene interferes with cell differentiation and induces neoplastic transformation in mammary and salivary glands, *Genes Dev.* **6**:345–355.

Jolicoeur, P., and Baltimore, D., 1976, Effect of Fv-1 gene product on proviral DNA formation and integration in cells infected with murine leukemia viruses, *Proc. Natl. Acad. Sci. USA* **73**:2236–2240.

Jolicoeur, P., and DesGroseillers, L., 1985, Neurotropic Cas-BR-E murine leukemia virus harbors several determinants of leukemogenicity mapping in different regions of the genome, *J. Virol.* **56**:639–643.

Jolicoeur, P., Rassart, E., Masse, G., and Paquette, Y., 1991, The specificity of the disease induced by defective murine retroviruses containing abl, fos, or Ha-ras is usually not determined by their LTR, *Virology* **180**:831–836.

Jorgensen, J. L., Reay, P. A., Ehrich, E. W., and Davis, M. M., 1992, Molecular components of T-cell recognition, *Annu. Rev. Immunol.* **10**:835–873.

Klarnet, J. P., Kern, D. E., Okuno, K., Holt, C., Lilly, F., and Greenberg, P. D., 1989, FBL-reactive CD8+ cytotoxic and CD4+ helper T lymphocytes recognize distinct Friend murine leukemia virus-encoded antigens, *J. Exp. Med.* **169**:457–467.

Klempnauer, K. H., Gonda, T. J., and Bishop, J. M., 1982, Nucleotide sequence of the retroviral leukemia gene v-myb and its cellular progenitor c-myb: The architecture of a transduced oncogene, *Cell* **31**:453–463.

Knepper, J. E., Kittrell, F. S., Medina, D., and Butel, J. S., 1989, Spontaneous progression of hyperplastic outgrowths of the D1 lineage to mammary tumors: Expression of mouse mammary tumor virus and cellular proto-oncogenes, *Mol. Carcinog.* **1**:229–238.

Koehne, C. F., Lazo, P. A., Alves, K., Lee, J. S., Tsichlis, P. N., and O'Donnell, P. V., 1989, The Mlvi-1 locus involved in the induction of rat T-cell lymphomas and the pvt-1/Mis-1 locus are identical, *J. Virol.* **63**:2366–2369.

Konopka, J. B., Watanabe, S. M., and Witte, O. N., 1984, An alteration of the human c-abl protein in K562 leukemia cells unmasks associated tyrosine kinase activity, *Cell* **37**:1035–1042.

Korostoff, J. M., Nakada, M. T., Faas, S. J., Blank, K. J., and Gaulton, G. N., 1990, Neonatal exposure to thymotropic Gross murine leukemia virus induces virus-specific immunologic nonresponsiveness, *J. Exp. Med.* **172**:1765–1775.

Kozak, C. A., and Ruscetti, S., 1992, Retroviruses in rodents, in: *The Retroviridae*, Vol. 1 (J. A. Levy, ed.), p. 405, Plenum Press, New York.

Lazo, P. A., Lee, J. S., and Tsichlis, P. N., 1990a, Long-distance activation of the Myc protooncogene by provirus insertion in Mlvi-1 or Mlvi-4 in rat T-cell lymphomas, *Proc. Natl. Acad. Sci. USA* **87**:170–173.

Lazo, P. A., Klein-Szanto, A. J., and Tsichlis, P. N., 1990b, T-cell lymphoma lines derived from rat thymomas induced by Moloney murine leukemia virus: Phenotypic diversity and its implications, *J. Virol.* **64**:3948–3959.

Leder, A., Pattengale, P. K., Kuo, A., Stewart, T. A., and Leder, P., 1986, Consequences of widespread deregulation of the c-myc gene in transgenic mice: Multiple neoplasms and normal development, *Cell* **45**:485–495.

Lee, J. C., and Ihle, J. N., 1981a, Increased responses to lymphokines are correlated with preleukemia in mice inoculated with Moloney leukemia virus, *Proc. Natl. Acad. Sci. USA* **78**:7712–7716.

Lee, J. C., and Ihle, J. N., 1981b, Chronic immune stimulation is required for Moloney leukaemia virus-induced lymphomas, *Nature* **289**:407–409.

Lee, J. C., Horak, I., and Ihle, J. N., 1981, Mechanisms in T cell leukemogenesis. II. T cell responses of preleukemic BALB/c mice to Moloney leukemia virus antigens, *J. Immunol.* **126**:715–722.

Lee, W. T., Prakash, O., Klein, D., and Sarkar, N. H., 1987, Structural alterations in the long terminal repeat of an acquired mouse mammary tumor virus provirus in a T-cell leukemia of DBA/2 mice, *Virology* **159**:39–48.

Lenz, J., and Haseltine, W. A., 1983, Localization of the leukemogenic determinants of SL3-3, an ecotropic, XC-positive murine leukemia virus of AKR mouse origin, *J. Virol.* **47**:317–328.

Lenz, J., Celander, D., Crowther, R. L., Patarca, R., Perkins, D. W., and Haseltine, W. A., 1984, Determination of the leukaemogenicity of a murine retrovirus by sequences within the long terminal repeat, *Nature* **308**:467–470.

Levy, L. S., Gardner, M. B., and Casey, J. W., 1984, Isolation of a feline leukaemia provirus containing the oncogene myc from a feline lymphosarcoma, *Nature* **308**:853–856.

Levy, L. S., Fish, R. E., and Baskin, G. B., 1988, Tumorigenic potential of a myc-containing strain of feline leukemia virus in vivo in domestic cats, *J. Virol.* **62**:4770–4773.

Li, J. P., and Baltimore, D., 1991, Mechanism of leukemogenesis induced by mink cell focus-forming murine leukemia viruses, *J. Virol.* **65**:2408–2414.

Li, J. P., D'Andrea, A. D., Lodish, H. F., and Baltimore, D., 1990, Activation of cell growth by binding of Friend spleen focus-forming virus gp55 glycoprotein to the erythropoietin receptor, *Nature* **343**:762–764.

Li, Q. X., and Fan, H., 1990, Combined infection by Moloney murine leukemia virus and a mink cell focus-forming virus recombinant induces cytopathic effects in fibroblasts or in long-term bone marrow cultures from preleukemic mice, *J. Virol.* **64**:3701–3711.

Li, Q. X., and Fan, H., 1991, Bone marrow depletion by 89Sr complements a preleukemic defect in a long terminal repeat variant of Moloney murine leukemia virus, *J. Virol.* **65**:4442–4448.

Li, Y., Golemis, E., Hartley, J. W., and Hopkins, N., 1987, Disease specificity of nondefective Friend and Moloney murine leukemia viruses is controlled by a small number of nucleotides, *J. Virol.* **61**:693–700.

Lilly, F., 1967, Susceptibility to two strains of Friend leukemia virus in mice, *Science* **155**:461–462.

Lilly, F., 1970, Fv-2: Identification and location of a second gene governing the spleen focus response to Friend leukemia virus in mice, *J. Natl. Cancer Inst.* **45**:163–169.

Lilly, F., Duran-Reynals, M. L., and Rowe, W. P., 1975, Correlation of early murine leukemia virus titer and H-2 type with spontaneous leukemia in mice of the BALB/c times AKR cross: A genetic analysis, *J. Exp. Med.* **141**:882–889.

Lonai, P., and Haran-Ghera, N., 1977, Resistance genes to murine leukemia in the I immune response gene region of the H-2 complex, *J. Exp. Med.* **146**:1164–1168.

Lu, X., Park, S. H., Thompson, T. C., and Lane, D. P., 1992, Ras-induced hyperplasia occurs with mutation of p53, but activated ras and myc together can induce carcinoma without p53 mutation, *Cell* **70**:153–161.

Luciw, P. A., and Leung, N. J., 1992, Mechanism of retrovirus replication, in: *The Retroviridae*, Vol. 1 (J. A. Levy, ed.), p. 159, Plenum Press, New York.

Lung, M. L., Hartley, J. W., Rowe, W. P., and Hopkins, N. H., 1983, Large RNase T1-resistant oligonucleotides encoding p15E and the U3 region of the long terminal repeat distinguish two biological classes of mink cell focus-forming type C viruses of inbred mice, *J. Virol.* **45**:275–290.

MacDonald, M. E., Reynolds, F. H., Jr., Van de Ven, W. J., Stephenson, J. R., Mak, T. W., and Bernstein, A., 1980, Anemia- and polycythemia-inducing isolates of Friend spleen focus-forming virus. Biological and molecular evidence for two distinct viral genomes, *J. Exp. Med.* **151**:1477–1492.

McEndy, D. P., Boon, M. C., and Furth, J., 1944, On the role of thymus, spleen and gonads in the development of leukemia in a high-leukemia stock of mice, *Cancer Res.* **4**:377–383.

McGrath, M. S., and Weissman, I. L., 1979, AKR leukemogenesis: Identification and biological significance of thymic lymphoma receptors for AKR retroviruses, *Cell* **17**:65–75.

McGrath, M. S., Pillemer, E., Kooistra, D. A., Jacobs, S., Jerabek, L., and Weissman, I. L., 1980, T-lymphoma retroviral receptors and control of T-lymphoma cell proliferation, *Cold Spring Harbor Symp. Quant. Biol.* **44**:1297–1304.

Marchetti, A., Robbins, J., Campbell, G., *et al.*, 1991, Host genetic background effect on the frequency of mouse mammary tumor virus-induced rearrangements of the int-1 and int-2 loci in mouse mammary tumors, *J. Virol.* **65**:4550–4554.

Marcu, K. B., Harris, L. J., Stanton, L. W., Erikson, J., Watt, R., and Croce, C. M., 1983, Transcriptionally active c-myc oncogene is contained within NIARD, a DNA sequence associated with chromosome translocations in B-cell neoplasia, *Proc. Natl. Acad. Sci. USA* **80**:519–523.

Marrack, P., Kushnir, E., and Kappler, J., 1991, A maternally inherited superantigen encoded by a mammary tumour virus [see comments], *Nature* **349**:524–526.

Marth, J. D., Peet, R., Krebs, E. G., and Perlmutter, R. M., 1985, A lymphocyte-specific protein-tyrosine kinase gene is rearranged and overexpressed in the murine T cell lymphoma LSTRA, *Cell* **43**:393–404.

Martinez, J., Georgoff, I., and Levine, A. J., 1991, Cellular localization and cell cycle regulation by a temperature-sensitive p53 protein, *Genes Dev.* **5**:151–159.

Meruelo, D., Leiberman, M., Ginzton, N., Deak, B., and McDevitt, H. O., 1977, Genetic control of radiation leukemia virus-induced tumorigenesis. I. Role of the major murine histocompatibility complex, H-2, *J. Exp. Med.* **146**:1079–1087.

Mester, J., Wagenaar, E., Sluyser, M., and Nusse, R., 1987, Activation of int-1 and int-2 mammary oncogenes in hormone-dependent and independent mammary tumors of GR mice, *J. Virol.* **61**:1073–1078.

Michalides, R., and Wagenaar, E., 1986, Site-specific rearrangements in the long terminal repeat of extra mouse mammary tumor proviruses in murine T-cell leukemias, *Virology* **154**:76–84.

Miyazawa, M., Nishio, J., and Chesebro, B., 1988, Genetic control of T cell responsiveness to the Friend murine leukemia virus envelope antigen. Identification of class II loci of the H-2 as immune response genes, *J. Exp. Med.* **168**:1587–1605.

Miyazawa, M., Nishio, J., Wehrly, K., and Chesebro, B., 1992, Influence of MHC genes on spontaneous recovery from Friend retrovirus-induced leukemia, *J. Immunol.* **148**:644–647.

Moreau-Gachelin, F., Tavitian, A., and Tambourin, P., 1988, Spi-1 is a putative oncogene in virally induced murine erythroleukaemias, *Nature* **331**:277–280.

Morishita, K., Parker, D. S., Mucenski, M. L., Jenkins, N. A., Copeland, N. G., and Ihle, J. N., 1988, Retroviral activation of a novel gene encoding a zinc finger protein in IL-3-dependent myeloid leukemia cell lines, *Cell* **54**:831–840.

Morris, D. W., Barry, P. A., Bradshaw, H. D., Jr., and Cardiff, R. D., 1990, Insertion mutation of the int-1 and int-2 loci by mouse mammary tumor virus in premalignant and malignant neoplasms from the GR mouse strain, *J. Virol.* **64**:1794–1802.

Mowat, M., Cheng, A., Kimura, N., Bernstein, A., and Benchimol, S., 1985, Rearrangements of the cellular p53 gene in erythroleukaemic cells transformed by Friend virus, *Nature* **314**:633–636.

Mucenski, M. L., Taylor, B. A., Ihle, J. N., *et al.*, 1988, Identification of a common ecotropic viral integration site, Evi-1, in the DNA of AKXD murine myeloid tumors, *Mol. Cell Biol.* **8**:301–308.

Mukhopadhyaya, R., and Wolff, L., 1992, New sites of proviral integration associated with murine promonocytic leukemias and evidence for alternate modes of c-myb activation, *J. Virol.* **66**:6035–6044.

Muller, W. J., Lee, F. S., Dickson, C., Peters, G., Pattengale, P., and Leder, P., 1990, The int-2 gene product acts as an epithelial growth factor in transgenic mice, *EMBO J.* **9**:907–913.

Nason-Burchenal, K., and Wolff, L., 1993, Activation of c-myb is an early bone-marrow event in a murine model for acute promonocytic leukemia, *Proc. Natl. Acad. Sci. USA* **90**:1619–1623.

Neil, J. C., Hughes, D., McFarlane, R., *et al.*, 1984, Transduction and rearrangement of the myc gene by feline leukaemia virus in naturally occurring T-cell leukaemias, *Nature* **308**:814–820.

Nusse, R., and Varmus, H. E., 1982, Many tumors induced by the mouse mammary tumor virus contain a provirus integrated in the same region of the host genome, *Cell* **31**:99–109.

Odaka, T., Ikeda, H., Yoshikura, H., Moriwaki, K., and Suzuki, S., 1981, Fv-4: Gene controlling resistance to NB-tropic Friend murine leukemia virus. Distribution in wild mice, introduction into genetic background of BALB/c mice, and mapping of chromosomes, *J. Natl. Cancer Inst.* **67**:1123–1127.

O'Donnell, P. V., Woller, R., and Chu, A., 1984, Stages in development of mink cell focus-inducing (MCF) virus-accelerated leukemia in AKR mice, *J. Exp. Med.* **160**:914–934.

O'Donnell, P. V., Fleissner, E., Lonial, H., Koehne, C. F., and Reicin, A., 1985, Early clonality and high-frequency proviral integration into the c-myc locus in AKR leukemias, *J. Virol.* **55**:500–503.

Oldstone, M. B., Aoki, T., and Dixon, F. J., 1972, The antibody response of mice to murine leukemia virus in spontaneous infection: Absence of classical immunologic tolerance (AKR mice–complement-fixing antibodies–lymphocytic choriomeningitis virus–immunofluorescence–glomerular deposits of antigen-antibody complexes), *Proc. Natl. Acad. Sci. USA* **69**:134–138.

Oliner, J. D., Kinzler, K. W., Meltzer, P. S., George, D. L., and Vogelstein, B., 1992, Amplification of a gene encoding a p53-associated protein in human sarcomas [see comments], *Nature* **358**:80–83.

Overhauser, J., and Fan, H., 1985, Generation of glucocorticoid-responsive Moloney murine leukemia virus by insertion of regulatory sequences from murine mammary tumor virus into the long terminal repeat, *J. Virol.* **54**:133–144.

Patriotis, C., Makris, A., Bear, S. E., and Tsichlis, P. N., 1993, Tumor progression locus 2 (Tpl-2) encodes a protein kinase involved in the progression of rodent T-cell lymphomas and in T-cell activation, *Proc. Natl. Acad. Sci. USA* **90**:2251–2255.

Paul, R., Schuetze, S., Kozak, S. L., Kozak, C. A., and Kabat, D., 1991, The Sfpi-1 proviral integration site of Friend erythroleukemia encodes the ets-related transcription factor Pu.1, *J. Virol.* **65**:464–467.

Payne, G. S., Bishop, J. M., and Varmus, H. E., 1982, Multiple arrangements of viral DNA and an activated host oncogene in bursal lymphomas, *Nature* **295**:209–214.

Pellegrini, S., and Basilico, C., 1987, Amplification and expression of foreign genes in cells producing polyoma virus large T-antigen, *Oncogene Res.* **1**:23–41.

Peters, G., Smith, R., Brookes, S., and Dickson, C., 1982, Conservation of protein coding potential in the long terminal repeats of exogenous and endogenous mouse mammary tumor viruses, *J. Virol.* **42**:880–888.

Peters, G., Brookes, S., Smith, R., and Dickson, C., 1983, Tumorigenesis by mouse mammary tumor virus: Evidence for a common region for provirus integration in mammary tumors, *Cell* **33**:369–377.

Pincus, T., Rowe, W. P., and Lilly, F., 1971, A major genetic locus affecting resistance to infection with murine leukemia viruses. II. Apparent identity to a major locus described for resistance to Friend murine leukemia virus, *J. Exp. Med.* **133**:1234–1241.

Poirier, Y., Kozak, C., and Jolicoeur, P., 1988, Identification of a common helper provirus integration site in Abelson murine leukemia virus-induced lymphoma DNA, *J. Virol.* **62**:3985–3992.

Polsky, D., and Lilly, F., 1991, Suppression of H-2b-associated resistance to Friend erythroleukemia virus by a class I gene from the H-2d major histocompatibility complex haplotype, *Proc. Natl. Acad. Sci. USA* **88**:9243–9247.

Prywes, R., Foulkes, J. G., Rosenberg, N., and Baltimore, D., 1983, Sequences of the A-MuLV protein needed for fibroblast and lymphoid cell transformation, *Cell* **34**:569–579.

Pullen, A. M., Choi, Y., Kushnir, E., Kappler, J., and Marrack, P., 1992, The open reading frames in the 3′ long terminal repeats of several mouse mammary tumor virus integrants encode V beta 3-specific superantigens, *J. Exp. Med.* **175**:41–47.

Quint, W., Quax, W., Van der Putten, H., and Berns, A., 1981, Characterization of AKR murine leukemia virus sequences in AKR mouse substrains and structure of integrated recombinant genomes in tumor tissues, *J. Virol.* **39**:1–10.

Radke, K., Beug, H., Kornfeld, S., and Graf, T., 1982, Transformation of both erythroid and myeloid cells by E26, an avian leukemia virus that contains the myb gene, *Cell* **31**:643–653.

Rasheed, S., Norman, G. L., and Heidecker, G., 1983, Nucleotide sequence of the Rasheed rat sarcoma virus oncogene: New mutations, *Science* **221**:155–157.

Reisner, Y., Sharon, N., and Haran-Ghera, N., 1980, Expression of peanut agglutinin receptors on virus-induced preleukemic cells in mice, *Proc. Natl. Acad. Sci. USA* **77**:2244–2246.

Rich, R. F., Fujii, T., and Green, W. R., 1992, CD4-CD8+ T lymphocytes mediate AKR/Gross murine leukemia virus nonresponsiveness in moderately aged AKR.H-2b:Fv-1b mice, *J. Immunol.* **148**:2961–2967.

Ringold, G. M., Yamamoto, K. R., Tomkins, G. M., Bishop, M., and Varmus, H. E., 1975, Dexamethasone-mediated induction of mouse mammary tumor virus RNA: A system for studying glucocorticoid action, *Cell* **6**:299–305.

Robertson, D. L., and Varmus, H. E., 1979, Structural analysis of the intracellular RNAs of murine mammary tumor virus, *J. Virol.* **30**:576–589.

Robertson, M. N., Spangrude, G. J., Hasenkrug, K., *et al.*, 1992, Role and specificity of T-cell subsets in spontaneous recovery from Friend virus-induced leukemia in mice, *J. Virol.* **66**:3271–3277.

Robinson, H. L., Blais, B. M., Tsichlis, P. N., and Coffin, J. M., 1982, At least two regions of the viral genome determine the oncogenic potential of avian leukosis viruses, *Proc. Natl. Acad. Sci. USA* **79**:1225–1229.

Roelink, H., Wagenaar, E., Lopes Da Silva, S., and Nusse, R., 1990, Wnt-3, a gene activated by proviral insertion in mouse mammary tumors, is homologous to int-1/Wnt-1 and is normally expressed in mouse embryos and adult brain, *Proc. Natl. Acad. Sci. USA* **87**:4519–4523.

Rosenberg, N., and Baltimore, D., 1978, The effect of helper virus on Abelson virus-induced transformation of lymphoid cells, *J. Exp. Med.* **147**:1126–1141.

Rous, P., 1911, A sarcoma of the fowl transmissible by an agent separable from the tumor cells, *J. Exp. Med.* **13**:397–411.

Ruscetti, S. K., Linemeyer, D., Feild, J., Troxler, D., and Scolnick, E. M., 1979, Characterization of a protein found in cells infected with the spleen focus-forming virus that shares immunological cross-reactivity with the gp70 found in mink cell focus-inducing virus particles, *J. Virol.* **30**:787–798.

Ruscetti, S., Davis, L., Feild, J., and Oliff, A., 1981, Friend murine leukemia virus-induced leukemia is associated with the formation of mink cell focus-inducing viruses and is blocked in mice expressing endogenous mink cell focus-inducing xenotropic viral envelope genes, *J. Exp. Med.* **154**:907–920.

Ruscetti, S., Matthai, R., and Potter, M., 1985, Susceptibility of BALB/c mice carrying various DBA/2 genes to development of Friend murine leukemia virus-induced erythroleukemia, *J. Exp. Med.* **162**:1579–1587.

Salmons, B., Erfle, V., Brem, G., and Gunzburg, W. H., 1990, naf, a trans-regulating negative-acting factor encoded within the mouse mammary tumor virus open reading frame region, *J. Virol.* **64**:6355–6359.

Saris, C. J., Domen, J., and Berns, A., 1991, The pim-1 oncogene encodes two related protein-serine/threonine kinases by alternative initiation at AUG and CUG, *EMBO J.* **10**:655–664.

Scheidereit, C., and Beato, M., 1984, Contacts between hormone receptor and DNA double helix within a glucocorticoid regulatory element of mouse mammary tumor virus, *Proc. Natl. Acad. Sci. USA* **81**:3029–3033.

Schwarz, H., Fischinger, P. J., Ihle, J. N., *et al.*, 1979, Properties of mouse leukemia viruses. XVI. Suppression of spontaneous fetal leukemias in AKR mice by treatment with broadly reacting antibody against the viral glycoprotein gp 71, *Virology* **93**:159–174.

Sels, F. T., Langer, S., Schulz, A. S., Silver, J., Sitbon, M., and Friedrich, R. W., 1992, Friend murine leukaemia virus is integrated at a common site in most primary spleen tumours of erythroleukaemic animals, *Oncogene* **7**:643–652.

Selten, G., Cuypers, H. T., Zijlstra, M., Melief, C., and Berns, A., 1984, Involvement of c-myc in MuLV-induced T cell lymphomas in mice: Frequency and mechanisms of activation, *EMBO J.* **3**:3215–3222.

Shen-Ong, G. L., and Wolff, L., 1987, Moloney murine leukemia virus-induced myeloid tumors in adult BALB/c mice: Requirement of c-myb activation but lack of v-abl involvement, *J. Virol.* **61**:3721–3725.

Shibuya, M., and Hanafusa, H., 1982, Nucleotide sequence of Fujinami sarcoma virus: Evolutionary relationship of its transforming gene with transforming genes of other sarcoma viruses, *Cell* **30**:787–795.

Sieweke, M. H., Stoker, A. W., and Bissell, M. J., 1989, Evaluation of the cocarcinogenic effect of wounding in Rous sarcoma virus tumorigenesis, *Cancer Res.* **49**:6419–6424.

Sieweke, M. H., Thompson, N. L., Sporn, M. B., and Bissell, M. J., 1990, Mediation of wound-related Rous sarcoma virus tumorigenesis by TGF-beta, *Science* **248**:1656–1660.

Silver, J., and Buckler, C. E., 1986, A preferred region for integration of Friend murine leukemia virus in hematopoietic neoplasms is closely linked to the Int-2 oncogene, *J. Virol.* **60**:1156–1158.

Simon, I., Lohler, J., and Jaenisch, R., 1982, Virus-specific transcription and translation in organs of BALB/Mo mice: Comparative study using quantitative hybridization, in situ hybridization, and immunocytochemistry, *Virology* **120**:106–121.

Sitbon, M., Sola, B., Evans, L., *et al.*, 1986, Hemolytic anemia and erythroleukemia, two distinct pathogenic effects of Friend MuLV: Mapping of the effects to different regions of the viral genome, *Cell* **47**:851–859.

Slamon, D. J., Clark, G. M., Wong, S. G., Levin, W. J., Ullrich, A., and McGuire, W. L., 1987, Human breast cancer: Correlation of relapse and survival with amplification of the HER-2/neu oncogene, *Science* **235**:177–182.

Slamon, D. J., Godolphin, W., Jones, L. A., *et al.*, 1989, Studies of the HER-2/neu proto-oncogene in human breast and ovarian cancer, *Science* **244**:707–712.

Smith, R. E., 1982, Avian osteopetrosis, *Curr. Top. Microbiol. Immunol.* **101**:75–94.

Sola, B., Fichelson, S., Bordereaux, D., Tambourin, P. E., and Gisselbrecht, S., 1986, fim-1 and fim-2: Two new integration regions of Friend murine leukemia virus in myeloblastic leukemias, *J. Virol.* **60**:718–725.

Sola, B., Simon, D., Mattei, M. G., *et al.*, 1988, Fim-1, Fim-2/c-fms, and Fim-3, three common integration sites of Friend murine leukemia virus in myeloblastic leukemias, map to mouse chromosomes 13, 18, and 3, respectively, *J. Virol.* **62**:3973–3978.

Speck, N. A., and Baltimore, D., 1987, Six distinct nuclear factors interact with the 75-base-pair repeat of the Moloney murine leukemia virus enhancer, *Mol. Cell Biol.* **7**:1101–1110.
Speck, N. A., Renjifo, B., Golemis, E., Fredrickson, T. N., Hartley, J. W., and Hopkins, N., 1990, Mutation of the core or adjacent LVb elements of the Moloney murine leukemia virus enhancer alters disease specificity, *Genes Dev.* **4**:233–242.
Steeves, R. A., Bennett, M., Mirand, E. A., and Cudkowicz, G., 1968, Genetic control by the W locus of susceptibility to (Friend) spleen focus-forming virus, *Nature* **218**:372–374.
Steffen, D., 1984, Proviruses are adjacent to c-myc in some murine leukemia virus-induced lymphomas, *Proc. Natl. Acad. Sci. USA* **81**:2097–2101.
Steffen, D., and Weinberg, R. A., 1978, The integrated genome of murine leukemia virus, *Cell* **15**:1003–1010.
Steffen, D., Bird, S., Rowe, W. P., and Weinberg, R. A., 1979, Identification of DNA fragments carrying ecotropic proviruses of AKR mice, *Proc. Natl. Acad. Sci. USA* **76**:4554–4558.
Stehelin, D., Varmus, H. E., Bishop, J. M., and Vogt, P. K., 1976, DNA related to the transforming gene(s) of avian sarcoma viruses is present in normal avian DNA, *Nature* **260**:170–173.
Stocking, C., Kollek, R., Bergholz, U., and Ostertag, W., 1985, Long terminal repeat sequences impart hematopoietic transformation properties to the myeloproliferative sarcoma virus, *Proc. Natl. Acad. Sci. USA* **82**:5746–5750.
Storch, T. G., Arnstein, P., Manohar, V., Leiserson, W. M., and Chused, T. M., 1985, Proliferation of infected lymphoid precursors before Moloney murine leukemia virus-induced T-cell lymphoma, *J. Natl. Cancer Inst.* **74**:137–143.
Stoye, J. P., Moroni, C., and Coffin, J. M., 1991, Virological events leading to spontaneous AKR thymomas, *J. Virol.* **65**:1273–1285.
Sugamura, K., and Hinuma, Y., 1993, Human retroviruses: HTLV-I and HTLV-II, in: *The Retroviridae*, Vol. 2 (J. A. Levy, ed.), pp. 399–435, Plenum Press, New York.
Suzuki, S., and Axelrad, A. A., 1980, Fv-2 locus controls the proportion of erythropoietic progenitor cells (BFU-E) synthesizing DNA in normal mice, *Cell* **19**:225–236.
Sveda, M. M., and Soeiro, R., 1976, Host restriction of Friend leukemia virus: Synthesis and integration of the provirus, *Proc. Natl. Acad. Sci. USA* **73**:2356–2360.
Tan, J. C., Nocka, K., Ray, P., Traktman, P., and Besmer, P., 1990, The dominant W42 spotting phenotype results from a missense mutation in the c-kit receptor kinase, *Science* **247**:209–212.
Taub, R., Kirsch, I., Morton, C., *et al.*, 1982, Translocation of the c-myc gene into the immunoglobulin heavy chain locus in human Burkitt lymphoma and murine plasmacytoma cells, *Proc. Natl. Acad. Sci. USA* **79**:7837–7841.
Teich, N., Wyke, J., Mak, T., Bernstein, A., and Hardy, W., 1984, Pathogenesis of retrovirus-induced disease, in: *RNA Tumor Viruses: Molecular Biology of Tumor Viruses*, 2nd ed. (R. Weiss, N. Teich, H. Varmus, and J. Coffin, eds.), pp. 785–998, Cold Spring Harbor Laboratory, Cold Spring Harbor, N.Y.
Thomas, C. Y., and Coffin, J. M., 1982, Genetic alterations of RNA leukemia viruses associated with the development of spontaneous thymic leukemia in AKR/J mice, *J. Virol.* **43**:416–426.
Thomas, C. Y., Khiroya, R., Schwartz, R. S., and Coffin, J. M., 1984, Role of recombinant ecotropic and polytropic viruses in the development of spontaneous thymic lymphomas in HRS/J mice, *J. Virol.* **50**:397–407.
Tikhonenko, A. T., and Linial, M. L., 1992, gag as well as myc sequences contribute to the transforming phenotype of the avian retrovirus FH3, *J. Virol.* **66**:946–955.
Traina, V. L., Taylor, B. A., and Cohen, J. C., 1981, Genetic mapping of endogenous mouse mammary tumor viruses: Locus characterization, segregation, and chromosomal distribution, *J. Virol.* **40**:735–744.
Tremblay, P. J., Pothier, F., Hoang, T., *et al.*, 1989, Transgenic mice carrying the mouse mammary tumor virus ras fusion gene: Distinct effects in various tissues, *Mol. Cell Biol.* **9**:854–859.
Tremblay, P. J., Kozak, C. A., and Jolicoeur, P., 1992, Identification of a novel gene, Vin-1, in murine leukemia virus-induced T-cell leukemias by provirus insertional mutagenesis [published erratum appears in *J. Virol.* 1992 **66**:5176], *J. Virol.* **66**:1344–1353.
Tsichlis, P. N., Strauss, P. G., and Hu, L. F., 1983, A common region for proviral DNA integration in MoMuLV-induced rat thymic lymphomas, *Nature* **302**:445–449.
Tsichlis, P. N., Lohse, M. A., Szpirer, C., Szpirer, J., and Levan, G., 1985, Cellular DNA regions involved in the induction of rat thymic lymphomas (Mlvi-1, Mlvi-2, Mlvi-3, and c-myc) represent independent loci as determined by their chromosomal map location in the rat, *J. Virol.* **56**:938–942.

Tsichlis, P. N., Shepherd, B. M., and Bear, S. E., 1989, Activation of the Mlvi-1/mis1/pvt-1 locus in Moloney murine leukemia virus-induced T-cell lymphomas, *Proc. Natl. Acad. Sci. USA* **86**:5487–5491.

Tsichlis, P. N., Lee, J. S., Bear, S. E., *et al.*, 1990, Activation of multiple genes by provirus integration in the Mlvi-4 locus in T-cell lymphomas induced by Moloney murine leukemia virus, *J. Virol.* **64**:2236–2244.

Van Klaveren, P., and Bentvelzen, P., 1988, Transactivating potential of the 3′ open reading frame of murine mammary tumor virus, *J. Virol.* **62**:4410–4413.

van Lohuizen, M., Verbeek, S., Krimpenfort, P., *et al.*, 1989, Predisposition to lymphomagenesis in pim-1 transgenic mice: Cooperation with c-myc and N-myc in murine leukemia virus-induced tumors, *Cell* **56**:673–682.

van Lohuizen, M., Verbeek, S., Scheijen, B., Wientjens, E., Van der Gulden, H., and Berns, A., 1991, Identification of cooperating oncogenes in E mu-myc transgenic mice by provirus tagging, *Cell* **65**:737–752.

Veillette, A., Bookman, M. A., Horak, E. M., and Bolen, J. B., 1988, The CD4 and CD8 T cell surface antigens are associated with the internal membrane tyrosine-protein kinase p56lck, *Cell* **55**:301–308.

Vijaya, S., Steffen, D. L., Kozak, C., and Robinson, H. L., 1987, Dsi-1, a region with frequent proviral insertions in Moloney murine leukemia virus-induced rat thymomas, *J. Virol.* **61**:1164–1170.

Villemur, R., Monczak, Y., Rassart, E., Kozak, C., and Jolicoeur, P., 1987, Identification of a new common provirus integration site in Gross passage A murine leukemia virus-induced mouse thymoma DNA, *Mol. Cell Biol.* **7**:512–522.

Villeneuve, L., Rassart, E., Jolicoeur, P., Graham, M., and Adams, J. M., 1986, Proviral integration site Mis-1 in rat thymomas corresponds to the pvt-1 translocation breakpoint in murine plasmacytomas, *Mol. Cell Biol.* **6**:1834–1837.

Voronova, A. F., and Sefton, B. M., 1986, Expression of a new tyrosine protein kinase is stimulated by retrovirus promoter insertion, *Nature* **319**:682–685.

Wegmann, K. W., McMaster, J. S., and Green, W. R., 1991, Mechanism of nonresponsiveness to AKR/Gross leukemia virus in AKR.H-2b:Fv-1b mice. An analysis of precursor cytotoxic T lymphocyte frequencies in young versus moderately aged mice, *J. Immunol.* **146**:2469–2477.

Weinberg, R. A., 1991, Tumor suppressor genes, *Science* **254**:1138–1146.

Weinhold, K. J., Huper, G., Matthews, T. J., *et al.*, 1984, Properties of mouse leukemia viruses. XIX. Effective antibody therapy of AKR leukemia occurs independently of virus neutralization and produces long-term changes in the virus status of the thymus, *Virology* **135**:105–117.

Wiener, F., Ohno, S., Spira, J., Haran-Ghera, N., and Klein, G., 1978, Chromosome changes (trisomies #15 and 17) associated with tumor progression in leukemias induced by radiation leukemia virus, *J. Natl. Cancer Inst.* **61**:227–237.

Wolff, L., and Ruscetti, S., 1985, Malignant transformation of erythroid cells in vivo by introduction of a nonreplicating retrovirus vector, *Science* **228**:1549–1552.

Wolff, L., Tambourin, P., and Ruscetti, S., 1986, Induction of the autonomous stage of transformation in erythroid cells infected with SFFV: Helper virus is not required, *Virology* **152**:272–276.

Wolff, L., Mushinski, J. F., Shen-Ong, G. L., and Morse, H. C., 3D, 1988, A chronic inflammatory response. Its role in supporting the development of c-myb and c-myc related promonocytic and monocytic tumors in BALB/c mice, *J. Immunol.* **141**:681–689.

Yang, W. K., Kiggans, J. O., Yang, D. M., *et al.*, 1980, Synthesis and circularization of N- and B-tropic retroviral DNA Fv-1 permissive and restrictive mouse cells, *Proc. Natl. Acad. Sci. USA* **77**:2994–2998.

Yoshida, T., Miyagawa, K., Odagiri, H., *et al.*, 1987, Genomic sequence of hst, a transforming gene encoding a protein homologous to fibroblast growth factors and the int-2-encoded protein [published erratum appears in *Proc. Natl. Acad. Sci. USA* 1988 **85**:1967], *Proc. Natl. Acad. Sci. USA* **84**:7305–7309.

Yoshimura, F. K., Davison, B., and Chaffin, K., 1985, Murine leukemia virus long terminal repeat sequences can enhance gene activity in a cell-type-specific manner, *Mol. Cell Biol.* **5**:2832–2835.

CHAPTER 8

Immune Response to Retroviral Infection

STEPHEN NORLEY AND REINHARD KURTH

I. INTRODUCTION

Since their discovery, retroviruses have been shown to be responsible for diseases in a wide range of hosts, from birds to humans. Indeed, the fact that most vertebrates contain within their genomes numerous (mostly noninfectious) retroviral sequences indicates just how long these viruses have maintained their intimate relationship with their host species as they evolved together (see Chapter 9; Temin, 1992). The recent appearance and relentless spread of the AIDS virus perhaps gives the impression that the host is helpless against retroviral infection. However, as with all infections with a foreign agent, the infected host will usually respond to active retroviral replication with the generation of a virus-specific immune response. This reaction is able in most cases, if not to eliminate the virus, then to suppress its replication to a harmless level. Identifying these protective mechanisms is therefore obviously of crucial importance in the design and production of prophylactic vaccines and immunotherapeutics.

What makes a retrovirus particularly interesting immunologically is the fact that most of these viruses interact with and disrupt in some way the function of the very cells responsible for generation of the immune response. One can therefore look at the immune response to retroviral infections from two sides: the influence the immune system has on the virus and the influence the virus has on the immune system.

In this review we will attempt to describe the general properties of the immune response to retroviral infections, pointing out the important similarities and, where they exist, the differences. Beneficial aspects of the immune response, particularly those important for vaccine development, will be described first,

STEPHEN NORLEY AND REINHARD KURTH • Paul-Ehrlich-Institute, 63225 Langen, Germany.

The Retroviridae, Volume 3, edited by Jay A. Levy. Plenum Press, New York, 1994.

followed by those which cause actual harm to the host. In each section we will endeavor to give examples of each topic under discussion, starting with avian retroviruses and moving through the murine, feline, ovine, and bovine viruses to those which infect primates, particularly the human immunodeficiency virus (HIV).

II. RETROVIRAL STRUCTURE AND INFECTIOUS CYCLE

To understand the relevance of the host's immune response to a particular component of the virus particle or one of its nonstructural gene products, it is important to keep in mind the function of each protein, if known, and how it is "seen" by the various arms of the immune system. Although there are differences in size and shape, all retroviruses share the same basic morphologic structure and mode of replication (Fig. 1). Two copies of the RNA genome are packaged together with the enzyme reverse transcriptase (RT) within a "core" of Gag proteins. This core is enclosed within a lipid bilayer derived from the host cell membrane and embedded in this membrane are the viral envelope glycoproteins. When an infectious retrovirus comes into contact with its target cell, these envelope glycoproteins interact with one or more cell-surface proteins and by a mechanism only now becoming partially understood the virus enters the cell (Weiss, 1993). Once inside, using RT, a double-stranded DNA copy of the RNA genome is made which can integrate into the host cell genome. This provirus can then remain silent within the cell for long periods of time. Once activated, however, the cell can transcribe and translate the information in the normal way to produce new viral particles that bud from the cell membrane. Before budding occurs, the newly synthesized envelope glycoproteins are arranged on the cell surface to be incorporated into the progeny virus. A detailed description of retroviral structure and infectious cycle are given in other volumes of this series (Coffin, 1992; Luciw and Leung, 1992; Temin, 1992).

Despite their characteristic association with oncogenesis, perhaps the most important feature of retroviruses is their intimate relationship with the cells which constitute the immune system. It is this relationship which gives retroviruses their ability to disrupt the normal immunologic regulation in the host leading to disease. In the next section we will briefly summarize the processes of immune stimulation and the different mechanisms by which the host immune system attempts to combat retroviral infection.

III. INDUCTION OF THE IMMUNE RESPONSE

The immune system has evolved to prevent or control replication in the body of infectious organisms and has therefore developed a number of immune mechanisms capable of eliminating foreign organisms or infected cells. The first stage in triggering these mechanisms is the recognition of nonself. Although the individual components of the immune system cannot distinguish between self and nonself, the system as a whole performs this function very well. During development, cells are produced which, by random gene rearrangement, have the capability of reacting with virtually any given protein. Any cells reacting with self

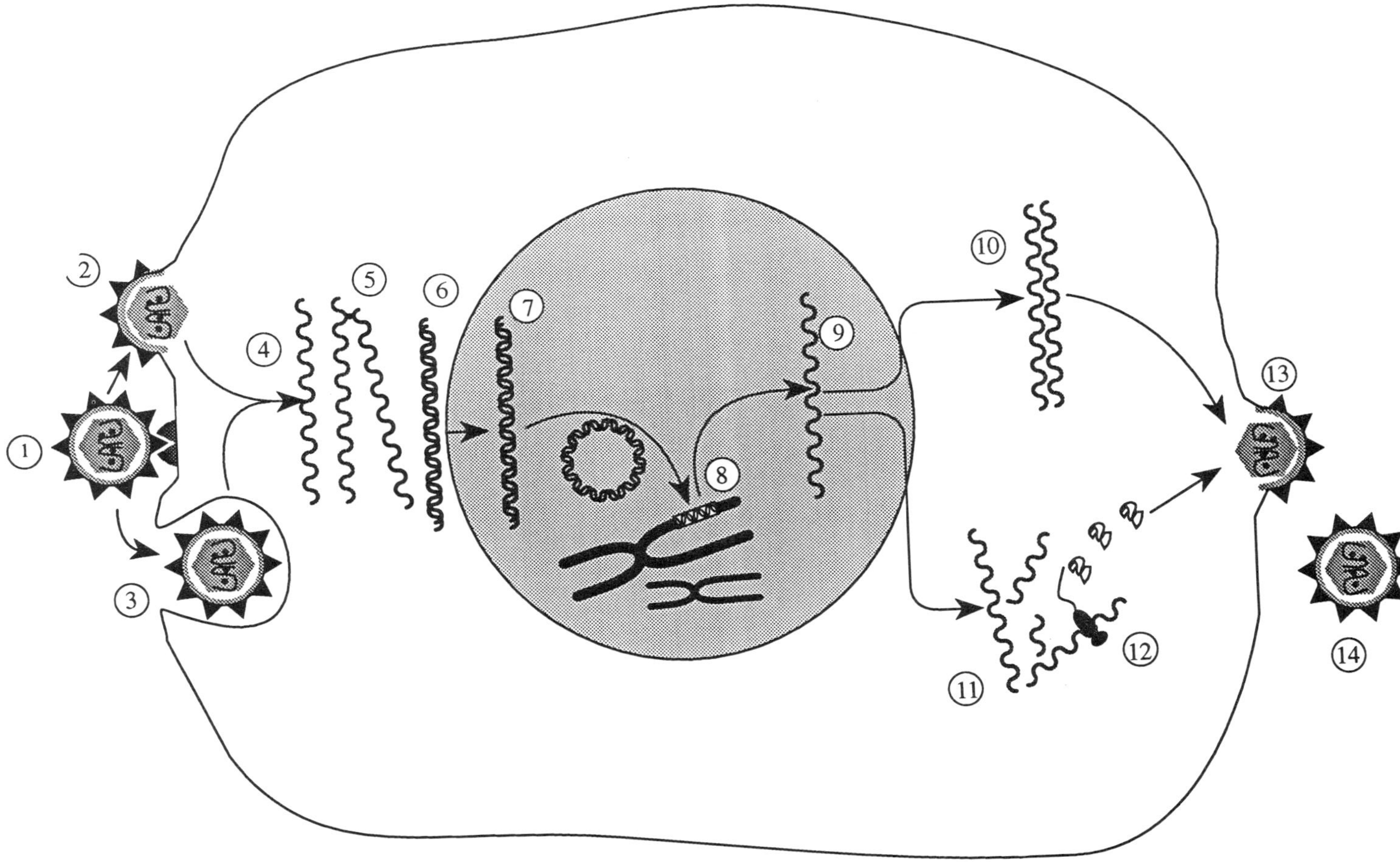

FIGURE 1. Retrovirus life cycle: (1) A free virus particle binds to its specific receptor on the cell surface and enters the cell by fusion of the viral and cell membranes (2) or by receptor-mediated endocytosis (3). Viral RNA released from the particle (4) is transcribed to DNA (5) by the reverse transcriptase carried in the virus. The double-stranded DNA (6) is transported to the nucleus (7) where it integrates in the host cell (via a circular form) chromosome (8) to form a provirus. This provirus can then be transcribed to make full-length viral RNA copies (9) which are transported to the cytoplasm (10) or spliced to make messenger RNA (11) for translation to viral proteins (12). Viral proteins and RNA are then assembled and bud from cell surface (13) to make free enveloped progeny virus (14).

proteins, however, are eliminated during the early stages of development so that only those with the potential to recognize nonself mature and become part of immune surveillance.

Perhaps the most important feature of the immune system is its capacity for immunologic memory. It is during the effort to overcome infection that the pool of cells reactive with that particular pathogen will be expanded and primed. Hence, any subsequent encounter with the same or closely related organism will result in a rapid reactivation of immunity and an inapparent infection. This process is the basis of vaccination.

The stimulation of an immune response to an antigen, retroviral or otherwise, is a complex series of events involving the interactions of many cell types (Fig. 2) and responses to a multitude of cytokines. The end product of such an immune response could be B cells producing antibodies able to bind antigen or T cells able to attack infected cells. These responses are conveniently classified as parts of the humoral or cellular arms of immunity, respectively. In this regard, T and B cells need to collaborate closely in the initial stages of immune activation. However, it is processing and presentation of antigen to T-helper (Th) cells that constitutes the vital first step of a functional immune response (Fig. 3).

A. Antigen Presentation

Immune stimulation naturally requires that a protein is recognized as "nonself" or foreign to the body. T- and B-cell precursors specific for a substantial proportion of all possible antigens are produced as a result of Ir-gene rearrangements, with those specific for "self" being quickly eliminated or suppressed. Lymphocytes are sequestered in the lymph nodes and other sites waiting for contact with specific antigen presented in the correct manner. T cells cannot themselves recognize free antigen, and the stimulation of Th cells requires that the antigen be presented as processed fragments in association with autologous major histocompatibility complex class II (MHC-II) molecules. A number of cells can act as antigen-presenting cells (APC) in this manner. Macrophages, for example, can phagocytose particulate or opsonized antigen, degrade it intracellularly, and reexpress the processed fragments on the cell surface in association with the MHC-II molecule. Other APC such as the class II-positive dendritic cells do not phagocytose antigen. It therefore seems likely that degradation and association with the MHC molecules occurs in this case at the cell surface.

The MHC/antigen complex is recognized by the specific Th cell via its T-cell receptor which is then primed for stimulation. However, the Th cell requires, in addition to the antigen trigger, a further signal from the APC, the cytokine interleukin-1 (IL-1). Binding of IL-1 to the IL-1 receptor of the Th cell stimulates both the expression of the receptor for IL-2 and the production of IL-2 itself. Some of the IL-2 will immediately bind to the IL-2 receptors of the same cell, and provide the final trigger for cell stimulation.

B. Stimulation of T and B Lymphocytes

As described in more detail in section V, cytotoxic T cells (Tc or CTL) recognize and are stimulated by antigen processed and presented in the context of the

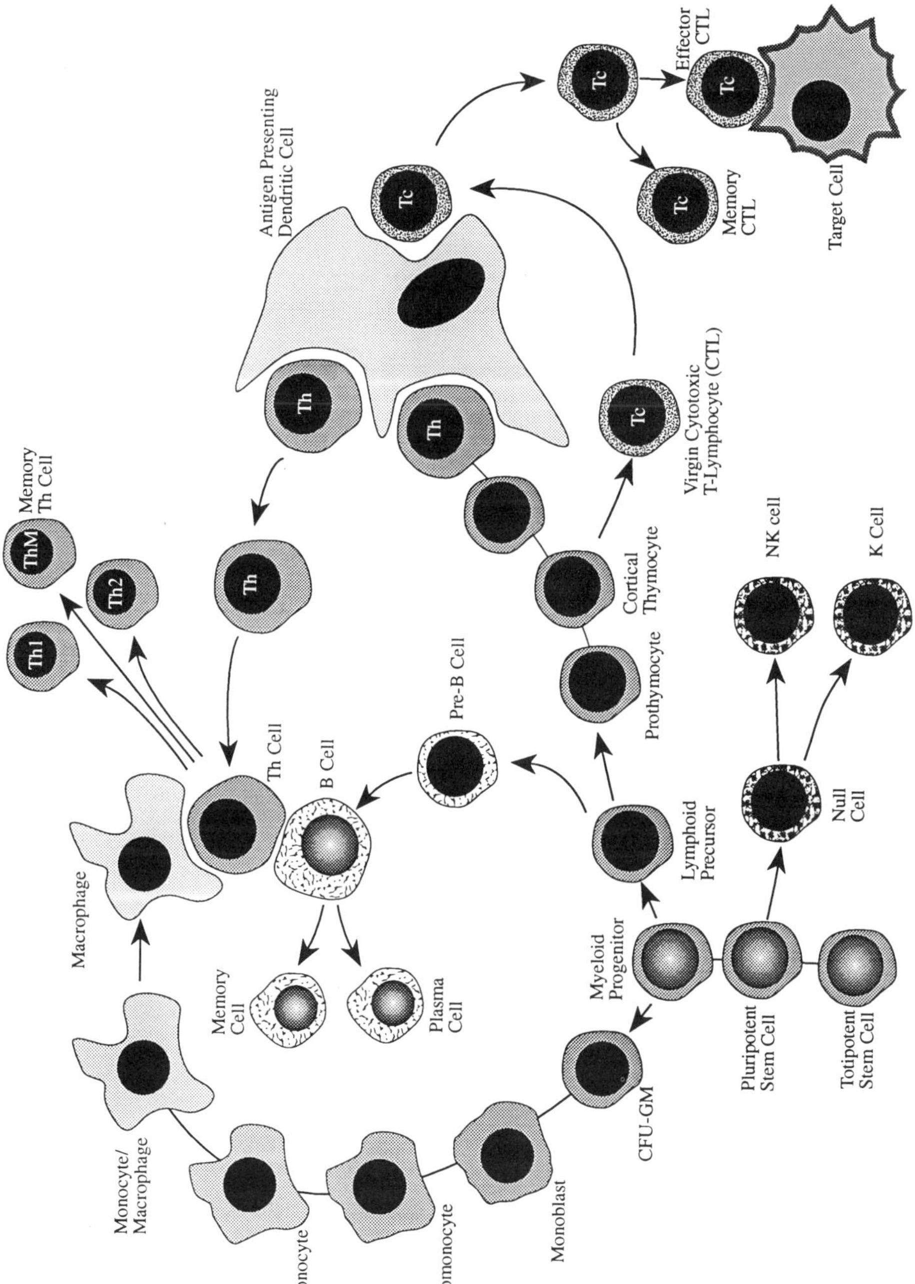

FIGURE 2. Generalized scheme for the development of cells involved in specific immune responses.

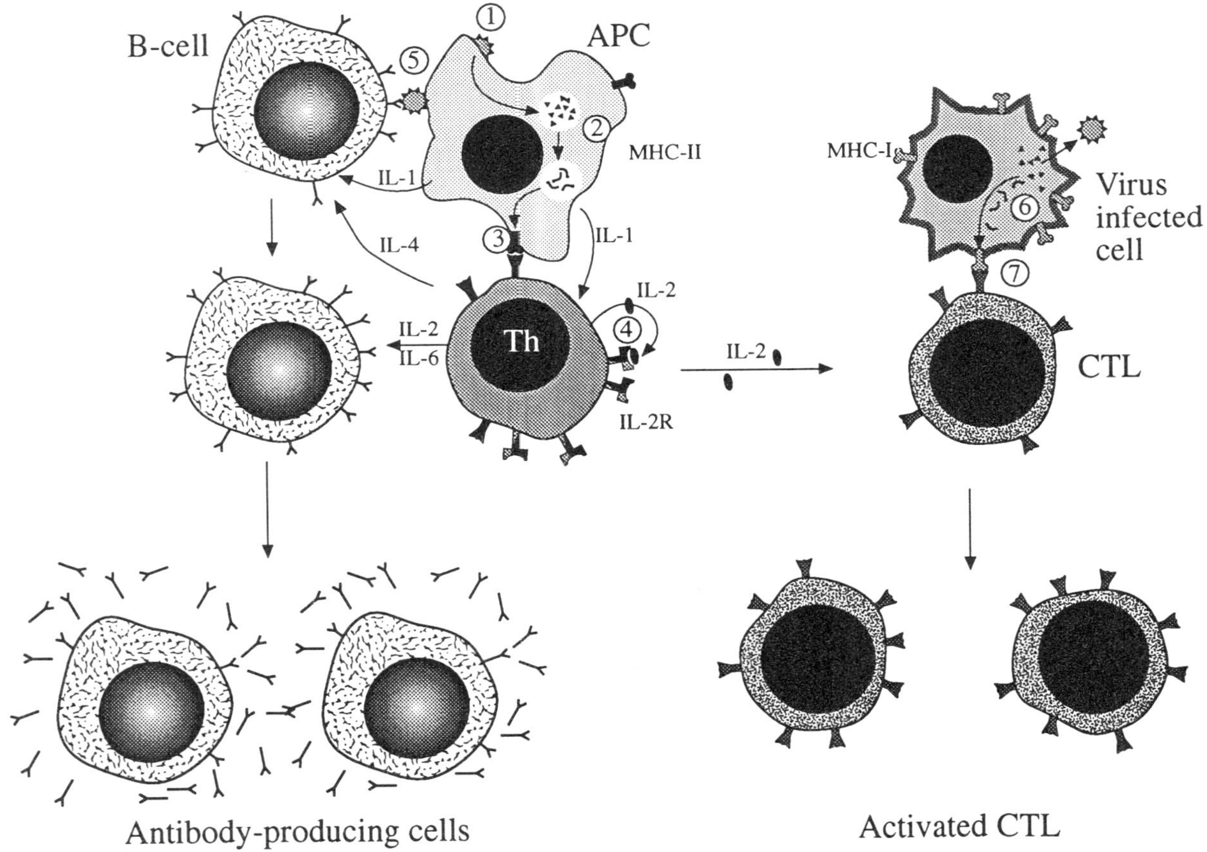

FIGURE 3. Simplified scheme of immune response stimulation. Antigen-presenting cells (APC) take up and process antigen (1) into short peptides which are processed (2) and presented with the MHC class II molecule (3) to antigen-specific T-helper cells (Th). This recognition, plus the stimulation by APC-produced IL-1 causes the Th cell to produce both IL-2 (4) and the IL-2 receptor (IL-2R). Antigen adsorbed to the surface of the APC may also be recognized by specific B cells (5) which are then stimulated, after receiving help in the form of IL-2, IL-4, and IL-6, to progress to antibody-producing cells. Cytotoxic T cells (CTL) are stimulated by specific viral antigen endogenously processed (6) by a virus-infected cell and presented in the context of MHC class I molecules (7). Also after receiving help in the form of IL-2, the CTL proliferates and differentiates to form mature CTL.

class I MHC molecule. As MHC-I is ubiquitous, this antigen presentation function can be fulfilled by any nucleated cell. However, even Tc cells require help in the form of IL-2 secreted by Th cells in order to proliferate and differentiate further.

B cells are triggered to replicate and differentiate into antibody-producing cells by recognition of intact antigen. The antigen may be "presented" to the B cell by class II-negative accessory cells (or indeed macrophages) after binding to the APC surface via Fc or C3 receptors. However, with the exception of T-independent antigens, this presentation in itself is not sufficient. B cells, like T cells, also require stimulus from a Th cell. Although the same APC could theoretically interact with both the T and the B cell, the APC is not in this case necessary. The B cell can itself endocytose bound antigen, particularly if attached to the surface Ig, process it, and express it with MHC-II. The Th cell, once activated, can then in return provide the necessary help for differentiation of the B cell. In addition, a Th cell stimulated in this way by a B cell acting as an APC could proceed to provide help for antigen-specific CTL.

C. The Role of Cytokines

The different cells of the immune system secrete and are susceptible to the effects of a multitude of soluble factors, the cytokines (Table I). The definition of these cellular factors is still a matter of some controversy, and efforts to standardize the nomenclature (e.g., the interleukins) are not always successful. It is increasingly evident, however, that the cytokines play a pivotal role in the regulation of the immune response, particularly in maintaining the fine balance of normal immune status.

Vilcek and Le (1991) proposed the following general definition of a cytokine: "Cytokines are regulatory proteins secreted by white blood cells and a variety of other cells of the body; the pleiotropic actions of cytokines include numerous effects on cells of the immune system and modulation of inflammatory responses." Under this definition one can include the lymphokines, monokines, interleukins, and interferons. Characteristically, cytokines are only synthesized in the cell to any great degree when the cell is stimulated by one or more external signals. The cytokine, a small protein or glycoprotein, is then secreted (usually transiently) to bind with high affinity to a specific receptor of the target cell. Usually this receptor is in close proximity to the cellular factor. This binding then induces a change in the behavior of the target cell, most often through alteration of the pattern of gene expression. For example, there could be increased growth or a changed differentiation state or secretion of another cytokine.

In the early years of cytokine research, it was often difficult to assign a function to a single cytokine because the molecules could not be purified to homogeneity. However, the advent of molecular biology has allowed the cloning (and sequencing), production, and characterization of homogenous cytokine preparations. In this way a wealth of information concerning the various actions of individual cytokines has been accumulated. Although of undoubted importance, such information can be misleading, for the action of one particular cytokine can be dramatically influenced by the presence of another. Cytokines often interact in a synergistic or antagonistic manner, and as a cell *in vivo* rarely comes

TABLE I. Examples of Sources and Effects of Cytokines on the Immune System

Cytokine[a]	Source[b]	Effect[c]
IL-1	Monocyte/macrophages Dendritic cells NK cells B and T cells	B-cell proliferation or differentiation Induces IL-2 (and IL-2R), IL-4, CSF, and IFN production in T cells Activates NK cells
IL-2	Activated T cells NK cells	B-cell proliferation or differentiation Growth of activated T cells Stimulated T-cell lymphokine production and CTL/NK activity
IL-3	T cells Mast cells	Growth and differentiation of pluripotential hemopoietic stem cells
IL-4	Activated T cells (Th2) Mast cells Transformed B cells	B-cell activation, differentiation, and growth Th2 growth CTL generation Stimulates proliferation of NK cell growth but inhibits IL-2-dependent growth
IL-5	T cells	Differentiation of eosinophils B-cell proliferation (mouse)
IL-6	T cells Monocyte/macrophages	Growth and antibody production of plasma cells Role in T-cell activation, growth, and differentiation Proliferation of pluripotential hemopoietic progenitors
IL-7	Bone marrow stromal cells	Proliferation of B- and T-cell precursors Stimulates thymocytes Co-stimulates mature T cells CTL generation
IL-8	Monocytes, lymphocytes, neutrophils, endothelial cells	Neutrophil activation
IL-9	Activated T2 cells	Erythroid colony stimulation Mast cell enhancement T-cell growth
IL-10	T2 cells	Inhibits Th1 cytokine synthesis
IL-11	Stromal cells	Lymphopoiesis Hematopoiesis (e.g., megakaryocytopoiesis)
IL-12	Transformed B cells Macrophages	NK cell activation Stimulation of Th1 cells
IFN-α/β	All nucleated cells	Antiviral Modulation of MHC expression Enhances expression of macrophage FcR Stimulates Ig synthesis
IFN-γ	T cells NK cells	Antiviral Modulation of MHC expression Enhances expression of macrophage FcR, antimicrobial and tumoricidal activities Stimulates NK activity Enhances TNF-α and β activity
GM-CSF	Macrophages Endothelial cells Activated T cells	Colony stimulation of numerous hemopoietic progenitors Granulocyte/macrophage differentiation
G-CSF	Macrophages Fibroblasts Endothelial cells T cells	Activation/differentiation of neutrophils
M-CSF	T cells Macrophage/monocytes	Macrophage colony growth
TNF-α	Monocyte/macrophages	Central role in inflammatory responses Upregulation of MHC-I Increased phagocytosis, degranulation, and oxygen radical release by polymorphonuclear cells
TNF-β	T cells, B cells	Activation of macrophages, granulocytes, and CTL

[a] The "modern" name of the cytokine is given here, although many of the cytokines are also known by other names (e.g., TCGF, BCGF)

[b] Only cells of the immune system are usually listed here, although many of the cytokines are also produced by a variety of other cell types (e.g., endothelial cells, fibroblasts)

[c] The effects on cells of the immune system are listed, not the multitude of influences cytokines have on other tissues of the body.

into contact with one single cytokine, the results of *in vitro* assays employing cell clones and recombinant cytokines may bear little resemblance to the normal effects in the living host.

1. IL-2

Many cytokines influence the growth and activities of T cells, but IL-2 appears to be by far the most important and is consequently the best characterized. The identification of T-cell growth factor (Morgan *et al.*, 1976), as it was initially known, facilitated the continuous growth of T cells in culture and the subsequent isolation of certain human retroviruses. IL-2 itself is produced by activated T cells and consists of a 15.5-kDa glycoprotein, the sequence and three-dimensional structure of which have been partially determined (Brandhuber *et al.*, 1987; Ciardelli *et al.*, 1988).

Of particular importance is the cell-surface receptor for IL-2 (IL-2R) which is made up of two subunits, IL-2Rα and IL-2Rβ with molecular masses of 55 and 75 kDa, respectively. Individually, these molecules can act as low- and intermediate-affinity receptors for IL-2, but when expressed as a heterodimer the affinity (K_d) increases to 10–50 pM. The IL-2Rβ molecule is responsible for the transmembrane signaling of activation following binding of IL-2 and is expressed on many cell types. It is the controlled transcription of the IL-2Rα subunit, and hence the formation of the high-affinity heterodimer, which renders a cell susceptible to physiologic levels of IL-2. One of the most important effects of IL-1 produced by the APC is to stimulate the expression of IL-2R on T cells and render them susceptible to IL-2. In addition, the cytokines IL-4, IL-6, IL-7, and TNF can upregulate IL-2R expression.

2. IL-4

Similar to the pivotal role played by IL-2 during T-cell growth is the effect of IL-4 on B cells. IL-4 is involved in many of the stages of B-cell activation, proliferation, and differentiation. For example, recombinant IL-4 can induce pre-B cells to mature into surface-IgM$^+$ B cells. The molecule is produced predominantly by T cells of the T-helper 2 subclass, although other populations such as mast cells and bone marrow stromal cells also secrete this cytokine. Like many cellular factors, the effects of IL-4 are not limited to one cell type. The protein is apparently important during T-cell ontogeny and can act synergistically or antagonistically with other cytokines.

3. Other Cytokines

The growth and differentiation of macrophages is also regulated by a family of cytokines, in particular the colony-stimulating factors (CSFs) named for their ability to promote the growth of colonies in soft agar, and IL-3. Macrophages, as APC, play a crucial role in the initiation of the immune response. As processed antigen is expressed to Th cells in the context of the class II MHC molecules, the levels of class II expression can profoundly influence the ensuing level of immune activation. A number of cytokines can upregulate class II MHC, most notably IFN-γ (Sztein *et al.*, 1984). This event results in an enhanced antigen-presenting capability (Zlotnik *et al.*, 1983). The presence of IL-4 also increases the expression of class II (and class I) MHC expression on macrophages (Crawford *et al.*, 1987; Stuart *et al.*, 1988), giving similar results (Zlotnik *et al.*, 1987).

As well as serving as APC, macrophages have a number of effector functions, most notably an array of tumoricidal and antimicrobial activities mediated predominantly by TNF and oxygen metabolites. The tumoricidal activity of macrophages is stimulated by factors such as GM-CSF, IL-4, and IFN-γ. M-CSF acts in synergy with these other cytokines.

With two or three new cytokines presently being identified and characterized each year, it is likely to be some time before the full impact of this diverse group of molecules is well understood.

D. Cross-Regulation by Th-Lymphocyte Subsets

The ability to produce Th lymphocyte clones has recently revealed that at least two distinct subsets of the $CD4^+$ population exist, based on the pattern of cytokine production. This finding has to some extent helped explain the multitudinous, yet often reciprocal functions of $CD4^+$ cell populations. T-helper 1 (Th1) clones, secreting IFN-γ, IL-2, and lymphotoxin, are involved in DTH inflammatory responses and provide help to cytotoxic T lymphocytes and monocytes. Th2 cells, producing IL-4, IL-5, IL-6, and IL-10, stimulate B-cell growth and Ig secretion.

The relationship between these two subsets is one of cross-regulation and mutually induced control: the cytokines produced by one can inhibit the functional activities of the other. IFN-γ, produced by Th1 cells, will suppress the activity of Th2 cells whereas IL-4 and IL-10 released from Th2 cells inhibit the cytokine production of Th1 cells. As will be discussed later, of particular importance for an infected host is the fact that stimulation of Th2 cells will result in suppression of Th1 cells, and hence many aspects of the cell-mediated immune response (Clerici and Shearer, 1993).

IV. HUMORAL RESPONSES TO RETROVIRUSES

For two basic reasons the humoral immune responses during viral infection, which involve antibody molecules, are often characterized in greater detail than cellular immune responses. First, antibodies are easier to handle and store in the laboratory. Second, the presence of a specific humoral immune response to a retrovirus or other infectious agent provides a convenient means of diagnosing an ongoing or previous infection. Most standard diagnostic methods are therefore based on the detection of antibodies to a retrovirus rather than detection of the retrovirus itself.

During the course of replication, structural and nonstructural retroviral proteins become exposed to the immune system, resulting in the production of antibodies specific for each of these foreign components. We will first describe how these antibodies can be used for diagnosis and then discuss the manner in which they fulfill their true role: elimination or control of the retrovirus.

A. Diagnostic Assays Using Antibodies

The humoral response to the various viral components can be visualized by a number of standard serologic techniques such as enzyme-linked immunosorbent assay (ELISA), radioimmunoassay (RIA), immunofluorescence assay (IFA), immunoblot, or radioimmune precipitation assay (RIPA). For a more detailed review of this subject, see Hardy (1991).

1. ELISA

The ELISA is the standard diagnostic method used for testing serum samples for the presence of antibodies specific for a given antigen. Its relative ease of use (readily automated), low cost, and high capacity make it an ideal test for screening large numbers of samples, such as those generated in a blood-donation clinic or hospital.

In its simplest form (Fig. 4), the ELISA is performed by allowing serum to react with virus antigen immobilized on a solid phase (microtiter plate or bead). Antibodies specific for the antigen will bind and be in turn recognized by an indicator antibody specific for the first antibody. This second antibody carries an enzyme that induces a color reaction to occur in the substrate/chromogen mixture added as the final step.

Besides using whole disrupted virus or virus-infected cells as antigen in the ELISA, many commercial kits use purified native or recombinant proteins or even synthetic peptides on the solid phase. Although having the advantages mentioned above, ELISA suffers from giving the occasional borderline result or even false-positive result. The latter is often caused by the presence of contaminating cellular proteins. Positive results are therefore usually confirmed by secondary methods such as immunoblot (see below).

ELISA is now used routinely for the detection of antibodies specific for many retroviruses (Fig. 5). Detection of antibodies to reticuloendotheliosis virus (REV)

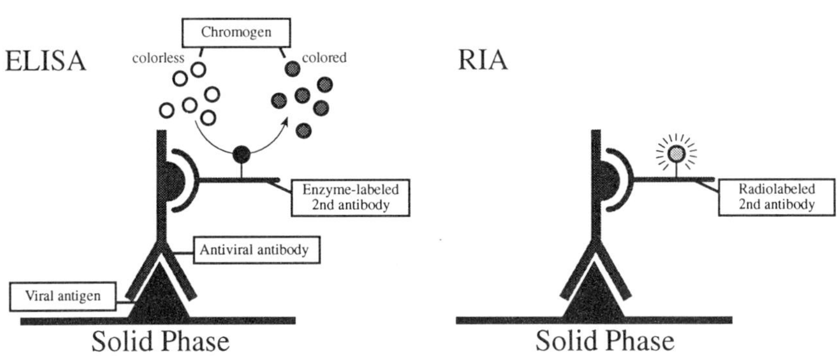

FIGURE 4. Enzyme-linked immunosorbent assay (ELISA) for antiretroviral antibodies. Viral antigen bound to the solid phase (e.g., microtiter plates) is incubated with the putative virus-specific antibody. The presence of bound antibody is detected by an enzyme-coupled second antibody specific for the first. The enzyme catalyzes a reaction in the final development solution resulting in a color change of the chromogen. The RIA is performed in essentially the same way with the second antibody being detected through a bound radiolabel.

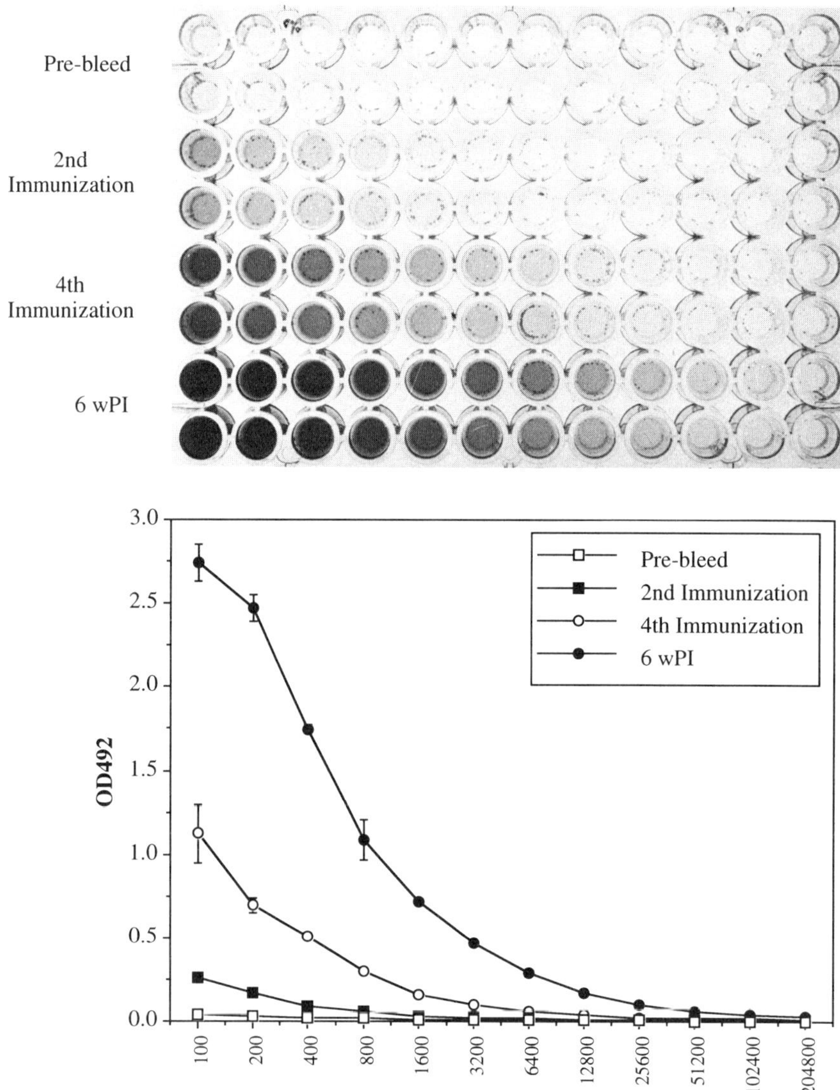

FIGURE 5. Example of a typical ELISA plate. A 96-well microtiter plate coated with retroviral antigen (in this case SIVmac) was exposed to different dilutions of sera from a monkey immunized with inactivated SIVmac antigen and challenged with live SIVmac (grown in rhesus PBMC). The sera were taken before immunization (pre-bleed), 2 weeks after the second and fourth immunizations, and 6 weeks after challenge (6 wPI). The intensity of the color reaction (OD_{492}), proportional to the amount of bound antibody, is also shown graphically as the average of duplicate tests (±standard deviation). The increase in antibody titer after the immunizations can be seen as can the anamnestic response to infection (the immunizations did not elicit a protective immunity).

of chickens using ELISA, for example, was shown to be consistently more sensitive than indirect immunofluorescent-antibody tests (Smith and Witter, 1983), as was an ELISA using virus-infected chick embryo fibroblasts as coating antigen (Nicholas and Thornton, 1987). The use of REV-specific monoclonal antibodies in ELISA allowed the differentiation between different strains of the virus (Cui *et al.*, 1986, 1988).

In the feline system, ELISA has been used to demonstrate the relatively high percentage of cats infected with feline immunodeficiency virus (FIV; O'Connor *et al.*, 1989), although a comparison of FIV ELISA with Western (immuno-) blot (Hosie and Jarrett, 1990) showed that even commercial ELISA kits can give false-positive results. Diagnosis of feline leukemia virus (FeLV) infection is normally carried out using an ELISA-based antigen-trapping test whereby monoclonal antibodies bound to the ELISA plate are used to trap FeLV p27 in the cat serum. The p27 is then in turn recognized by a second virus-specific antibody (Lutz *et al.*, 1983).

Using purified caprine arthritis–encephalitis virus (CAEV) and a biotin/avidin amplification system, Archambault *et al.* (1988) showed the ELISA to be a simpler and more accurate method of diagnosing infection than the more commonly used agar gel immunodiffusion test and RIA. Although Western blot analysis remains the definitive test for anti-CAEV antibodies, replacing the second (detection) antibody with protein G was shown to enhance the specificity of the ELISA (Zanoni *et al.*, 1989).

In 1978, Ressang *et al.* (1978a,b) developed an ELISA-based system for diagnosing infection of cattle with bovine leukemia virus (BLV) and compared the test with other assays commonly used at that time. Since then, the assays have been refined and adapted to employ, for example, monoclonal antibodies (Portetelle *et al.*, 1983) or purified viral proteins (Molloy *et al.*, 1990) and have greatly aided the control of this infection.

The ELISA technique is also useful for detecting complexes of antigen and antibody in serum. Immune complexes of antibodies with retroviruslike particles were demonstrated using ELISA in the blood of patients with psoriatic arthritis (Iversen and Rodahl, 1985). Similarly, Schetters *et al.* (1990) used an adaptation of the ELISA method to determine levels of murine leukemia virus (MuLV) gp70-specific immune complexes in mouse sera.

The need to routinely screen the sera of blood donors for antibodies to HIV-1 and HIV-2 has led to the development of numerous commercial ELISA kits. Intense competition between the various manufacturers has resulted in ELISA tests that are highly sensitive and reliable. Early tests depended on the use of disrupted HIV to coat the wells of the ELISA plates (Weiss *et al.*, 1985), but many tests now use recombinant proteins or even synthetic peptides as antigen to detect HIV-specific antibodies. Peptides in particular allow kits to distinguish between HIV-1 and HIV-2 infection (Broliden *et al.*, 1991; Fenouillet *et al.*, 1990; Modrow *et al.*, 1989).

2. RIA

RIA is performed in much the same way as the ELISA described above except that in place of an enzyme to label the second antibody a radioactive marker is used (usually ^{125}I) and binding is measured using a γ- or β-counter or exposure to X-ray film.

The difficulties involved in handling radioactive substances on a routine basis have led to RIA being replaced by ELISA in most situations. However, RIA was for many years a standard method for screening sera specific for antibodies specific for a given antigen. A liquid-phase RIA using radiolabeled core antigen was shown to be a very reliable means of diagnosing BLV infection of cattle (Levy *et al.*, 1977), whereas Gupta and Ferrer (1981) found RIA using BLV envelope glycoprotein to be more sensitive, and used the assay to help eradicate the disease from an infected herd (Ferrer, 1982). Jacobsen *et al.* (1985) showed radioimmune precipitation assays to be more sensitive than commercial agarose gel immunodiffusion assays in diagnosing BLV infection.

The plethora and simplicity of commercial ELISA kits have decreased the importance of RIA in the diagnosis of HIV infection, although the method was initially used in some laboratories. Competition RIAs have, however, proved helpful for the detection of HIV-antigen and immune complexes (Caruso *et al.*, 1987; Lucas *et al.*, 1990; Neurath *et al.*, 1985).

3. IFA

IFA usually detects antibodies binding to infected cells (either the cell surface or the cytoplasm, depending on the method used to fix the cells). Antibodies produced in response to the envelope glycoproteins expressed on the infected cell surface are often more virus strain-specific than those reacting with the viral core or Gag protein. The IFA test can therefore be useful where cross-reactivity is a problem. Using infected cells (usually fixed by acetone and/or methanol) on slides as the solid phase, the assay is performed much like the ELISA or RIA. However, with IFA the second antibody is labeled with a fluorescent dye that lights up the infected cell when exposed to the correct wavelength of excitatory light (Figs. 6 and 7). Despite its long history, certain technical difficulties, some degree of subjective evaluation, and the advent of recombinant or peptide antigens have now made the method less popular. In the avian system, Nicholas and Thornton (1983) showed IFA to be superior to other methods for identifying birds infected with REV. However, this assay was soon superseded by ELISA (Nicholas and Thornton, 1987; Smith and Witter, 1983).

Varnier *et al.* (1983) found an adapted IFA-based system to be superior to other assays for detecting murine xenotropic retroviruses. Use of monoclonal antibodies in such a system allowed Sitbon *et al.* (1985) to detect, differentiate, and biologically clone C-type murine leukemia viruses of different tropism.

In the 1970s, diagnosis of BLV infection was simplified by the development of a sensitive IFA test employing fixed infected cells (Ferrer *et al.*, 1975), an assay still being used to screen herds for antibodies to BLV, bovine visna virus, bovine syncytial virus (Amborski *et al.*, 1989), and even bovine immunodeficiency virus (Whetstone *et al.*, 1990).

Reciprocal cross-immunofluorescence assays helped Nemo *et al.* (1978) determine that a foamy virus isolated from a human was closely related to a simian foamy virus.

Although not routinely employed for diagnosing HIV infection, IFA is often used in place of immunoblot as a confirmatory test or to distinguish between

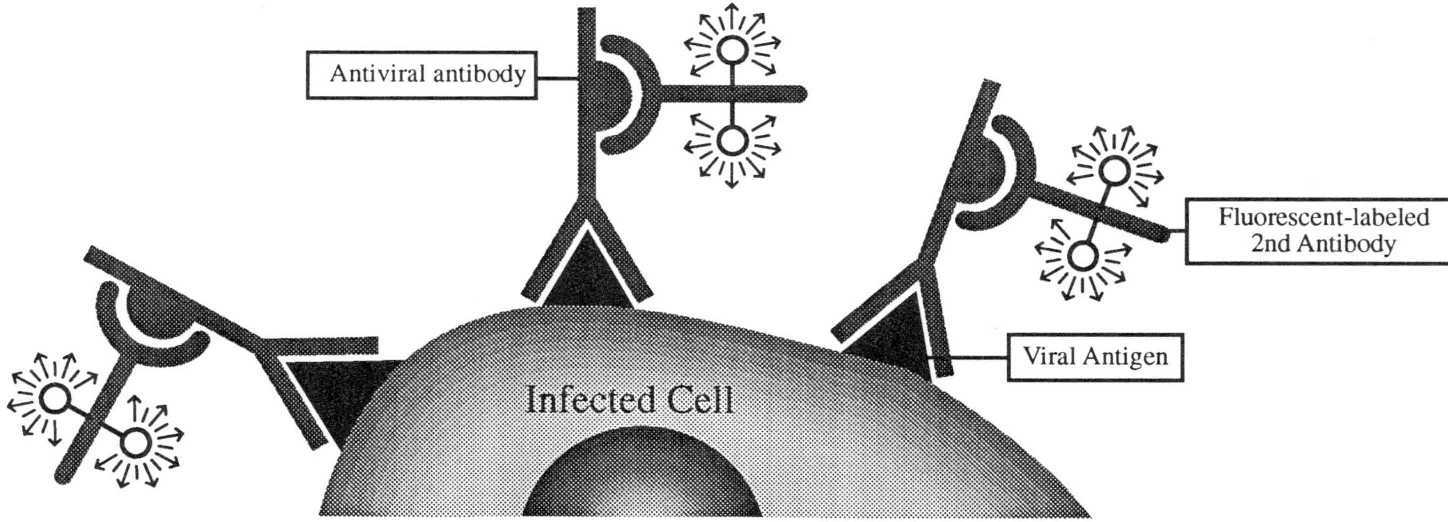

FIGURE 6. Immunofluorescence assay (IFA) for antiretroviral antibodies. Infected cells are incubated with the virus-specific antibody followed by a fluorescent-labeled antibody specific for the first. Infected cells can then be seen using a fluorescence microscope.

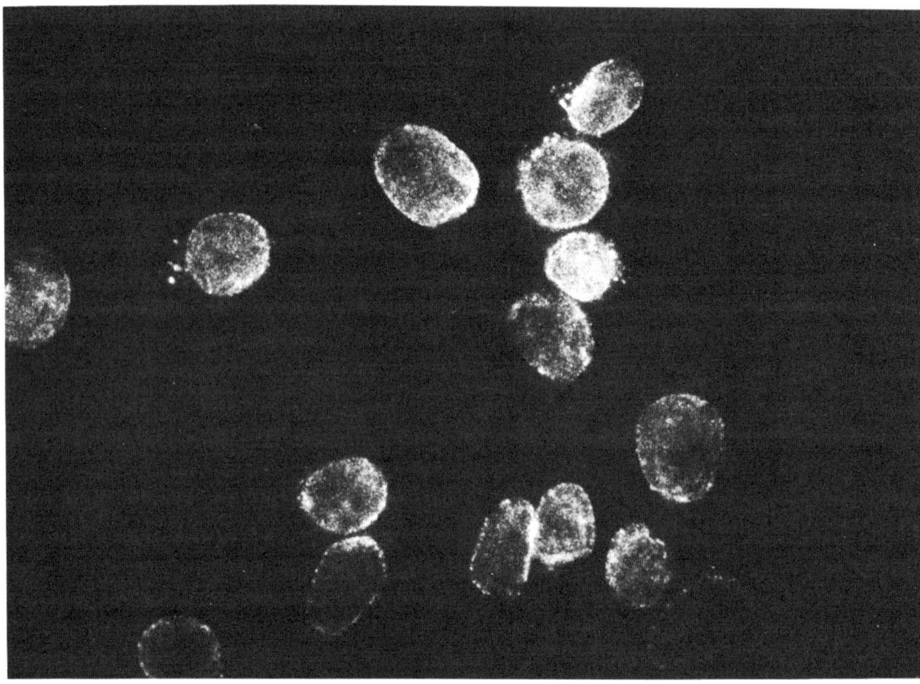

FIGURE 7. Example of immunofluorescent staining of cells expressing retroviral neoantigens. Monoclonal antibodies specific for the avian sarcoma virus transforming protein pp60^{src} are shown bound to the protein expressed in Schmidt–Ruppin virus (SRV-D) transformed cells. This binding is visualized by subsequent binding of a fluorescein-labeled goat anti-mouse IgG antibody which is visible under UV light. (From Tanaka and Kurth, 1984.)

HIV-1 and HIV-2 infection (Gallo *et al.*, 1992; Hedenskog *et al.*, 1986; Imrie *et al.*, 1986; Kvinesdal *et al.*, 1989; Lennette *et al.*, 1987; Sandstrom *et al.*, 1985). It is also very helpful in developing countries which may not be able to afford the more expensive tests for antiviral antibodies.

4. Immunoblot

When purified virus is disrupted and separated by gel electrophoresis, it is possible, after transfer to a suitable membrane, to immunochemically stain the different proteins using sera from an infected individual or animal (Fig. 8). A typical retrovirus immunoblot will demonstrate the presence of antibodies to the outer envelope and transmembrane envelope glycoproteins, to the core (Gag) proteins, and often to the reverse transcriptase. Reactivity to a number of precursor molecules might also be detected.

Although the general pattern is similar for most retroviruses, the specificities of the reactions differ enormously. Reactivity to the envelope glycoproteins is often virus-specific whereas reactivity to the core proteins is usually characterized by wide cross-reactions. For example, reaction of sera with FIV envelope glycoproteins is virtually diagnostic for infection. No cross-reactivity is seen (Steinman *et al.*, 1990) with sera from cats infected with FeLV or feline syncytium-forming virus (FeSFV). However, because of the relatedness of equine

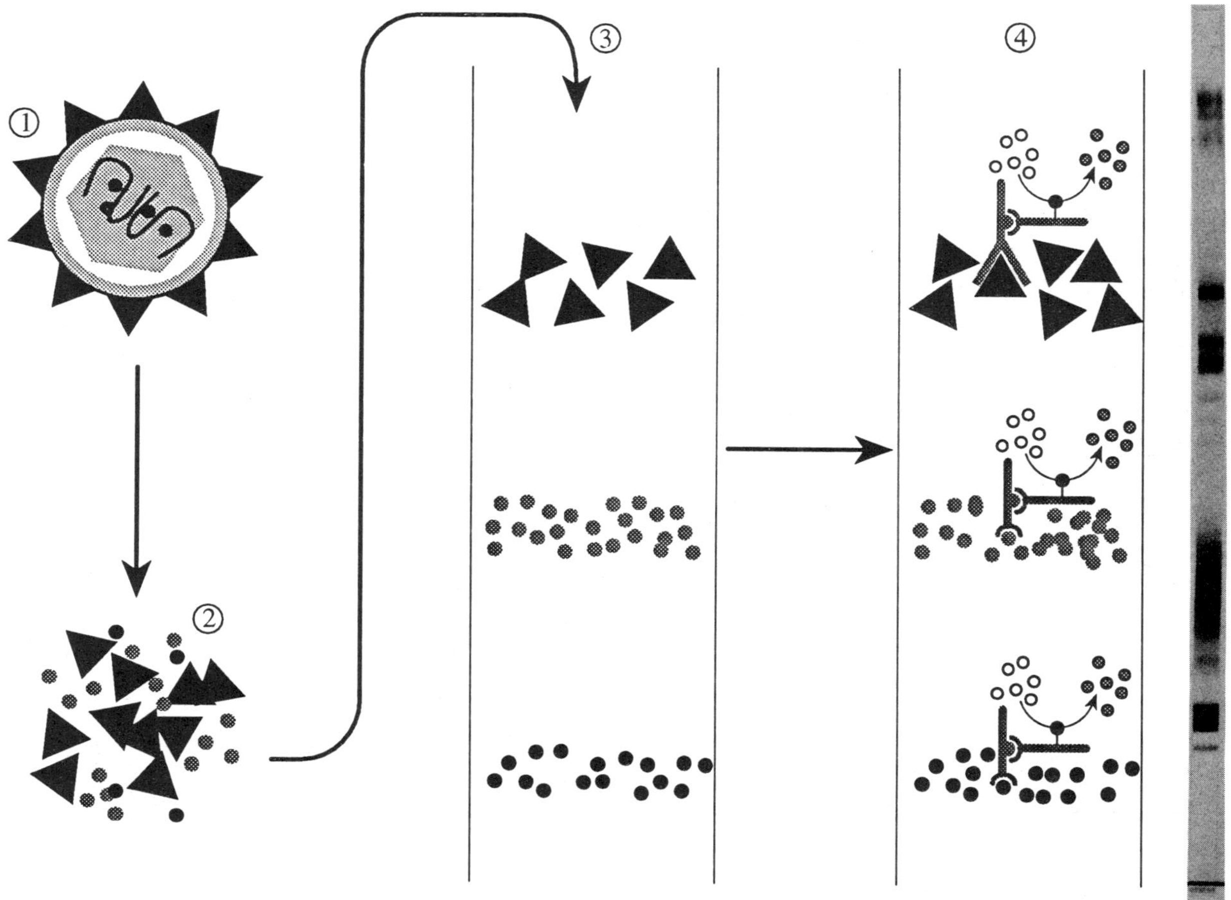

FIGURE 8. Immunoblot assay for antiretroviral antibodies. Purified virus (1) is disrupted (2) and the individual proteins separated by gel electrophoresis (3). Proteins are "blotted" to nitrocellulose paper which is then allowed to react with virus-specific antibody, enzyme-labeled antibody, and chromogen containing development buffer (4). Antibodies specific for the individual viral proteins can then be identified.

infectious anemia virus (EIAV) to FIV (both are lentiviruses), the core proteins of FIV can be recognized by sera specific for EIAV (Egberink *et al.*, 1990). This cross-reaction, however, should be readily recognized.

Despite the occasional problems with core-protein cross-reactivity, immunoblot remains the standard method for confirming and characterizing an antibody response to most retroviruses including simian retrovirus D (Kwang *et al.*, 1988), FIV (Hosie and Jarrett, 1990), BLV and bovine immunodeficiency virus (BIV; P. J. Walker *et al.*, 1987; Whetstone *et al.*, 1990), CAEV (Zanoni *et al.*, 1989), ovine lentivirus (Kajikawa *et al.*, 1990), and HIV-1 and HIV-2 (Dodd and Fang, 1990). The technique is also very helpful in distinguishing HIV-1 from HIV-2 (Zuber *et al.*, 1990).

5. RIPA

Immunoblot gives information about the reactivity to different viral proteins. However, as purified virus is normally used to prepare the strips, only structural proteins are included. During the replication of retroviruses (particularly lentiviruses), various important regulatory proteins will be expressed but not packaged within the progeny virus particles and are therefore not detectable by immunoblot. In addition, the proteins in an immunoblot strip, having been separated by gel electrophoresis, are in a denatured form with many of the possibly crucial discontinuous or conformational epitopes disrupted.

By metabolically labeling the proteins being newly synthesized in the infected cells (e.g., by pulsing with [^{35}S]methionine), it is potentially possible to detect antibodies to all viral proteins, including regulatory proteins. RIPA is performed by mixing the labeled cell lysate with serum before immunoprecipitating, denaturing, and running on a gel (Fig. 9). Under the right conditions, it is of course possible for IFA to detect antibodies to all viral proteins. Similarly, if the source of antigen is selected correctly, ELISA can be used to assay any antibody/antigen interaction.

Due to its complexity and requirement for radioisotopes, RIPA is not usually used to diagnose infection. However, many laboratories have chosen the technique as an alternative to immunoblot for confirmation, finding it to be the more sensitive and specific of the two. In addition, RIPA has been used to distinguish between HIV-1 and HIV-2 infection and to quantitate the antibody response to one particular viral protein.

6. Reactivity to Retrovirus-Transformed Cells

In the 1970s, the confirmation that lymphoma or sarcoma cells transformed by retroviral infection expressed unique antigens on the cell surface prompted an intense characterization of these antigens in the hope that they would serve as suitable targets for immunization and therapy (for review see Kurth *et al.*, 1979). Part of the host's immune response to retrovirus-infected cells was directed to modified (glycosylated) viral structural proteins expressed on the cell surface (i.e., the envelope glycoproteins). However, resistance to tumors was found mainly to be the result of responses against tumor cell surface antigens not found in the virus *per se*.

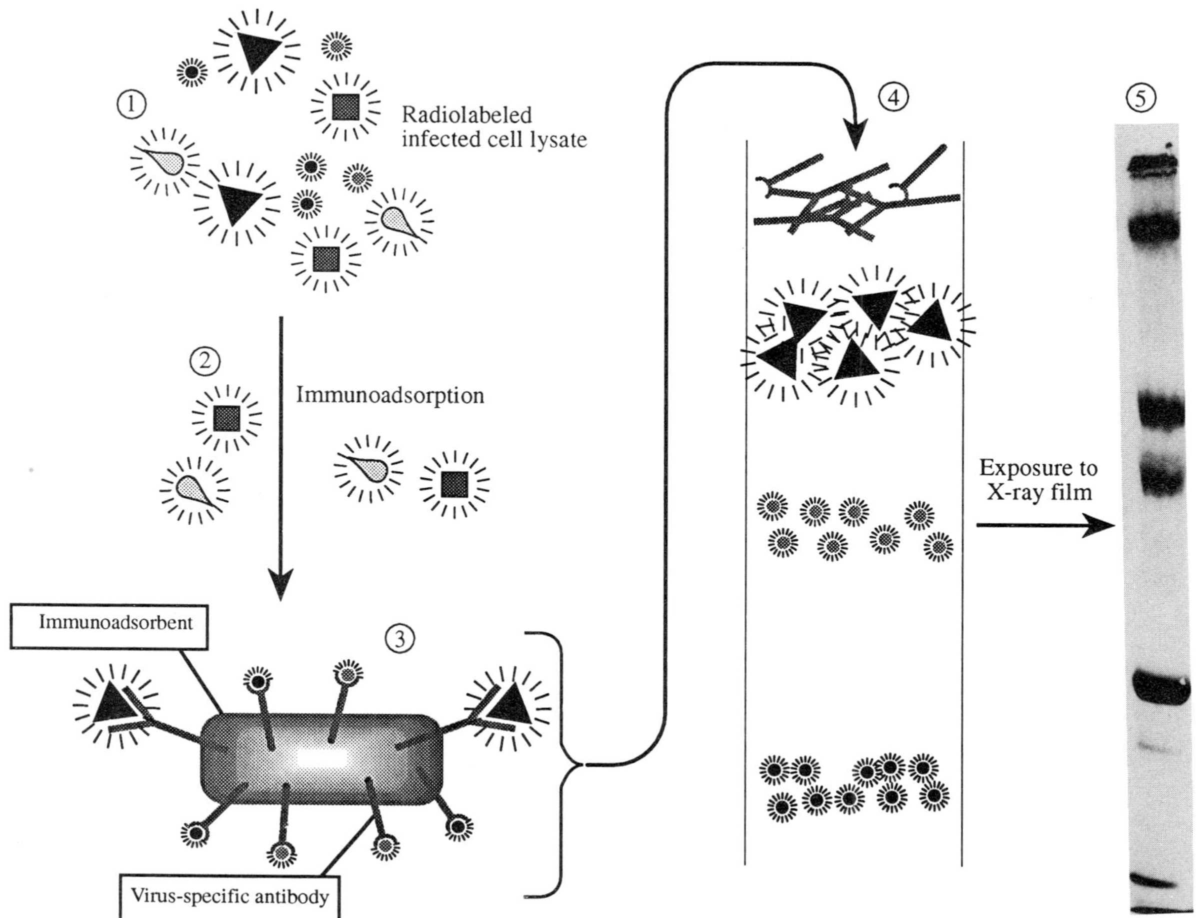

FIGURE 9. Radioimmune precipitation assay (RIPA) for antiretroviral antibodies. Infected cells grown in the presence of radioactive amino acids are lysed to give a mixture of radiolabeled cellular and viral proteins (1). The proteins are allowed to react with virus-specific antibodies and the antigen/antibody complexes are separated (2) from free proteins using an immunosorbent (3). These complexes are then disrupted and the individual proteins separated by gel electrophoresis (4). The radioactive protein bands are visualized by exposure to X-ray film (5).

Perhaps the best known example of this phenomenon involves the response to the feline oncornavirus-associated cell membrane antigen (FOCMA) expressed (Essex *et al.*, 1971) on all tumors transformed by FeLV and feline sarcoma virus (FeSV). Evidence for FOCMA being distinct from viral structural genes accumulated rapidly: Adsorption of FOCMA-positive antisera with FeLV core and envelope proteins failed to remove FOCMA reactivity and the titers of antibody to FOCMA did not correlate with titers to Gag or Env proteins (Hardy *et al.*, 1976; Sherr *et al.*, 1978; Stephenson *et al.*, 1977). Furthermore, peptide maps of FOCMA were shown to be distinct from those of FeLV $gp70^{env}$.

FOCMA was long considered to be a precursor of the viral *gag* gene product although the evidence now is that it is related to, but distinct from, the envelope glycoprotein of FeLV-C (Snyder *et al.*, 1983). The fact that immunization of cats with a genetically engineered FeLV $gp70^{env}$ vaccine induces antibodies able to recognize FOCMA (Marciani *et al.*, 1991) supports the latter possibility. Immune responses to such molecules, although ineffective in clearing virus itself, appear to play an important role in combating retrovirus-induced lymphoma (Essex, 1975; Essex *et al.*, 1971).

B. Functional Humoral Immune Responses

Although a useful marker for infection, the immune system obviously did not evolve to help diagnosis. Its function is to eliminate or control an established infection or to prevent reinfection by the same or similar agent. A number of different "effector" mechanisms operate in the host which have the ability to abrogate the infectivity of free viral particles or to eliminate infected cells (Tables II and III). Although the functional immune response mechanisms generally work in a similar fashion against the different retroviruses, there are important differences that could influence the outcome of the infection. A description of each of the known major effector mechanisms follows with examples of their efficacy against retroviral infections.

1. Neutralizing Antibody

This response is perhaps the simplest, operating by preventing a free virus particle from infecting the host cell. In order for a retroviral infection to begin, an interaction between the envelope glycoprotein and one or more cell surface receptors is usually required. If an antibody is able to bind to the virus and interfere with this interaction, then the virus infectivity is neutralized (Fig. 10).

a. Examples of Retrovirus Neutralization

In many retroviral systems, regions on one or both envelope glycoproteins can be targeted for neutralization. For example, inoculation of chickens with purified Rous sarcoma virus (RSV) gp85 and gp37 stimulates a neutralizing antibody response (Bauer *et al.*, 1979) and even protection from RSV-induced tumor production (Bennett and Wright, 1987). Similarly, birds immunized with cells infected with an avian erythroblastosis virus vector bearing the Rous-associated virus *env* gene developed neutralizing antibodies and were protected against tu-

TABLE II. Potential Benefits and Disadvantages of Specific Antiretroviral Immunologic Mechanisms

Mechanism	Specificity through:	Effector mechanism	Benefits	Disadvantages
Neutralization of virus	Antibody	Antibody	Blocks infection	Enhancement Autoantibodies (molecular mimicry)
Virolysis	Antibody	Complement	Destroys virus	
Complement-mediated cell lysis	Antibody	Complement	Destroys infected cell → no progeny virus	Autoimmunity?
ADCC	Antibody	Killer cell-mediated cytolysis	Destroys infected cell → no progeny virus	Immunopathology? i.e., killing of uninfected cells carrying bound viral proteins
Natural killer cells	Unknown	Natural killer cell-mediated cytolysis	Destroys infected/ transformed cell → no progeny virus	Release of progeny virus
Cytotoxic T lymphocytes	T-cell receptor	CTL-mediated cytolysis	Destroys infected cell → no progeny virus	Autoimmunity? Release of progeny virus

mor production after RSV challenge (Chebloune *et al.*, 1991). Molecular and biologic analysis of different avian retrovirus envelope genes demonstrated that type-specific neutralization was associated with the gp85-coding domain of the virus (Bova *et al.*, 1988). In ducks artificially infected with RSV, a decline in viremia was found to correlate with increasing titers of neutralizing antibodies, although the high levels of antibodies were not able to completely clear the virus from duck tissues (Nehyba *et al.*, 1990).

Similarly, a comparison of the immune response in tumor progressor and tumor regressor chicken lines following RSV challenge revealed a strong correlation between the early, increasing and sustained presence of "anti-RSV factors," which were presumably neutralizing antibodies, in the blood and regression of the sarcoma (Bower and Gyles, 1982). Infection of chicks with recombinant fowlpox viruses expressing the reticuloendotheliosis virus (REV) envelope glycoprotein resulted in more rapid generation of REV-specific neutralizing antibodies and protection from subsequent disease after challenge (Calvert *et al.*, 1993).

In the murine system, gp70 of Friend murine leukemia virus (F-MuLV) also elicits antibodies able to neutralize the virus itself (Fischinger *et al.*, 1976; Hunsmann *et al.*, 1974) and again protects from virus-induced disease (Hunsmann *et al.*, 1975a,b). Indeed, passive transfer of gp70-specific serum or purified IgG was able to prevent erythroleukemia after challenge with F-MuLV (Hunsmann *et al.*, 1975a,b). Similarly, AKR mice were shown to be protected from retrovirus-induced spontaneous leukemia by regular injection of goat antiviral immune globulin (Fish *et al.*, 1979; Huebner *et al.*, 1976b) or by type-specific immunization regimes (Huebner *et al.*, 1976a).

TABLE III. Antiretroviral Immune Mechanisms

Mechanism	Retrovirus	Target protein[c]
Elimination of free virus		
Virus neutralization	HIV	gp120, gp41, p17
	SIV	gp130, gp36
	Visna virus	
	EIAV	gp90
	CAEV[a]	
	FeLV	gp70
	BLV	gp51
	F-MuLV	gp70
	RSV	gp85, gp37
Virolysis via complement	M-MuLV	p15E
	Feline	
	Avian	
	Simian	
	HIV-1[a]	
Elimination of virus-infected cells		
Antibody + complement	F-MuLV	gp70
	REV-A	
	FeLV	gp70
	SIV	
Antibody-dependent cellular cytotoxicity	MuLV	
	FeLV, FeSV[b]	
	EIAV	
	HTLV-I	
	HIV-1, HIV-2, SIVmac, SIVagm	gp120, gp41
Natural killer cells (transformed target cells)	FeLV	
Cytotoxic T lymphocytes	MuLV	Env, Gag
	M-MuLV	gp70, p15E
	F-MuLV	Env, Gag
	REV	
	FeSV	
	HIV, SIV	Gag, Pol, Env, Nef, RT, Vif

[a] Virolysis only in the presence of specific antibody.
[b] Conflicting evidence of FeLV and FeSV specific ADCC (see text).
[c] Where known.

Cats immunized with an FeLV recombinant $gp70^{env}$ immunogen also develop a strong neutralizing antibody response and are protected from viremia upon challenge (Marciani *et al.*, 1991). However, passive transfer of a neutralizing monoclonal antibody did not afford protection against FeLV infection of kittens (Weijer *et al.*, 1989). Flower *et al.* (1985) used neutralization assays to serotype feline syncytial virus (a spumavirus) and to show the dissemination of one subtype in the cat population over a period of 5 years.

Visna virus envelope glycoproteins are known to stimulate in sheep neutralizing antibodies against a number of different epitopes (Crane *et al.*, 1988). However, the virus persists in cells of the macrophage lineage despite the presence of these antibodies. Studies comparing the mechanisms of neutralization in different cells suggested that whereas antibodies prevent binding of visna to fibro-

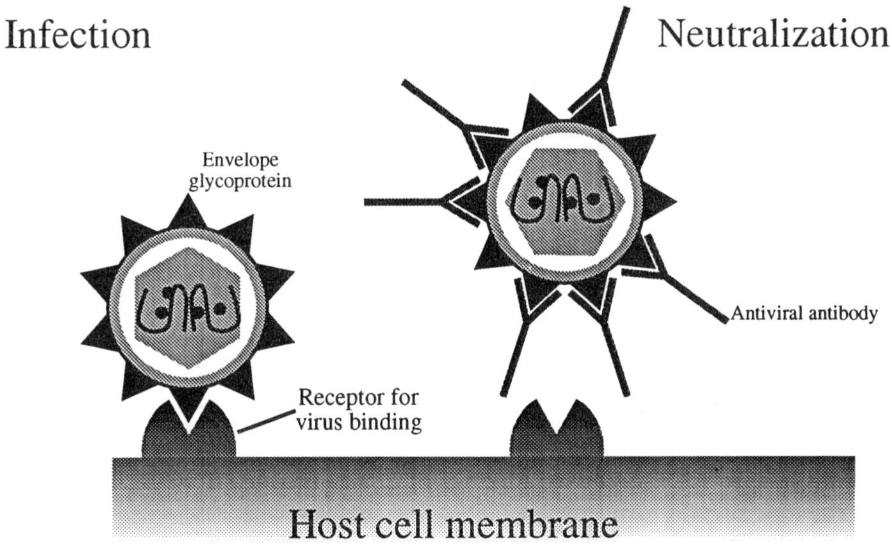

FIGURE 10. Blocking of infection by neutralizing antibody. Antibodies binding to the viral surface prevent subsequent recognition of the cellular receptor for viral binding and entry. Antibodies might also be effective in preventing entry of virus in a postbinding event.

blasts, they can also facilitate rapid phagocytosis and uncoating of the virus leading to infection (Crane *et al.*, 1988; Jolly *et al.*, 1989b; Kennedy-Stoskopf and Narayan, 1986). In some cases an enhancement of infection takes place (see below). However, the fact that the kinetics of virus binding to the macrophages was much faster than that of antibody to virus suggests that the virus might persist by macrophage-to-macrophage spread before neutralization can occur.

Antibodies specific for the major envelope glycoprotein of bovine leukemia virus appeared to be responsible for preventing the release of viral particles in cultures of peripheral blood lymphocytes from BLV-infected cattle (Driscoll *et al.*, 1977). BLV neutralizing antibodies were further characterized by Portetelle *et al.* (1989b) who raised antibodies to synthetic peptides corresponding to the BLV gp51.

b. Neutralization of Equine Infectious Anemia Virus

The neutralizing antibody response in horses to EIAV has been studied in considerable detail because this process appears to routinely bring viral replication and disease under control, a somewhat unique situation for lentiviruses. Persistent infection of horses with EIAV is characterized in the first year or so by periodic episodes of viremia associated with fever, hemolytic anemia, weight loss, and other symptoms (Cheevers and McGuire, 1985). These episodes coincide with the emergence of antigenically distinct virus variants followed by the appearance of neutralizing antibodies specific for the new variant. The bouts of disease decrease in severity and frequency, however, and the animal, if it survives, enters an asymptomatic carrier state. The fact that such animals can be induced to relapse by treatments which suppress the immune system indicates that the inapparent state is the result of efficient immune control, presumably through neutralizing antibody (Kono *et al.*, 1973).

Further proof of the essential role played by the immune system in controlling EIAV comes from experiments in which normal horses and horses with combined immunodeficiency were infected by a biologic clone of EIAV (Perryman *et al.*, 1988). In the normal horses the viremia which appeared by day 7 postinfection was cleared 2 to 4 weeks later whereas the immunodeficient horses remained viremic (and anemic) until the time of death some weeks later. Stimulation of an antiviral immune response able to control infection is therefore essential in this infection. As with other retroviruses, the outer envelope gp90 and gp45 of EIAV have been shown to be the dominant targets for neutralization (Montelaro *et al.*, 1984; Rwambo *et al.*, 1990b).

c. *Neutralization of Primate and Human Retroviruses*

Neutralizing antibodies are also considered to be an important response to primate (and human) retroviral infections. For example, type D simian retroviruses (SRV-D) which induce an AIDS-like disease in monkeys can be grouped into five distinct neutralization serotypes with differing pathogenic potential. The long-term survival upon SRV-D infection appears to be correlated with the production of high-titer neutralizing antibodies (Kwang *et al.*, 1987). Immunization with a formalin-inactivated whole SRV-D vaccine or with a vaccinia-*env* hybrid virus induced substantial levels of neutralizing antibody (Marx *et al.*, 1986; Hu *et al.*, 1989) and completely protected macaques against infection with an otherwise lethal live virus challenge.

Like many retroviruses, infection of cell cultures with human T-cell leukemia virus I (HTLV-I) induces formation of syncytia, giant cells resulting from cell fusion. This phenomenon can be used in assays to detect antibodies reacting with biologically relevant areas of the envelope glycoprotein. For example, antibodies from patients with adult T-cell lymphoma-leukemia were shown to inhibit syncytium formation in a virus-specific manner (Nagy *et al.*, 1983). The same sera failed to inhibit syncytium formation by bovine leukosis virus, Moloney murine leukemia virus, or simian sarcoma-associated virus.

The intensive study of the mechanisms of HIV-1 infection has revealed that the recognition, binding, and entry stages of retroviral infection constitute a complex process. Being instrumental in binding and virus entry, the envelope glycoproteins gp120 and gp41 of HIV-1 are, as expected, the major targets for neutralizing antibodies. There is indeed an initial high-affinity interaction between the viral gp120 and the cellular CD4 molecule (Lasky *et al.*, 1987; McDougal *et al.*, 1986). However, this interaction appears to trigger a series of conformational changes, including exposure of the envelope glycoprotein V3 loop to proteolytic cleavage (Werner and Levy, 1993). These changes culminate in the attachment of the gp41 amino-terminus to the host cell membrane (Gallaher, 1987) and fusion of the virus and cell membranes (McClure *et al.*, 1988; Stein *et al.*, 1987). Therefore, antibodies which bind and interfere with any of these stages can potentially neutralize virus infectivity.

Although the so-called V3 principal neutralizing determinant (PND) of HIV-1 gp120 has been the subject of intensive study (Rusche *et al.*, 1988), neutralizing antibodies directed against the linear epitopes of this domain are notoriously type-specific and the mechanism by which neutralization occurs is still unclear. A broader specificity is seen with antibodies directed against the discontinuous, conformational regions of gp120 responsible for binding to CD4 (Haigwood *et al.*, 1992; Ho *et al.*, 1991; J. E. Robinson *et al.*, 1990). When both antibodies are present, they can act in synergy to neutralize the virus as might be expected (Tilley *et al.*, 1991).

Anti-core protein antibodies, although often predominant, should theoretically be unable to play a role in neutralization because the protein is not directly expressed on the surface of the virus particle. It is noteworthy, however, that antibodies directed at the p17 of HIV-1 were reported to neutralize infectivity (Boucher *et al.*, 1990). This finding suggests that p17 is at least partially exposed on the surface of the virus. The relative importance of anti-p17 neutralizing antibodies remains to be determined, however.

d. Role of Glycosylation

Serum neutralizing antibodies are detectable during most retroviral infections. The response during infection by the lentivirus group is, however, surprisingly poor. This observation is particularly true for CAEV (Jolly *et al.*, 1989a). It seems likely that the intensity or pattern of glycosylation in the lentivirus envelope glycoprotein is responsible (Steinman *et al.*, 1990). Carbohydrate residues can mask partially or completely the neutralizing targets on the intact protein. However, carbohydrates could in some cases form part of or even entirely constitute the sensitive epitope. HIV gp120 is, for example, one of the most heavily glycosylated proteins known in nature with over 30 predicted glycosylation sites.

Even retroviruses such as avian leukosis sarcoma virus (ALSV) possess envelope glycoproteins whose mass is over 50% carbohydrate (Steinman *et al.*, 1990). Masking of neutralization by carbohydrate residues has been shown for other retroviruses such as Rauscher murine leukemia virus (R-MuLV; Alexander and Elder, 1984). In contrast, the reactivity of healthy human sera with simian sarcoma virus–simian sarcoma-associated virus (SSV-SSAV) complex in radioimmunoprecipitation assays was shown to be likely the result of cross-reactivity with the carbohydrate moiety of the viral gp70 molecules (Snyder and Felissner, 1980).

2. Virolysis

As is discussed below, activation of the complement cascade at the infected cell surface can result in lysis of that cell. In addition, it is possible for the complement cascade to be activated against enveloped virus particles themselves (Fig. 11). Many members of the retrovirus family can be lysed directly by complement in the absence of specific antibody. The transmembrane glycoprotein p15(E) of some murine retroviruses binds the complement component Clq which then activates the classical complement cascade resulting in viral lysis (Bartholomew *et al.*, 1978). However, virolysis in the absence of specific antibodies was difficult to demonstrate against HIV, simian immunodeficiency virus (SIV), or even HTLV-I (Banapour *et al.*, 1986; Hoshino *et al.*, 1984).

Specific antibodies binding to a virus, although possibly not neutralizing, may activate the complement cascade resulting in virolysis. Antibodies raised to CAEV after immunization of rabbits were shown to prime virions for neutralization upon addition of complement (Klevjer-Anderson and McGuire, 1982). No

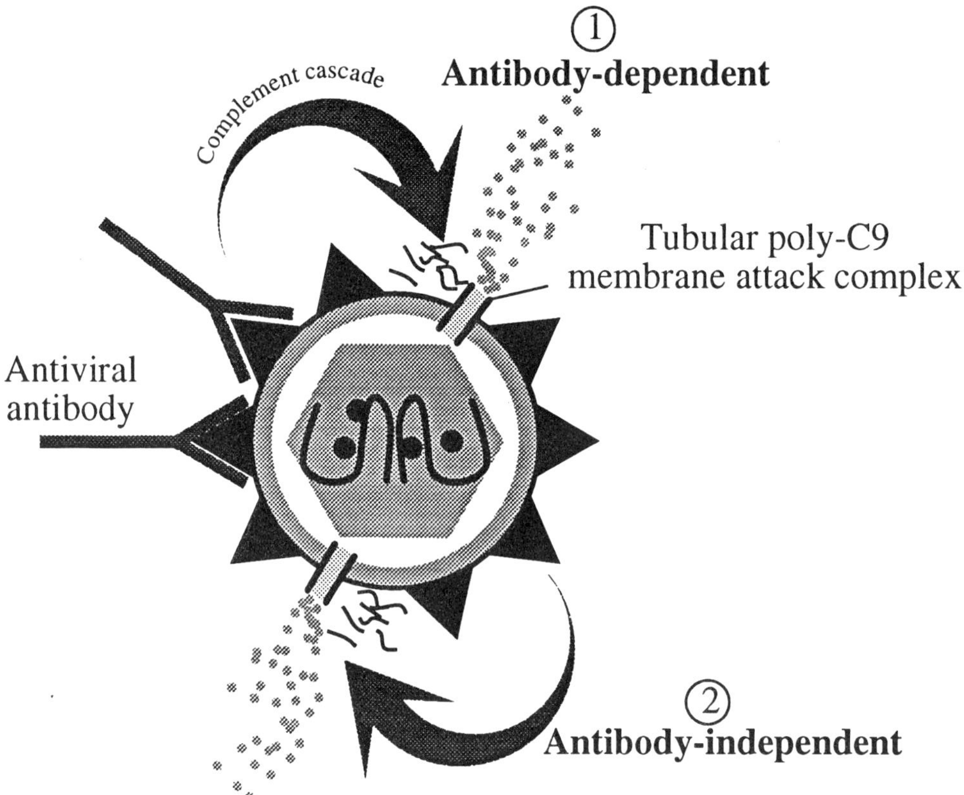

FIGURE 11. Elimination of free virus particles by virolysis. (1) Antibodies binding to the virus may activate the complement cascade resulting in disruption of the virus by the tubular membrane attack complex. (2) In some cases the viral proteins may activate complement directly.

such activity could be detected in the sera of goats infected or hyperimmunized with the virus. Presumably, the transmembrane glycoproteins of other lysable retroviruses stimulate the same effect. Recently, virolysis of HIV-1 has been demonstrated (Spear *et al.*, 1990b, 1991). Indeed, using a panel of human monoclonal antibodies, it was possible to identify the V3 loop of HIV-1 as the major target for antibodies able to initiate virolysis (Spear *et al.*, 1993b).

As will be discussed below, binding of complement by a retrovirus or retrovirus/antibody complex, rather than acting to eliminate the virus, may in fact result in an increase of infectivity by the mechanism of complement-mediated enhancement.

Retroviruses appear to take with them in their envelopes human cell proteins from the host cell membrane (Arthur *et al.*, 1992; Azocar *et al.*, 1983; Gelderblom *et al.*, 1987). This phenomenon has caused the possible misinterpretation of many successful SIVmac vaccine experiments although the mechanism of protection via host cell proteins is unknown (see Section XI). It may therefore be relevant that in the 1970s, Azocar and Essex (1979) demonstrated virolysis of FeLV by anti-HLA or anticell antibodies in the presence of complement. Indeed, recent results from Spear *et al.* (1993a) show that in vaccinated macaques, antibodies specific for the human cell components incorporated in HIV and SIV can induce complement-mediated virolysis of the virions.

3. Complement-Mediated Lysis of Infected Cells

a. Mechanism

As mentioned in Section I, prior to retroviral budding from the infected cell, the envelope glycoproteins are expressed on the cell surface. An infected cell can therefore be recognized as such by antibodies. Using FACS analysis or simple immunofluorescence, it is possible to identify infected cells using antibodies specific for the retroviral envelope glycoprotein. In addition, binding of antibodies to the envelope glycoproteins can activate two effector mechanisms capable of lysing the infected cell. The first mechanism, complement-mediated lysis (Fig. 12), involves activation of the complement cascade, a series of serum proteins that interact in a "chain-reaction" fashion. The process ultimately forms large numbers of tubular protein complexes inserted in the cell membrane. Unable to maintain intracellular hypotonicity, the cell swells and dies (for review see Sissons and Oldstone, 1980).

b. Complement-Mediated Lysis of Retrovirus-Infected Cells

A number of studies have suggested that the activation of complement by antibodies might be an important defense mechanism against retroviral infec-

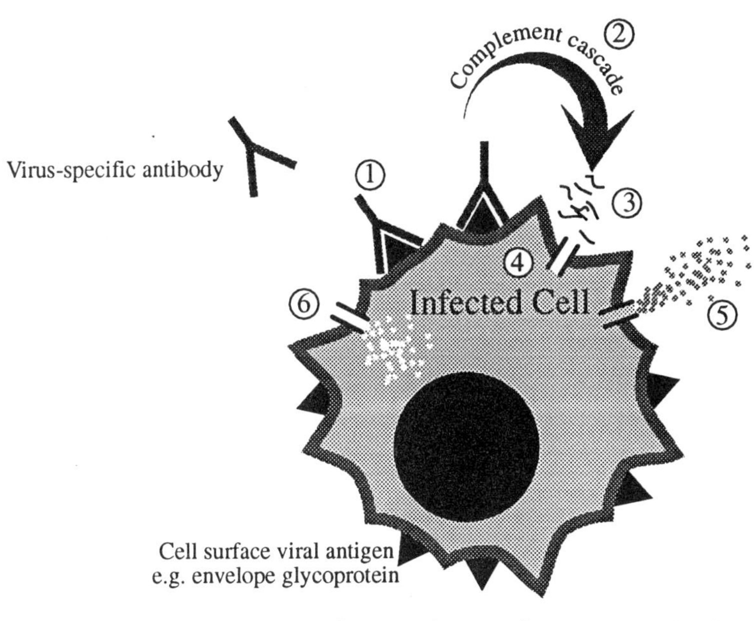

FIGURE 12. Complement-mediated lysis of virus-infected cells. Antibodies binding to envelope glycoproteins expressed on the surface of infected cells (1) activate the complement cascade (2) resulting in the polymerization (3) and insertion (4) of tubular poly-C9 membrane attack complexes. The cell, unable to maintain osmotic control, is subsequently lysed by loss of cellular proteins (5) and influx of surrounding fluid (6).

tion. In the case of avian reticuloendotheliosis virus for example, gp90-specific antibodies raised in response to peptides were shown to induce lysis of REV-A-infected cells (Tsai and Oroszlan, 1988). *In vivo* experiments involving infection of chickens with a recombinant retroviral vector expressing the subgroup A envelope glycoprotein resulted in smaller tumor formation after challenge with avian sarcoma virus-A. This protection appeared to correlate with the appearance of both neutralizing and complement activating antibodies (Wright and Bennett, 1992).

Many other retroviruses stimulate the production of antibodies able to mediate lysis of virus-infected cells in the presence of complement. Purified gp70 of F-MuLV induced both neutralizing and cytotoxic antibodies and when used as an immunogen protected mice from lethal F-MuLV challenge. Passive transfer of immune IgG also had a protective effect (Hunsmann *et al.*, 1975a). Passive transfer of monoclonal IgM antibodies specific for Moloney murine sarcoma virus/Moloney murine leukemia virus (M-MuSV/M-MuLV) was also shown to substantially inhibit tumor growth in mice infected with M-MuSV (Lamon *et al.*, 1987). Although maximum inhibition required an intact T-cell system, the fact that the monoclonal exhibited no virus neutralizing activity but induced complement and cell-mediated cytotoxicity suggested a beneficial role for these mechanisms in tumor inhibition. In addition, antibodies specific for tumor cells induced by the radiation leukemia virus were shown to induce lysis of target leukemic cells (Haran-Ghera and Rubio, 1977).

The role of complement-activating antibodies has been especially studied in some detail in the FeLV system because of their apparent involvement in protection and recovery from FeLV-induced leukemia in cats. Grant and co-workers (Grant *et al.*, 1980c) showed that a humoral immune response to FeLV-induced antigens (FOCMA) protected against development of leukemia. They concluded that complement-activating antibodies were able to eliminate the virus-positive lymphomas. Similarly, immunization with feline leukemia cells induced cytotoxic antibodies specific for FOCMA and prevented lethal disease upon challenge with feline sarcoma virus (Grant *et al.*, 1980a). However, despite the common target protein, cross-reactivity between strains of FeLV was not seen in the complement-dependent antibody populations (Grant and Michalek, 1981).

If cats suffering from FeLV-induced lymphoma are subjected to immunoglobulin depletion by injection of staphylococcus protein A or by *ex vivo* immunoadsorption, the immune system responds initially by replenishing the body with specific antibodies able to initiate complement-mediated lysis of lymphoma cells. The activity of these antibodies is thought to correlate with remission from disease and disappearance of viremia (Liu *et al.*, 1984a,b). Snyder *et al.* (1984) showed that, following extracorporeal immunoadsorption, the one serologic parameter correlating with clearance of FeLV was the replenishment of free antibodies specific for FeLV gp70. Removal of circulating immune complex "blocking factors" might possibly stimulate existing antibody responses (Snyder *et al.*, 1989). However, others have suggested that it is the level of FeLV-specific neutralizing antibody, directed against gp70, which correlates with natural recovery from disease (de Noronha *et al.*, 1978). In terms of protection from initial infection, there have been successful FeLV vaccine preparations that did not seem to stimulate neutralizing antibodies (Essex, 1988), and others that did (Marciani *et al.*, 1991; Osterhaus *et al.*, 1985).

One study compared the appearance of antibodies as measured by immunoblot, neutralization assay, and complement fixation in sheep sera throughout the course of natural and experimental infection with visna virus. Complement-activating antibodies appeared relatively early, coincident with first appearance of antibodies specific for the major envelope glycoprotein (Torfason *et al.*, 1992).

Sera from cattle infected with BLV contain antibodies able to initiate lysis of BLV-producing cells but not noncultured malignant lymphocytes in the presence of complement (Portetelle *et al.*, 1978). It was concluded that these antibodies, although ineffective at controlling the growth of tumors *per se,* were able to reduce viremia and hence limit the spread of the virus. Honma *et al.* (1980) also found antibodies cytotoxic for BLV-infected cells in infected cattle and sheep and showed them to be specific for the viral envelope glycoprotein.

Complement-activating antibodies have in fact been implicated as a crucial protective mechanism against retroviral diseases in mice, cats, chimpanzees, and man (Fischinger *et al.*, 1976; Grant *et al.*, 1980b; Nara *et al.*, 1987a; Tam *et al.*, 1980). Given the apparent importance of complement-activating antibodies in combating several retroviral infections, it is disappointing that it is difficult to demonstrate complement-mediated antiviral activity in the abundant envelope-glycoprotein-specific antibodies of sera from HIV-infected humans or SIV-infected monkeys. Depending on the nature of the assay used, infected cells have been shown to be resistant (Hartung *et al.*, 1992; Nara *et al.*, 1987b; Norley *et al.*, 1990a; Posner *et al.*, 1992; Solder *et al.*, 1989; Spear *et al.*, 1990a; Yefenof *et al.*, 1991) or susceptible (Fultz *et al.*, 1990; Gregersen *et al.*, 1990; Toth *et al.*, 1989) to lysis by complement in the presence of antibodies. The *in vivo* relevance of the mechanism therefore remains unknown.

4. Antibody-Dependent Cellular Cytotoxicity

a. Mechanism

Like complement, the antibody-dependent cellular cytotoxicity (ADCC) reaction depends on binding of antibodies to the envelope glycoproteins expressed on the surface of infected cells. In this case, however, the lethal hit is delivered not by a series of serum proteins but by a "killer" cell (Fig. 13). These cells, usually large granular lymphocytes, bear receptors for the Fc portion of antibody. When the antibody/antigen complex at the cell surface is recognized, the effector cell secretes proteins (perforins) which polymerize to form polyperforin channels in the target cell membrane, similar to complement (for reviews see Yagita *et al.*, 1992).

ADCC is considered to be important for the control of a number of (non-retroviral) infections (Shore *et al.*, 1974), particularly herpes simplex virus (HSV). Because of the high sensitivity of the mechanism, ADCC is probably the first antibody-directed immune mechanism to become functional after infection and represents the first line of defense against dissemination of virus.

b. Antibody-Dependent Cellular Cytolysis of Retrovirus-Infected Cells

Antibodies capable of initiating ADCC have been demonstrated in a number of retroviral systems, although the *in vivo* relevance of the mechanism remains

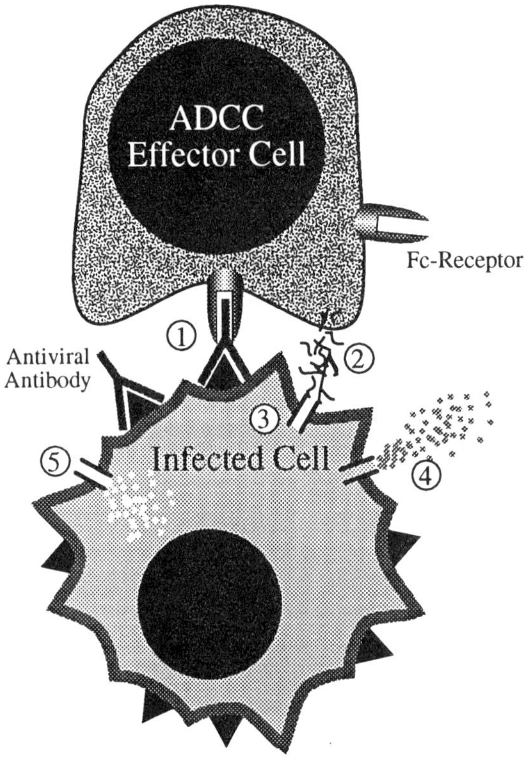

FIGURE 13. Antibody-dependent cellular cytolysis (ADCC) of virus-infected cells. The Fc portion of antibodies binding to envelope glycoproteins expressed on the surface of infected cells is recognized by killer cells bearing Fc receptors (1). This recognition stimulates the secretion and deposition of perforin molecules (2) which, similar to the complement C9 component, polymerize to form a lytic, tubular polyperforin attack complex (3). The infected cell is then lysed by osmotic shock involving loss of cellular proteins (4) and influx of surrounding fluid (5).

unclear. For example, there is evidence that ADCC could be important for resistance to MuLV infection (Peters and Theofilopoulos, 1977). In the FeLV system, conflicting evidence for ADCC activity exists, with groups able to show lysis of FeLV- or FeSV-infected cat cells (McCarty and Grant, 1983b) and others able to demonstrate ADCC only against erythrocytes pulsed with FeLV and not against FeLV-infected cells (Kooistra and Splitter, 1985). Even sera containing high levels of antibodies able to initiate complement attack of FeLV-infected cells were unable to activate ADCC effectors (McCarty and Grant, 1983b). ADCC is therefore unlikely to play an important role in recovery from this infection.

ADCC activity mediated by antibodies in horse serum using EIAV-infected cells was demonstrated during the acute phase of infection, although this response disappeared as the animals entered the chronic state (Gerencer *et al.*, 1989).

HTLV-I- or -II-infected cells are readily lysed by specific antibody in the presence of human PBMC effector cells with considerable cross-reactivity between the two (Miyakoshi *et al.*, 1984; Zhang *et al.*, 1992). The level of activity in the HTLV-I system appears to be generally higher than that seen for HIV-1 (Sinclair *et al.*, 1988). As expected, the use of peptide-specific antibodies indicated that the critical epitopes are located primarily in hydrophilic regions of the exterior part of the envelope glycoprotein.

ADCC has in fact been described and in some cases finely characterized for HIV-1, HIV-2, SIVmac, and SIVagm (Ljunggren *et al.*, 1987a,b, 1989; Norley *et al.*, 1990a,b; Tyler *et al.*, 1989). During these infections the process appears to be generally more broadly reactive than neutralization. Antibodies specific for HIV-

2 readily initiate lysis of SIVmac-infected cells, for example (Ljunggren *et al.*, 1989a) and seem to have a broader specificity than HIV-1 specific sera (Von Gegerfelt *et al.*, 1993). However, a small percentage of HIV-1-specific sera can act equally well against HIV-1, HIV-2, and SIVmac (Norley *et al.*, 1990b).

In adults, the effector cell primarily responsible for ADCC has been shown to be the $CD16^+$ (FcRIII) NK-like lymphocyte (Murayama *et al.*, 1990; Norley *et al.*, 1990b). Normally, macrophages from seronegative donors are unable to mediate HIV-specific ADCC, despite their abundance of Fc receptors. However, studies have shown that monocytes from HIV-seropositive (but not seronegative) patients can act as ADCC effectors using autologous or heterologous HIV-positive serum (Jewett and Bonavida, 1990; Jewett *et al.*, 1990).

c. *In Vivo Relevance of ADCC*

The *in vivo* relevance of ADCC has been difficult to prove. An antibody population which mediates ADCC will usually also neutralize virus or stimulate complement-mediated lysis. In the HIV system, many groups have failed to find a correlation between the levels of antibody mediating ADCC and the clinical stage of the subject (Brenner *et al.*, 1991; Dalgleish *et al.*, 1990; Norley *et al.*, 1990b; Tyler *et al.*, 1990a). Some smaller studies did, however, find reduced levels of ADCC antibody in asymptomatic individuals compared with AIDS or ARC patients (Goudsmit *et al.*, 1988b). ADCC requires both antibody and an effector cell to function. It is therefore noteworthy that later stages of AIDS appear in some studies to be associated with a substantial drop in the number or activity of these effector cells (Bender *et al.*, 1986; Brenner *et al.*, 1991; Ljunggren *et al.*, 1989b). This was true even in seropositive patients prior to the onset of disease (Ljunggren *et al.*, 1989b; Tyler *et al.*, 1990a). Others, however, observed no such impairment in AIDS patients (Katz *et al.*, 1987, 1988; Ojo-Amaize *et al.*, 1989; Schroff *et al.*, 1983). In this regard, polymorphonuclear leukocytes (PMNL) are able to mediate ADCC and the activity of these cells has been shown to be reduced in AIDS/ARC patients compared with healthy seropositives (Kinne and Gupta, 1989).

In addition, HIV-positive children appear to have a defect in both PMNL effector and antibody functions compared with adults (Szelc *et al.*, 1992). By comparing sera from infants infected perinatally, Ljunggren *et al.* (1990) found a correlation between the level of ADCC activity and lack of disease, although disease status also correlated with neutralizing antibody levels. In this study, HIV-specific ADCC and neutralizing antibodies did not seem to protect against transmission of HIV from mother to child.

In a multicenter AIDS cohort study employing serial samples taken from patients before and after seroconversion, HIV-1 Env-specific ADCC-mediating antibodies were detected much earlier than neutralizing antibodies and these early ADCC responses were associated with a high mean $CD4^+$ count and absence of lymphadenopathy during the 2-year observation period (Sawyer *et al.*, 1990). Although such data might suggest a beneficial role for ADCC soon after infection, they could reflect differences in the assays used. ADCC has a much broader specificity (Ljunggren *et al.*, 1989a; Norley *et al.*, 1990b) than the first appearing neutralizing antibodies and the sensitivity is usually much higher. The later appearance of neutralizing antibodies might therefore reflect the broadening of the response to include the laboratory HIV strains used in the assay. In support of this hypothesis, Koup *et al.* (1991a) showed broad ADCC activity in early sera with a strain-specific neutralizing antibody response. Only later did the sera acquire the ability to neutralize a broad spectrum of HIV-1 isolates.

V. CELL-MEDIATED IMMUNE RESPONSE TO RETROVIRUSES

A. Diagnostic Assays Using Lymphocytes

Lymphocytes primed by infection or vaccination of the host will be stimulated *in vitro* by the presence of specific antigen. This stimulation results in lymphoproliferation which can be quantified by measuring the incorporation of radiolabeled thymidine into the cellular DNA. Although not commonly used for diagnosis because of its need for cells rather than serum, the lymphoproliferation assay is often used during vaccine trials as a crude marker for stimulation of the cellular immune response. In some cases, however, it appears that this assay can detect subclinical infections in situations where even the highly sensitive polymerase chain reaction fails. For example, Clerici *et al.* (1991a) recently demonstrated a positive lymphoproliferative response to HIV-1 envelope glycoprotein synthetic peptides in the seronegative partners of some HIV-1-infected patients. This observation could reflect the induction of cell-mediated responses in the absence of the later onset of antibody production.

B. Functional Cell-Mediated Immune Responses

Antibodies are not the only weapon available to the immune system in its efforts to recognize and control a retroviral infection. Macrophages and natural killer cells may also be involved in removing virus or virus-infected cells. The T lymphocytes, one population of which (T-helper) is instrumental during the stimulation of the immune response, also have highly evolved mechanisms (cytotoxic T lymphocytes) for recognizing and rapidly eliminating virus-infected cells.

1. T-Lymphocyte Responses

a. $CD8^+$ Cytotoxic T Cells

i. Mechanism. The CTL response is considered the most important antiviral defense mechanism involved in clearing a viral infection (Rouse *et al.*, 1988). It is therefore perhaps worthwhile to describe this system in some detail.

The other effector mechanisms described above depend on recognition of viral proteins (i.e., usually the envelope glycoproteins) expressed on the virus or cell surface as a result of virus-driven processes. Antibodies specific for regulatory or internal proteins probably have no antiviral effect. During the course of synthesis in the cell, however, some proteins are diverted away from the normal pathways and are processed into short peptide fragments (six to ten amino acids). These fragments then become associated with the nascent MHC class I molecules (Fig. 14). The peptide/MHC complex, when expressed on the cell surface, is recognized by the CTL, which stimulates a lytic attack (Fig. 15).

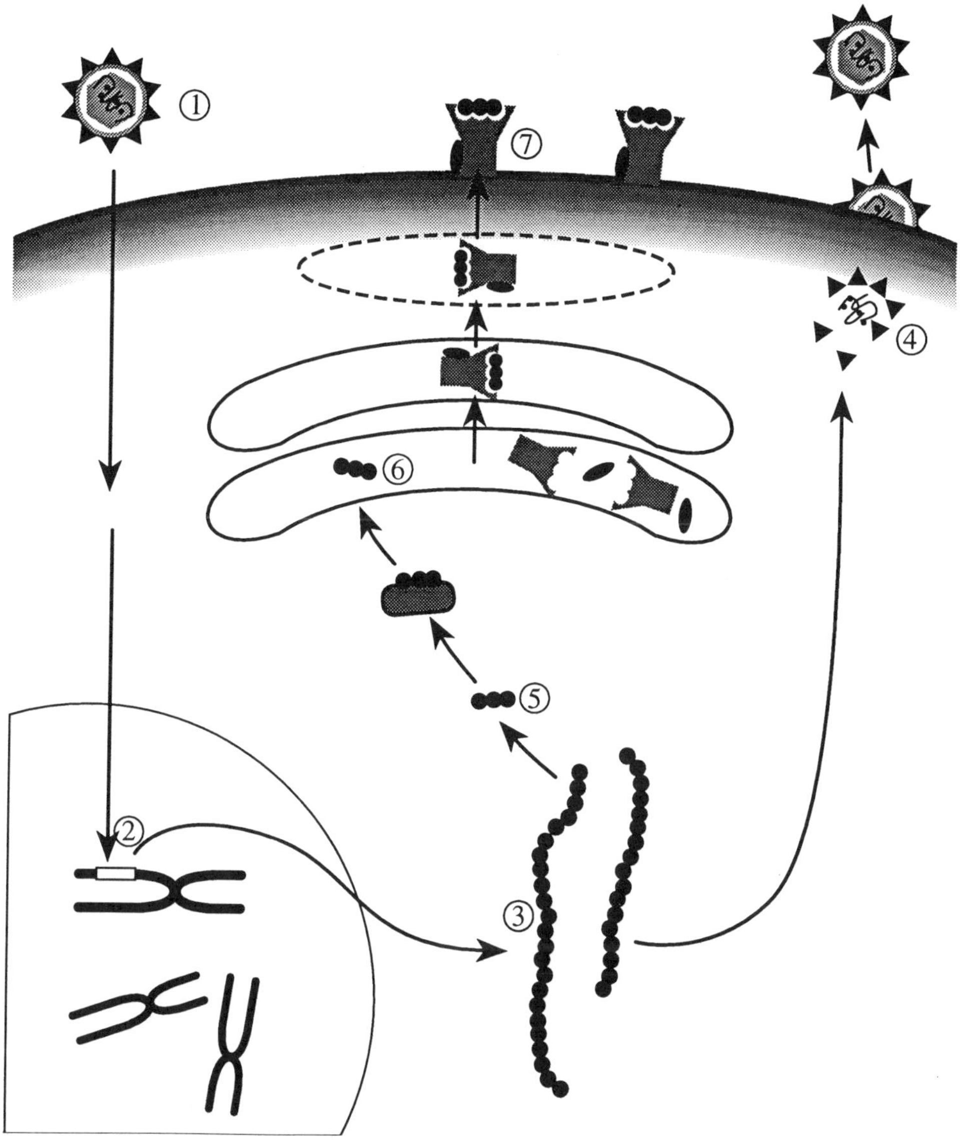

FIGURE 14. Processing and expression of CTL epitopes. Retrovirus entering the host cell (1) is uncoated and the proviral DNA is integrated into the cell genome (2). Transcription and translation of virus sequences lead to the synthesis of viral proteins (3) which form progeny virus (4). However, some of the new proteins are processed into small peptide fragments (5) and carried to the endoplasmic reticulum (6) where association with the MHC class I/β_2-microglobulin complex occurs. The complex is then transported to the surface of the infected cell (7). (Adapted from Nixon and McMichael, 1991.)

The mechanism of cell lysis is similar to that observed during activation of complement, with the secretion and deposition of perforin molecules which polymerize to form a tubular polyperforin attack complex spanning the cell membrane. In addition to the osmotic shock induced by these complexes, CTL attack can also induce degradation of the target cell DNA, presumably by activation of

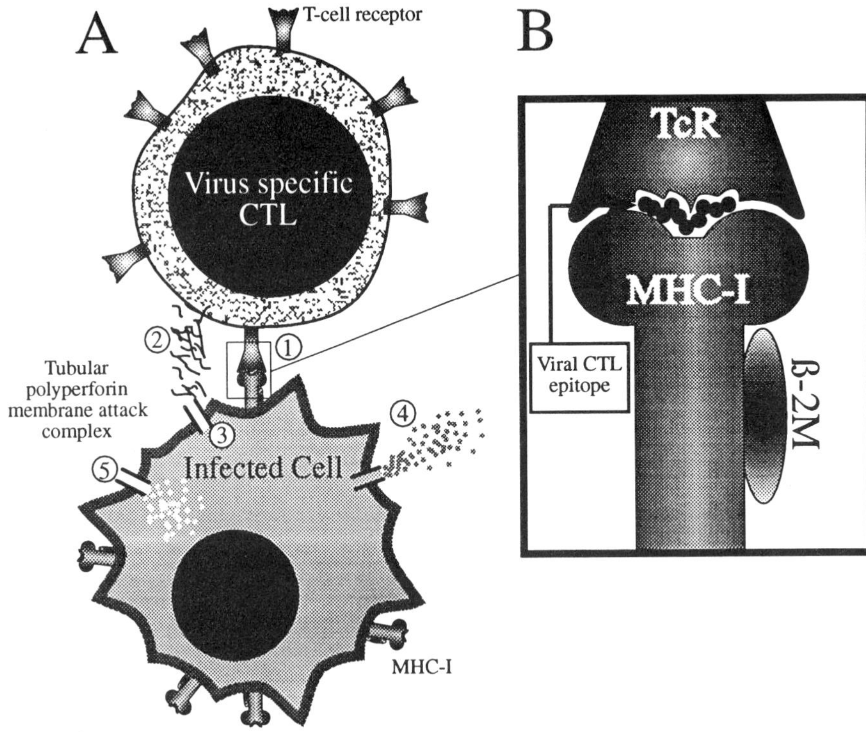

FIGURE 15. Cytotoxic T-lymphocyte (CTL) attack of infected cells. (A) Short viral protein segments are presented on the surface of infected cells by the MHC class I molecule and this epitope/MHC complex is recognized by specific CTLs (1). The CTLs are stimulated to initiate a lytic attack and secrete perforin molecules (2) which polymerize to form the polyperforin attack complex (3). The infected cell is then lysed by efflux of cellular proteins (4) and influx of surrounding fluid (5).

apoptosis (Vaux, 1993). In recent years, the mechanisms by which proteins are processed and presented at the cell surface in association with MHC molecules to be recognized by the T-cell receptor have been partially elucidated. This finding allows strategies for artificial CTL stimulation to be developed.

The MHC class I molecule consists of a membrane-spanning heavy chain and a light chain (β_2-microglobulin) joined noncovalently. The polymorphic $\alpha 1$ and $\alpha 2$ regions are arranged so that a 25-Å-long cleft is formed between the two and it is within this peptide-binding groove that the processed peptide lies (Bjorkman *et al.*, 1987a,b). Evidence that it is the association of a suitable peptide within the groove which stabilizes the complex and allows expression on the cell surface to occur is accumulating. The reasons why a certain protein region is processed to create the correct peptide for MHC association are still largely unknown. It appears that the MHC molecule itself is involved in the selection.

Analysis of peptides binding to a particular MHC class I molecule reveals a common motif. The amino acids at critical "anchor" positions are highly conserved, although high variability is seen at other positions (Falk *et al.*, 1991). Whether the peptides are derived from cellular proteins or from viral proteins is irrelevant for the processing, selection, and expression mechanisms. CTL epitopes can therefore be broadly specific among viruses, but are strongly restricted to the MHC type of the target cell (for review see Rammensee *et al.*, 1993).

As mentioned above, the fact that any suitable protein can be diverted into the endogenous pathway of processing and MHC-I association implies that any viral protein is potentially a CTL target. During replication, expression of the intact envelope glycoproteins, which renders the cell immunologically visible, may occur immediately before release of progeny virus. This event gives only a short "window of opportunity" for the immune system to act. However, as soon as the viral genome is transcribed, the endogenous processing pathway ensures that suitable target peptides are visible on the cell surface, even peptides derived from early regulatory proteins. The virus-specific CTL therefore has time to come into contact with, recognize, and destroy the infected cell before assembly of progeny virus occurs.

ii. Examples of Antiretroviral CTL. The difficulties imposed by the requirement for MHC class I matching between the target and CTL effector cells mean that the bulk of knowledge concerning antiretroviral CTL activity comes from models in which inbred hosts are available (e.g., mice) or where specific reagents are common (e.g., human).

Avian, Murine, and Feline CTL. Despite the difficulties associated with investigations of CTL in the avian system, MHC-restricted CTL specific for reticuloendotheliosis virus have been demonstrated in the spleens of infected chickens (Maccubbin and Schierman, 1986) and further characterized as Ia^+ cells (Weinstock *et al.*, 1989). However, it is noteworthy that Bose (1984) showed that the REV-A helper virus induces a splenic suppressor cell which inhibits the proliferation of CTL specific for REV-T tumor cells.

In the murine system, Plata *et al.* (1986) showed selective CTL recognition of MuLV gene products (Env and Gag) which, perhaps expectedly, varied according to the origin of the CTL. This group (Plata, 1985) also presented evidence that CTL are actively involved in rejection of Gross MuLV-induced tumors. In this study, chronic infection of mice by *Trypanosoma cruzi* was shown to suppress CTL differentiation at the level of CTL precursor activation, and this suppression facilitated tumor growth.

The importance of the immune response in controlling virus-induced lymphoma was also reported by Ronchese *et al.* (1984) using C57BL/6 Mov-13 mice. Mice of this strain carry the M-MuLV genome integrated into their germ line and express the virus early in lymphoid and nonlymphoid tissues. In accordance with the theories of clonal deletion during ontogeny, the viral proteins are recognized as "self" and lymphocyte precursors specific for the virus are eliminated, leaving a "hole" in the immunologic repertoire. Consequently, when such mice are challenged with M-MuSV, virtually no antibody or CTL response is seen and a high incidence of T-cell lymphoma ensues.

Using target cells transfected with $gp70^{env}$ and $p15(E)^{env}$ of M-MuLV, Env-specific CTL can be demonstrated in infected mice (Flyer *et al.*, 1983). Holt *et al.* (1986) used similar techniques to demonstrate effector cells able to lyse target cells expressing both the syngeneic HLA-I and *gag* or *env* genes of F-MuLV. Indeed, F-MuLV-specific CTL were suggested to be at least partially responsible for protection against leukemia following their induction by infection with an F-MuLV*env*/vaccinia construct prior to challenge (Earl *et al.*, 1986).

The development of CTL also correlates with protection of mice from neurologic disease induced by Cas-Br-M MuLV infection (Robbins and Hoffman, 1991). Indeed, adoptive transfer of virus-specific CTL was sufficient to prevent the progression of gliosis, vacuolation, and cell death in infected recipient mice (Hoffman *et al.*, 1991). Moreover, adoptive transfer of an M-MuSV-specific CTL clone was sufficient to protect T-cell-deficient mice against M-MuSV-induced tumors (Cerundolo *et al.*, 1987).

In an unusual approach to vaccination, Ruprecht *et al.* (1990) stimulated a protective immune response against Rauscher MuLV (R-MuLV) by infecting mice with the lethal virus and then using a pharmacological blockade of AZT and IFN-α to prevent viremia. Mice immunized by this abortive infection were subsequently resistant to challenge, and adoptive transfer of immune T cells indicated that the cellular arm of the immune system was responsible.

The ability of CTL clones to protect from disease can be augmented by IL-2. Matis *et al.* (1986) stimulated the expression of IL-2 receptors in cultures of a murine tumor-specific CTL prior to adoptive transfer into syngeneic mice suffering from F-MuLV-induced lymphoma. If the mice were then treated daily by systemic administration of IL-2, half of the mice underwent complete tumor regression and survived. Nonstimulated (resting) CTL plus IL-2 or stimulated CTL alone had little or no effect on tumor progression.

Identification of important effector mechanisms can be achieved using strategies which impair a particular aspect of the immune response. For example, a transient functional impairment of $CD4^+$ cells in mice by treatment with anti-CD4 monoclonal antibody immediately prior to infection with M-MuLV induced a state of tolerance. This phenomenon allowed, on challenge, the uninhibited spread of Moloney murine sarcoma virus with associated development of tumors (Biasi *et al.*, 1991). In contrast, untreated mice would undergo regression because of antitumor T-cell activity (Chieco-Bianchi *et al.*, 1988; Collavo *et al.*, 1976). Similarly, comparing RSV-induced tumor development in inbred chickens, McMahon *et al.* (1986) suggested that the susceptibility of one chicken line was related to immunologic tolerance induced by expression of a cross-reacting endogenous retroviral envelope glycoprotein.

As mentioned above, lysis of infected cells by CTL requires recognition of antigen expressed in association with the MHC class I complex. Interference with MHC expression would therefore inhibit recognition and lysis and some retroviruses might use this mechanism to avoid CTL attack. Reduced H-2K expression in cells transformed by Gross/AKR MuLV appears to render the cells resistant to CTL attack and allows uninhibited progression in the host (Goodenow *et al.*, 1985). Even tumors expressing H-2K were resistant to lysis (Schaefer and Schmidt, 1986), indicating that qualitative changes in the molecule or other factors help the cells resist immune control. However, M-MuLV-infected cells were shown to exhibit a tenfold increase in surface MHC expression, although this increase was not seen if cells were infected with the M-MuSV/M-MuLV complex (Flyer *et al.*, 1985).

In the feline system an apparently MHC-restricted cytolysis of feline fibrosarcoma cells after priming *in vivo* (McCarty and Grant, 1983a) or *in vitro* (Kooistra and Splitter, 1985) with the same cells could be demonstrated. It was suggested that whereas antibody and complement represent the major defense mechanism against FeLV-induced lymphoma, the CTL response protects against fibroblastic tumors induced by FeSV (McCarty and Grant, 1983b). Cats infected with canarypox recombinant viruses expressing the FeLV *env* and *gag* genes were shown to resist challenge with FeLV despite the absence of neutralizing antibodies (Tartaglia *et al.*, 1993).

Human CTL. CTL lines specific for autologous HTLV-I-infected cells have been generated from the blood of patients infected with adult T-cell leukemia virus (Kannagi *et al.*, 1983). In addition, one of the lines could lyse fresh adult T-cell leukemia cells that were not expressing the ATLV surface antigen as detected by serology. Similarly, Mitsuya *et al.* (1983) generated a CTL line able to kill autologous cultured tumor cells from a patient who had a history of unusually long survival with HTLV-associated T-cell lymphoma.

The need to develop an effective vaccine against HIV has prompted an extensive characterization of the human CTL response to HIV. One of the surprising aspects of HIV infection is the abnormally high level of CTL activity induced. Normally, restimulation of PBMC *in vitro* is necessary in order to expand the antigen-specific CTL, but demonstration of anti-HIV CTL activity in fresh PBMC from HIV-seropositive patients is relatively easy (Walker and Plata, 1990). This activity appears to be related to a high number of fully differentiated CTL effectors in the circulation (Hoffenbach *et al.*, 1989; Koenig *et al.*, 1988; Nixon *et al.*, 1988; Plata *et al.*, 1987; Riviere *et al.*, 1989; Sethi *et al.*, 1988; B. D. Walker *et al.*, 1987). This phenomenon could be a result of continuous viral replication and therefore continuous antigenic stimulation in the immunologically active tissues. There is indeed evidence that HIV-1 replication is episodic throughout the course of the infection (Coombs *et al.*, 1989).

In support of the fact that all viral proteins are potential CTL targets, CTLs specific for viral accessory proteins Nef and reverse transcriptase, as well as for Env and Gag have been demonstrated in infected humans (Claverie *et al.*, 1988; Clerici *et al.*, 1991b; Culmann *et al.*, 1989, 1991; Hammond *et al.*, 1991; Hosmalin *et al.*, 1990; Koenig *et al.*, 1990; Littaua *et al.*, 1991; Walker *et al.*, 1989).

Despite the abnormally high level of HIV-specific CTL during the asymptomatic phase of HIV infection, the virus eventually replicates out of control and the patient progresses to AIDS. A decline in the virus-specific CTL population preceding the onset of symptoms has been shown although it is not yet known whether this reduction is a cause or an effect of disease progression (Autran *et al.*, 1988; Joly *et al.*, 1989; Sharma and Gupta, 1985). The mechanism for this degradation of the CTL response is not fully understood. One hypothesis is that the general impairment of the immune system as a whole, involving the diminishment of macrophage and $CD4^+$ cell function, would leave the CTL precursor without the necessary help and cytokines needed for correct maturation. However, there is evidence that as the patient progresses toward disease a population of $CD8^+$ $CD57^+$ T-suppressor cells develops with the ability to nonspecifically inhibit the effector phase of anti-HIV CTL lysis (Joly *et al.*, 1989).

b. Antiviral Suppressor Factor Produced by $CD8^+$ T Cells

i. Suppression of HIV-1. The adoptive transfer experiments in mice which so precisely demonstrate an *in vivo* role for CTL in protection from disease cannot, of course, be performed in humans for HIV. However, increasing *in vitro* evidence exists that $CD8^+$ lymphocytes have the ability to abrogate the replication of HIV in PBMC. If $CD8^+$ cells are removed from a culture of PBMC from an

HIV-infected patient, the virus replicates to high levels with increased kinetics (Walker *et al.*, 1986). One assumption (Tsubota *et al.*, 1989) was that virus-specific $CD8^+$ CTL were removing the infected cells by classic cytolytic mechanisms. The somewhat circumstantial evidence for this conclusion was the fact that (1) contact between the $CD8^+$ cells and the infected cells was required, (2) the effector cell bore classic CTL, not suppressor, cell markers, and (3) the inhibition was MHC-restricted.

However, as early as 1986, Walker and co-workers demonstrated that infected cells were apparently not being eliminated from the culture, as removal of the $CD8^+$ cells after 3 weeks of co-culture allowed the subsequent growth of the virus. The same group subsequently showed that the inhibition of HIV replication also occurred when the infected $CD4^+$ cells and the $CD8^+$ cells were separated by a 0.45-μm filter (Walker and Levy, 1989). A soluble, as yet unidentified, factor produced by the $CD8^+$ cells therefore appears to be at least partly responsible for inhibiting the replication of HIV.

The fact that suppression of virus production, not elimination of infected cells, is involved in this process was later supported (Walker *et al.*, 1991) by the lack of correlation between CTL activity as measured by ^{51}Cr release assay and anti-HIV inhibitory activity. Moreover, these studies showed that the number of HIV-infected $CD4^+$ cells increased in the culture despite the absence of detectable free virus production. This activity was broadly specific as exogenously added HIV-1, HIV-2, or SIV was prevented from replicating by $CD8^+$ cells from HIV-1-infected patients. In contrast to these other reports, Kannagi *et al.* (1990) found that PBMC from seropositive, asymptomatic patients were resistant to exogenous HIV-1 infection in the presence of $CD8^+$ cells, but this resistance did not extend to HIV-2. Resolution of these differences requires further study.

The nature of the factor(s) responsible for this antiviral effect has yet to be resolved and there is still disagreement as to whether the factor or its production is virus-specific and MHC-restricted. Nevertheless, the therapeutic potential of a molecule able to dramatically inhibit viral production without toxic side effects is obvious.

ii. Suppression of Other Primate Retroviruses. This inhibitory effect of $CD8^+$ cells is also demonstrable in many of the animal models for AIDS. Depletion of cells bearing this marker from PBMC culture facilitates the isolation of SIVmac (Kannagi *et al.*, 1988), SIVsmm (Powell *et al.*, 1990), SIVsyk (Emau *et al.*, 1991), and SIVagm (Ennen *et al.*, 1994). In the latter case, the use of double-chamber culture plates with $CD4^+$ and $CD8^+$ cells separated by a porous membrane confirmed that SIVagm infection can be also suppressed by a soluble factor.

It is noteworthy that in contrast to seronegative human or rhesus PBMC, $CD8^+$ cells from seronegative chimpanzees have the ability *in vitro* to suppress HIV-1 replication in $CD4^+$ cells after addition of the virus to the culture (Castro *et al.*, 1991). This "inherent" anti-HIV activity could possibly be one reason for the chimpanzees' natural resistance to AIDS development. However, others (Husch *et al.*, 1993) only detected this activity in the PBMC of HIV-1 infected chimps. In what might be a related phenomenon, cell-free cytotoxic mediators secreted by H-2-restricted CTL clones specific for tumor cells induced by Rous sarcoma virus were shown to eliminate syngeneic but not allogeneic tumor cells (Haraguchi *et al.*, 1988).

c. $CD4^+$ Cytotoxic T Cells

In addition to the numerous reports of "classical" $CD8^+$ CTL induction during retroviral infection, several groups have demonstrated the generation of $CD4^+$ CTL (Curiel *et al.*, 1993). For example, chimpanzees immunized with a live HIV-1 gp160 vaccinia recombinant develop $CD4^+$ CTL (Zarling *et al.*, 1987). However, $CD4^+$ CTL have not been detected in PBMC taken directly from infected individuals and are only demonstrable after extensive *in vitro* culturing or cloning. It is therefore unclear whether such cells play a role in the host.

2. Macrophage/Monocytes

As described in Section III, monocyte/macrophages are intimately involved in the induction and regulation of the immune response to antigen, in relation to both their role in antigen presentation and their secretion of cytokines. Concerning a functional antiviral immune response, however, these cells have no mechanism for specific recognition of foreign proteins.

Macrophages, as the name suggests, are effective in engulfing and breaking down large particulate material, including bacteria, although viruses are too small to be recognized in this fashion. However, because of their possession of Fc receptors and complement receptors on their cell surface, macrophages do have the ability to recognize viral particles opsonized by virus-specific antibodies or coated in complement components. This recognition can then lead to phagocytosis and destruction of the viral particle, although as described in Section IX, it can also lead to enhancement of infection. Macrophages activated by a variety of stimuli (e.g., γ-interferon) may also inactivate nearby free virus particles by the release of reactive oxygen intermediates (Chase and Klebanoff, 1992; Ennen and Kurth, 1993).

3. Natural Killer Cells

Natural killer (NK) cells can also act as an antiviral effector mechanism, although their major role is thought to be immune surveillance against tumors. NK cells generally have a high level of cytolytic activity against transformed cells. The exact nature and lineage of the NK effector cell, large granular lymphocytes often bearing Fc receptors, is a matter of some debate. Although sharing some markers with both T cells and myeloid cells, NK cells appear to have a different origin. If NK cells and T-cell precursors have a common progenitor, divergence must occur very early. In most cases the effectors for NK lysis and ADCC belong to identical or largely overlapping populations (Abo and Balch, 1981; Bradley and Bonavida, 1982; Kay *et al.*, 1977; Timonen *et al.*, 1981) with only the method of target recognition distinguishing the two mechanisms.

The mechanism of lysis is known to be essentially the same as that used by ADCC and CTL effectors, i.e., deposition of perforins on the target cell surface which polymerize into lytic polyperforin transmembrane pores. How an NK cell recognizes its target is unfortunately less well understood, but it is not the inducible immunologically specific recognition typified by antibodies or T-cell receptors.

NK cells have been shown to be active against certain virus-infected cells, and are often thought to be an additional first-line defense against infection. They may help keep the level of infection in check while the more specific and aggressive immune responses develop.

The importance of NK activity against retrovirus-infected cells in limiting viral spread is unknown. However, NK cells probably play an essential role in eliminating cells transformed to a malignant state by retroviruses. Tumor cells which normally spread rapidly in mice after transplantation were rejected if infected *in vitro* by an attenuated strain of Friend leukemia virus (Thiel *et al.*, 1988). This rejection was shown to be the result of induction of increased NK activity at the sites of tumor inoculation. Conversely, infection of mice with F-MuLV itself was shown to result in a steady decline of splenocytic NK activity (Moody *et al.*, 1984), although the ability of effector cells to bind targets was unaltered. Others (Hertenstein *et al.*, 1984) found no difference between the susceptibility of NK-competent and -deficient mice to F-MuLV infection and no differences in the levels of NK activity in F-MuLV-resistant and -susceptible mice. However, the induction of interferon production by retrovirus-infected cells was thought by Silagi and Calvelli (1982) to stimulate the *in vivo* antitumor activity of murine NK cells.

There are reports of NK activity in feline PBMC against FeLV-infected lymphoma lines (McCarty and Grant, 1983b) which others (Kooistra and Splitter, 1985) failed to demonstrate. This response against lymphoma lines was easier to show in the spleen cells of cats, and was activated nonspecifically by infection (Tompkins and Tompkins, 1985).

As might be expected from the fact that the responsible cells are often the same, the decline in ADCC effector function in HIV-1-positive humans has been shown to be accompanied by a similar loss of NK activity (Ljunggren *et al.*, 1989b; Wisecarver *et al.*, 1983). Nevertheless, some studies have found a reduction in the level of NK activity without loss of ADCC capability (Katz *et al.*, 1987).

VI. POTENTIALLY PROTECTIVE MECHANISMS

We have just described many examples where infection with a retrovirus or immunization with retroviral components has been shown to stimulate one or more arms of the immune system. As can be seen, attempts to correlate levels of a particular antibody or the activity of a T-cell response with protection from infection or disease manifestation are a common practice. However, even when positive correlations are shown, care should be taken before concluding that a "cause-and-effect" relationship exists. High levels of neutralizing antibodies, for example, might reflect efficient stimulation of the immune response in general, and unless all parameters are measured (i.e., ADCC, complement-mediated lysis, CTL), it is inappropriate to assign a protective role to the one mechanism evaluated. During natural infection and following many vaccine regimes, two or more effector mechanisms may well operate in tandem to control the virus and its effects.

In the animal model systems, however, it is often possible to stimulate or confer one response with the complete exclusion of others. Protection from infection following passive transfer of antibodies, particularly when monospecific or monoclonal in nature, can strongly indicate that the humoral response alone is sufficient. If the antibodies are characterized in terms of their ability to neutralize, mediate ADCC, or stimulate complement, the effector mechanism can be narrowed down even further.

Adoptive transfer of virus-specific CTLs in the absence of antibodies can similarly be used to ascertain the role of the cellular arm of the immune response in protection. Because of the restrictions imposed by the requirement for MHC-matching, such experiments have been performed almost exclusively in inbred syngeneic laboratory animals (e.g., mice) and therefore with viruses that replicate in these species. However, the development of the severe combined immunodeficiency (SCID) mouse model has recently allowed, for example, HIV-specific adoptive transfer experiments to be performed. SCID mice, lacking a functional immune system of their own, will accept human lymphoid cells as a substitute and this "human immune system" can then be challenged by HIV infection.

Using such a system, adoptive transfer of HIV-1 Nef-specific CTL clones has been shown to protect mice against HIV challenge as has passive transfer of V3-specific monoclonal antibodies (Safrit *et al.*, 1992; van Kuyk *et al.*, 1992). In the former experiment, however, the lack of HLA restriction and the very high numbers of transferred cells make interpretation of the results difficult. Possibly, secretion by the $CD8^+$ cells of the as yet poorly characterized HIV-suppressive factor described in Section V is the key mechanism in this case.

VII. VIRUS EPITOPES RECOGNIZED BY THE IMMUNE SYSTEM

The advent and widespread use of molecular biology, T-cell cloning, and monoclonal antibodies has revolutionized our understanding of how the molecules involved in immune recognition interact with their targets. It is known that antibodies and T cells recognize only short stretches of a foreign protein, typically six to ten amino acids, although in the case of antibodies these need not be continuous. The development of a vaccine through recombinant DNA or synthetic peptide technology relies in part on identifying these epitopes and understanding the functional activity of the regions. Cloning and sequencing of a viral gene allows the amino acid sequence to be derived. By a variety of techniques the precise epitopes recognized by the immune system can then be predicted or determined experimentally (Table IV).

A. B-Cell Epitopes

A large molecule such as the retrovirus outer envelope glycoprotein has a complex three-dimensional structure, but only a limited number of regions on this molecule are accessible to antibody. Epitope mapping is obviously an area of intensive study for HIV-1 and indeed most information concerning retrovirus immunogenic epitopes has come from HIV studies.

1. Computer Analysis

A number of different approaches have been used to identify epitopes on a given molecule. Often, the first step employs one of the many computer algo-

TABLE IV. Tools for Identifying Retroviral Targets of Immune Recognition

Tool	Procedure	Uses
Monoclonal/monospecific antibodies	Prepare monoclonal or monospecific antibodies reacting with individual viral proteins or epitopes	In assays for antibody activity in neutralization, ADCC, CML, or enhancement tests
Purified recombinant or native proteins	Purify viral proteins from virus or virus-infected cells or produce pure viral proteins using recombinant DNA technology	As antigen in ELISA, immunoblot, etc. to detect protein-specific antibodies As antigen in lymphoproliferation assays to detect virus-specific cellular responses To compete with or adsorb protein-specific antibodies before assay
Recombinant virus vectors/transfections	Induce expression of single viral protein in a cell by transfection or by infection with a genetically engineered vector virus	To identify targets for ADCC or complement-mediated lysis In autologous/syngeneic cells to identify targets for CTL
Synthetic peptides	Knowing the nucleic acid sequence of a viral protein, chemically synthesize short fragments of the corresponding proteins	As antigen in ELISA to precisely define linear B-cell epitopes As antigen in lymphoproliferation assay to precisely define T-cell epitopes To pulse autologous/syngeneic cells for use as targets in CTL assay
Immunizations	Immunize animals/humans with purified proteins, synthetic peptides, or recombinant viral vectors	To identify proteins or epitopes able to stimulate a functional or protective immune response

rithms available to analyze the known amino acid sequence of the protein. B-cell epitopes are characteristically *hydrophilic* in nature, as they are expressed on the outside surface of the molecule where antibodies have access. Knowing the hydrophilic index of each amino acid (see Table V), it is possible to simply determine the average hydrophilicity of small region (e.g., 15 amino acids) centered at each point along the molecule. Such algorithms, particularly the more recent manifestations which take into account many other properties of the proteins, often reveal domains with a high probability of antigenicity (Fig. 16).

Many antibodies, however, perhaps the most important, do not recognize linear epitopes but rather domains of the molecule which are conformational in nature. Such domains can comprise discontinuous, discrete regions brought into close proximity by the complex folding of the protein. Without knowledge of this folding pattern, usually determined using X-ray crystallography, such epitopes are virtually impossible to predict. Further complications for such a procedure include the role of carbohydrates in forming an epitope, which is particularly true for HIV gp120, and the vagaries of the host immune system in terms of processing and recognition. The biologic rules which govern protein synthesis are continually being elucidated and the prediction algorithms are hence being constantly refined.

TABLE V. Amino Acids

Amino acid	Three-letter code	One-letter code	Hydrophilicity[a]
Alanine	Ala	A	−1.8
Arginine	Arg	R	4.5
Asparagine	Asn	N	3.5
Aspartic acid	Asp	D	3.5
Cysteine	Cys	C	−2.5
Glutamine	Gln	Q	3.5
Glutamic acid	Glu	E	3.5
Glycine	Gly	G	0.4
Histidine	His	H	3.2
Isoleucine	Ile	I	−4.5
Leucine	Leu	L	−3.8
Lysine	Lys	K	3.9
Methionine	Met	M	−1.9
Phenylalanine	Phe	F	−2.8
Proline	Pro	P	1.6
Serine	Ser	S	0.8
Threonine	Thr	T	0.7
Tryptophan	Trp	W	0.9
Tyrosine	Tyr	Y	1.3
Valine	Val	V	−4.2

[a] Values used in the calculation of the hydrophilicity plot of a protein (see Fig. 16).

2. Epitope Mapping

Given their limitations, it is perhaps not surprising that the computer-derived predictions of antigenicity often do not reflect the actual situation. For example, the major immunodominant epitope of the HIV envelope residing in the gp41 is not predicted to be immunogenic by computer analysis (Gnann *et al.*, 1987). Direct measurement of immune reactivity is therefore commonly used to identify the critical regions.

Using recombinant DNA technology, small fragments of a given protein can be produced and used to determine which areas of the whole are recognized by sera from infected individuals. Alternatively, which fragment, when injected, can stimulate antibodies able to recognize or neutralize the native virus can be evaluated.

Synthetic peptides are being increasingly used to identify linear B-cell and T-cell epitopes. Thus, with modern equipment a laboratory can quickly produce and test many hundreds of short peptides (10–20 amino acids) spanning the entire protein in question. Each individual peptide preparation bound to ELISA plates can be incubated with the sera from infected patients or animals in order to identify sequences which react with antibody. Addition of overlapping synthetic peptide panels to microcultures of patients' PBMC and subsequent measurement of cell proliferation can also help pinpoint T-cell epitopes.

Although the information derived from such studies is often invaluable, the use of linear synthetic peptides suffers from many of the same limitations noted above which affect the use of computer predictions. Many important epitopes,

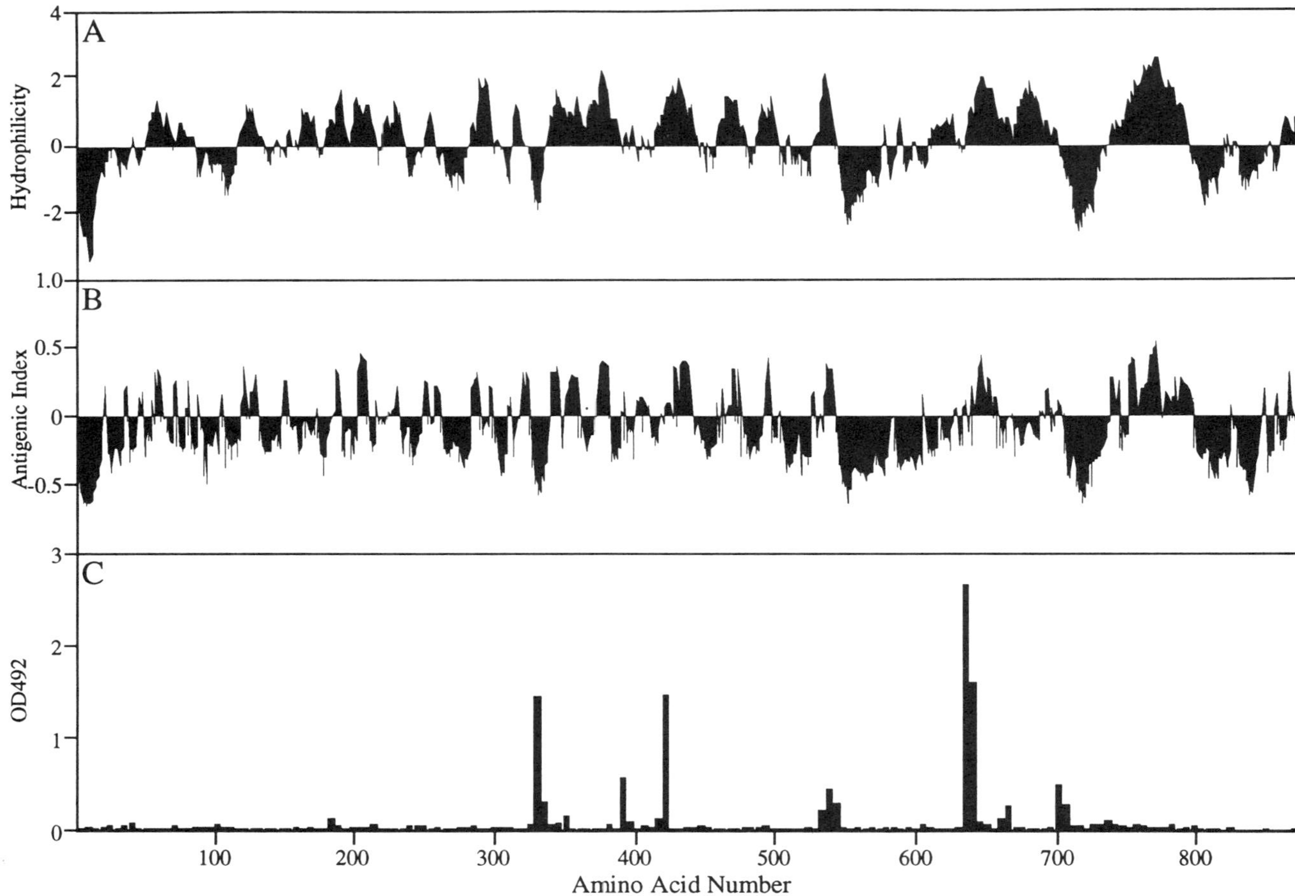

FIGURE 16. Comparison of the B-cell epitopes of SIVagm Env determined by different methods. (A) Hydrophilic regions of the protein predicted by computer analysis using the method of Kyte–Doolittle (15-amino-acid moving window). (B) Computer prediction of antigenic index. (C) Actual antibody reactivities determined by incubating sera from infected monkeys with a panel of overlapping synthetic 15-mer peptides corresponding to the entire SIVagm envelope glycoprotein.

particularly the discontinuous ones, will not be identified. For identifying these epitopes, monoclonal antibodies are usually required.

3. Monoclonal Antibodies

The use of monoclonal antibodies permeates most virology and immunology research, including the study of retroviruses. Apart from their value in identifying and purifying distinct cell subsets, monoclonal antibodies raised to viral components allow the identification of the all-important conformational and discontinuous B-cell epitopes. For viruses such as HIV, monoclonal antibodies produced by immunizing a mouse with the complete virus are often useful for diagnosis. However, the pattern of reactivity may be completely different from that seen in infected humans. Thus, the most important monoclonal antibodies to help understand the natural immune response are those produced by human hybridomas.

By fusing B cells from an infected human with a myeloma cell line or by transforming B cells with Epstein–Barr virus (Teeuwsen *et al.*, 1990), monoclonal antibodies which reflect precisely those produced in the natural situation can be produced in large quantity (Zolla-Pazner and Gorny, 1992). Monoclonal antibodies are impossible to map using linear synthetic peptides if they are specific for discontinuous epitopes and it is often necessary to determine the important regions by site-directed mutagenesis of the gene coding for the protein in question and seeing which changes abolish the ability of the antibody to bind.

4. Retroviral B-Cell Epitopes

The B-cell epitopes of a number of retroviral proteins have now been mapped. For example, Tsai and Oroszlan (1988) used antibodies raised in response to a COOH-terminal tridecapeptide of REV gp90 to demonstrate that this domain is expressed on the surface of infected cells and is the target for complement-activating antibodies.

A comparison of the known immunogenic domains from different retroviruses reveals both striking similarities and profound differences. Two of the best characterized retrovirus envelope glycoproteins, EIAV and HIV-1, share major structural and functional similarities despite the lack of substantial sequence homology (Gallaher *et al.*, 1989; Rushlow *et al.*, 1986). DNA sequence analysis of the EIAV envelope genes from sequential isolates revealed that base substitutions are clustered, as with HIV, in discrete variable and conserved regions (Payne *et al.*, 1987). These changes result in variants resistant to the present immune control. When B-cell epitope mapping was performed on the EIAV envelope glycoprotein gp90 (Ball *et al.*, 1992), a strikingly similar pattern of reactivity was revealed, with even a structure resembling the HIV-1 V3 loop apparently acting as the principal neutralizing determinant.

The envelope proteins of SIVagm have linear B-cell epitopes recognized by sera from infected monkeys in regions similar to those of HIV-1. Reactivity is seen in the area corresponding to the V3 loop of HIV-1, at the COOH-terminal end of the outer envelope protein, in a region of the transmembrane glycoprotein (reactivity with which is a reliable marker for infection) and in a second TMP domain (Siegel *et al.*, 1992). As well as correlating with HIV-1, the corresponding areas also appear to be dominant in HIV-2 and SIVmac (Goudsmit *et al.*, 1989; Mannervik *et al.*, 1992; Zvelebil *et al.*, 1988; Norley *et al.*, unpublished data).

5. Neutralizing B-Cell Epitopes

Many epitopes recognized by antibody appear to be immunologically irrelevant, i.e., the antibodies do not neutralize the virus, nor do they activate complement attack or ADCC. Identifying those epitopes whose immunologic recognition can help eliminate a virus or virus-infected cells, can enhance infectivity or induce a pathogenic effect is important in the design of synthetic or recombinant vaccines.

Using the techniques described above, the neutralizing determinants of the FeLV outer envelope gp70 have been identified (Elder *et al.*, 1987; Nunberg *et al.*, 1984) as well as one in the transmembrane protein p15(E) (Nick *et al.*, 1990). As type-specific neutralizing antibodies were directed in some cases at sequences common with other subtypes, it seems possible that conformation plays an important role (Elder *et al.*, 1987).

Epitopes in the bovine leukemia virus envelope glycoprotein have also been localized (Bruck *et al.*, 1982a,b, 1983; Portetelle *et al.*, 1989b), with the major neutralizing domain being identified in the BLV gp51. This domain, comprising three epitopes, was shown to be conformational in nature and to be the site of variability defining BLV antigenic variants (Portetelle *et al.*, 1989a).

Comparing the susceptibility of visna virus variants to antifusion and neutralizing antibodies suggested that the neutralizing and fusion domains are distinct from each other, with only the former showing variability (Crane *et al.*, 1988).

Drawing predominantly from the comparatively well-characterized HIV field, it seems that the most important neutralizing epitopes are conformational or even discontinuous in nature and are therefore difficult to mimic artificially (Steimer *et al.*, 1991). However, linear regions are known which can, when presented as a synthetic peptide, induce an antibody capable of neutralizing the virus itself (Chanh *et al.*, 1986; Dalgleish *et al.*, 1988; Durda *et al.*, 1990; Hart *et al.*, 1990). Human monoclonal antibodies were used to identify four neutralizing antibody epitopes in the HTLV-1 envelope glycoprotein (Baba *et al.*, 1993).

The so-called principal neutralizing determinant (PND) in the V3 loop of HIV-1 (Goudsmit *et al.*, 1988a; Putney *et al.*, 1986; Rusche *et al.*, 1988) has been used to construct multivalent synthetic immunogens which stimulate a strong and persistent neutralizing antibody response (Wang *et al.*, 1991). However, the strain specificity of this region probably limits the general usefulness of this approach. Using HIV-1 V3 loop multivalent synthetic peptides, Vogel *et al.* (1994) showed that the vast majority of neutralizing antibodies in patients' sera do in fact recognize the V3 loop.

By immunizing monkeys with synthetic peptides corresponding to the HIV-1 gp120, Vahlne *et al.* (1991) identified multiple linear epitopes which can act as targets for neutralizing antibody. SIVmac possesses a region similar in sequence and structure to the HIV-1 V3 loop (Regier and Desrosiers, 1990), a region which can be shown to be immunogenic in infected monkeys. In contrast to HIV-1, antibodies specific for this region do not appear to neutralize the virus itself (Javaherian *et al.*, 1992). In fact, other regions of the SIVmac outer envelope glycoprotein seem to be the targets for neutralizing antibody (Javaherian *et al.*, 1992; Kent *et al.*, 1992; Torres *et al.*, 1993). Such a lack of immunologic selection pressure might explain why the SIVmac V3 is not particularly variable.

The V3 loop of HIV-2, known to be relatively conserved among different isolates, also does not appear to be a major target for neutralization (Robert-Guroff *et al.*, 1992). Using neutralizing monoclonal antibodies specific for the SIVmac transmembrane envelope glycoprotein, Kodama *et al.* (1991) were able to identify a strain-specific neutralizing determinant, corresponding to a weak neutralizing HIV-1 determinant, which showed progressive variability *in vivo*. This phenomenon suggests selective pressure.

As mentioned previously, the B-cell epitopes of the retroviral envelope proteins are of prime interest because of their potential use in vaccine development. Mapping the linear B-cell epitopes of other retroviral proteins, such as Gag, is more of interest when one is designing diagnostics or examining protein structure. For example, Langedijk *et al.* (1990) used synthetic peptides and a panel of mono- and polyclonal antibodies to validate a proposed structure of the HIV-1 p24 Gag, whereas others (Hinkula *et al.*, 1990) used similar tools to identify seven immunogenic regions coded for by the HIV-1 *gag* gene. In what is perhaps the only exception to the rule that neutralizing antibodies are directed to retroviral envelope glycoproteins, Boucher *et al.* (1990) described a neutralizing epitope in the p17 Gag molecule of HIV-1 which is related to an epitope on thymosin $\alpha 1$ (Sarin *et al.*, 1986). These results, which require confirmation, suggest that part of p17 is expressed on the virus particle.

6. ADCC Epitopes

Target epitopes for ADCC on the HIV envelope glycoproteins have also been identified. A human monoclonal antibody recognizing a conformational epitope independent of the V3 loop was shown to mediate ADCC of HIV-1-infected cells in a type-specific manner (Koup *et al.*, 1991b). Human monoclonal antibodies were also used to identify ADCC epitopes in the HIV-1 transmembrane gp41 (Tyler *et al.*, 1990b). Antibodies specific for peptides corresponding to the HIV-2 V3 loop and two other linear epitopes of gp125 were shown to initiate lysis of HIV-2-infected cells. However, with the exception of the V3-specific site, no correlation was found between the linear neutralizing and ADCC epitopes (Bjoerling *et al.*, 1991).

B. T-Cell Epitopes

Epitope mapping of retroviral molecules other than the envelope glycoproteins becomes important for vaccine development when T-cell reactivity is considered. Selection, processing, and expression of short linear CTL epitopes is a cell-driven process, and suitable epitopes can potentially be found in any structural or regulatory protein. In practical terms, conformational and discontinuous T-cell epitopes do not exist and one would therefore suppose that the prediction of important regions by computer analysis of amino acid sequence would be much easier. As described in Section V, however, selection of the correct peptide fragment is determined predominantly by the characteristics of the MHC molecule with which it is expressed and by as yet poorly understood intracellular mechanisms.

Although all CTL epitopes presented by a cell of a given MHC type probably conform to the correct motif (Falk *et al.*, 1991), not all regions which conform are CTL epitopes. Berzofsky (1988) noticed that many T-cell epitopes are from regions of the protein which are α-helical and amphipathic (adjacent regions of hydrophobicity and hydrophilicity) in nature. Again, whereas computer predictions based on this fact are often successful in identifying T-cell epitopes, they often fail. As with B-cell epitopes, elucidation over the next decade of the biologic rules which direct selection should facilitate accurate computer prediction of T-cell epitopes.

1. Use of Synthetic Peptides

Synthetic peptides are particularly useful for identifying CTL epitopes. Given a population of effector cells containing virus-specific CTL, one must only pulse the ^{51}Cr-labeled autologous, syngeneic, or MHC-matched target cell with the different peptides before washing and mixing with the effector population. If a certain peptide is of the correct sequence to bind the target cell MHC-I and if it represents the target epitope for some of the CTLs in the effector population, the cells will be lysed and release their radioactive marker. Many CTL epitopes have been mapped using derivations of this system, although the peptides in question must have sufficient affinity for the peptide-binding groove to displace the resident peptide derived from the cell.

Synthetic peptides can be used to rapidly screen for putative CTL epitopes by binding them to a microtiter plate and incubating with recombinant, ^{125}I-labeled MHC-I in an assay similar in principle to the RIA described in Section IV (Choppin *et al.*, 1990; Frelinger *et al.*, 1990). If the peptide conforms to the correct motif necessary for fitting in the peptide-binding groove, the MHC molecule will bind and be detected via its radiolabel. Another method (Townsend *et al.*, 1989) is to add exogenous peptide to a cell line defective in its antigen-processing pathway. If the peptide fulfills the requirements for MHC binding, the MHC complex is stabilized and only then does cell surface expression occur. Although these last two methods are very useful for prescreening, they will yield many false-positives. Many peptides which fit the peptide-binding groove will have no corresponding CTL effectors due possibly to the holes in the CTL repertoire necessary to prevent recognition of self.

2. Retroviral CTL Epitopes

CTL epitopes have been identified in the F-MuLV *env* gene product (Ruan and Lilly, 1991). Epitopes recognized by sheep CTL and CD4$^+$ cells were identified in the bovine leukemia virus gp51 envelope protein (Gatei *et al.*, 1993). The *gag*-encoded protein of SIVmac has been characterized in some detail in terms of CTL reactivity by Letvin's group (Miller *et al.*, 1990, 1991; Yamamoto *et al.*, 1990) who first showed, using a vaccinia/*gag* recombinant to infect target cells, that PBMC from SIVmac-infected rhesus recognized and lysed cells expressing *gag* gene products. Indeed, CTLs specific for SIVmac *gag* and *nef* gene products are detectable in monkeys as early as 4–6 days after infection (Yasutomi *et al.*,

1993c). The same group then demonstrated that in all infected monkeys of the Mamu-A1 MHC haplotype, this recognition was directed solely to a particular epitope in the protein. Immunization of macaques with a peptide corresponding to this epitope was shown to elicit SIVmac-specific CTLs (Yasutomi *et al.*, 1993b).

Also using a combination of recombinant vaccinia viruses and synthetic peptides, it has been possible to identify three individual CTL epitopes in the Nef protein of SIVmac (Bourgault *et al.*, 1992). In the HIV-1 and -2 systems, CTL specific for Env, Gag, Pol, Nef (Buseyae *et al.*, 1993; Clerici *et al.*, 1991b; Culmann *et al.*, 1991; Gotch *et al.*, 1993; Johnson *et al.*, 1993; Nixon *et al.*, 1988; Walker *et al.*, 1989) have been demonstrated (for reviews see Autran *et al.*, 1991; Plata *et al.*, 1989; Walker and Plata, 1990).

VIII. ESCAPE MUTATION

One characteristic feature of retroviruses, particularly HIV, is their relatively high rate of variability *in vivo.* This feature is predominantly a result of the low fidelity of the reverse transcriptase enzyme which has no proofreading mechanisms. Obviously, if mutation occurs in an area essential for the virus infection cycle the progeny virus will be defective and fail to replicate. If, on the other hand, the mutation occurs in a nonessential epitope which serves as a target for a controlling immune response, the virus could escape the immune control (Fig. 17).

In the murine system, for example, inoculation of mice with wild-type or biologically cloned Gross murine leukemia virus resulted in the generation *in vivo* of variants with changes in specific epitopes of the *env* gene product gp70 (Pozsgay *et al.*, 1989). These changes rendered the variants resistant to neutralizing antibodies specific for the original virus.

There are many examples of conflict between retrovirus variability and the host immune response. Because of the problems posed by this phenomenon to the development of vaccines, the topic is of immense importance. The classic exam-

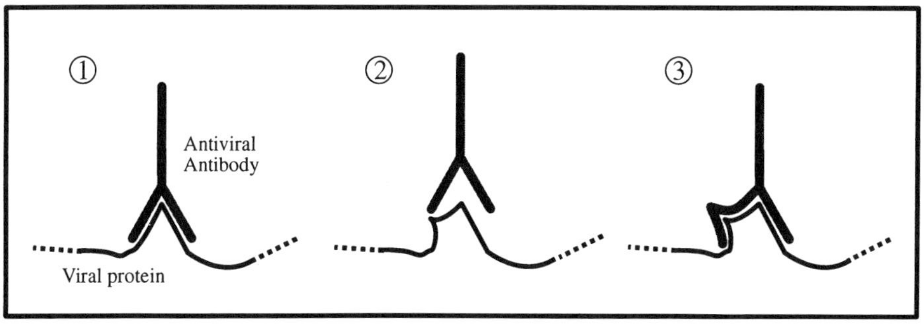

FIGURE 17. Mutational escape from neutralizing antibody. (1) Epitope on viral protein recognized by antibody (e.g., neutralizing antibody). (2) Mutation within the epitope prevents binding of the antibody or reduces its affinity. The mutant therefore has a growth advantage over the original virus. (3) Stimulation of the immune response by the mutant epitope results in the generation of new antibodies able to bind the virus once more.

ple of a retrovirus temporarily escaping immune control is in fact the primary infection of horses with EIAV.

A. Escape Mutation by EIAV

As mentioned in Section IV, during the first year following infection, horses undergo periodic episodes of viremia associated with hemolytic crisis. The viruses responsible for the viremia have been shown to be variants resistant to the neutralizing antibodies which have developed in response to (and control) all previously occurring variants (Montelaro *et al.*, 1984). Analysis of sequential variant isolates revealed changes in the amino acid sequence and glycosylation patterns of the surface glycoproteins gp90 and gp45, whereas internal proteins such as p15, p26, and p9 remained unchanged (Montelaro *et al.*, 1984; Salinovich *et al.*, 1986). Soon after the viremic episode, neutralizing antibodies to this escape mutant develop and the virus is brought under control, that is, until the next variant able to escape these (and the previous) neutralizing antibodies appears. The generation of neutralizing escape mutants has been simulated *in vitro* by passaging EIAV in culture in the presence of serum containing antibodies to group- and type-specific determinants of EIAV envelope glycoproteins (Rwambo *et al.*, 1990a).

Infected horses enter a state of low-level persistence after about a year and three to six episodes of viremia, possibly because the repertoire of possible variants able to completely escape neutralization is exhausted. Other mechanisms such as CTL might of course play a role in suppressing viral replication to a subacute state. In any case, the fact that compromising the immune system during the subacute persistent phase allows reoccurrence of the viremia and associated symptoms (Kono *et al.*, 1973) suggests some form of immune control.

B. Escape Mutation by Other Retroviruses

During visna virus infection of sheep, animals that develop neutralizing antibodies with a restricted range of specificity show development of antigenic variants whereas those that respond to infection with broadly neutralizing antibodies do not (Narayan *et al.*, 1981). As with EIAV (and indeed HIV; Masuda *et al.*, 1990), the development *in vitro* of visna virus variants can be forced by including immune sera in the culture medium (Narayan *et al.*, 1981). In this case the antigenic profiles of the emerging variants reflect the neutralizing spectrum of the selecting antibody. Only minor changes are necessary to escape neutralization by sera taken early after infection whereas extensive changes are required to escape "late" sera with a broader neutralization range.

Unlike the EIAV or HIV systems, there is no evidence that the development of visna virus variants *in vivo* plays any role in disease progression. In one study involving isolation of variants from sheep over a period of 7 years, emerging variants resistant to neutralization did not replace the earlier phenotypes and in fact isolates taken at the time of death were susceptible to autologous neutralization (Lutley *et al.*, 1983).

Despite the fact that the neutralizing antibody response to CAEV infection is very low, neutralization-resistant variants can be demonstrated *in vivo*. Ellis *et al.* (1987) isolated a variant of a cloned CAEV inoculum which although being susceptible to neutralization by antisera from the host animal was resistant to neutralization by sera from animals which did not develop the variant. Cheevers *et al.* (1991) infected newborn goats with a biologic clone of CAEV and in the majority of animals which developed neutralizing antibody found arthritic swelling of carpal joints from which antigenic variants with type-specific neutralization epitopes could be isolated. Rather than control the development of inflammatory lesions, neutralizing antibody was apparently associated with the clonal expansion of antigenic variants within the carpal joints with the resulting pathology. The fact that recombinant envelope glycoproteins from both the original and variant isolates could inhibit the type-specific neutralizing activity of sera against the variant was interpreted as evidence that CAEV variation involves amino acid changes outside of conserved neutralizing epitopes (Lichtensteiger *et al.*, 1991).

Although the dramatic cycling of escape mutation followed by a controlling immune response is not usually seen in other retrovirus systems, episodes of antigenemia coincident with diminished antibody in the SIV system (Zhang *et al.*, 1988) suggests a similar phenomenon. Indeed, the development *in vivo* of SIVmac variants resistant to neutralization has been observed (Burns *et al.*, 1993; Joag *et al.*, 1993).

C. Escape Mutation by HIV-1

The exceptionally high variation of HIV in humans is probably related, at least in part, to the low fidelity of the HIV reverse transcriptase. HIV RT has been shown to be more error prone than the RT of ALSV or MuLV, for example (Roberts *et al.*, 1988; Takeuchi *et al.*, 1988). Most mutations arising from random errors are of course lethal to the virus. However, the fact that over 80% of the observed (and therefore surviving) HIV mutations at the RNA level are nonsynonymous (i.e., result in an amino acid change) is strong evidence that the changes are imparting a survival advantage (Cichutek *et al.*, 1992).

Some of the most important areas of the viral envelope glycoprotein in terms of targets for neutralizing antibody or ADCC are highly variable, and this diversity probably helps the virus to overcome repeatedly the host's immune response. The fact that the V3 loop, also known as the "principal neutralizing domain," of the HIV-1 envelope glycoprotein is also one of the most variable areas of the viral genome supports the idea of an ongoing contest between the ever-changing virus and an immune system which is constantly being forced to respond, perhaps to the point of immunologic exhaustion (Nowak *et al.*, 1991). If AIDS is indeed the result of *in vivo* variation, the high mortality rate associated with HIV infection reveals which of the two usually prevails.

As noted in Section VII, targets for neutralizing antibody are often conformational and the virus has many options for change. This fact is demonstrated by infection of chimpanzees with HIV-1 strains of known sequence. In one case (Nara *et al.*, 1990), the dominant variant which arose was no longer neutralized by a monoclonal antibody specific for the V3 loop, although no mutations had occurred in this area. Rather, changes in the gp120 outside of the V3 loop seemed to have exerted a conformational change in the molecule. This change prevented antibody binding and presumably allowed the virus to replicate, for a short while at least, without immune control. The generation of neutralization escape mutants of HIV has also been observed *in vitro* (McKeating *et al.*, 1993; Robert-Guroff *et al.*, 1986; Watkins *et al.*, 1993) and *in vivo* (Arendrup *et al.*, 1992; Montefiori *et al.*, 1991b; Wahlberg *et al.*, 1991).

D. Mutational Escape from CTL

All of the above examples of variation leading to escape from immune control involved sensitivity to neutralizing antibody. The first description during natural infection of escape mutation in an epitope serving as a target for CTL recognition came from the HIV system (Phillips *et al.*, 1991). In a longitudinal study of HIV-infected patients, variants arose over time that were not recognized by autologous CTL, and this failure was shown to result from changes in a critical Gag CTL epitope. In contrast, Meyerhans *et al.* (1991) found that over a 14-month period a predominant form of HIV in terms of *gag* sequences persisted and no change in the sequence specificities of the CTL was seen.

The fact that in this study only 2 out of the 17 variants could theoretically escape CTL recognition indicates that CTL escape mutation is not necessarily a general property of HIV infection. However, when it does occur the worrisome aspect of this phenomenon is that CTL recognition is strongly dependent on the MHC background of the host. Thus, the accumulation of changes in CTL epitopes could theoretically render a virus (i.e., virus-infected cell) totally invisible to CTLs with no possibility of a response developing to the new variants. Although CTL escape mutations have not been formally shown in other natural retrovirus infections, it is reasonable to assume that they do occur, albeit probably with a low frequency.

IX. NEGATIVE CONSEQUENCES OF THE IMMUNE RESPONSE

We have so far concentrated on the positive aspects of the immune response to retroviruses: (1) antibodies which neutralize the virus or eliminate infected cells through activation of complement or ADCC effector cells and (2) the development of cytotoxic T cells. However, the generation of a specific immune response can have its negative side, sometimes enhancing the infectivity of the virus or even mediating pathogenesis.

A. Enhancement of Infection

Virus-specific antibodies can enhance, rather than neutralize, the infectivity of free virus particles via two known mechanisms (Fig. 18). First, antibodies bound to the virus particle can possibly be recognized by Fc receptor-bearing cells (e.g., macrophages) resulting in rapid uptake and infection. This phenomenon is referred to as Fc receptor antibody-dependent enhancement (FcR-ADE). Second,

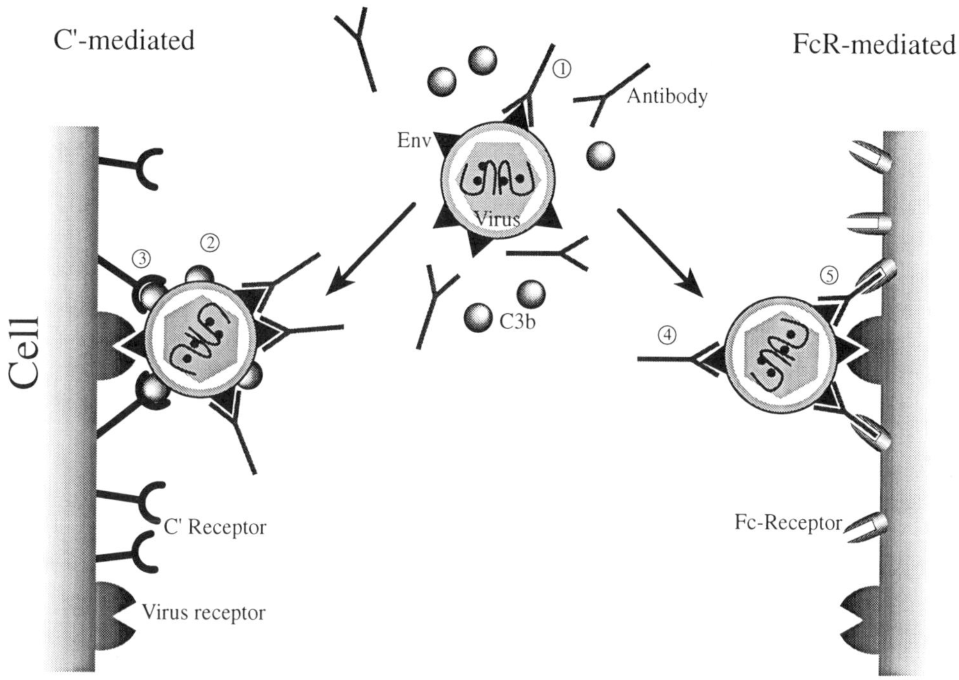

FIGURE 18. Enhancement of viral infectivity. Virus-specific antibodies bind the free virus (1) which stimulates the complement pathway, resulting in deposition of C3b on the viral surface (2). Activation of alternative pathway by virus alone may have the same effect. In C′-ADE, binding of the C3b to complement receptors (3) stabilizes the interaction of the virus with its normal receptor (e.g., CD4), facilitating infection. Similarly, binding of the Fc portion of bound antibody (4) with cellular Fc receptors (5) can stabilize the virus/receptor interaction with the same consequences.

complement binding to opsonized virus particles can lead to enhanced infection by interaction of the complement with cell-surface complement receptors (C′-ADE).

1. Fc Receptor-Mediated Enhancement

a. Examples of Fc Receptor-Mediated Enhancement of Retroviral Infection

FcR-ADE, commonly associated with dengue virus infection (Halstead, 1988), has been shown to operate (*in vitro* at least) in a number of retroviral systems. Monoclonal antibodies specific for murine retroviruses were shown, for example, to directly enhance the infectivity of both ecotropic and amphotropic retroviruses *in vitro* (Legrain *et al.*, 1986). Antibodies specific for epitopes in both gp70 and p15(E) which mediate enhancement of feline leukemia virus infectivity through an uncharacterized mechanism have also been described (Nick *et al.*, 1990). In fact, vaccination of kittens with native FeLV gp70 or gp85 vaccines, although resulting in the development of smaller tumors after challenge with FeSV, had the effect of enhancing both FeLV and FeSV viremia.

Sheep immunized with inactivated CAEV particles prior to challenge showed an increase in the severity of subsequent lesions. However, *in vitro* assays indicated that whereas non-neutralizing antibodies facilitate the binding, internalization, and uncoating of CAEV in primary macrophages, the enhanced production of virus particles does not occur (Jolly *et al.*, 1989a).

One of the most dramatic examples of the possible dangers associated with vaccine induction of enhancing antibodies was the response of horses to a purified EIAV envelope glycoprotein subunit vaccine. The vaccine induced protection against infection by the homologous challenge virus strain but enhanced the disease following challenge with a heterologous isolate (Issel *et al.*, 1992), even to the point of causing death (Montelaro *et al.*, 1992).

Similarly, enhancement of HIV-1 infection of human macrophage/monocytes via Fc receptors is of extreme interest because of the potential hazards associated with immunization. Antibodies could be induced which increase the chances of infection upon exposure or which would precipitate disease development in people already HIV-positive. At this point there is no clear evidence that enhancement occurs *in vivo*, although the demonstration by Homsy *et al.* (1990) that levels of enhancing antibodies in the sera of HIV-1-infected patients correlated with the progression to AIDS does indicate a possible *in vivo* role. As the regions of the viral glycoprotein responsible for binding enhancing antibodies are now identified, it might be considered judicious to exclude them from a vaccine preparation if possible.

b. Mechanism of HIV-1 Enhancement

The mechanism of Fc receptor-directed enhancement of HIV infection is not entirely clear. It was originally suggested (Takeda *et al.*, 1988) that enhancement is via Fc receptor-mediated entry of HIV into the cell. Indeed, results from Homsy *et al.* (1989) showed that the presence of antibodies specific for the CD4 molecule or soluble CD4 itself failed to prevent enhancement. This result suggested that an infection pathway independent of CD4 operates. The same group also found that the third Fc receptor (FcRIII) and not the FcRI or FcRII was apparently involved in augmenting infection, although the use of highly specific monoclonal antibodies renders overall conclusions difficult.

In contrast, others using different cells (Takeda *et al.*, 1990) found that the CD4 molecule was indeed necessary for infection in the presence of enhancing antibodies. Infectivity was blocked by soluble CD4 or monoclonal antibodies to CD4. In this study the FcRI was implicated as the responsible element. Finally, studies using bispecific antibodies to target HIV-1 to various cell surface molecules (Connor *et al.*, 1991) suggested that the FcRs are not infectivity receptors on human monocyte/macrophages. However, cytomegalovirus-induced expression of FcRs was shown to render human fibroblasts susceptible to infection by HIV/antibody complexes (McKeating *et al.*, 1990). These results indeed suggest a role for FcR-mediated enhancement of virus spread.

Whether or not an antibody enhances infection or neutralizes the virus might in fact depend on the affinity and quantity of the antibody as well as the cell type involved. The multivalent nature of highly opsonized HIV-1 has been suggested to allow high-affinity interactions with the FcR, triggering endocytosis and particle degradation. However, if insufficient interactions occur between the cellular FcR and a less-opsonized virus/antibody complex, endocytosis might not be triggered. Instead, stabilization of the complex at the cell surface could facilitate virus/CD4 interaction and thus enhance infectivity.

2. Complement-Mediated Enhancement

The second mechanism by which the immune response can enhance rather than restrict infection involves complement binding to the antibody/virus complex followed by interaction with complement receptors on target cells. C′-ADE requires, on the side of the virus, the envelope glycoproteins and, on the side of the host, the alternative complement pathway, CD4, and the complement receptor CR2. As can be expected, the retrovirus systems which have been most closely examined for this phenomenon are HIV and its SIV model. C′-ADE has been demonstrated using sera from HIV-infected humans and chimpanzees (Robinson *et al.*, 1987, 1988, 1989, 1990c) and SIVmac-infected monkeys (Montefiori *et al.*, 1990a,b).

In addition to C′-ADE, HIV-1 infection has also been shown to be enhanced by complement in the absence of antibody. Boyer *et al.* (1991) demonstrated enhanced HIV infection of C3 receptor-bearing T cells via a CD4-independent mechanism involving the alternative pathway of complement. As the levels of complement required were higher than those needed for C′-ADE, the antibody-independent mechanism would presumably only play a significant *in vivo* role before seroconversion. Similarly, Montefiori *et al.* (1989) found that infectivity in T cells of HIV grown in the presence of mannosidase inhibitors was enhanced by the alternative complement pathway. Infection of B-cell lines was also enhanced by complement via antibody-dependent and -independent mechanisms (Gras and Dormont, 1991).

3. Envelope Regions Responsible for Antibody-Dependent Enhancement

Identification of the HIV envelope glycoprotein domains responsible for binding enhancing antibodies is important as many believe that any putative vaccine should exclude epitopes likely to stimulate such antibodies. It is known that sera from animals immunized with whole inactivated virus immunogen (Takeda *et al.*, 1988) often contain enhancing antibodies.

Using a variety of methods involving recombinant proteins, monoclonal antibodies, and synthetic peptides, the domains targeted by enhancing antibodies have been identified. However, the results are not always in complete agreement. Using different human monoclonal antibodies, Takeda *et al.* (1992) showed that neutralizing and FcR-antibody-dependent enhancing epitopes are at distinct sites in the gp120. Characterization of the binding to recombinant HIV-1 proteins revealed that the enhancing antibody binds to a conformational domain located in the carboxyl half of the gp120. As mentioned in Section IV, the HIV-1 gp120 is heavily glycosylated, and the pattern of carbohydrates seems to profoundly affect susceptibility to neutralization. Nevertheless, whereas monoclonal antibodies specific for carbohydrate epitopes themselves can neutralize HIV infection of lymphocytes, some can also enhance infection of monocytic cells (Hansen *et al.*, 1991).

Several studies have shown that epitopes for C′-ADE are located in the HIV-1 transmembrane gp41 (Robinson *et al.*, 1990a,b, 1991). Using human monoclonal antibodies, the specificity for enhancement was shown to be located in two gp41 domains. The first, corresponding to the immunodominant area (amino acids 579–613), also appeared to be the major target for enhancing antibodies in patients' sera. Antibodies to this epitope could act synergistically with antibodies specific for the second, "weaker" epitope (amino acids 644–663) to enhance infection (Tilley *et al.*, 1992).

In addition to the epitopes in gp41, Jiang *et al.* (1991) found that antibodies specific for different V3-loop peptides (the principal neutralizing determinant) had the ability to enhance infection in relation to their serologic cross-reactivity with the V3 peptide of the virus strain used in the C'-ADE assay. This result suggests that V3-loop immunogens might stimulate enhancing rather than neutralizing antibodies depending on the level of cross-reactivity between the immunogen and the wild-type virus.

4. *In Vivo* Relevance of Antibody-Dependent Enhancement

Despite the concerns surrounding enhancement, the phenomena of C'-ADE and FcR-ADE in HIV infection have only so far been observed *in vitro* using established cell lines or cultured PBMC. The *in vivo* relevance thus remains to be established. In one study, antibody-induced enhancement of HIV infection by infected patients' sera could not be observed using fresh human blood monocytes or peritoneal macrophages (Shadduck *et al.*, 1991), whereas others demonstrated enhancement of homotypic HIV-1 isolates in monocytes (Homsy *et al.*, 1990).

In a study of a large cohort of HIV-1-seropositive patients, the levels of antibodies mediating C'-ADE at the time of disease onset did not correlate with the subsequent severity of disease progression (Montefiori *et al.*, 1991a). Yet, results comparing levels of neutralizing and enhancing antibodies at different clinical stages suggested that the appearance of enhancing antibodies coincident with the loss of neutralizing antibodies might indicate a poor prognosis (Homsy *et al.*, 1990; Toth *et al.*, 1991). In the SIVmac animal model, the levels of antibodies mediating C'-ADE increased over time and peaked just prior to death from AIDS (Montefiori *et al.*, 1990a). In addition, plasma from monkeys infected with isolates or clones of SIVsm, a natural isolate from sooty mangabey monkeys, was able to enhance infection of SIVmac in T-cell lines, indicating a broad specificity (Montefiori *et al.*, 1990b).

B. Immunopathology and Autoimmunity

Another manner by which the immune response to a retroviral infection might prove detrimental is by directly causing pathogenesis. For all its complexity the immune system appears "blind" with each component responding to the relevant stimuli regardless of the outcome for the host. The lysis of virus-infected cells by CTL, ADCC, or complement-mediated lysis, for example, although obviously necessary to prevent continued production of virus, would be detrimental if a substantial proportion of a particular tissue were involved. Opsonization of virus particles by specific antibody can initiate neutralization or phagocytosis but can also result in the formation of immune complexes. Trapping such complexes in the glomeruli, for example, can stimulate the infiltration of lymphocytes and macrophages with a corresponding inflammatory response. These and other immunopathologic mechanisms demonstrate that the immune response is often a "double-edged sword" particularly if the detrimental effects occur without successful clearance or control of the virus.

1. Examples of Immunopathology

The mechanisms of immunopathology are often difficult to determine. Indeed, it seems likely that both the humoral and cellular arms of the immune system are often involved, with antibodies causing the formation of immune complexes which then activate inflammation, for example. Experiments showing a correlation between the levels of one particular component and the severity of immunopathology may therefore be partially misleading. For this reason we will not attempt to separate antibody-mediated immunopathology from that apparently mediated by cells.

Examples of immunopathology are abundant in retroviral infections. In mice carrying an endogenous MuLV (see Kozak and Ruscetti, 1992) whose expression is often associated with the spontaneous development of necrotizing polyarteritis (NPA), only those animals which developed anti-gp70 antibodies able to activate complement-mediated lysis of MuLV-infected cells were affected (Miyazawa *et al.*, 1987). This result supports the notion that NPA in this system is a result of antibody/complement lytic attack of arterial smooth muscle expressing the endogenous MuLV gp70.

Animals suffering from retroviral infections which establish a chronic state are particularly prone to the development of immunopathology. Visna infection of sheep is, for example, typically associated with severe inflammatory lesions (Narayan and Cork, 1985) which appear to result from activation of infected monocytes and infiltration or proliferation of other lymphoid cells (see Narayan *et al.*, 1993). This activation might also result in excessive cytokine secretion. Excessive local immune responses indeed seem to play a role in pathogenesis, as the early lesions in experimental infections could be abolished by immunosuppressive treatment (Nathanson *et al.*, 1976). Similarly, hyperimmunization followed by infection resulted in more severe disease. In fact, lymphadenopathy and follicular hyperplasia in lymph nodes and spleen are common features of most lentivirus infections.

CAEV infects primarily mature macrophage and reticuloendothelial phagocytic cells of the goat synovial lining membrane (Michaels *et al.*, 1991) leading occasionally to a chronic and slowly progressing oligoarticular arthritis (see Narayan *et al.*, 1993). In these cases, there is strong evidence that the immune response is itself responsible for the development of arthritis. The subsynovial accumulation of mononuclear cells suggests that a chronic immune stimulus relating to persistent infection leads to local tissue damage (Cheevers *et al.*, 1988). The antibody titers to gp135 in the synovial fluid and serum correlate with the severity of the carpal arthritis in chronically infected goats (Knowles *et al.*, 1990). Artificial immunosuppression of goats prevents the development of lesions, but previous immunization with inactivated CAEV enhances the arthritis (McGuire, 1987).

The enhanced expression of MHC class II antigens, known to increase substantially the immune response to limiting quantities of antigen (Matis *et al.*, 1983), together with the intermittent expression of viral antigen on the surface of infected cells would trigger a cell-mediated immune response. Similarly, the production of an ill-defined interferon (Narayan *et al.*, 1985) by infected macrophages that could restrict visna replication but at the same time induce MHC class II antigen expression might augment such an intra-articular antiviral immune response. As in the case of primary or early visna infection, immunosuppressive intervention might help to prevent development of pathology resulting from these excessive but localized immune responses (McGuire, 1987). Rheumatoid symptoms such as Reiter's syndrome reminiscent of CAEV-induced immunopathology can also be associated with HIV infection of humans (Berman *et al.*, 1988; Winchester *et al.*, 1987).

EIAV infection of horses is characterized during the phases of acute disease by anorexia, lethargy, fever, and—as the name implies—anemia (see Montelaro *et al.*, 1993). Part of the anemia, erythrocyte destruction, appears to be caused by the immune response. Although the mechanisms are not fully understood, infection of macrophages results in dysfunction of iron metabolism (McGuire *et al.*, 1990). Deposition of viral antigen/antibody/C3 complexes results in both a severe glomerulitis and a reduction in plasma C3 levels. The fact that lesions in the liver and other organs are infiltrated with lymphocytes and monocyte/macrophages in the interstitial areas suggests an immunopathologic mechanism. Immune complexes comprised at least in part of HIV and HIV-specific antibody have also been demonstrated in the majority of AIDS patients (Morrow *et al.*, 1986).

In humans, tropical spastic paraparesis (TSP) induced by HTLV-I infection has clinical and pathological features similar to those seen with visna in sheep (see Sugamura and Hinuma, 1993). It has been suggested that TSP is the result of an inflammatory response in the spinal cord mediated by HTLV-I-infected lymphocyte/macrophages (Cruickshank *et al.*, 1989). With HIV infection, the fact that a $CD8^+$ infiltration of lungs is frequently observed in patients prompted Autran *et al.* (1990) to examine the HIV-specific cytolytic activity of the cells. Although the levels of alveolar CTL activity decreased with progression toward AIDS, there was a correlation between this activity and functional abnormalities of the lung epithelium, indicating a potentially immunopathogenic mechanism involving the CTL.

2. Sensitization for Lytic Attack

Although not autoimmunity in the classic sense, there is another possible mechanism of "self-destruction" active during HIV-1 infection. The number of HIV-infected $CD4^+$ cells *in vivo* is usually too low to account for the loss of these cells by virus-induced cytolysis. However, free virus fragments or envelope glycoproteins are probably shed during virus production that can bind to the CD4 of uninfected cells. Uninfected cells coated with exogenous gp120 have been shown *in vitro* to be excellent targets for ADCC attack. This finding suggests that uninfected cells primed by binding of viral proteins could be erroneously eliminated by the host immune system (Lyerly *et al.*, 1987). This process might be one mechanism by which the progressive depletion of $CD4^+$ cells occurs despite a relatively low level of peripheral viral infection.

3. Autoimmunity Induced by Molecular Mimicry

A more sinister aspect of many retroviral infections is the fact that areas of the viral proteins are able to mimic self antigens, in particular those involved in immune regulation itself. Apart from the possibility that these proteins could cause dysfunction of the immune system by direct interaction with the relevant receptors, these self-like domains could also break tolerance and stimulate a vigorous immune response to self proteins, i.e., induce a state of autoimmunity.

It is noteworthy that the development of retrovirus-induced immunodeficiency in the murine AIDS model, characterized by phenotypic and functional abnormalities in the T-cell population, requires the presence of mature B cells: high-level virus expression in mice depleted from birth of B cells has little or no effect on T-cell function and prevents disease development (Cerny *et al.*, 1990).

Autoimmune New Zealand Black/New Zealand White (F1) mice develop spontaneously retroviral gp70 immune complexes and antibodies to double-stranded DNA, concurrent with the development of glomerulonephritis (Shirai *et al.*, 1986). The renal disease was shown to be associated strongly with the levels of immune complexes and not free antibodies to DNA (Izui *et al.*, 1981b). It seems that the prevalence of glomerulonephritis in susceptible animals is linked to their ability to make antibody to their own xenotropic gp70 as a result of immunologic dysfunction (Izui *et al.*, 1981a; Kozak and Ruscetti, 1992).

In humans there is some evidence that endogenous retroviral sequences responsible for molecular mimicry and immunosuppression are important in the etiology and pathogenesis of Sjögren's syndrome and systemic lupus erythematosus (Garry *et al.*, 1990; Talal *et al.*, 1992). Patients with connective tissue disease carry autoantibodies which as well as reacting with U1 small nuclear ribonucleoproteins also cross-react with a shared epitope in the type C retrovirus p30gag (Query and Keene, 1987), suggesting a role for retroviruses in the initiation of autoimmunity. Virtually the same epitope is recognized by autoantibodies to DNA topoisomerase I, the characteristic marker for diffuse systemic sclerosis (Maul *et al.*, 1989).

Antibodies to HIV-1 envelope glycoproteins have been shown to recognize antigens expressed on activated monocytes (Beretta *et al.*, 1987). The homology between regions of MHC class II and HIV-1 gp41 (Golding *et al.*, 1988) could explain the presence of antilymphocyte antibodies in the sera of HIV-infected humans (Kiprov *et al.*, 1985). Kion and Hoffmann (1991) suggested an anti-idiotypic network model to explain HIV pathogenesis based on the fact that alloimmune mice react with HIV gp120. A peptide from the carboxy terminus of gp120 which has homology to MHC class I and II alpha-helices was shown to sensitize syngeneic target cells for lysis by alloactivated cytotoxic T effector cells (Clerici *et al.*, 1993). The envelope gene of HTLV-1 also shows some homology with class I HLA (Clarke *et al.*, 1983) which could account for the extra HLA-A and -B specificities detected using alloantisera on the surface of HTLV-1-infected cells (Gallo *et al.*, 1982).

There is some controversy surrounding the possible role for immune responses to retroviral proteins in the development of autoimmune connective tissue disorders, although it is known that the sera of patients suffering from such disorders often contain antibodies able to recognize retroviral (e.g., MuLV) Gag proteins (Rucheton *et al.*, 1985).

C. Activation of HIV-1 by the Immune System: Cytokine Production

The fact that HIV can remain dormant in immune cells as an integrated provirus until those cells are activated suggests that immune stimulation resulting from other infections could increase the levels of HIV present in the body. For example, IL-6, particularly in synergy with TNF-α, is known to upregulate production of HIV in acutely and chronically infected monocytes (Poli *et al.*, 1990). In addition, physiologic levels of heat acting synergistically with IL-6 and GM-CSF have been shown to productively stimulate integrated HIV-provirus expression *in vitro* (Stanley *et al.*, 1990). This suggests that fever and immune responses resulting from opportunistic infections may help induce HIV expression. Indeed, increased levels of the IL-6 receptor able to suppress the proliferative responses of PBMC can be detected in the blood of HIV-seropositive patients (Honda *et al.*, 1992). Conversely, pretreatment of macrophages with recombinant IFN-α, -β, and -γ (or bacterial LPS) prevents HIV replication in macrophages, apparently acting at a late stage in the viral infection cycle (Kornbluth *et al.*, 1989). In some studies, cytokine production itself resulting from a retroviral infection can have pathologic consequences for the host (for reviews see Matsuyama *et al.*, 1991; Merrill, 1992; Rosenberg and Fauci, 1990).

X. EFFECTS OF RETROVIRUSES ON THE IMMUNE SYSTEM

A. Direct Mechanisms

Infection of a cell by a retrovirus and the recruitment of the cellular machinery for the production of progeny virus particles can of course lead to the eventual death of that cell. If enough cells with a crucial function are involved, the consequences for the host can be severe. In addition, binding of the viral envelope glycoproteins expressed on the surface of the cell to their target molecules on the surface of neighboring cells can lead to multiple cell fusion and the formation of giant cells (syncytia). For viruses such as HIV, however, the number of infected cells appears insufficient to account for the profound disruption of the immune system through such direct effects. Retroviruses achieve this disruption through more subtle, indirect mechanisms (Table VI).

B. Indirect Mechanisms

The immune system is an immensely complex and finely regulated system. Minute quantities of different cytokines can have a profound effect on the balance of this system. It is therefore perhaps not surprising that retroviruses, which make their home predominantly in the cells of the immune system, often suppress or chronically activate this system. Indeed, examples of immune dysfunc-

TABLE VI. Detrimental Effects of Retroviral Infection

Direct killing of infected cells
Induction of syncytium formation
Immunopathogenesis
Autoimmunity
Induction of apoptosis
Inappropriate immune attack of uninfected cells
Stimulation/depletion of T cells by superantigen
Disruption of immune homeostasis
Oncogenesis

tion induced by infection with animal retroviruses are plentiful (for review see Kurth and Norley, 1993).

Infection of mice with Friend leukemia virus has long been known to induce a transient suppression, concurrent with the time of viremia, of the primary humoral immune response (Ceglowski *et al.*, 1975; Ceglowski and Friedman, 1968). Similarly, inoculation of mice with radiation leukemia virus variants reduced the immune response to sheep erythrocytes (Haran-Ghera *et al.*, 1977; Haran-Ghera and Rubio, 1977). Suppression of humoral and cellular immune responses in mice inoculated with visna virus appeared to depend on active virus replication as heat treatment abolished the immunosuppressive effect (Svennerholm *et al.*, 1978). Immune evasion by maedi-visna virus within the lymphoid tissue was shown to already occur during the initial stages of infection (Bird *et al.*, 1993).

Denner *et al.* (1980) showed that disrupted type C and D retroviruses inhibited the mitogen-driven blastogenic response of human lymphocytes, and that Rauscher MuLV had the same effect on murine cells. In addition, inactivated avian leukosis and sarcoma viruses, Sendai virus, Friend leukemia virus, and murine mammary tumor virus inhibited the mitogenic responses of avian lymphocytes (Wainberg *et al.*, 1980). The persistence of Gross murine leukemia virus infection of the murine central nervous system appears to depend, in part, on the impairment of the antiviral immune response and APC function in the CNS (Korostoff *et al.*, 1991).

One intriguing piece of evidence that endogenous retroviruses might play a role in regulation of the immune response comes from the use of specific antisense oligonucleotides complementary to type C murine retroviral sequences. Antisense constructs corresponding to the initiation regions of the endogenous mink cell focus-forming envelope gene dramatically enhanced murine spleen cell RNA synthesis and surface expression of Ia and Ig proteins (Krieg *et al.*, 1989). This finding suggests that expression of endogenous retroviral genes, occurring as a result of immune stimuli, might play a role in the regulation of immune homeostasis.

The mechanisms by which so many of these retroviruses cause some form of immune dysfunction are not clearly understood, and indeed, the effects may vary from one virus to another (Chirmule *et al.*, 1989). A number of possible categories can be suggested, however. The virus or parts thereof might interact directly with components of the immune system, blocking or stimulating normal responses. The immune response to the virus itself might be damaging, particularly if viral proteins bear homology to immune cell components. In addition, the nature of the viral genome integration into the host (immune) cell chromosomes might activate or inactivate crucial genes involved in immune regulation.

1. Immunosuppressive Domains

It is attractive to think that a section of a retroviral protein might simply inhibit in some way a crucial stage of immune stimulation as this might then serve as a target for therapeutic intervention. Indeed, candidate retroviral sequences have been minutely characterized, particularly the putative immunosuppressive domain of the retroviral envelope transmembrane glycoprotein. This short region (6–20 amino acids) is highly conserved among all retroviruses and is considered by some to be responsible for many of the immunosuppressive activities of retroviruses *in vitro* and *in vivo*.

For example, a peptide (CKS-17) corresponding to this region of the murine leukemia virus p15E has been shown to have a multitude of detrimental effects on the immune system. As well as inhibiting mitogen- and antigen-driven proliferation of lymphocytes *in vitro* (Cianciolo *et al.*, 1985), the presence of this peptide: (1) blocks the response to IL-1 (Kleinerman *et al.*, 1987), apparently by interfering with signal transduction through a protein kinase C pathway (Gottlieb *et al.*, 1989); (2) inhibits IL-2 (Nelson and Nelson, 1990), IFN-γ (Ogasawara *et al.*, 1991), and immunoglobulin (Mitani *et al.*, 1987) production; and (3) interferes with the lytic phase of NK cytolysis (Harris *et al.*, 1987) and respiratory burst of monocytes (Harrell *et al.*, 1986).

The corresponding region of the HIV-1 gp41 has been shown to exhibit, at least in part, many of these activities against human cells. However, it is not clear how the relatively low amount of viral protein which exists *in vivo* could initiate the profound disturbances observed. Possibly, the disturbances initiated by this region are enough to shift the delicate balance of the immune system into a state of immunosuppression or chronic stimulation.

2. Disruption of the Th1/Th2 Balance

It has long been known that immune responses tend to polarize during infection with a number of different parasites, and this has provided insight into the roles of the different lymphocyte subsets and their cytokines. For example, recovery from leishmaniasis infection in mice depends on the activity of the Th1 subset whereas activation of the Th2 subset is detrimental, leading to uncontrolled disease (Heinzel *et al.*, 1989, 1991; Reiner *et al.*, 1993; Scott *et al.*, 1988). In addition, during murine schistosomiasis or infections with *T. cruzi* and *T. gondii*, the Th2 cells are artificially stimulated, to the detriment of the Th1 cells whose activities are suppressed (Sher *et al.*, 1992). The benefit to the parasite is that the macrophage effector function, a major component of the antiparasite defense system, is compromised, both directly by the Th2 cytokines IL-4 and IL-10 and indirectly by suppression of IFN-γ secretion by Th1 cells (Fig. 19).

Parasites are of course not the only organisms which stand to benefit from indirectly downregulating cellular immunity via stimulation of Th2 activity. For example, C57BL/6 mice infected with the LP-BM5 mixture of murine leukemia

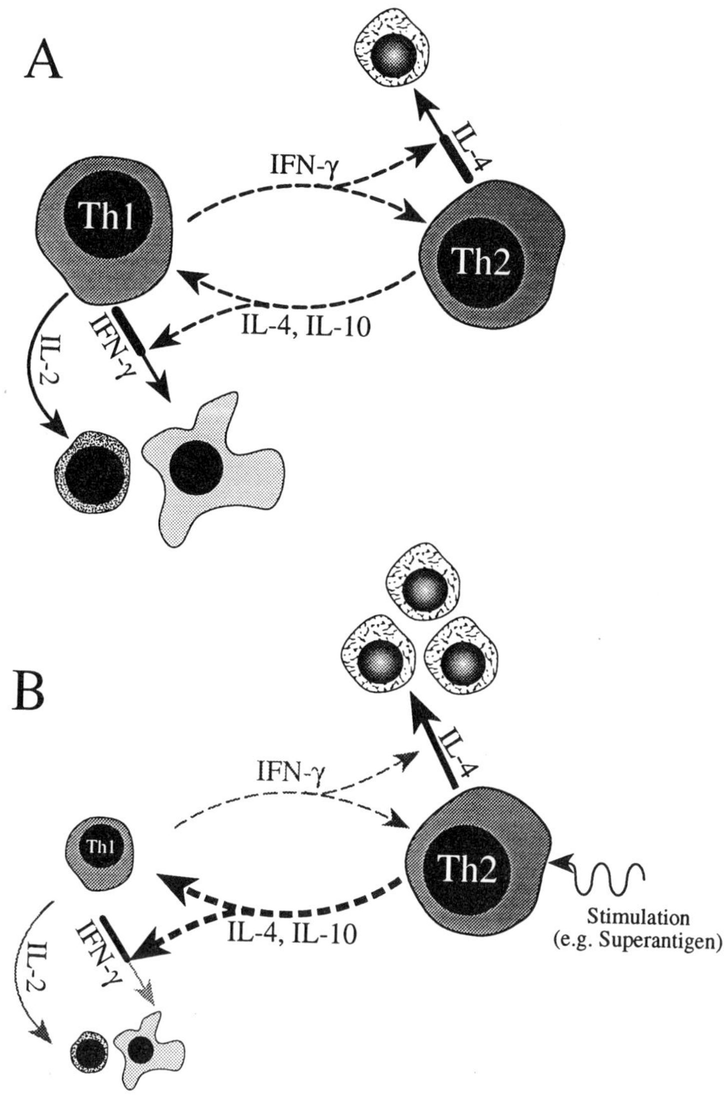

FIGURE 19. Mutual regulation of T-helper cell subsets and possible results of dysregulation. (A) Th1 cells, secreting IL-2 and IFN-γ, help maintain (solid arrows) macrophage and T-cell-mediated immune responses while Th2 cells stimulate B-cell activity via IL-4. Each subset can control the activity of the other by downregulation and inhibition of cytokine activity (dashed arrows). (B) Preferential stimulation of the Th2 subset by superantigen, for example, would lead to both an increase in B-cell activity and a strong suppression of Th1 activity, and hence a reduction of macrophage and T-cell-mediated immune responses.

viruses, which induces a severe immune deficiency disease, exhibit an activation and expansion of T and B cells, rendering the cells unresponsive to normal immune stimuli. This effect seems to be caused primarily by an overstimulation of the Th2 subset of $CD4^+$ cells (Gazzinelli *et al.*, 1992b). The stimulation could account for the suppression of cell-mediated immunity, particularly the loss of $CD8^+$ CTL activity, as well as the symptoms of polyclonal B-cell activation, hypergammaglobulinemia (including excessive IgE production), lymphadenopathy, and splenomegaly (Buller *et al.*, 1987; Gazzinelli *et al.*, 1992b; Mosier *et al.*, 1985; Pitha *et al.*, 1988; Yetter *et al.*, 1988). This phenomenon leaves the host susceptible to malignancy (Klinken *et al.*, 1988), as well as infection by other viruses (Buller *et al.*, 1987), yeasts (Darban *et al.*, 1991), and in particular by the parasites described above (Gazzinelli *et al.*, 1992a). Indeed, mice defective in Th2 (IL-4) cytokine responses have a delayed onset of murine AIDS with no lethality following infection with the LP-BM5 mixture of murine leukemia viruses (Kanagawa *et al.*, 1993). At the molecular level, the $CD4^+$ cell activation appears to be mediated by defective activation of protein kinase C, not by normal TcR signaling (Fitzpatrick *et al.*, 1992).

Th1/Th2 Imbalance in AIDS

The symptoms described above resulting from the inappropriate activation of the Th2 $CD4^+$ lymphocyte subset closely resemble those seen in AIDS. The loss of CD4-immune functions such as IL-2 and IFN-γ production concurrently with elevated production of B-cell growth and differentiation factors (Clerici *et al.*, 1989a,b; Lane *et al.*, 1983; Lucey *et al.*, 1990; Maggi *et al.*, 1987) is in accordance with abnormal Th2 activation.

Activation of Th2 cells with the resulting suppression of Th1 activities may be more than an accidental by-product of HIV infection. The fact that a large proportion of seronegative individuals who participate in "high-risk" behavior appear to have an established anti-HIV Th1 immune response (Clerici *et al.*, 1991a, 1992) suggests that it is possible for infection to occur but be suppressed to a level below that needed for activation of the humoral immune system. This observation, coupled with the fact that progression to AIDS is often preceded by a decrease in the Th1:Th2 ratio (Clerici and Shearer, 1993) suggests that Th1 cytokines help stimulate an HIV-1-specific cellular immunity able to control the virus. Stimulation of the Th2 response during the production of HIV-specific antibodies would downregulate this "protective" Th1 response. If so, anything which affects the Th balance in favor of the Th2 population, e.g., opportunistic infection, could act as a cofactor for increased risk for infection or disease. Coinfection of mice with a vaccinia/HIV-1 gp160 construct and the helminth *S. mansoni* has in fact been shown to prevent Th1 cytokine responses and hence severely delay the clearance of the virus (Actor *et al.*, 1993).

3. Cytokine Dysregulation in Retroviral Infections

Cytokines also appear to be important for the pattern of infection seen during CAEV infection of goats. IFN produced by PBMC from CAEV-infected goats inhibits monocyte proliferation and maturation to macrophages, thereby indirectly downregulating viral replication. In addition, this cytokine lowered viral gene expression in mature macrophages at the level of transcription (Zink and Narayan, 1989). Lentivirus-induced IFN might therefore inadvertently help maintain CAEV persistence.

In humans, HTLV-I infection often results in disruption of the normal pattern of cytokine expression. Indeed, the pattern of this retrovirus-induced cytokine expression may determine the nature of the ensuing disease. Elevated levels of TGF-β1 produced by the leukemic cells during adult T-cell leukemia would impair cellular and humoral immune responses. Upregulation of IFN-γ, TNF-α, and IL-1β could stimulate the immune-mediated destruction of the central nervous system which leads to HTLV-associated myelopathy/tropical spastic paraparesis (Tendler *et al.*, 1991). Like HIV, HTLV-I possesses a number of regulatory genes. As well as affecting the expression of viral proteins, these genes may influence the production of normal cellular proteins. For example, the HTLV-I *trans*-activating protein *tax* increases the expression of IL-2 and IL-2R genes, as well as those coding for GM-CSF and IL-3 (Leung and Nabel, 1988; Miyatake *et al.*, 1988; Siekevitz *et al.*, 1987). The end result can be oncogenic changes in the cell (see Gardner *et al.*, this volume).

The profound effects of HIV infection on various aspects of human immune regulation are becoming increasingly evident. Many of these effects are almost to be expected when one considers the cells (Th cells, macrophages) in which the virus replicates. HIV-1 infection of macrophages, for example, compromises their antigen presentation ability (Ennen *et al.*, 1990). However, other mechanisms of immune disruption appear so far to be unique to HIV. For example, adherent cells from HIV-infected patients spontaneously produce a 29-kDa molecule distinct from IFN-γ and TNF-α which inhibits IL-2 production by normal activated T cells and is possibly partially responsible for the immune dysfunctions seen in AIDS (Ammar *et al.*, 1991). The HIV envelope glycoprotein has a region of homology to the area of IL-2 responsible for binding to the IL-2 receptor (Reiher *et al.*, 1986; Weigent *et al.*, 1986). This fact suggests that direct competition for binding to the IL-2 receptor or generation of antibodies which cross-react with and neutralize IL-2 could be contributory mechanisms of immune suppression.

4. Apoptosis

a. Natural Role of Apoptosis

In addition to the immunologic mechanisms for eliminating virus-infected cells (e.g., ADCC, CML, CTL), it is important, in a variety of situations, for the body to have the means of killing unwanted but otherwise healthy cells (Fig. 20). Programmed cell death, or apoptosis, plays an essential role during embryonic development, particularly in the formation of the nervous system, gut, limbs, and bones (Antalikova *et al.*, 1989; Furtwaengler *et al.*, 1985; Kimura and Truman, 1990; Williams and Bell, 1991). Programmed cell death of cells in the gut (Duncan and Heddle, 1984) and skin (Young, 1987) helps reduce the chances of cancer development, as does the induction of apoptosis in transformed cells by TNF (Heller *et al.*, 1992). In the immune system, lymphocytes specific for self antigen can be removed during the early stages of differentiation by apoptosis, thereby preventing the development of autoimmunity (Golstein *et al.*, 1991). Apoptosis may also play an important role in controlling lymphotropic viral infections such as EBV and measles (McChesney and Oldstone, 1989; Uehara *et al.*, 1992).

b. Mechanism of Apoptosis

Although the physiological effects of apoptosis have been known for some time, the mechanisms and triggers for programmed cell death are only now being characterized. Incomplete signal transduction can induce cell death, which is an active process requiring initiation of protein synthesis. In particular, an endoge-

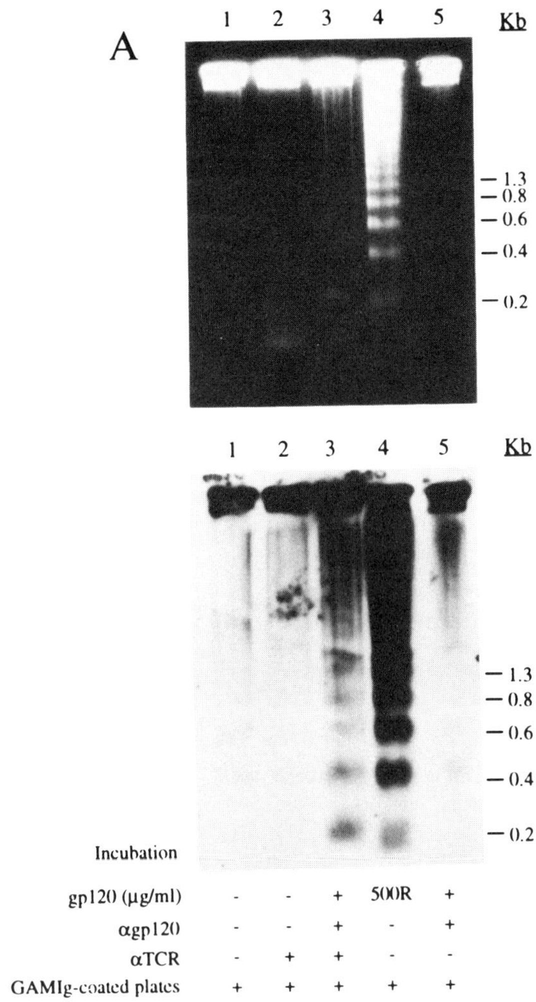

FIGURE 20. Induction of cell apoptosis by cross-linking of cell surface CD4. (A) Agarose gel electrophoresis of total DNA from: untreated cells (lane 1); cells treated with anti-TCR (lane 2); cells treated with gp120, anti-gp120 antibody, and anti-TCR (lane 3); irradiated cells (lane 4); or cells treated with gp120 and anti-gp120 antibody (lane 5). The characteristic DNA ladder is observed only after cross-linking of the CD4 followed by stimulation with anti-TCR (lane 3) or after irradiation (lane 4). (B) Electron micrographs of normal cells (i) or cells undergoing apoptosis after treatment with gp120, anti-gp120 antibody, and anti-TCR (ii). ×10,140. Bar = 1 μm. (Reproduced with kind permission from Banda *et al.*, 1992.)

nous endonuclease is involved. Indeed, the most striking feature of programmed cell death is the fragmentation of the genomic DNA (Fig. 20) into pieces 200 bp in length or multiples thereof, shown by the characteristic DNA "ladder" when run in an agarose gel (McConkey *et al.*, 1990).

c. *Inappropriate Activation of Apoptosis*

Apoptosis, although an important mechanism for general physiologic control, would obviously be harmful if activated on a large scale under inappropriate

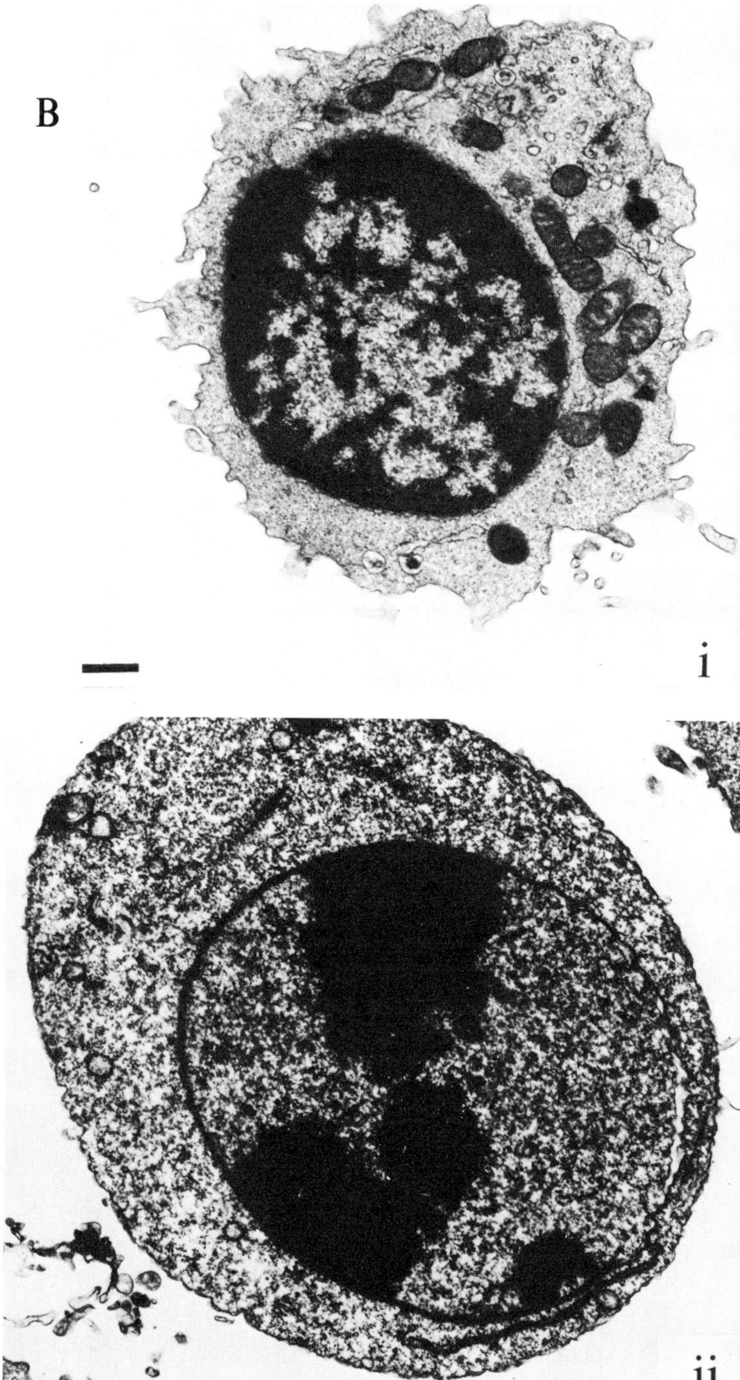

FIGURE 20. (*Continued*)

conditions. This is precisely what is presently being postulated to occur during HIV-1 infection. One of the puzzles surrounding AIDS is that the progressive and eventually fatal decline in the CD4$^+$ cell population cannot be accounted for entirely by direct infection of the cells. A number of alternative explanations have been suggested, some of which are described above, such as immunosuppression by viral proteins (Ruegg *et al.*, 1989), or exhaustive stimulation (Ascher and Sheppard, 1990).

Evidence is now accumulating that T cells are primed as a result of the initial HIV-1 infection of the host to undergo apoptosis upon later stimulation. Groux *et al.* (1992) showed that stimulation of PBMC from HIV-1-infected patients by mitogen or superantigen resulted in programmed cell death, while others (Laurent-Crawford *et al.*, 1991; Terai *et al.*, 1991) demonstrated that HIV-1-induced cell death *in vitro* is the result of apoptosis. In addition, cells infected with the feline AIDS virus, FIV, have been shown to undergo apoptosis in the presence of TNF-α (Ohno *et al.*, 1993). This cytokine can be produced in elevated levels in AIDS patients (Lähdevirta *et al.*, 1988; Matsuyama *et al.*, 1991).

Exactly how HIV-1 can prime cells for activation-induced death is a matter of debate. The recent demonstration (Banda *et al.*, 1992) that cross-linking of CD4 by HIV-1 gp120/anti-gp120 antibody complexes at physiological (picomolar) concentrations renders CD4$^+$ cells susceptible to apoptosis on subsequent stimulation of the TcR may explain why CD4$^+$ cells and not CD8$^+$ cells are depleted *in vivo* in the absence of high viral loads (Fig. 21). However, others (Meyaard *et al.*, 1992) have shown that both CD4$^+$ and CD8$^+$ T cells from asymptomatic HIV-1-infected patients are primed for apoptosis upon *in vitro* stimulation. It is perhaps important that the pathogenic and apathogenic systems appear to differ with regard to apoptosis. Whereas PBMC from HIV-1 infected humans and SIV_{mac} infected rhesus monkeys appear to be primed *in vivo* for subsequent apoptosis, this is not true for cells from HIV-1 infected chimpanzees or SIVagm infected African green monkeys (Gougeon *et al.*, 1993; Norley, personal observation).

5. Superantigens

In recent years a new class of molecules able to cause profound disruption of the immune system have been described in detail. These "superantigens" are so called because of their ability to almost indiscriminately activate T cells (up to 20% of the total population) regardless of their specificity. Superantigens are produced by microorganisms as diverse as bacteria (Janeway *et al.*, 1989; Marrack and Kappler, 1990; White *et al.*, 1989), mycoplasmas (Cole *et al.*, 1990), and viruses (Lafon *et al.*, 1992) and can be severely pathogenic. For example, the acute symptoms of food poisoning caused by *Staphylococcus aureus* enterotoxins are related to the superantigenic nature of the molecules (Johnson *et al.*, 1991).

a. Retroviral Superantigens

It has been known for some time that the murine genome carries genes coding for proteins with the properties of superantigens (Festenstein, 1973). The fact that these proteins stimulated T cells neither on the basis of their antigenic specificity nor their MHC type led to the term "minor lymphocyte stimulating" (Mls) antigen being used. Mls genes were at first presumed to be native rodent genes of

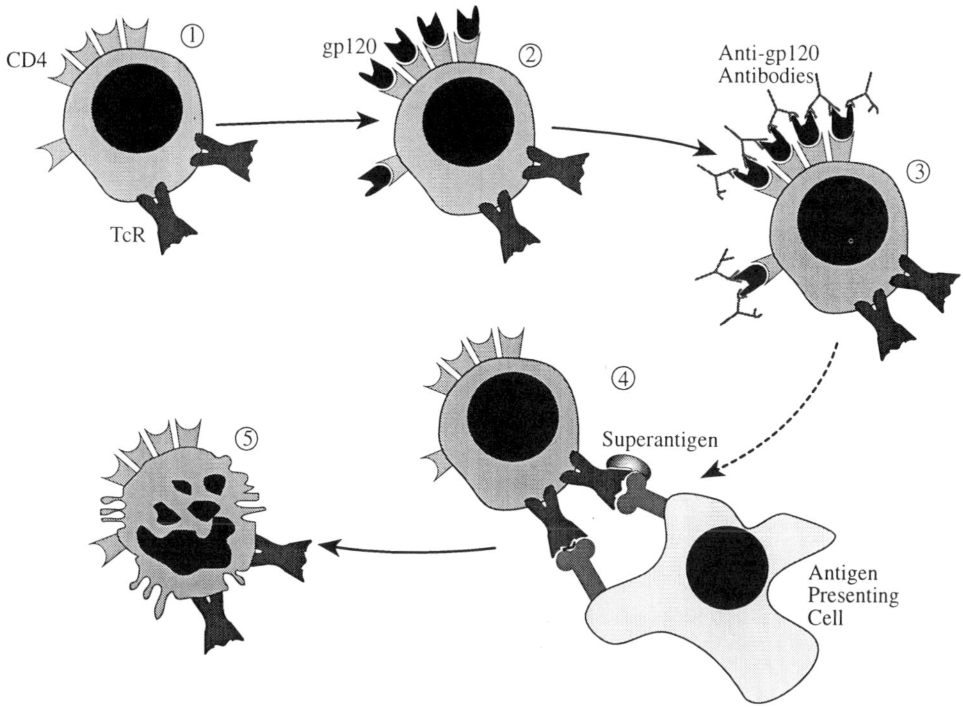

FIGURE 21. Hypothetical mechanism for depletion of uninfected $CD4^+$ cells in HIV-1-infected individuals by induction of apoptosis. T-helper cells (1) bearing CD4 and T-cell receptors (TcR) become coated with free gp120 (2) via specific interaction with the CD4 receptors. These gp120 molecules are then cross-linked by gp120-specific antibody (3), priming the cell for apoptosis. Later, a second signal resulting from antigen-specific recognition or superantigen activation (4) triggers the cell to undergo apoptosis (5).

unknown function (self superantigens). However, it is now known that the Mls genes are in fact part of integrated endogenous retroviral proviruses (Acha-Orbea *et al.*, 1991; Choi *et al.*, 1991; Dyson *et al.*, 1991; Frankel *et al.*, 1991; Marrack *et al.*, 1991; Woodland *et al.*, 1991). These mouse mammary tumor proviruses have, over time, scattered themselves throughout the mouse genome so that a selection can be found in all modern strains of mice (Lee and Eicher, 1990; Peters *et al.*, 1986), most of which encode a superantigen in their 3′ long terminal repeat.

b. Mechanism and Consequences of T-Cell Activation by Superantigen

The molecular actions of superantigens which cause T-cell activation have recently been elucidated (Janeway *et al.*, 1989; Pontzer *et al.*, 1991; Russell *et al.*, 1991). Basically, superantigens cross-link the T-cell receptor and the MHC molecule by binding simultaneously to the outer face of the variable region of the TcR β-chain (the Vβ region) and the MHC molecule (Fig. 22). This circumvents the need for processed antigen recognition and results in uncontrolled stimulation of the T cell.

Each superantigen may recognize a different Vβ region, and therefore stimulate a distinct population of T cells, leading to expansion and possibly eventual elimination of these cells. Organisms exposed to superantigens therefore display

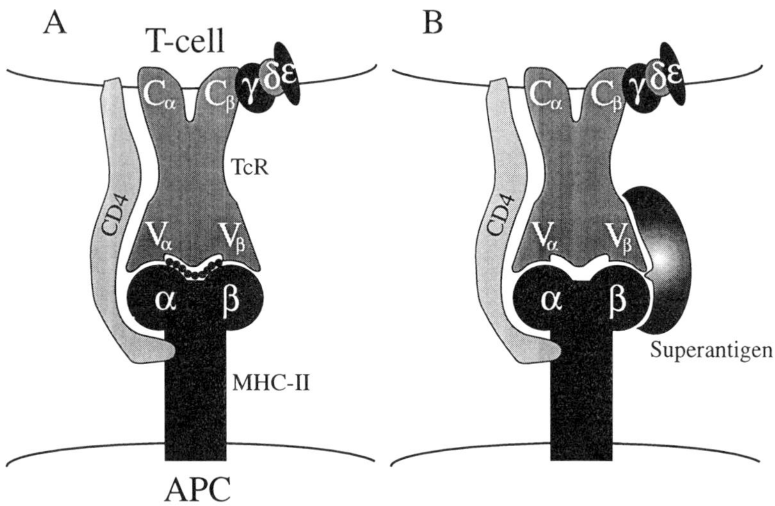

FIGURE 22. Mechanism of T-cell activation by superantigen. (A) Normal activation of T cell by processed antigen presented to the specific T-cell receptor (TcR) by the MHC-II molecule of an antigen-presenting cell. The different variable (V) and constant (C) domains of the TcR are shown. (B) Inappropriate activation of T cells by superantigen following cross-linking of the MHC to the TcR, regardless of specificity.

an unusual and distinctive pattern of Vβ-gene expression. The recent observation that HIV-infected humans display these characteristic perturbations (Dalgleish *et al.*, 1992; Imberti *et al.*, 1991; Laurence *et al.*, 1992; Pantaleo *et al.*, 1992) raises the reasonable possibility that HIV itself codes for or induces a superantigen. Indeed, a virus-encoded superantigen has been demonstrated to be active during murine AIDS (Huegin *et al.*, 1991).

A superantigen effect would support some of the other proposed mechanisms of CD4 cell depletion and dysfunction, i.e., indiscriminate activation of cells primed to undergo apoptosis and perhaps even Th2 stimulation. In addition, the fact that *in vitro* stimulation of T cells increases both the susceptibility to HIV-1 infection and the likelihood of provirus activation suggests a detrimental role for unrelated superantigens *in vivo*. It may therefore be important that the exotoxin of *S. aureus*, frequently isolated from HIV-infected individuals (Levine *et al.*, 1990), has been shown to efficiently activate *in vitro* HIV-1 production in patients' $CD4^+$ cells (Brinchmann *et al.*, 1992).

Despite our increasing knowledge of the disturbances HIV exerts on the normal functioning of the immune system, the search for a mechanism of AIDS induction by HIV is continuing. In fact, AIDS induction will almost certainly turn out to be the result of a number of inhibitory and stimulatory interactions of this virus and the immune system.

XI. VACCINES

The study of the immune response to a retrovirus is primarily concerned with understanding pathogenesis and with the development of an effective vaccine. This is of course particularly true for HIV, and a startling array of putative vaccines have been developed and tested for efficacy against HIV infection in chimpanzees or in the SIV nonhuman primate model. These experiments have been reviewed elsewhere (Kurth *et al.*, 1991), but it is relevant to describe here the properties in terms of immune responses of the successful HIV/SIV vaccines described to date.

A. Whole Inactivated Virus Immunogen

One of the simplest forms of a vaccine consists of the complete inactivated virus particle (Table VII). Although, for a number of safety reasons, there is resistance to the idea of using a whole inactivated HIV vaccine in humans, the success of such a preparation in reducing the extent of viremia in chimpanzees (Gibbs *et al.*, 1991) and of laboratories worldwide in achieving complete protection against SIVmac infection using inactivated SIVmac immunogen (Desrosiers *et al.*, 1989; Hartung *et al.*, 1992; Johnson *et al.*, 1992; Murphey-Corb *et al.*, 1989; Stott *et al.*, 1990) suggested that a form of inactivated virus might be usable (Table VIII). However, the realization that, in the SIVmac system at least, much of the protection was a result of human T-cell proteins contained in both the immunogen and challenge virus (Stott, 1991) and that no protection was seen against virus grown in macaque PBMC (Le Grand *et al.*, 1992; Norley, unpublished observation) has called these results into question. This phenomenon represents either a discouraging artifact of the SIVmac animal model or an opportunity to produce a vaccine by entirely novel means, i.e., immunization against host cell components.

In the HIV/SIV field, only a handful of animals have been successfully protected by whole inactivated immunogen against infection in a situation where no involvement of human cell proteins is possible (Berman *et al.*, 1990; Girard *et al.*, 1991; Putkonen *et al.*, 1991b) and even then only when a very low challenge dose was used. Against other lentivirus infections or disease, protection using whole inactivated virus immunogen has been achieved with visna in sheep and goats (Narayan *et al.*, 1981, 1984) and EIAV (Issel *et al.*, 1992) in horses but not against CAEV (Narayan *et al.*, 1984). Protection against FIV infection of cats has also been achieved using whole inactivated virus or fixed infected cells (Yamamoto *et al.*, 1991), but once again the system used does not totally exclude the possibility of protection mediated by response to cellular components in the immunogen.

TABLE VII. Approaches to Antiretroviral Vaccines

Immunization with whole inactivated virus
Immunization with empty virus particles
Immunization with purified native subunit
Immunization with genetically engineered subunit
Immunization with synthetic peptide
Infection with live attenuated virus
Infection with live recombinant virus vector
Infection with live recombinant virus followed by boost with subunit or peptide immunogen
Immunization with anti-idiotypic antibodies
Injection of naked DNA corresponding to retroviral genes

TABLE VIII. Examples of Successful Protection against Retroviral Infection or Disease

Immunization protocol	Retrovirus	Specificity	References
Passive antibody transfer	F-MuLV	gp70	Hunsmann *et al.* (1975a,b)
	FeLV	Broad	de Noronha *et al.* (1978), Haley *et al.* (1985)
	SIV	Broad	Putkonen *et al.* (1991a)
	HIV-1	gp120(V3)	Emini *et al.* (1992)
	HIV-2	Broad	Putkonen *et al.* (1991a)
Whole inactivated virus	MuLV	?	Ishihara *et al.* (1992)
	Visna	?	Narayan *et al.* (1981, 1984)
	FIV	?	Yamamoto *et al.* (1991)
	EIAV	?	Issel *et al.* (1992)
	SIV[a]	?[a]	Desrosiers *et al.* (1989), Hartung *et al.* (1992), Johnson *et al.* (1992), Murphey-Corb *et al.* (1989), Stott *et al.* (1990)
Subunit vaccine	MuLV	gp70	Hunsmann *et al.* (1975a,b)
	RSV	gp85, gp37	Bennett and Wright (1987)
	Prague ASV	Env	Bennett and Wright (1987)
	FeLV	gp70	Marciani *et al.* (1991)
	EIAV	Env	Issel *et al.* (1992)
	HIV	gp120	Berman *et al.* (1990)
Live, attenuated virus	SIV		Desrosiers (1992)

[a] The exact nature (antiviral or anticellular) of the protective immune response in most studies is currently unknown.

B. Subunit Vaccines

Subunit vaccines based on the retroviral envelope glycoproteins have generally been successful. This fact prompted most of the initial HIV vaccine efforts to be centered on the production of recombinant gp120. Despite a number of disappointing failures, protection of a small number of chimps against HIV infection has been achieved using recombinant gp120, gp160, or a mixture of envelope-based immunogens (Berman *et al.*, 1990; Girard *et al.*, 1991). Macaques have also been protected against SIVmne infection by infection with a vaccinia recombinant expressing the envelope gene followed by a boost with the purified protein itself (Hu *et al.*, 1992).

That mice can be protected against MuLV infection and disease by immunization with the purified gp70 has long been known (Hunsmann, 1985; Hunsmann *et al.*, 1975a). Infection with a vaccinia vector expressing F-MuLV envelope glycoprotein primed mice so that challenge with Friend virus complex was followed by a rapid anamnestic response in terms of neutralizing antibody and CTL which protected against leukemia (Earl *et al.*, 1986). Similarly, chickens are immune to Rous sarcoma virus after receiving cells infected with a RSV^{env}/vaccinia construct (Chebloune *et al.*, 1991). Subgroup-specific protection against Prague avian sarcoma virus was also achieved by immunization of envelope glycoproteins (Bennett and Wright, 1987).

However, the greatest successes have been made in immunizing against FeLV. Commercial vaccines based on a crude FOCMA and FeLV envelope glycoprotein preparation have been available since 1985 (Mastro *et al.*, 1986), and a recombinant gp70-based vaccine is even more effective (Marciani *et al.*, 1991). Presenting the gp70 in the form of immunostimulatory complexes (ISCOMS) increases further the immunogenicity in terms of neutralizing antibody and protective capacity of the molecule (Jarrett, 1990; Osterhaus *et al.*, 1985, 1987).

In addition, a variety of novel immunogens have been tested for their ability to stimulate a (protective) immune response to HIV or SIV. Stimulation of an immune response by segments of the HIV envelope glycoprotein expressed on yeast Ty particle (Griffiths *et al.*, 1991; Griffiths *et al.*, 1993; Layton *et al.*, 1993) or hepatitis B surface antigen (Michel *et al.*, 1990) holds promise, as does the use of live vectors to deliver HIV proteins to the immune system (Aldovini and Young, 1991; Chanda *et al.*, 1990; Cooney *et al.*, 1991; Waug *et al.*, 1993; Yasutomi *et al.*, 1993a). Synthetic peptides have been shown to protect against SIVmne-induced disease (Shafferman *et al.*, 1991) and the stimulation of specific CTL by peptides linked to lipids (Deres *et al.*, 1989) or by envelope glycoprotein presented as ISCOMS (Takahashi *et al.*, 1990) is promising. In what is perhaps the most novel approach, Letvin's group stimulated the production in macaques of SIVmac "neutralizing antibodies" by breaking tolerance to self-CD4 using injections of soluble human and macaque CD4 (Watanabe *et al.*, 1991).

C. Live Attenuated Vaccines

Even though the risks in using a replication-competent attenuated HIV mutant as a vaccine are even higher than those associated with whole inactivated immunogens, the use of live attenuated vaccines cannot be completely ruled out. Such vaccines should stimulate an immune response most closely matching that produced to the pathogen itself and are historically the most efficient vaccines. In the case of retroviruses, horses which recover from infection with a mild strain of EIAV have been shown to be resistant to infection with a highly pathogenic variant. In addition, Desrosiers's group (Daniel *et al.*, 1992; Desrosiers, 1992) showed that preinfection of macaques with a slow-replicating mutant of SIVmac attenuated by removal of the *nef* gene of SIVmac, prevented infection and therefore disease upon challenge by the pathogenic wild-type virus.

D. Passive Immunization

Although not vaccination in the classical sense, passive transfer of antibodies to chimps immediately prior to challenge can protect against infection (Emini *et al.*, 1992), showing the potential of antibody-mediated immune mechanisms (e.g., neutralizing antibody, ADCC). Similarly, passive transfer of immune serum to macaques protected against HIV-2 or SIVsm infection (Putkonen *et al.*, 1991a), although there was no clear evidence that neutralizing antibodies were responsible. Successful immunotherapy of cats infected with FeLV has been achieved by passive transfer of FeLV-hyperimmune serum (de Noronha *et al.*, 1978; Haley *et al.*, 1985) or even F-MuLV gp71-specific serum (de Noronha *et al.*, 1978). Moreover, mice can be protected against the effects of F-MuLV and R-MuLV infection by administration of F-MuLV-specific serum (Hunsmann *et al.*, 1975a; Schaefer *et al.*, 1975).

Successful vaccination against retroviral infection and disease appears then to depend on the virus. The relative ease with which protection can be achieved against murine and feline retroviruses has not been seen for the lentiviruses. Problems of high variability and mode of transmission (e.g., infected cells) complicate further the task of producing successful vaccines against HIV. Nevertheless, the numerous approaches which are now yielding some results hold promise for the future.

XII. CONCLUSIONS

This chapter summarizes what is known regarding the interactions of retroviruses with the immune system. The induction and control of the immune response is a complex system of cooperation involving many cell types and soluble factors. This system has evolved to combat and prevent infection by microorganisms such as retroviruses, and indeed this is the result in most cases. Presentation of foreign retroviral antigen to the immune system by macrophages and infected cells brings into motion a cascade of interactive events culminating in the production of specific antibodies and immune cells able to eliminate virus or virus-infected cells. Some of these responses are incidentally invaluable tools for diagnosing infection.

Most retroviruses are, however, not "helpless" against the onslaught of the immune system. Many have evolved properties or mechanisms that allow them to survive or even negate the immune control. For example, the relatively high rate of variability of crucial immunogenic regions often allows the virus to escape the controlling efforts of the immune system (for a while at least). Moreover, the immune response may inadvertently help the virus by enhancing infectivity or cause damage to the host by immunopathologic or autoimmune mechanisms. Finally, the predilection of retroviruses for the cells of the immune system allows the normal, delicate balance of immune control to be disrupted, often resulting in the development of immunodeficiency or chronic immune stimulation. All of these factors appear to conspire against the development of effective vaccines or therapeutics to combat retroviral infections.

Determining how a retrovirus interacts with the immune system is therefore not just an interesting academic exercise: identifying the precise cause of a retrovirus-induced disease may allow the development of successful intervention therapies. Achievements in terms of vaccination and therapy have been made in several of the retrovirus/host systems (e.g., FeLV, simian type D). It can only be hoped that the lessons learned from studying this diverse and intriguing family of viruses will ultimately provide the necessary insight into combating the problem of HIV and AIDS.

XIII. REFERENCES

Abo, T., and Balch, C. M., 1981, A differentiation antigen of human NK and K cells identified by a monoclonal antibody, *J. Immunol.* **127**:1024.

Acha-Orbea, H., Shakhov, A. N., Scarpellino, L., Kolb, E., Mueller, V., Vessaz-Shaw, A., Fuchs, R., Bloechlinger, K., Rollini, P., Billotte, J., Sarafidou, M., MacDonald, H. R., and Diggelmann, H., 1991, Clonal deletion of V beta 14-bearing T cells in mice transgenic for mammary tumour virus, *Nature* **350**:207.

Actor, J. K., Shirai, M., Kullberg, M. C., Buller, R. M., Sher, A., and Berzofsky, J. A., 1993, Helminth infection results in decreased virus-specific CD8$^+$ cytotoxic T-cell and Th1 cytokine responses as well as delayed virus clearance, *Proc. Natl. Acad. Sci. USA* **90**:948.

Aldovini, A., and Young, R. A., 1991, Humoral and cell-mediated immune responses to live recombinant BCG-HIV vaccines, *Nature* **351**:479.

Alexander, S., and Elder, J. H., 1984, Carbohydrate dramatically influences immune reactivity of antisera to viral glycoprotein antigens, *Science* **226**:1328.

Amborski, G. F., Lo, J. L., and Seger, C. L., 1989, Serological detection of multiple retroviral infections in cattle: Bovine leukemia virus, bovine syncytial virus and bovine visna virus, *Vet. Microbiol.* **20**:247.

Ammar, A., Cibert, C., Bertoli, A. M., Tsilivakos, V., Jasmin, C., and Georgoulias, V., 1991, Biological and biochemical characterization of a factor produced spontaneously by adherent cells of human immunodeficiency virus-infected patients inhibiting interleukin-2 receptor alpha chain (Tac) expression on normal T cells, *J. Clin. Invest.* **87**:2048.

Antalikova, L., Kren, V., Kasparek, R., and Bila, V., 1989, Patterns of physiological cell death and mitoses in the apical ectodermal ridge in normodactylous and polydactylous rat limb buds. A quantitative evaluation, *Folia Biol. (Prague)* **35**:339.

Archambault, D., East, N., Perk, K., and Dahlberg, J. E., 1988, Development of an enzyme-linked immunosorbent assay for caprine arthritis–encephalitis virus, *J. Clin. Microbiol.* **26**:971.

Arendrup, M., Nielsen, C., Hansen, J. E., Pedersen, C., Mathiesen, L., and Nielsen, J. O., 1992, Autologous HIV-1 neutralizing antibodies: Emergence of neutralization-resistant escape virus and subsequent development of escape virus neutralizing antibodies, *J. AIDS* **5**:303.

Arthur, L. O., Bess, J. W., Jr., Sowder, R. C., II, Benveniste, R. E., Mann, D. L., Chermann, J. C., and Henderson, L. E., 1992, Cellular proteins bound to immunodeficiency viruses: Implications for pathogenesis and vaccines, *Science* **258**:1935.

Ascher, M. S., and Sheppard, H. W., 1990, AIDS as immune system activation. II. The panergic imnesia hypothesis, *J. AIDS* **3**:177.

Autran, B., Mayaud, C. M., Raphael, M., Plata, F., Denis, M., Bourguin, A., Guillon, J. M., Debre, P., and Akoun, G., 1988, Evidence for a cytotoxic T-lymphocyte alveolitis in human immunodeficiency virus-infected patients, *AIDS* **2**:179.

Autran, B., Plata, F., Guillon, J. M., Joly, P., Mayaud, C., and Debre, P., 1990, HIV-specific cytotoxic T lymphocytes directed against alveolar macrophages in HIV-infected patients, *Res. Virol. (France)* **141**:131.

Autran, B., Plata, F., and Debre, P., 1991, MHC-restricted cytotoxicity against HIV, *J. AIDS* **4**:361.

Azocar, J., and Essex, M., 1979, Incorporation of HLA antigens into the envelope of RNA tumor viruses grown in human cells, *Cancer Res.* **39**:3388.

Azocar, J., Essex, M., and Yunis, E. J., 1983, Cell culture density modulates the incorporation of HLA antigens by enveloped viruses, *Hum. Immunol.* **7**:59.

Baba, E., Nakamura, M., Tanaka, Y., Kuroki, M., Itoyama, Y., Nakano, S., and Niho, Y., 1993, Multiple neutralizing B-cell epitopes of human T-cell leukemia virus type 1 (HTLV-1) identified by human monoclonal antibodies. A basis for the design of an HTLV-1 peptide vaccine, *J. Immunol.* **151**:1013.

Ball, J. M., Rushlow, K. E., Issel, C. J., and Montelaro, R. C., 1992, Detailed mapping of the antigenicity of the surface unit glycoprotein of equine infectious anemia virus by using synthetic peptide strategies, *J. Virol.* **66**:732.

Banapour, B., Sernatinger, J., and Levy, J. A., 1986, The AIDS-associated retrovirus is not sensitive to lysis or inactivation by human serum, *Virology* **152**:268.

Banda, N. K., Bernier, J., Kurahara, D. K., Kurrle, R., Haigwood, N., Sekaly, R. P., and Finkel, T. H., 1992, Crosslinking CD4 by human immunodeficiency virus gp120 primes T cells for activation-induced apoptosis, *J. Exp. Med.* **176**:1099.

Bartholomew, R. M., Esser, A. F., and Mueller-Eberhard, H. J., 1978, Lysis of oncornaviruses by human serum. Isolation of the viral complement (C1) receptor and identification as p15E, *J. Exp. Med.* **147**:844.

Bauer, H., Hayami, M., and Stehfen-Gervinus, J. C., 1979, Influence of different routes of anti-tumor immunization: Alternative induction of tumor immunity and tumor enhancement, *J. Immunol.* **122**:806.

Bender, B. S., Auger, F. A., Quinn, T. C., Redfield, R., Gold, J., and Folks, T. M., 1986, Impaired antibody-dependent cell-mediated cytotoxic activity in patients with the acquired immunodeficiency syndrome, *Clin. Exp. Immunol.* **64**:166.

Bennett, D. D., and Wright, S. E., 1987, Immunization with envelope glycoprotein of an avian RNA tumor virus protects against sarcoma virus tumor induction: Role of subgroup, *Virus Res.* **8**:73.

Beretta, A., Grassi, F., Pelagi, M., Clivio, A., Parravicini, C., Giovinazzo, G., Andronico, F., Lopalco, L., Verani, P., Butto, S., Titti, F., Rossi, G. B., Viale, G., Ginelli, E., and Siccardi, A. G., 1987, HIV env glycoprotein shares a cross-reacting epitope with a surface protein present on activated human monocytes and involved in antigen presentation, *Eur. J. Immunol.* **17**:1793.

Berman, A., Espinoza, L. R., Diaz, J. D., Aguilar, J. L., Rolando, T., Vasey, F. B., Germain, B. F., and Lockey, R. F., 1988, Rheumatic manifestations of human immunodeficiency virus infection, *Am. J. Med.* **85**:59.

Berman, P. W., Gregory, T. J., Riddle, L., Nakamura, G. R., Champe, M. A., Porter, J. P., Wurm, F. M., Hershberg, R. D., Cobb, E. K., and Eichberg, J. W., 1990, Protection of chimpanzees from infection by HIV-1 after vaccination with recombinant glycoprotein gp120 but not gp160, *Nature* **345**:622.

Berzofsky, J. A., 1988, Features of T-cell recognition and antigen structure useful in the design of vaccines to elicit T-cell immunity, *Vaccine* **6**:89.

Biasi, G., Facchinetti, A., Panozzo, M., Zanovello, P., Chieco-Bianchi, L., and Collavo, D., 1991, Moloney murine leukemia virus tolerance in anti-CD4 monoclonal antibody-treated adult mice, *J. Immunol.* **147**:2284.

Bjoerling, E., Broliden, K., Bernardi, D., Utter, G., Thorstensson, R., Chiodi, F., and Norrby, E., 1991, Hyperimmune antisera against synthetic peptides representing the glycoprotein of human immunodeficiency virus type 2 can mediate neutralization and antibody-dependent cytotoxic activity, *Proc. Natl. Acad. Sci. USA* **88**:6082.

Bird, P., Blacklaws, B., Reyburn, H. T., Allen, D., Hopkins, J., Sargan, D., and McConnell, I., 1993, Early events in immune evasion by the lentivirus maedi-visna occurring within infected lymphoid tissue, *J. Virol.* **67**:5187.

Bjorkman, P. J., Saper, M. A., Samraoui, B., Bennett, W. S., Strominger, J. L., and Wiley, D. C., 1987a, The foreign antigen binding site and T cell recognition regions of class I histocompatibility antigens, *Nature* **329**:512.

Bjorkman, P. J., Saper, M. A., Samraoui, B., Bennett, W. S., Strominger, J. L., and Wiley, D. C., 1987b, Structure of the human class I histocompatibility antigen, HLA-A2, *Nature* **329**:506.

Bose, H. R., 1984, Reticuloendotheliosis virus and disturbance in immune regulation, *Microbiol. Sci.* **1**:107.

Boucher, C. A., Krone, W. J., Goudsmit, J., Meloen, R. H., Naylor, P. H., Goldstein, A. L., Sun, D. K., and Sarin, P. S., 1990, Immune response and epitope mapping of a candidate HIV-1 p17 vaccine HGP30, *J. Clin. Lab. Anal.* **4**:43.

Bourgault, I., Venet, A., and Levy, J. P., 1992, Three epitopic peptides of the simian immunodeficiency virus Nef protein recognized by macaque cytolytic T lymphocytes, *J. Virol.* **66**:750.

Bova, C. A., Olsen, J. C., and Swanstrom, R., 1988, The avian retrovirus env gene family: Molecular analysis of host range and antigenic variants, *J. Virol.* **62**:75.

Bower, R. K., and Gyles, N. R., 1982, Immunologic response in chickens from different genetic lines to Rous sarcoma virus, *Cancer Detect. Prev.* **5**:221.

Boyer, V., Desgranges, C., Trabaud, M. A., Fischer, E., and Kazatchkine, M. D., 1991, Complement mediates human immunodeficiency virus type 1 infection of a human T cell line in a CD4- and antibody-independent fashion, *J. Exp. Med.* **173**:1151.

Bradley, T. P., and Bonavida, B., 1982, Mechanism of cell-mediated cytotoxicity at the single cell level. V. The importance of target cell structure in cytotoxic T lymphocyte-mediated antigen nonspecific lectin-dependent cellular cytotoxicity, *J. Immunol.* **129**:2352.

Brandhuber, B. J., Boone, T., Kenney, W. C., and McKay, D. B., 1987, Crystals and a low resolution structure of interleukin-2, *J. Biol. Chem.* **262**:12306.

Brenner, B. G., Gryllis, C., and Wainberg, M. A., 1991, Role of antibody-dependent cellular cytotoxicity and lymphokine-activated killer cells in AIDS and related diseases, *J. Leukocyte Biol.* **50**:628.

Brinchmann, J. E., Gaudernack, G., Thorsby, E., and Vartdal, F., 1992, Staphylococcal exotoxin superantigens induce human immunodeficiency virus type 1 expression in naturally infected CD4$^+$ T cells, *J. Virol.* **66**:5924.

Broliden, P. A., Ruden, U., Ouattara, A. S., de The, G., Solver, E., Trojnar, J., and Wahren, B., 1991, Specific synthetic peptides for detection of and discrimination between HIV-1 and HIV-2 infection, *J. AIDS* **4**:952.

Bruck, C., Mathot, S., Portetelle, D., Berte, C., Franssen, J. D., Herion, P., and Burny, A., 1982a, Monoclonal antibodies define eight independent antigenic regions on the bovine leukemia virus (BLV) envelope glycoprotein gp51, *Virology* **122**:342.

Bruck, C., Portetelle, D., Burny, A., and Zavada, J., 1982b, Topographical analysis by monoclonal antibodies of BLV-gp51 epitope involved in viral functions, *Virology* **122**:353.

Bruck, C., Portetelle, D., Zavada, J., and Burny, A., 1983, Molecular dissection of the bovine leukemia virus envelope glycoprotein (BLV gp51) by a monoclonal antibody study, *Haematol. Bluttransfus.* **28**:222.

Buller, R. M., Yetter, R. A., Fredrickson, T. N., and Morse, H. C., 1987, Abrogation of resistance to severe mousepox in C57BL/6 mice infected with LP-BM5 murine leukemia viruses, *J. Virol.* **61**:383.

Burns, D. P., Collignon, C., and Desrosiers, R. C., 1993, Simian immunodeficiency virus mutants resistant to serum neutralization arise during persistent infection of rhesus monkeys, *J. Virol.* **67**:4104.

Buseyne, F., McChesney, M., Porrot, F., Kovarik, S., Guy, B., and Riviere, Y., 1993, Gag-specific cytotoxic T lymphocytes from human immunodeficiency virus type 1-infected individuals: Gag epitopes are clustered in three regions of the p24gag protein, *J. Virol.* **67**:694.

Calvert, J. G., Nazerian, K., Witter, R. L., and Yanagida, N., 1993, Fowlpox virus recombinants expressing the envelope glycoprotein of an avian reticuloendotheliosis retrovirus induce neutralizing antibodies and reduce viremia in chickens, *J. Virol.* **67**:3069.

Caruso, A., Terlenghi, L., Ceccarelli, R., Verardi, R., Foresti, I., Scura, G., Manca, N., Bonfanti, C., and Turano, A., 1987, Liquid competition radioimmunoassay for the detection and quantitation of the HIV p24, *J. Virol. Methods* **17**:199.

Castro, B. A., Walker, C. M., Eichberg, J. W., and Levy, J. A., 1991, Suppression of human immunodeficiency virus replication by CD8$^+$ cells from infected and uninfected chimpanzees, *Cell. Immunol.* **132**:246.

Ceglowski, W. S., and Friedman, H., 1968, Immunosuppression by leukemia viruses. I. Effect of Friend disease virus on cellular and humoral hemolysin responses of mice to a primary immunization with sheep erythrocytes, *J. Immunol.* **101**:594.

Ceglowski, W. S., Campbell, B. P., and Friedman, H., 1975, Immunosuppression by leukemia viruses. Effect of Friend leukemia virus on humoral immune competence of leukemia-resistant C57BL/6 mice, *J. Immunol.* **114**:231.

Cerny, A., Huegin, A. W., Hardy, R. R., Hayakawa, K., Zinkernagel, R. M., Makino, M., and Morse, H. C., 1990, B cells are required for induction of T cell abnormalities in a murine retrovirus-induced immunodeficiency syndrome, *J. Exp. Med.* **171**:315.

Cerundolo, V., Lahaye, T., Horvath, C., Zanovello, P., Collavo, D., and Engers, H. D., 1987, Functional activity in vivo of effector T cell populations. III. Protection against Moloney murine sarcoma virus (M-MSV)-induced tumors in T cell deficient mice by the adoptive transfer of a M-MSV-specific cytolytic T lymphocyte clone, *Eur. J. Immunol.* **17**:173.

Chanda, P. K., Natuk, R. J., Mason, B. B., Bhat, B. M., Greenberg, L., Dheer, S. K., Molnar-Kimber, K. L., Mizutani, S., Lubeck, M. D., Davis, A. R., 1990, High level expression of the envelope glycoproteins of the human immunodeficiency virus type I in presence of rev gene using helper-independent adenovirus type 7 recombinants, *Virology* **175**:535.

Chanh, T. C., Dreesman, G. R., Kanda, P., Linette, G. P., Sparrow, J. T., Ho, D. D., and Kennedy, R. C., 1986, Induction of anti-HIV neutralizing antibodies by synthetic peptides, *EMBO J.* **5**:3065.

Chase, M. J., and Klebanoff, S. J., 1992, Viricidal effect of stimulated human mononuclear phagocytes on human immunodeficiency virus type 1, *Proc. Natl. Acad. Sci. USA* **89**:5582.

Chebloune, Y., Rulka, J., Cosset, F. L., Valsesia, S., Ronfort, C., Legras, C., Drynda, A., Kuzmak, J., Nigon, V. M., and Verdier, G., 1991, Immune response and resistance to Rous sarcoma virus challenge of chickens immunized with cell-associated glycoproteins provided with a recombinant avian leukosis virus, *J. Virol.* **65**:5374.

Cheevers, W. P., Stem, T. A., Knowles, D. P., and McGuire, T. C., 1988, Precursor polypeptides of caprine arthritis–encephalitis lentivirus structural proteins, *J. Gen. Virol.* **69**:675.

Cheevers, W. P., and McGuire, T. C., 1985, Equine infectious anemia virus: Immunopathogenesis and persistence, *Rev. Infect. Dis.* **7**:83.

Cheevers, W. P., Knowles, D. P., and Norton, L. K., 1991, Neutralization-resistant antigenic variants of caprine arthritis–encephalitis lentivirus associated with progressive arthritis, *J. Infect. Dis.* **164**:679.

Chieco-Bianchi, L., Collavo, D., and Biasi, G., 1988, Immunologic unresponsiveness to murine leukemia virus antigens: Mechanisms and role in tumor development, *Adv. Cancer Res.* **51**:277.

Chirmule, N., Saxinger, C., and Pahwa, S., 1989, Influences of related retroviruses on lymphocyte functions, *FEMS Microbiol. Immunol.* **1**:271.

Choi, Y., Kappler, J. W., and Marrack, P., 1991, A superantigen encoded in the open reading frame of the 3′ long terminal repeat of mouse mammary tumour virus, *Nature* **350**:203.

Choppin, J., Martinon, F., Gomard, E., Bahraoui, E., Connan, F., Bouillot, M., and Levy, J. P., 1990, Analysis of physical interactions between peptides and HLA molecules and application to the detection of human immunodeficiency virus 1 antigenic peptides, *J. Exp. Med.* **172**:889.

Cianciolo, G. J., Copeland, T. D., Oroszlan, S., and Snyderman, R., 1985, Inhibition of lymphocyte proliferation by a synthetic peptide homologous to retroviral envelope proteins, *Science* **230**:453.

Ciardelli, T. L., Landgraf, B., Gadski, R., Strnad, J., Cohen, F. E., and Smith, K. A., 1988, A design approach to the structural analysis of interleukin-2, *J. Mol. Recogn.* **1**:42.

Cichutek, K., Merget, H., Norley, S., Linde, R., Kreuz, W., Gahr, M., and Kurth, R., 1992, Development of a quasispecies of human immunodeficiency virus type 1 in vivo, *Proc. Natl. Acad. Sci. USA* **89**:7365.

Clarke, M. F., Gelmann, E. P., and Reitz, M. S., 1983, Homology of human T-cell leukaemia virus envelope gene with class I HLA gene, *Nature* **305**:60.

Claverie, J. M., Kourilsky, P., Langlade-Demoyen, P., Chalufour-Prochnicka, A., Dadaglio, G., Tekaia, F., Plata, F., and Bougueleret, L., 1988, T-immunogenic peptides are constituted of rare sequence patterns. Use in the identification of T epitopes in the human immunodeficiency virus gag protein, *Eur. J. Immunol.* **18**:1547.

Clerici, M., and Shearer, G. M., 1993, A Th1 → Th2 switch is a critical step in the etiology of HIV infection, *Immunol. Today* **14**:107.

Clerici, M., Stocks, N. I., Zajac, R. A., Boswell, R. N., Bernstein, D. C., Mann, D. L., Shearer, G. M., and Berzofsky, J. A., 1989a, Interleukin-2 production used to detect antigenic peptide recognition by T-helper lymphocytes from asymptomatic HIV-seropositive individuals, *Nature* **339**:383.

Clerici, M., Stocks, N. I., Zajac, R. A., Boswell, R. N., Lucey, D. R., Via, C. S., and Shearer, G. M., 1989b, Detection of three distinct patterns of T helper cell dysfunction in asymptomatic, human immunodeficiency virus-seropositive patients. Independence of $CD4^+$ cell numbers and clinical staging, *J. Clin. Invest.* **84**:1892.

Clerici, M., Berzofsky, J. A., Shearer, G. M., and Tacket, C. O., 1991a, Exposure to human immunodeficiency virus (HIV) type I indicated by HIV-specific T helper cell responses before detection of infection by polymerase chain reaction and serum antibodies, *J. Infect. Dis.* **164**:178.

Clerici, M., Lucey, D. R., Zajac, R. A., Boswell, R. N., Gebel, H. M., Takahashi, H., Berzofsky, J. A., and Shearer, G. M., 1991b, Detection of cytotoxic T lymphocytes specific for synthetic peptides of gp160 in HIV-seropositive individuals, *J. Immunol.* **146**:2214.

Clerici, M., Giorgi, J. V., Chou, C. C., Gudeman, V. K., Zack, J. A., Gupta, P., Ho, H. N., Nishanian, P. G., Berzofsky, J. A., and Shearer, G. M., 1992, Cell-mediated immune response to human immunodeficiency virus (HIV) type 1 in seronegative homosexual men with recent sexual exposure to HIV-1, *J. Infect. Dis.* **165**:1012.

Clerici, M., Shearer, G., Hounsell, E. F., Jameson, B., Habeshaw, J., and Dalgleish, A. G., 1993, Alloactivated cytotoxic T cells recognize the carboxy-terminal domain of human immunodeficiency virus-1 gp120 envelope glycoprotein, *Eur. J. Immunol.* **23**:2022.

Coffin, J. M., 1992, Structure and classification of retroviruses, in: *The Retroviridae*, Vol. 1 (J. A. Levy, ed.), p. 19, Plenum Press, New York.

Cole, B. C., Kartchner, D. R., and Wells, D. J., 1990, Stimulation of mouse lymphocytes by a mitogen derived from Mycoplasma arthritidis (MAM). VIII. Selective activation of T cells expressing distinct V beta T cell receptors from various strains of mice by the "superantigen" MAM, *J. Immunol.* **144**:425.

Collavo, D., Colombatti, A., Biasi, G., Chieco-Bianchi, L., and Davies, A. J., 1976, Immune reactivity in the Moloney strain of murine sarcoma virus oncogenesis: Requirement of thymus-derived lymphocytes for in vivo protection, *J. Natl. Cancer Inst.* **56**:603.

Connor, R. I., Dinces, N. B., Howell, A. L., Romet-Lemonne, J. L., Pasquali, J. L., and Fanger, M. W., 1991, Fc receptors for IgG (Fc gamma Rs) on human monocytes and macrophages are not infectivity receptors for human immunodeficiency virus type 1 (HIV-1): Studies using bispecific antibodies to target HIV-1 to various myeloid cell surface molecules, including the Fc gamma R, *Proc. Natl. Acad. Sci. USA* **88**:9593.

Coombs, R. W., Collier, A. C., Allain, J. P., Nikora, B., Leuther, M., Gjerset, G. F., and Corey, L., 1989, Plasma viremia in human immunodeficiency virus infection, *N. Engl. J. Med.* **321**:1626.

Cooney, E. L., Collier, A. C., Greenberg, P. D., Coombs, R. W., Zarling, J., Arditti, D. E., Hoffman, M. C., Hu, S. L., and Corey, L., 1991, Safety of and immunological response to a recombinant vaccinia virus vaccine expressing HIV envelope glycoprotein, *Lancet* **337**:567.

Crane, S. E., Clements, J. E., and Narayan, O., 1988, Separate epitopes in the envelope of visna virus are responsible for fusion and neutralization: Biological implications for anti-fusion antibodies in limiting virus replication, *J. Virol.* **62**:2680.

Crawford, R. M., Finbloom, D. S., Ohara, J., Paul, W. E., and Meltzer, M. S., 1987, B cell stimulatory factor-1 (interleukin 4) activates macrophages for increased tumoricidal activity and expression of Ia antigens, *J. Immunol.* **139**:135.

Cruickshank, J. K., Rudge, P., Dalgleish, A. G., Newton, M., McLean, B. N., Barnard, R. O., Kendall, B. E., and Miller, D. H., 1989, Tropical spastic paraparesis and human T cell lymphotropic virus type 1 in the United Kingdom, *Brain* **112**:1057.

Cui, Z. Z., Lee, L. F., Silva, R. F., and Witter, R. L., 1986, Monoclonal antibodies against avian reticuloendotheliosis virus: Identification of strain-specific and strain-common epitopes, *J. Immunol.* **136**:4237.

Cui, Z. Z., Lee, L. F., Smith, E. J., Witter, R. L., and Chang, T. S., 1988, Monoclonal-antibody-mediated enzyme-linked immunosorbent assay for detection of reticuloendotheliosis viruses, *Avian Dis.* **32**:32.

Culmann, B., Gomard, E., Kieny, M. P., Guy, B., Dreyfus, F., Saimot, A. G., Sereni, D., and Levy, J. P., 1989, An antigenic peptide of the HIV-1 NEF protein recognized by cytotoxic T lymphocytes of seropositive individuals in association with different HLA-B molecules, *Eur. J. Immunol.* **19**:2383.

Culmann, B., Gomard, E., Kieny, M. P., Guy, B., Dreyfus, F., Saimot, A. G., Sereni, D., Sicard, D., and Levy, J. P., 1991, Six epitopes reacting with human cytotoxic $CD8^+$ T cells in the central region of the HIV-1 NEF protein, *J. Immunol.* **146**:1560.

Curiel, T. J., Wong, J. T., Gorczyca, P. F., Schooley, R. T., and Walker, B. D., 1993, $CD4^+$ human immunodeficiency virus type 1 (HIV-1) envelope-specific cytotoxic T lymphocytes derived from the peripheral blood cells of an HIV-1-infected individual, *AIDS Res. Hum. Retroviruses* **9**:61.

Dalgleish, A. G., Chanh, T. C., Kennedy, R. C., Kanda, P., Clapham, P. R., and Weiss, R. A., 1988, Neutralization of diverse HIV-1 strains by monoclonal antibodies raised against a gp41 synthetic peptide, *Virology* **165**:209.

Dalgleish, A., Sinclair, A., Steel, M., Beatson, D., Ludlam, C., and Habeshaw, J., 1990, Failure of ADCC to predict HIV-associated disease progression or outcome in a haemophiliac cohort, *Clin. Exp. Immunol.* **81**:5.

Dalgleish, A. G., Wilson, S., Gompels, M., Ludlam, C., Gazzard, B., Coates, A. M., and Habeshaw, J., 1992, T-cell receptor variable gene products and early HIV-1 infection, *Lancet* **339**:824.

Daniel, M. D., Kirchhoff, F., Czajak, S. C., Sehgal, P. K., and Desrosiers, R. C., 1992, Protective effects of a live attenuated SIV vaccine with a deletion in the nef gene, *Science* **258**:1938.

Darban, H., Enriquez, J., Sterling, C. R., Lopez, M. C., Chen, G., Abbaszadegan, M., and Watson, R. R., 1991, Cryptosporidiosis facilitated by murine retroviral infection with LP-BM5, *J. Infect. Dis.* **164**:741.

Denner, J., Wunderlich, V., and Bierwolf, D., 1980, Suppression of human lymphocyte mitogen response by disrupted primate retroviruses of type C (baboon endogenous virus) and type D (PMFV), *Acta Biol. Med. Ger.* **39**:19.

de Noronha, F., Schaefer, W., Essex, M., and Bolognesi, D. P., 1978, Influence of antisera to oncornavirus glycoprotein (gp71) on infections of cats with feline leukemia virus, *Virology* **85**:617.

Deres, K., Schild, H., Wiesmueller, K. H., Jung, G., and Rammensee, H. G., 1989, In vivo priming of virus-specific cytotoxic T lymphocytes with synthetic lipopeptide vaccine, *Nature* **342**:561.

Desrosiers, R. C., 1992, HIV with multiple gene deletions as a live attenuated vaccine for AIDS, *AIDS Res. Hum. Retroviruses* **8**:411.

Desrosiers, R. C., Wyand, M. S., Kodama, T., Ringler, D. J., Arthur, L. O., Sehgal, P. K., Letvin, N. L., King, N. W., and Daniel, M. D., 1989, Vaccine protection against simian immunodeficiency virus infection, *Proc. Natl. Acad. Sci. USA* **86**:6353.

Dodd, R. Y., and Fang, C. T., 1990, The Western immunoblot procedure for HIV antibodies and its interpretation, *Arch. Pathol. Lab. Med.* **114**:240.

Driscoll, D. M., Onuma, M., and Olson, C., 1977, Inhibition of bovine leukemia virus release by antiviral antibodies, *Arch. Virol.* **55**:139.

Duncan, A. M., and Heddle, J. A., 1984, The frequency and distribution of apoptosis induced by three non-carcinogenic agents in mouse colonic crypts, *Cancer Lett.* **23**:307.

Durda, P. J., Bacheler, L., Clapham, P., Jenoski, A. M., Leece, B., Matthews, T. J., McKnight, A., Pomerantz, R., Rayner, M., and Weinhold, K. J., 1990, HIV-1 neutralizing monoclonal antibodies induced by a synthetic peptide, *AIDS Res. Hum. Retroviruses* **6**:1115.

Dyson, P. J., Knight, A. M., Fairchild, S., Simpson, E., and Tomonari, K., 1991, Genes encoding ligands for deletion of V beta 11 T cells cosegregate with mammary tumour virus genomes, *Nature* **349**:531.

Earl, P. L., Moss, B., Morrison, R. P., Wehrly, K., Nishio, J., and Chesebro, B., 1986, T-lymphocyte priming and protection against Friend leukemia by vaccinia–retrovirus env gene recombinant, *Science* **234**:728.

Egberink, H. F., Ederveen, J., Montelaro, R. C., Pedersen, N. C., Horzinek, M. C., and Koolen, M. J., 1990, Intracellular proteins of feline immunodeficiency virus and their antigenic relationship with equine infectious anaemia virus proteins, *J. Gen. Virol.* **71**:739.

Elder, J. H., McGee, J. S., Munson, M., Houghten, R. A., Kloetzer, W., Bittle, J. L., and Grant, C. K., 1987, Localization of neutralizing regions of the envelope gene of feline leukemia virus by using anti-synthetic peptide antibodies, *J. Virol.* **61**:8.

Ellis, T. M., Wilcox, G. E., and Robinson, W. F., 1987, Antigenic variation of caprine arthritis–encephalitis virus during persistent infection of goats, *J. Gen. Virol.* **68**:3145.

Emau, P., McClure, H. M., Isahakia, M., Else, J. G., and Fultz, P. N., 1991, Isolation from African Sykes' monkeys (Cercopithecus mitis) of a lentivirus related to human and simian immunodeficiency viruses, *J. Virol.* **65**:2135.

Emini, E. A., Schleif, W. A., Nunberg, J. H., Conley, A. J., Eda, Y., Tokiyoshi, S., Putney, S. D., Matsushita, S., Cobb, K. E., Jett, C. M., Eicherg, J. W., and Murthy, K. K., 1992, Prevention of HIV-1 infection in chimpanzees by gp120 V3 domain-specific monoclonal antibody, *Nature* **355**:728.

Ennen, J., and Kurth, R., 1993, Interferon gamma-activated monocytes impair infectivity of HIV particles by an oxygen metabolite-dependent reaction, *Immunology* **78**:171.

Ennen, J., Seipp, I., Norley, S. G., and Kurth, R., 1990, Decreased accessory cell function of macrophages after infection with human immunodeficiency virus type 1 in vitro, *Eur. J. Immunol.* **20**:2451.

Ennen, J., Findeklee, H., Dittmar, M., Werner, K., Norley, S., and Kurth, R., 1993, $CD8^+$ T-lymphocytes of African green monkeys secrete an immunodeficiency virus suppressing lymphokine, *Proc. Natl. Acad. Sci. USA, in press.*

Essex, M., 1975, Horizontally and vertically transmitted oncornaviruses of cats, *Adv. Cancer Res.* **21**:175.

Essex, M., 1988, Retroviruses: Leukemia and immunosuppression. The Yohei Ito memorial lecture, *Leukemia* **2**:3.

Essex, M., Klein, G., Snyder, S. P., and Harrold, J. B., 1971, Correlation between humoral antibody and regression of tumours induced by feline sarcoma virus, *Nature* **233**:195.

Falk, K., Roetzschke, O., Stevanovic, S., Jung, G., and Rammensee, H. G., 1991, Allele-specific motifs revealed by sequencing of self-peptides eluted from MHC molecules, *Nature* **351**:290.

Fenouillet, E., Sorensen, A. M., Lacroix, M., Coutellier, A., Herson, S., Fretz-Foucault, C., and Gluckman, J. C., 1990, Early and specific diagnosis of seropositivity to HIVs by an enzyme-linked immunosorbent assay using env-derived synthetic peptides, *AIDS* **4**:1137.

Ferrer, J. F., 1982, Eradication of bovine leukemia virus infection from a high-prevalence herd, using radioimmunoassay for identification of infected animals, *J. Am. Vet. Med. Assoc.* **180**:890.

Ferrer, J. F., Bhatt, D. M., Abt, D. A., Marshak, R. R., and Baliga, V. L., 1975, Serological diagnosis of infection with the putative bovine leukemia virus, *Cornell Vet.* **65**:527.

Festenstein, H., 1973, Immunogenetic and biological aspects of in vitro lymphocyte allotransformation (MLR) in the mouse, *Transplant. Rev.* **15**:62.

Fischinger, P. J., Schäfer, W., and Bolognesi, D. P., 1976, Neutralization of homologous and heterologous oncornaviruses by antisera against the p15(E) and gp71 polypeptides of Friend murine leukemia virus, *Virology* **71**:169.

Fish, D. C., Gilden, R. V., Bare, R. M., Trimmer, R. M., and Huebner, R. J., 1979, Prevention of spontaneous leukemia in AKR mice by passive immunization and type specificity of this protection, *J. Natl. Cancer Inst.* **62**:943.

Fitzpatrick, E. A., Bryson, J. S., Rhoads, C., Kaplan, A. M., and Cohen, D. A., 1992, T-deficient transmembrane signaling in $CD4^+$ T cells of retroviral-induced immune-deficient mice, *J. Immunol.* **148**:3377.

Flower, R. L., Wilcox, G. E., Cook, R. D., and Ellis, T. M., 1985, Detection and prevalence of serotypes of feline syncytial spumaviruses, *Arch. Virol.* **83**:53.

Flyer, D. C., Burakoff, S. J., and Faller, D. V., 1983, Cytotoxic T lymphocyte recognition of transfected cells expressing a cloned retroviral gene, *Nature* **305**:815.

Flyer, D. C., Burakoff, S. J., and Faller, D. V., 1985, Retrovirus-induced changes in major histocompatibility complex antigen expression influence susceptibility to lysis by cytotoxic T lymphocytes, *J. Immunol.* **135**:2287.

Frankel, W. N., Rudy, C., Coffin, J. M., and Huber, B. T., 1991, Linkage of Mls genes to endogenous mammary tumour viruses of inbred mice, *Nature* **349**:526.

Frelinger, J. A., Gotch, F. M., Zweerink, H., Wain, E., and McMichael, A. J., 1990, Evidence of widespread binding of HLA class I molecules to peptides, *J. Exp. Med.* **172**:827.

Fultz, P. N., Stricker, R. B., McClure, H. M., Anderson, D. C., Switzer, W. M., and Horaist, C., 1990, Humoral response to SIV/SMM infection in macaque and mangabey monkeys, *J. AIDS* **3**:319.

Furtwaengler, J. A., Hall, S. H., and Koskinen-Moffett, L. K., 1985, Sutural morphogenesis in the mouse calvaria: The role of apoptosis, *Acta Anat. (Basel)* **124**:74.

Gallaher, W. R., 1987, Detection of a fusion peptide sequence in the transmembrane protein of human immunodeficiency virus, *Cell* **50**:327.

Gallaher, W. R., Ball, J. M., Garry, R. F., Griffin, M. C., and Montelaro, R. C., 1989, A general model for the transmembrane proteins of HIV and other retroviruses, *AIDS Res. Hum. Retroviruses* **5**:431.

Gallo, D., Hoffman, M. N., Yeh, E. T., George, J. R., and Hanson, C. V., 1992, Comparison of indirect immunofluorescence and membrane fluorescence assays for the differentiation of antibodies to human immunodeficiency virus types 1 and 2, *J. Clin. Microbiol.* **30**:2275.

Gallo, R. C., Mann, D., Broder, S., Ruscetti, F. W., Maeda, M., Kalyanaraman, V. S., Robert-Guroff, M., and Reitz, M. S., 1982, Human T-cell leukemia–lymphoma virus (HTLV) is in T but not B lymphocytes from a patient with cutaneous T-cell lymphoma, *Proc. Natl. Acad. Sci. USA* **79**:5680.

Garry, R. F., Fermin, C. D., Hart, D. J., Alexander, S. S., Donehower, L. A., and Hong, L. Z., 1990, Detection of a human intracisternal A-type retroviral particle antigenically related to HIV, *Science* **250**:1127.

Gatei, M. H., Good, M. F., Daniel, R. C., and Lavin, M. F., 1993, T-cell responses to highly conserved CD4 and CD8 epitopes on the outer membrane protein of bovine leukemia virus: relevance to vaccine development, *J. Virol.* **67**:1796.

Gazzinelli, R. T., Hartley, J. W., Fredrickson, T. N., Chattopadhyay, S. K., Sher, A., and Morse, H. C., 1992a, Opportunistic infections and retrovirus-induced immunodeficiency: Studies of acute and chronic infections with Toxoplasma gondii in mice infected with LP-BM5 murine leukemia viruses, *Infect. Immun.* **60**:4394.

Gazzinelli, R. T., Makino, M., Chattopadhyay, S. K., Snapper, C. M., Sher, A., Huegin, A. W., and Morse, H. C., 1992b, $CD4^+$ subset regulation in viral infection. Preferential activation of Th2 cells during progression of retrovirus-induced immunodeficiency in mice, *J. Immunol.* **148**:182.

Gelderblom, H., Reupke, H., Winkel, T., Kunze, R., and Pauli, G., 1987, MHC-antigens: Constituents of the envelopes of human and simian immunodeficiency viruses, *Z. Naturforsch.* **42**:1328.

Gerencer, M., Valpotic, I., Jukic, B., Tomaskovic, M., and Basic, I., 1989, Qualitative analyses of cellular immune functions in equine infectious anemia show homology with AIDS, *Arch. Virol.* **104**:249.

Gibbs, C. J., Peters, R., Gravell, M., Johnson, B. K., Jensen, F. C., Carlo, D. J., and Salk, J., 1991, Observations after human immunodeficiency virus immunization and challenge of human immunodeficiency virus seropositive and seronegative chimpanzees, *Proc. Natl. Acad. Sci. USA* **88**:3348.

Girard, M., Kieny, M. P., Pinter, A., Barre-Sinoussi, F., Nara, P., Kolbe, H., Kusumi, K., Chaput, A., Reinhart, T., Muchmore, E., Ronco, J., Kaczorek, M., Gomad, E., Gluckman, J., and Fultz, P. N., 1991, Immunization of chimpanzees confers protection against challenge with human immunodeficiency virus, *Proc. Natl. Acad. Sci. USA* **88**:542.

Gnann, J. W., Nelson, J. A., and Oldstone, M. B., 1987, Fine mapping of an immunodominant domain in the transmembrane glycoprotein of human immunodeficiency virus, *J. Virol.* **61**:2639.

Golding, H., Robey, F. A., Gates, F. T., Linder, W., Beining, P. R., Hoffman, T., and Golding, B., 1988, Identification of homologous regions in human immunodeficiency virus I gp41 and human MHC class II beta 1 domain. I. Monoclonal antibodies against the gp41-derived peptide and patients' sera react with native HLA class II antigens, suggesting a role for autoimmunity in the pathogenesis of acquired immune deficiency syndrome, *J. Exp. Med.* **167**:914.

Golstein, P., Ojcius, D. M., and Young, D. E., 1991, Cell death mechanisms and the immune system, *Immunol. Rev.* **121**:29.

Goodenow, R. S., Vogel, J. M., and Linsk, R. L., 1985, Histocompatibility antigens on murine tumors, *Science* **230**:777.

Gotch, F., McAdam, S. N., Allsopp, C. E., Gallimore, A., Elvin, J., Kieny, M. P., Hill, A. V., McMichael, A. J., and Whittle, H. C., 1993, Cytotoxic T cells in HIV2 seropositive Gambians. Identification of a virus-specific MHC-restricted peptide epitope, *J. Immunol.* **151**:3361.

Gottlieb, R. A., Lennarz, W. J., Knowles, R. D., Cianciolo, G. J., Dinarello, C. A., Lachman, L. B., and Kleinerman, E. S., 1989, Synthetic peptide corresponding to a conserved domain of the retroviral protein p15E blocks IL-1-mediated signal transduction, *J. Immunol.* **142**:4321.

Goudsmit, J., Debouck, C., Meloen, R. H., Smit, L., Bakker, M., Asher, D. M., Wolff, A. V., Gibbs, C. J., and Gajdusek, D. C., 1988a, Human immunodeficiency virus type 1 neutralization epitope with conserved architecture elicits early type-specific antibodies in experimentally infected chimpanzees, *Proc. Natl. Acad. Sci. USA* **85**:4478.

Goudsmit, J., Ljunggren, K., Smit, L., Jondal, M., and Fenyoe, E. M., 1988b, Biological significance of the antibody response to HIV antigens expressed on the cell surface, *Arch. Virol.* **103**:189.

Goudsmit, J., Meloen, R. H., Brasseur, R., and Barin, F., 1989, Human B-cell epitopes of HIV-2 transmembrane protein are similarly spaced as in HIV-1, *J. AIDS* **2**:297.

Gougeon, M. L., Garcia, S., Heeney, J., Tschopp, R., Lecoeur, H., Guetard, D., Rame, V., Dauguet, C., and Montagnier, L., 1993, Programmed cell death in AIDS-related HIV and SIV infections, *AIDS Res. Hum. Retroviruses* **9**:553.

Grant, C. K., and Michalek, M. T., 1981, Feline leukemia—Unique and cross-relating antigens on individual virus-producing tumors identified by complement-dependent antibody, *Int. J. Cancer* **28**:209.

Grant, C. K., de Noronha, F., Tusch, C., Michalek, M. T., and McLane, M. F., 1980a, Protection of cats against progressive fibrosarcomas and persistent leukemia virus infection by vaccination with feline leukemia cells, *J. Natl. Cancer Inst.* **65**:1285.

Grant, C. K., Essex, M., Gardner, M. B., and Hardy, W. D., 1980b, Natural feline leukemia virus infection and the immune response of cats of different ages, *Cancer Res.* **40**:823.

Grant, C. K., Harris, D., Essex, M. E., Pickard, D. K., Hardy, W. D., and de Noronha, F., 1980c, Protection of cats against feline leukemia virus-positive and virus-negative tumors by complement-dependent antibody, *J. Natl. Cancer Inst.* **64**:1527.

Gras, G. S., and Dormont, D., 1991, Antibody-dependent and antibody-independent complement-mediated enhancement of human immunodeficiency virus type 1 infection in a human, Epstein–Barr virus-transformed B-lymphocytic cell line, *J. Virol.* **65**:541.

Gregersen, J. P., Mehdi, S., Baur, A., and Hilfenhaus, J., 1990, Antibody- and complement-mediated lysis of HIV-infected cells and inhibition of viral replication, *J. Med. Virol.* **30**:287.

Griffiths, J. C., Berrie, E. L., Holdsworth, L. N., Moore, J. P., Harris, S. J., Senior, J. M., Kingsman, S. M., Kingsman, A. J., and Adams, S. E., 1991, Induction of high-titer neutralizing antibodies, using hybrid human immunodeficiency virus V3-Ty viruslike particles in a clinically relevant adjuvant, *J. Virol.* **65**:450.

Griffiths, J. C., Harris, S. J., Layton, G. T., Berrie, E. L., French, T. J., Burns, N. R., Adams, S. E., and Kingsman, A. J., 1993, Hybrid human immunodeficiency virus Gag particles as an antigen carrier system: induction of cytotoxic T-cell and humoral responses by a Gag:V3 fusion, *J. Virol.* **67**:3191.

Groux, H., Torpier, G., Monte, D., Mouton, Y., Capron, A., and Ameisen, J. C., 1992, Activation-induced death by apoptosis in $CD4^+$ T cells from human immunodeficiency virus-infected asymptomatic individuals, *J. Exp. Med.* **175**:331.

Gupta, P., and Ferrer, J. F., 1981, Comparison of various serological and direct methods for the diagnosis of BLV infection in cattle, *Int. J. Cancer* **28**:179.

Haigwood, N. L., Nara, P. L., Brooks, E., Van Nest, G. A., Ott, G., Higgins, K. W., Dunlop, N., Scandella, C. J., Eichberg, J. W., and Steimer, K. S., 1992, Native but not denatured recombinant human immunodeficiency virus type 1 gp120 generates broad-spectrum neutralizing antibodies in baboons, *J. Virol.* **66**:172.

Haley, P. J., Hoover, E. A., Quackenbush, S. L., Gasper, P. W., and Macy, D. W., 1985, Influence of antibody infusion on pathogenesis of experimental feline leukemia virus infection, *J. Natl. Cancer Inst.* **74**:821.

Halstead, S. B., 1988, Pathogenesis of dengue: Challenges to molecular biology, *Science* **239**:476.

Hammond, S. A., Obah, E., Stanhope, P., Monell, C. R., Strand, M., Robbins, F. M., Bias, W. B., Karr, R. W., Koenig, S., and Siliciano, R. F., 1991, Characterization of a conserved T cell epitope in HIV-1 gp41 recognized by vaccine-induced human cytolytic T cells, *J. Immunol.* **146**:1470.

Hansen, J. E., Nielsen, C., Clausen, H., Mathiesen, L. R., and Nielsen, J. O., 1991, Effect of anti-carbohydrate antibodies on HIV infection in a monocytic cell line (U937), *Antiviral Res.* **16**:233.

Haraguchi, S., Fujii, T., Matsuo, T., and Yoshida, T. O., 1988, Mechanism of target cell lysis by cytotoxic T lymphocytes (CTL) employing RSV-induced H-2 congenic and recombinant mouse tumor cells: Demonstration of soluble mediators released from H-2-restricted CTL clones, *Cell. Immunol.* **112**:279.

Haran-Ghera, N., and Rubio, N., 1977, Immunologic characteristics in relation to high and low leukemogenic activity of radiation leukemia virus variants. II. Analysis of the immune response, *J. Immunol.* **118**:607.

Haran-Ghera, N., Ben-Yaakov, M., and Peled, A., 1977, Immunologic characteristics in relation to high and low leukemogenic activity of radiation leukemia virus variants. I. Cellular analysis of immunosuppression, *J. Immunol.* **118**:600.

Hardy, W. D., 1991, General principles of retrovirus immunodetection tests, *J. Am. Vet. Med. Assoc.* **199**:1282.

Hardy, W. D., Hess, P. W., MacEwen, E. G., McClelland, A. J., Zuckerman, E. E., Essex, M., Cotter, S. M., and Jarrett, O., 1976, Biology of feline leukemia virus in the natural environment, *Cancer Res.* **36**:582.

Harrell, R. A., Cianciolo, G. J., Copeland, T. D., Oroszlan, S., and Snyderman, R., 1986, Suppression of the respiratory burst of human monocytes by a synthetic peptide homologous to envelope proteins of human and animal retroviruses, *J. Immunol.* **136**:3517.

Harris, D. T., Cianciolo, G. J., Snyderman, R., Argov, S., and Koren, H. S., 1987, Inhibition of human natural killer cell activity by a synthetic peptide homologous to a conserved region in the retroviral protein, p15E, *J. Immunol.* **138**:889.

Hart, M. K., Palker, T. J., Matthews, T. J., Langlois, A. J., Lerche, N. W., Martin, M. E., Scearce, R. M., McDanal, C., Bolognesi, D. P., and Haynes, B. F., 1990, Synthetic peptides containing T and B cell epitopes from human immunodeficiency virus envelope gp120 induce anti-HIV proliferative responses and high titers of neutralizing antibodies in rhesus monkeys, *J. Immunol.* **145**:2677.

Hartung, S., Norley, S. G., Ennen, J., Cichutek, K., Plesker, R., and Kurth, R., 1992, Vaccine protection against SIVmac infection by high- but not low-dose whole inactivated virus immunogen, *J. AIDS* **5**:461.

Hedenskog, M., Dewhurst, S., Ludvigsen, C., Sinangil, F., Rodriguez, L., Wu, Y. T., and Volsky, D. J., 1986, Testing for antibodies to AIDS-associated retrovirus (HTLV-III/LAV) by indirect fixed cell immunofluorescence: Specificity, sensitivity, and applications, *J. Med. Virol.* **19**:325.

Heinzel, F. P., Sadick, M. D., Holaday, B. J., Coffman, R. L., and Locksley, R. M., 1989, Reciprocal expression of interferon gamma or interleukin 4 during the resolution or progression of murine leishmaniasis. Evidence for expansion of distinct helper T cell subsets, *J. Exp. Med.* **169**:59.

Heinzel, F. P., Sadick, M. D., Mutha, S. S., and Locksley, R. M., 1991, Production of interferon gamma, interleukin 2, interleukin 4, and interleukin 10 by CD4+ lymphocytes in vivo during healing and progressive murine leishmaniasis, *Proc. Natl. Acad. Sci. USA* **88**:7011.

Heller, R. A., Song, K., Fan, N., and Chang, D. J., 1992, The p70 tumor necrosis factor receptor mediates cytotoxicity, *Cell* **70**:47.

Hertenstein, B., Kreja, L., and Seidel, H. J., 1984, Natural killer (NK) cells and interferon in Friend-virus-infected susceptible and resistant mice, *Exp. Hematol.* **12**:603.

Hinkula, J., Rosen, J., Sundqvist, V. A., Stigbrand, T., and Wahren, B., 1990, Epitope mapping of the HIV-1 gag region with monoclonal antibodies, *Mol. Immunol.* **27**:395.

Ho, D. D., McKeating, J. A., Li, X. L., Moudgil, T., Daar, E. S., Sun, N. C., and Robinson, J. E., 1991, Conformational epitope on gp120 important in CD4 binding and human immunodeficiency virus type 1 neutralization identified by a human monoclonal antibody, *J. Virol.* **65**:489.

Hoffenbach, A., Langlade-Demoyen, P., Dadaglio, G., Vilmer, E., Michel, F., Mayaud, C., Autran, B., and Plata, F., 1989, Unusually high frequencies of HIV-specific cytotoxic T lymphocytes in humans, *J. Immunol.* **142**:452.

Hoffman, P. M., Cimino, E. F., and Robbins, D. S., 1991, Effects of viral specific cytotoxic lymphocytes on the expression of murine leukemia virus induced neurologic disease, *J. Neuroimmunol.* **33**:157.

Holt, C. A., Osorio, K., and Lilly, F., 1986, Friend virus-specific cytotoxic T lymphocytes recognize both gag and env gene-encoded specificities, *J. Exp. Med.* **164**:211.

Homsy, J., Meyer, M., Tateno, M., Clarkson, S., and Levy, J. A., 1989, The Fc and not CD4 receptor mediates antibody enhancement of HIV infection in human cells, *Science* **244**:1357.

Homsy, J., Meyer, M., and Levy, J. A., 1990, Serum enhancement of human immunodeficiency virus (HIV) infection correlates with disease in HIV-infected individuals, *J. Virol.* **64**:1437.

Honda, M., Yamamoto, S., Cheng, M., Yasukawa, K., Suzuki, H., Saito, T., Osugi, Y., Tokunaga, T., and Kishimoto, T., 1992, Human soluble IL-6 receptor: Its detection and enhanced release by HIV infection, *J. Immunol.* **148**:2175.

Honma, T., Onuma, M., Suneya, M., Mikami, T., and Izawa, H., 1980, Cytotoxic antibody in cattle and sheep exposed to bovine leukemia virus, *Arch. Virol.* **66**:293.

Hoshino, H., Tanaka, H., Miwa, M., and Okada, H., 1984, Human T-cell leukaemia virus is not lysed by human serum, *Nature* **310**:324.

Hosie, M. J., and Jarrett, O., 1990, Serological responses of cats to feline immunodeficiency virus, *AIDS* **4**:215.

Hosmalin, A., Clerici, M., Houghten, R., Pendleton, C. D., Flexner, C., Lucey, D. R., Moss, B., Germain, R. N., Shearer, G. M., and Berzofsky, J. A., 1990, An epitope in human immunodeficiency virus 1 reverse transcriptase recognized by both mouse and human cytotoxic T lymphocytes, *Proc. Natl. Acad. Sci. USA* **87**:2344.

Hu, S. L., Zarling, J. M., Chinn, J., Travis, B. M., Moran, P. A., Sias, J., Kuller, L., Morton, W. R., Heidecker, G., and Benveniste, R. E., 1989, Protection of macaques against simian AIDS by immunization with a recombinant vaccinia virus expressing the envelope glycoproteins of simian type D retrovirus, *Proc. Natl. Acad. Sci. USA* **86**:7213.

Hu, S. L., Abrams, K., Barber, G. N., Moran, P., Zarling, J. M., Langlois, A. J., Kuller, L., Morton, W. R., and Benveniste, R. E., 1992, Protection of macaques against SIV infection by subunit vaccines of SIV envelope glycoprotein gp160, *Science* **255**:456.

Huebner, R. J., Gilden, R. V., Lane, W. T., Toni, R., Trimmer, R. W., and Hill, P. R., 1976a, Suppression of murine type-C RNA virogenes by type-specific oncornavirus vaccines: Prospects for prevention of cancer, *Proc. Natl. Acad. Sci. USA* **73**:620.

Huebner, R. J., Gilden, R. V., Toni, R., Hill, R. W., Trimmer, R. W., Fish, D. C., and Sass, B., 1976b, Prevention of spontaneous leukemia in AKR mice by type-specific immunosuppression of endogenous ecotropic virogenes, *Proc. Natl. Acad. Sci. USA* **73**:4633.

Huegin, A. W., Vacchio, M. S., and Morse, H. C., 1991, A virus-encoded "superantigen" in a retrovirus-induced immunodeficiency syndrome of mice, *Science* **252**:424.

Hunsmann, G., 1985, Subunit vaccines against exogenous retroviruses: Overview and perspectives, *Cancer Res.* **45**(Suppl.):4691.

Hunsmann, G., Moennig, V., Pister, L., Seifert, E., and Schaefer, W., 1974, Properties of mouse leukemia viruses. VIII. The major viral glycoprotein of Friend leukemia virus. Seroimmunological, interfering and hemagglutinating capacities, *Virology* **62**:307.

Hunsmann, G., Moennig, V., and Schaefer, W., 1975a, Properties of mouse leukemia viruses. IX. Active and passive immunization of mice against Friend leukemia with isolated viral GP71 glycoprotein and its corresponding antiserum, *Virology* **66**:327.

Hunsmann, G., Moennig, V., Seifert, E., and Schaefer, W., 1975b, Protection of mice against Friend leukemia by active and passive immunization with isolated viral glycoprotein and its antiserum respectively, *Z. Naturforsch.* **30c**:309.

Husch, B., Eibl, M. M., and Mannhalter, J. W., 1993, CD3, CD8 double-positive cells from HIV-1-infected chimpanzees show group-specific inhibition of HIV-1 replication, *AIDS Res. Hum. Retroviruses* **9**:405.

Imberti, L., Sottini, A., Bettinardi, A., Puoti, M., and Primi, D., 1991, Selective depletion in HIV infection of T cells that bear specific T cell receptor V beta sequences, *Science* **254**:860.

Imrie, A. A., Kehrer, S., Smith, G. W., Penny, R., and Cooper, D. A., 1986, Seroimmunology of AIDS retrovirus infection. I. Use of immunofluorescence assay to confirm sera with ELISA reactivity, *Pathology* **18**:438.

Ishihara, C., Miyazawa, M., Nishio, J., Azuma, I., and Chesebro, B., 1992, Use of low toxicity adjuvants and killed virus to induce protective immunity against the Friend murine leukaemia retrovirus-induced disease, *Vaccine* **10**:353.

Issel, C. J., Horohov, D. W., Lea, D. F., Adams, W. V., Hagius, S. D., McManus, J. M., Allison, A. C., and Montelaro, R. C., 1992, Efficacy of inactivated whole-virus and subunit vaccines in preventing infection and disease caused by equine infectious anemia virus, *J. Virol.* **66**:3398.

Iversen, O. J., and Rodahl, E., 1985, The major internal protein, p27, of a retrovirus-like particle participates in immune complex formation in psoriasis, *Arch. Virol.* **86**:37.

Izui, S., Elder, J. H., McConahey, P. J., and Dixon, F. J., 1981a, Identification of retroviral gp70 and anti-gp70 antibodies involved in circulating immune complexes in NZB × NZW mice, *J. Exp. Med.* **153**:1151.

Izui, S., McConahey, P. J., Clark, J. P., Hang, L. M., Hara, I., and Dixon, F. J., 1981b, Retroviral gp70 immune complexes in NZB × NZW F2 mice with murine lupus nephritis, *J. Exp. Med.* **154**:517.

Joag, S. V., Anderson, M. G., Clements, J. E., McEntee, M. F., Sharma, D. P., Adams, R. J., and Narayan, O., 1993, Antigenic variation of molecularly cloned SIVmac239 during persistent infection in a rhesus macaque, *Virology* **195**:406.

Jacobsen, K. L., Kaneene, J. B., Miller, J. M., and Bull, R. W., 1985, Comparison of the commercial agar-gel immunodiffusion test and radioimmunoprecipitation assay for detection of antibodies to bovine leukemia virus, *Am. J. Vet. Res.* **46**:1430.

Janeway, C. A., Yagi, J., Conrad, P. J., Katz, M. E., Jones, B., Vroegop, S., and Buxser, S., 1989, T-cell responses to Mls and to bacterial proteins that mimic its behavior, *Immunol. Rev.* **107**:61.

Jarrett, O., 1990, Vaccination against feline retroviruses, *Dev. Biol. Stand.* **72**:185.

Javaherian, K., Langlois, A. J., Schmidt, S., Kaufmann, M., Cates, N., Langedijk, J. P., Meloen, R. H., Desrosiers, R. C., Burns, D. P., Bolognesi, D. P., LaRosa, G. J., and Putney, S. D., 1992, The principal neutralization determinant of simian immunodeficiency virus differs from that of human immunodeficiency virus type 1, *Proc. Natl. Acad. Sci. USA* **89**:1418.

Jewett, A., and Bonavida, B., 1990, Peripheral blood monocytes derived from HIV$^+$ individuals mediate antibody-dependent cellular cytotoxicity (ADCC), *Clin. Immunol. Immunopathol.* **54**:192.

Jewett, A., Giorgi, J. V., and Bonavida, B., 1990, Antibody-dependent cellular cytotoxicity against HIV-coated target cells by peripheral blood monocytes from HIV seropositive asymptomatic patients, *J. Immunol.* **145**:4065.

Jiang, S. B., Lin, K., and Neurath, A. R., 1991, Enhancement of human immunodeficiency virus type 1 infection by antisera to peptides from the envelope glycoproteins gp120/gp41, *J. Exp. Med.* **174**:1557.

Johnson, H. M., Russell, J. K., and Pontzer, C. H., 1991, Staphylococcal enterotoxin superantigens, *Proc. Soc. Exp. Biol. Med.* **198**:765.

Johnson, P. R., Montefiori, D. C., Goldstein, S., Hamm, T. E., Zhou, J., Kitov, S., Haigwood, N. L., Misher, L., London, W. T., Gerin, J. L., Allison, A., Purcell, R. H., Chanock, R. M., and Hirsch, V. M., 1992, Inactivated whole-virus vaccine derived from a proviral DNA clone of simian immunoeficiency virus induces high levels of neutralizing antibodies and confers protection against heterologous challenge, *Proc. Natl. Acad. Sci. USA* **89**:2175.

Johnson, R. P., Trocha, A., Buchanan, T. M., and Walker, B. D., 1993, Recognition of a highly conserved region of human immunodeficiency virus type 1 gp120 by an HLA-Cw4-restricted cytotoxic T-lymphocyte clone, *J. Virol.* **67**:438.

Jolly, P. E., Huso, D., Hart, G., and Narayan, O., 1989a, Modulation of lentivirus replication by antibodies. Non-neutralizing antibodies to caprine arthritis–encephalitis virus enhance early stages of infection in macrophages, but do not cause increased production of virions, *J. Gen. Virol.* **70**:2221.

Jolly, P. E., Huso, D. L., Sheffer, D., and Narayan, O., 1989b, Modulation of lentivirus replication by antibodies: Fc portion of immunoglobulin molecule is essential for enhancement of binding, internalization, and neutralization of visna virus in macrophages, *J. Virol.* **63**:1811.

Joly, P., Guillon, J. M., Mayaud, C., Plata, F., Theodorou, I., Denis, M., Debre, P., and Autran, B., 1989, Cell-mediated suppression of HIV-specific cytotoxic T lymphocytes, *J. Immunol.* **143**:2193.

Kajikawa, O., Lairmore, M. D., and DeMartini, J. C., 1990, Analysis of antibody responses to phenotypically distinct lentiviruses, *J. Clin. Microbiol.* **28**:764.

Kannagi, M., Sugamura, K., Sato, H., Okochi, K., Uchino, H., and Hinuma, Y., 1983, Establishment of human cytotoxic T cell lines specific for human adult T cell leukemia virus-bearing cells, *J. Immunol.* **130**:2942.

Kannagi, M., Chalifoux, L. V., Lord, C. I., and Letvin, N. L., 1988, Suppression of simian immunodeficiency virus replication in vitro by $CD8^+$ lymphocytes, *J. Immunol.* **140**:2237.

Kannagi, M., Masuda, T., Hattori, T., Kanoh, T., Nasu, K., Yamamoto, N., and Harada, S., 1990, Interference with human immunodeficiency virus (HIV) replication by $CD8^+$ T cells in peripheral blood leukocytes of asymptomatic HIV carriers in vitro, *J. Virol.* **64**:3399.

Katz, J. D., Mitsuyasu, R., Gottlieb, M. S., Lebow, L. T., and Bonavida, B., 1987, Mechanism of defective NK cell activity in patients with acquired immunodeficiency syndrome (AIDS) and AIDS-related complex. II. Normal antibody-dependent cellular cytotoxicity (ADCC) mediated by effector cells defective in natural killer (NK) cytotoxicity, *J. Immunol.* **139**:55.

Katz, J. D., Nishanian, P., Mitsuyasu, R., and Bonavida, B., 1988, Antibody-dependent cellular cytotoxicity (ADCC)-mediated destruction of human immunodeficiency virus (HIV)-coated $CD4^+$ T lymphocytes by acquired immunodeficiency syndrome (AIDS) effector cells, *J. Clin. Immunol.* **8**:453.

Kay, H. D., Bonnard, G. D., West, W. H., and Herberman, R. B., 1977, A functional comparison of human Fc-receptor-bearing lymphocytes active in natural cytotoxicity and antibody-dependent cellular cytotoxicity, *J. Immunol.* **118**:2058.

Kennedy-Stoskopf, S., and Narayan, O., 1986, Neutralizing antibodies to visna lentivirus: Mechanism of action and possible role in virus persistence, *J. Virol.* **59**:37.

Kanagawa, O., Vaupel, B. A., Gayama, S., Koehler, G., and Kopf, M., 1993, Resistance of mice deficient in IL-4 to retrovirus-induced immunodeficiency syndrome (MAIDS), *Science* **262**:240.

Kent, K. A., Rud, E., Corcoran, T., Powell, C., Thiriart, C., Collignon, C., and Stott, E. J., 1992, Identification of two neutralizing and 8 non-neutralizing epitopes on simian immunodeficiency virus envelope using monoclonal antibodies, *AIDS Res. Hum. Retroviruses* **8**:1147.

Kimura, K. I., and Truman, J. W., 1990, Postmetamorphic cell death in the nervous and muscular systems of Drosophila melanogaster, *J. Neurosci.* **10**:403.

Kinne, T. J., and Gupta, S., 1989, Antibody-dependent cellular cytotoxicity by polymorphonuclear leucocytes in patients with AIDS and AIDS-related complex, *J. Clin. Lab. Immunol.* **30**:153.

Kion, T. A., and Hoffmann, G. W., 1991, Anti-HIV and anti-anti-MHC antibodies in alloimmune and autoimmune mice, *Science* **253**:1138.

Kiprov, D. D., Anderson, R. E., Morand, P. R., Simpson, D. M., Chermann, J. C., Levy, J. A., and Ross, A. R., 1985, Antilymphocyte antibodies and seropositivity for retroviruses in groups at high risk for AIDS, *N. Engl. J. Med.* **312**:1517.

Kleinerman, E. S., Lachman, L. B., Knowles, R. D., Snyderman, R., and Cianciolo, G. J., 1987, A synthetic peptide homologous to the envelope proteins of retroviruses inhibits monocyte-mediated killing by inactivating interleukin 1, *J. Immunol.* **139**:2329.

Klevjer-Anderson, P., and McGuire, T. C., 1982, Neutralizing antibody response of rabbits and goats to caprine arthritis–encephalitis virus, *Infect. Immun.* **38**:455.

Klinken, S. P., Fredrickson, T. N., Hartley, J. W., Yetter, R. A., and Morse, H. C., 1988, Evolution of B cell lineage lymphomas in mice with a retrovirus-induced immunodeficiency syndrome, MAIDS, *J. Immunol.* **140**:1123.

Knowles, D., Jr., Cheevers, W., McGuire, T., Stem, T., and Gorham, J., 1990, Severity of arthritis is predicted by antibody response to gp135 in chronic infection with caprine arthritis–encephalitis virus, *J. Virol.* **64**:2396.

Kodama, T., Burns, D. P., Silva, D. P., Veronese, F. D., and Desrosiers, R. C., 1991, Strain-specific neutralizing determinant in the transmembrane protein of simian immunodeficiency virus, *J. Virol.* **65**:2010.

Koenig, S., Earl, P., Powell, D., Pantaleo, G., Merli, S., Moss, B., and Fauci, A. S., 1988, Group-specific, major histocompatibility complex class I-restricted cytotoxic responses to human immunodeficiency virus 1 (HIV-1) envelope proteins by cloned peripheral blood T cells from an HIV-1-infected individual, *Proc. Natl. Acad. Sci. USA* **85**:8638.

Koenig, S., Fuerst, T. R., Wood, L. V., Woods, R. M., Suzich, J. A., Jones, G. M., de la Cruz, V. F., Davey, R. T., Venkatesan, S., Moss, B., Biddison, W. E., and Fauci, A. S., 1990, Mapping the fine specificity of a cytolytic T cell response to HIV-1 nef protein, *J. Immunol.* **145**:127.

Kono, Y., Hirasawa, K., Fukunaga, Y., and Taniguchi, T., 1973, Recrudescence of equine infectious anemia by treatment with immunosuppressive drugs, *NIAH Q.* **16**:8.

Kooistra, L. H., and Splitter, G. A., 1985, Killer cells of feline leukemia virus- and feline sarcoma virus-infected transformed cells: The role of NK, ADCC, and in vitro generated cytotoxic cells, *Cell. Immunol.* **94**:466.

Kornbluth, R. S., Oh, P. S., Munis, J. R., Cleveland, P. H., and Richman, D. D., 1989, Interferons and bacterial lipopolysaccharide protect macrophages from productive infection by human immunodeficiency virus in vitro, *J. Exp. Med.* **169**:1137.

Korostoff, J. M., Nakada, M. T., Markman, J. F., and Gaulton, G. N., 1991, Impaired immune responsiveness is an essential component in persistent central nervous system infection with Gross murine leukemia virus, *J. Neuroimmunol.* **34**:1.

Koup, R. A., Pikora, C. A., Mazzara, G., Panicali, D., and Sullivan, J. L., 1991a, Broadly reactive antibody-dependent cellular cytotoxic response to HIV-1 envelope glycoproteins precedes broad neutralizing response in human infection, *Viral Immunol.* **4**:215.

Koup, R. A., Robinson, J. E., Nguyen, Q. V., Pikora, C. A., Blais, B., Roskey, A., Panicali, D., and Sullivan, J. L., 1991b, Antibody-dependent cell-mediated cytotoxicity directed by a human monoclonal antibody reactive with gp120 of HIV-1, *AIDS* **5**:1309.

Kozak, C. A., and Ruscetti, S., 1992, Retroviruses in rodents, in: *The Retroviridae,* Vol. 1 (J. A. Levy, ed.), p. 405, Plenum Press, New York.

Krieg, A. M., Gause, W. C., Gourley, M. F., and Steinberg, A. D., 1989, A role for endogenous retroviral sequences in the regulation of lymphocyte activation, *J. Immunol.* **143**:2448.

Kurth, R., and Norley, S., 1993, Animal immunodeficiency viruses, in: *Focus on HIV: Proceedings of an International Symposium,* Churchill Livingstone, Edinburgh.

Kurth, R., Fenyoe, E. M., Klein, E., and Essex, M., 1979, Cell-surface antigens induced by RNA tumour viruses, *Nature* **279**:197.

Kurth, R., Binninger, D., Ennen, J., Denner, J., Hartung, S., and Norley, S., 1991, The quest for an AIDS vaccine: The state of the art and current challenges, *AIDS Res. Hum. Retroviruses* **7**:425.

Kvinesdal, B. B., Nielsen, C. M., Poulsen, A. G., and Hojlyng, N., 1989, Immunofluorescence assay for detection of antibodies to human immunodeficiency virus type 2, *J. Clin. Microbiol.* **27**:2502.

Kwang, H. S., Pedersen, N. C., Lerche, N. W., Osborn, K. G., Marx, P. A., and Gardner, M. B., 1987, Viremia, antigenemia, and serum antibodies in rhesus macaques infected with simian retrovirus type 1 and their relationship to disease course, *Lab. Invest.* **56**:591.

Kwang, H. S., Barr, P. J., Sabin, E. A., Sujipto, S., Marx, P. A., Power, M. D., Bathurst, I. C., and Pedersen, N. C., 1988, Simian retrovirus-D serotype 1 (SRV-1) envelope glycoproteins gp70 and gp20: Expression in yeast cells and identification of specific antibodies in sera from monkeys that recovered from SRV-1 infection, *J. Virol.* **62**:1774.

Lafon, M., Lafage, M., Martinez-Arends, A., Ramirez, R., Vuillier, F., Charron, D., Lotteau, V., and Scott-Algara, D., 1992, Evidence for a viral superantigen in humans, *Nature* **358**:507.

Lähdevirta, J., Maury, C. P., Teppo, A. M., and Repo, H., 1988, Elevated levels of circulating cachectin/tumor necrosis factor in patients with acquired immunodeficiency syndrome, *Am. J. Med.* **85**:289.

Lamon, E. W., Powell, T. J., Walia, A. S., Lidin, B. I., Srinivas, R. V., Baskin, J. G., and Kearney, J. F., 1987, Monoclonal IgM antibodies that inhibit primary Moloney murine sarcoma growth, *J. Natl. Cancer Inst.* **78**:547.

Lane, H. C., Masur, H., Edgar, L. C., Whalen, G., Rook, A. H., and Fauci, A. S., 1983, Abnormalities of B-cell activation and immunoregulation in patients with the acquired immunodeficiency syndrome, *N. Engl. J. Med.* **309**:453.

Langedijk, J. P., Schalken, J. J., Tersmette, M., Huisman, J. G., and Meloen, R. H., 1990, Location of epitopes on the major core protein p24 of human immunodeficiency virus, *J. Gen. Virol.* **71**:2609.

Lasky, L. A., Nakamura, G., Smith, D. H., Fennie, C., Shimasaki, C., Patzer, E., Berman, P., Gregory, T., and Capon, D. J., 1987, Delineation of a region of the human immunodeficiency virus type 1 gp120 glycoprotein critical for interaction with the CD4 receptor, *Cell* **50**:975.

Laurence, J., Hodtsev, A. S., and Posnett, D. N., 1992, Superantigen implicated in dependence of HIV-1 replication in T cells on TCR V beta expression, *Nature* **358**:255.

Laurent-Crawford, A. G., Krust, B., Muller, S., Riviere, Y., Rey-Cuille, M. A., Bechet, J. M., Montagnier, L., and Hovanessian, A. G., 1991, The cytopathic effect of HIV is associated with apoptosis, *Virology* **185**:829.

Layton, G. T., Harris, S. J., Gearing, A. J., Hill, P. M., Cole, J. S., Griffiths, J. C., Burns, N. R., Kingsman, A. J., and Adams, S. E., 1993, Induction of HIV-specific cytotoxic T lymphocytes in vivo with hybrid HIV-1 V3:Ty-virus-like particles, *J. Immunol.* **151**:1097.

Lee, B. K., and Eicher, E. M., 1990, Segregation patterns of endogenous mouse mammary tumor viruses in five recombinant inbred strain sets, *J. Virol.* **64**:4568.

Legrain, P., Goud, B., and Buttin, G., 1986, Increase of retroviral infection in vitro by the binding of antiretroviral antibodies, *J. Virol.* **60**:1141.

Le Grand, R., Vaslin, B., Vogt, G., Roques, P., Humbert, M., and Dormont, D., 1992, AIDS vaccine developments, *Nature* **355**:684.

Lennette, E. T., Karpatkin, S., and Levy, J. A., 1987, Indirect immunofluorescence assay for antibodies to human immunodeficiency virus, *J. Clin. Microbiol.* **25**:199.

Leung, K., and Nabel, G. J., 1988, HTLV-1 transactivator induces interleukin-2 receptor expression through an NF-kappa B-like factor, *Nature* **333**:776.

Levine, S. J., White, D. A., and Fels, A. O., 1990, The incidence and significance of Staphylococcus aureus in respiratory cultures from patients infected with the human immunodeficiency virus, *Am. Rev. Respir. Dis.* **141**:89.

Levy, D., Deshayes, L., Parodi, A. L., Levy, J. P., Stephenson, J. R., Devare, S. G., and Gilden, R. V., 1977, Bovine leukemia virus specific antibodies among French cattle. II. Radioimmunoassay with the major structural protein (BLV p24), *Int. J. Cancer* **20**:543.

Lichtensteiger, C. A., Knowles, D. P., McGuire, T. C., and Cheevers, W. P., 1991, Recombinant gp135 envelope glycoproteins of caprine arthritis–encephalitis lentivirus variants inhibit homologous and heterologous variant-specific neutralizing antibodies, *Virology* **185**:2.

Littaua, R. A., Oldstone, M. B., Takeda, A., Debouck, C., Wong, J. T., Tuazon, C. U., Moss, B., Kievits, F., and Ennis, F. A., 1991, An HLA-C-restricted $CD8^+$ cytotoxic T-lymphocyte clone recognizes a highly conserved epitope on human immunodeficiency virus type 1 gag, *J. Virol.* **65**:4051.

Liu, W. T., Engelman, R. W., Trang, L. Q., Hau, K., Good, R. A., and Day, N. K., 1984a, Appearance of cytotoxic antibody to viral gp70 on feline lymphoma cells (FL-74) in cats during ex vivo immunoadsorption therapy: Quantitation, characterization, and association with remission of disease and disappearance of viremia, *Proc. Natl. Acad. Sci. USA* **81**:3516.

Liu, W. T., Good, R. A., Trang, L. Q., Engelman, R. W., and Day, N. K., 1984b, Remission of leukemia and loss of feline leukemia virus in cats injected with Staphylococcus protein A: Association with increased circulating interferon and complement-dependent cytotoxic antibody, *Proc. Natl. Acad. Sci. USA* **81**:6471.

Ljunggren, K., Boettiger, B., Biberfeld, G., Karlson, A., Fenyoe, E. M., and Jondal, M., 1987a, Antibody-dependent cellular cytotoxicity-inducing antibodies against human immunodeficiency virus. Presence at different clinical stages, *J. Immunol.* **139**:2263.

Ljunggren, K., Fenyoe, E. M., Biberfeld, G., and Jondal, M., 1987b, Detection of antibodies which mediate human immunodeficiency virus-specific cellular cytotoxicity (ADCC) in vitro, *J. Immunol. Methods* **104**:7.

Ljunggren, K., Biberfeld, G., Jondal, M., and Fenyoe, E. M., 1989a, Antibody-dependent cellular cytotoxicity detects type- and strain-specific antigens among human immunodeficiency virus types 1 and 2 and simian immunodeficiency virus SIVmac isolates, *J. Virol.* **63**:3376.

Ljunggren, K., Karlson, A., Fenyoe, E. M., and Jondal, M., 1989b, Natural and antibody-dependent cytotoxicity in different clinical stages of human immunodeficiency virus type 1 infection, *Clin. Exp. Immunol.* **75**:184.

Ljunggren, K., Moschese, V., Broliden, P. A., Giaquinto, C., Quinti, I., Fenyoe, E. M., Wahren, B., Rossi, P., and Jondal, M., 1990, Antibodies mediating cellular cytotoxicity and neutralization correlate with a better clinical stage in children born to human immunodeficiency virus-infected mothers, *J. Infect. Dis.* **161**:198.

Lucas, C., Peterson, M. L., Bennett, G. L., Frie, S. W., Berman, P. W., and Chen, A. B., 1990, A sensitive radioimmunoprecipitation assay for the detection and quantitation of antibodies to the envelope glycoprotein gp120 of the human immunodeficiency virus (HIV-1), *AIDS Res. Hum. Retroviruses* **6**:357.

Lucey, D. R., Zajac, R. A., Melcher, G. P., Butzin, C. A., and Boswell, R. N., 1990, Serum IgE levels in 622 persons with human immunodeficiency virus infection: IgE elevation with marked depletion of $CD4^+$ T-cells, *AIDS Res. Hum. Retroviruses* **6**:427.

Luciw, P. A., and Leung, N. J., 1992, Mechanism of retrovirus replication, in: *The Retroviridae*, Vol. 1 (J. A. Levy, ed.), Plenum Press, New York.

Lutley, R., Petursson, G., Palsson, P. A., Georgsson, G., Klein, J., and Nathanson, N., 1983, Antigenic drift in visna: Virus variation during long-term infection of Icelandic sheep, *J. Gen. Virol.* **64**:1433.

Lutz, H., Pederson, N. C., Durbin, R., and Theilen, G. H., 1983, Monoclonal antibodies to three epitopic regions of feline leukemia virus p27 and their use in enzyme-linked immunosorbent assay of p27, *J. Immunol. Methods* **56**:209.

Lyerly, H. K., Matthews, T. J., Langlois, A. J., Bolognesi, D. P., and Weinhold, K. J., 1987, Human T-cell lymphotropic virus IIIB glycoprotein (gp120) bound to CD4 determinants on normal lymphocytes and expressed by infected cells serves as target for immune attack, *Proc. Natl. Acad. Sci. USA* **84**:4601.

McCarty, J. M., and Grant, C. K., 1983a, Cellular immune response in the blood of cats is restricted to autochthonous feline sarcoma virus-transformed cells, *Int. J. Cancer* **31**:627.

McCarty, J. M., and Grant, C. K., 1983b, Feline cytotoxic immune mechanisms against virus-associated leukemia and fibrosarcoma, *Cell. Immunol.* **81**:157.

McChesney, M. B., and Oldstone, M. B., 1989, Virus-induced immunosuppression: Infections with measles virus and human immunodeficiency virus, *Adv. Immunol.* **45**:335.

McClure, M. O., Marsh, M., and Weiss, R. A., 1988, Human immunodeficiency virus infection of CD4-bearing cells occurs by a pH-independent mechanism, *EMBO J.* **7**:513.

McConkey, D. J., Orrenius, S., and Jondal, M., 1990, Cellular signalling in programmed cell death (apoptosis), *Immunol. Today* **11**:120.

Maccubbin, D. L., and Schierman, L. W., 1986, MHC-restricted cytotoxic response of chicken T cells: Expression, augmentation, and clonal characterization, *J. Immunol.* **136**:12.

McDougal, J. S., Kennedy, M. S., Sligh, J. M., Cort, S. P., Mawle, A., and Nicholson, J. K., 1986, Binding of HTLV-III/LAV to $T4^+$ T cells by a complex of the 110K viral protein and the T4 molecule, *Science* **231**:382.

McGuire, T. C., 1987, The immune response to viral antigens as a determinant of arthritis in caprine arthritis–encephalitis virus infection, *Vet. Immunol. Immunopathol.* **17**:465.

McGuire, T. C., O'Rourke, K. I., and Perryman, L. E., 1990, Immunopathogenesis of equine infectious anemia lentivirus disease, *Dev. Biol. Stand.* **72**:31.

McKeating, J. A., Griffiths, P. D., and Weiss, R. A., 1990, HIV susceptibility conferred to human fibroblasts by cytomegalovirus-induced Fc receptor, *Nature* **343**:659.

McKeating, J. A., Bennett, J., Zolla, P. S., Schutten, M., Ashelford, S., Brown, A. L., and Balfe, P., 1993, Resistance of a human serum-selected human immunodeficiency virus type 1 escape mutant to neutralization by CD4 binding site monoclonal antibodies is conferred by a single amino acid change in gp120, *J. Virol.* **67**:5216.

McMahon, S. B., Ewert, D., Carroll, R., and Halpern, M. S., 1986, Expression of endogenous retroviral envelope glycoprotein as a determinant of immunity to Rous sarcoma, *Virology* **155**:737.

Maggi, E., Macchia, D., Parronchi, P., Mazzetti, M., Ravina, A., Milo, D., and Romagnani, S., 1987, Reduced production of interleukin 2 and interferon-gamma and enhanced helper activity for IgG synthesis by cloned CD4+ T cells from patients with AIDS, *Eur. J. Immunol.* **17**:1685.

Mannervik, M., Putkonen, P., Ruden, U., Kent, K. A., Norrby, E., Wahren, B., and Broliden, P. A., 1992, Identification of B-cell antigenic sites on HIV-2 gp125, *J. AIDS* **5**:177.

Marciani, D. J., Kensil, C. R., Beltz, G. A., Hung, C. H., Cronier, J., and Aubert, A., 1991, Genetically-engineered subunit vaccine against feline leukaemia virus: Protective immune response in cats, *Vaccine* **9**:89.

Marrack, P., and Kappler, J., 1990, The staphylococcal enterotoxins and their relatives, *Science* **248**:705.

Marrack, P., Kushnir, E., and Kappler, J., 1991, A maternally inherited superantigen encoded by a mammary tumour virus, *Nature* **349**:524.

Marx, P. A., Pedersen, N. C., Lerche, N. W., Osborn, K. G., Lowenstine, L. J., Lackner, A. A., Maul, D. H., Kwang, H. S., Kluge, J. D., Zaiss, C. P., Sharpe, V., Spinner, A. P., Allison, A. C., and Gardner, M. B., 1986, Prevention of simian acquired immune deficiency syndrome with a formalin-inactivated type D retrovirus vaccine, *J. Virol.* **60**:431.

Mastro, J. M., Lewis, M. G., Mathes, L. E., Sharpee, R., Tarr, M. J., and Olsen, R. G., 1986, Feline leukemia vaccine: Efficacy, contents and probable mechanism, *Vet. Immunol. Immunopathol.* **11**:205.

Masuda, T., Matsushita, S., Kuroda, M. J., Kannagi, M., Takatsuki, K., and Harada, S., 1990, Generation of neutralization-resistant HIV-1 in vitro due to amino acid interchanges of third hypervariable env region, *J. Immunol.* **145**:3240.

Matis, L. A., Glimcher, L. H., Paul, W. E., and Schwartz, R. H., 1983, Magnitude of response of histocompatibility-restricted T-cell clones is a function of the product of the concentrations of antigen and Ia molecules, *Proc. Natl. Acad. Sci. USA* **80**:6019.

Matis, L. A., Shu, S., Groves, E. S., Zinn, S., Chou, T., Kruisbeek, A. M., Rosenstein, M., and Rosenberg, S. A., 1986, Adoptive immunotherapy of a syngeneic murine leukemia with a tumor-specific cytotoxic T cell clone and recombinant human interleukin 2: Correlation with clonal IL 2 receptor expression, *J. Immunol.* **136**:3496.

Matsuyama, T., Kobayashi, N., and Yamamoto, N., 1991, Cytokines and HIV infection: Is AIDS a tumor necrosis factor disease? *AIDS* **5**:1405.

Maul, G. G., Jimenez, S. A., Riggs, E., and Ziemnicka-Kotula, D., 1989, Determination of an epitope of the diffuse systemic sclerosis marker antigen DNA topoisomerase I: Sequence similarity with retroviral p30gag protein suggests a possible cause for autoimmunity in systemic sclerosis, *Proc. Natl. Acad. Sci. USA* **86**:8492.

Merrill, J. E., 1992, Cytokines and retroviruses, *Clin. Immunol. Immunopathol.* **64**:23.

Meyaard, L., Otto, S. A., Jonker, R. R., Mijnster, M. J., Keet, R. P., and Miedema, F., 1992, Programmed death of T cells in HIV-1 infection, *Science* **257**:217.

Meyerhans, A., Dadaglio, G., Vartanian, J. P., Langlade-Demoyen, P., Frank, R., Asjoe, B., Plata, F., and Wain-Hobson, S., 1991, In vivo persistence of a HIV-1-encoded HLA-B27-restricted cytotoxic T lymphocyte epitope despite specific in vitro reactivity, *Eur. J. Immunol.* **21**:2637.

Michaels, F. H., Banks, K. L., and Reitz, M. S., 1991, Lessons from caprine and ovine retrovirus infections, *Rheum. Dis. Clin. North Am.* **17**:5.

Michel, M. L., Mancini, M., Riviere, Y., Dormont, D., and Tiollais, P., 1990, T- and B-lymphocyte responses to human immunodeficiency virus (HIV) type 1 in macaques immunized with hybrid HIV/hepatitis B surface antigen particles, *J. Virol.* **64**:2452.

Miller, M. D., Lord, C. I., Stallard, V., Mazzara, G. P., and Letvin, N. L., 1990, The gag-specific cytotoxic T lymphocytes in rhesus monkeys infected with the simian immunodeficiency virus of macaques, *J. Immunol.* **144**:122.

Miller, M. D., Yamamoto, H., Hughes, A. L., Watkins, D. I., and Letvin, N. L., 1991, Definition of an epitope and MHC class I molecule recognized by gag-specific cytotoxic T lymphocytes in SIVmac-infected rhesus monkeys, *J. Immunol.* **147**:320.

Mitani, M., Cianciolo, G. J., Snyderman, R., Yasuda, M., Good, R. A., and Day, N. K., 1987, Suppressive effect on polyclonal B-cell activation of a synthetic peptide homologous to a transmembrane component of oncogenic retroviruses, *Proc. Natl. Acad. Sci. USA* **84**:237.

Mitsuya, H., Matis, L. A., Megson, M., Bunn, P. A., Murray, C., Mann, D. L., Gallo, R. C., and Broder, S., 1983, Generation of an HLA-restricted cytotoxic T cell line reactive against cultured tumor cells from a patient infected with human T cell leukemia/lymphoma virus, *J. Exp. Med.* **158**:994.

Miyakoshi, H., Koide, H., and Aoki, T., 1984, In vitro antibody-dependent cellular cytotoxicity against human T-cell leukemia/lymphoma virus (HTLV)-producing cells, *Int. J. Cancer* **33**:287.

Miyatake, S., Seiki, M., Malefijt, R. D., Heike, T., Fujisawa, J., Takebe, Y., Nishida, J., Shlomai, J., Yokota, T., Yoshida, M., and Arai, N., 1988, Activation of T cell-derived lymphokine genes in T cells and fibroblasts: Effects of human T cell leukemia virus type I p40x protein and bovine papilloma virus encoded E2 protein, *Nucleic Acids Res.* **16**:6547.

Miyazawa, M., Nose, M., Kawashima, M., and Kyogoku, M., 1987, Pathogenesis of arteritis of SL/Ni mice. Possible lytic effect of anti-gp70 antibodies on vascular smooth muscle cells, *J. Exp. Med.* **166**:890.

Modrow, S., Hoeflacher, B., Mertz, R., and Wolf, H., 1989, Carrier-bound synthetic peptides. Use as antigen in HIV-1 ELISA tests and in antiserum production, *J. Immunol. Methods* **118**:1.

Molloy, J. B., Walker, P. J., Baldock, F. C., Rodwell, B. J., and Cowley, J. A., 1990, An enzyme-linked immunosorbent assay for detection of bovine leukaemia virus p24 antibody in cattle, *J. Virol. Methods* **28**:47.

Montefiori, D. C., Robinson, W. E., and Mitchell, W. M., 1989, Antibody-independent, complement-mediated enhancement of HIV-1 infection by mannosidase I and II inhibitors, *Antiviral Res.* **11**:137.

Montefiori, D. C., Robinson, W. E., Hirsch, V. M., Modliszewski, A., Mitchell, W. M., and Johnson, P. R., 1990a, Antibody-dependent enhancement of simian immunodeficiency virus (SIV) infection in vitro by plasma from SIV-infected rhesus macaques, *J. Virol.* **64**:113.

Montefiori, D. C., Robinson, W. E., Hirsch, V. M., Modliszewski, A., Mitchell, W. M., and Johnson, P. R., 1990b, Antibody-dependent enhancement of SIV infection: Further characterization and cross reactivity between macaque and sooty mangabey isolates, *J. Med. Primatol.* **19**:269.

Montefiori, D. C., Lefkowitz, L. B., Keller, R. E., Holmberg, V., Sandstrom, E., and Phair, J. P., 1991a, Absence of a clinical correlation for complement-mediated, infection-enhancing antibodies in plasma or sera from HIV-1-infected individuals. Multicenter AIDS Cohort Study Group, *AIDS* **5**:513.

Montefiori, D. C., Zhou, I. Y., Barnes, B., Lake, D., Hersh, E. M., Masuho, Y., and Lefkowitz, L. B., 1991b, Homotypic antibody responses to fresh clinical isolates of human immunodeficiency virus, *Virology* **182**:635.

Montelaro, R. C., Parekh, B., Orrego, A., and Issel, C. J., 1984, Antigenic variation during persistent infection by equine infectious anemia virus, a retrovirus, *J. Biol. Chem.* **259**:10539.

Montelaro, R. C., Rushlow, K. E., Wang, S., Issel, C. J., and Horohov, D. W., 1992, Pony tales of the immunologic management and mismanagement of infection and disease by experimental vaccines to equine infectious anemia virus, in: *Fifth Annual Meeting of the National Cooperative Vaccine Development Groups for AIDS.*

Montelaro, R., Ball, J. M., and Rushlow, K. E., 1993, Equine retroviruses, in: *The Retroviridae,* Vol. 2 (J. A. Levy, ed.), pp. 257–360, Plenum Press, New York.

Moody, D. J., Specter, S., Bendinelli, M., and Friedman, H., 1984, Suppression of natural killer cell activity by Friend murine leukemia virus, *J. Natl. Cancer Inst.* **72**:1349.

Morgan, D. A., Ruscetti, F. W., and Gallo, R., 1976, Selective in vitro growth of T lymphocytes from normal human bone marrows, *Science* **193**:1007.

Morrow, W. J., Wharton, M., Stricker, R. B., and Levy, J. A., 1986, Circulating immune complexes in patients with acquired immune deficiency syndrome contain the AIDS-associated retrovirus, *Clin. Immunol. Immunopathol.* **40**:515.

Mosier, D. E., Yetter, R. A., and Morse, H. C., 1985, Retroviral induction of acute lymphoproliferative disease and profound immunosuppression in adult C57BL/6 mice, *J. Exp. Med.* **161**:766.

Murayama, T., Cai, Q., and Rineldo, C. R., Jr., 1990, Antibody-dependent cellular cytotoxicity mediated by $CD16^+$ lymphocytes from HIV-seropositive homosexual men, *Clin. Immunol. Immunopathol.* **55**:297.

Murphey-Corb, M., Martin, L. N., Davison-Fairburn, B., Montelaro, R. C., Miller, M., West, M., Ohkawa, S., Baskin, G. B., Zhang, J. Y., Putney, S. D., Allison, A. C., and Eppstein, D. A., 1989, A formalin-inactivated whole SIV vaccine confers protection in macaques, *Science* **246**:1293.

Nagy, K., Clapham, P., Cheingsong-Popov, R., and Weiss, R. A., 1983, Human T-cell leukemia virus type I: Induction of syncytia and inhibition by patients' sera, *Int. J. Cancer* **32**:321.

Nara, P. L., Robey, W. G., Arthur, L. O., Asher, D. M., Wolff, A. V., Gibbs, C. J., Gajdusek, D. C., and Fischinger, P. J., 1987a, Persistent infection of chimpanzees with human immunodeficiency virus: Serological responses and properties of reisolated viruses, *J. Virol.* **61**:3173.

Nara, P. L., Robey, W. G., Gonda, M. A., Carter, S. G., and Fischinger, P. J., 1987b, Absence of cytotoxic antibody to human immunodeficiency virus-infected cells in humans and its induction in animals after infection or immunization with purified envelope glycoprotein gp120, *Proc. Natl. Acad. Sci. USA* **84**:3797.

Nara, P. L., Smit, L., Dunlop, N., Hatch, W., Merges, M., Waters, D., Kelliher, J., Gallo, R. C., Fischinger, P. J., and Goudsmit, J., 1990, Emergence of viruses resistant to neutralization by V3-specific antibodies in experimental human immunodeficiency virus type 1 IIIB infection of chimpanzees, *J. Virol.* **64**:3779.

Narayan, O., and Cork, L. C., 1985, Lentiviral diseases of sheep and goats: Chronic pneumonia leukoencephalomyelitis and arthritis, *Rev. Infect. Dis.* **7**:89.

Narayan, O., Clements, J. E., Griffin, D. E., and Wolinsky, J. S., 1981, Neutralizing antibody spectrum determines the antigenic profiles of emerging mutants of visna virus, *Infect. Immun.* **32**:1045.

Narayan, O., Sheffer, D., Griffin, D. E., Clements, J., and Hess, J., 1984, Lack of neutralizing antibodies to caprine arthritis–encephalitis lentivirus in persistently infected goats can be overcome by immunization with inactivated Mycobacterium tuberculosis, *J. Virol.* **49**:349.

Narayan, O., Sheffer, D., Clements, J. E., and Tennekoon, G., 1985, Restricted replication of lentiviruses. Visna viruses induce a unique interferon during interaction between lymphocytes and infected macrophages, *J. Exp. Med.* **162**:1954.

Narayan, O., Zink, M. C., Gorrell, M., Crane, S., Huso, D., Jolly, P., Saltarelli, M., Adams, R. J., and Clements, J. E., 1993, The lentiviruses of sheep and goats, in: *The Retroviridae*, Vol. 2 (J. A. Levy, ed.), pp. 229–255, Plenum Press, New York.

Nathanson, N., Panitch, H., Palsson, P. A., Petursson, G., and Georgsson, G., 1976, Pathogenesis of visna. II. Effect of immunosuppression upon early central nervous system lesions, *Lab. Invest.* **35**:444.

Nehyba, J., Svoboda, J., Karakoz, I., Geryk, J., and Hejnar, J., 1990, Ducks: A new experimental host system for studying persistent infection with avian leukaemia retroviruses, *J. Gen. Virol.* **71**:1937.

Nelson, M., and Nelson, D., 1990, Inhibition of interleukin-2 production by tumor cell products and by CKS-17, a synthetic retroviral envelope peptide, *Cancer Immunol. Immunother.* **30**:331.

Nemo, G. J., Brown, P. W., Gibbs, C. J., Jr., and Gajdusek, D. C., 1978, Antigenic relationship of human foamy virus to the simian foamy viruses, *Infect. Immun.* **20**:69.

Neurath, A. R., Strick, N., Sproul, P., Baker, L., Rubinstein, P., Stevens, C. E., Taylor, P., Gallo, R. C., Gold, J. W., Lee, Y. S., and Nilsen, T., 1985, Radioimmunoassay and enzyme-linked immunoassay of antibodies to the core protein (P24) of human T-lymphotropic virus (HTLV III), *J. Virol. Methods* **11**:75.

Nicholas, R. A., and Thornton, D. H., 1983, Relative efficiency of techniques for detecting avian reticuloendotheliosis virus as a vaccine contaminant, *Res. Vet. Sci.* **34**:377.

Nicholas, R. A., and Thornton, D. H., 1987, An enzyme-linked immunosorbent assay for the detection of antibodies to avian reticuloendotheliosis virus using whole cell antigen, *Res. Vet. Sci.* **43**:403.

Nick, S., Klaws, J., Friebel, K., Birr, C., Hunsmann, G., and Bayer, H., 1990, Virus neutralizing and enhancing epitopes characterized by synthetic oligopeptides derived from the feline leukaemia virus glycoprotein sequence, *J. Gen. Virol.* **71**:77.

Nixon, D. F., and McMichael, A. J., 1991, Cytotoxic T-cell recognition of HIV proteins and peptides, *AIDS* **5**:1049.

Nixon, D. F., Townsend, A. R., Elvin, J. G., Rizza, C. R., Gallwey, J., and McMichael, A. J., 1988, HIV-1 gag-specific cytotoxic T lymphocytes defined with recombinant vaccinia virus and synthetic peptides, *Nature* **336**:484.

Norley, S. G., Kraus, G., Ennen, J., Bonilla, J., König, H., and Kurth, R., 1990a, Immunological studies of the basis for the apathogenicity of simian immunodeficiency virus from African green monkeys, *Proc. Natl. Acad. Sci. USA* **87**:9067.

Norley, S. G., Mikschy, U., Werner, A., Staszewski, S., Helm, E. B., and Kurth, R., 1990b, Demonstration of cross-reactive antibodies able to elicit lysis of both HIV-1- and HIV-2-infected cells, *J. Immunol.* **145**:1700.

Nowak, M. A., Anderson, R. M., McLean, A. R., Wolfs, T. F., Goudsmit, J., and May, R. M., 1991, Antigenic diversity thresholds and the development of AIDS, *Science* **254**:963.

Nunberg, J. H., Rodgers, G., Gilbert, J. H., and Snead, R. M., 1984, Method to map antigenic determinants recognized by monoclonal antibodies: Localization of a determinant of virus neutralization on the feline leukemia virus envelope protein gp70, *Proc. Natl. Acad. Sci. USA* **81**:3675.

O'Connor, T. P., Tanguay, S., Steinman, R., Smith, R., Barr, M. C., Yamamoto, J. K., Pedersen, N. C., Andersen, P. R., and Tonelli, Q. J., 1989, Development and evaluation of immunoassay for detection of antibodies to the feline T-lymphotropic lentivirus (feline immunodeficiency virus), *J. Clin. Microbiol.* **27**:474.

Ogasawara, M., Cianciolo, G. J., Mitani, M., Kizaki, T., Good, R. A., and Day, N. K., 1991, The suppressive effect of a synthetic retroviral peptide on the human IFN gamma production is abrogated by the combined stimulation with IL-1 and IL-2, *Cancer Detect. Prev.* **15**:205.

Ohno, K., Nakano, T., Matsumoto, Y., Watari, T., Goitsuka, R., Nakayama, H., Tsujimoto, H., and Hasegawa, A., 1993, Apoptosis induced by tumor necrosis factor in cells chronically infected with feline immunodeficiency virus, *J. Virol.* **67**:2429.

Ojo-Amaize, E., Nishanian, P. G., Heitjan, D. F., Rezai, A., Esmail, I., Korns, E., Detels, R., Fahey, J., and Giorgi, J. V., 1989, Serum and effector-cell antibody-dependent cellular cytotoxicity (ADCC) activity remains high during human immunodeficiency virus (HIV) disease progression, *J. Clin. Immunol.* **9**:454.

Osterhaus, A., Weijer, K., Uytdehaag, F., Jarrett, O., Sundquist, B., and Morein, B., 1985, Induction of protective immune response in cats by vaccination with feline leukemia virus iscom, *J. Immunol.* **135**:591.

Osterhaus, A., Weijer, K., Uytdehaag, F., Knell, P., Jarrett, O., and Morein, B., 1987, Comparison of serological responses in cats vaccinated with two different FeLV vaccine preparations, *Vet. Rec.* **121**:260.

Pantaleo, G., Rebai, N., Graziosi, C., Lane, H. C., Sekaly, R. P., and Fauci, A. S., 1992, Perturbation of the T cell receptor Vβ repertoire in peripheral blood T cells of HIV infected individuals, *Clin. Res.* **40**:253.

Payne, S. L., Fang, F. D., Liu, C. P., Dhruva, B. R., Rwambo, P., Issel, C. J., and Montelaro, R. C., 1987, Antigenic variation and lentivirus persistence: Variations in envelope gene sequences during EIAV infection resemble changes reported for sequential isolates of HIV, *Virology* **161**:321.

Perryman, L. E., O'Rourke, K. I., and McGuire, T. C., 1988, Immune responses are required to terminate viremia in equine infectious anemia lentivirus infection, *J. Virol.* **62**:3073.

Peters, C. J., and Theofilopoulos, A. N., 1977, Antibody-dependent cellular cytotoxicity against murine leukemia viral antigens: Studies with human lymphoblastoid cell lines and human peripheral lymphocytes as effector cells comparing rabbit, goat, and mouse antisera, *J. Immunol.* **119**:1089.

Peters, G., Placzek, M., Brookes, S., Kozak, C., Smith, R., and Dickson, C., 1986, Characterization, chromosome assignment, and segregation analysis of endogenous proviral units of mouse mammary tumor virus, *J. Virol.* **59**:535.

Phillips, R. E., Rowland-Jones, S., Nixon, D. F., Gotch, F. M., Edwards, J. P., Ogunlesi, A. O., Elvin, J. G., Rothbard, J. A., Bangham, C. R., Rizza, C. R., and McMichael, A. J., 1991, Human immunodeficiency virus genetic variation that can escape cytotoxic T cell recognition, *Nature* **354**:453.

Pitha, P. M., Biegel, D., Yetter, R. A., and Morse, H. C., 1988, Abnormal regulation of IFN-alpha, -beta, and -gamma expression in MAIDS, a murine retrovirus-induced immunodeficiency syndrome, *J. Immunol.* **141**:3611.

Plata, F., 1985, Enhancement of tumor growth correlates with suppression of the tumor-specific cytolytic T lymphocyte response in mice chronically infected by Trypanosoma cruzi, *J. Immunol.* **134**:1312.

Plata, F., Langlade-Demoyen, P., Abastado, J. P., Berbar, T., and Kourilsky, P., 1986, Molecular definition of retrovirus-induced antigens recognized by tumour-specific H-2-restricted cytolytic T lymphocytes, *J. Immunogenet.* **13**:263.

Plata, F., Autran, B., Martins, L. P., Wain-Hobson, S., Raphael, M., Mayaud, C., Denis, M., Guillon, J. M., and Debre, P., 1987, AIDS virus-specific cytotoxic T lymphocytes in lung disorders, *Nature* **328**:348.

Plata, F., Dadaglio, G., Chenciner, N., Hoffenbach, A., Wain-Hobson, S., Michel, F., and Langlade-Demoyen, P., 1989, Cytotoxic T lymphocytes in HIV-induced disease: Implications for therapy and vaccination, *Immunodefic. Rev.* **1**:227.

Poli, G., Bressler, P., Kinter, A., Duh, E., Timmer, W. C., Rabson, A., Justement, J. S., Stanley, S., and Fauci, A. S., 1990, Interleukin 6 induces human immunodeficiency virus expression in infected monocytic cells alone and in synergy with tumor necrosis factor alpha by transcriptional and post-transcriptional mechanisms, *J. Exp. Med.* **172**:151.

Pontzer, C. H., Russell, J. K., and Johnson, H. M., 1991, Structural basis for differential binding of staphylococcal enterotoxin A and toxic shock syndrome toxin 1 to class II major histocompatibility molecules, *Proc. Natl. Acad. Sci. USA* **88**:125.

Portetelle, D., Bruck, C., Burny, A., Dekegel, D., Mammerickx, M., and Urbain, J., 1978, Detection of complement-dependent lytic antibodies in sera from bovine leukemia virus-infected animals, *Ann. Rech. Vet.* **9**:667.

Portetelle, D., Bruck, C., Mammerickx, M., and Burny, A., 1983, Use of monoclonal antibody in an ELISA test for the detection of antibodies to bovine leukaemia virus, *J. Virol. Methods* **6**:19.

Portetelle, D., Couez, D., Bruck, C., Kettmann, R., Mammerickx, M., Van Der Maaten, M., Brasseur, R., and Burny, A., 1989a, Antigenic variants of bovine leukemia virus (BLV) are defined by amino acid substitutions in the NH_2 part of the envelope glycoprotein gp51, *Virology* **169**:27.

Portetelle, D., Dandoy, C., Burny, A., Zavada, J., Siakkou, H., Gras-Masse, H., Drobecq, H., and Tartar, A., 1989b, Synthetic peptides approach to identification of epitopes on bovine leukemia virus envelope glycoprotein gp51, *Virology* **169**:34.

Posner, M. R., Elboim, H. S., Cannon, T., Cavacini, L., and Hideshima, T., 1992, Functional activity of an HIV-1 neutralizing IgG human monoclonal antibody: ADCC and complement-mediated lysis, *AIDS Res. Hum. Retroviruses* **8**:553.

Powell, J. D., Villinger, F., Yehuda-Cohen, T., Vuchetich, M., McClure, H. M., Sell, K. W., and Ahmed-Ansari, A., 1990, Identification of SIV/SMM viral proteins that induce T cell response in experimentally infected rhesus macaques and naturally infected sooty mangabeys by the cellular Western blot assay, *J. Med. Primatol.* **19**:227.

Pozsgay, J. M., Klyczek, K. K., and Blank, K. J., 1989, In vivo generation of antigenic variants of murine retroviruses, *Virology* **173**:330.

Putkonen, P., Thorstensson, R., Ghavamzadeh, L., Albert, J., Hild, K., Biberfeld, G., and Norrby, E., 1991a, Prevention of HIV-2 and SIVsm infection by passive immunization in cynomolgus monkeys, *Nature* **352**:436.

Putkonen, P., Thorstensson, R., Walther, L., Albert, J., Akerblom, L., Granquist, O., Wadell, G., Norrby, E., and Biberfeld, G., 1991b, Vaccine protection against HIV-2 infection in cynomolgus monkeys, *AIDS Res. Hum. Retroviruses* **7**:271.

Putney, S. D., Matthews, T. J., Robey, W. G., Lynn, D. L., Robert-Guroff, M., Mueller, W. T., Langlois, A. J., Ghrayeb, J., Petteway, S. R., Weinhold, K. J., Fischinger, P. J., Wong-Staal, F., Gallo, R. C., and Bolognese, D. P., 1986, HTLV-III/LAV-neutralizing antibodies to an E. coli-produced fragment of the virus envelope, *Science* **234**:1392.

Query, C. C., and Keene, J. D., 1987, A human autoimmune protein associated with U1 RNA contains a region of homology that is cross-reactive with retroviral p30gag antigen, *Cell* **51**:211.

Rammensee, H. G., Falk, K., and Roetzschke, O., 1993, Peptides naturally presented by MHC class I molecules, *Annu. Rev. Immunol.* **11**:213.

Regier, D. A., and Desrosiers, R. C., 1990, The complete nucleotide sequence of a pathogenic molecular clone of simian immunodeficiency virus, *AIDS Res. Hum. Retroviruses* **6**:1221.

Reiher, W. E., 3d, Blalock, J. E., and Brunck, T. K., 1986, Sequence homology between acquired immunodeficiency syndrome virus envelope protein and interleukin 2, *Proc. Natl. Acad. Sci. USA* **83**:9188.

Reiner, S. L., Wang, Z. E., Hatam, F., Scott, P., and Locksley, R. M., 1993, TH1 and TH2 cell antigen receptors in experimental leishmaniasis, *Science* **259**:1457.

Ressang, A. A., Gielkens, A. L., Quak, S., Mastenbroek, N., Tuppert, C., and De Castro, A., 1978a, Studies on bovine leukosis. VI. Enzyme linked immunosorbent assay for the detection of antibodies to bovine leukosis virus, *Ann. Rech. Vet.* **9**:663.

Ressang, A. A., Mastenbroek, N., and Quak, J., 1978b, Studies on bovine leucosis V. A comparative study on the practical value of the agar gel immunodiffusion test, the indirect fluorescent antibody technique and the micro complement fixation test for the detection of antibodies to bovine leucosis virus, *Tijdschr. Diergeneeskd.* **103**:758.

Riviere, Y., Tanneau-Salvadori, F., Regnault, A., Lopez, O., Sansonetti, P., Guy, B., Kieny, M. P., Fournel, J. J., and Montagnier, L., 1989, Human immunodeficiency virus-specific cytotoxic responses of seropositive individuals: Distinct types of effector cells mediate killing of targets expressing gag and env proteins, *J. Virol.* **63**:2270.

Robbins, D. S., and Hoffman, P. M., 1991, Virus-specific cytotoxic lymphocyte response in a neurotropic murine leukemia virus infection, *J. Neuroimmunol.* **31**:9.

Robert-Guroff, M., Reitz, M. S., Jr., Robey, W. G., and Gallo, R. C., 1986, In vitro generation of an HTLV-III variant by neutralizing antibody, *J. Immunol.* **137**:3306.

Robert-Guroff, M., Aldrich, K., Muldoon, R., Stern, T. L., Bansal, G. P., Matthews, T. J., Markham, P. D., Gallo, R. C., and Franchini, G., 1992, Cross-neutralization of human immunodeficiency virus type 1 and 2 and simian immunodeficiency virus isolates, *J. Virol.* **66**:3602.

Roberts, J. D., Bebenek, K., and Kunkel, T. A., 1988, The accuracy of reverse transcriptase from HIV-1, *Science* **242**:1171.

Robinson, J. E., Holton, D., Pacheco-Morell, S., Liu, J., and McMurdo, H., 1990, Identification of conserved and variant epitopes of human immunodeficiency virus type 1 (HIV-1) gp120 by human monoclonal antibodies produced by EBV-transformed cell lines, *AIDS Res. Hum. Retroviruses* **6**:567.

Robinson, W. E., Montefiori, D. C., and Mitchell, W. M., 1987, A human immunodeficiency virus type 1 (HIV-1) infection-enhancing factor in seropositive sera, *Biochem. Biophys. Res. Commun.* **149**:693.

Robinson, W. E., Montefiori, D. C., and Mitchell, W. M., 1988, Antibody-dependent enhancement of human immunodeficiency virus type 1 infection, *Lancet* **1**:790.

Robinson, W. E., Montefiori, D. C., Gillespie, D. H., and Mitchell, W. M., 1989, Complement-mediated, antibody-dependent enhancement of HIV-1 infection in vitro is characterized by increased protein and RNA syntheses and infectious virus release, *J. AIDS* **2**:33.

Robinson, W. E., Kawamura, T., Gorny, M. K., Lake, D., Xu, J. Y., Matsumoto, Y., Sugano, T., Masuho, Y., Mitchell, W. M., Hersh, E., and Zolla-Pazner, S., 1990a, Human monoclonal antibodies to the human immunodeficiency virus type 1 (HIV-1) transmembrane glycoprotein gp41 enhance HIV-1 infection in vitro, *Proc. Natl. Acad. Sci. USA* **87**:3185.

Robinson, W. E., Kawamura, T., Lake, D., Masuho, Y., Mitchell, W. M., and Hersh, E. M., 1990b, Antibodies to the primary immunodominant domain of human immunodeficiency virus type 1 (HIV-1) glycoprotein gp41 enhance HIV-1 infection in vitro, *J. Virol.* **64**:5301.

Robinson, W. E., Montefiori, D. C., and Mitchell, W. M., 1990c, Complement-mediated antibody-dependent enhancement of HIV-1 infection requires CD4 and complement receptors, *Virology* **175**:600.

Robinson, W. E., Gorny, M. K., Xu, J. Y., Mitchell, W. M., and Zolla-Pazner, S., 1991, Two immunodominant domains of gp41 bind antibodies which enhance human immunodeficiency virus type 1 infection in vitro, *J. Virol.* **65**:4169.

Ronchese, F., D'Andrea, E., Collavo, D., De Rossi, A., Zanovello, P., and Chieco-Bianchi, L., 1984, Tolerance to viral antigens in Mov-13 mice carrying endogenized Moloney-murine leukemia virus, *Cell. Immunol.* **83**:379.

Rosenberg, Z. F., and Fauci, A. S., 1990, Immunopathogenic mechanisms of HIV infection: Cytokine induction of HIV expression, *Immunol. Today* **11**:176.

Rouse, B. T., Norley, S., and Martin, S., 1988, Antiviral cytotoxic T lymphocyte induction and vaccination, *Rev. Infect. Dis.* **10**:16.

Ruan, K. S., and Lilly, F., 1991, Identification of an epitope encoded in the env gene of Friend murine leukemia virus recognized by anti-Friend virus cytotoxic T lymphocytes, *Virology* **181**:91.

Rucheton, M., Graafland, H., Fanton, H., Ursule, L., Ferrier, P., and Larsen, C. J., 1985, Presence of circulating antibodies against gag-gene MuLV proteins in patients with autoimmune connective tissue disorders, *Virology* **144**:468.

Ruegg, C. L., Monell, C. R., and Strand, M., 1989, Inhibition of lymphoproliferation by a synthetic peptide with sequence identity to gp41 of human immunodeficiency virus type 1, *J. Virol.* **63**:3257.

Ruprecht, R. M., Mullaney, S., Bernard, L. D., Gama Sosa, M. A., Hom, R. C., and Finberg, R. W., 1990, Vaccination with a live retrovirus: The nature of the protective immune response, *Proc. Natl. Acad. Sci. USA* **87**:5558.

Rusche, J. R., Javaherian, K., McDanal, C., Petro, J., Lynn, D. L., Grimaila, R., Langlois, A., Gallo, R. C., Arthur, L. O., Fischinger, P. J., Bolognese, D. P., Putney, S. D., and Matthews, T. J., 1988, Antibodies that inhibit fusion of human immunodeficiency virus-infected cells bind a 24-amino acid sequence of the viral envelope, gp120, *Proc. Natl. Acad. Sci. USA* **85**:3198.

Rushlow, K., Olsen, K., Stiegler, G., Payne, S. L., Montelaro, R. C., and Issel, C. J., 1986, Lentivirus genomic organization: The complete nucleotide sequence of the env gene region of equine infectious anemia virus, *Virology* **155**:309.

Russell, J. K., Pontzer, C. H., and Johnson, H. M., 1991, Both alpha-helices along the major histocompatibility complex binding cleft are required for staphylococcal enterotoxin A function, *Proc. Natl. Acad. Sci. USA* **88**:7228.

Rwambo, P. M., Issel, C. J., Hussain, K. A., and Montelaro, R. C., 1990a, In vitro isolation of a neutralization escape mutant of equine infectious anemia virus, *Arch. Virol.* **111**:275.

Rwambo, P. M., Issel, C. J., Adams, W. V., Hussain, K. A., Miller, M., and Montelaro, R. C., 1990b, Equine infectious anemia virus (EIAV): Humoral responses of recipient ponies and antigenic variation during persistent infection, *Arch. Virol.* **111**:199.

Safrit, J. T., Fung, M. S., Andrews, C. A., Braun, D. G., Sun, W. N., Chang, T. W., and Koup, R. A., 1992, Protection of Hu-PBL-SCID mice from infection with human immunodeficiency virus type 1 through passive transfer of a monoclonal antibody directed against the third variable region of the envelope gp120, in: *Fifth Annual Meeting of the National Cooperative Vaccine Development Groups for AIDS.*

Salinovich, O., Payne, S. L., Montelaro, R. C., Hussain, K. A., Issel, C. J., and Schnorr, K. L., 1986, Rapid emergence of novel antigenic and genetic variants of equine infectious anemia virus during persistent infection, *J. Virol.* **57**:71.

Sandstrom, E. G., Schooley, R. T., Ho, D. D., Byington, R., Sarngadharan, M. G., MacLane, M. E., Essex, M., Gallo, R. C., and Hirsch, M. S., 1985, Detection of human anti-HTLV-III antibodies by indirect immunofluorescence using fixed cells, *Transfusion* **25**:308.

Sarin, P. S., Sun, D. K., Thornton, A. H., Naylor, P. H., and Goldstein, A. L., 1986, Neutralization of HTLV-III/LAV replication by antiserum to thymosin alpha 1, *Science* **232**:1135.

Sawyer, L. A., Katzenstein, D. A., Hendry, R. M., Boone, E. J., Vujcic, L. K., Williams, C. C., Zeger, S. L., Saah, A. J., Rinaldo, C. R., Phair, J. P., Giorgi, J. V., and Quinnan, G. V., 1990, Possible beneficial effects of neutralizing antibodies and antibody-dependent, cell-mediated cytotoxicity in human immunodeficiency virus infection, *AIDS Res. Hum. Retroviruses* **6**:341.

Schaefer, A., and Schmidt, W., 1986, Resistance to cellular immune response in AKR leukemias, *Eur. J. Immunol.* **16**:753.

Schaefer, W., Claviez, M., Frank, H., Hunsmann, G., Moennig, V., Schwarz, H., Thiel, H. J., Bolognesi, D. P., Green, R. W., Langlois, A. J., Fischinger, P. J., and De Noronha, F., 1975, Mammalian C-type oncorna viruses. Relationships between structural virus and cell surface antigens and their possible significance in immunological defense mechanisms, *Biblio. Haematol.* **43**:88.

Schetters, H., Rohmer, H., Hehlmann, R., and Erfle, V., 1990, Quantitative determination of retrovirus-specific immune complexes, *J. Immunol. Methods* **134**:113.

Schroff, R. W., Gottlieb, M. S., Prince, H. E., Chai, L. L., and Fahey, J. L., 1983, Immunological studies of homosexual men with immunodeficiency and Kaposi's sarcoma, *Clin. Immunol. Immunopathol.* **27**:300.

Scott, P., Natovitz, P., Coffman, R. L., Pearce, E., and Sher, A., 1988, Immunoregulation of cutaneous leishmaniasis. T cell lines that transfer protective immunity or exacerbation belong to different T helper subsets and respond to distinct parasite antigens, *J. Exp. Med.* **168**:1675.

Sethi, K. K., Naeher, H., and Stroehmann, I., 1988, Phenotypic heterogeneity of cerebrospinal fluid-derived HIV-specific and HLA-restricted cytotoxic T-cell clones, *Nature* **335**:178.

Shadduck, P. P., Weinberg, J. B., Haney, A. F., Bartlett, J. A., Langlois, A. J., Bolognesi, D. P., and Matthews, T. J., 1991, Lack of enhancing effect of human anti-human immunodeficiency virus type 1 (HIV-1) antibody on HIV-1 infection of human blood monocytes and peritoneal macrophages, *J. Virol.* **65**:4309.

Shafferman, A., Jahrling, P. B., Benveniste, R. E., Lewis, M. G., Phipps, T. J., Eden-McCutchan, F., Sadoff, J., Eddy, G. A., and Burke, D. S., 1991, Protection of macaques with a simian immunodeficiency virus envelope peptide vaccine based on conserved human immunodeficiency virus type 1 sequences, *Proc. Natl. Acad. Sci. USA* **88**:7126.

Sharma, B., and Gupta, S., 1985, Antigen-specific primary cytotoxic T lymphocyte (CTL) responses in acquired immune deficiency syndrome (AIDS) and AIDS-related complexes (ARC), *Clin. Exp. Immunol.* **62**:296.

Sher, A., Gazzinelli, R. T., Oswald, I. P., Clerici, M., Kullberg, M., Pearce, E. J., Berzofsky, J. A., Mosmann, T. R., James, S. L., Morse, H. C., and Shearer, G. M., 1992, Role of T-cell derived cytokines in the downregulation of immune responses in parasitic and retroviral infection, *Immunol. Rev.* **127**:183.

Sherr, C. J., Todaro, G. J., Sliski, A., and Essex, M., 1978, Characterization of a feline sarcoma virus-coded antigen (FOCMA-S) by radioimmunoassay, *Proc. Natl. Acad. Sci. USA* **75**:4489.

Shirai, T., Ohta, K., Kohno, A., Furukawa, F., Yoshida, H., Maruyama, N., and Hirose, S., 1986, Naturally occurring antibody response to DNA is associated with the response to retroviral gp70 in autoimmune New Zealand mice, *Arthritis Rheum.* **29**:242.

Shore, S. L., Nahmias, A. J., Starr, S. E., Wood, P. A., and McFarlin, D. E., 1974, Detection of cell-dependent cytotoxic antibody to cells infected with herpes simplex virus, *Nature* **251**:350.

Siegel, F., Norley, S., Hartung, S., and Kurth, R., 1992, B-cell epitope mapping of the entire SIVagm envelope glycoprotein including fine mapping of immunogenic regions, *J. AIDS* **5**:583.

Siekevitz, M., Feinberg, M. B., Holbrook, N., Wong-Staal, F., and Greene, W. C., 1987, Activation of interleukin 2 and interleukin 2 receptor (Tac) promoter expression by the trans-activator (tat) gene product of human T-cell leukemia virus, type I, *Proc. Natl. Acad. Sci. USA* **84**:5389.

Silagi, S., and Calvelli, T. A., 1982, Relationship between rejection of several syngeneic tumors and retrovirus production by 5-bromodeoxyuridine-grown melanoma cells: Lack of protection in natural killer-deficient beige mice, *Cancer Res* **42**:2562.

Sinclair, A. L., Habeshaw, J. A., Muir, L., Chandler, P., Forster, S., Cruickshank, K., and Dalgleish, A. G., 1988, Antibody-dependent cell-mediated cytotoxicity: Comparison between HTLV-I and HIV-1 assays, *AIDS* **2**:465.

Sissons, J. G., and Oldstone, M. B., 1980, Antibody-mediated destruction of virus-infected cells, *Adv. Immunol.* **29**:209.

Sitbon, M., Nishio, J., Wehrly, K., Lodmell, D., and Chesebro, B., 1985, Use of a focal immunofluorescence assay on live cells for quantitation of retroviruses: Distinction of host range classes in virus mixtures and biological cloning of dual-tropic murine leukemia viruses, *Virology* **141**:110.

Smith, E. J., and Witter, R. L., 1983, Detection of antibodies against reticuloendotheliosis viruses by an enzyme-linked immunosorbent assay, *Avian Dis.* **27**:225.

Snyder, H. W., and Fleissner, E., 1980, Specificity of human antibodies to oncovirus glycoproteins: Recognition of antigen by natural antibodies directed against carbohydrate structures, *Proc. Natl. Acad. Sci. USA* **77**:1622.

Snyder, H. W., Singhal, M. C., Zuckerman, E. E., Jones, F. R., and Hardy, W. D., 1983, The feline oncornavirus-associated cell membrane antigen (FOCMA) is related to, but distinguishable from, FeLV-C gp70, *Virology* **131**:315.

Snyder, H. W., Singhal, M. C., Hardy, W. D., and Jones, F. R., 1984, Clearance of feline leukemia virus from persistently infected pet cats treated by extracorporeal immunoadsorption is correlated with an enhanced antibody response to FeLV gp 70, *J. Immunol.* **132**:1538.

Snyder, H. W., Reed, D. E., and Jones, F. R., 1989, Remission of FeLV-associated lymphosarcoma and persistent viral infection after extracorporeal immunoadsorption of plasma using staphylococcal protein A columns: Details of immune response, *Semin. Hematol.* **26**:25.

Solder, B. M., Schulz, T. F., Hengster, P., Lower, J., Larcher, C., Bitterlich, G., Kurth, R., Wachter, H., and Dierich, M. P., 1989, HIV and HIV-infected cells differentially activate the human complement system independent of antibody, *Immunol. Lett.* **22**:135.

Spear, G. T., Landay, A. L., Sullivan, B. L., Dittel, B., and Lint, T. F., 1990a, Activation of complement on the surface of cells infected by human immunodeficiency virus, *J. Immunol.* **144**:1490.

Spear, G. T., Sullivan, B. L., Landay, A. L., and Lint, T. F., 1990b, Neutralization of human immunodeficiency virus type 1 by complement occurs by viral lysis, *J. Virol.* **64**:5869.

Spear, G. T., Sullivan, B. L., Takefman, D. M., Landay, A. L., and Lint, T. F., 1991, Human immunodeficiency virus (HIV)-infected cells and free virus directly activate the classical complement pathway in rabbit, mouse and guinea-pig sera; activation results in virus neutralization by virolysis, *Immunology* **73**:377.

Spear, G. T., Takefman, D. M., Sullivan, B. L., Landay, A. L., Jennings, M. B., and Carlson, J. R., 1993a, Anti-cellular antibodies in sera from vaccinated macaques can induce complement-mediated virolysis of human immunodeficiency virus and simian immunodeficiency virus, *Virology* **195**:475.

Spear, G. T., Takefman, D. M., Sullivan, B. L., Landay, A. L., and Zolla-Pazner, S., 1993b, Complement activation by human monoclonal antibodies to human immunodeficiency virus, *J. Virol.* **67**:53.

Stanley, S. K., Bressler, P. B., Poli, G., and Fauci, A. S., 1990, Heat shock induction of HIV production from chronically infected promonocytic and T cell lines, *J. Immunol.* **145**:1120.

Steimer, K. S., Scandella, C. J., Skiles, P. V., and Haigwood, N. L., 1991, Neutralization of divergent HIV-1 isolates by conformation-dependent human antibodies to Gp120, *Science* **254**:105.

Stein, B. S., Gowda, S. D., Lifson, J. D., Penhallow, R. C., Bensch, K. G., and Engleman, E. G., 1987, pH-independent HIV entry into CD4-positive T cells via virus envelope fusion to the plasma membrane, *Cell* **49**:659.

Steinman, R., Dombrowski, J., O'Connor, T., Montelaro, R. C., Tonelli, Q., Lawrence, K., Seymour, C., Goodness, J., Pedersen, N. C., and Andersen, P. R., 1990, Biochemical and immunological characterization of the major structural proteins of feline immunodeficiency virus, *J. Gen. Virol.* **71**:701.

Stephenson, J. R., Essex, M., Hino, S., Hardy, W. D., and Aaronson, S. A., 1977, Feline oncornavirus-associated cell-membrane antigen (FOCMA): Distinction between FOCMA and the major virion glycoprotein, *Proc. Natl. Acad. Sci. USA* **74**:1219.

Stott, E. J., 1991, Anti-cell antibody in macaques, *Nature* **353**:393.

Stott, E. J., Chan, W. L., Mills, K. H., Page, M., Taffs, F., Cranage, M., Greenaway, P., and Kitchin, P., 1990, Preliminary report: Protection of cynomolgus macaques against simian immunodeficiency virus by fixed infected-cell vaccine, *Lancet* **336**:1538.

Stuart, P. M., Zlotnik, A., and Woodward, J. G., 1988, Induction of class I and class II MHC antigen expression on murine bone marrow-derived macrophages by IL-4 (B cell stimulatory factor 1), *J. Immunol.* **140**:1542.

Sugamura, K., and Hinuma, Y., 1993, Human retroviruses: HTLV-I and HTLV-II, in: *The Retroviridae,* Vol. 2 (J. A. Levy, ed.), pp. 399–435, Plenum Press, New York.

Svennerholm, B., Strannegard, O., and Lycke, E., 1978, Immune reactivity of visna virus-inoculated mice, *Infect. Immun.* **20**:412.

Szelc, C. M., Mitcheltree, C., Roberts, R. L., and Stiehm, E. R., 1992, Deficient polymorphonuclear cell and mononuclear cell antibody-dependent cellular cytotoxicity in pediatric and adult human immunodeficiency virus infection, *J. Infect. Dis.* **166**:486.

Sztein, M. B., Steeg, P. S., Johnson, H. M., and Oppenheim, J. J., 1984, Regulation of human peripheral blood monocyte DR antigen expression in vitro by lymphokines and recombinant interferons, *J. Clin. Invest.* **73**:556.

Takahashi, H., Takeshita, T., Morein, B., Putney, S., Germain, R. N., and Berzofsky, J. A., 1990, Induction of $CD8^+$ cytotoxic T cells by immunization with purified HIV-1 envelope protein in ISCOMs, *Nature* **344**:873.

Takeda, A., Tuazon, C. U., and Ennis, F. A., 1988, Antibody-enhanced infection by HIV-1 via Fc receptor-mediated entry, *Science* **242**:580.

Takeda, A., Sweet, R. W., and Ennis, F. A., 1990, Two receptors are required for antibody-dependent enhancement of human immunodeficiency virus type 1 infection: CD4 and Fc gamma R, *J. Virol.* **64**:5605.

Takeda, A., Robinson, J. E., Ho, D. D., Debouck, C., Haigwood, N. L., and Ennis, F. A., 1992, Distinction of human immunodeficiency virus type 1 neutralization and infection enhancement by human monoclonal antibodies to glycoprotein 120, *J. Clin. Invest.* **89**:1952.

Takeuchi, Y., Nagumo, T., and Hoshino, H., 1988, Low fidelity of cell-free DNA synthesis by reverse transcriptase of human immunodeficiency virus, *J. Virol.* **62**:3900.

Talal, N., Flescher, E., and Dang, H., 1992, Are endogenous retroviruses involved in human autoimmune disease? *J. Autoimmun.* (Suppl. A) 61.

Tam, M. R., Green, W. R., and Nowinski, R. C., 1980, Cytotoxic activities of monoclonal antibodies against the envelope proteins of murine leukemia virus, *Cancer Res.* **40**:3850.

Tanaka, T., and Kurth, R., 1984, Monoclonal antibodies specific for the avian sarcoma virus transforming protein pp60src, *Virology* **133**:202.

Tartaglia, J., Jarrett, O., Neil, J. C., Desmettre, P., and Paoletti, E., 1993, Protection of cats against feline leukemia virus by vaccination with a canarypox virus recombinant, ALVAC-FL, *J. Virol.* **67**:2370.

Teeuwsen, V. J., Siebelink, K. H., Crush-Stanton, S., Swerdlow, B., Schalken, J. J., Goudsmit, J., van de Akker, R., Stukart, M. J., Uytdehaag, F. G., and Osterhaus, A. D., 1990, Production and characterization of a human monoclonal antibody, reactive with a conserved epitope on gp41 of human immunodeficiency virus type I, *AIDS Res. Hum. Retroviruses* **6**:381.

Temin, H., 1992, Origin and general nature of retroviruses, in: *The Retroviridae*, Vol. 1 (J. A. Levy, ed.), p. 1, Plenum Press, New York.

Tendler, C. L., Greenberg, S. J., Burton, J. D., Danielpour, D., Kim, S. J., Blattner, W. A., Manns, A., and Waldmann, T. A., 1991, Cytokine induction in HTLV-I associated myelopathy and adult T-cell leukemia: Alternate molecular mechanisms underlying retroviral pathogenesis, *J. Cell. Biochem.* **46**:302.

Terai, C., Kornbluth, R. S., Pauza, C. D., Richman, D. D., and Carson, D. A., 1991, Apoptosis as a mechanism of cell death in cultured T lymphoblasts acutely infected with HIV-1, *J. Clin. Invest.* **87**:1710.

Thiel, K., Genovesi, E. V., Iglehart, J. D., Bolognesi, D. P., and Weinhold, K. J., 1988, Loss of tumorigenicity following in vitro MuLV infection is associated with induction of peritoneal natural killer cell activity, *Adv. Exp. Med. Biol.* **239**:169.

Tilley, S. A., Honnen, W. J., Racho, M. E., Hilgartner, M., and Pinter, A., 1991, Human monoclonal antibodies against the putative CD4 binding site and the V3 loop of HIV gp120 act in concert to neutralize virus, *VII International Conference on AIDS.*

Tilley, S. A., Honnen, W. J., Racho, M. E., Chou, T. C., and Pinter, A., 1992, Synergistic neutralization of HIV-1 by human monoclonal antibodies against the V3 loop and the CD4-binding site of gp120, *AIDS Res. Hum. Retroviruses* **8**:461.

Timonen, T., Ortaldo, J. R., and Herberman, R. B., 1981, Characteristics of human large granular lymphocytes and relationship to natural killer and K cells, *J. Exp. Med.* **153**:569.

Tompkins, M. B., and Tompkins, W. A., 1985, Stimulation of a cell-mediated cytotoxic response to FeLV-induced T cell lymphomas in the cat, *J. Immunol.* **135**:2817.

Torfason, E. G., Gudnadottir, M., and Loeve, A., 1992, Comparison of immunoblots with neutralizing and complement fixing antibodies in experimental and natural cases of visna-maedi, *Arch. Virol.* **123**:47.

Torres, J. V., Malley, A., Banapour, B., Anderson, D. E., Axthelm, M. K., Gardner, M. B., and Benjamini, E., 1993, An epitope on the surface envelope glycoprotein (gp130) of simian immunodeficiency virus (SIVmac) involved in viral neutralization and T cell activation, *AIDS Res. Hum. Retroviruses* **9**:423.

Toth, F. D., Kiss, J., Szabo, B., Fuest, G., Ujhelyi, E., Hollan, S. R., and Vaczi, L., 1989, Complement-dependent cytotoxicity of antibodies reactive with HIV-induced cell surface antigens in HIV-carrying haemophiliacs, *Acta Virol.* **33**:521.

Toth, F. D., Szabo, B., Ujhelyi, E., Paloczi, K., Horvath, A., Fuest, G., Kiss, J., Banhegyi, D., and Hollan, S. R., 1991, Neutralizing and complement-dependent enhancing antibodies in different stages of HIV infection, *AIDS* **5**:263.

Townsend, A., Ohlen, C., Bastin, J., Ljunggren, H. G., Foster, L., and Kaerre, K., 1989, Association of class I major histocompatibility heavy and light chains induced by viral peptides, *Nature* **340**:443.

Tsai, W. P., and Oroszlan, S., 1988, Site-directed cytotoxic antibody against the C-terminal segment of the surface glycoprotein gp90 of avian reticuloendotheliosis virus, *Virology* **166**:608.

Tsubota, H., Lord, C. I., Watkins, D. I., Morimoto, C., and Letvin, N. L., 1989, A cytotoxic T lymphocyte inhibits acquired immunodeficiency syndrome virus replication in peripheral blood lymphocytes, *J. Exp. Med.* **169**:1421.

Tyler, D. S., Lyerly, H. K., and Weinhold, K. J., 1989, Anti-HIV-1 ADCC, *AIDS Res. Hum. Retroviruses* **5**:557.

Tyler, D. S., Stanley, S. D., Nastala, C. A., Austin, A. A., Bartlett, J. A., Stine, K. C., Lyerly, H. K., Bolognesi, D. P., and Weinhold, K. J., 1990a, Alterations in antibody-dependent cellular cytotoxicity during the course of HIV-1 infection. Humoral and cellular defects, *J. Immunol.* **144**:3375.

Tyler, D. S., Stanley, S. D., Zolla-Pazner, S., Gorny, M. K., Shadduck, P. P., Langlois, A. J., Matthews, T. J., Bolognesi, D. P., Palker, T. J., and Weinhold, K. J., 1990b, Identification of sites within gp41 that serve as targets for antibody-dependent cellular cytotoxicity by using human monoclonal antibodies, *J. Immunol.* **145**:3276.

Uehara, T., Miyawaki, T., Ohta, K., Tamaru, Y., Yokoi, T., Nakamura, S., and Taniguchi, N., 1992, Apoptotic cell death of primed $CD45RO^+$ T lymphocytes in Epstein–Barr virus-induced infectious mononucleosis, *Blood* **80**:452.

Vahlne, A., Horal, P., Eriksson, K., Jeansson, S., Rymo, L., Hedstroem, K. G., Czerkinsky, C., Holmgren, J., and Svennerholm, B., 1991, Immunizations of monkeys with synthetic peptides disclose conserved areas on gp120 of human immunodeficiency virus type 1 associated with cross-neutralizing antibodies and T-cell recognition, *Proc. Natl. Acad. Sci. USA* **88**:10744.

Von Gegerfelt, A., Diaz-Pohl, C., Fenyoe, E. M., and Broliden, K., 1993, Specificity of antibody-dependent cellular cytotoxicity in sera from human immunodeficiency virus type 2-infected individuals, *AIDS Res. Hum. Retroviruses* **9**:883.

van Kuyk, R., Torbett, B., Gulizia, R., MacIsaac, P., Leath, S., Mosier, D., and Koenig, S., 1992, Human CTL specific for the Nef protein of HIV protect Hu-PBL-SCID mice from HIV infection, in: *Fifth Annual Meeting of the National Cooperative Vaccine Development Groups for AIDS.*

Varnier, O. E., Melioli, G., Repetto, C. M., Merli, A., and Raffanti, S. P., 1983, Sensitive micro in situ fluorescent focus assay for detection of murine xenotropic retroviruses, *Intervirology* **20**:159.

Vaux, D. L., 1993, Toward an understanding of the molecular mechanisms of physiological cell death, *Proc. Natl. Acad. Sci. USA* **90**:786.

Vilcek, J., and Le, J., 1991, Immunology of cytokines: An introduction, in: *The Cytokine Handbook,* (A. Thomson, ed.), pp. 1-17, Academic Press, New York.

Vogel, T., Kurth, R., and Norley, S., 1993, The majority of neutralizing antibodies in HIV-1 infected patients recognize linear V3 loop sequences: Studies using HIV-1MN multiple antigenic peptides, *J. Immunol.* Submitted for publication.

Wahlberg, J., Albert, J., Lundeberg, J., Von Gegerfelt, A., Broliden, K., Utter, G., Fenyoe, E. M., and Uhlen, M., 1991, Analysis of the V3 loop in neutralization-resistant human immunodeficiency virus type 1 variants by direct solid-phase DNA sequencing, *AIDS Res. Hum. Retroviruses* **7**:983.

Wainberg, M. A., Beiss, B., and Israel, E., 1980, Virus-mediated abrogation of chicken lymphocyte responsiveness to mitogenic stimulus, *Avian Dis.* **24**:580.

Walker, B. D., and Plata, F., 1990, Cytotoxic T lymphocytes against HIV, *AIDS* **4**:177.

Walker, B. D., Chakrabarti, S., Moss, B., Paradis, T. J., Flynn, T., Durno, A. G., Blumberg, R. S., Kaplan, J. C., Hirsch, M. S., and Schooley, R. T., 1987, HIV-specific cytotoxic T lymphocytes in seropositive individuals, *Nature* **328**:345.

Walker, B. D., Flexner, C., Birch-Limberger, K., Fisher, L., Paradis, T. J., Aldovini, A., Young, R., Moss, B., and Schooley, R. T., 1989, Long-term culture and fine specificity of human cytotoxic T-lymphocyte clones reactive with human immunodeficiency virus type 1, *Proc. Natl. Acad. Sci. USA* **86**:9514.

Walker, C. M., and Levy, J. A., 1989, A diffusible lymphokine produced by $CD8^+$ T lymphocytes suppresses HIV replication, *Immunology* **66**:628.

Walker, C. M., Moody, D. J., Stites, D. P., and Levy, J. A., 1986, $CD8^+$ lymphocytes can control HIV infection in vitro by suppressing virus replication, *Science* **234**:1563.

Walker, C. M., Erickson, A. L., Hsueh, F. C., and Levy, J. A., 1991, Inhibition of human immunodeficiency virus replication in acutely infected $CD4^+$ cells by $CD8^+$ cells involves a noncytotoxic mechanism, *J. Virol.* **65**:5921.

Walker, P. J., Molloy, J. B., and Rodwell, B. J., 1987, A protein immunoblot test for detection of bovine leukemia virus p24 antibody in cattle and experimentally infected sheep, *J. Virol. Methods* **15**:201.

Wang, B., Ugen, K. E., Srikantan, V., Agadjanyan, M. G., Dang, K., Refaeli, Y., Sato, A. I., Boyer, J., Williams, W. V., and Weiner, D. B., 1993, Gene inoculation generates immune responses against human immunodeficiency virus type 1, *Proc. Natl. Acad. Sci. USA* **90**:4156.

Wang, C. Y., Looney, D. J., Li, M. L., Walfield, A. M., Ye, J., Hosein, B., Tam, J. P., and Wong-Staal, F., 1991, Long-term high-titer neutralizing activity induced by octameric synthetic HIV-1 antigen, *Science* **254**:285.

Watanabe, M., Chen, Z. W., Tsubota, H., Lord, C. I., Levine, C. G., and Letvin, N. L., 1991, Soluble human CD4 elicits an antibody response in rhesus monkeys that inhibits simian immunodeficiency virus replication, *Proc. Natl. Acad. Sci. USA* **88**:120.

Weigent, D. A., Hoeprich, P. D., Bost, K. L., Brunck, T. K., Reiher, W. E., 3d, and Blalock, J. E., 1986, The HTLV-III envelope protein contains a hexapeptide homologous to a region of interleukin-2 that binds to the interleukin-2 receptor, *Biochem. Biophys. Res. Commun.* **139**:367.

Weijer, K., Uytdehaag, F. G., and Osterhaus, A. D., 1989, Control of feline leukaemia virus, *Vet. Immunol. Immunopathol.* **21**:69.

Weinstock, D., Schat, K. A., and Calnek, B. W., 1989, Cytotoxic T lymphocytes in reticuloendotheliosis virus-infected chickens, *Eur. J. Immunol.* **19**:267.

Weiss, R., 1993, Cellular receptors and viral glycoproteins involved in retrovirus entry, in: *The Retroviridae,* Vol. 2 (J. A. Levy, ed.), p. 1, Plenum Press, New York.

Weiss, S. H., Goedert, J. J., Sarngadharan, M. G., Bodner, A. J., Gallo, R. C., and Blattner, W. A., 1985, Screening test for HTLV-III (AIDS agent) antibodies. Specificity, sensitivity, and applications, *J. Am. Med. Assoc.* **253**:221.

Werner, A., and Levy, J., 1993, HIV-1 envelope gp120 is cleaved at the V3 loop after incubation with recombinant soluble CD4, *J. Virol.* **67**:2566.

Whetstone, C. A., Van Der Maaten, M. J., and Black, J. W., 1990, Humoral immune response to the bovine immunodeficiency-like virus in experimentally and naturally infected cattle, *J. Virol.* **64**:3557.

White, J., Herman, A., Pullen, A. M., Kubo, R., Kappler, J. W., and Marrack, P., 1989, The V beta-specific superantigen staphylococcal enterotoxin B: Stimulation of mature T cells and clonal deletion in neonatal mice, *Cell* **56**:27.

Williams, L., and Bell, L., 1991, Asynchronous development of the rat colon, *Anat. Embryol.* **183**:573.

Watkins, B. A., Reitz, M. S., Wilson, C. A., Aldrich, K., Davis, A. E., and Robert-Guroff, M., 1993, Immune escape by human immunodeficiency virus type 1 from neutralizing antibodies: evidence for multiple pathways, *J. Virol.* **67**:7493.

Winchester, R., Bernstein, D. H., Fischer, H. D., Enlow, R., and Solomon, G., 1987, The co-occurrence of Reiter's syndrome and acquired immunodeficiency, *Ann. Intern. Med.* **106**:19.

Wisecarver, J., Bechtold, T., Lipscomb, H., Davis, J., Collins, M., Purtilo, D., and Sonnabend, J. A., 1983, Comparison of ADCC and NK activities of peripheral blood mononuclear cells from patients at risk for acquired immune deficiency syndrome, *AIDS Res.* **1**:347.

Woodland, D. L., Happ, M. P., Gollob, K. J., and Palmer, E., 1991, An endogenous retrovirus mediating deletion of alpha beta T cells? *Nature* **349**:529.

Wright, S. E., and Bennett, D. D., 1992, Avian retroviral recombinant expressing foreign envelope delays tumour formation of ASV-A-induced sarcoma, *Vaccine* **10**:375.

Yagita, H., Nakata, M., Kawasaki, A., Shinkai, Y., and Okumura, K., 1992, Role of perforin in lymphocyte-mediated cytolysis, *Adv. Immunol.* **51**:215.

Yamamoto, H., Miller, M. D., Tsubota, H., Watkins, D. I., Mazzara, G. P., Stallard, V., Panicali, D. L., Aldovini, A., Young, R. A., and Letvin, N. L., 1990, Studies of cloned simian immunodeficiency virus-specific T lymphocytes. gag-specific cytotoxic T lymphocytes exhibit a restricted epitope specificity, *J. Immunol.* **144**:3385.

Yamamoto, J. K., Okuda, T., Ackley, C. D., Louie, H., Pembroke, E., Zochlinski, H., Munn, R. J., and Gardner, M. B., 1991, Experimental vaccine protection against feline immunodeficiency virus, *AIDS Res. Hum. Retroviruses* **7**:911.

Yasutomi, Y., Koenig, S., Haun, S. S., Stover, C. K., Jackson, R. K., Conard, P., Conley, A. J., Emini, E. A., Fuerst, T. R., and Letvin, N. L., 1993a, Immunization with recombinant BCG-SIV elicits SIV-specific cytotoxic T lymphocytes in rhesus monkeys, *J. Immunol.* **150**:3101.

Yasutomi, Y., Palker, T. J., Gardner, M. B., Haynes, B. F., and Letvin, N. L., 1993b, Synthetic peptide in mineral oil adjuvant elicits simian immunodeficiency virus-specific $CD8^+$ cytotoxic T lymphocytes in rhesus monkeys, *J. Immunol.* **151**:5096.

Yasutomi, Y., Reimann, K. A., Lord, C. I., Miller, M. D., and Letvin, N. L., 1993c, Simian immunodeficiency virus-specific $CD8^+$ lymphocyte response in acutely infected rhesus monkeys, *J. Virol.* **67**:1707.

Yefenof, E., Asjo, B., and Klein, E., 1991, Alternative complement pathway activation by HIV infected cells: C3 fixation does not lead to complement lysis but enhances NK sensitivity, *Int. Immunol.* **3**:395.

Yetter, R. A., Buller, R. M., Lee, J. S., Elkins, K. L., Mosier, D. E., Fredrickson, T. N., and Morse, H. C., 1988, $CD4^+$ T cells are required for development of a murine retrovirus-induced immunodeficiency syndrome (MAIDS), *J. Exp. Med.* **168**:623.

Young, A. R., 1987, The sunburn cell, *Photodermatology* **4**:127.

Zanoni, R., Krieg, A., and Peterhans, E., 1989, Detection of antibodies to caprine arthritis–encephalitis virus by protein G enzyme-linked immunosorbent assay and immunoblotting, *J. Clin. Microbiol.* **27**:580.

Zarling, J. M., Eichberg, J. W., Moran, P. A., McClure, J., Sridhar, P., and Hu, S. L., 1987, Proliferative and cytotoxic T cells to AIDS virus glycoproteins in chimpanzees immunized with a recombinant vaccinia virus expressing AIDS virus envelope glycoproteins, *J. Immunol.* **139**:988.

Zhang, J. Y., Martin, L. N., Watson, E. A., Montelaro, R. C., West, M., Epstein, L., and Murphey-Corb, M., 1988, Simian immunodeficiency virus/delta-induced immunodeficiency disease in rhesus monkeys: Relation of antibody response and antigenemia, *J. Infect. Dis.* **158**:1277.

Zhang, X. Q., Yang, L., Ho, D. D., Kuritzkes, D. R., Chen, I. S., Ching, W. T., Chen, Y. M., and Schooley, R. T., 1992, Human T lymphotropic virus types I- and II-specific antibody-dependent cellular cytotoxicity: Strain specificity and epitope mapping, *J. Infect. Dis.* **165**:805.

Zink, M. C., and Narayan, O., 1989, Lentivirus-induced interferon inhibits maturation and proliferation of monocytes and restricts the replication of caprine arthritis–encephalitis virus, *J. Virol.* **63**:2578.

Zlotnik, A., Shimonkevitz, R. P., Gefter, M. L., Kappler, J., and Marrack, P., 1983, Characterization of the gamma-interferon-mediated induction of antigen-presenting ability in P388D1 cells, *J. Immunol.* **131**:2814.

Zlotnik, A., Fischer, M., Roehm, N., and Zipori, D., 1987, Evidence for effects of interleukin 4 (B cell stimulatory factor 1) on macrophages: Enhancement of antigen presenting ability of bone marrow-derived macrophages, *J. Immunol.* **138**:4275.

Zolla-Pazner, S., and Gorny, M. K., 1992, Passive immunization for the prevention and treatment of HIV infection, *AIDS* **6**:1235.

Zuber, M., Samuel, K. P., Lautenberger, J. A., Kanki, P. J., and Papas, T. S., 1990, Bacterially produced HIV-2 env polypeptides specific for distinguishing HIV-2 from HIV-1 infections, *AIDS Res. Hum. Retroviruses* **6**:525.

Zvelebil, M. J., Sternberg, M. J., Cookson, J., and Coates, A. R., 1988, Predictions of linear T-cell and B-cell epitopes in proteins encoded by HIV-1, HIV-2 and SIVMAC and the conservation of these sites between strains, *FEBS Lett.* **242**:9.

CHAPTER 9

Endogenous Human Retroviruses

DAVID A. WILKINSON, DIXIE L. MAGER, AND JO-ANN C. LEONG

I. INTRODUCTION

A. Historical Background on Endogenous Retroviruses

The genomes of all eukaryotes contain multiple copies of DNA sequences that are related to sequences found in infectious retroviruses (for review, see Coffin, 1984; Garfinkel, 1992). These elements are transmitted through the germ line as stable Mendelian genes, yet they exhibit structural and sequence similarities to infectious exogenous retroviruses. It is these similarities that have led investigators to speculate that endogenous retroviruses are remnants of prior infections with exogenous retroviral agents and, with evolutionary time, changes have occurred to make them no longer infectious or pathogenic. These speculations have been supported with experimental studies that show that the genomes of infectious, exogenous retroviruses can integrate into the host chromosome and be inherited through the germ line. Since retroviruses are thought to have evolved from retrotransposons (Temin, 1980, 1992), it is also possible that some endogenous retrovirus-related sequences are actually precursors of infectious forms. In either case, once they are part of the host genome, these proviruses can serve as a pool of genetic material that exogenous viruses can use to produce variants with altered host specificities and phenotypes; they can encode gene products that compete for or complement in *trans* retrovirus function(s); and they can, themselves, act as insertional mutagens to change the regulation of host genes.

Much of the early work on endogenous viruses was carried out on chickens as a model. The chicken genome carries at least 16 endogenous Rous-associated virus-0-related proviral copies (ev-1 through ev-16), each located at a unique chromosomal site (Astrin, 1978; Rovigatti and Astrin, 1983). Some endogenous

DAVID A. WILKINSON AND DIXIE L. MAGER • The Terry Fox Laboratory, B.C. Cancer Agency, Vancouver, and Department of Medical Genetics, University of British Columbia, Vancouver, British Columbia V5Z 1L3, Canada. JO-ANN C. LEONG • Department of Microbiology, Oregon State University, Corvallis, Oregon 97331-3804.

The Retroviridae, Volume 3, edited by Jay A. Levy. Plenum Press, New York, 1994.

proviral loci were able to specify the synthesis of infectious virus (Crittenden *et al.*, 1983), and others represented deleted genomes capable of coding for only a subset of retroviral proteins (Astrin, 1978; Crittenden and Astrin, 1981). The presence of these multiple copies in the chicken genome led to the proposal that these genes were involved in ontogeny. However, the breeding of healthy, fertile chickens lacking all avian leukosis virus (ALV)-related endogenous viral genes (ev−) was considered strong evidence against this hypothesis (Astrin *et al.*, 1979). It was also suggested that endogenous viruses conferred a selective advantage on a host animal exposed to infectious virus particles, as evidenced by the unusual susceptibility of ev$^-$ chickens to infection with exogenous retroviruses (Robinson *et al.*, 1981; Crittenden *et al.*, 1982). However, the role of endogenous proviruses in the development or evolution of vertebrate species was made less clear by the finding that ev$^-$ chickens contain *gag-*, *pol-*, *env-*, and LTR-related sequences linked on a single continuous piece of DNA in the order expected for a proviral structure (Dunwiddie and Faras, 1985; Dunwiddie *et al.*, 1986). Under less stringent hybridization conditions, cloned viral DNA probes from the avian virus genome detected related sequences in the ev$^-$ chicken. The patterns obtained with the various probes suggested the presence of between 40 and 100 copies of retrovirus-related sequences in ev$^-$ chickens, suggesting that more distantly evolved families of endogenous retrovirus genes are present.

The genomes of mice have also been shown to contain at least four major families of endogenous sequences that are distinguished by relatedness to different classes of exogenous viruses (Gelb *et al.*, 1973; Steffen *et al.*, 1979, 1982; Hoggan *et al.*, 1983; Risser and Horowitz, 1983; Coffin, 1984; Stoye and Coffin, 1985, 1987). For most of these endogenous retroviral elements, the sequences are defective so that no functional gene products are encoded by them. However, some strains of mice do carry complete proviral DNAs in the germ line, and the induction of these DNAs can result in viral spread and ultimately leukemia (Chattopadhyay *et al.*, 1975; Breindl *et al.*, 1979; Steffen *et al.*, 1979; Chumakov *et al.*, 1982). Most mouse strains also contain partially defective endogenous proviruses that can contribute sequences to replicating viruses by recombination to generate viruses with new host ranges and phenotypes. The mink cell focus-forming (MCF) viruses are recombinant viruses with expanded host range. They have been shown to have arisen by recombination events with the host genome presumably after infection of many mouse strains (Stephenson *et al.*, 1974; Fischinger *et al.*, 1975; Hartley *et al.*, 1977; Chattopadhyay *et al.*, 1981; Famulari, 1983; Khan, 1984; Evans and Cloyd, 1984, 1985). More recently, endogenous retroviral sequences have been shown to provide functional DNA sequences to deletion-defective exogenous viruses during replication in NIH/3T3 cells. Analysis of these revertant viruses showed that the original deletion had been repaired by recombination with weakly homologous sequences derived from the mouse genome to re-create a fully infectious virus (Schwartzberg *et al.*, 1985; Colicelli and Goff, 1986; Martinelli and Goff, 1990).

Similar studies characterizing endogenous retroviral genes in many different vertebrate species have emphasized the ubiquitous nature of these elements. The studies conducted in chickens and mice suggest that these retroviral elements may be important in the biology of their host and in the generation of new phenotypes of exogenous retroviruses. In light of these findings, it is important to consider the role of endogenous retroviruses in the human genome.

B. Detection of Human Endogenous Retroviral Elements

It has long been recognized that human DNA contains sequences related to retroviruses (Benveniste and Todaro, 1974). Since the first such sequences were isolated by Martin and co-workers in 1981, it has become apparent that the human genome harbors a wide variety of endogenous retroviruslike elements (ERV) present in numbers ranging from a single copy to thousands of dispersed copies. These elements have been identified in several different ways, such as:

1. Screening of the genome by low-stringency hybridization with probes derived from nonhuman endogenous and exogenous retroviruses (Martin *et al.*, 1981; Bonner *et al.*, 1982; Callahan *et al.*, 1982; O'Connell *et al.*, 1984; Perl *et al.*, 1989)
2. Screening of the genome by hybridization to oligonucleotide probes with the 3'-terminal sequence of select tRNAs (Harada *et al.*, 1987; Kroger and Horak, 1987)
3. Direct amplification of genomic DNA by polymerase chain reaction (PCR) using redundant primers specific for conserved reverse transcriptase sequences (Bangham *et al.*, 1988; Shih *et al.*, 1989)
4. Analyzing the flanking regions of unrelated cellular genes (Mager and Henthorn, 1984; Maeda, 1985)

The human ERVs (HERVs) that have been described have the same fundamental structure as the integrated proviral form of exogenous retroviruses with flanking long terminal repeats (LTRs) of several hundred nucleotides. Full-length HERVs contain sequences homologous to the *gag*, *pol*, and *env* genes of infectious retroviruses. However, despite the similarities to infectious retroviruses, most HERVs are defective and contain multiple termination signals throughout those regions corresponding to *gag*, *pol*, and *env*. Although no infectious HERV has been recovered to date, the presence of such large numbers of HERVs and other retroviruslike elements in the human genome has led scientists to speculate on their origin and function(s).

Several possibilities for the function of endogenous retroviruses and retroviruslike elements in human DNA have been suggested (Temin, 1985). They may have played a role in evolution of the human genome by serving as a source of genetic variation, by increasing the effective target size for possible lethal mutations, or by acting as insertional mutagens. Their contributions to the origins of pathogenic retroviruses have been suggested by the endogenous–exogenous retrovirus recombinations which produce hybrid viruses with new tropisms and pathology in other mammals. For example, Harvey and Kirsten murine leukemia viruses resulted from the recombination of exogenous murine leukemia virus with endogenous VL30 elements (Chien *et al.*, 1979; Ellis *et al.*, 1980). Endogenous retroviruses may also be associated with human neoplasias by producing gene products that act as transcriptional activators; ERVs in other mammalian species have been shown to encode gene products that compete in *trans* for retroviral functions. In addition, endogenous retroviruses have been implicated in the development of autoimmune diseases (Krieg *et al.*, 1992). Thus, it would seem that ERVs could have an important role in the biology of the human species. Their prevalence and the maintenance of these elements in the human genome would seem to support this hypothesis.

This review will describe the known families of human endogenous retroviruslike elements and their relationship to the exogenous/endogenous retroviruses of other mammals, their classification in the context of other retroelements, and the possible biological significance of these elements.

II. CLASSIFICATION OF RETROELEMENTS

A. An Overview of Retroelements

A retroelement is a transposable or mobile genetic element which utilizes an RNA intermediate in its replicative cycle. These elements are widespread in nature and have been identified in both vertebrate and invertebrate animals, plants, fungi, and possibly prokaryotes (Finnegan, 1989; Temin, 1989, 1992; Garfinkel, 1992). They utilize a mechanism of transposition that involves the synthesis of an RNA intermediate which then acts as a template for an RNA-dependent DNA polymerase, commonly referred to as reverse transcriptase. All retroelements use reverse transcriptase to generate a DNA copy of their sequence from an RNA transcript. The new copy is then able to integrate back into the genome and complete the transposition process. This process appears to be very important in nature since it has been conserved throughout evolution; reverse transcriptase genes have been identified in bacteria and in mitochondrial plasmids present in fungi (Garfinkel, 1992). In myxobacteria and some strains of *Escherichia coli*, this reverse transcriptase-encoding element exists as a single-copy gene which has been termed a retron (Temin, 1989). The only known source of reverse transcriptase in eukaryotes are retroelements themselves; no cellular gene for reverse transcriptase has been identified. However, because reverse transcription is involved in the maintenance of telomeres, it is likely that a cellular gene coding for reverse transcriptase exists in higher organisms (Lundblad and Blackburn, 1990). It has been speculated that retroelements encoding reverse transcriptase may have evolved by co-opting the cellular enzyme for their own replicative purpose in a far distant eukaryotic ancestor (Temin, 1980). This possibility is suggested by two observations: (1) the broad distribution of retroelements among all eukaryotic kingdoms and (2) the conserved amino acid homology among all known reverse transcriptases, including that encoded by retrons.

As a consequence of their common transposition mechanism, all retroelements share certain features. First, because reverse transcriptase generates a new copy of the element, transposition increases the element's copy number within the genome. Thus, retroelements are present in multiple copies per genome, ranging from 10 to 10^5. Second, because the site of insertion is generally not restricted, these multiple copies are dispersed throughout the genome. Finally, during the integration step the target site is opened by staggered nicks to each DNA strand. This creates a short direct repeat of the target DNA sequence which immediately flanks the element on each end. These three features are characteristic of many transposable elements, including both transposons (DNA intermediate) and retroelements (RNA intermediate). There are also several hallmarks unique to retroelements, although not ubiquitous among them. Sequence homology with reverse transcriptase is one example. In addition, many retroelements contain an indication of their RNA intermediate, such as the presence of a 3′ polyadenosine tract downstream of a polyadenylation signal.

Retroelements have been subdivided into two categories based on the sequence structure of the element. The primary distinction made among retroelements is whether the element is bound by long terminal direct repeats (LTRs) or not (see Fig. 1). Elements that contain LTRs are designated as retrotransposons, while those that do not contain LTRs comprise a more heterogeneous group that has been termed retroposons (Hull and Will, 1989). Within each of these groups a further distinction can be made: both groups contain elements that encode reverse transcriptase and those that do not. Those elements that do not encode reverse transcriptase are thought to be able to serve as an efficient template for it when it is supplied by another element.

B. Retroposons

In humans and other mammals, retroposons were originally identified as repetitive sequences present in very high copy numbers within the genome. Originally, these elements were classified on the basis of conserved restriction endonuclease sites that they contained. Two examples in humans are *Kpn*I elements (now generally referred to as L1 elements) and Alu elements. Repetitive sequences have also been categorized on the basis of length into long interspersed elements (LINEs) and short interspersed elements (SINEs) (Singer, 1982a). At the time that the LINE nomenclature was suggested, provision for multiple LINE families within a species was made by identifying different families numerically by the abbreviation L1 and so on. However, in mammals (the only group of species for which this nomenclature has been adopted), only one example of such a retroelement has been identified per species. Consequently, the term LINE element has become associated with a specific structure of retroposon and is used interchangeably with the term L1. Sometimes a species designation is added, derived from the first letter of the genus and species (L1Hs for L1 in *Homo sapiens*, for example). Coincidentally, the designations of LINE and SINE also correspond with the capacity of the element to encode reverse transcriptase (LINEs) or not (SINEs).

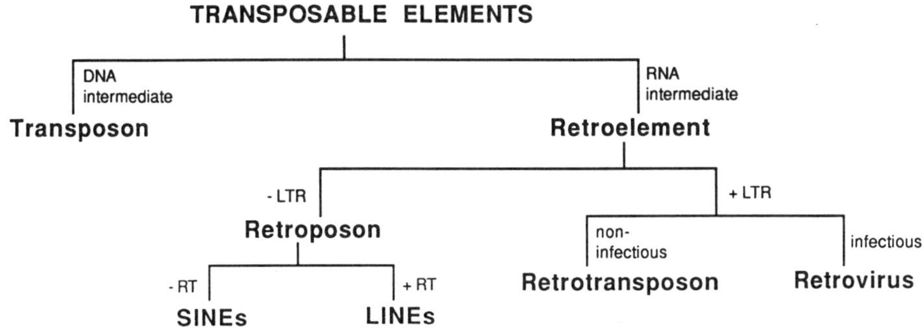

FIGURE 1. Classification of retroelements. Elements that transpose via an RNA intermediate are termed retroelements. They can be subdivided into LTR-containing elements (retrotransposons and retroviruses) and non-LTR-containing elements or retroposons (LINEs and SINEs). See text for details. RT, reverse transcriptase.

1. LINEs

In those mammals that have been studied, the L1 family of elements is comprised of approximately 10^4 randomly integrated, dispersed copies (Singer and Skowronski, 1985). Full-length L1 elements are approximately 6.5 kb in size and are terminated at their 3′ end by a polyadenosine tract resembling an mRNA. Characterization of randomly cloned genomic copies of L1 elements has indicated that the vast majority of L1 sequences are disrupted by deletions and point mutations. Commonly, the elements are truncated at the 5′ end (Singer and Skowronski, 1985). However, coding-competent L1 elements containing two large open reading frames (ORFs) have been identified in mice and humans (Martin *et al.*, 1984; Dombroski *et al.*, 1991). Similar elements have also been found in species of insects, protozoans, plants, and fungi (Eickbush, 1992). The larger, 3′ ORF encodes a protein with deduced amino acid homology to reverse transcriptase (Hattori *et al.*, 1986). In fact, the product of that gene has recently been demonstrated to possess reverse transcriptase activity (Mathias *et al.*, 1991). Deragon *et al.* (1990) have demonstrated that full-length L1 transcripts and reverse transcriptase are associated together in cytoplasmic macromolecular complexes. The complexes also contain a protein which is the predicted size for the product of the 5′ L1 ORF (Deragon *et al.*, 1990). Similar observations have been made in mice (Martin, 1991). It is possible that such complexes represent a reaction center in which the L1 sequence is reverse transcribed.

All genes that encode proteins in eukaryotes are transcribed by RNA polymerase II. Normally the promoter sequences which allow RNA polymerase II to initiate transcription are located 5′ to the start site and are therefore not included in the transcript. Consequently, retroposons are confronted with the problem of conserving their promoter sequences within their transcript. If this event did not happen, new copies of the element could not be transcribed and there would be no further genetic movement of the element. A novel feature of L1 elements is that they possess an internal RNA polymerase II promoter, i.e. 3′ to the transcription start site (Swergold, 1990; Minakami *et al.*, 1992). Promoters of this type have also been characterized in analogous elements in *Drosophila melanogaster* (Mizrokhi *et al.*, 1988). With the internal transcription start site, new copies of the element retain the ability to be transcribed. A simplified diagram of an L1 or LINE-1 element is shown in Fig. 2.

2. SINEs

The term SINE does not exclusively refer to retroposons; however, the most abundant and best characterized example of this class of repetitive sequence is indeed a retroposon. These sequences, called Alu elements, are present in 10^5–10^6 copies in the human genome. They are approximately 300 bp in length and terminate in a polyadenosine tract at their 3′ end (Weiner *et al.*, 1986) (see Fig. 2). They do not contain an ORF; rather, their nucleotide sequence indicates that they were derived from the gene coding for the 7SL structural RNA (Ullu and Tschudi, 1984). This small (~300 nucleotide) RNA is a component of the signal recognition particle located on the rough endoplasmic reticulum. Actually, Alu elements in humans are a unit consisting of a dimeric repeat of internally deleted 7SL RNAs, each ending in an adenosine-rich tract. Because a cellular gene has been

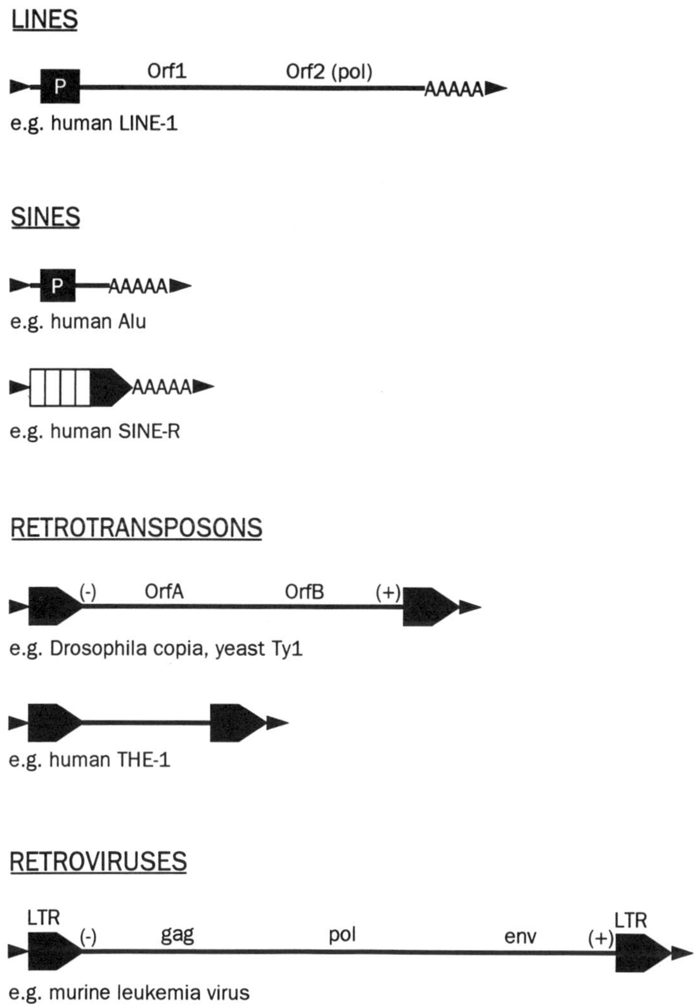

FIGURE 2. Structural features of retroelements. The general structures of different classes of retroelements are depicted with an emphasis on the human elements discussed in this review. Small arrows indicate short direct repeats of cellular DNA and large arrows are LTRs. P denotes the internal promoters in LINEs and SINEs. The SINE-R elements consist of variable numbers of a GC-rich repeat (open rectangles) followed by HERV-K LTR sequences. The (−) and (+) symbols show locations of primer binding sites for minus- and plus-strand DNA synthesis in retrotransposons and retroviruses.

identified as the source of Alu elements, they can be regarded as a form of processed pseudogene. Processed pseudogene is the term used to designate a dispersed copy of a cellular gene, which by the absence of introns or the presence of a polyadenosine tract and flanking short direct repeats, appears to be a reversed transcribed and reintegrated copy of that gene.

Several important differences set Alu elements apart from most processed pseudogenes. The extremely high number of Alu elements indicates that they serve as efficient templates for reverse transcriptase. In contrast, the copy number of most processed pseudogenes is in the range of a single copy to no more than a few hundred (Vanin, 1985). Also, some Alu sequences can be autonomously transcribed *in vivo* (Matera *et al.*, 1990); most pseudogenes do not contain the necessary promoter sequences for reexpression after integration. The 7SL RNA progenitor of Alu elements is an RNA polymerase III transcript and contains important regulatory sequences internally. Another hypothesis is that the deletions or other point mutations present in Alu elements may have enhanced the function of these internal promoters to the point that transcription of the Alu element is no longer dependent on upstream flanking sequences as is the case for the 7SL RNA gene (Ullu and Weiner, 1985; Weiner *et al.*, 1986). Thus, new copies of Alu elements may produce transcripts which in turn may serve to generate more genomic copies and so on resulting in the high copy number that is currently present within the human genome.

The fact that Alu elements are transcribed by RNA polymerase III has also been used as an explanation for their efficient template utilization by reverse transcriptase. The enzyme depends on a base-paired nucleotide primer for the initiation of polymerization. The termination signal for RNA polymerase III is a tract of thymidine bases in the flanking DNA. Hence, Alu elements might self-prime via pairing of the polyuridine tract at the end of the transcript with the polyadenosine tract at the end of the Alu element (Weiner *et al.*, 1986). Alternatively, Alu transcripts may preferentially associate with a source of reverse transcriptase, such as the L1 macromolecular complexes. Finding L1–Alu fusion elements flanked by short direct repeats does, indeed, suggest that this may occur (Miyaki *et al.*, 1983). The two possibilities are not mutually exclusive.

Alu elements are not the only example of retroposons within the SINE class of repetitive sequences. The human genome contains 4000 to 5000 copies of another sequence, called SINE-R elements (Ono *et al.*, 1987a). These elements are actually derivatives of the LTR-containing endogenous retroviruslike element, HERV-K (see below). They are approximately 630 bp in length and are comprised of a short stretch of HERV-K 3′ internal sequence followed by the LTR which contains a 370-bp deletion and terminates in an adenosine-rich tract immediately downstream of the LTR's polyadenylation signal (see Fig. 2). The 5′ end of these elements was not derived from the HERV-K sequence, and consists of multiple tandemly repeated copies of a 40-bp GC-rich repeat which may be an internal RNA polymerase II promoter (Ono *et al.*, 1987a). Distinct transcripts from these elements have not been detected; however, there has been a proliferation of SINE-R elements with their deleted HERV-K LTR sequences within the human genome. These LTRs normally contain transcriptional regulatory sequences that may or may not retain their function in SINE-R. These elements, like the Alu family, appear to have no coding capacity. Until function has been determined, the possible impact of the amplification of the HERV-K LTR sequences in the human genome remains to be evaluated.

C. Retrotransposons

Retrotransposons have been identified in a wide range of eukaryotes, including species of fungi, plants, protozoans, and animals (Boeke and Corces, 1989; Garfinkel, 1992). Typically, they are 6–7 kb in length and usually contain two ORFs (which often overlap) bounded by 300 to 600-bp LTRs (Fig. 2). They commonly occur in approximately 10–50 copies per genome, although in some mammals, retrotransposon families with thousands of members have been identified.

Virtually all retrotransposons encode homologues of the same two genes, arranged in the same order, although in some elements these genes may be disrupted by mutations (Boeke and Corces, 1989). The gene products of retrotransposons have been best characterized for the Ty1 family in *Saccharomyces cerevisiae* and the copia family in *Drosophila melanogaster.* Expression of these retrotransposons leads to the formation of ribonucleoprotein particles within the cell (Shiba and Saigo, 1983; Garfinkel *et al.*, 1985; Garfinkel, 1992). The 5′ ORF in these elements, ORFA, encodes the major protein component of these particles. The particles have also been demonstrated to contain reverse transcriptase, encoded by the second ORF, ORFB, as well as the full-length mRNA of the element (Adams *et al.*, 1987). These intracellular particles appear to represent a transposition intermediate in which the reverse transcript is generated (Garfinkel, 1992). It should be noted, however, that elements rendered unable to encode proteins by mutations are often still able to transpose. That is, their transcripts can be reverse transcribed into DNA and integrated into the host genome. Therefore, the products of coding-competent elements can act in *trans* on transcripts from defective elements to facilitate their transposition. The replicative cycle of retrotransposons such as Ty1 is shown in Fig. 3 where it is compared with that of retroviruses.

Unlike retroposons, the transcription of retrotransposons is driven by a clas-

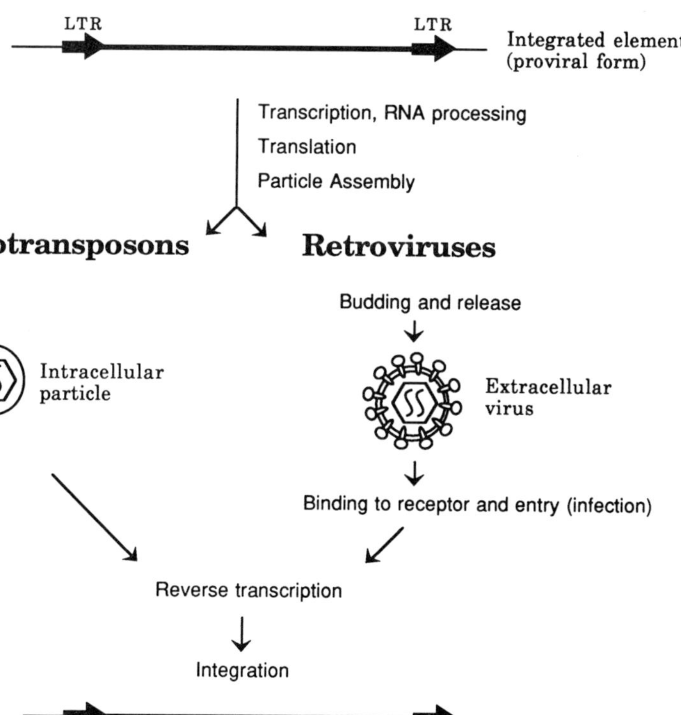

FIGURE 3. Comparison of the life cycles of retrotransposons and retroviruses. Although the general replicative schemes are similar, expression of *env* proteins (shown as spiked ovals) allows retroviral virions to recognize and infect new cells.

sical RNA polymerase II promoter contained within the LTR. As a consequence, the transcript is not a complete copy of the integrated form of the element. To compensate for the loss of transcriptional control sequences at the 5′ end of the transcript, these elements have developed an elaborate mechanism of reverse transcription which regenerates these sequences at both ends of the DNA form of the element, resulting in LTRs. In addition to LTRs, the replication and integration processes of retrotransposons require several important sequence features which distinguish this class of elements (Boeke and Corces, 1989). These features include: (1) an 8- to 18-bp stretch of sequence identity with a tRNA species, usually to its 3′ terminus, located just downstream of the 5′ LTR; this sequence serves as a primer for minus-strand synthesis; (2) a short purine-rich tract, just 5′ to the 3′ LTR, which primes plus-strand synthesis; (3) 2- to 5-bp inverted repeats, ending in TG . . . CA, which define the ends of the LTRs and are important to the integration process. Many of the structural and genetic features of retrotransposons are shared with retroviruses, as shown schematically in Fig. 2.

THE-1 Elements

One family of putative retrotransposons has been recognized in the human genome. Elements in this family have been termed THE-1 sequences, for *t*ransposon-like *h*uman *e*lements. Limited structural similarities between THE-1 elements and retrotransposons have promoted the speculation that they have dispersed within the genome by a mechanism of retrotransposition (Deka *et al.*, 1988a). The elements are bounded by ~350-bp LTRs, and are typically flanked by 5-bp direct repeats. However, the consensus THE-1 element (see Fig. 2) is only 2.3 kb in length, and lacks any discernible ORFs, nor is there any detectable sequence homology with either retrotransposons or retroviruses (Paulson *et al.*, 1985). In addition, the elements also lack discernible primer binding sites for both plus- and minus-strand synthesis by reverse transcriptase.

The haploid human genome has been estimated to contain approximately 10,000 THE-1 sequences as well as an additional 10,000 solitary LTRs (Paulson *et al.*, 1985). Thus, together these sequences comprise 1% of the total genome. Because of their relatively high copy number, these elements and their solitary LTRs were identified independently by several groups of workers interested in repetitive sequences in the human genome (Law *et al.*, 1982; Sun *et al.*, 1984; Mermer *et al.*, 1987). Other names that have been applied to these related sequences are: O family repeats (solitary LTRs; Sun *et al.*, 1984), LRS, for *l*ow *r*epeat *s*equence (Law *et al.*, 1987), and *Mst*II repeats, after a conserved *Mst*II site in the sequence (solitary LTRs; Mermer *et al.*, 1987). For simplicity, only the term THE-1 will be used in this discussion. It should be noted that two of these groups have estimated the human genome to contain approximately 5000 copies of the element (Law *et al.*, 1982; Mermer *et al.*, 1987). The reason for the discrepancy between these estimates and that by Paulson and co-workers has not been established, but it could be that the elements isolated by those groups detect a subpopulation within the total range of related sequences (Fields *et al.*, 1992).

THE-1 sequences are also present in the genomes of other primates, but not in more distantly related species such as rodents (Paulson *et al.*, 1985). Normally, the elements are found in high copy number in primates but the prosimian genus *Galago* has only a few copies in its genome (Schmid *et al.*, 1990). Isolation and characterization of two of the *Galago* genomic loci has revealed a striking finding. Each of these loci are flanked on their 5′ end by a common repeat sequence that is related to the R region of the THE-1 LTR. The sequences immediately 3′ to this repeat contain a region of homology to different parts of the THE-1 internal sequence (Schmid *et al.*, 1990). Thus, it appears that THE-1 sequences are actually a composite derived from two preexisting genomic loci. The mechanism by which these sequences were fused to create the current THE-1 structure is not understood.

During the retrotransposition process, a DNA copy of the element is reverse transcribed from an RNA template. Therefore, it is of particular interest that extrachromosomal circular DNAs containing THE-1 sequences have been detected in several cell lines (Paulson *et al.*, 1985; Misra *et al.*, 1987). A fraction of these DNA molecules, estimated to represent approximately 10% of the unintegrated THE-1 DNA forms present in Hela cells, has been cloned and characterized (Misra *et al.*, 1987, 1989). Analysis of these clones has revealed that they were derived from a single genomic locus by a recombination event involving the adjacent cellular sequence and consequently do not represent typical retrotransposition intermediates (Misra *et al.*, 1989). It remains to be determined, however, if the other THE-1 DNA forms present in Hela cells are derived from similar recombination events at other unstable loci or if they are true retrotransposition intermediates.

The association of THE-1 sequences with recombination events has been detected in several other instances. For example, two different *v*ariable *n*umber of *t*andem *r*epeat (VNTR) sequences have been found to be associated with THE-1 solitary LTRs. One of these VNTRs is comprised of a tandem repeat derived from the internal sequence of the THE-1 LTR, while the other VNTR was generated from the junction region with flanking DNA (Mermer *et al.*, 1987). It has been suggested that VNTRs may arise in regions of the genome that are particularly prone to rearrangements (Jeffreys *et al.*, 1985). Thus, the derivation of two different VNTRs from THE-1 LTRs suggests that THE-1 sequences might predispose the locus in which they reside to rearrange. In this regard, it is notable that the breakpoints of two independent deletions within a deletion-prone region of the dystrophin gene have been localized within a THE-1 sequence (Pizzuti *et al.*, 1992). It has also been observed that 673 bp of two 893-bp regions identified as recombinational hotspots in the immunoglobulin heavy chain gene are actually THE-1 sequences (Keyeux *et al.*, 1989; Fields *et al.*, 1992). THE-1 sequences have also been identified close to the integration site of a DNA duplication event which transposed more than 15 kb of sequence from chromosome 6p25-ter, to chromosome 16p11 (Wong *et al.*, 1990), and close to the integration site of papillomavirus DNA in a cervical carcinoma cell line (Baker *et al.*, 1987). It appears, therefore, that THE-1 elements are involved in recombination events that may sometimes incur pathological consequences.

Because THE-1 sequences lack a clearly defined coding strand and do not have a 5′ LTR as defined by a primer binding site homologous to a known tRNA sequence, it is difficult to predict the expected transcriptional orientation of the element. In addition, although RNAs containing THE-1 sequences have been detected in several cell lines and tissues, including a putative unit-length 2-kb transcript, a THE-1 LTR-promoted transcript has not been identified (Paulson *et al.*, 1985, 1987). Indeed, it appears that most transcripts containing THE-1 sequences are initiated by external cellular promoters (Paulson *et al.*, 1987). This finding is supported by the analysis of several cDNAs isolated from a fibroblast cell line, as well as Northern blot data that show that both strands of THE-1 sequences are incorporated into the polyadenylated fraction of cellular RNA (Paulson *et al.*, 1987). A putative promoter sequence has been identified in one LTR (Misra *et al.*, 1987) and is generally conserved (Fields *et al.*, 1992). However, this promoter sequence is in the opposite orientation to a similarly conserved, functional polyadenylation signal identified in two different cDNAs with THE-1 sequences. These cDNAs are initiated in unique cellular sequence and contain full-length THE-1 elements at their 3′ ends. They are both polyadenylated by the same signal within the downstream THE-1 LTR (Paulson *et al.*, 1987). A sequence that possibly represents a THE-1 pseudogene has also been identified (Deka *et al.*, 1988a). It, too, is polyadenylated by the same signal sequence. The 5′ end of the THE-1 pseudogene is in the upstream LTR and might correspond to the transcription start site.

One THE-1-containing transcript has been studied in detail. This transcript was originally identified as a cDNA clone that contained a full-length THE-1 embedded in a unique cellular sequence (Paulson *et al.*, 1987). Characterization of the cDNA clone and its genomic locus has revealed that the THE-1 sequence is in the 3′ untranslated region of an mRNA that codes for a member of the calmodulin superfamily of calcium-binding proteins (Deka *et al.*, 1988b). The transcripts of this gene are alternatively processed into four different forms, three of which contain the THE-1 sequence. It has been postulated that the THE-1 sequence may play a role in regulating the expression of the gene, most likely at the posttranscriptional level (Deka *et al.*, 1988b). However, any such effect has not yet been demonstrated. THE-1 sequences have also been speculated to be associated with other cellular functions, such as origins of replication (Misra *et al.*, 1989), and attachment points for the nuclear matrix (Law *et al.*, 1989). Although these suggestions have not been rigorously substantiated, it is interesting that a 66-kDa nuclear protein has been identified that specifically binds to THE-1 sequences *in vitro* (Law *et al.*, 1989). This finding supports the hypothesis that THE-1 sequences may be involved in as yet unrecognized processes within the cell.

D. Retroviruses

It has been postulated that retrotransposons have evolved from retroposons through a process in which LTRs were acquired by an ancestral element (Temin, 1980). This suggestion is consistent with an analysis of reverse transcriptase sequence homologies (Doolittle *et al.*, 1989), as well as the concept that LTRs represent an evolutionary refinement providing a flexible means to conserve promoter and enhancer sequences during the reverse transcription process. In turn, retrotransposons appear to be the evolutionary precursors of closely related infectious agents, called retroviruses.

Several lines of evidence suggest that retroviruses have evolved from retrotransposons. First, while retroviruses are very similar to retrotransposons, retroviruses are more complex. As shown in Fig. 2, retroviruses contain the same sequence features as retrotransposons, and homologues of the two ORFs, but they also contain at least one other ORF. This additional ORF accounts for the key biological property that distinguishes retroviruses from retrotransposons, i.e., infectivity. In retroviruses, the third ORF is called the envelope gene (*env*) and is always located between the 3′ LTR and the ORF encoding reverse transcriptase. The products of the envelope gene are essential for the infection of new cells by the retrovirus. The life cycle of retroviruses is also more complex than that of retrotransposons. Figure 3 illustrates that, although the general features of replication are very similar, the retroviral life cycle involves an extracellular phase. It is also of note that infectious retroviruses have only been identified in vertebrate species, including fish and snakes. The limited distribution of retroviruses throughout phylogeny, relative to that of retrotransposons, suggests that it is retroviruses which have evolved more recently, by a mechanism which involved the acquisition of an envelope gene (Temin, 1980, 1992). In this regard, it is noteworthy that some retrotransposons, such as the 17.6, 297, and gypsy families of elements in *Drosophila,* contain a third ORF in the same position as the envelope gene, and among retrotransposons, they have the reverse transcriptase genes most closely related to those of retroviruses (Boeke and Corces, 1989; Xiong and Eickbush, 1990). However, the possible product of the third ORF has not been characterized and its function is unknown. In addition, these elements do not appear to be infectious.

Because retroviruses are an agent of infectious disease among mammals, including humans, they have been intensively studied during the last 25 years. In fact, reverse transcriptase was first discovered through the study of retroviruses (Baltimore, 1970; Temin and Mizutani, 1970). It and the other gene products of retroviruses have been more thoroughly characterized than those of any other retroelement. In retroviruses, the homologues of the two ORFs of retrotransposons are called the *gag* gene, which encodes the structural proteins of the virion, and the *pol* gene, which encodes the enzymes necessary for reverse transcription and integration. The three definitive genes of simple infectious retroviruses always occur in the same order, giving integrated retroviruses the general structure: LTR–*gag*–*pol*–*env*–LTR (see Fig. 2). The term provirus is used to distinguish the integrated DNA form of the retrovirus from the genomic RNA form which is transcribed from it and packaged into virions. Various means of classifying retroviruses have been devised, based on either pathological, morphological, or structural characteristics, or sequence relatedness of the *pol* genes. [See Coffin (1992b) for the current classification scheme approved by the International Committee on Taxonomy of Viruses (ICTV).] Initially, retroviruses were classified on the basis of the morphology of their virion when visualized by electron microscopy. This led to the designation of type A, B, C, and D virion particles (see Teich, 1984, for details). Phylogenetic trees constructed from comparisons of the amino acid sequence of reverse transcriptase have also been used to divide retroviruses into four major groups (Doolittle *et al.*, 1989; Xiong and Eickbush, 1990) (see Fig. 4). It has been suggested that these groups provide a useful alternative system for classifying retroviruses, indicative of some of the molecular characteristics of the different viruses (Doolittle *et al.*, 1989).

The idea that reverse transcription of retrotransposons takes place in assembled particles was actually preceded by experiments which demonstrated that retroviral reverse transcription occurs within a nucleoprotein complex derived from the infecting virion. Reverse transcriptase activity of defective mutants can

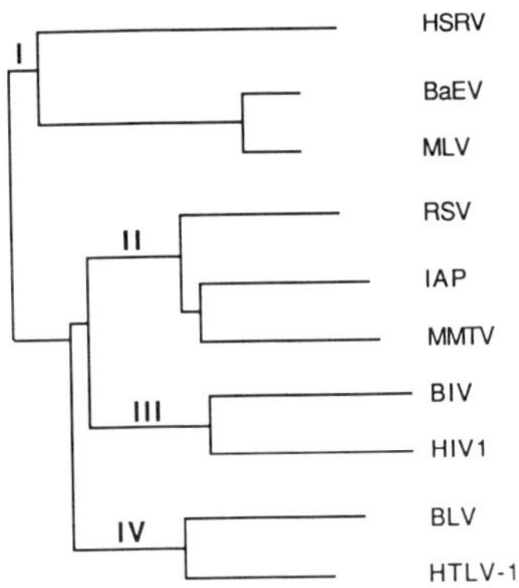

FIGURE 4. Phylogenetic tree of retroviruses based on reverse transcriptase homology (simplified from Xiong and Eickbush, 1990). Some representative viruses from each group are shown. HSRV, human spumaretrovirus; BaEV, baboon endogenous retrovirus; MLV, Moloney murine leukemia virus; RSV, Rous sarcoma virus; IAP, intracisternal A-particle (rodent endogenous retroviral element); MMTV, mouse mammary tumor virus; BIV, bovine immunodeficiency virus; HIV1, human immunodeficiency virus 1; BLV, bovine leukemia virus; HTLV-1, human T-cell leukemia virus 1.

be complemented by a functional gene from another provirus, but only at the time of virus assembly, and not in the infected cell (Varmus and Brown, 1991). Like retrotransposons, the LTRs of the provirus contain RNA polymerase II promoters and associated enhancer sequences. Another similarity between retroviruses and retrotransposons is the fact that most species of vertebrates contain proviral sequences as a normal component of their genome. These endogenous proviruses are presumably the result of retroviral infection of the germ cells or early embryo (and hence the germ line) which must periodically become fixed within the population. They are a stable component of the genomes of these organisms and consequently are genetically inherited from parent to offspring. Typically, 1 to 100 copies of a particular provirus are present within an organism's genome; however, like retrotransposons, most copies are disrupted by mutations and unable to encode infectious virions (Coffin, 1984). Some examples include the baboon endogenous retrovirus, the endogenous feline leukemia virus, the endogenous murine leukemia viruses, and the avian endogenous retroviruses (Coffin, 1984; Kozak and Ruscetti, 1992; Payne, 1992). In addition to endogenous retroviruses that are able to encode infectious virions, the genomes of many mammals contain dispersed, multicopy families with varying degrees of nucleotide sequence similarity to retroviruses. Because of the historical focus on retroviruses as a cause of disease, these sequence families are commonly referred to as endogenous retroviruslike elements, although the sequence structure of some families indicates that they might be more correctly viewed as possible retrotransposons. Because THE-1 elements lack any sequence homology with retroviruses, they are an exception. Only after characterization of their structural features have they become viewed as possible retrotransposons. In any case, the term retroviruslike designates that members coding for infectious retroviruses have not been detected. Several types of retroviruslike elements have been described in mice, of which the IAP (*i*ntracisternal *A* *p*article) and VL30 (*v*irus-*l*ike 30) families are the most well known. Because IAP elements are known to transpose intracellularly (Kuff and Lueders, 1988; Heidmann and Heidmann, 1991), they are also frequently termed retrotransposons. Indeed, when considering these types of sequences, the distinction between endogenous retroviruses and retrotransposons is somewhat arbitrary. IAPs and other murine elements have been extensively discussed in several recent reviews (Kuff and Lueders, 1988; Keshet *et al.*, 1991; Kozak and Ruscetti, 1992). The primary aim of this review is to present the current state of knowledge on the endogenous retroviruslike sequences found in the human genome. Some of the material discussed here has also been presented in earlier reviews on this subject (Cohen and Larsson, 1988; Larsson *et al.*, 1989; Brack-Werner *et al.*, 1989b; Leib-Mösch *et al.*, 1990, 1992a; Abraham and Khan, 1990; Johnson *et al.*, 1990).

III. HUMAN ENDOGENOUS RETROVIRUSLIKE ELEMENTS

There is ample evidence in the literature that the human genome contains a wide variety of endogenous retroviruslike elements. Those elements that have been sequenced in their *pol* region fall into two of the primary phylogenetic retroviral groups shown in Fig. 4: class I or class II. This observation has prompted the suggestion that *pol* sequence homology provides a useful criterion by which HERVs may be classified (Callahan, 1988). HERVs with homology to mammalian type C retroviruses such as murine leukemia virus (MLV) have been termed class I, and those that share sequence homology to type A, type B, and avian type C retroviruses have been termed class II. A dendrogram of a *pol* region sequence comparison of different HERV elements for which sequence is available is illustrated in Fig. 5.

Another system that has been used to classify HERVs is based on the identity of the tRNA which has homology to the minus-strand primer binding site (PBS) located immediately downstream of the 5′ LTR (Larsson *et al.*, 1989). However, this classification scheme is complicated by the fact that some distinctly different families (based on internal sequence) share homology with the same tRNA, and that, as a result of deletions or partial cloning, not all elements contain this very short sequence (18 bp). In addition, definitive assignment of a tRNA counterpart to a primer binding site can sometimes be hindered by the lack of sequence information for all human tRNAs. There may also be point mutations within the primer binding site since available evidence indicates that most elements were fixed in the primate genome for 40 million years or more. Despite these shortcomings, this classification scheme remains valuable for distinguishing the various families of HERV which belong to the two classes and so will be employed where possible in the following discussion of the different HERV classes.

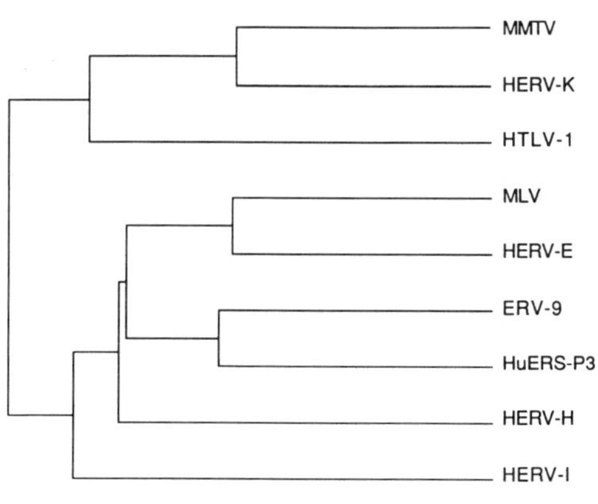

FIGURE 5. Comparisons of HERV *pol* sequences. The deduced amino acid sequences corresponding to a 90- to 97-amino-acid region of the reverse transcriptase gene (domains 3–5 of Xiong and Eickbush, 1990) were compared with the Genetics Computer Group (GCG) program Pileup. (The dendrogram is based on sequence differences and is not a true phylogenetic tree.) Three known retroviruses, MMTV, MLV, and HTLV-1, are included for reference. HERV-K, Ono *et al.* (1986); HERV-E, Repaske *et al.* (1985); ERV-9, La Mantia *et al.* (1991); HuERS-P3, Kroger and Horak (1987); HERV-H, Mager and Freeman (1987), Wilkinson *et al.* (1993); HERV-I, Maeda and Kim (1990). See following sections for complete descriptions of these elements.

Neither the use of the primer binding site designation or a standard acronym has been universally adopted by researchers in this field. Therefore, the original appellations assigned by the researcher(s) who discovered each family have also been included in this review. As an aid to the reader, Table I lists the various terms that have been used to describe the human retroelements discussed here.

A. Class I HERV

1. The HERV-ERI Superfamily

At the time when most HERV clones were isolated, they were routinely hybridized back to human genomic Southern blots to identify other cross-hybridizing sequences within the genome and obtain an estimate of their relative copy number. Under stringent conditions, some clones hybridized to single DNA fragments and could be distinguished as single elements; other clones would hybridize to multiple DNA fragments even under high stringency conditions and were representatives of a family of closely related elements. Subsequently, the nucleotide sequence of all or part of most clones has been determined. Comparison of the *env* and LTR regions of HERV clones indicates that several of the families and single elements identified so far are significantly related, despite the fact that they have different primer binding sites. Nucleotide similarity among the

TABLE I. Terms Used for Human Retroelements

Term	Previously used or other terms	Description
LINE		Long interspersed element
L1	LINE-1, *Kpn*I repeat	Major family of human LINEs
SINE		Short interspersed element
Alu		Major family of human SINEs
SINE-R		Retrovirus-derived SINE family
THE-1	O family, LRS, *Mst*II repeats	Transposon-like human element
ERV		Endogenous retroviruses
HERV		Human endogenous retrovirus-like elements
HERV-ERI		Superfamily of related HERVs with Glu, Arg, or Ile tRNA PBS
HERV-E	4.1 family	HERV family with a Glu tRNA PBS
Hs5		Member of HERV-ERI superfamily
NP-2		Member of HERV-ERI superfamily
ERV1		Member of HERV-ERI superfamily
HERV-R	ERV3	Member of HERV-ERI superfamily
RRHERV-I		Member of HERV-ERI superfamily
HERV-I	RTVL-I	HERV family with Ile tRNA PBS
S71		HERV related to simian sarcoma virus
HRES-1		HTLV-1-related HERV
HERV-P		HERV family with a Pro tRNA PBS
HuERS-P1, 2		Different HERV-P families
-P3	HuRRS-P	Different HERV-P family
ERV-9		HERV family
HERV-H	RTVL-H	HERV family with His tRNA PBS
HERV-K	HLM, HM, NMWV	Class II HERV family with Lys tRNA PBS
HML		MMTV-like (class II) HERV

env and LTR regions is particularly sensitive for detecting closely related elements since it has been shown that these regions diverge faster than other parts of the genome for infectious retroviruses (Shimotohno *et al.*, 1985; Desai *et al.*, 1986; Doolittle *et al.*, 1989). Therefore, this observation suggests that these HERVs might be regarded as a large group which we propose to call the HERV-ERI superfamily. This designation reflects the fact that several HERV families with different primer binding sites (homologous to Glu, Arg, or Ile tRNAs) are among the different isolates which are included in the superfamily. Table II shows the HERV clones isolated to date which comprise the HERV-ERI superfamily.

a. 4.1 (HERV-E)

The first clones to be isolated in this superfamily were designated 51-1 and 4-1 and 4-14. These sequences were isolated by an approach which involved two cross-species genomic library screenings. First, a genomic clone of African green monkey DNA was isolated on the basis of cross hybridization to a probe containing the MLV genome. A portion of this monkey clone was then recognized to be highly related to the *gag–pol* region of baboon endogenous virus (BaEV), and was subsequently used to screen a human genomic library (Martin *et al.*, 1981). This approach followed previous experiments which had indicated that both Old World monkey and human DNA contained sequences which hybridized to BaEV (Benveniste and Todaro, 1974).

The analysis of these clones revealed 4-1 to contain a full-length HERV (8.8 kb), having *gag*, *pol*, and *env*-related regions bounded by ~490-bp LTRs (Steele *et*

TABLE II. Isolates of the Class I HERV-ERI Superfamily

Isolate	Sequence structure	Sequence determined	Reference
a. HERV-E			
51-1	RPT–*gag–pol*–RPT[a]	RPT–*gag*–(*pol*)[b]	Repaske *et al.* (1983)
4-1	Provirus	Complete	Repaske *et al.* (1985)
4-14	Provirus	LTRs	Repaske *et al.* (1983)
ERVA1A	Provirus	5′ LTR	Emi *et al.* (1988)
		3′ LTR	Samuelson *et al.* (1990)
ERVA1B	Provirus	—[c]	Samuelson *et al.* (1990)
ERVA1C	Provirus	(*gag*)–LTR	Ting *et al.* (1992)
ERVA2A	Solitary LTR	LTR	Samuelson *et al.* (1990)
pL1, cDNA	*env*–LTR	Complete	Rabson *et al.* (1985)
jnLTR8, cDNA	Solitary LTR	Complete	Tomita *et al.* (1990)
jnLTR22, cDNA	Solitary LTR	Complete	Tomita *et al.* (1990)
LTRpdhB	Solitary LTR	LTR	GAN: D90086[d]
b. Hs5	Partial *pol–env*	*pol–env*	Levy *et al.* (1990)
c. NP-2	*gag–pol–env*–LTR	(*env*)	Silver *et al.* (1987)
d. ERV1	*gag–pol–env*–LTR	3′ LTR	Bonner *et al.* (1982)
e. ERV3	Provirus	(*gag*)–(*pol*)	O'Connell *et al.* (1984)
		env, 3′ LTR	Cohen *et al.* (1985)
f. RRHERV-I, cDNA	Spliced *env* transcript	Complete	Kannan *et al.* (1991)

[a] RPT: tandem repeat structure at the termini of clone 51-1.
[b] Parentheses in the "Sequence determined" column indicate that those regions were only partially sequenced.
[c] No reported sequence.
[d] GAN, GenBank accession number is provided for those clones with no published reference.

al., 1984; Repaske *et al.*, 1985). The primer binding site inside the 5′ LTR was homologous to the 3′ terminus of the rat glutamic acid tRNA (Steele *et al.*, 1984) and hence, this family has been termed HERV-E, the E denoting glutamic acid (Larsson *et al.*, 1989). In addition to full-length elements, solitary HERV-E-type LTR sequences (i.e., not associated with *gag*, *pol* or *env* sequence) have been identified in the human genome (Steele *et al.*, 1984). Solitary LTR sequences are thought to result from a recombination event between the two LTRs of a full-length element.

The element in clone 51-1 was shown to contain almost complete *gag* and *pol* regions having 85–93% nucleotide homology to the 4-1 sequence (Repaske *et al.*, 1983, 1985). However, the ~6-kb element in clone 51-1 did not contain sequences homologous to an *env* region and was bounded by unique repeat structures consisting of a 72- to 76-bp sequence tandemly repeated 8–13 times (Steele *et al.*, 1984). These repeat structures did not bear any recognizable similarity to the LTRs of retroviruses. Rather, they appeared to be unique to this family of truncated elements (Steele *et al.*, 1984).

The copy number of both the full-length and truncated elements has been estimated at 35–50 copies of each family within the human genome (Steele *et al.*, 1984). Both families are associated with regions of amplified genomic DNA. This association was discovered by the detection of conserved restriction endonuclease sites in the 3′ DNA flanking full-length elements (Steele *et al.*, 1984, 1986) and the use of DNA 5′ to the truncated elements as a probe to isolate multiple, distinct clones of that family (Repaske *et al.*, 1983). Many of the conserved bands observed on a Southern blot could not be correlated with a single chromosome suggesting that these elements were dispersed throughout the genome (Steele *et al.*, 1986). It is striking, therefore, that these two families were found in marked clusters in the telomeric and, to a lesser extent, the centromeric regions of chromosomes when mapped by *in situ* hybridization (Taruscio and Manuelidis, 1991).

Characterization of the amylase gene family has identified one cluster of HERV-E elements. At this locus, a region of DNA spanning a primordial amylase gene has undergone a series of tandem duplications, which in humans has created a cluster of five functional genes. Three of these genes are expressed in the salivary gland and two in the pancreas. Analysis of the complete amylase genomic region identified full-length HERV-E elements, termed ERVA1A–1C (Table II), situated approximately 800 bp 5′ to each salivary gland gene in the opposite transcriptional orientation, and a solo LTR, termed ERVA2A, located 5′ to one of the pancreatic genes (Emi *et al.*, 1988; Samuelson *et al.*, 1990). A comparison of the genomic structure and transcriptional start sites of the salivary gland genes versus the pancreatic gene suggested that the insertion of the full-length HERV-E element activated a cryptic promoter which directed tissue-specific expression of the salivary gland genes in the parotid gland. Indeed, further investigation of the promoter region of the salivary gland genes identified parotid-specific regulatory sequences in the 5′ internal regions of the HERV-E elements at this locus (Ting *et al.*, 1992). This finding aptly explained why deletion of the proviral sequence via LTR–LTR recombination resulted in a reversion back to pancreatic expression (Samuelson *et al.*, 1990). Thus, the ancient insertion of a HERV-E element appears to have played an important role in the evolution of this gene family. It is also noteworthy that the salivary amylase gene region in human DNA appears to have undergone additional duplication events compared with the same region in the chimpanzee genome (Samuelson *et al.*, 1990). This finding suggests that amplification events involving endogenous retroviral elements could possibly contribute to important species differences.

Members of the HERV-E family have been detected in the genomes of Old World monkeys, apes, and humans (cited in Repaske *et al.*, 1985). In fact, PCR analysis of the genomic locus containing the element in clone 4-14 (Repaske *et al.*, 1983) has determined that it is present in the same location in the genomes of Old World monkeys, apes, and humans (Shih *et al.*, 1991). Therefore, the element must have inserted into the primate lineage at a time when these species shared a common ancestor, estimated to be at least 25 million years ago (Gingerich, 1984). In addition, some full-length elements have been amplified within a similar sequence of DNA in the genomes of both chimpanzees and humans (Steele *et al.*, 1986). This finding effectively dates this amplification event to at least 5 million years ago (Sibley and Ahlquist, 1987). Polymorphisms associated with these elements within the human population have not been detected (Steele *et al.*, 1984; Taruscio and Manuelidis, 1991; Nakamura *et al.*, 1991; Sugino *et al.*, 1992). Thus, these elements appear to be an ancient and comparatively stable part of the genome. At one time, however, they were a component of DNA that underwent dramatic amplification and dispersal within the primate genome.

In addition to the dispersal of HERV-E throughout the genome by amplification events, successive integrations into the primate germ line have probably occurred. This possibility is suggested by evidence that some HERV-E sequences are dispersed as single elements within the genome (Taruscio and Manuelidis, 1991) and from the degree of divergence of the LTR sequences of several HERV-E clones which display a marked continuum of divergence. The contribution of genetic drift among the integrated elements can be estimated to be approximately 5–10% by comparison of the 5′ and 3′ LTRs within a single full-length element (Repaske *et al.*, 1985; Samuelson *et al.*, 1990). However, the observed divergence among the LTRs of different isolates can be substantially greater than this, as shown in Fig. 6 (lightly shaded boxes). This finding is consistent with the hypothesis that the divergent HERV-E isolates represent stages along the evolutionary lineage of an ancient retrovirus which has been preserved by integration into the germ line of a common primate ancestor.

The complete nucleotide sequence of the HERV-E clone 4-1 has been reported (Repaske *et al.*, 1985). In-frame stop codons occur in each of the retroviral coding regions rendering that element incapable of encoding an infectious virion. However, an approximately 1.3-kb ORF was detected in the *env* region which spanned almost the entire surface glycoprotein domain. This observation is notable, since the primary transcripts from this family appear to be analogous to retroviral spliced *env* mRNA. Polyadenylated 3.0- and 1.7-kb transcripts have been detected in term placenta, normal colon mucosa, spleen, and liver (Rabson *et al.*, 1983; Gattoni-Celli *et al.*, 1986). In addition, lower levels of 3.6- and 2.2-kb transcripts were also expressed in placenta (Rabson *et al.*, 1983, 1985) and suggest the use of an alternative splice site ~500 bp upstream of the one which generates the 3.0- and 1.7-kb transcripts. Analysis of partial cDNA clones isolated from a placental cDNA library was consistent with the interpretation of the 3.0 and 3.6 kb RNAs as spliced *env* transcripts (Rabson *et al.*, 1985). It also revealed that the 1.7- and 2.2-kb RNAs most likely result from splicing events analogous to their

	jnLTR8	jnLTR22	ERVA1A	LTRpdhB	4-14	4-1	RRHERVI	HERV-R
jnLTR22	90% 445bp							
ERVA1A	85% 438bp	86% 431bp						
LTRpdhB	75% 419bp	72% 444bp	79% 302bp					
4-14	75% 291bp	71% 437bp	72% 419bp	72% 462bp				
4-1	68% 247bp	72% 260bp	70% 305bp	72% 422bp	77% 426bp			
RRHERVI	65% 58bp	73% 22bp	68% 31bp	70% 44bp	62% 117bp	79% 19bp		
HERV-R	69% 50bp	78% 27bp	74% 53bp	83% 18bp	76% 25bp	85% 20bp	74% 303bp	
ERV1	79% 19bp	59% 61bp	75% 33bp	67% 36bp	79% 19bp	78% 23bp	71% 223bp	76% 253bp

FIGURE 6. Comparison of the nucleotide sequence similarities among the LTRs of the HERV-ERI superfamily isolates. Percent homology, determined by the GCG program Bestfit (gap weight = 5.0; gap length weight = 0.3), is shown in the upper left corner, and the length of the homology is shown in the bottom right corner of each square. Comparisons among members of the HERV-E family are lightly shaded. Significant homologies among other members of the superfamily are darkly shaded. All LTR sequences were as defined by the workers who first reported the sequence, as presented in Table II. The 3′ LTR of full-length elements was used in each case, except for ERVA1A, for which only the 5′ LTR sequence was available. LTRpdhB is a previously unrecognized solo LTR present in the 5′ flanking region of the pyruvate dehydrogenase gene. The sequence is deposited in Genbank under the accession number D90086, in which the LTR sequence lies between nucleotide coordinates 1227 through 1676 inclusively.

larger counterparts in transcripts of elements that contain two deletions. One deletion removes 222 bp in the surface glycoprotein domain of *env*, the other removes 938 bp including all of the transmembrane domain as well as part of the U3 domain of the 3′ LTR (Rabson *et al.*, 1985). The nucleotide sequence of these clones revealed in frame termination codons precluding them from representing possible mRNAs for a HERV-E envelope protein.

Expression of the 3.0- and 1.7-kb transcripts has been detected in colon and breast carcinoma cell lines (Rabson *et al.*, 1985) and increased expression of the 1.7-kb transcript has been reported in primary colon tumors and colon cell lines (Gattoni-Celli *et al.*, 1986). Other researchers have reported detecting full-length HERV-E transcripts, as well as the 1.7-kb RNA, in human brain (unpublished results cited in Taruscio and Manuelidis, 1991; Yeh *et al.*, 1991). In addition to transcripts containing the *env* region, LTR-specific transcripts, including a 3.6-kb species, can be readily detected in normal colon mucosa and colon carcinoma cell lines (Rabson *et al.*, 1985; Gattoni-Celli *et al.*, 1986). The hybridization pattern of this transcript indicates that it is distinct from the other 3.6-kb transcript mentioned above, and probably contains transcribed cellular sequences polyadenylated by a HERV-E LTR (Rabson *et al.*, 1985). Similarly, two partial cDNAs, termed jnLTR8 and jnLTR22, polyadenylated by HERV-E LTRs have been isolated from a lung adenocarcinoma cDNA library (Tomita *et al.*, 1990). PCR analysis has also been used to detect HERV-E expression in peripheral blood mononuclear cells (Medstrand *et al.*, 1992). Thus far, the identification of possible

HERV-E protein products has not been reported. Consequently, the significance of these transcripts is difficult to assess.

b. Hs5

A clone called Hs5 was isolated from a human genomic library by hybridization with a DNA fragment containing the feline leukemia virus (FeLV) 3′ *pol, env,* and LTR sequence (Levy *et al.*, 1990) (see Table II). Many of the other clones isolated by this screen were identified by the similarity of their restriction maps to HERV-E clone 4-1 to be members of the HERV-E family. However, clone Hs5 had a distinctive map. Nucleotide sequence analysis revealed that the element in Hs5 has largely been deleted since only 3′ *pol* and *env*-related sequences could be identified (Levy *et al.*, 1990). These sequences were 77% homologous overall to HERV-E clone 4-1 (Levy *et al.*, 1990) and contained a fragment of an LTR 80% homologous to LTRpdhB, a solitary HERV-E LTR in the 5′ flanking region of the pyruvate dehydrogenase beta chain gene (D. Wilkinson, unpublished observation). The potential retroviral coding regions of this clone were disrupted by mutations that would prohibit their possible expression. Whether the element in Hs5 is actually a member of the HERV-E family cannot be assessed because of the deletion of the 5′ sequences, including the primer binding site. Those sequences that remain, however, are clearly related.

c. NP-2

HERV-E *pol* region fragments were used to search for related endogenous sequences by screening a genomic library using hybridization at low stringency but not at high stringency as the criterion by which clones were selected and purified (Silver *et al.*, 1987). The clone NP-2 was isolated by this strategy. This clone contained a 3′ LTR plus *gag, pol,* and *env*-related regions (Table II). Unfortunately, the cloning site at one end excluded the 5′ LTR and part of the *gag* region. A 1.1-kb fragment of this clone was sequenced and shown to have 73% nucleotide similarity with the *env* region of clone 4-1 (Silver *et al.*, 1987). Further analysis, shown in Fig. 7, indicates that NP-2 has decreasing homology with other members of the HERV-ERI superfamily for which *env* sequence is known, placing it at one end of the spectrum of elements comprising the superfamily. Genomic Southern blots of male and female DNA indicate that there are only two such elements in the genome and that they are located on the Y chromosome (Silver *et al.*, 1987). Furthermore, like some HERV-E, these two elements are associated with highly similar flanking sequences, indicating that the genomic region in

	NP-2	Hs5	4-1	HERV-R
Hs5	76%			
4-1	75%	76%		
HERV-R	55%	58%	59%	
RRHERVI	51%	57%	56%	68%

FIGURE 7. Comparisons of *env*-region nucleotide sequences among members of the HERV-ERI superfamily. Percent homology was calculated by the GCG program Gap (gap weight = 5.0; gap length weight = 0.3) over regions limited to the equivalent positions of the sequence available for NP-2.

which they reside underwent a duplication (Silver *et al.*, 1987). No NP-2-derived transcripts were detected in polyadenylated RNA from normal testis, or in lung, colon, and pancreatic carcinomas (Silver *et al.*, 1987).

d. ERV1

This clone was isolated using the approach of cross-species hybridization. A fragment from a previously isolated chimpanzee genomic clone that had been determined to have homology with the *pol* region of BaEV was used to screen a human genomic library (Bonner *et al.*, 1982). A group of overlapping clones was isolated that constituted a single genomic locus which was subsequently mapped to 18q22-23 (Bonner *et al.*, 1982; O'Brien *et al.*, 1983; Renan and Reeves, 1987). The locus was homologous to the sequence in the original chimpanzee clone and contained within it an *e*ndogenous *r*etro*v*iral element, designated ERV1 (Table II). Thus, this element has been in the primate genome at least 5 million years, based on an estimate of the time at which chimpanzees and humans last shared a common ancestor (Sibley and Ahlquist, 1987). However, the element could be much older, since sequences hybridizing to ERV1 have also been detected in the genomes of Old World monkeys (unpublished results cited in O'Connell *et al.*, 1984). ERV1 contains *gag, pol, env*-related regions as well as a putative 3′ LTR sequence; however, the 5′ LTR of the element is deleted and the primer binding site, if it remains, has not been characterized (Bonner *et al.*, 1982). Short nucleotide stretches within the *gag* and *pol* regions as well as the 3′ LTR of ERV1 have been sequenced and found to be most closely related to homologous sequences in BaEV, 4-1, ERV3, and RRHERV-I (unpublished observations) (ERV3 and RRHERV-I are discussed below). Although ERV1 can be distinguished as a single element under high-stringency conditions, hybridization of a fragment containing 3′ *pol, env* and 3′ LTR sequences under low-stringency conditions revealed that there may be as many as 15 related elements in human DNA (O'Brien *et al.*, 1983). It is not known if ERV1 is expressed, but deletion of the 5′ LTR makes this doubtful.

e. ERV3 (HERV-R)

ERV3 is a single-copy HERV element that was isolated by hybridization with a fragment containing 3′ *pol* and *env* sequence from the same chimpanzee clone used for ERV1 as well as a probe containing BaEV LTR sequence (O'Connell *et al.*, 1984) (Table II). This 9.9-kb element has been mapped to human chromosome 7 and is comprised of *gag, pol,* and *env*-related internal regions flanked by ~590-bp LTRs (O'Connell *et al.*, 1984; O'Connell and Cohen, 1984). The primer binding site of ERV3 most closely matches mouse arginine tRNA, hence this element is now usually termed HERV-R (Larsson *et al.*, 1989). Like ERV1, hybridization with a HERV-R *env* region probe clearly identified a single band on human genomic Southern blots under high-stringency conditions. However, ten or more additional bands were detected if the stringency of hybridization was lowered (O'Connell *et al.*, 1984). HERV-R can also be detected in the genomes of Old World monkeys and apes (but not gorillas) and related sequences have been detected in New World monkeys (cited in Cohen and Larsson, 1988). This result suggested that HERV-R represents an ancient proviral integration into the germ line of a common primate ancestor (O'Connell *et al.*, 1984; Shih *et al.*, 1991). The LTRs and *env* region of HERV-R have been sequenced, as have parts of the *gag* and *pol* regions. Analysis of the *gag* and *pol* sequences revealed the close relationship between HERV-R, ERV1, and clone 51-1 of the HERV-E family (O'Connell *et al.*, 1984). More recently, the *env* region of HERV-R has been related to the RRHERV-I elements with a 68% nucleotide sequence similarity stretching over more than 2 kb (Kannan *et al.*, 1991; see also Fig. 7) (RRHERV-I is discussed below).

The nucleotide sequence determined in the *gag* and *pol* regions of HERV-R indicated that the ORFs homologous to retroviral proteins in both of these regions are disrupted by mutations creating stop codons (O'Connell *et al.*, 1984). The *env* region was deduced to contain a 1.9-kb ORF spanning the surface glycoprotein domain and most of the transmembrane domain (Cohen *et al.*, 1985). However, similar to the *env*-related ORF identified in clone 4-1 of the HERV-E family, the predicted protein product would not be anchored in the cell membrane like typical retroviral *env* proteins are. The *env*-related ORF is especially noteworthy since the major transcripts from HERV-R are all spliced in a manner similar to a retroviral subgenomic mRNA (Kato *et al.*, 1987). These transcripts are polyadenylated and are 3.5, 7.3, and 9 kb in size. They are all initiated in the 5′ LTR and are spliced identically to remove 5.9 kb of *gag–pol* sequence. The 3.5-kb transcript terminates as expected in the 3′ LTR to generate a typical subgenomic *env* transcript, but the 7.3- and 9-kb transcripts extend into the 3′ flanking cellular DNA sequence (Kato *et al.*, 1987). Analysis of partial cDNAs corresponding to the 7.3- and 9-kb transcripts revealed that the associated cellular DNA contains a 1.3-kb ORF deduced to encode a zinc finger protein related to the Kruppel gene of *Drosophila* (Kato *et al.*, 1990). In addition, the 7.3- and 9-kb transcripts contain approximately 1 and 2.7 kb, respectively, of uncharacterized cellular sequence. The 5′ two-thirds of the Kruppel-related ORF as well as some of the proximal 5′ noncoding sequence is reported to be identical to the 5′ end of an independently isolated Kruppel-related cDNA (Kato *et al.*, 1990). Together, these observations suggest that HERV-R is situated in the proximal 5′ region (or within) of a Kruppel-related gene, named H-*plk*, through which the 7.3- and 9-kb RNAs are transcribed. It should be noted that the Kruppel-related ORF in the 7.3- and 9-kb transcripts is downstream of the 1.9-kb ORF coding for a putative *env* protein and its stop codon might prevent the subsequent translation of the Kruppel-related ORF.

A wide survey of 170 tissue samples and cell lines has been conducted to screen for HERV-R RNA expression (Cohen *et al.*, 1988; Kato *et al.*, 1988). Samples of placental chorionic villi were found to express 3- to 50-fold higher levels of HERV-R than the other samples tested and to be the only tissue in which the 7.3-kb transcript was expressed. Although the 3.5- and 9-kb RNAs were widely expressed, in some samples only the 3.5-kb transcript could be detected (Kato *et al.*, 1987; Cohen *et al.*, 1988). More recently, Larsson *et al.* (1993) have used *in situ* hybridization to localize HERV-R *env* transcripts to syncytiotrophoblasts in placenta. Antibodies raised against *env*-derived synthetic peptides also detect putative *env*-encoded proteins at the same site suggesting that HERV-R is translationally competent.

In contrast to the generally broad transcription pattern of HERV-R, it is of note that little or no HERV-R RNA could be detected in six choriocarcinoma cell lines (Cohen *et al.*, 1988; Kato *et al.*, 1988). This was very surprising because choriocarcinomas are derived from the placental trophoblastic epithelium, the outermost cells of the chorionic villi. The same lack of HERV-R mRNA expression was found for a tissue biopsy from an invasive hydatidiform mole, a cystic outgrowth from the chorionic villi of the placenta. Subsequent Southern blots of the chromosomal DNA from the choriocarinomas showed that the lack of HERV-R expression in these cells was not the result of translocations or deletions of the HERV-R locus. In addition, the methylation pattern of the HERV-R genome in the choriocarcinoma cells was identical to that found in normal human placenta. Thus, the reason for abrogation of HERV-R *env* gene expression in these cells is not known. Nonetheless, this observation was particularly intriguing considering the association of HERV-R with the Kruppel-related zinc finger gene. The susceptibility gene for Wilms's tumor has been identified as a zinc finger protein (Call *et al.*, 1990). Thus, it has been speculated that the downregulation of HERV-R expression in choriocarcinomas is associated with the loss of expression of the H-*plk* gene, a potential tumor suppressor gene (Kato *et al.*, 1988). However, this possibility is unlikely, since recent work has shown that HERV-R is in fact not expressed in the normal cytotrophoblast from which choriocarcinomas are derived (Boyd *et al.*, 1993).

Thus far, HERV-R is the only element that has been characterized to have LTR homology with BaEV. Another group of clones were isolated from a human genomic library using the BaEV LTR as a probe, but were not fully characterized (Noda *et al.*, 1982). Preliminary estimates of copy number produced conflicting results. Approximately ten bands could be detected on a genomic Southern blot under nonstringent conditions (similar to HERV-R). However, the frequency of positive plaques detected in the library corresponded to a copy number of approximately 300 LTR-related sequences per haploid genome (Noda *et al.*, 1982). Considering that the library used had been amplified (Bonner *et al.*, 1982), the estimate obtained from it may be less accurate. These results suggest that the human genome contains a small family (~ten members) of divergent HERVs that have BaEV-related LTRs, and that HERV-R is a member of this family.

f. RRHERV-I

This family of sequences was inadvertently discovered by a PCR strategy designed to clone developmentally regulated transcription factors (Kannan *et al.*, 1991). A PCR fragment with homology to the HERV-R *env* region was isolated and, when used as a probe, detected an RNA transcript in teratocarcinoma cells (subline 9117 from line PA-1) induced by retinoic acid treatment. Subsequently, the PCR fragment was used to isolate several clones from a cDNA library of retinoic acid-treated teratocarcinoma cells. A 3.3-kb composite cDNA structure of this element, designated RRHERV-I (Table II), was assembled from several cDNA clones (Kannan *et al.*, 1991). Homology with the *env* regions of HERV-R and 4-1 and with the LTR of ERV1 was detected (see Figs. 6 and 7). The sequence of this composite strongly suggested that RRHERV-I, like HERV-E and HERV-R, was primarily expressed as a spliced subgenomic RNA. Only a single transcript, approximately 4 kb in size, hybridized to an RRHERV-I probe in a Northern blot of RNA from the retinoic acid-treated teratocarcinoma cells (Kannan *et al.*, 1991). Clones of the transcript were found to contain termination codons in the putative *env* ORF rendering them incapable of encoding a protein product (Kannan *et al.*, 1991). Hybridization of RRHERV-I probes to Southern blots has been used to estimate that there are approximately 20 copies of this family in humans. A similar number of elements can also be detected in the genomes of Old World monkeys and apes, indicating that, like other HERVs, RRHERV-I have been present in the primate germ line for 25 million years or more (Kannan *et al.*, 1991).

Further investigation of the different members of the HERV-ERI superfamily promises to provide interesting insights into the dynamics of retroviral and host evolution. For example, the predominant expression of spliced subgenomic RNAs from several members of this superfamily might be indicative of a common mechanism by which a host species adapts to limit the release of infectious virions from an endogenous retrovirus. Alternatively, these *env* transcripts may possibly have been adopted to serve a cellular function. Detection of putative HERV-R *env* proteins specifically in the syncytiotrophoblast layer of placenta supports this hypothesis (Larsson *et al.*, 1993). Other questions such as the relative roles of recombination, LTR divergence, and primer binding site adaptations during retroviral evolution might possibly be addressed. The relatively low degree of sequence homology in the LTR and *env* regions among different isolates (Figs. 6 and 7) suggests that individual members of this family represent stages in the evolutionary lineage(s) of an ancient exogenous retrovirus that was periodically preserved by fixation of the provirus, or "endogenization," within the much slower evolving host genome. It will also be important to determine if this group of HERVs contains the closest human equivalents to what appears to be a functionally conserved family of elements related to BaEV present in the genomes of chimpanzees, gorillas, and Old World monkeys. For unknown reasons, conservation of that family was lost during the evolution of the human, gibbon, and orangutan lineages (Shih *et al.*, 1991). Unfortunately, at the present time it is difficult to more precisely evaluate the relationships of the different isolates in this superfamily because of a lack of sequence information from the *gag* and *pol*-related regions from many of the existing clones. However, comparisons done with the small amount of sequence data available from these regions support the conclusion that the members of the HERV-ERI superfamily are all closely related (Werner *et al.*, 1990).

2. Other Class I HERV Families

Several other families of HERV elements have been detected that also have amino acid homology to mammalian type C or class I retroviruses such as MLV. However, the level or extent of this homology is less than with the HERV-ERI superfamily. Primary isolates from these different families are shown in Table III.

a. RTVL-I (HERV-I)

The prototype of this family of elements was discovered during the analysis of genomic sequence data from the haptoglobin-related locus on chromosome 16 (Maeda, 1985). It is a 9-kb element containing ~500-bp LTRs, which flank internal sequences related to *gag*, *pol*, and *env* (Maeda, 1985; Maeda and Kim, 1990). The primer binding site in this element is most closely homologous to the mouse

TABLE III. Other Class I HERV Isolates

Isolate	Sequence structure	Sequence determined	Reference
a. HERV-I[a]:			
RTVL-Ia	Provirus	Complete	Maeda (1985)
RTVL-Ib	Provirus	Complete	Maeda and Kim (1990)
RTVL-Ic	*pol–env*–LTR	Complete	Maeda and Kim (1990)
b. S71	*gag*–Δ*pol*–LTR	Complete	Werner *et al.* (1990)
c. HRES-1	LTR–*gag*–?	5′ LTR, *gag*	Perl *et al.* (1989)
d. HERV-P			
HuERS-P1-1	Provirus	LTRs	Harada *et al.* (1987)
HuERS-P1-2	Provirus	LTRs	Harada *et al.* (1987)
HuERS-P2	Deleted provirus?	LTRs	Harada *et al.* (1987)
HuERS-P3-1	Deleted provirus?	3′ LTR	Harada *et al.* (1987)
HuERS-P3-2[b]	Provirus	LTRs, (*pol*)	Kroger and Horak (1987)
e. ERV-9			
pHE.1 cDNA	Δ*gag–pol–env*–LTR	Complete	La Mantia *et al.* (1991)
λFix 1.1	Provirus	LTRs	La Mantia *et al.* (1991)
λFix Ť3	Provirus	5′ LTR	Lania *et al.* (1992)
pTR2 cDNA[c]	Spliced transcript	Complete	La Mantia *et al.* (1989)
pTR7 cDNA[c]	Spliced transcript	Complete	La Mantia *et al.* (1989)
f. HERV-H[d]			
RTVL-H1	Provirus (Δ form)[e]	LTRs	Mager and Henthorn (1984)
RTVL-H2	Provirus (Δ form)	Complete	Mager and Freeman (1987)
N10-13 cDNA	Unit-length transcript (Δ form)	Complete	Wilkinson *et al.* (1990)
N10-14 cDNA	Spliced transcript (Δ form)	Complete	D. Mager (unpublished)
RGH2	Provirus (intact form)	Complete	Hirose *et al.* (1993)
RGH1	Partial *pol–env*–LTR	Complete	Hirose *et al.* (1993)

[a] HERV-I solitary LTRs have also been isolated (Armour *et al.*, 1989; Suzuki *et al.*, 1990).
[b] This element is also termed HuRRS-P.
[c] The finding that these two cDNAs were actually ERV-9 spliced transcripts was reported by Lania *et al.* (1992).
[d] Numerous other HERV-H partial and full-length genomic and cDNA clones and solitary LTRs have been characterized (e.g., see Mager, 1989; Johansen *et al.*, 1989; Wilkinson *et al.*, 1990, 1993; Hirose *et al.*, 1993).
[e] Δ form indicates the common HERV-H type with no *env* and a partially deleted *pol*.

tRNA for isoleucine, hence it was called RTVL-I (for *r*e*t*ro*v*irus*l*ike element with an *i*soleucine tRNA primer binding site) (Maeda, 1985) and has also been termed HERV-I.

Characterization of the haptoglobin gene cluster has revealed that three separate HERV-I insertions have occurred during the course of primate evolution. Subsequent to the divergence of humans and chimpanzees, a region of DNA containing one of these elements and a haptoglobin-related gene was deleted from the human genome (McEvoy and Maeda, 1988; Maeda and Kim, 1990). The elements that remain in humans are located within the first intron of the haptoglobin-related gene and in the 3′ flanking region of the haptoglobin-related gene. They have been designated RTVL-Ia and RTVL-Ic, respectively. The RTVL-Ia element has been determined to be integrated at the same position in the genomes of Old World monkeys and apes, indicating that it inserted into the primate germ line more than 25 million years ago (Shih *et al.*, 1991). The complete nucleotide sequences of these two elements in humans, as well as the additional element in chimpanzees (designated RTVL-Ib) have been determined (Maeda and Kim, 1990). The data reveal that all three elements are frequently disrupted by mutations that would prevent them from encoding retroviruslike proteins. However, the deduced amino acid sequence of these clones reveals detectable homology with the *gag* and *pol* genes of MuLV, ranging from 31 to 41% in the *pol* region and 40% in the *gag* region (Maeda and Kim, 1990). Investigation of HERV-I expression at the RNA level has not been reported.

In addition to termination codons and short deletions, the 3′ internal region of RTVL-Ia contains an inserted Alu sequence, while the 5′ LTR and *gag* region of RTVL-Ic are deleted and the remaining sequence is interrupted by two Alu insertions (Maeda and Kim, 1990). Deletion of the 5′ region of HERV-I elements may be common since the *gag* region is estimated to be present in 8 copies in the human genome while probes specific for the *env*-related region detect approximately 25 copies (Maeda and Kim, 1990). Alternatively, the difference in copy number between these two regions might indicate that recombinant elements containing the HERV-I *env* region and different 5′ sequences exist within the genome.

An interesting feature of the haptoglobin gene cluster is the relative predisposition of this region to homologous but unequal recombination events. Such events are thought to have created the third haptoglobin-related gene in primates and explain its deletion in the human genome (Maeda and Kim, 1990). In addition, different arrangements of the gene cluster in humans have been discovered which vary in the number of copies of the haptoglobin-related gene. Up to six tandemly repeated copies, thought to result from homologous but unequal recombination, have been described (Maeda *et al.*, 1986). It is not known, however, if the HERV-I elements in this region are directly involved in such events.

A solitary HERV-I LTR has been discovered to be associated with another region of genomic variation among humans. In this case, a minisatellite sequence mapped to 1q42-43 is located within the HERV-I LTR and appears to have been derived from its sequence (Armour *et al.*, 1989). Another partial HERV-I LTR has been identified in the 5′ regulatory region of the cytochrome c_1 gene (Suzuki *et al.*, 1990). The LTR-related sequence associated with this gene was found to contain three sequence motifs conserved among mitochondrion-related nuclear genes and to bind nuclear proteins in *in vitro* binding assays (Suzuki *et al.*, 1990). Based on these observations, it was suggested that the HERV-I LTR sequence is involved in the regulation of transcription of the cytochrome c_1 gene.

b. S71

Following preliminary experiments which detected sequences related to the simian sarcoma-associated virus (SSAV) in the human genome, the entire genome of SSAV was used to screen a human genomic library under reduced stringency conditions (Leib-Mösch *et al.*, 1986). The resulting clone, designated S71 (Table III), can be distinguished under high-stringency conditions as a single element within the human genome, and has been mapped to chromosomal position 18q21 (Brack-Werner *et al.*, 1989a). The element appears to be fixed at this locus within the human population, since it is present in the DNA of all individuals tested thus far (Leib-Mösch *et al.*, 1986; Brack-Werner *et al.*, 1989a). Under low-stringency hybridization conditions, fragments specific to the *gag*, *pol*, and LTR regions of S71 detect approximately 20 related sequences per haploid human genome (Leib-Mösch *et al.*, 1992b). A recent phylogentic study suggests that S71 is present in Old World but not New World monkeys (Brack-Werner *et al.*, 1993).

The nucleotide sequence of the entire region of clone S71 containing SSAV homology has been determined. Short direct repeats within the sequence were used to delineate an element that is a composite of retrovirus-related and unrelated sequences. Together they constitute a 5.4-kb HERV that contains *gag* and *pol*-related sequences and ends in a 535-bp sequence with the hallmarks of a 3′ LTR. The first 400 bp at the 5′ end, plus another region of 1.1 kb inserted between the *gag* and *pol*-related regions were initially reported to have no detectable homology with known retroviral sequences (Werner *et al.*, 1990). However, it has recently been found that almost all of the segment between the *gag* and *pol*-related regions is actually a solitary HERV-K LTR (see below) oriented in the antisense direction (Leib-Mösch *et al.*, 1993). Analysis of the S71 sequence showed that, like other HERVs, the *gag* and *pol* regions were disrupted by mutations, including a deletion that removed the protease and reverse transcriptase domains of the *pol* region. In addition, the element did not contain a 5′ LTR or a putative *env* sequence. Sequence comparisons made with the hypothetical retrovirus-related amino acid sequences in the *gag* and *pol* regions confirmed the relatedness of S71 and SSAV, and also indicated it was about equally related to BaEV and MuLV. The overall homology within the RNase H domain of the *pol* region to the homologous domain in SSAV was determined to be 60% over 156 residues (Werner *et al.*, 1990). A short ORF capable of coding for 167 amino acids was identified at the 5′ end of the *gag* region, but the significance of this ORF is not known.

Transcripts that hybridize to S71 sequences have been detected in the leukemic cell line K562. This cell line expressed a prominent transcript of 2.9 kb, as well as two minor transcripts of 2.5 and 3.6 kb (Leib-Mösch *et al.*, 1992b). These transcripts could not be detected in normal peripheral lymphocytes or in several other human cell lines derived from various cell types. In addition, a 1.1-kb cDNA which contains a 350-bp region related to the 3′ end of S71 has been isolated from a placental library (Leib-Mösch *et al.*, 1992b). The structure of this cDNA has not been fully characterized nor has its possible relationship to the transcripts detected in cell line K562 been determined. Thus, the potential biological significance of these transcripts remains to be established.

c. *HRES-1*

In an attempt to identify HERV related to infectious retroviruses of humans, a human genomic library was screened under low-stringency conditions with a probe containing HTLV-I LTR and *gag* sequence (Perl *et al.*, 1989). Eight recombinant clones were isolated by this approach, and one of these has been characterized in detail. This clone contains a HERV which has been called HRES-I, denoting *H*TLV-1-*r*elated *e*ndogenous *s*equence-I (Table III). This element can be distinguished by high-stringency hybridization to represent a single locus fixed within the human genome, and has been mapped to chromosome band 1q42 (Perl *et al.*, 1989, 1991a). A homologous element can also be detected in the genomes of Old and New World monkeys, but not other mammals, indicating that this element was assimilated by the genome of a common primate ancestor 40 million years ago or more (Gingerich, 1984). One other clone, called HRES-2, has been isolated by the same researchers, but characterization of that clone has not been reported (Perl *et al.*, 1991b).

A 5′ LTR and *gag*-related sequence have been identified in HRES-1 by sequence analysis of a 2151-bp fragment of the genomic clone (Perl *et al.*, 1989). The 684-bp LTR region is followed by a primer binding site that most closely matches (12/18 nucleotides) chimpanzee histidine tRNA. The *gag*-related region contains two overlapping, alternative ORFs, coding for possible proteins. However, only one of these ORFs has detectable sequence homology with retroviral *gag* proteins. This homology is low and is limited to two short stretches. An amino-terminal region of 31 amino acids was shown to have 32 and 39% homology to the p19 matrix protein of HTLV-I and HTLV-II, respectively. Another 36 amino acids in the middle of the 223-residue ORF have 25 and 28% homology to the p24 capsid protein of HTLV-I and HTLV-II, respectively, as well as 33% homology with a region of the HIV-2 *gag* gene. Thus far, HRES-1 sequences homologous to other retroviral genes have not been reported. Thus, it is not known at this time whether HRES-1 is a truncated element containing only 5′ sequences or if the 3′ sequences remain to be characterized. Lacking information from the *pol* region, HRES-1 has been included among class I elements because other class I elements, such as HERV-H (see below), have also been determined to have limited homology with the *gag* proteins of HTLV-I.

A 6-kb transcript containing the *gag*-related ORF was detected in the polyadenylated RNA of several cell lines including MA-T cells, HL60 cells, W7A and W7B melanoma cells, but not K562 or U937 cells (Perl *et al.*, 1989). In other experiments, two antisera were generated against polypeptides derived from each of the domains of HTLV homology mentioned above. These two antisera both recognized a 28-kDa protein on Western blots of H9 T cell lysates (Banki *et al.*, 1992). This size corresponds precisely to the predicted molecular mass of the protein encoded by the HTLV-related ORF. In addition, both antisera react with *in vitro* translated HRES-1 protein (A. Perl, personal communication). The protein recognized by these antisera was further characterized as an intracellular cytoplasmic protein with some localization to nuclear bodies (Banki *et al.*, 1992). However, its function has not been determined.

d. HERV-P Families

Two different groups of researchers have used a similar approach to attempt to identify endogenous elements related to infectious human retroviruses. Their strategy was based on the knowledge that both HTLV-I and HTLV-II have a primer binding site corresponding to the tRNA for proline, and that tRNAs are processed by the posttranscriptional addition of the three nucleotides CCA to their 3′ terminus. One group screened a genomic library with oligonucleotide probes that extended past the primer binding site to the conserved CA dinucleotide which defines the 3′ end of the LTR, adding several additional bases of homology to a putative proviral sequence as opposed to a tRNA gene. Thus, a close match to a retroviral sequence could be distinguished from possible cross hybridization with a tRNA gene by adjusting hybridization stringency (Kroger and Horak, 1987). The other group used 3′ end-labeled, 3′ fragments of purified proline tRNA to screen a genomic library. They identified possible retroviral hybrids by digesting with RNase A during the screening procedure to remove the unbound CCA tail (and consequently the radiolabel) of probe fragments bound to tRNA clones (Harada *et al.*, 1987).

In all, three different families of HERV containing a primer binding site that matched proline tRNA were identified and distinguished by their LTR sequences. They have been designated HuERS-P1, -P2, and -P3 where P designates proline (Harada *et al.*, 1987) (Table III). Two elements representing HuERS-P1 were isolated, one about 8 kb in size (P1-1), and the other about 7.5 kb in size (P1-2) (Harada *et al.*, 1987). This length difference suggests that P1-2 likely contains an internal deletion(s) not present in P1-1. The nucleotide sequence of the LTRs of the two elements has been determined. These data showed that the HuERS-P1 LTR is about 690 bp, and that an Alu sequence has inserted into the 5′ LTR of P1-2 (Harada *et al.*, 1987). An internal fragment of P1-1 detects homologous elements in the genome of an Old World monkey, but not in mice, and has been used to estimate the copy number of HuERS-P1 at 10–20 copies in the human genome (Harada *et al.*, 1987).

A single element representing HuERS-P2 was isolated (Harada *et al.*, 1987). This element consists of 890-bp LTRs that flank approximately 3.2 kb of internal sequence. The two LTRs of this element are not related to the LTRs of HuERS-P1 or HuERS-P3. The short internal part of this element suggests that large regions of the putative retrovirus-related sequences are deleted. Like HuERS-P1, an internal fragment of HuERS-P2 detects related sequences in an Old World monkey genome, but not in mice. Hybridization of the internal fragment to a Southern blot was also used to estimate that between 20 and 30 elements related to HuERS-P2 are present in the human genome (Harada *et al.*, 1987).

Both groups of researchers isolated members of the third family of elements, and consequently it has been called both HuERS-P3 (Harada *et al.*, 1987) and HuRRS-P (Kroger and Horak, 1987). For purposes of clarity, the designation HuERS-P3 will be used in this discussion. Only one complete element has been reported (Kroger and Horak, 1987), along with two other partial clones containing primarily LTR sequence (Kroger and Horak, 1987; Harada *et al.*, 1987). The full-length element was 8.1 kb in length and contained ~630-bp LTRs. A small fragment from the internal region of the full-length element was sequenced. Translation of this sequence revealed 62% amino acid homology over 145 residues with the reverse transcriptase domain of MuLV. However, the HuERS-P3 sequence was interrupted by an in-frame termination codon. An HuERS-P3 LTR fragment detected related sequences in the genomes of apes, Old World and New World monkeys, but not mice (Kroger and Horak, 1987; Harada *et al.*, 1987). The number of bands on Southern blots that hybridized to fragments from the putative *gag* and *env* regions indicated that the human genome contains between 20 and 40 HuERS-P3 elements (Kroger and Horak, 1987).

To date, the internal sequences of these HERV isolates are largely uncharacterized. Further analysis of these elements will be necessary to determine if the different HERV-P families distinguished by LTR sequence have any similarity in their internal regions and to establish their relatedness to infectious retroviruses. In addition, more isolates of HuERS-P2 will need to be studied to determine if the 3.2-kb internal sequence in the one reported element is representative of that family, or a deletion mutant. It will also be of interest to investigate the possible expression of these three families of HERV elements.

e. ERV-9

A novel sequence consisting of three 85-bp tandem repeats was discovered at the 5′ end of a cDNA clone coding for a protein with zinc-finger-related domains (Pannuti *et al.*, 1988). Further investigations revealed that these repeats were commonly associated with other short repeated sequences. Each type of repeat was found to be highly reiterated within the human genome and, with one exception, unrelated to previously described repetitive sequences. Expression of these repeat sequences was detected in the teratocarcinoma cell line NTera2D1, and the transcripts containing them appeared to be downregulated by retinoic acid-induced differentiation of the cell line (La Mantia *et al.*, 1989). During the investigation of cDNAs containing these repeats isolated from an NTera2D1 library, one cDNA was discovered that also contained retrovirus-related sequences (La Mantia *et al.*, 1991). This was the first isolate of a family of elements called ERV-9.

The complete nucleotide sequence of this cDNA, termed pHE.1, was determined. Regions of *gag*, *pol*, and *env*-related sequence were identified, although the *env* region appears to be partially deleted and the 5′ terminus of the clone did not extend through a complete *gag* region. In addition, the predicted amino acid sequence of the retrovirus-related regions revealed that they were disrupted by mutations and therefore could not code for the characteristic retroviral proteins (La Mantia *et al.*, 1991). Analysis of the reported sequence shows that ERV-9 is most closely related to BaEV; with nucleotide homologies in the *pol* and *env* regions of 54% (1340 bp) and 62% (300 bp), respectively (D. Wilkinson, unpublished observations).

The pattern of bands detected on a Southern blot with a *pol* region fragment indicates that there are about 40 related elements in the human genome comprising the ERV-9 family. Other primate species, but not rodents, have also been reported to contain ERV-9 elements in their genome (La Mantia *et al.*, 1991). Furthermore, it has been estimated that 3000–4000 solitary LTRs related to this family are present in the genome (Zucchi and Schlessinger, 1992). The *pol* region fragment of the original clone was also used to isolate genomic clones of other members of the sequence family. Analysis of one such clone, λ Fix 1.1, revealed that the ERV-9 element contained exceptionally long, 1.8 kb, LTRs that flanked approximately 6 kb of internal sequence and contained a primer binding site for arginine tRNA (La Mantia *et al.*, 1991). The 5′ and 3′ LTRs of the element showed 90% sequence similarity, and contained two types of the novel repeats. Interestingly, the genomic LTR differed from the corresponding LTR region of the initially isolated cDNA clone in the number of repeats each sequence contained. This observation, as well as a preliminary analysis of additional genomic clones, indicates that the repeats are an important source of heterogeneity among ERV-9 LTRs (La Mantia *et al.*, 1991).

Expression of an 8-kb ERV-9 transcript corresponding to a full-length RNA has been detected in undifferentiated NTera2D1 cells and is downregulated when the cells are induced to differentiate by retinoic acid (La Mantia *et al.*, 1991). Other ERV-9 transcripts were detected in placenta, but not in several leukemic and carcinoma cell lines tested. More sensitive PCR methods have also recently been used to amplify ERV-9 segments from a glioma RNA sample corresponding to the conserved immunosuppressive region of the *env* protein (Lindeskog *et al.*, 1993). Small NTera2D1 transcripts of 1.5 and 2 kb detected with an ERV-9 LTR probe have been shown by cDNA analysis to be spliced RNAs which remove most of the internal sequences (Lania *et al.*, 1992). It has also been demonstrated that the ERV-9 LTR sequences in both the cDNA and genomic clones direct expression of a reporter gene only in undifferentiated NTera2D1 cells; expression did not occur in differentiated NTera2D1 cells (La Mantia *et al.*, 1991). The consistent association of the novel repeat sequences with the transcripts expressed in undifferentiated NTera2D1 cells, and their discovery in ERV-9 LTRs present an intriguing correlation; however, any direct involvement of these sequences in the regulation of tissue-specific transcription remains to be tested. In fact, deletion analysis of the promoter sequences of ERV-9 LTRs indicates that the repeat sequences are not essential for transcription from the LTR (La Mantia *et al.*, 1992). Characterization of the ERV-9 LTR sequences revealed that they lack the TATA sequence found in most LTR promoters and contain an unusually short R region of only 6 bp (La Mantia *et al.*, 1992). Functional analysis of the LTR minimal promoter region indicates that it may be regulated by Sp1 binding and that another protein factor interacts with an initiatorlike element (Inr) located in the region from −7 to +6 with respect to the transcription start site (La Mantia *et al.*, 1992).

f. RTVL-H (HERV-H)

The first member of this repetitive family of elements was identified by virtue of the fact that it is located near the 3′ breakpoints of three large deletions involving the β-globin gene cluster (Mager and Henthorn, 1984; Mager *et al.*, 1985). Sequence analysis of that element revealed that it contained a primer binding site that matched human histidine tRNA, hence the family was originally designated RTVL-H. However, the term HERV-H has also been used, and will be employed in the following section. Analysis of several genomic isolates of HERV-H elements has revealed that the majority of elements are approximately 5.8 kb in length and contain *gag* and *pol*-related sequences that are flanked by 400- to 450-bp LTRs (Mager and Henthorn, 1984; Mager and Freeman, 1987). Hence, the structure of most HERV-H elements is similar to that of retrotransposons, in that they lack an *env*-related region. However, it has recently been recognized that these retrotransposonlike elements (800–900 copies per haploid genome) are actually deleted derivatives of less abundant elements that have the classic retroviruslike structure of *gag–pol–env* (100 or fewer per haploid genome) (Hirose *et al.*, 1993; Wilkinson *et al.*, 1993; Wilkinson and Mager, 1993). In humans, the HERV-H family was determined to consist of approximately 1000 elements per haploid genome, as well as a similar number of solitary LTRs (Mager and Henthorn, 1984). Approximately the same number of HERV-H elements are present in the genomes of apes and Old World monkeys, but they are not detectable in rodents (Fraser *et al.*, 1988; Johansen *et al.*, 1989; Goodchild *et al.*, 1993). *In situ* hybridization has shown that HERV-H sequences are widely dispersed within the human genome, although possible clusters exist at chromosomal positions 1p31 and 7q31 (Fraser *et al.*, 1988). For the most part, HERV-H elements appear to be fixed within the genome, although analysis of one chromosomal HERV-H locus among 70 individuals detected two siblings who carried an allele at which the element had deleted via LTR–LTR recombination, leaving a solitary LTR (Mager and Goodchild, 1989).

Sequence comparisons with known retroviruses were originally made with the nucleotide sequence of a genomic isolate having the common deleted form (Mager and Freeman, 1987). The *gag* and *pol* regions of that element were disrupted by mutations that would preclude it from expressing retroviruslike proteins. Comparisons of the hypothetical retroviral reading frames to known retroviral protein sequences revealed that sections of the *gag* region of HERV-H elements are ~50% similar to BaEV p30 capsid protein and 55–61% similar to the nucleocapsid protein of the retroviruses in the HTLV-I phylogenetic group. In the *pol* region, the element was determined to be most closely related to MLV, with amino acid sequence homologies of 40–50% in the different *pol* domains (Mager and Freeman, 1987). However, the *pol* region of the majority of HERV-H elements contains four shared deletions, termed A, B, C, and D (Wilkinson *et al.*, 1993; Wilkinson and Mager, 1993), which remove substantial portions of the functional domains within *pol*. Undeleted HERV-H elements have recently been detected and characterized (Hirose *et al.*, 1993; Wilkinson *et al.*, 1993). Among the elements that are intact in their *pol* region, a fraction of these have been identified that also contain an *env*-related region (Hirose *et al.*, 1993; Wilkinson and Mager, 1993). The *env*-related region of HERV-H elements shows the highest homology with the transmembrane proteins of type D retroviral *env* genes, sharing approximately 40% amino acid identity over a range of 240 residues (Hirose *et al.*, 1993; D. Wilkinson, unpublished results). Recent PCR experiments using primers from highly conserved domains of the immunosuppressive region of the *env* transmembrane protein have also amplified endogenous retroviral *env* segments (Lindeskog *et al.*, 1993) that appear to be derived from HERV-H (D. Mager and D. Wilkinson, unpublished observation).

HERV-H sequences are transcribed in a variety of cell lines with the highest level of expression occurring in teratocarcinoma cell lines and others derived from germ cell tumors (Wilkinson *et al.*, 1990, 1993; Hirose *et al.*, 1993). Expression in normal tissues is readily detectable in placentas (Johansen *et al.*, 1989), most notably in amniotic and chorionic membranes (Wilkinson *et al.*, 1990). Much lower levels of HERV-H RNAs have also been detected by PCR in peripheral blood mononuclear cells (Medstrand *et al.*, 1992) and by Northern blot analysis in some other normal tissues (Wilkinson *et al.*, 1990; unpublished data). Consistent with the high copy number of the family, complex patterns of expression are observed among cell lines (Wilkinson *et al.*, 1990). The most abundant transcript is a 5.6-kb species corresponding to unit-length RNAs from the deleted 5.8-kb elements. In addition, a spliced 3.7-kb derivative of this transcript is also commonly detected (Wilkinson *et al.*, 1990). Analysis of cDNA clones indicates that the region removed from this spliced 3.7-kb RNA extends from a common splice donor site 140 bp internal to the 5′ LTR to one of a cluster of splice acceptor sites located just upstream of the *pol*-related domain (Wilkinson *et al.*, 1990). This type of splicing event, which removes *gag* while leaving *pol* intact, has not been observed for other retroviruses or retroviruslike elements. Several other transcripts ranging in size from less than 1 kb to greater than 10 kb are also detected by HERV-H probes but only a fraction of these have been characterized further (Wilkinson *et al.*, 1990). It is noteworthy that the expression of HERV-H sequences, like that of ERV-9 (La Mantia *et al.*, 1989) and L1 elements (Skowronski and Singer, 1985), is greatly reduced in NTera2D1 cells induced to differentiate by retinoic acid (Wilkinson, 1993).

Transcription of the much less abundant structurally intact HERV-H sequences has also recently been detected in some teratocarcinoma and testicular tumor cell lines (Hirose *et al.*, 1993; Wilkinson *et al.*, 1993). Probes from the *pol* or *env* regions that are deleted in the majority of HERV-H elements detect several RNAs including a putative unit-length 8.6-kb species. A 6.5-kb transcript derived from elements that have *pol* regions A, B, and C intact but are deleted in regions D and the *env* domain is also observed in Tera1 cells (Wilkinson *et al.*, 1993). Smaller transcripts, which may represent spliced RNAs, and transcripts of 10 kb or larger, which may correspond to HERV-H-cellular fusion transcripts, are also detected but their structures have not yet been elucidated (Wilkinson *et al.*, 1993; Hirose *et al.*, 1993). No HERV-H element capable of encoding proteins homologous to those of retroviruses has yet been isolated but the great expansion of clearly defective elements within the genome suggests that their retrotransposition has been facilitated by complemention, possibly by coding-competent HERV-H. Since defective elements have continued to expand in relatively recent evolutionary times (Goodchild *et al.*, 1993), the coding potential of this family of elements remains an intriguing possibility. Indeed, some of the structurally intact HERV-H elements that have recently been isolated contain ORFs of greater than 1 kb in the *pol* and/or *env* domains (Hirose *et al.*, 1993; Wilkinson *et al.*, 1993).

One unique feature of the HERV-H family is that three subtypes of LTRs have been found to be associated with these elements (Mager, 1989; Wilkinson, 1993; Goodchild *et al.*, 1993). The different LTR subtypes are closely related in the R and U5 regions, as well as the first 130 bp of U3. The remainder of the U3 region, however, contains different tandemly repeated sequences that distinguish each type of LTR (Mager and Freeman, 1987; Mager, 1989; Wilkinson, 1993). The two most abundant LTR subtypes, types I and II, are ancient and are each present in several hundred copies within the genomes of humans, apes, and Old World monkeys. In contrast, the third subtype, type Ia, is not present in Old World monkeys but has expanded within the genomes of higher primates, including humans, to a copy number of 100 (Goodchild *et al.*, 1993). Sequence comparisons suggest that the more recently expanded LTR subtype is a recombinant derived from the other two more ancient subtypes (Wilkinson, 1993; Goodchild *et al.*, 1993). The different LTR subtypes have also been determined to correspond to functional differences of promoter activity as assayed by transfection experiments with chloramphenicol acetyltransferase (CAT) constructs (Feuchter and Mager, 1990). In general, HERV-H type I and II LTRs showed little or no detectable promoter activity in most cell lines when assayed in CAT construct transfection experiments (Feuchter and Mager, 1990). However, two type Ia LTRs tested showed much stronger promoter activity in a variety of cell lines (Feuchter and Mager, 1990; unpublished observations). Endogenous expression of the different LTR subtypes also differs among different cell lines. Expression of elements associated with the ancestral LTR types I and II is either very low, or restricted to teratocarcinoma cell lines, whereas expression of elements with the more recent recombinant LTR type Ia is readily detected in many different cell lines (Wilkinson, 1993; Goodchild *et al.*, 1993). These findings suggest that sequence elements within the LTR U3 repeat region are involved in governing HERV-H transcription. However, such regulatory elements have not yet been characterized. All types of HERV-H LTRs have also been discovered to be activated by simian virus 40 (SV40) large T antigen (T Ag) in transfection experiments. Levels of LTR promoter activity are increased 5- to 30-fold in response to T Ag, depending on the LTR, although not all LTRs are responsive (Feuchter and Mager, 1992). It has been suggested that this LTR activation is likely caused by pleiotropic effects of T Ag on transcription factor activities rather than a direct interaction between T Ag and the LTRs (Feuchter and Mager, 1992).

Several cDNAs containing unrelated cellular sequences polyadenylated by HERV-H LTRs have been reported (Mager, 1989). One such cDNA, isolated from a placental library, contains an ORF for an unknown protein product. The transcripts of this gene, termed *PLT*, are apparently alternatively spliced at their 3′ end, creating one transcript which is polyadenylated by a HERV-H LTR (Goodchild *et al.*, 1992). Other fusion cDNAs have been identified that initiate within a HERV-H LTR. In one case, the corresponding transcript initiates in a 5′ HERV-H LTR and splices from the donor site just upstream of the HERV-H *gag* region into downstream sequences encoding a cellular gene. This gene, termed *PLA2L*, was discovered by virtue of this novel promoter arrangement (Feuchter-Murthy *et al.*, 1993). Thus far, detection of the *PLA2L* transcripts has been limited to two teratocarcinoma cell lines. Analysis of the putative coding frame of one cDNA revealed that the gene encodes a protein with two phospholipase A_2 (PLA_2)-related domains (Feuchter-Murthy *et al.*, 1993). However, the protein products of the gene have not yet been identified, and its function remains to be determined.

Other cDNAs containing cellular sequences apparently promoted by HERV-H LTRs have been isolated. In one example, a cDNA isolated from an NTera2D1 cell library appears to represent a transcript that initiates in the LTR and proceeds directly into flanking cellular sequence. This sequence contains a long ORF and a possible initiation methionine. Analysis of the predicted protein sequence identified sequence motifs related to those found in the yeast CDC4 protein and β transducin, a G protein in photoreceptor rod cells (Feuchter *et al.*, 1992). Expression of the fusion transcript was undetectable, however, and a normal genomic association between a HERV-H element and the gene encoding the protein could not be established. Thus, it has been suggested that this cDNA clone represents the novel insertion of a HERV-H element into that gene in a clone of NTera2D1 cells that subsequently expired during passaging of the cell line (Feuchter *et al.*, 1992). If true, it is the first insertion event involving a HERV to be detected. Another fusion cDNA between HERV-H LTR sequences and adjacent cellular sequences has been isolated (Feuchter *et al.*, 1992). In this case, the adjacent sequences have the characteristics of a CpG island, which strongly suggests the presence of a nearby gene (Larsen *et al.*, 1992). However, this putative gene has not been isolated and the potential role that the LTR may play in expression of that gene has not yet been explored.

In another example, a chimeric cDNA was isolated from a subtraction library constructed from two different prostate metastasis-derived cell lines (Liu and Abraham, 1991). It represents a spliced fusion between a HERV-H element and the second exon of the calbindin gene, which is apparently expressed at the protein level (Liu and Abraham, 1991). It has been postulated that deregulation of calbindin expression via this fusion transcript contributed to the malignant phenotype of the tumor from which the cell line was derived. The genomic structure of the calbindin gene has not been characterized, and therefore it is not known if the HERV-H/calbindin fusion transcript is the result of mutations that activate a HERV-H element normally quiescent within the calbindin locus, or if the insertion of a HERV-H element has deregulated expression of the calbindin gene. Together, these findings involving HERV-H sequences illustrate that some HERV elements have become associated with normal cellular functions, while other members of the same family may be linked to aberrant cellular processes leading to disease.

B. Class II HERV

1. HERV-K

HERV related to class II retroviruses also exist in the genome. These class II HERV were originally isolated because they cross-hybridize to murine mammary tumour virus (MMTV) *pol*-region containing probes. Initial investigations, which employed an MMTV *gag-pol*-containing fragment to probe human genomic Southern blots, revealed that the human genome contained many related sequences (Callahan *et al.*, 1982; May *et al.*, 1983). Genomic clones were subsequently isolated by several groups (Callahan *et al.*, 1982; Westley and May, 1984; Deen and Sweet, 1986; Ono, 1986).

Analysis of these clones indicates that the prototypic element is 9.5 kb in length, has ~970 bp LTRs, and contains a primer binding site for lysine tRNA (Callahan *et al.*, 1985; Ono *et al.*, 1986). This family of elements has hence been termed HERV-K with K denoting lysine. Homology with MMTV is restricted primarily to *pol*-related sequences, where amino acid sequence identity is as high as 70% in some regions (Callahan *et al.*, 1985; Deen and Sweet, 1986; May and Westley, 1986; Ono *et al.*, 1986). Homology with the class II squirrel monkey retrovirus (SMRV) LTRs was also detected in cross-hybridization experiments (Callahan *et al.*, 1985). Other regions of HERV-K homology to class II retroviruses, avian Rous sarcoma virus (RSV), murine Intracisternal A particle elements (IAPs), and SMRV, have also been detected (Callahan *et al.*, 1985; Ono *et al.*, 1986). This finding suggests that HERV-K or related elements account for the reported hybridization of SMRV sequences with sequences in the human genome (Chiu *et al.*, 1983). Estimates of copy numbers suggest that approximately 50 HERV-K are present in the haploid human genome (Westley and May, 1984; Callahan *et al.*, 1985; Deen and Sweet 1986; Ono 1986).

Initially, studies on the RNA expression of HERV-K elements detected transcripts, including a full length 8.8 kb RNA, in several cell lines (Ono *et al.*, 1987b). In the breast carcinoma cell line T47D, which expresses estrogen and progesterone receptors, HERV-K mRNA was induced by treatment with estradiol followed by progesterone (Ono *et al.*, 1987b). Because the HERV-K elements were initially identified by hybridization to mouse mammary tumor virus (MMTV) *pol*-gene specific probes, it was interesting to note that the HERV-K LTR region contained a glucocorticoid- or protesterone-responsive sequence (Ono *et al.*, 1986). Probes specific for the different class II HERV subgroups identified by cross-hybridization (see below) also detected different transcripts in several cell lines and placenta (Franklin *et al.*, 1988). These included both large RNAs of 7.0, 8.0, 8.3, 10, and 12 kb in size, as well as a variety of smaller transcripts. In general, placenta was found to express the highest levels of HERV-K transcripts (Franklin *et al.*, 1988). HERV-K RNAs have also been detected in colon tumors as well as in normal colon mucosa by Northern blot analysis (Moisher *et al.*, 1986) and in peripheral blood mononuclear cells and leukocytes by using PCR (Medstrand *et al.*, 1992; Brodsky *et al.*, 1993). Recently, HERV-K RNA expression has been detected and analyzed in teratocarcinoma cell lines (Löwer *et al.*, 1993 a,b). A full-length transcript of 8.6 kb and smaller spliced RNAs of 3.3, 1.8, and 1.5 kb were detected by Northern blot analysis in the GH teratocarcinoma cell line (Löwer *et al.*, 1993 a, b), which produces retrovirus-like particles in culture (Boller *et al.*, 1983). The 3.3 kb transcript corresponds to a subgenomic *env* RNA while the spliced smaller transcripts remove most of the interior retroviral sequences (Löwer *et al.*, 1993a). The 1.8 kb RNA is doubly spliced and generates an ORF, termed cORF (for *c*entral), that could potentially encode a 12 kD protein. Interestingly, the predicted cORF protein sequence contains a region of basic amino acids with similarity to the RNA binding domains of the *rex, rev* and *tat* proteins of HTLV-1 and HIV (Löwer *et al.*, 1993a).

Unlike many other sequenced HERV elements, complete sequencing of the first full-length HERV-K element isolated, HERV-K10, revealed that it contained very few mutations interrupting its putative coding regions (Ono *et al.*, 1986). The *gag*-related region contained one frameshift mutation and the *env*-related region contained one nonsense mutation. The protease related region was intact, containing an ORF coding for 334 amino acids. The *pol*-related region was also uninterrupted by mutations over 937 amino acids, however a 290 bp deletion removed part of the carboxyl terminal region of *pol* as well as part of the amino terminal region of the *env* region. In addition to the paucity of mutations in this element, the 5′ and 3′ LTRs were nearly identical, containing just 2 nucleotide differences over 968 bp. Together, these sequence data suggest that this element represents a relatively new proviral insertion into the human germ line.

Recently, it has become clear that some HERV-K elements are indeed coding competent. HERV-K clones containing intact *gag* and *env* ORFs have been isolated using RNA PCR from GH cells (Löwer *et al.*, 1993a). Furthermore, a HERV-K *gag* gene cloned by PCR from genomic DNA has been expressed in *E. coli* as a fusion protein (Müeller-Lantzsch *et al.*, 1993). The size of this fusion product indicates that the HERV-K portion is approximately 73,000 kD, which is the molecular weight predicted from the *gag* ORF. In addition, expression in *E. coli* of the entire *gag* gene, including the protease homologous region at the 3′ end, resulted in the autoproteolytic generation of smaller proteins, which are detected by anti-*gag* antiserum (Müeller-Lantzsch *et al.*, 1993). These proteins were not detected when constructs deleted in the protease domain were used. Thus, at least some HERV-K elements appear to encode an enzymatically functional protease. The possibility that HERV-K envelope proteins may be expressed is supported by another recent study that has detected antibodies that recognize a recombinant HERV-K *env* protein in human sera from 12% of normal individuals (Vogetseder *et al.*, 1993). There is now also substantial evidence that HERV-K encodes the retrovirus-like particles, termed HTDV (*h*uman *t*eratocarcinoma *d*erived *v*iruses), expressed in relatively high amounts by the GH cell line (Figure 8). Besides the HERV-K RNA expression data discussed above, HTDV producing cells are stained in immunofluorescence tests by an anti-HERV-K *gag* antiserum (Boller *et al.*, 1993). Furthermore, immunoelection microscopy of ultrathin sections demonstrates that this antiserum reacts specifically with the HTDV particles (Figure 8, inset) (Boller *et al.*, 1993). In addition, Western blotting experiments of GH cell membrane preparations show that the antiserum reacts primarily with a protein of 30 kD (Boller *et al.*, 1993), the size of the putative major core protein (p30) of HERV-K (Müeller-Lantzsch *et al.*, 1993). Finally, Korbmacher *et al.*, 1993, showed that expression of the HERV-K *gag* gene is sufficient for particle formation when transfected into mammalian cells. The HTDV particles appear as immature virions that seem to be arrested in budding stages and lack the electron lucent space between the core and envelope typical of mature retroviral virions (Boller *et al.*, 1983; Löwer *et al.*, 1993a; Boller *et al.*, 1993; see Figure 8). Collapsed or condensed cores, which result from final proteolytic steps (Luciw and Leung, 1992), have not been observed. This finding sug-

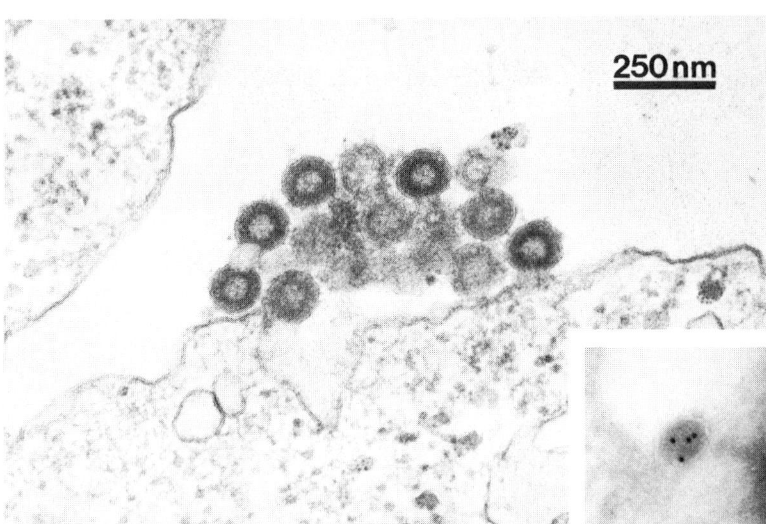

FIGURE 8. Group of HTDV particles from the human teratocarcinoma cell line GH in an ultrathin section. Note that no particles have a condensed core and that virtually all are in contact with the cell or other virus membranes. The inset shows an ultrathin frozen section of an HTDV particle immunogold labeled with anti-HERV-K *gag* antiserum. (Figure courtesy of Klaus Boller.)

gests that proper cleavage of the HERV-K *gag* precursor protein does not occur, although some HERV-K elements encode an apparently functional protease (Müeller-Lantzsch *et al.*, 1993).

Sequences related to HERV-K have also been detected in the genomes of great apes and Old World monkeys, but not in more distantly related species (Mariani-Costantini *et al.*, 1989). These elements are dispersed throughout the human genome, and their positions appear to be fixed within the population and largely conserved among humans and apes (Horn *et al.*, 1986; Mariani-Costantini *et al.*, 1989). However, identification of an insertion in the chimpanzee genome not present in humans is an indication that this family of elements has been transpositionally active since the two species diverged (Craig *et al.*, 1991). Although the primary mechanism for the dispersal of these elements was probably via retroviral integration, there is evidence that some elements are associated with tandemly repeated genomic sequences (Horn *et al.*, 1986; May and Westley, 1989). One of these tandem repeats appears to be specifically rearranged in some breast carcinoma cell lines (May and Westley, 1989).

Recent studies suggest that as many as 25,000 solitary HERV-K LTRs are present in the genome and that these LTRs are expressed as parts of cellular transcripts in a variety of human tissues (Leib-Mösch *et al.*, 1993). This estimate of LTR copy number does not include the repetitive SINE-R family discussed above (Section II.B.2) since the hybridization probe used in these studies distinguished between intact LTRs and the SINE-R elements, which contain only a portion of the HERV-K LTR sequence (Ono *et al.*, 1987a). The few solitary LTRs that have been isolated appear to be the result of LTR-LTR recombination (Kambhu *et al.*, 1990; Liu and Chan, 1990; Leib-Mösch *et al.*, 1993). However, given the very large number of these sequences in the genome, it is possible that some or most may have been generated not by HERV-K retrotransposition and subsequent recombination but by a DNA amplification mechanism such as has apparently occurred with the HERV-E family (Steele *et al.*, 1984; 1986). The observation that the chromosomal distribution of HERV-K LTRs is non random supports this possibility (Leib-Mösch *et al.*, 1993). Another finding suggesting a non-random distribution of HERV-K sequences is that two HERV-K related loci have been identified on chromosome Xq28 within 160 kb of each other (Sedlacek *et al.*, 1993).

Both HERV-K solitary LTRs and SINE-R sequences have been found to be associated with genomic variation among humans. In the case of solitary LTRs, two solo LTRs only 5 kb apart in the major histocompatibility locus HLA-DQ are associated with a polymorphism at that locus in which the intervening sequence is deleted (Kambhu *et al.*, 1990). This finding could possibly be due to recombination between the two LTRs. In the case of the SINE-R family, an element in the third intron of the complement C2 gene has been found to be associated with a variable number of tandem repeats (VNTR) locus (Zhu *et al.*, 1992). The 40 bp GC-rich repeats at the 5′ end of the element are reiterated 17 times in one allele and 23 times in another allele. This type of structure is similar to the VNTRs associated with THE-1 LTRs (Mermer *et al.*, 1987) and the minisatellite associated with a HERV-I LTR (Armour *et al.*, 1989). Such observations suggest that HERV may tend to be associated with recombinational hot spots (Jeffreys *et al.*, 1985).

2. Other Class II HERV

Although HERV-K elements have received the most attention, other less well characterized families or groups of class II HERV also exist. In one study, 100 genomic clones were subdivided into nine groups, in which their *pol*-related sequences hybridized to each other, but not to the members of other groups. The largest of these groups, which contained 64 of the 100 clones, included elements closely related to HERV-K (Franklin *et al.*, 1988). In general, the clones that have been sequenced in the HERV-K group are approximately 10% divergent in their nucleotide sequences (Ono *et al.*, 1986). The isolation of several other groups indicates that, while they are all related to MMTV at reduced stringency, appreciable divergence exists among different class II elements. Recently, PCR has been used to isolate 245 bp fragments from the reverse transcriptase domain of class II HERVs (Medstrand and Blomberg, 1993). In this study, six groups of elements, termed HML-1 through 6, were identified based on sequence comparisons of the PCR fragments. The HML-2 group corresponds to the HERV-K family with the sequences within this group being approximately 90% identical. However, the most divergent group, HML-6, is only 55–60% identical to the HERV-K group or to MMTV and is slightly more related to IAP elements (Medstrand and Blomberg, 1993). Southern blots using group-specific probes indicate that HML-1, 2, 3 and 6 are most abundant with the total copy number being 60–80 per haploid genome. RNA expression of the different groups was also detected in several tissues (Medstrand and Blomberg, 1993). Further analysis of the structures and organization of these new class II families will require the isolation of full length elements corresponding to each group.

Rearrangements within different class II HERV also contribute to their heterogeneity. Several of the clones that have been characterized were found to contain rearrangements, commonly deletions (May and Westley, 1986; Deen and Sweet, 1986; Ono *et al.*, 1986). Another difference among class II elements exists in their LTRs, as revealed by the isolation of a 6 kb element containing ~430 bp LTRs (May and Westley, 1986). This element, termed NMWV4, also contains a lysine tRNA primer binding site, but it corresponds to a different codon (UUU) than the PBS found in other HERV-K elements (CUU) (May and Westley, 1986).

C. Other HERV

As alluded to above, retroviruslike particles have been detected from several human sources not known to be infected by virus (see Section IV.C.2 below). These sources include normal term placentas (Kalter *et al.*, 1973; Dalton *et al.*, 1974; Vernon *et al.*, 1974; Imamura *et al.*, 1976; Nelson *et al.*, 1978; for reviews, see Dirksen and Levy, 1977; Johnson *et al.*, 1990), human milk (Moore *et al.*, 1971; Vaidya, 1973), breast cancer biopsies (Axel *et al.*, 1972), monocytes from breast cancer patients (Al-Sumidaie *et al.*, 1988), lymphocytes of patients with myeloproliferative disease (Boyd *et al.*, 1989), and muscular dystrophy patients (see Haahr, 1992), as well as several breast cancer (McGrath *et al.*, 1972; Keydar *et al.*, 1984) and teratocarcinoma cell lines (Boller *et al.*, 1983; Bronson *et al.*, 1984; Löwer *et al.*, 1987). In addition, retrovirus-related antigens have also been detected in many of the same sources (see Section IV.C.1). The failure to demonstrate an infectious retrovirus associated with these particles, together with the recent findings discussed in the previous section, suggests that at least some of these particles are encoded by HERV sequences. The particles could be encoded by a single functional element, or via complementation between partly defective elements. Furthermore, the particles could be encoded by one or more of the HERV families that have been described. Rare functional ERV elements within a family of mostly defective sequences have been discovered in the genomes of other organisms (Reeves and O'Brien, 1984), and among other retroelements in humans (Matera *et al.*, 1990; Dombroski *et al.*, 1991; Deininger *et al.*, 1992). Alternatively, in some cases the characterization of these particles and the identification of their source(s) could provide a novel approach for identifying HERV sequences that have thus far eluded detection.

As discussed in the preceding section, the possibility that the genome contains yet uncharacterized HERV families has been indicated by a cloning procedure in which PCR primers were designed against sequences conserved among reverse transcriptases and used to isolate short fragments of different reverse transcriptase-related sequences (Shih *et al.*, 1989; Bangham *et al.*, 1988; Medstrand and Blomberg, 1993). Analysis of the clones revealed that the genome contains a great variety of reverse transcriptase-related sequences, not all of which could be attributed to previously identified HERV families. However, except for the class II HML groups (Medstrand and Blomberg, 1993), no new HERV families have yet been established as a result of the analysis of reverse transcriptase PCR fragments (Shih *et al.*, 1989; Wichman and Van Den Bussche, 1992). Other investigations have been reported that may also involve uncharacterized HERV families. For example, hybridization studies with probes containing LTR, *gag*, and *env* regions of Mason–Pfizer monkey virus detected distantly related sequences in the genomes of humans and apes (Barker *et al.*, 1985). These sequences have not been isolated, hence their possible relation to characterized HERV families is difficult to assess. In another case, the intriguing report of the possible rescue of a HERV element by pseudotyping with MLV could also lead to the discovery of a new HERV family, but has not been confirmed by further experimentation (Zavada *et al.*, 1986).

Unlike many other organisms, HERV elements that are closely related to the exogenous retroviruses of humans have not been discovered. Attempts to identify HERV sequences related to HIV by low-stringency hybridization library screening have been unsuccessful (Horwitz *et al.*, 1992), and only small segments related to HTLV-I have been found (Mager and Freeman, 1987; Perl *et al.*, 1989).

A HERV sequence proposed to have a structure similar to murine sarcoma virus (i.e., MLV plus associated *v-mos* sequence) has been reported on the basis of hybridization studies (Zabarovsky *et al.*, 1983; Chumakov *et al.*, 1985). However, independent analysis of the published LTR–U5 sequence reveals that it is derived from an Alu element (D. Wilkinson, unpublished observation). Therefore, further characterization of the clone will be required to confirm the initial hypothesis.

A summary of the known human endogenous retroviruslike elements or families of elements that have now been characterized is shown in Table IV along with some of their properties. As discussed above, it is unlikely that this list is complete since the use of techniques such as PCR has begun to uncover new families that have escaped detection by other methods. Many of these families are

TABLE IV. Summary of Known HERV Families

HERV family	Subfamily	Copy number	Method of detection	Some sites of transcription	Evidence for protein expression
HERV-ERI	HERV-E	35–50	rtv[a] homology	Placenta, colon, brain, colon & breast tumor lines	
	Hs5	1?	rtv homology	?	
	NP-2	2	HERV-E homology	None detected	
	ERV1	1 (~15 related)	rtv homology	Transcription unlikely	
	HERV-R (ERV3)	1 (~10 related)	rtv homology	Placenta, low levels in most cells	*env*
	RRHERV-I	~20	Chance	Retinoic acid-treated TC cells[b]	
HERV-I (RTVL-I)		~25	Chance	?	
S71		1 (~20 related)	rtv homology	Placenta, K562 cells	
HRES-1		1	rtv homology	Tumor cell lines	28 kDa (*gag*?)
HERV-P	HuERS-P1	10–20	tRNA homology	?	
	HuERS-P2	20–30	tRNA homology	?	
	HuERS-P3	20–40	tRNA homology	?	
ERV-9		~40	Chance	TC cells, placenta	
HERV-H (RTVL-H)		~900 (deleted) 50–100 (intact)	Chance	TC cells, placenta, tumor cell lines	
HERV-K		~50	rtv homology	TC cells, placenta, tumor cell lines	*gag, prt*

[a] rtv, retroviral.
[b] TC, teratocarcinoma cell lines.

multicopy and are expressed at the RNA level. In some cases, there is also evidence for HERV protein expression. In the following section, the potential biological functions of HERVs and other retroelements will be considered.

IV. ENDOGENOUS RETROVIRUSES: IMPLICATIONS FOR THE ORGANISM

By the early 1980s, it was recognized that interspersed repetitive DNA sequences constituted a significant fraction of the genome. Major sequence families, such as LINEs and Alu elements, had been identified and their dispersal via a mechanism involving reverse transcriptase had become appreciated (see Singer, 1982b; Rogers, 1985, for reviews). At that time, however, the biological significance of repetitive interspersed sequences created by the reverse flow of genetic information could not be fully understood because data supporting the role of these sequences in the biology of an organism were lacking. This problem led to the suggestion that such sequences could be regarded as "junk DNA," or as molecular parasites that were able to self-replicate and disperse within the genome, but did not directly contribute to biological functions within the organism (Doolittle and Sapienza, 1980; Orgel and Crick, 1980). While that analysis may be partly true, it is becoming clear that retroelements contribute to the biology of the organism in which they reside in several different ways. The following discussion will elaborate on these interactions, with an emphasis on the human context as affected by HERV. Examples from other human retroelements or other organisms will be included in cases where HERV data are lacking, but by analogy, similar events may be possible.

A. Genome Structure

It has been estimated that between 5 and 10% of the human genome is comprised of sequences that were created by the reverse transcription of RNA followed by integration back into the genome (Baltimore, 1985). As discussed in earlier sections of this chapter, reverse transcribed sequences include those that encode reverse transcriptase as well as those that do not. Because endogenous retroviruses and related sequences are a source of reverse transcriptase activity, it is possible that they can facilitate the retroposition of genetic information in addition to their own. Thus, they may contribute to the reverse flow of genetic information in several ways: (1) retroposition of the sequence coding for a functional ERV; (2) complementation of mutated ERV to facilitate their retroposition; (3) providing the reverse transcriptase required for the retroposition of unrelated sequences. The plausibility of the latter two events has been confirmed by their demonstration in experimental systems (Linial, 1987; Tchenio and Heidmann, 1991). Although the rate of retroposition is so slow that it is most appreciable by genomic differences accrued over an evolutionary time scale, evidence indicates that it is an ongoing process in all species. The characterization of insertion events involving L1 and Alu elements (see below), along with the detection of polymorphic L1 and Alu elements and a polymorphic dihydrofolate reductase pseudogene, indicate that humans are no exception (Anagnou *et al.*, 1988; Woods-Samuels *et al.*, 1989; Batzer *et al.*, 1991).

The consequence of an insertion event greatly depends on its location. The distribution of retroelements within different eukaryotic genomes demonstrates that the target sites of insertions are very flexible, but not completely random. In general, retroelements preferentially insert into adenine-rich nucleotide stretches (Furano *et al.*, 1986; Deka *et al.*, 1988a; Deininger, 1989; Sandmeyer *et al.*, 1990). It has also been observed that retroelements frequently occur in clusters (Rogers, 1985; Kuff and Lueders, 1988). This finding is also observed for HERVs, as indicated in Table V, which shows that many of the HERV sequences that have been isolated are either flanked or interrupted by other retroelements, or in close proximity to them. The observed clustering of retroelements of differ-

TABLE V. Clustering of HERV with Other Retroelements

HERV element	Associated sequence	Reference
jnLTR8	Alu adjacent	Tomita *et al.* (1990)
4-1	L1 adjacent	Steele *et al.* (1984)
ERVA1A	Inserted into γ-actin pseudogene	Samuelson *et al.* (1990)
LTRpdhB	Alu 500 bp flanking	Unpublished
HERV-R	Alu 200 bp flanking	Kato *et al.* (1990)
RTVL-Ia	Alu inserted within, 2 Alu's within 1 kb	Maeda and Kim (1990)
RTVL-Ic	2 Alu's inserted within	Maeda and Kim (1990)
S71	HERV-K LTR within	Leib-Mösch *et al.* (1993)
HuERS-P1-2	Alu inserted within	Harada *et al.* (1987)
RTVL-H1	Inserted into L1	Mager and Henthorn (1984)
RTVL-H LTR in proenkephalin locus	Alu adjacent	Comb *et al.* (1983)
RTVL-H3	Alu 200 bp flanking	D. Mager (unpublished data)
RTVL-H Sol 1 LTR	Inserted into Alu	D. Mager (unpublished data)
RTVL-H *env* clone	Alu inserted within	D. Mager (unpublished data)
RTVL-H cDNA clone	Alu juxtaposed	Hirose *et al.* (1993)
HERV-K LTR in HLA-DQ locus	Alu adjacent	Kambhu *et al.* (1990)
HERV-K LTR	Inserted into L1 4 Alu's flanking, 750 bp or closer	Liu and Chan (1990)
Chimpanzee class II LTR	Inserted into TPI[a] pseudogene	Craig *et al.* (1991)

[a] TPI, triose phosphate isomerase.

ent classes suggests that they, at least in part, share some integration specificities. It might also represent a co-adaptive feature of the elements, since insertion into a previously inserted sequence is not likely to have a deleterious effect on the organism. Targeting of a retroelement to a specific region of the genome in the absence of insertion site sequence specificity has been observed in several cases. In humans, an example of this phenomenon are the three separate HERV-I insertions into the haptoglobin-related gene (Maeda and Kim, 1990). Such specificity may be the result of an interaction between the integration complex of the retroelement and cellular proteins that are targeted to the site. For example, an interaction between the yeast retrotransposon Ty3 and the RNA polymerase III transcription factor TFIIIB appears to direct Ty3 insertions to the 5′ region of yeast tRNA genes (Sandmeyer *et al.*, 1990). Retrotransposons and retroviruses also display a bias for inserting into transcribed regions of the genome, with a particular affinity for regions close to DNase I hypersensitive sites (Sandmeyer *et al.*, 1990; Craigie, 1992). Such insertions often affect the regulation of transcription of the target gene. In some organisms, the insertion of ERV into the 5′ regulatory region of a gene has introduced novel regulatory sequences that have evolved to serve a normal role in the regulation of the gene (discussed below).

The reverse flow of genetic information is responsible for a novel mechanism of gene duplication. On rare occasions, mRNAs are reverse transcribed and integrated back into the genome in a structure that either retains or acquires promoter sequences that allow the new copy of the gene to be expressed. Such genes have been termed retrogenes. As a consequence of the retroposition pathway, retrogenes typically lose all or most of the introns present in the original gene, acquire polyadenosine tails, and are flanked by short direct repeats as described for pseudogenes. Although retrogenes are not common, examples exist in several different organisms and are listed in Table VI. They include the autosomal human phosphoglycerate kinase gene (Boer *et al.*, 1987; McCarrey and Thomas, 1987), the human prointerleukin-1 beta gene (Clark *et al.*, 1986), the mouse autosomal pyruvate dehydrogenase E1a gene (Dahl *et al.*, 1990), the mouse *Zfa* gene (Ashworth *et al.*, 1990), the rat preproinsulin I gene (Soares *et al.*, 1985), the chicken muscle-specific calmodulin gene (Gruskin *et al.*, 1987), and the woodchuck *N-myc2* gene (Fourel *et al.*, 1990). In fact, it has been suggested that the primarily intronless genomic structure of *Saccharomyces cerevisiae* could have been created by a similar mechanism (Fink, 1988). ERV could contribute to the creation of retrogenes in several different ways. As mentioned, ERV may be one source of reverse transcriptase activity which is essential for the process. In addition, ERV could play a role either in the retention of original promoter sequences or in the acquisition of new promoter activity. Original promoter sequences could be retained if the transcript that was reverse transcribed initiated in an upstream ERV and read through the gene. Alternatively, promoter sequences from an ERV LTR could be donated to drive the expression of the newly created gene copy at the insertion site. The proclivity of HERV to integrate into the 5′ flanking regions of genes and in clusters with other retroelements would increase the likelihood of events such as these. An involvement in gene duplication, regarded as a primary component of the evolutionary process, is just one of the ways in which ERV may contribute to the evolution of an organism.

Because their sequences are repeated throughout the genome, ERV and other retroelements may serve as foci for recombination events. The effects of these rearrangements are varied: some appear to be neutral, creating polymorphisms within a population; some are deleterious, resulting in disease (discussed below); and some have created alterations in genomic structure that have become fixed during evolution. Rearrangements that appear to be neutral include the several VNTRs that are associated with LTR sequences (Mermer *et al.*, 1987; Armour *et al.*, 1989; Zhu *et al.*, 1992), and a deletion in the HLA-DQ locus that possibly involves two solo HERV-K LTRs (Kambhu *et al.*, 1990). Although HERVs are not requisite for homologous but unequal crossing over, their association with two gene clusters in the human genome created in that way (Maeda, 1985; Samuelson *et al.*, 1990) suggests that HERVs may enhance the likelihood of such an event. In

TABLE VI. Retrogenes[a]

Gene	Species	Reference
Phosphoglycerate kinase	Human	Boer *et al.* (1987), McCarrey and Thomas (1987)
Prointerleukin-1 beta	Human	Clark *et al.* (1986)
Pyruvate dehydrogenase E1a	Mouse	Dahl *et al.* (1990)
Zfa	Mouse	Ashworth *et al.* (1990)
Preproinsulin	Rat	Soares *et al.* (1985)
Calmodulin	Chicken	Gruskin *et al.* (1987)
N-myc2	Woodchuck	Fourel *et al.* (1990)

[a] Retrogenes are genes derived from mRNAs that have retained or acquired promoter sequences which permit the newly reintegrated gene copy to be expressed.

fact, duplication of the γ-globin gene has been demonstrated to have been caused by homologous but unequal crossing over between two L1 sequences that flanked the original gene (Fitch *et al.*, 1991). Similarly, the human growth hormone gene and the haptoglobin-related locus have been duplicated by recombination events mediated by Alu sequences (Barsh *et al.*, 1983; Erickson *et al.*, 1992). Several clusters of HERV elements have been detected within the genome, comprised of HERV-H and HERV-E family members (Fraser *et al.*, 1988; Taruscio and Manuelidis, 1991). It would be of interest to investigate those clusters to determine if they also are associated with gene families.

B. Effects on Adjacent Genes

Like their infectious counterparts, ERVs contain transcriptional regulatory sequences within their LTRs that control the expression of the element. The distribution of multiple ERV copies throughout an organism's genome, combined with their affinity for transcriptionally active targets, creates the possibility that some ERVs may modify cellular gene expression. Some general ways in which ERVs can affect adjacent gene expression are shown in Fig. 9. Such effects are best documented in cases of somatic or tumor-specific retroviral insertions that have altered expression of a proto-oncogene (see Favor and Morawetz, 1992, for review). However, studies conducted in several organisms reveal that during the course of evolution, ERV sequences have on occasion become assimilated by the organism to provide functions to the cellular gene in which they are integrated. As discussed in Section III, HERV elements contribute promoter or enhancer elements to at least four different genes in humans. The best characterized of these examples are the salivary amylase genes which acquired a parotid-specific enhancer element by the insertion of a HERV-E element in the 5′ flanking

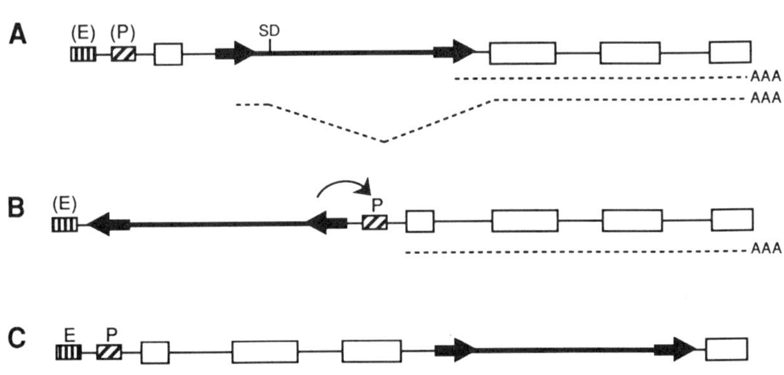

FIGURE 9. Potential effects of proviral sequences on gene expression. A hypothetical gene with four exons (white boxes) and enhancer and promoter elements (striped boxes) is shown. Dashed lines represent possible transcripts (splicing events between exons are not shown). (A) Promoter insertion: In this case, the enhancer and promoter sequences in the 5′ LTR (via splicing) or the 3′ LTR of an inserted retroviral element now regulate expression of the gene. (B) Enhancer insertion: In this situation, retroviral enhancer elements act in conjunction with the gene's natural promoter. (C) Differential polyadenylation: Insertion of a retroviral element in this instance provides an alternative polyadenylation signal. Numerous variations on these effects are possible depending on the orientation and structure of the retroviral element and on the gene involved.

region of a duplicated pancreatic amylase gene (Samuelson *et al.*, 1990; Ting *et al.*, 1992). Enhancer sequences present in the 5′ internal region of the HERV-E element created an amylase gene with new tissue specificity, a mutation that was positively selected for and subsequently fixed within the population (Ting *et al.*, 1992). The human cytochrome c_1 gene also appears to have derived part of its transcriptional regulatory sequences from a HERV element. In the case of that gene, however, the HERV sequences are limited to a short region having homology with a HERV-I LTR (Suzuki *et al.*, 1990). Transcripts from two other genes have been found to be driven by HERV promoters. In one case, two of the three transcripts initiated in the HERV-R promoter have been found to read through the 3′ LTR and splice into a Kruppel-related gene, called *H-plk,* that encodes a zinc-finger protein (Kato *et al.*, 1990). The fusion transcripts are readily detectable in placental tissue, but characterization of the protein product has not yet been reported. In the other example, transcripts initiated in the 5′ LTR of a HERV-H element splice into a gene, designated *PLA2L,* that contains two phospholipase A_2-related domains (Feuchter-Murthy *et al.*, 1993). Transcripts of *PLA2L* have been detected in teratocarcinoma cell lines but the putative protein products of this alternatively spliced gene have not been characterized. Several other examples of ERV promoted/regulated genes have been discovered in other organisms, such as the rat oncomodulin gene (Banville and Boie, 1989; Furter *et al.*, 1989), the mouse sex-limited gene (Stavenhagen and Robins, 1988; Adler *et al.*, 1991), and the mouse *MIPP* gene (Chang-Yeh *et al.*, 1991), demonstrating that the assimilation of ERV promoter/enhancer elements has been a recurrent event throughout evolution.

In addition to sequences that control transcription initiation, ERV LTRs also contain a polyadenylation signal. During the course of investigating HERV expression, several cDNAs comprised of cellular sequences polyadenylated by a HERV have been isolated. In one case, a new gene, termed *PLT,* was discovered by virtue of the fact that one of its alternatively processed transcripts is polyadenylated by a HERV-H LTR (Goodchild *et al.*, 1992). Other cDNAs polyadenylated by HERV-H (Mager, 1989), HERV-E LTRs (Tomita *et al.*, 1990), and THE-1 LTRs (Paulson *et al.*, 1987) have been identified, but it has not been determined if they too are associated with genes.

As discussed below (Section IV.D), the possible deleterious consequences of ERV can be readily appreciated. Hence, ERVs were once regarded as parasitic DNAs that were able to propagate their sequences within a species despite negative selective pressures that could be perceived to operate against such expansion. However, as the elements have become better understood, the possible beneficial contributions to a species can now be appreciated. Examples discussed above illustrate that ERVs have provided an important source of biological diversity during the course of evolution. In general, transposition events are a rare occurrence, thus minimizing the negative effects on the population as a whole. Exceptions to this tight regulation of transposition may occur in times of stress to the organism. Many conditions found to induce the transcription of ERV and retrotransposons correspond to conditions that also induce stress responses in the organism. For example, murine ERVs have been found to be induced by ultraviolet radiation and nucleotide analogues (both treatments that would also activate DNA repair mechanisms), anoxia, and various immunological factors involved in an immune response (Coffin, 1984; Anderson *et al.*, 1988; Mietz *et al.*, 1992; Itoh *et al.*, 1992). Retrotransposons in other organisms are similarly activated by conditions that stress the organism (Echalier, 1989; Arnault *et al.*, 1991). This coordination of ERV induction with stress responses may have evolved as a mechanism to enhance the biological diversity within a species during times of extreme selective pressure and thereby enhance the possibility of beneficial mutations generated by ERVs being fixed within the population.

C. HERV Protein Expression

As discussed above in Sections III.A and B, the expression of most HERV families has been detected at the RNA level. However, most elements are disrupted by mutations that would prevent them from encoding proteins. Nonetheless, the finding of some HERVs with long ORFs and the fact that families of mostly defective retroviral elements containing a few coding-competent members have been documented in other species (Reeves and O'Brien, 1984; Kuff and Lueders, 1988; Fredholm *et al.*, 1991) suggest that protein-encoding HERVs exist. In this section, we summarize the evidence for HERV protein expression in the context of retroviruslike antigens and particles that have been described in human tissues and cell lines.

1. HERV Expression and Retroviral Antigens in Human Tissues

The *env* RNA expression of the single-copy HERV-R has been well documented and appears to be widespread (Section III.A.1.e). The *env* ORF (1940 bp) of HERV-R is sufficient to code for a polypeptide of 577 amino acids. The deduced amino acid sequence of the potential polypeptide indicates that it has several features characteristic of retroviral glycoproteins including seven *N*-glycosylation sites and a hydrophobic transmembrane domain (Cohen *et al.*, 1985). Indeed, as discussed in the above section on HERV-R, evidence for protein expression of HERV-R *env* or a highly related polypeptide at the syncytiotrophoblast layer of the placenta has recently been obtained (Larsson *et al.*, 1993).

For the element, ERV1, no specific mRNA was detected in placental tissue. The lack of expression was not unexpected because the element does not have a 5′ LTR that normally contains the transcriptional promoter sites. Yet, antibodies raised against a synthetic undecapeptide derived from the p30 *gag*-related region of ERV1 reacted with a protein of 75 kDa in the syncytiotrophoblast layer of normal first-trimester placentas. The antibody detected the protein in blighted ova and benign and malignant trophoblastic tumors as well as in three different lines of cultured choriocarinoma cells (Suni *et al.*, 1984; Wahlstrom *et al.*, 1984). The undecapeptide blocked the reactivity of the antibody suggesting specificity in the reaction. Other evidence for the presence of a baboon endogenous virus (BaEV) *gag* p30-like protein in humans has been reported by Suni *et al.* (1981). In this study, 1540 human cord-blood sera were screened for antibodies reactive to *gag* p30 of RD114 and BaEV; 118 (8%) were found to have reactive antibody. They also found that an antigen related to the *gag* p30 protein of RD114/BaEV was localized in the placental syncytiotrophoblast and not the cytotrophoblast layer. Whether this cross-reactive placental protein was derived from an uncharacterized ERV related to ERV1 or HERV-R was not determined. When low-stringency hybridization studies were carried out with ERV1 or HERV-R probes, 10–15 related elements were detected in the human genome. It is possible that the expression of *gag* p30-related proteins in placenta results from these HERVs.

Antibodies against the structural protein p30 *gag* of SSV/SSAV and BaEV recognize proteins (M_r 30,000) in human leukemic sera (Derks *et al.*, 1982; Hehlmann *et al.*, 1983). With immunohistochemical techniques, a placental antigen related to the SSAV and GALV (gibbon ape leukemia virus) gag p30 protein was localized to the basal aspect of the syncytiotrophoblast (Maeda *et al.*, 1983; Jerabek *et al.*, 1984). These investigators also found a 30-kDa protein by immunoblotting with anti-SSAV p30 in one- and two-dimensional electrophoretic gels of lysates from 12 of 14 human placentas. When sera from autoimmune patients were screened for antibodies to human placenta, the same region of reactivity was observed at the blast layer of the syncytiotrophoblast; no reactive antibodies were detected in sera of normal patients. The binding of the autoimmune sera was blocked with disrupted SSAV or by extracts of human placenta. Antiplacental antibodies were also detected in the serum of leukemic patients and reacted with 19- and 32-kDa proteins in Western immunoblots of placenta protein lysates (S. Maeda *et al.*, 1986). These same proteins were also reactive with anti-SSAV-specific antisera. Antisera raised to SSAV or to HEL-12, a retrovirus which was spontaneously expressed in a diploid strain of human embryonic lung fibroblast cells (Panem *et al.*, 1975), produced positive immunofluorescence in 24 of 24 human term placentas. The antibody reaction could be specifically blocked by purified SSAV but not by fetal calf serum, human immunoglobulin, or serum albumin. An extract of placentas which was enriched in 70-kDa glycoprotein was found to block the reaction (Sawyer *et al.*, 1978). There was no difference in expression between normal human placentas and placentas from patients with systemic lupus erythematosus (Panem *et al.*, 1976). Furthermore, serum proteins related to the SSAV gp70 protein are found in patients in blast crisis with acute leukemias or chronic myelogenous leukemia (Hehlmann *et al.*, 1984). These are interesting findings since the SSAV-related HERV element S71 contains no complete ORF for p30 *gag* and no *env* region. More likely these results may be accounted for by the expression of the SSAV *gag* p30 and gp70 cross-reactive proteins derived from retroviral elements related to S71 of which there are approximately 15 to 20 copies (Leib-Mösch *et al.*, 1986).

The HTLV-1-related endogenous retroviral element, HRES-1, is transcriptionally active and contains ORFs with some homology to the HTLV-1 *gag* p19 and *gag* p24 genes (Perl *et al.*, 1989). The calculated molecular mass of the protein encoded by the HRES-1 *gag* p24-like sequence was 28 kDa, much larger than the 24 kDa of the HTLV-1 capsid protein. Antibodies made to two nonoverlapping peptides taken from the deduced amino acid sequence of this protein did detect a 28-kDa protein in lysates of H9 human T cells by Western blotting (Banki *et al.*, 1992). Furthermore, HRES-1 protein translated *in vitro* is recognized by both antisera (A. Perl, personal communication). This protein was also found in lymphoblastoid T-cell lines, normal peripheral blood T cells, and in normal brain and liver by Western blot analysis and in cutaneous T-cell lymphoma tissues by immunohistochemistry. When cells were examined by confocal immunofluorescence microscopy for the p28 HRES-1 protein, specific staining was localized to the cytoplasm and to nuclear bodies, possibly nucleoli. No surface expression of p28 was detected and this observation suggested that it was relatively unexposed to the immune system. However, antibodies to HRES-1 p28 have been detected by serum reactivity with the p28-derived peptides in 65 normal blood donors and 146 patients with immunological disorders. Sera of patients with multiple sclerosis (19 out of 65, 29%), progressive systemic sclerosis (4 out of 17, 23%), systemic lupus erythematosus (4 out of 19, 21%), and Sjögren's syndrome (2 out of 19, 11%) contained significantly higher HRES-1 peptide binding activity. When 30 HRES-1 seropositive patients were tested for immunoreactivity to HTLV-1 *gag* p24, 9 showed reactivity. These findings have led investigators to speculate that the autoantibodies associated with HTLV-1 infection result from the cross-reactivity of HTLV-1 p24 antigens with HRES-1 p28.

Any discussion of the significance of ERV proteins in human tissues presents an interesting dilemma because there is no direct evidence of a functional role for these proteins. There are the correlations that can be made after surveys of different tissues and cell lines for these proteins are made, but experimental studies introducing these genes into cells and assessing the effect of gene expression on cell/animal phenotype have not been carried out. Nevertheless, some discussion of the interactions of ERV proteins with the host and ERV elements in other mammalian systems is possible. For example, ERV-encoded proteins have been implicated to play a role in immunoregulatory pathways in the mouse. Immunoglobulin E binding factors (IgE-BF) induced by IgE and antigenic stimuli have been identified that function to either potentiate or suppress IgE production. Expression cloning of the cDNAs encoding IgE-BF via a functional assay revealed that the proteins were actually encoded by a subset of intracisternal A particle elements (IAPs), a family of ERV in mice (Toh *et al.*, 1985; Moore *et al.*, 1986; Kuff *et al.*, 1986). The IgE-BF activity was determined to be encoded by the *gag* region of the mouse IAP element, which is partly comprised of sequences that are significantly divergent from those in IAP elements in other rodent species (Kuff *et al.*, 1986). Consequently, the evolutionary development of this interaction between IAPs and the immune system is thought to be specific to mouse species (Kuff and Lueders, 1988).

When considering the possible interactions between ERV and the organism, the known biological properties of retroviral proteins are often used as paradigms to suggest ERV activities. For example, the transmembrane protein encoded by the *env* gene has been found to exert immunosuppressive effects *in vivo* and *in vitro* (see Krieg and Steinberg, 1990, for a review). Inhibition of the expression of the *env* gene of one class of MLV-related ERV by antisense oligonucleotides has been found to stimulate murine lymphocytes (Krieg *et al.*, 1989). This finding has prompted the suggestion that these ERV proteins may be involved in normal murine immunoregulation, but further studies are clearly required to confirm the hypothesis.

Because HERV proteins remain poorly characterized, the possibility that they contribute important functions to human biology has not been investigated. However, the detection of retroviruslike particles and reverse transcriptase activity in several human tissues and cell lines (see below) suggests that HERV proteins are expressed, and hence may have become functionally integrated into the organism. Several possibilities are suggested by the frequent detection of HERV transcripts and retrovirus-related proteins in normal placenta (for review see Johnson *et al.*, 1990). It has been postulated that HERV-encoded *env* proteins may play a role in the creation of the syncytiotrophoblast layer of the placenta, since retroviral envelope proteins have fusogenic properties (Harris, 1991). The recent finding of putative HERV-R *env* protein expression at this site (Larsson *et al.*, 1993) is particularly intriguing in this regard. Another possibility is that HERV-encoded *env* proteins may be involved in suppressing possible maternal immunological reactions to the fetal allograft, again based on the immunosuppressive activity of retroviral *env* proteins. A third role that HERV may play is to block infectious retroviruses from infecting the organism. Such an effect could result from receptor interference, as seen in the case of the mouse locus Fv-4 which encodes an ERV *env* gene that confers resistance to retroviral infection (Ikeda and Sugimura, 1989), or by other mechanisms, such as target deletion, as seen in the case of murine ERV related to MMTV (Golovkina *et al.*, 1992).

2. Retroviruslike Particles in Human Tissues

Particles resembling complete retroviruses have been observed in normal human placentas, oocytes, teratocarcinoma cells, and mammary carcinoma tissues. Perhaps the most well-documented evidence for endogenous human retroviral particles may be found in electron microscopic studies of human placenta (for review, see Kalter *et al.*, 1973; Vernon *et al.*, 1974; Dalton *et al.*, 1974; Imamura *et al.*, 1976; Dirksen and Levy, 1977; Panem, 1979). These particles were first observed in 1973 at the basal surface of the placental syncytiotrophoblast (Kalter *et al.*, 1973) and appeared to be type D retroviruses since they were missing the distinct space between the nucleocapsid and the envelope. When these particles were purified by differential density centrifugation in sucrose, retroviruslike particles banding at 1.15–1.16 g/ml were isolated with reverse transcriptase (RT) activity. The RT preferred Mg^{2+} as the divalent cation, a characteristic of class II retroviruses (avian type C and mammalian types B and D). After mild detergent treatment of the placental particles, the RT activity was associated with particles having the density of retroviral cores at 1.24–1.26 g/ml (Nelson *et al.*, 1978). The RT activity was detected in 80% of over 100 human placentas examined and there appeared to be a positive correlation between the number of previous pregnancies and a higher level of RT activity. In the process of characterizing these retroviruslike particles, an RT inhibitor was discovered and postulated to account for the lack of RT activity in some human placentas. This inhibitor (60–65 kDa) was very selective for mammalian retroviral enzymes. In fact, it had its greatest effect on purified BaEV RT among the retroviral enzymes tested. It had no effect on avian retroviral RT or calf thymus DNA polymerases α and γ (Nelson *et al.*, 1981; Leong *et al.*, 1984). The frequency of finding these particles in placenta has suggested to some investigators that there is a role for these particles in placental function. As mentioned above, they may be involved in the formation of the syncytiotrophoblast by virtue of the fusogenic properties of retroviral glycoproteins, or they may act as immunosuppressive agents to prevent the fetus-derived placenta from being rejected. These particles appear to be mature retroviral particles and attempts have been made to find a cell line suitable for the replication of these particles. However, despite numerous attempts on different cell lines including many human, canine, rabbit, and mink lung cells, no direct evidence for their replication has been observed. It was not possible, at the time, to distinguish between replication of the human placenta-derived retrovirus and the induction of endogenous host cell-derived viruses (Levy *et al.*, 1982; J. Levy, and J. Leong, unpublished observations).

Electron microscopy has also demonstrated retroviruslike particles in teratocarcinoma cell lines (Bronson *et al.*, 1979, 1984; Kurth *et al.*, 1980) and in a human mammary tumor line (Keydar *et al.*, 1979). This evidence was associated with numerous reports of RT activity in supernatant fluids of these cells and the detection of retroviral cross-reactive proteins in these cell lines. Treating these cells with agents such as 5-azacytidine, iododeoxyuridine, dexamethasone, and dimethylsulfoxide would induce the transient release of RT-containing particles into the supernatant fluid (Bronson *et al.*, 1979; Kurth *et al.*, 1980). Kurth and his co-workers had established that the human teratocarcinoma-derived viruses (HTDV) contain a high-molecular-weight (60 S) RNA and an associated RT activity (Löwer *et al.*, 1987). However, further characterization of the HTDV was hampered by the very low amounts of virus produced in these cells; it was not possible to harvest sufficient quantities of virus particles to clone the HTDV genome. These investigators also found that, as in placenta, an inhibitor of RT copurified with the particles (Löwer *et al.*, 1987). More recently, DNA primers complementary to the tRNA binding site of several different retroviruses were used successfully to generate DNA probes from the 5′ end of polyadenylated RNA isolated from teratocarcinoma cells by PCR. As discussed in the section on HERV-K elements, these probes detected four major HERV-K mRNA species of 8.9, 3.3, 1.8, and 1.5 kb in teratocarcinoma cells (Löwer *et al.*, 1993a,b). The size distribution of these mRNA species, which included a full-length mRNA, was reminiscent of the complex expression pattern seen with exogenous lenti- or spumaviruses (Löwer *et al.*, 1993a). Sequence analysis of the expressed HERV-K mRNAs revealed complete *gag* ORFs. In fact, antiserum raised against a recombinant expressed HERV-K *gag* protein specifically stained the HTDV particles in GH teratocarcinoma cells under immunoelectron microscopy (Boller *et al.*, 1993) (see Fig. 8). These findings were further supported by immunoblot experiments which showed that the antiserum reacted with a 30-kDa protein from teratocarcinoma cell lysates (Boller *et al.*, 1993).

Small (60- to 80-nm diameter) retroviruslike particles have also been detected at the surface of human preovulatory oocytes. These particles, termed viruslike particles (VLPs), were associated with RT activity in the follicular fluid of three of nine oocytes examined (Larsson *et al.*, 1981). The VLPs were only observed in oocytes taken from women treated with the ovulation-inducing drugs HCG (human chorionic gonadotropin) and clomiphene. They were not present in oocytes recovered from untreated women. The follicular fluid was also shown to contain protein reactive with antibody to RD114 *gag* p30 protein (Nilsson *et al.*, 1992).

Taken together, these studies suggest that genomes of endogenous retroviruses can encode sufficient information to produce budding retrovirus particles. No single HERV element capable of autonomous replication has yet been identified among the vast majority of elements in the genome that are translationally defective. However, since only a tiny fraction of HERV genomes have been characterized in detail, it is quite feasible that fully functional elements do exist. While no evidence of virus infectivity has been obtained, it remains possible that, with the appropriate cell lines and conditions to detect their replication, infectious endogenous human retroviruses will be isolated. Indeed, HERVs may resemble the murine xenotropic retroviruses that replicate best in cells from species far removed from their host.

D. Pathological Conditions

One obvious harmful consequence that ERV might have on the host organism is the insertional mutagenesis of cellular genes resulting from ERV transposition. Several mutations in mice have been identified that are caused by the insertion of retroviruslike elements into or adjacent to a gene (Jenkins *et al.*, 1981; Stoye *et al.*, 1988; Adachi *et al.*, 1993). In addition, numerous examples of retrotransposition of such elements resulting in either the loss of gene function or the upregulation of gene expression have been detected in mouse cell lines (Hawley *et al.*, 1982; Shell *et al.*, 1987; Blatt *et al.*, 1988; Kongsuwan *et al.*, 1989; Blankenstein *et al.*, 1990) (reviewed by Kuff and Lueders, 1988; Favor and Morawetz, 1992). Although the transposition of a HERV element has not yet been detected, insertions of other retroelements have been found in humans. Two cases of germ line insertions of a LINE-1 element into exon 14 of the factor VIII gene have been identified in unrelated patients with hemophilia A (Kazazian *et al.*, 1988). Furthermore, a germ line insertion of an Alu sequence into an intron of the NF-1 gene has been described in a patient with neurofibromatosis (Wallace *et al.*, 1991). Another *de novo* germ line Alu insertion has also been found in the coding region of the factor IX gene in a patient with hemophilia B (Vidaud *et al.*, 1993). In addition, somatic insertions of LINE-1 elements into intron 2 of the *myc* gene and the last exon of the adenomatous polyposis coli (*APC*) gene have been reported (Morse *et al.*, 1988; Miki *et al.*, 1992). In these latter two cases, the inserted element was found only in tumor samples but not in normal tissue from the same individual. Therefore, it is possible that the transpositional insertion of a retroelement played a role in the etiology of the cancers in those patients.

As previously mentioned, recombination events mediated by repetitive sequences such as ERV can also result in disease. For example, several deletions and duplications in the low-density lipoprotein receptor gene result from recombination between Alu sequences normally present in the loci (Hobbs *et al.*, 1986; Lehrman *et al.*, 1987). A deletion in the adenosine deaminase (ADA) gene has also occurred via homologous recombination between Alu elements (Markert *et al.*, 1988). Some HERV may recombine with nonhomologous sequences, as is the case for deletions involving a THE-1 element in the dystrophin locus (Pizzuti *et al.*, 1992), or may possibly predispose a genomic region to delete, as postulated for the HERV-H element downstream of the β-globin locus (Mager *et al.*, 1985).

Recombination between ERV and infectious retroviruses is also possible and can generate infectious recombinant virus with altered receptor specificity, tissue tropism, and pathogenicity (Weiss *et al.*, 1973; Katz and Skalka, 1990; Coffin, 1992a). Recombination between virion sequences usually requires copackaging of each genome into a heterozygous particle. Because of this, as well as a requirement for homologous pairing at the point of recombination, recombination of infectious retroviruses with ERV is most likely in species that contain ERV very closely related to the infectious retroviruses of that species (e.g., mice, cats, and domestic chickens) (Weiss *et al.*, 1973; Schwartzberg *et al.*, 1985; Martinelli and Goff, 1990; Pandey *et al.*, 1991). In contrast to those species, none of the sequences characterized in the human genome are as closely related to the presently recognized human infectious retroviruses, reducing the likelihood of recombination between HERV and human exogenous retroviruses. However, nonhomologous recombination between retroviruses and cellular sequences is a well-documented, albeit rare event and has occurred in at least one instance between an endogenous and an exogenous retrovirus (see Katz and Skalka, 1990; Swain and Coffin, 1992). Therefore, it is conceivable that HERV sequences containing a long ORF could recombine with infectious human retroviruses, to create a virus with modified replicative and pathogenic properties. A recombinant virus would need to possess a strong selective advantage for it to establish itself in the face of overwhelming competition from the precursor infectious form, and no such recombinants have been detected among humans.

One school of thought regarding ERVs postulates that they have evolved relatively benign interactions with the host species, thereby ensuring their continued propagation by the success of the species (Doolittle *et al.*, 1989). In this light, the activation of ERVs under conditions of stress might be regarded as a "bailout" mechanism, whereby the virus seeks self-preservation by abandoning its present host for a better-adapted individual within the same species, or even of a different species. Cross-species transmission of ERVs appears to have occurred several times during the course of evolution and is also thought to have occurred among contemporary infectious retroviruses (Doolittle *et al.*, 1989). The result of a cross-species infection might be expected to be pathogenic since the virus and the new host species would not have had time to evolve a benign symbiotic relationship. Thus, ERVs in general may constitute a reservoir of potential infectious agents that may on rare occasions be triggered to infect and cause disease in other organisms. The fate of ERVs that reside quiescent for too long would be to fall victim to random genetic mutations inactivating their essential gene functions. It appears that most HERVs fall into this latter category.

The causal relationship between infectious retroviruses and tumorgenesis in their hosts suggests that some ERVs may also be associated with cancer. This hypothesis has been confirmed in certain strains of mice, bred for their genetic predisposition to develop tumors. In those mice, replication-competent (i.e., infectious) endogenous retroviruses have been shown to play a central role in tumor induction via the insertional mutagenesis of proto-oncogenes (see Stoye and Coffin, 1985; Favor and Morawetz, 1992). In humans however, data on the importance of HERV to cancer etiology remain circumstantial. Retroviruslike particles and retrovirus-related antigens have frequently been detected in primary tumor samples and in tumor-derived cell lines in the absence of detectable infectious virus (see Sections IV.C.1 and 2) (reviewed in: Weiss, 1984; Faff *et al.*, 1992). In addition, antibodies reactive with retroviral proteins have been demonstrated in the sera of some cancer patients (see Section IV.C.1) (Weiss, 1984). Those findings correlate with reports that HERV sequences are expressed most strongly in certain tumor-derived cell lines, in particular teratocarcinoma cell lines (reviewed in Section III), and may be activated in some cancers (McClain and Wilkowski, 1985; Gattoni-Celli *et al.*, 1986). At the molecular level, two HERVs have been implicated in cancer-causing events. It is possible that the HERV-R sequence is involved in the tumorigenesis of choriocarcinomas via a mechanism that involves downregulation of an adjacent cellular gene (Cohen *et al.*, 1988; Kato *et al.*, 1988, 1990). In the other example, a HERV-H element has been found to drive the expression of a calbindin-related gene in a cell line derived from a prostate metastasis. It has been suggested that the aberrant expression of that gene may have contributed to the progression of the original tumor (Liu and Abraham, 1991). Yet no HERV sequences capable of producing infectious virions have been discovered, nor have any HERV transpositions been characterized, leaving the possible mechanism of HERV involvement in cancer open to speculation and further investigation.

A possible role for ERV in autoimmune diseases has also been suggested, but again, evidence in humans is circumstantial, whereas mouse models are better characterized (for reviews see Abraham and Khan, 1990; Krieg *et al.*, 1992). Autoimmune diseases are, in part, characterized by the spectrum of autoantibodies that are formed by patients afflicted with different diseases. Two characteristic antigens recognized by autoantibodies in systemic lupus erythematosus (SLE) and Sjögren's syndrome have been found to share homology and cross-reactivity with retroviral *gag* proteins. The Sm antigen recognized by sera from 50% of SLE patients shares epitope determinants with p24 of HIV (Garry, 1990; Krieg *et al.*, 1992). Likewise, the SS-B/La antigen, recognized by more than 60% of Sjögren's syndrome patients, has homology with feline sarcoma virus capsid protein in an autoepitope sequence (Garry, 1990). Other cellular proteins have been shown to immunologically cross-react with retroviral proteins via regions of short amino acid homology. For example, the 70-kDa protein associated with U1 small nuclear ribonucleoproteins (snRNPs) cross-reacts with the p30 *gag* protein of MLV (Query and Keene, 1987), and the heterogeneous nuclear ribonucleoparticle protein A1 cross-reacts with gp70 *env* protein of MLV (Trauger *et al.*, 1990). In addition, several other cellular proteins that contain regions of sequence homology with retroviral proteins have been identified, such as the rat transcription factor FE65 and retroviral integrases (Duilio *et al.*, 1991), the erythroid-potentiating activity factor (EPA) and Rous sarcoma virus p27 *gag* protein (Patarca and Haseltine, 1985), and topoisomerase-I and feline sarcoma virus capsid protein (Garry, 1990). Therefore, it has been suggested that an immune response initiated against the aberrant expression of a HERV element, could trigger an autoimmune response by cross-reacting with immunologically related cellular proteins. Expression of HERV at the protein level in Sjögren's syndrome patients has been suggested by the detection of intracisternal A-type particles in cells of a T-lymphoblastoid cell line cocultivated with salivary gland tissue samples from two of six patients (Garry *et al.*, 1990). Other studies have found that a specific RFLP allele of HERV-R is disproportionately common among rheumatoid arthritis patients (Rubin *et al.*, 1991) but the significance of that finding remains to be determined. In addition, it has been suggested that other diseases, such as muscular dystrophy, psoriasis, and possibly Creutzfeldt–Jakob disease, may have HERV-associated etiologies (Iversen, 1990; Murdoch *et al.*, 1990; Haahr, 1992).

V. CONCLUDING REMARKS

Research on HERVs, which began 20 years ago with observations of retroviruslike particles associated with uninfected human tissues or cells, entered an accelerated phase in the 1980s when retroviruslike elements were detected and cloned from normal human DNA. Indeed, there is now strong evidence (discussed in Section III.B) that at least some of the particles are in fact encoded by an endogenous source, namely the HERV-K family. It is thus satisfying to see the observational and molecular genetic studies on HERVs begin to merge. A number of potential HERV biological roles, which were discussed above, are shown in Table VII. Such a list is useful to illustrate that the DNA sequences themselves or their encoded proteins may be of functional significance. At the DNA level, a growing number of studies implicate HERVs and other retroelements in gene

TABLE VII. Possible Biological Roles of HERV and Other Human Retroelements

Function	Example/evidence	Reference
DNA Sequences		
Promoter of cellular gene	HERV-H, *PLA2L* gene	Feuchter-Murthy *et al.* (1993)
	HERV-R, *plk* gene	Kato *et al.* (1990)
Polyadenylate cellular gene transcripts	HERV-H, *PLT* gene	Goodchild *et al.* (1992)
Regulatory region of cellular gene	HERV-E, salivary amylase genes	Samuelson *et al.* (1990), Ting *et al.* (1992)
	HERV-I, cytochrome c_1 gene	Suzuki *et al.* (1990)
Cancer etiology/progression	HERV-H dysregulation of calbindin gene	Liu and Abraham (1991)
	LINE-1 disruption of *myc* gene	Morse *et al.* (1988)
	LINE-1 disruption of *APC* gene	Miki *et al.* (1992)
Promote genomic recombination	HERV-K solo LTRs, HLA-DQ deletion	Kambhu *et al.* (1990)
	THE-1, dystrophin deletions	Pizzuti *et al.* (1992)
	Alu-mediated gene duplications	Barsh *et al.* (1983), Erickson *et al.* (1992)
	LINE-1-mediated γ-globin gene duplication	Fitch *et al.* (1991)
	Alu-mediated deletion mutations	Hobbs *et al.* (1986), Markert *et al.* (1988)
Recombination with exogenous retroviruses	Analogous events in mice and chickens	Coffin (1992a)
Insertional mutagen	Alu insertions	Wallace *et al.* (1991), Vidaud *et al.* (1993)
	LINE-1 insertions	Kazazian *et al.* (1988), Morse *et al.* (1988), Miki *et al.* (1992)
Proteins		
Syncytiotrophoblast development	HERV-R expression, fusion activity of envelope protein	Larsson *et al.* (1993)
Fetal allograft tolerance	HERV-R expression	Larsson *et al.* (1993)
	Immunosuppressive activity of envelope protein	Johnson *et al.* (1990)
Autoimmunity	HRES-1 antigens in patients	Banki *et al.* (1992)
	Retroviruslike particles in Sjögren's syndrome patients	Garry *et al.* (1990)
Reverse flow of genetic information	Existence of retrogenes and processed pseudogenes	See Section IV.A, Table VI

regulatory and genetic recombination events (Table VII). These studies, along with the ancient origins and abundance of human retroelements, suggest that these sequences have an evolutionary role in the generation of genetic diversity. Furthermore, as in other species, human retroelements can also act as insertional mutagens resulting in disease.

Possible functional roles of HERV proteins are more speculative at this time. The RNA expression of HERV elements is common, being in general most readily detectable in placenta, teratocarcinoma cells, and other tumor cell lines (see Table IV). High levels of HERV transcription in fetal-derived cells and cells of germ line origin is not unexpected since germ line expression is a prerequisite for the inheritance and subsequent fixation of new insertional events within a population. This expression pattern is also suggestive of potential roles that HERVs may play in early development (Table VII). While most elements are translationally defective, recent evidence indicates that some of them are coding competent. The elucidation of potential HERV protein involvement in normal biological processes, as well as in diseases such as cancer and autoimmune disorders, will depend upon the characterization of these proteins and the particular elements that encode them. As work towards this end is still in its early stages, the coming years promise to yield important and intriguing results from the investigation of HERV.

ACKNOWLEDGMENTS. We are grateful to numerous colleagues who provided "in press" manuscripts and other information relevant to this review. Special thanks to Dr. Klaus Boller for providing an electron micrograph of HTDV particles (Fig. 8). We also thank Christine Kelly and Rebecca Day for help with manuscript preparation. This review was modified and expanded from the introductory chapter of the Ph.D. thesis written by D.A.W. at the University of British Columbia. D.A.W. was a Research Student and D.L.M. is a Research Scholar of the Medical Research Council (MRC) of Canada. Work from the laboratory of D.L.M. was supported by grants from the MRC of Canada, the British Columbia Health Research Foundation, and the National Cancer Institute of Canada, with core support provided by the British Columbia Cancer Agency. J.-A.C.L. acknowledges the support of Bonneville Power Administration Contract DE-FG79-88BP92431, Project No. 88-152; United States Department of Agriculture to the Western Regional Aquaculture Consortium under Grant No. 92-38500-7195, Project No. 9208044-1; USDA Grant Nos. 91-34123-6124 and 92-37204-7997; and Oregon Sea Grant with funds from the National Oceanic and Atmospheric Administration, Office of Sea Grant, Department of Commerce, under Grant No. NA89AA-D-SG108, Project No. R/FSD-16.

VI. REFERENCES

Abraham, H. N., and Khan, A. S., 1990, Human endogenous retroviruses and immune disease, *Clin. Immunol. Immunopathol.* **56**:1.

Adachi, M., Watanabe-Fukunaga, R., and Nagata, S., 1993, Aberrant transcription caused by the insertion of an early transposable element in an intron of the Fas antigen gene of *lpr* mice, *Proc. Natl. Acad. Sci. USA* **90**:1756.

Adams, S., Mellor, J., Gull, K., Sim, R., and Tuite, M., 1987, The functions and relationships of Ty-VLP proteins in yeast reflect those of mammalian proteins, *Cell* **49**:111.

Adler, A. J., Scheller, A., Hoffman, Y., and Robins, D. M., 1991, Multiple components of a complex androgen-dependent enhancer, *Mol. Endocrinol.* **5**:1587.

Al-Sumidaie, A. M., Hart, C. A., Leinster, S. J., Green, C. D., and McCarthy, K., 1988, Particles with properties of retroviruses in monocytes from patients with breast cancer, *Lancet* **2**:5.

Anagnou, N., Antonarakis, S., O'Brien, S., Modi, W., and Nienhuis, A., 1988, Chromosomal localization and racial distribution of the polymorphic human dihydrofolate reductase pseudogene (DHFRPI), *Am. J. Hum. Genet.* **42**:345.

Anderson, G. R., Stoler, D. L., Scott, J. P., and Farkas, B. K., 1988, Induction of VL30 element expression as a response to anoxic stress, in: *Branbury Reports 30: Eukaryotic Transposable Elements as Mutagenic Agents* (M. Lambert, J. McDonald, and I. Weinstein, eds.), pp. 265–274, Cold Spring Harbor Laboratory, Cold Spring Harbor, N.Y.

Armour, J., Wong, Z., Wilson, V., Royle, N., and Jeffreys, A., 1989, Sequences flanking the repeat arrays of human minisatellites: Association with tandem and dispersed repeat elements, *Nucleic Acids Res.* **17**:4925.

Arnault, C., Heizmann, A., Loevenbruck, C., and Biémont, C., 1991, Environmental stresses and mobilization of transposable elements in inbred lines of *Drosophila melanogaster*, *Mutat. Res.* **248**:51.

Ashworth, A., Skene, B., Swift, S., and Lovell-Badge, R., 1990, Zfa is an expressed retroposon derived from an alternative transcript of the *Zfx* gene, *EMBO J.* **9**:1065.

Astrin, S. M., 1978, Endogenous viral genes of White Leghorn chickens: Common site of residence and sites associated with specific phenotypes of viral gene expression, *Proc. Natl. Acad. Sci. USA* **75**:5941.

Astrin, S. M., Buss, E. G., and Hayward, W. S., 1979, Endogenous viral genes are non-essential in the chicken, *Nature* **282**:339.

Axel, R., Gulati, S. C., and Spiegelman, S., 1972, Particles containing RNA-instructed DNA polymerase and virus-related RNA in human breast cancer, *Proc. Natl. Acad. Sci. USA* **69**:3133.

Baker, C., Phelps, W., Lindgren, V., Braun, M., Gonda, M., and Howley, P., 1987, Structural and transcriptional analysis of human papillomavirus type 16 sequences in cervical carcinoma cell lines, *J. Virol.* **61**:962.

Baltimore, D., 1970, RNA-dependent DNA polymerase in virions of RNA tumour viruses, *Nature* **226**:1209.

Baltimore, D., 1985, Retroviruses and retrotransposons: The role of reverse transcription in shaping the eukaryotic genome, *Cell* **40**:481.

Bangham, C., Daenke, S., Phillips, R., Cruickshank, J., and Bell, J., 1988, Enzymatic amplification of exogenous and endogenous retroviral sequences from DNA of patients with tropical spastic paraparesis, *EMBO J.* **7**:4179.

Banki, K., Maceda, J., Hurley, E., Ablonczy, E., Mattson, D., Szegedy, L., Hung, C., and Perl, A., 1992, Human T-cell lymphotropic virus (HTLV)-related endogenous sequence, HRES-1, encodes a 28-kDa protein: A possible autoantigen for HTLV-I *gag*-reactive autoantibodies, *Proc. Natl. Acad. Sci. USA* **89**:1939.

Banville, D., and Boie, Y., 1989, Retroviral long terminal repeat is the promoter of the gene encoding the tumor-associated calcium-binding protein oncomodulin in the rat, *J. Mol. Biol.* **207**:481.

Barker, C. S., Wills, J. W., Bradac, J. A., and Hunter, E., 1985, Molecular cloning of the Mason–Pfizer monkey virus genome: Characterization and cloning of subgenomic fragments, *Virology* **142**:223.

Barsh, G. S., Seeburg, P. H., and Gelinas, R. E., 1983, The human growth hormone gene family: Structure and evolution of the chromosomal locus, *Nucleic Acids Res.* **11**:3939.

Batzer, M. A., Gudi, V. A., Mena, J. C., Foltz, D. W., Herrera, R. V., and Deininger, P. L., 1991, Amplification dynamics of human-specific (HS) alu family members, *Nucleic Acids Res.* **19**:3619.

Benveniste, R., and Todaro, G., 1974, Evolution of type C viral genes: I. Nucleic acid from baboon type C virus as a measure of divergence among primate species, *Proc. Natl. Acad. Sci. USA* **71**:4513.

Blankenstein, T., Qin, Z., Li, W., and Diamantstein, T., 1990, DNA rearrangement and constitutive expression of the interleukin 6 gene in a mouse plasmacytoma, *J. Exp. Med.* **171**:965.

Blatt, C., Aberdam, D., Schwartz, R., and Sachs, L., 1988, DNA rearrangement of a homeobox gene in myeloid leukaemic cells, *EMBO J.* **7**:4283.

Boeke, J. D., and Corces, V., 1989, Transcription and reverse transcription of retrotransposons, *Annu. Rev. Microbiol.* **43**:403.

Boer, P. H., Adra, C. N., Lau, Y.-F., and McBurney, M. W., 1987, The testis-specific phosphoglycerate kinase gene *pgk*-2 is a recruited retroposon, *Mol. Cell. Biol.* **7**:3107.

Boller, K., Frank, H., Löwer, J., Löwer, R., and Kurth, R., 1983, Structural organization of unique retrovirus-like particles budding from human teratocarcinoma cell lines, *J. Gen. Virol.* **64**:2549.

Boller, K., Konig, H., Sauter, M., Mueller-Lantzsch, N., Löwer, R., Löwer, J., and Kurth, R., 1993, Evidence that HERV-K is the endogenous retrovirus sequence that codes for the human teratocarcinoma derived retrovirus HTDV, *Virology* **196**:349.

Bonner, T., O'Connell, C., and Cohen, M., 1982, Cloned endogenous retroviral sequences from human DNA, *Proc. Natl. Acad. Sci. USA* **79**:4709.

Boyd, M. T., Bax, C. M. R., Bax, B. E., Bloxam, D. L., and Weiss, R., 1993, The human endogenous retrovirus ERV-3 is upregulated in differentiating placental trophoblast cells. *Virology* **196**:905.

Boyd, M. T., Maclain, N., and Oscier, D. G., 1989, Detection of retrovirus in patients with myeloproliferative disease, *Lancet* **1**:814.

Brack-Werner, R., Barton, D., Werner, T., Foellmer, B., Leib-Mösch, C., Francke, U., Erfle, V., and Hehlmann, R., 1989a, Human SSAV-related endogenous retroviral element: LTR-like sequence and chromosomal localization to 18q21, *Genomics* **4**:68.

Brack-Werner, R., Leib-Mösch, C., Werner, T., Erfle, V., and Hehlmann, R., 1989b, Human endogenous retrovirus-like sequences. in: *Modern Trends in Human Leukemia VIII* (R. Neth, ed.), pp. 464–477, Springer-Verlag, Berlin.

Brack-Werner, R., Sander, I., Leib-Mösch, C., Ohlmann, M., Erfle, V., and Werner, T., 1993, Endogenous retroviral elements as tools for phylogenetic analysis, *J. Cancer Res. Clin. Oncol.* **119**(Suppl. 1):S4.

Breindl, M., Doehmer, J., Willecke, K., Dausman, J., and Jaenisch, R., 1979, Germline integration of Moloney leukemia virus: Identification of the chromosomal integration site, *Proc. Natl. Acad. Sci. USA* **76**:1938.

Brodsky, I., Foley, B., Haines, D., Johnston, J., Cuddy, K., and Gillespie, D., 1993, Expression of HERV-K proviruses in human leukocytes. *Blood* **81**:2369.

Bronson, D., Fraley, E., Fogh, J., and Kalter, S., 1979, Induction of retrovirus particles in human testicular tumor (Tera-1) cell cultures: An electron microscopic study, *J. Natl. Cancer Inst.* **63**:337.

Bronson, D. L., Saxinger, W. C., Ritzi, D. M., and Fraley, E. E., 1984, Production of virions with retrovirus morphology by human embryonal carcinoma cells in vitro, *J. Gen. Virol.* **65**:1043.

Call, K. M., Glaser, T., Ito, C. Y., Buckler, A. J., Pelletier, J., Haber, D. A., Rose, E. A., Kral, A., Yeger, H., Lewis, W. H., Jones, C., and Houseman, D. E., 1990, Isolation and characterization of a zinc finger polypeptide gene at the human chromosome 11 Wilms' tumor locus, *Cell* **60**:509.

Callahan, R., 1988, Two families of human endogenous retroviral genomes, in: *Branbury Reports 30: Eukaryotic Transposable Elements as Mutagenic Agents* (M. Lambert, J. McDonald, and I. Weinstein, eds.), pp. 91–100, Cold Spring Harbor Laboratory, Cold Spring Harbor, N.Y.

Callahan, R., Drohan, W., Tronick, S., and Schlom, J., 1982, Detection and cloning of human DNA sequences related to the mouse mammary tumor virus genome, *Proc. Natl. Acad. Sci. USA* **79**:5503.

Callahan, R., Chiu, I.-M., Tronick, S., Roe, B., Aaronson, S., and Schlom, J., 1985, A new class of endogenous human retroviral genomes, *Science* **228**:1208.

Chang-Yeh, A., Mold, D. E., and Huang, R. C. C., 1991, Identification of a novel murine IAP-promoted placenta-expressed gene, *Nucleic Acids Res.* **19**:3667.

Chattopadhyay, S. K., Rowe, W. P., Teich, N. M., and Lowy, D. R., 1975, Definitive evidence that the murine C-type virus inducing locus Akv-1 is viral genetic material, *Proc. Natl. Acad. Sci. USA* **72**:906.

Chattopadhyay, S. K., Lander, M. R., Gupta, S., Rands, E., and Lowy, D., 1981, Origin of mink cytopathic focus-forming (MCF) viruses: Comparison with ecotropic and xenotropic murine leukemia virus genomes, *Virology* **113**:465.

Chien, Y., Lai, M., Shih, T. Y., Verma, I. M., Scolnick, E. M., Roy-Burman, P., and Davidson, N., 1979, Heteroduplex analysis of the sequence relationships between the genomes of Kirsten and Harvey sarcoma viruses, their respective parental murine leukemia viruses and the rat endogenous 30S RNA, *J. Virol.* **31**:752.

Chiu, I.-M., Andersen, P. R., Aaronson, S. A., and Tronick, S. R., 1983, Molecular cloning of the unintegrated squirrel monkey retrovirus genome: Organization and distribution of related sequences in primate DNAs, *J. Virol.* **47**:434.

Chumakov, I., Stuhlmann, H., Harbers, K., and Jaenisch, R., 1982, Cloning of two genetically transmitted Moloney leukemia proviral genomes: Correlation between biological activity of the cloned DNA and viral genome activation in the animal, *J. Virol.* **42**:1088.

Chumakov, I. M., Zabarovskii, E. R., Prasolov, V. S., Mett, V. L., and Kiselev, L. L., 1985, Sections of human genome containing analogues of oncogenes and retroviral genes: A new class of retroviral-like elements, *Mol. Biol.* **19**:351.

Clark, B. D., Collins, K. L., Gandy, M. S., Webb, A. C., and Auron, P. E., 1986, Genomic sequence for human prointerleukin 1 beta: Possible evolution from a reverse transcribed prointerleukin 1 alpha gene, *Nucleic Acids Res.* **14**:7897.

Coffin, J., 1984, Endogenous retroviruses, in: *RNA Tumor Viruses,* 2nd ed., Vol. 1 (R. Weiss, N. Teich, H. Varmus, and J. Coffin, eds.), pp. 1109–1203, Cold Spring Harbor Laboratory, Cold Spring Harbor, N.Y.

Coffin, J., 1992a, Genetic diversity and evolution of retroviruses, *Curr. Top. Microbiol. Immunol.* **176**:143.

Coffin, J., 1992b, Structure and classification of retroviruses, in: *The Retroviridae,* Vol. 1 (J. A. Levy, ed.), pp. 19–49, Plenum Press, New York.

Cohen, M., and Larsson, E., 1988, Human endogenous retroviruses, *Bioessays* **9**:191.

Cohen, M., Powers, M., O'Connell, C., and Kato, N., 1985, The nucleotide sequence of the *env* gene from the human provirus ERV3 and isolation and characterization of an ERV3-specific cDNA, *Virology* **147**:449.

Cohen, M., Kato, N., and Larsson, E., 1988, ERV3 human endogenous provirus mRNAs are expressed in normal and malignant tissues and cells, but not in choriocarcinoma tumor cells, *J. Cell. Biochem.* **36**:121.

Colicelli, J., and Goff, S. P., 1986, Isolation of a recombinant murine leukemia virus utilizing a new primer tRNA, *J. Virol.* **57**:37.

Comb, M., Rosen, H., Seeburg, P., Adelman, J., and Herbert, E., 1983, Primary structure of the human proenkephalin gene, *DNA* **3**:213.

Craig, L. C., Pirtle, I. L., Gracy, R. W., and Pirtle, R. M., 1991, Characterization of the transcription unit and two processed pseudogenes of chimpanzee triosephosphate isomerase (TPI), *Gene* **99**:217.

Craigie, R., 1992, Hotspots and warm spots: Integration specificity of retroelements, *Trends Genet.* **8**:187.

Crittenden, L. B., and Astrin, S. M., 1981, Genes, viruses, and avian leukosis, *BioScience* **31**:305.

Crittenden, L. B., Fadly, A. M., and Smith, E. J., 1982, Effect of endogenous leukosis virus genes on response to infection with avian leukosis and reticuloendotheliosis viruses, *Avian Dis.* **26**:279.

Crittenden, L. B., Astrin, S. M., and Smith, E. J., 1983, Independent segregation of ev-10 and ev-11, genetic loci for spontaneous production of endogenous avian retroviruses, *Virology* **129**:514.

Dahl, H.-H. M., Brown, R. M., Hutchison, W. M., Maragos, C., and Brown, G. K., 1990, A testis-specific form of the human pyruvate dehydrogenase E1a subunit is coded for by an intronless gene on chromosome 4, *Genomics* **8**:225.

Dalton, A. J., Hellman, A., Kalter, S. S., and Helmke, R. J., 1974, Ultrastructural comparison of placental virus with several type-C oncogenic viruses, *J. Natl. Cancer Inst.* **52**:1379.

Deen, K., and Sweet, R., 1986, Murine mammary tumor virus *pol*-related sequences in human DNA: Characterization and sequence comparison with the complete murine mammary tumor virus *pol* gene, *J. Virol.* **57**:422.

Deininger, P., 1989, SINEs: Short interspersed repeated DNA elements in higher eucaryotes, in: *Mobile DNA* (D. Berg and M. Howe, eds.), pp. 619–636, American Society for Microbiology, Washington, D.C.

Deininger, P. L., Batzer, M. A., Hutchison, C. A., III, and Edgell, M. H., 1992, Master genes in mammalian repetitive DNA amplification, *Trends Genet.* **8**:307.

Deka, N., Willard, C., Wong, E., and Schmid, C., 1988a, Human transposon-like elements insert at a preferred target site: Evidence for a retrovirally mediated process, *Nucleic Acids Res.* **16**:1143.

Deka, N., Wong, E., Matera, A., Kraft, R., Leinwand, L., and Schmid, C., 1988b, Repetitive nucleotide sequence insertions into a novel calmodulin-related gene and its processed pseudogene, *Gene* **71**:123.

Deragon, J., Sinnett, D., and Labuba, D., 1990, Reverse transcriptase activity from embryonal carcinoma cells, *EMBO J.* **9**:3363.

Derks, J. P. A., Homans, L., Bruning, H. W., and von Rood, J. J., 1982, Synthesis of a viral protein with molecular weight of 30,000 (p30) by leukemic cells and antibodies cross-reacting with simian sarcoma virus p30 in serum of a chronic myeloid leukemia patient, *Cancer Res.* **42**:618.

Desai, S., Kalyanaraman, V., Casey, J., Srinivasan, A., Andersen, P., and Devare, S., 1986, Molecular cloning and primary nucleotide sequence analysis of a distinct human immunodeficiency virus isolate reveal significant divergence in its genomic sequences, *Proc. Natl. Acad. Sci. USA* **83**:380.

Dirksen, E. R., and Levy, J. A., 1977, Virus-like particles in placentas from normal individuals and patients with systemic lupus erythematosus, *J. Natl. Cancer Inst.* **59**:1187.

Dombroski, B., Mathias, S., Nanthakumar, E., Scott, A., and Kazazian, H., 1991, Isolation of an active human transposable element, *Science* **254**:1805.

Doolittle, R., Feng, D.-F., Johnson, M., and McClure, M., 1989, Origins and evolutionary relationships of retroviruses, *Q. Rev. Biol.* **64**:1.

Doolittle, W. F., and Sapienza, C., 1980, Selfish genes, the phenotype paradigm and genome evolution, *Nature* **284**:601.

Duilio, A., Zambrano, N., Mogavero, A. R., Ammendola, R., Cimino, F., and Russo, R., 1991, A rat brain mRNA encoding a transcriptional activator homologous to the DNA binding domain of retroviral integrases, *Nucleic Acids Res.* **19**:5269.

Dunwiddie, C. T., and Faras, A. J., 1985, Presence of retrovirus reverse transcriptase-related gene sequences in avian cells lacking endogenous leukosis viruses, *Proc. Natl. Acad. Sci. USA* **82**:5097.

Dunwiddie, C. T., Resnick, R., Boyce-Jacino, M., Alegre, J. N., and Faras, A. J., 1986, Molecular cloning and characterization of *gag-*, *pol-*, and *env*-related gene sequences in the ev^- chicken, *J. Virol.* **59**:669.

Echalier, G., 1989, *Drosophila* retrotransposons: Interactions with genome, *Adv. Virus Res.* **36**:33.

Eickbush, T. H., 1992, Transposing without ends: The non-LTR retrotransposable elements, *New Biol.* **4**:430.

Ellis, R. W., DeFeo, D., Maryak, J. M., Young, H. A., Shih, T. Y., Chang, E. H., Lowy, D. R., and Scolnick, E. M., 1980, Dual evolutionary origin for the rate genetic sequences of Harvey murine sarcoma virus, *J. Virol.* **36**:408.

Emi, M., Horii, A., Tomita, N., Nishide, T., Ogawa, M., Mori, T., and Matsubara, K., 1988, Overlapping two genes in human DNA: A salivary amylase gene overlaps with a gamma-actin pseudogene that carries an integrated human endogenous retroviral DNA, *Gene* **62**:229.

Erickson, L. M., Kim, H. S., and Maeda, N., 1992, Junctions between genes in the haptoglobin gene cluster in primates, *Genomics* **14**:948.

Evans, L. H., and Cloyd, M. W., 1984, Generation of mink cell focus forming viruses by Friend murine leukemia virus: Recombination with specific endogenous proviral sequences, *J. Virol.* **49**:772.

Evans, L. H., and Cloyd, M. W., 1985, Friend and Moloney murine leukemia viruses specifically recombine with different endogenous retroviral sequences to generate mink cell focus-forming viruses, *Proc. Natl. Acad. Sci. USA* **82**:459.

Faff, O., Murray, A. B., Schmidt, J., Leib-Mösch, C., Erfle, V., and Hehlmann, R., 1992, Retrovirus-like particles from the human T47D cell line are related to mouse mammary tumour virus and are of human endogenous origin, *J. Gen. Virol.* **73**:1087.

Famulari, N. G., 1983, Murine leukemia viruses with recombinant *env* genes: A discussion of their role in leukemogenesis, *Curr. Top. Microbiol. Immunol.* **103**:75.

Favor, J., and Morawetz, C., 1992, Insertional mutations in mammals and mammalian cells, *Mutat. Res.* **284**:53.

Feuchter, A., and Mager, D., 1990, Functional heterogeneity of a large family of human LTR-like promoters and enhancers, *Nucleic Acids Res.* **18**:1261.

Feuchter, A., and Mager, D., 1992, SV40 large T antigen trans-activates the long terminal repeats of a large family of human endogenous retrovirus-like sequences, *Virology* **187**:242.

Feuchter, A., Freeman, D., and Mager, D., 1992, Strategy for detecting cellular transcripts promoted by human endogenous long terminal repeats: Identification of a novel gene (CDC4L) with homology to yeast CDC4, *Genomics* **13**:1237.

Feuchter-Murthy, A. E., Freeman, J. D., and Mager, D. L., 1993, Splicing of a human endogenous retrovirus to a novel phospholipase A_2 related gene, *Nucleic Acids Res.* **21**:135.

Fields, C., Grady, D., and Moyzis, R., 1992, The human THE-LTR(O) and MstII interspersed repeats are subfamilies of a single widely distributed highly variable repeat family, *Genomics* **13**:431.

Fink, G. R., 1988, Pseudogenes in yeast? *Cell* **49**:5.

Finnegan, D., 1989, Eukaryotic transposable elements and genome evolution, *Trends Genet.* **5**:103.

Fischinger, P. J., Nomura, S., and Bolognesi, D. P., 1975, A novel murine oncornavirus with dual eco- and xenotropic properties, *Proc. Natl. Acad. Sci. USA* **72**:5150.

Fitch, D., Bailey, W., Tagle, D., Goodman, M., Sieu, L., and Slightom, J., 1991, Duplication of the γ-globin gene mediated by L1 long interspersed repetitive elements in an early ancestor of simian primates, *Proc. Natl. Acad. Sci. USA* **88**:7396.

Fourel, G., Trepo, C., Bougueleret, L., Henglein, B., Ponzetto, A., Tiollais, P., and Buendia, M.-A., 1990, Frequent activation of N-*myc* genes by hepadnavirus insertion in woodchuck liver tumours, *Nature* **347**:294.

Franklin, G., Chretien, S., Hanson, I., Rochefort, H., May, F., and Westley, B., 1988, Expression of human sequences related to those of mouse mammary tumor virus, *J. Virol.* **62**:1203.

Fraser, C., Humphries, R., and Mager, D., 1988, Chromosomal distribution of the RTVL-H family of human endogenous retrovirus-like sequences, *Genomics* **2**:280.

Fredholm, M., Policastro, P. F., and Wilson, M., 1991, The dispersion of defective endogenous murine retroviral elements suggests retrotransposition-mediated amplification, *DNA Cell Biol.* **10**:713.

Furano, A., Somerville, C., Tsichlis, P., and D'Ambrosio, E., 1986, Target sites for the transposition of rat long interspersed repeated DNA elements (LINEs) are not random, *Nucleic Acids Res.* **14**:3717.

Furter, C. A., Heizmann, C., and Berchtold, M. W., 1989, Isolation and analysis of a rat genomic clone containing a long terminal repeat with high similarity to the oncomodulin mRNA leader sequence, *J. Biol. Chem.* **264**:18276.

Garfinkel, D., 1992, Retroelements in microorganisms, in: *The Retroviridae*, Vol. 1 (J. A. Levy, ed.), pp. 107–158, Plenum Press, New York.

Garfinkel, D., Boeke, J., and Fink, G., 1985, Ty element transposition: Reverse transcriptase and virus-like particles, *Cell* **42**:507.

Garry, R., 1990, Extensive antigenic mimicry by retrovirus capsid proteins, *AIDS Res. Hum. Retroviruses* **6**:1361.

Garry, R. F., Fermin, C. D., Hart, D. J., Alexander, S. S., Donehower, L. A., and Lou-Zhang, H., 1990, Detection of a human intracisternal A-type retroviral particle antigenically related to HIV, *Science* **250**:1127.

Gattoni-Celli, S., Kirsch, K., Kalled, S., and Isselbacher, K., 1986, Expression of type C-related endogenous retroviral sequences in human colon tumors and colon cancer cell lines, *Proc. Natl. Acad. Sci. USA* **83**:6127.

Gelb, L. D., Milstein, J. B., Martin, M. A., and Aaronson, S. A., 1973, Characterization of murine leukemia virus-specific DNA present in normal mouse cells. *Nature New Biol.* **244**:76.

Gingerich, P. D., 1984, Primate evolution: Evidence from the fossil record, comparative morphology, and molecular biology, *Yearbk. Phys. Anthropol.* **27**:57.

Golovkina, T., Chervonsky, A., Dudley, J., and Ross, S., 1992, Transgenic mouse mammary tumor virus superantigen expression prevents viral infection, *Cell* **69**:637.

Goodchild, N., Wilkinson, D., and Mager, D., 1992, A human endogenous long terminal repeat polyadenylates a novel alternatively spliced transcript in normal placenta, *Gene* **121**:287.

Goodchild, N., Wilkinson, D., and Mager, D., 1993, Recent evolutionary expansion of a subfamily of RTVL-H human endogenous retrovirus-like elements, *Virology* **196**:778.

Gruskin, K. S., Smith, T. F., and Goodman, M., 1987, Possible origin of a calmodulin gene that lacks intervening sequences, *Proc. Natl. Acad. Sci. USA* **84**:1605.

Haahr, S., 1992, Second international symposium on retrovirus in multiple sclerosis and related diseases, *J. Neuroimmunol.* **38**:163.

Harada, F., Tsukada, N., and Kato, N., 1987, Isolation of three kinds of human endogenous retrovirus-like sequences using $tRNA^{Pro}$ as a probe, *Nucleic Acids Res.* **15**:9153.

Harris, J., 1991, The evolution of placental mammals, *Fed. Eur. Biol. Soc.* **295**:3.

Hartley, J. W., Wolford, N. K., Old, L. J., and Rowe, W. P., 1977, A new class of murine leukemia virus associated with development of spontaneous leukemias, *Proc. Natl. Acad. Sci. USA* **74**:789.

Hattori, M., Kuhara, S., Takenaka, O., and Sakaki, Y., 1986, L1 family of repetitive DNA sequences in primates may be derived from a sequence encoding a reverse transcriptase-related protein, *Nature* **321**:625.

Hawley, R. G., Shulman, M. J., Murialdo, H. M., Gibson, D. M., and Hozumi, N., 1982, Mutant immunoglobulin genes have repetitive DNA elements inserted into their intervening sequences, *Proc. Natl. Acad. Sci. USA* **79**:7425.

Hehlmann, R., Schetters, H., Erfle, V., and Leib-Mösch, C., 1983, Detection and biochemical characterization of antigens in human leukemic sera that cross-react with primate C-type viral proteins (Mr 30,000), *Cancer Res.* **43**:392.

Hehlmann, R., Erfle, V., Schetters, H., Luz, A., Rohmer, H., Schreiber, M., Pralle, H., Essers, U., and Weber, E., 1984, Antigens and circulating immune complexes related to the primate retroviral glycoprotein SiSV gp70: Indicators of early mortality in human acute leukemias and chronic myelogenous leukemias in blast crisis, *Cancer* **54**:2927.

Heidmann, O., and Heidmann, T., 1991, Retrotransposition of a mouse IAP sequence tagged with an indicator gene, *Cell* **64**:159.

Hirose, Y., Takamatsu, M., and Harada, F., 1993, Presence of *env* genes in members of the RTVL-H family of human endogenous retrovirus-like elements, *Virology* **192**:52.

Hobbs, H. H., Brown, M. S., Goldstein, J. L., and Russell, S. W., 1986, Deletion of exon encoding cysteine rich repeat of low density lipoprotein receptor alters its binding specificity in a subject with familial hypercholesterolemia, *J. Biol. Chem.* **261**:13144.

Hoggan, M. D., Buckler, C. E., Sears, J. F., Rowe, W. P., and Martin, M. A., 1983, Organization and stability of endogenous xenotropic murine leukemia virus proviral DNA in mouse genomes, *J. Virol.* **45**:473.

Horn, T., Huebner, K., Croce, C., and Callahan, R., 1986, Chromosomal locations of members of a family of novel endogenous human retroviral genomes, *J. Virol.* **58**:955.

Horwitz, M. S., Boyce-Jacino, M. T., and Faras, A. J., 1992, Novel human endogenous sequences related to human immunodeficiency virus type I, *J. Virol.* **66**:2170.

Hull, R., and Will, H., 1989, Molecular biology of viral and nonviral retroelements, *Trends Genet.* **5**:357.

Ikeda, H., and Sugimura, H., 1989, *Fv-4* resistance gene: A truncated endogenous murine leukemia virus with ecotropic interference properties, *J. Virol.* **63**:5405.

Imamura, M., Phillips, P. E., and Mellors, R. C., 1976, The occurrence and frequency of type-C virus like particles in placentas from patients with systemic lupus erythematosus and from normal subjects, *Am. J. Pathol.* **83**:383.

Itoh, Y., Maruyama, N., Kitamura, M., Shirasawa, T., Shigemoto, K., and Koike, T., 1992, Induction of endogenous retroviral gene product (SU) as an acute-phase protein by IL-6 in murine hepatocytes, *Clin. Exp. Immunol.* **88**:356.

Iversen, O.-J., 1990, The expression of retrovirus-like particles in psoriasis, *J. Invest. Dermatol.* **95**:41S.

Jeffreys, A., Wilson, V., and Thein, S., 1985, Hypervariable minisatellite regions in human DNA, *Nature* **314**:67.

Jenkins, N. A., Copeland, N. G., Taylor, B. A., and Lee, B. K., 1981, Dilute (d) coat colour mutation of DBA/2J mice is associated with the site of integration of an ecotropic MuLV genome, *Nature* **293**:370.

Jerabek, L., Mellors, R., Elkon, K., and Mellors, J., 1984, Detection and immunochemical characterization of a primate type C retrovirus-related p30 protein in normal human placentas, *Proc. Natl. Acad. Sci. USA* **81**:6501.

Johansen, T., Holm, T., and Bjorklid, R., 1989, Members of the RTVL-H family of human endogenous retrovirus-like elements are expressed in placenta, *Gene* **79**:259.

Johnson, P. M., Lyden, T. W., and Mwenda, J. M., 1990, Endogenous retroviral expression in the human placenta, *Am. J. Reprod. Immunol.* **23**:115.

Kalter, S. S., Helmke, R. J., Heberling, R. L., Panigel, M., Fowler, A. K., Strickland, J. E., and Hellman, A., 1973, C-type particles in normal human placentas, *J. Natl. Cancer Inst.* **50**:1081.

Kambhu, S., Falldorf, P., and Lee, J., 1990, Endogenous retroviral long terminal repeats within the HLA-DQ locus, *Proc. Natl. Acad. Sci. USA* **87**:4927.

Kannan, P., Buettner, R., Pratt, D., and Tainsky, M., 1991, Identification of a retinoic acid-inducible endogenous retroviral transcript in the human teratocarcinoma-derived cell line PA-1, *J. Virol.* **65**:6343.

Kato, N., Pfeifer-Ohlsson, S., Kato, M., Larsson, E., Rydnert, J., Ohlsson, R., and Cohen, M., 1987, Tissue-specific expression of human provirus ERV3 mRNA in human placenta: Two of the three ERV3 mRNAs contain human cellular sequences, *J. Virol.* **61**:2182.

Kato, N., Larsson, E., and Cohen, M., 1988, Absence of expression of a human endogenous retrovirus is correlated with choriocarcinoma, *Int. J. Cancer* **41**:380.

Kato, N., Shimotohno, K., VanLeeuwen, D., and Cohen, M., 1990, Human proviral mRNAs down regulated in choriocarcinoma encode a zinc finger protein related to Kruppel, *Mol. Cell. Biol.* **10**:4401.

Katz, R., and Skalka, A., 1990, Generation of diversity in retroviruses, *Annu. Rev. Genet.* **24**:409.

Kazazian, H., Jr., Wong, C., Youssoufian, H., Scott, A., Phillips, D., and Antonarakis, S., 1988, Haemophilia A resulting from de novo insertion of L1 sequences represents a novel mechanism for mutation in man, *Nature* **332**:164.

Keshet, E., Schiff, R., and Itin, A., 1991, Mouse retrotransposons: A cellular reservoir of long terminal repeat (LTR) elements with diverse transcriptional specificities, *Adv. Cancer Res.* **56**:215.

Keydar, I., Chen, L., Karby, S., Weiss, F., Delarea, J., Radu, M., Chaitcik, S., and Brenner, H., 1979, Establishment and characterization of a cell line of human breast carcinoma origin, *Eur. J. Cancer* **15**:659.

Keydar, I., Ohno, T., Nayak, R., Sweet, R., Simoni, F., Weiss, F., Karby, S., Mesa-Tejada, R., and Spiegelman, S., 1984, Properties of retrovirus-like particles produced by a human breast carcinoma cell line: Immunological relationship with mouse mammary tumor virus proteins, *Proc. Natl. Acad. Sci. USA* **81**:4188.

Keyeux, G., Lefranc, G., and Lefranc, M.-P., 1989, A multigene deletion in the human IGH constant region locus involves highly homologous hot spots of recombination, *Genomics* **5**:431.

Khan, A. S., 1984, Nucleotide sequence analysis establishes the role of endogenous murine leukemia virus DNA segments in the formation of recombinant mink cell focus forming murine leukemia viruses, *J. Virol.* **50**:864.

Kongsuwan, K., Allen, J., and Adams, J. M., 1989, Expression of Hox-2.4 homeobox gene directed by proviral insertion in a myeloid leukemia, *Nucleic Acids Res.* **17**:1881.

Korbmacher, C., Konig, H., Boller, K., Löwer, R., Kurth, R., and Löwer, J., 1993, The *gag* gene of the endogenous retrovirus HERV-K is sufficient for particle production, *J. Cancer Res. Clin. Oncol.* **119**(Suppl. 1):S5.

Kozak, C. A., and Ruscetti, S., 1992, Retroviruses in rodents, in: *The Retroviridae*, Vol. 1 (J. A. Levy, ed.), pp. 405–481, Plenum Press, New York.

Krieg, A., and Steinberg, A., 1990, Retroviruses and autoimmunity, *J. Autoimmunity* **3**:137.

Krieg, A., Gause, W., Gourley, M., and Steinberg, A., 1989, A role for endogenous retroviral sequences in the regulation of lymphocyte activation, *J. Immunol.* **143**:2448.

Krieg, A. M., Gourley, M. F., and Perl, A., 1992, Endogenous retroviruses: Potential etiologic agents in autoimmunity, *FASEB J.* **6**:2537.

Kroger, B., and Horak, I., 1987, Isolation of novel human retrovirus-related sequences by hybridization to synthetic oligonucleotides complementary to the $tRNA^{Pro}$ primer-binding site, *J. Virol.* **61**:2071.

Kuff, E., and Lueders, K., 1988, The intracisternal A-particle gene family: Structure and functional aspects, *Adv. Cancer Res.* **51**:183.

Kuff, E., Mietz, J., Trounstine, M., Moore, K., and Martens, C., 1986, cDNA clones encoding murine IgE-binding factors represent multiple structural variants of intracisternal A-particle genes, *Proc. Natl. Acad. Sci. USA* **83**:6583.

Kurth, R., Löwer, J., Löwer, R., Harzman, R., Pfeiffer, R., Schmidt, R., Fogh, C., and Frank, H., 1980, Oncornavirus synthesis in human teratocarcinoma cultures and an increased antiviral immune reactivity in corresponding patients, in *Cold Spring Harbor Conferences on Cell Proliferation*, Vol. 7, pp. 835–846, Cold Spring Harbor Laboratory, Cold Spring Harbor, N.Y.

La Mantia, G., Pengue, G., Maglione, D., Pannuti, A., Pascucci, A., and Lania, L., 1989, Identification of new human repetitive sequences: Characterization of the corresponding cDNAs and their expression in embryonal carcinoma cells, *Nucleic Acids Res.* **17**:5913.

La Mantia, G., Maglione, D., Pengue, G., Di Cristofano, A., Simeone, A., Lanfrancone, L., and Lania, L., 1991, Identification and characterization of novel human endogenous retroviral sequences preferentially expressed in undifferentiated embryonal carcinoma cells, *Nucleic Acids Res.* **19**:1513.

La Mantia, G., Majello, B., Di Cristofano, A., Strazzullo, M., Minchiotti, G., and Lania, L., 1992, Identification of regulatory elements within the minimal promoter region of the human endogenous ERV9 proviruses: Accurate transcription initiation is controlled by an Inr-like element, *Nucleic Acids Res.* **20**:4129.

Lania, L., Di Cristofano, A., Strazzullo, M., Pengue, G., Majello, B., and La Mantia, G., 1992, Structure and functional organization of the human endogenous retroviral ERV9 sequences, *Virology* **191**:464.

Larsen, F., Gundersen, G., Lopez, R., and Prydz, H., 1992, CpG islands as gene markers in the human genome, *Genomics* **13**:1095.

Larsson, E., Nilsson, B. O., Sundstrom, P., and Widehn, S., 1981, Morphological and microbiological signs of endogenous C-virus in human oocytes, *Int. J. Cancer* **18**:551.

Larsson, E., Kato, N., and Cohen, M., 1989, Human endogenous proviruses, *Curr. Top. Microbiol. Immunol.* **148**:115.

Larsson, E., Anderson, A. C., Holmberg, L., Ohlsson, R., Kato, N., Callacio, J., and Cohen, M., 1993, Expression of an endogenous retrovirus, HERV-R, in human tissues, *J. Cancer Res. Clin. Oncol.* **119**(Suppl. 1):S6.

Law, M., Davidson, J., and Kao, F.-T., 1982, Isolation of a human repetitive sequence and its application to regional chromosome mapping, *Proc. Natl. Acad. Sci. USA* **79**:7390.

Law, M. L., Xu, Y., Berger, R., and Tung, L., 1987, Molecular analysis of human repetitive sequence family and its use as genetic marker, *Somatic Cell Mol. Genet.* **13**:381.

Law, M., Gao, J., and Puck, T., 1989, A nuclear protein associated with human cancer cells binds preferentially to a human repetitive DNA sequence, *Proc. Natl. Acad. Sci. USA* **86**:8472.

Lehrman, M. A., Goldstein, J. L., Russell, D. W., and Brown, M. S., 1987, Duplication of seven exons in LDL receptor gene caused by Alu–Alu recombination in a subject with familial hypercholesterolemia, *Cell* **48**:827.

Leib-Mösch, C., Brack, R., Werner, T., Erfle, V., and Hehlmann, R., 1986, Isolation of an SSAV-related endogenous sequence from human DNA, *Virology* **155**:666.

Leib-Mösch, C., Brack-Werner, R., Werner, T., Bachmann, M., Faff, O., Erfle, V., and Hehlmann, R., 1990, Endogenous retroviral elements in human DNA, *Cancer Res.*(Suppl. 1) **50**:5636s.

Leib-Mösch, C., Bachmann, M., Brack-Werner, R., Werner, T., Erfle, V., and Hehlmann, R., 1992a, Expression and biological significance of human endogenous retroviral sequences, *Leukemia* **6**:72S.

Leib-Mösch, C., Bachmann, M., Geigl, E.-M., Brack-Werner, R., Werner, T., Erfle, V., and Hehlmann, R., 1992b, Expression of S71-related sequences in human cells, *Haematol. Blood Transfus.* **35**:256.

Leib-Mösch, C., Haltmeier, M., Werner, T., Geigl, E.-M., Brack-Werner, R., Francke, U., Erfle, V., and Hehlmann, R., 1993, Genomic distribution and transcription of solitary HERV-K LTRs, *Genomics* **18**:261.

Leong, J., Wood, S., Lyford, A., and Levy, J., 1984, Purification of a specific inhibitor of reverse transcriptase from human placenta, *Int. J. Cancer* **33**:435.

Levy, J. A., Volberding, P., Oppermann, H., and Leong, J., 1982, Retroviruses and differentiation, in: *Expression of Differentiated Functions in Cancer Cells* (R. F. Revoltella *et al.*, eds.), pp. 451–469, Raven Press, New York.

Levy, L., Lobelle-Rich, P., Elder, J., Payne, S., and Montelaro, R., 1990, An unusual retrovirus-like sequence identified in human DNA, *J. Gen. Virol.* **71**:1613.

Lindeskog, M., Medstrand, P., and Blomberg, J., 1993, Sequence variation of human endogenous retrovirus ERV9-related elements in an *env* region corresponding to an immunosuppressive peptide: Transcription in normal and neoplastic cells, *J. Virol.* **67**:1122.

Linial, M., 1987, Creation of a processed pseudogene by retroviral infection, *Cell* **49**:93.

Liu, A., and Abraham, B., 1991, Expression of a hybrid human endogenous retrovirus and calbindin gene in a prostate cell line, *Cancer Res.* **51**:4107.

Liu, Q.-R., and Chan, P., 1990, Identification of a long stretch of homopurine–homopyrimidine sequence in a cluster of retroposons in the human genome, *J. Mol. Biol.* **212**:453.

Löwer, J., Wondrak, E. M., and Kurth, R., 1987, Genome analysis and reverse transcriptase activity of human teratocarcinoma-derived retroviruses, *J. Gen. Virol.* **68**:2807.

Löwer, R., Boller, K., Hasenmaier, B., Korbmacher, C., Müller-Lantzsch, N., Löwer, J., and Kurth, R., 1993a, Identification of human endogenous retroviruses with complex mRNA expression and particle formation, *Proc. Natl. Acad. Sci. USA* **90**:4480.

Löwer, R., Löwer, J., Tondera-Koch, C., and Kurth, R., 1993b, A general method for the identification of transcribed retrovirus sequences (R-U5 PCR) reveals the expression of the human endogenous retrovirus loci HERV-H and HERV-K in teratocarcinoma cells, *Virology* **192**:501.

Luciw, P. A., and Leung, N. J., 1992, Mechanisms of retrovirus replication, in: *The Retroviridae*, Vol. 1 (J. A. Levy, ed.), pp. 159–298, Plenum Press, New York.

Lundblad, V., and Blackburn, E. H., 1990, RNA-dependent polymerase motifs in EST1: Tentative identification of a protein component of an essential yeast telomerase, *Cell* **60**:529.

McCarrey, J. R., and Thomas, K., 1987, Human testis-specific PGK gene lacks introns and possesses characteristics of a processed gene, *Nature* **326**:501.

McClain, K., and Wilkowski, C., 1985, Activation of endogenous retroviral sequences in human leukemia, *Biochem. Biophys. Res. Commun.* **133**:945.

McEvoy, S., and Maeda, M., 1988, Complex events in the evolution of the haptoglobin gene cluster in primates, *J. Biol. Chem.* **263**:15740.

McGrath, C. M., Grant, P. H., Sonle, H. D., Glancy, T., and Rich, M. A., 1972, Replication of oncornavirus-like particle in human breast carcinoma cell line MCF7, *Nature* **252**:247.

Maeda, N., 1985, Nucleotide sequence of the haptoglobin and haptoglobin-related gene pair, *J. Biol. Chem.* **260**:6698.

Maeda, N., and Kim, H.-S., 1990, Three independent insertions of retrovirus-like sequences in the haptoglobin gene cluster of primates, *Genomics* **8**:671.

Maeda, N., McEvoy, S., Harris, H., Huisman, T., and Smithies, O., 1986, Polymorphisms in the human haptoglobin gene cluster: Chromosomes with multiple haptoglobin-related (Hpr) genes, *Proc. Natl. Acad. Sci. USA* **83**:7395.

Maeda, S., Mellors, R., Mellors, J., Jerabek, L., and Zervoudakis, I., 1983, Immunohistochemical detection of antigen related to primate type C retrovirus p30 in normal human placentas, *Am. J. Pathol.* **112**:347.

Maeda, S., Yonezawa, K., Yoshizaki, H., Mori, M., Kobayashi, T., Akahonai, Y., Yachi, A., and Mellors, R. C., 1986, Leukemia serum reactive with retrovirus-related antigen in normal human placenta, *Int. J. Cancer* **38**:309.

Mager, D., 1989, Polyadenylation function and sequence variability of the long terminal repeats of the human endogenous retrovirus-like family RTVL-H, *Virology* **173**:591.

Mager, D. L., and Freeman, D., 1987, Human endogenous retroviruslike genome with type C *pol* sequences and *gag* sequences related to human T-cell lymphotropic viruses, *J. Virol.* **61**:4060.

Mager, D., and Goodchild, N., 1989, Homologous recombination between the LTRs of a human retrovirus-like element causes a 5-kb deletion in two siblings, *Am. J. Hum. Genet.* **45**:848.

Mager, D. L., and Henthorn, P. S., 1984, Identification of a retrovirus-like repetitive element in human DNA, *Proc. Natl. Acad. Sci. USA* **81**:7510.

Mager, D., Henthorn, P., and Smithies, O., 1985, A Chinese $^{G}\gamma+(^{A}\gamma\delta\beta)^{0}$ thalassemia deletion: Comparison to other deletions in the human β-globin gene cluster and sequence analysis of the breakpoints, *Nucleic Acids Res.* **13**:6559.

Mariani-Costantini, R., Horn, T., and Callahan, R., 1989, Ancestry of a human endogenous retrovirus family, *J. Virol.* **63**:4982.

Markert, M. L., Hutton, J. J., Wiginton, D. A., States, J. C., and Kaufman, R. E., 1988, Adenosine deaminase (ADA) deficiency due to deletion of the ADA gene promoter and first exon by homologous recombination between two Alu elements, *J. Clin. Invest.* **81**:1323.

Martin, M., Bryan, T., Rasheed, S., and Khan, A., 1981, Identification and cloning of endogenous retroviral sequences present in human DNA, *Proc. Natl. Acad. Sci. USA* **78**:4892.

Martin, S., 1991, Ribonucleoprotein particles with LINE-1 RNA in mouse embryonal carcinoma cells, *Mol. Cell. Biol.* **11**:4804.

Martin, S., Voliva, C., Burton, F., Edgell, M., and Hutchison, C., III, 1984, A large interspersed repeat found in mouse DNA contains a long open reading frame that evolves as if it encodes a protein, *Proc. Natl. Acad. Sci. USA* **81**:2308.

Martinelli, S., and Goff, S., 1990, Rapid reversion of a deletion mutation in Moloney murine leukemia virus by recombination with a closely related endogenous provirus, *Virology* **174**:135.

Matera, A., Hellmann, U., and Schmid, C., 1990, A transpositionally and transcriptionally competent Alu subfamily, *Mol. Cell. Biol.* **10**:5424.

Mathias, S., Scott, A., Kazazian, H., Jr., Boeke, J., and Gabriel, A., 1991, Reverse transcriptase encoded by a human transposable element, *Science* **254**:1808.

May, F., and Westley, B., 1986, Structure of a human retroviral sequence related to mouse mammary tumor virus, *J. Virol.* **60**:743.

May, F., and Westley, B., 1989, Characterization of sequences related to mouse mammary tumor virus that are specific to MCF-7 breast cancer cells, *Cancer Res.* **49**:3879.

May, F., Westley, B., Rochefort, H., Buetti, E., and Diggelmann, H., 1983, Mouse mammary tumour virus related sequences are present in human DNA, *Nucleic Acids Res.* **11**:4127.

Medstrand, P., and Blomberg, J., 1993, Characterization of novel reverse transcriptase encoding human endogenous retroviral sequences similar to type A and type B retroviruses: differential transcription in normal human tissues. *J. Virol.* **67**:6778.

Medstrand, P., Linderskog, M., and Blomberg, J., 1992, Expression of human endogenous retroviral sequences in peripheral blood mononuclear cells of healthy individuals, *J. Gen. Virol.* **73**:2463.

Mermer, B., Colb, M., and Krontiris, T., 1987, A family of short, interspersed repeats is associated with tandemly repetitive DNA in the human genome, *Proc. Natl. Acad. Sci. USA* **84**:3320.

Mietz, J. A., Fewell, J. W., and Kuff, E. L., 1992, Selective activation of a discrete family of endogenous proviral elements in normal BALB/c lymphocytes, *Mol. Cell. Biol.* **12**:220.

Miki, Y., Nishisho, I., Horii, A., Miyoshi, Y., Utsunomiya, J., Kinzler, K., Vogelstein, B., and Nakamura, Y., 1992, Disruption of the APC gene by a retrotransposal insertion of L1 sequence in a colon cancer, *Cancer Res.* **52**:643.

Minakami, R., Kurose, K., Etoh, K., Furuhata, Y., Hattori, M., and Sakaki, Y., 1992, Identification of an internal *cis*-element essential for the human L1 transcription and a nuclear factor(s) binding to the element, *Nucleic Acids Res.* **20**:3139.

Misra, R., Shih, A., Rush, M., Wong, E., and Schmid, C., 1987, Cloned extrachomosomal circular DNA copies of the human transposable element THE-1 are related predominantly to a single type of family member, *J. Mol. Biol.* **196**:233.

Misra, R., Matera, A., Schmid, C., and Rush, M., 1989, Recombination mediates production of an extrachromosomal circular DNA containing a transposon-like human element, THE-1, *Nucleic Acids Res.* **17**:8327.

Miyaki, T., Migita, K., and Sakaki, Y., 1983, Some KpnI family members are associated with the Alu family in the human genome, *Nucleic Acids Res.* **11**:6837.

Mizrokhi, L., Georgieva, S., and Ilyin, Y., 1988, Jockey, a mobile *Drosophila* element similar to mammalian LINEs, is transcribed from the internal promoter by RNA polymerase II, *Cell* **54**:685.

Moisher, J. A., Luk, G. D., and Huang, R. C. C., 1986, mRNA from human colon tumor and mucosa related to the *pol* gene of an endogenous A-type retrovirus, *Biochem. Biophys. Res. Commun.* **139**:1071.

Moore, D. H., Charney, J., Kramarsky, B., Lasfargues, E. Y., Sarkar, N. H., Brennan, M. J., Burrows, J. H., Sirsat, S. M., Paymaster, J. C., and Vaidya, A. B., 1971, Search for a human breast cancer virus, *Nature* **229**:611.

Moore, K., Jardieu, P., Mietz, J., Trounstine, M., Kuff, E., Ishizaka, K., and Martens, C., 1986, Rodent IgE-binding factor genes are members of an endogenous retrovirus-like gene family, *J. Immunol.* **136**:4283.

Morse, B., Rotherg, P., South, V., Spandorfer, J., and Astrin, S., 1988, Insertional mutagenesis of the myc locus by a LINE-1 sequence in a human breast carcinoma, *Nature* **333**:87.

Müeller-Lantzsch, N., Sauter, M., Weiskircher, A., Kramer, K., Best, B., Buck, M., and Grässer, F., 1993, The human endogenous retroviral element K10 (HERV-K10) encodes for a full length *gag* homologous 73 kD protein and a functional protease, *AIDS Res. Hum. Retroviruses* **9**:343.

Murdoch, G. H., Sklaviadis, T., Manuelidis, E. E., and Manuelidis, L., 1990, Potential retroviral RNAs in Creutzfeldt–Jakob disease, *J. Virol.* **64**:1477.

Nakamura, N., Sugino, H., Takahara, K., Jin, C., Fukushige, S., and Matsubara, K., 1991, Endogenous retroviral LTR DNA sequences as markers for individual human chromosomes, *Cytogenet. Cell Genet.* **57**:18.

Nelson, J., Leong, J., and Levy, J., 1978, Normal human placentas contain RNA-directed DNA polymerase activity like that in viruses, *Proc. Natl. Acad. Sci. USA* **75**:6263.

Nelson, J., Levy, J., and Leong, J., 1981, Human placentas contain a specific inhibitor of RNA-directed DNA polymerase, *Proc. Natl. Acad. Sci. USA* **78**:1670.

Nilsson, B. O., Kattstrom, P. O., Sundstrom, P., Jaquemin, P., and Larsson, E., 1992, Human oocytes express murine retroviral equivalents, *Virus Genes Six* **3**:221.

Noda, M., Kurihara, M., and Takano, T., 1982, Retrovirus-related sequences in human DNA: Detection and cloning of sequences which hybridize with the long terminal repeat of baboon endogenous virus, *Nucleic Acids Res.* **10**:2865.

O'Brien, S., Bonner, T., Cohen, M., O'Connell, C., and Nash, W., 1983, Mapping of an endogenous retroviral sequence to human chromosome 18, *Nature* **303**:74.

O'Connell, C., and Cohen, M., 1984, The long terminal repeat sequences of a novel human endogenous retrovirus, *Science* **226**:1204.

O'Connell, C., O'Brien, S., Nash, W., and Cohen, N., 1984, ERV3, a full-length human endogenous provirus: Chromosomal localization and evolutionary relationships, *Virology* **138**:225.

Ono, M., 1986, Molecular cloning and long terminal repeat sequences of human endogenous retrovirus genes related to types A and B retrovirus genes, *J. Virol.* **58**:937.
Ono, M., Yasunaga, T., Miyata, T., and Ushikubo, H., 1986, Nucleotide sequence of human endogenous retrovirus genome related to the mouse mammary tumor virus genome, *J. Virol.* **60**:589.
Ono, M., Kawakami, M., and Takezawa, T., 1987a, A novel human nonviral retroposon derived from an endogenous retrovirus, *Nucleic Acids Res.* **15**:8725.
Ono, M., Kawakami, M., and Ushikubo, H., 1987b, Stimulation of expression of the human endogenous retrovirus genome by female steroid hormones in the human breast cancer cell line T47D, *J. Virol.* **61**:2059.
Orgel, L. E., and Crick, F. H., 1980, Selfish DNA: The ultimate parasite, *Nature* **284**:604.
Pandey, R., Ghosh, A., Kumar, D., Bachman, B., Shibata, D., and Roy-Burman, P., 1991, Recombination between feline leukemia virus subgroup B or C and endogenous *env* elements alters the in vitro biological activities of the viruses, *J. Virol.* **65**:6495.
Panem, S., 1979, C-type virus expression in the placenta, in: *Current Topics in Pathology* (E. Grundmann and W. H. Kirsten, eds.), pp. 175–189, Springer-Verlag, Berlin.
Panem, S., Prochownik, E. V., Reale, F. R., and Kirsten, W. H., 1975, Isolation of type C virions from a normal human fibroblast strain, *Science* **189**:297.
Panem, S., Ordonez, N., Kirsten, W. H., Katz, A., and Spargo, B. H., 1976, C-type virus expression in systemic lupus erythematosus, *N. Engl. J. Med.* **295**:470.
Pannuti, A., Lanfrancone, L., Pascucci, A., Pelicci, P., La Mantia, G., and Lania, L., 1988, Isolation of cDNAs encoding finger proteins and measurement of the corresponding mRNA levels during myeloid terminal differentiation, *Nucleic Acids Res.* **16**:4227.
Patarca, R., and Haseltine, W. A., 1985, A major retroviral core protein related to EPA and TIMP, *Nature* **318**:390.
Paulson, K., Deka, N., Schmid, C., Misra, R., Schindler, C., Rush, M., Kadyk, L., and Leinwand, L., 1985, A transposon-like element in human DNA, *Nature* **316**:359.
Paulson, K., Matera, A., Deka, N., and Schmid, C., 1987, Transcription of a human transposon-like sequence is usually directed by other promoters, *Nucleic Acids Res.* **15**:5199.
Payne, L. N., 1992, Biology of avian retroviruses, in: *The Retroviridae*, Vol. 1 (J. Levy, ed.), pp. 299–404, Plenum Press, New York.
Perl, A., Rosenblatt, J., Chen, I., DiVincenzo, J., Bever, R., Poiesz, J., and Abraham, G., 1989, Detection and cloning of new HTLV-related endogenous sequences in man, *Nucleic Acids Res.* **17**:6841.
Perl, A., Isaacs, C., Eddy, R., Byers, M., Sait, S., and Shows, T., 1991a, The human T-cell leukemia virus-related endogenous sequence (HRES1) is located on chromosome 1 at q42, *Genomics* **11**:1172.
Perl, A., Pazmany, T., Hung, C., Mattson, D., Baraczka, K., Ablonczy, E., and Perl-Banki, K., 1991b, Cloning and expression of an HTLV *gag*-related autoantigen, *Arthritis Rheum.* **34**:S135.
Pizzuti, A., Pieretti, M., Fenwick, R., Gibbs, R., and Caskey, C., 1992, A transposon-like element in the deletion-prone region of the dystrophin gene, *Genomics* **13**:594.
Query, C. C., and Keene, J. D., 1987, A human autoimmune protein associated with U1 RNA contains a region of homology that is cross-reactive with retroviral p30 *gag* antigen, *Cell* **51**:211.
Rabson, A., Steele, P., Garon, C., and Martin, M., 1983, mRNA transcripts related to full-length endogenous retroviral DNA in human cells, *Nature* **306**:604.
Rabson, A., Hamagishi, Y., Steele, P., Tykocinske, M., and Martin, M., 1985, Characterization of human endogenous retroviral envelope RNA transcripts, *J. Virol.* **56**:176.
Reeves, R. H., and O'Brien, S. J., 1984, Molecular genetic characterization of the RD-114 gene family of endogenous feline retroviral sequences, *J. Virol.* **52**:164.
Renan, M. J., and Reeves, B. R., 1987, Chromosomal localization of human endogenous retroviral element ERV1 to 18q22-q23 by in situ hybridization, *Cytogenet. Cell Genet.* **44**:167.
Repaske, R., O'Neill, R., Steele, P., and Martin, M., 1983, Characterization and partial nucleotide sequence of endogenous type C retrovirus segments in human chromosomal DNA, *Proc. Natl. Acad. Sci. USA* **80**:678.
Repaske, R., Steele, P., O'Neill, R., Rabson, A., and Martin, M., 1985, Nucleotide sequence of a full-length human endogenous retroviral segment, *J. Virol.* **54**:764.
Risser, R., and Horowitz, J. M., 1983, Endogenous murine leukemia viruses, *Annu. Rev. Genet.* **17**:85.
Robinson, H. L., Astrin, S. M., Senior, A. M., and Salazar, F. H., 1981, Host susceptibility to endogenous viruses: Defective glycoprotein-expressing proviruses interfere with infections, *J. Virol.* **40**:745.

Rogers, J., 1985, The origin and evolution of retroposons, *Int. Rev. Cytol.* **93**:187.

Rovigatti, U. G., and Astrin, S. M., 1983, Avian endogenous viral genes, *Curr. Top. Microbiol. Immunol.* **103**:1.

Rubin, L. A., Siminovitch, K. A., Shi, M. H., and Cohen, M., 1991, A novel retroviral gene association with rheumatoid arthritis, *Arthritis Rheum.* **34**:S60.

Samuelson, L., Wiebauer, K., Snow, C., and Meisler, M., 1990, Retroviral and pseudogene insertion sites reveal the lineage of human salivary and pancreatic amylase genes from a single gene during primate evolution, *Mol. Cell. Biol.* **10**:2513.

Sandmeyer, S., Hansen, L., and Chalker, D., 1990, Integration specificity of retrotransposons and retroviruses, *Annu. Rev. Genet.* **24**:491.

Sawyer, M. H., Nachlas, N. E., Jr., and Panem, S., 1978, C-type virus antigen expression in normal human placenta, *Nature* **275**:62.

Schmid, C., Wong, E., and Deka, N., 1990, Single copy sequences in galago DNA resemble a repetitive human retrotransposon-like family, *J. Mol. Evol.* **31**:92.

Schwartzberg, P., Colicelli, J., and Goff, S., 1985, Recombination between a defective retrovirus and homologous sequences in host DNA: Reversion by patch repair, *J. Virol.* **53**:719.

Sedlacek, Z., Korn, B., Konecki, D. S., Siebenhaar, R., Coy, J. F., Kioschis, P., and Poustka, A., 1993, Construction of a transcription map of a 300 kb region around the human G6PD locus by direct cDNA selection, *Hum. Mol. Genet.* **2**:1865.

Shell, B., Szurek, P., and Dunnick, W., 1987, Interruption of two immunoglobulin heavy-chain switch regions in murine plasmacytoma P3.26Bu4 by insertion of retrovirus-like elements ETn, *Mol. Cell. Biol.* **7**:1364.

Shiba, T., and Saigo, K., 1983, Retrovirus-like particles containing RNA homologous to the transposable element *copia* in *Drosophila melanogaster*, *Nature* **302**:119.

Shih, A., Misra, R., and Rush, M. G., 1989, Detection of multiple, novel reverse transcriptase coding sequences in human nucleic acids: Relation to primate retroviruses, *J. Virol.* **63**:64.

Shih, A., Coutavas, E. E., and Rush, M. G., 1991, Evolutionary implications of primate endogenous retroviruses, *Virology* **182**:495.

Shimotohno, K., Takahashi, Y., Shimizu, N., Gojobori, T., Golde, D., Chen, I., Miwa, M., and Sugimura, T., 1985, Complete nucleotide sequence of an infectious clone of human T-cell leukemia virus type II: An open reading frame for the protease gene, *Proc. Natl. Acad. Sci. USA* **82**:3101.

Sibley, C. G., and Ahlquist, J. E., 1987, DNA hybridization evidence of hominoid phylogeny: Results from an expanded data set, *J. Mol. Evol.* **26**:99.

Silver, J., Rabson, A., Bryan, T., Willey, R., and Martin, M., 1987, Human retroviral sequences on the Y chromosome, *Mol. Cell. Biol.* **7**:1559.

Singer, M., 1982a, SINEs and LINEs: Highly repeated short and long interspersed sequences in mammalian genomes, *Cell* **28**:433.

Singer, M., 1982b, Highly repeated sequences in mammalian genomes, *Int. Rev. Cytol.* **76**:67.

Singer, M., and Skowronski, J., 1985, Making sense out of LINEs: Long interspersed repeat sequences in mammalian genomes, *Trends Biol. Sci.* **10**:119.

Skowronski, J., and Singer, M., 1985, Expression of a cytoplasmic LINE-1 transcript is regulated in a human teratocarcinoma cell line, *Proc. Natl. Acad. Sci. USA* **82**:6050.

Soares, M. B., Schon, E., Henderson, A., Karathanasis, S. K., Sate, R., Zeitlin, S., Chirgwin, J., and Efstratiadis, S., 1985, RNA-mediated gene duplication: The rat preproinsulin I gene is a functional retroposon, *Mol. Cell. Biol.* **5**:2090.

Stavenhagen, J. B., and Robins, D. M., 1988, An ancient provirus has imposed androgen regulation on the adjacent mouse sex-limited protein gene, *Cell* **55**:247.

Steele, P., Rabson, A., Bryan, T., and Martin, M., 1984, Distinctive termini characterize two families of human endogenous retroviral sequences, *Science* **225**:943.

Steele, P., Martin, M., Rabson, A., Bryan, T., and O'Brien, S., 1986, Amplification and chromosomal dispersion of human endogenous retroviral sequences, *J. Virol.* **59**:545.

Steffen, D., Bird, S., Rowe, W. P., and Weinberg, R. A., 1979, Identification of DNA fragments carrying ecotropic proviruses of AKR mice, *Proc. Natl. Acad. Sci. USA* **76**:4554.

Steffen, D. L., Mural, R., Cowling, D., Mielcarz, J., Young, J., and Roblin, R., 1982, Most of the murine leukemia virus sequences in the DNA of NIH/Swiss mice consist of two closely related proviruses, each repeated several times, *J. Virol.* **43**:127.

Stephenson, J. R., Anderson, G. R., Tronick, S. R., and Aaronson, S. A., 1974, Evidence for genetic recombination between endogenous and exogenous mouse RNA type C viruses, *Cell* **2**:87.

Stoye, J., and Coffin, J., 1985, Endogenous retroviruses, in: *RNA Tumor Viruses*, 2nd ed., Vol. 2 (R.

Weiss, N. Teich, H. Varmus, and J. Coffin, eds.), pp. 357–404, Cold Spring Harbor Laboratory, Cold Spring Harbor, N.Y.

Stoye, J., and Coffin, J. M., 1987, The four classes of endogenous murine leukemia virus: Structural relationships and potential for recombination, *J. Virol.* **61**:2659.

Stoye, J. P., Fenner, S., Greenoak, G. E., Moran, C., and Coffin, J. M., 1988, Role of endogenous retroviruses as mutagens: The hairless mutation of mice, *Cell* **54**:383.

Sugino, H., Oshimura, M., and Matsubara, K., 1992, Banding profiles of LTR of human endogenous retrovirus HERV-A in 24 chromosomes in somatic cell hybrids, *Genomics* **13**:461.

Sun, L., Paulson, K. E., Schmid, C. W., Kadyk, L., and Leinwand, L., 1984, Non-Alu family interspersed repeats in human DNA and their transcriptional activity, *Nucleic Acids Res.* **12**:2669.

Suni, J., Wahlstrom, T., and Vaheri, A., 1981, Retrovirus p30-related antigen in human syncytiotrophoblasts and IgG antibodies in cord-blood sera, *Int. J. Cancer* **29**:559.

Suni, J., Narvanen, A., Wahlstrom, T., Aho, M., Pakkanen, R., Vaheri, A., Copeland, T., Cohen, M., and Oroszlan, S., 1984, Human placental syncytiotrophoblast Mr 75,000 polypeptide defined by antibodies to a synthetic peptide based on a cloned human endogenous retroviral DNA sequence, *Proc. Natl. Acad. Sci. USA* **81**:6197.

Suzuki, H., Hosokawa, Y., Toda, H., Nishikimi, M., and Ozawa, T., 1990, Common protein-binding sites in the 5′-flanking regions of human genes for cytochrome c_1 and ubiquinone-binding protein, *J. Biol. Chem.* **265**:8159.

Swain, A., and Coffin, J., 1992, Mechanism of transduction by retroviruses, *Science* **255**:841.

Swergold, G. D., 1990, Identification, characterization and cell specificity of a human LINE-1 promoter, *Mol. Cell. Biol.* **10**:6718.

Taruscio, D., and Manuelidis, L., 1991, Integration site preferences of endogenous retroviruses, *Chromosoma* **101**:141.

Tchenio, T., and Heidmann, T., 1991, Defective retroviruses can disperse in the human genome by intracellular transposition, *J. Virol.* **65**:2113.

Teich, N., 1984, Taxonomy of retroviruses, in: *RNA Tumor Viruses*, 2nd ed., Vol. 1 (R. Weiss, N. Teich, H. Varmus, and J. Coffin, eds.), pp. 25–207, Cold Spring Harbor Laboratory, Cold Spring Harbor, N.Y.

Temin, H., 1980, Origin of retroviruses from cellular moveable genetic elements, *Cell* **21**:599.

Temin, H. M., 1985, Reverse transcriptase in the eukaryotic genome: Retroviruses, pararetroviruses, retrotransposons, and retrotranscripts, *Mol. Biol. Evol.* **2**:455.

Temin, H., 1989, Retrons in bacteria, *Nature* **339**:254.

Temin, H., 1992, Origin and general nature of retroviruses, in: *The Retroviridae*, Vol. 1 (J. A. Levy, ed.), pp. 1–18, Plenum Press, New York.

Temin, H., and Mizutani, S., 1970, RNA-dependent DNA polymerase in virions of Rous sarcoma virus, *Nature* **226**:1211.

Ting, C.-N., Rosenberg, M. P., Snow, C. M., Samuelson, L. C., and Meisler, M. H., 1992, Endogenous retroviral sequences are required for tissue-specific expression of a human salivary amylase gene, *Genes Dev.* **6**:1457.

Toh, H., Ono, M., and Miyata, T., 1985, Retroviral *gag* and DNA endonuclease coding sequences in IgE-binding factor gene, *Nature* **318**:388.

Tomita, N., Horii, A., Doi, S., Yokouchi, H., Ogawa, M., Mori, T., and Matsubara, K., 1990, Transcription of human endogenous retroviral long terminal repeat (LTR) sequence in a lung cancer cell line, *Biochem. Biophys. Res. Commun.* **166**:1.

Trauger, R. J., Talbott, R., Wilson, S. H., Karpel, R. L., and Elder, J. H., 1990, A single-stranded nucleic acid binding sequence common to the heterogeneous nuclear ribonucleoparticle protein A1 and murine recombinant virus GP70, *J. Biol. Chem.* **265**:3674.

Ullu, E., and Tschudi, C., 1984, Alu sequences are processed 7SL RNA genes, *Nature* **312**:171.

Ullu, E., and Weiner, A., 1985, Upstream sequences modulate the internal promoter of the human 7SL RNA gene, *Nature* **318**:371.

Vaidya, A. B., 1973, Molecular biology of human milk, *Science* **180**:776.

Vanin, E., 1985, Processed pseudogenes: Characteristics and evolution, *Annu. Rev. Genet.* **19**:253.

Varmus, H., and Brown, P., 1991, Retroviruses, in: *Mobile DNA* (D. Berg and M. Howe, eds.), pp. 53–108, American Society for Microbiology, Washington, D.C.

Vernon, M. L., McMahon, J. M., and Hackett, J. J., 1974, Additional evidence of type-C particles in human placentas, *J. Natl. Cancer Inst.* **42**:987.

Vidaud, D., Vidaud, M., Bahnak, B., Siguret, V., Sanchez, S., Laurian, Y., Meyer, D., Goossens, M., and Lavergne, J., 1993, Haemophilia B due to a de novo insertion of a human-specific alu subfamily member within the coding region of the factor IX gene, *Eur. J. Hum. Genet.* **1**:30.

Vogetseder, W., Dumfahrt, A., Mayersbach, P., Schonitzer, D., and Dierich, M. P., 1993, Antibodies in human sera recognizing a recombinant outer membrane protein encoded by the envelope gene of the human endogenous retrovirus K. *AIDS Res. Hum. Retroviruses* **9**:687.

Wahlstrom, T., Nieminen, P., Narvanen, A., Suni, J., Lehtovirta, P., Saksela, E., and Vaheri, A., 1984, Monoclonal antibody defining a human syncytiotrophoblastic polypeptide immunologically related to mammalian structural protein p30, *Placenta* **5**:465.

Wallace, M., Andersen, L., Saulino, A., Gregory, P., Glover, T., and Collins, F., 1991, A *de novo* Alu insertion results in neurofibromatosis type 1, *Nature* **353**:864.

Weiner, A., Deininger, P., and Efstratiadis, A., 1986, Nonviral retroposons: Genes, pseudogenes, and transposable elements generated by the reverse flow of genetic information, *Annu. Rev. Biochem.* **55**:631.

Weiss, R., 1984, The search for human RNA tumor viruses, in: *RNA Tumor Viruses*, 2nd ed., Vol. 1 (R. Weiss, N. Teich, H. Varmus, and J. Coffin, eds.), pp. 1205–1281, Cold Spring Harbor Laboratory, Cold Spring Harbor, N.Y.

Weiss, R. A., Mason, W. S., and Vogt, P. K., 1973, Genetic recombinants and heterozygotes derived from endogenous and exogenous avian RNA tumor viruses, *Virology* **52**:535.

Werner, T., Brack-Werner, R., Leib-Mösch, C., Backhaus, H., Erfle, V., and Hehlmann, R., 1990, S71 is a phylogenetically distinct human endogenous retroviral element with structural and sequence homology to simian sarcoma virus (SSV), *Virology* **174**:225.

Westley, B., and May, F., 1984, The human genome contains multiple sequences of varying homology to mouse mammary tumour virus DNA, *Gene* **28**:221.

Wichman, H. A., and Van Den Bussche, R. A., 1992, In search of retrotransposons: Exploring the potential of the PCR, *BioTechniques* **13**:258.

Wilkinson, D. A., 1993, Expression of the RTVL-H family of human endogenous retrovirus-like sequences, Ph.D. thesis, University of British Columbia.

Wilkinson, D. A., Freeman, J. D., Goodchild, N. L., Kelleher, C. A., and Mager, D. L., 1990, Autonomous expression of RTVL-H endogenous retroviruslike elements in human cells, *J. Virol.* **64**:2157.

Wilkinson, D. A., Goodchild, N. L., Saxton, T. M., Wood, S., and Mager, D. L., 1993, Evidence for a functional subclass of the RTVL-H family of human endogenous retrovirus-like sequences, *J. Virol.* **67**:2981.

Wilkinson, D. A., and Mager, D. L., 1993, RTVL-H human endogenous retrovirus-like elements with an intact pol and a potential env region, *J. Cancer Res. Clin. Oncology* **119**(Suppl. 1):S8.

Wong, Z., Royle, N., and Jeffreys, A., 1990, A novel human DNA polymorphism resulting from transfer of DNA from chromosome 6 to chromosome 16, *Genomics* **7**:222.

Woods-Samuels, P., Wong, C., Mathias, S. L., Scott, A. F., Kazazian, H. H., Jr., and Antonarakis, S. E., 1989, Characterization of a nondeleterious L1 insertion in an intron of the human factor VIII gene and further evidence of open reading frames in functional L1 elements, *Genomics* **4**:290.

Xiong, Y., and Eickbush, T., 1990, Origin and evolution of retroelements based upon their reverse transcriptase sequences, *EMBO J.* **9**:3353.

Yeh, K.-W., Yen, C.-P., Liu, J.-C., Feng, Y.-N., Wu, F. Y.-H., Yang, W.-K., and Wu, C.-W., 1991, Isolation of a cDNA clone of human endogenous retrovirus (HERV) from human cancer cell line, *FASEB J.* **5**:A884.

Zabarovsky, E. R., Chumakov, I., and Kisselev, L. L., 1983, Tight linkage of retroviral-like sequences to a variant human *c-mos* gene in the human genome, *Gene* **23**:379.

Zavada, J., Zavadova, Z., and Russ, G., 1986, Rescue of presumptive viral information from human cells by a helper oncovirus, *J. Gen. Virol.* **67**:1561.

Zhu, Z., Hsieh, S.-L., Bentley, D., Campbell, R., and Volanakis, J., 1992, A variable number of tandem repeats locus within the human complement C2 gene is associated with a retroposon derived from a human endogenous retrovirus, *J. Exp. Med.* **175**:1783.

Zucchi, I., and Schlessinger, D., 1992, Distribution of moderately repetitive sequences pTR5 and LF1 in $X_q24\text{-}_q28$ human DNA and their use in assembly YAC contigs, *Genomics* **12**:264.

Index